Computational Music Science

Series editors

Guerino Mazzola
Moreno Andreatta

More information about this series at http://www.springer.com/series/8349

Guerino Mazzola • René Guitart • Jocelyn Ho
Alex Lubet • Maria Mannone • Matt Rahaim
Florian Thalmann

The Topos of Music III: Gestures

Musical Multiverse Ontologies

Second Edition

Springer

Guerino Mazzola
School of Music
University of Minnesota
Minneapolis, MN, USA

René Guitart
Université Diderot Paris 7
Institut de Mathématiques de Jussieu
Paris, France

Jocelyn Ho
Department of Music
UCLA Herb Alpert School of Music
Los Angeles, CA, USA

Alex Lubet
School of Music
University of Minnesota
Minneapolis, MN, USA

Maria Mannone
School of Music
University of Minnesota
Minneapolis, MN, USA

Matt Rahaim
School of Music
University of Minnesota
Minneapolis, MN, USA

Florian Thalmann
School of Electronic Engineering
and Computer Science
Queen Mary University of London
London, United Kingdom

This is the third of four volumes in the set "The Topos of Music" in the series Computational Music Science. Volume I is ISBN 978-3-319-64363-2, Volume II is ISBN 978-3-319-64443-1, Volume III is ISBN 978-3-319-64479-0, Volume IV is ISBN 978-3-319-64494-3. The first edition of The Topos of Music appeared in one volume in the Birkhäuser program ISBN 978-3-7643-5731-3.

ISSN 1868-0305 ISSN 1868-0313 (electronic)
Computational Music Science
ISBN 978-3-030-09720-2 ISBN 978-3-319-64481-3 (eBook)
https://doi.org/10.1007/978-3-319-64481-3

Mathematics Subject Classification (2010): 14K99, 18A99, 57P99, 18F20, 20C99

Printed on acid-free paper

This Springer imprint is published by the registered company Springer International Publishing AG part of Springer Nature.
The registered company address is: Gewerbestrasse 11, 6330 Cham, Switzerland

Preface to the Second Edition

Comprendre, c'est
attraper le geste
et pouvoir continuer
Jean Cavaillès [181, p. 186]

A major reason for a second edition of *The Topos of Music*—besides the simple fact that the first edition is now sold out—goes back to spring 2002, when I was completing its first edition, published in fall 2002. I was asked to give a talk in the MaMuX seminar of the IRCAM in Paris, to explain how I applied the mathematics of *The Topos of Music* to my free jazz improvisations.

While preparing my talk I realized that despite the presence of mathematical music theory the decisive generator of my instant compositions was the gestural deployment of formulas, the "action painting" of musical thoughts, not the abstract formulas in their static facticity. First and foremost this was a shocking insight in view of the forthcoming publication of the formulaic setup in *The Topos of Music.*

Fortunately, I knew from Hermann Hesse that "every end is a beginning"[1], which meant in my case that the end of a scientific development as traced in the book's first edition initiated the next step: a music theory of gestures. It goes without saying that this new phase would not destroy the previous research, but incorporate it as the stratum of facticity in an extended ontology of embodiment, where facts are the output of processes and their gestural generators.

In the sequel, I discovered that I was far from being the first scholar and artist to discover the crucial role of gestures in music. For instance, free jazz pianist Cecil Tayor, music philosopher Theodor Wiesengrund Adorno, or lateral thinker Paul Valéry had clearly stressed the dancing essence of art, an insight that I had embodied in my own pianist's art, but never understood on an intellectual level.

Of course, I could not be satisfied by the very existence of gesture philosophy or gestural practice, just as I could not accept traditional music and performance theory when I started my enterprise of mathematical music theory in 1978. The gesturally colored thoughts and actions needed a rigorous conceptualization in the same vein as my efforts before the first edition of *The Topos of Music.*

In 2002, I was in the privileged position to work in the multimedia division of Peter Stucki at the Institut für Informatik of the University of Zurich. I had excellent PhD students, and we could, with one of them, Stefan Müller, realize a first experimental software for the gestural representation of a pianist's hand, a work presented at the ICMC conference in 2003 [772].

This experimental preliminary work was then taken as a point of departure for a mathematical theory of musical gestures. I presented this theory in a course in spring 2005 at École normale supérieure in Paris, a course that was later in 2007 taken as the material basis of my French book *La vérité de beau dans la musique* [718]. The first publication of a formally more evolved mathematical theory of musical gestures was

[1] Actually, he says that "Jeder Anfang ist ein Ende," but the reverse is immediate.

written with co-author Moreno Andreatta in 2007 [720]. This date could be called the birthday of a valid mathematical theory of musical gestures.

Until the publication of this second edition of *The Topos of Music*, several important conceptual extensions of the mathematical theory of musical gestures, models of musical gestural processes, as well as a number of theorems have been published. The decade from date of birth to the presence proved that the mathematical theory of musical gestures is an important added value to the theory described in the first edition of *The Topos of Music*.

We can however not state that this theory of gestures is in a complete state, quite the opposite is true: The coming years will reveal important news of theoretical as well as practical nature. So why did we make the decision to publish the present state of the art? The first argument is that the present state is rich enough to define concrete new directions, be it in music theory, such as harmony or counterpoint, be it in performance theory, or be it in the understanding of embodiment in the making of music. The second argument is that the present material, roughly 500 pages of new material, is ample enough to present a book's stature. And the third and very important argument is that we would like to communicate the state of the art in the spirit of Cavaillès: *Understanding is catching the gesture and being able to continue.* The co-authors of the gesture theory part, René Guitart, Jocelyn Ho, Alex Lubet, Maria Mannone, Matt Rahaim, and Florian Thalmann, are a wonderful confirmation of this philosophy. So let us continue!

Here is a summary of the new material and its authorship. Whenever I don't mention the author, it is my own contribution, all others are mentioned explicitly.

Until Part XIV, nearly everything is as in the book's first edition, refer to the preface of that edition (also included in this edition) for detailed summaries. The only new content—besides errata corrections—is Chapter 45 in Part XI, which is a shortened version of a paper [110] on a statistical analysis of Chopin's Prélude op. 28, No. 4, written with Jan Beran, Robert Goswitz, and Patrizio Mazzola.

Gesture theory starts with Part XV: Gesture Philosophy for Music. Chapter 57 gives an overview of philosophical aspects of gestures, including works by Jean-Claude Schmitt, Vilém Flusser, Michel Guérin, Adam Kendon, David McNeill, Juhani Pallasmaa, André Chastel, Émile Benveniste, Marie-Dominique Popelard, and Anthony Wall. In Chapter 58, we discuss the presemiotic approch to gestures in the French perspective of Maurice Merleau-Ponty, Gilles Deleuze, Jean Cavaillès, Charles Alunni, and Gilles Châtelet. Paul Valéry is also referenced in Section 59.4.

Chapter 59 deals with gestural aspects in cognitive science. After a review of Embodied AI and anthropology, Alex Lubet in Section 59.5 introduces us to gestural disability studies, focusing on two famous disabled pianists: Horace Parlan and Oscar Peterson (in his last years). Then in Section 59.6 Lubet reflects on perception of musical gesture as being inherently synaesthetic.

Chapter 60 concludes this part with a review of musical models of gesturality as proposed by Wolfgang Graeser, Theodor W. Adorno, Neil P. McAngus Todd, David Lewin, Robert Hatten, Marcelo Wanderley, Claude Cadoz, and Marc Leman.

Part XVI introduces the mathematics of gestures. Chapter 61 presents the mathematical concept of a gesture in a topological space and states the Diamond Conjecture, which deals with a hypothetical big space that unites algebraic and topological categories. Chapter 62 extends the theory from gestures in topological spaces to gestures in topological categories and introduces functorial gestures, i.e., functors on topological categories with values in the category of gestures, similar to functorial compositions in the previous theory.

Chapter 63 presents a generalized singular homology, where cubes are replaced by general hypergestures. Hypergesture homology applies to a gestural model of counterpoint and to a gestural refinement of performance stemma theory.

Chapter 64 presents—similar to Chapter 63—Stokes' Theorem for hypergestures. This theorem applies to problems in gestural modulation theory.

Chapters 65 and 66 discuss categories of local and global compositions, processes/networks, and gestures, together with their functorial relationships. This triple typology composition/process/gesture corresponds to the ontological dimension of embodiment with its three coordinates facts/processes/gestures.

In Sections 67.1–67.7 of Chapter 67, René Guitart develops a fascinating and demanding mathematical model of mathematical creativity, where thought is viewed as an algebra of gestures.

In Section 61.14, we, Maria Mannone and Guerino Mazzola, present a group-theoretical model of Georg Wilhelm Friedrich Hegel's initial discourse in his *Wissenschaft der Logik*, a model that applies to the Yoneda concept of creativity [726, Chapter 19.2]. We illustrate the method with an experimental composition by Mannone. This discussion extends over Sections 67.8–67.15 and completes Chapter 67.

Part XVII deals with concept architectures and software for musical gesture theory. Chapter 68 explains the denotator formalism for gestures over topological categories. Chapter 69 is a summary of the Java-based RUBATO® Composer software [739], written to sketch the framework of Chapters 70–74, where Florian Thalmann presents his gesture-oriented software component, the BigBang rubette. This discourse again follows the coordinates of the dimension of embodiment: facts, processes, gestures, which in this situation specify to: visualization and sonification of denotators (Chapter 71), BigBang's operation graph (Chapter 72), and gestural interaction and gesturalization (Chapter 73). In the final Chapter 74 of this part, Thalmann discusses musical examples.

Part XVIII is entitled *The Multiverse Perspective* because it opens up the relationship of gesture theory with string theory in theoretical physics. After a critical review of Hermann Hesse's *Glasperlenspiel* with regard to its gestural deficiencies, we, Mazzola and Mannone, develop the Euler-Lagrange formalism of world-sheets for musical gestures. This theory extends to functorial global gestures over global topological categories.

Part XIX is dedicated to applications of gesture theory to a number of musical themes.

Chapter 79 deals with singular gesture homology being applied to counterpoint.

Chapter 80 introduces a gestural restatement of modulation theory, applying in particular Stokes' Theorem for hypergestures.

Chapter 81 applies gesture theory to a gestural performance stemma theory.

Chapter 82 is written by Jocelyn Ho as a creative presentation of composition and analysis as embodied gestures in an inter-corporeal world. She presents two compositions, Toru Takemitsu's *Rain Tree Sketch II* and her own composition *Sheng* for piano, smartphones, and fixed playback.

Chapter 83 is Mannone's analysis and classification of a conductor's movements from the viewpoint of gesture theory.

Chapter 84 is a review of gestural aspects that were developed in *Flow, Gesture, and Spaces in Free Jazz* [721].

Chapter 85 is written by Matt Rahaim and presents the gestural approach to understanding Hindustani music in its vocal gesturality.

Chapter 86 is a first approach, written by Mannone, to a future theory of vocal gestures. The short addendum was written by Mazzola.

The Appendix has been enriched by additional complements on mathematics (Chapter J) plus complements on physics (Chapter K).

The Leitfaden III has been added to the original Leitfaden I & II for the gestural chapters.

The ToM_CD has been updated, containing now the present book's pdf ToposOfMusic.pdf. However the original CD is no longer added to the book, instead the ToM_CD can be downloaded from

`www.encyclospace.org/special/ToM_CD.zip`.

Concerning the division of the now very large book into parts, this is the split:

- Volume I: *Theory*, Prefaces and Table of Contents, Parts I to VII
- Volume II: *Performance*, Parts VIII to XIV
- Volume III: *Gestures*, Parts XV to XIX
- Volume IV: *Roots*, Appendices

My sincere acknowledgments go to my co-authors and to Springer's Ronan Nugent and Frank Holzwarth as well as to Birkhäuser's Thomas Hempfling.

Minneapolis, October 2017 Guerino Mazzola

Preface

Man kann
einen jeden Begriff,
einen jeden Titel,
darunter viele Erkenntnisse gehören,
einen logischen Ort nennen.
Immanuel Kant [519, p. B 324]

This book's title subject, *The Topos of Music*, has been chosen to communicate a double message: First, the Greek word "topos" ($\tau \acute{o} \pi o \varsigma$ = location, site) alludes to the logical and transcendental location of the concept of music in the sense of Aristotle's [40, 1154] and Kant's [519, p. B 324] topic. This view deals with the question of *where music is situated* ***as*** *a concept*—and hence with the underlying ontological problem: *What is the type of being and existence of music?* The second message is a more technical understanding insofar as the system of musical signs can be associated with the mathematical theory of *topoi*, which realizes a powerful synthesis of geometric and logical theories. It laid the foundation of a thorough geometrization of logic and has been successful in central issues of algebraic geometry (Grothendieck, Deligne), independence proofs and intuitionistic logic (Cohen, Lawvere, Kripke).

But this second message is intimately entwined with the first since the present concept framework of the musical sign system is technically based on topos theory, so the topos of music receives its topos-theoretic foundation. In this perspective, the double message of the book's title in fact condenses to a unified intention: to unite philosophical insight with mathematical explicitness.

According to Birkhäuser's initial plan in 1996, this book was first conceived as an English translation of my former book *Geometrie der Töne* [682], since the German original had suffered from its restricted access to the international public. However, the scientific progress since 1989, when it was written, has been considerable in theory and technology. We have known new subjects, such as the denotator concept framework, performance theory, and new software platforms for composition, analysis, and performance, such as RUBATO® or OpenMusic. Modeling concepts via the denotator approach in fact results from an intense collaboration of mathematicians and computer scientists in the object-oriented programming paradigm and supported by several international research grants.

Also, the scientific acceptance of mathematical music theory has grown since its beginnings in the late 1970s. As the first acceptance of mathematical music theory was testified to by von Karajan's legendary Ostersymposium "Musik und Mathematik" in 1984 in Salzburg [383], so is the significantly improved present status of acceptance testified to by the Fourth Diderot Forum on Mathematics and Music [711] in Paris, Vienna, and Lisbon 1999, which was organized by the European Mathematical Society. The corresponding extension of collaborative efforts in particular entail the inclusion of works by other research groups in this book, such as the "American Set Theory", the Swedish school of performance research at Stockholm's KTH, or the research on computer-aided composition at the IRCAM in Paris.

Therefore, as a result of these revised conditions, *The Topos of Music* appears as a vastly extended English update of the original work. The extension is visibly traced in the following parts which are new with respect to [682]: Part II exposes the theory of denotators and forms, part V introduces the topological theories of rhythms and motives, part VIII introduces the structure theory of performance, part IX deals with the expressive semantics of performance in the language of performance operators and stemmata (genealogical trees of successively refined performance), part X is devoted to the description of the RUBATO® software platform for representation, analysis, composition, and performance, part XI presents a statistical analysis of musical analysis, part XII concludes the subject of performance with an inverse performance theory, in fact a first formalization of the problem of music criticism.

This does however not mean that the other parts are just translations of the German text. Considerable progress has been made in most fields, except the last part XIV which reproduces the status quo in [682]. In particular, the local and global theories have been thoroughly functorialized and thereby introduce an ontological depth and variability of concepts, techniques, and results, which by far transcend the semiotically naive geometric approach in [682]. The present theory is as different from the traditional geometric conceptualization as is Grothendieck's topos theoretic algebraic geometry from classical algebraic geometry in the spirit of Segre, van der Waerden, or Zariski.

Beyond this topos-theoretic generalization, the denotator language also introduces a fairly exceptional technique of circular concept constructions. This more precisely is rooted in Finsler's pioneering work in foundations of set theory [322], a thread which has been rediscovered in modern theoretical computer sciences [5]. The present state of denotator theory rightly could be termed a Galois theory of concepts in the sense that circular definitions of concepts play the role of conceptual equations (corresponding to algebraic equations in algebraic Galois theory), the solutions of which are concepts instead of algebraic numbers.

Accordingly, the mathematical apparatus has been vastly extended, not only in the field of topos theory and its intuitionistic logic, but also with regard to general and algebraic topology, ordinary and partial differential equations, Pólya theory, statistics, multiaffine algebra and functorial algebraic geometry. It is mandatory that these technicalities had to be placed in a more elaborate semiotic perspective. However, this book does not cover the full range of music semiotics, for which the reader is referred to [703]. Of course, such an extension on the technical level has consequences for the readability of the theory. In view of the present volume of over 1300 pages, we could however not even make the attempt to approach a non-technical presentation. This subject is left to subsequent efforts. The critical reader may put the question whether music is really that complex. The answer is yes, and the reason is straightforward: We cannot pretend that Bach, Haydn, Mozart, or Beethoven, just to name some of the most prominent composers, are outstanding geniuses and have elaborated masterworks of eternal value, without trying to understand such singular creations with adequate tools, and this means: of adequate depth and power. After all, understanding God's 'composition', the material universe, cannot be approached without the most sophisticated tools as they have been elaborated in physics, chemistry, and molecular biology.

So who is recommended to read this book? A first category of readers is evidently the working scientist in the fields of mathematical music theory, the soft- and hardware engineer in music informatics, but also the mathematician who is interested in new applications from the above fields of pure mathematics. A second category are those theoretical mathematicians or computer scientists interested in the Galois theory of concepts; they may discover interesting unsolved problems. A third category of potential readers are all those who really want to get an idea of what music is about, of how one may conceptualize and turn into language the "ineffable" in music for the common language. Those who insist on the dogma that precision and beauty contradict each other, and that mathematics only produces tautologies and therefore must fail when aiming at substantial knowledge, should not read such a book.

Despite the technical character of *The Topos of Music*, there are at least four different approaches to its reading. To begin with, one may read it as a philosophical text, concentrating on the qualitative passages, surfing over technical portions and leaving those paragraphs to others. One may also take the book as a dictionary for computational musicology, including its concept framework and the lists of musical objects and processes (such as modulation degrees, contrapuntal steps) in the appendices. Observe however, that not all existing important lists have been included. For example, the list of all-interval series and the list of self-addressed chords are omitted, the reader may find these lists in other publications. Thirdly, the working scientist will have to read the full-fledged technicalities. And last, but not least, one may take the book as a source for ideas of how to go on with the whole subject of music. The GPL (General Public License[2]) software sources in the appended CD-ROM may support further development.

The prerequisites to a more in-depth reading of this book are these. Generally speaking, a good acquaintance with formal reasoning as mathematics (including formal logic) preconizes, is a conditio sine qua non. As to musicology and music theory, the familiarity with elementary concepts, like chords, motives, rhythm, and also musical notation, as well as a real interest in understanding music and not simply (ab)using it, are recommended. For the more computer-oriented passages, familiarity with the paradigm of object-oriented programming is profitable. We have not included the appendix on mathematical basics because it should help the reader get familiar with mathematics, but as an orientation in fields where the specialized mathematician possibly needs a specification of concepts and notation. The appendix was also included to expose the spectrum of mathematics which is needed to tackle the formal problems of computational musicology. It is by no means an overkill of mathematization: We have even omitted some non-trivial fields, such as statistics or Lambda calculus, for which we have to apologize.

There are different supporting instances to facilitate orientation in this book. To begin with, the table of contents and an extensive subject and name index may help find one's key-words. Further, following the list of contents, a leitfaden (on page xlix) is included for a generic navigation. Each chapter and section is headed by a summary that offers a first orientation about specific contents. Finally, the book is also available as a file `ToposOfMusic.pdf` with bookmarks and active cross-references in the appended CD-ROM (see page li for its contents). This version is also attractive because the figures' colors are visible only in this version.

In order to obtain a consistent first reading, we recommend chapters 1 to 5, and then appendix A: Common Parameter Spaces (appendix B is not mandatory here, though it gives a good and not so technical overview of auditory physiology). After that, the reader may go on with chapter 6 on denotators and then follow the outline of the leitfaden (see page xlix).

This book could not have been realized without the engaged support of nineteen collaborators and contributors. Above all, my PhD students Stefan Göller and Stefan Müller at the MultiMedia Laboratory of the Department of Information Technology at the University of Zurich have collaborated in the production of this book on the levels of the LaTeX installation, the final production of hundreds of figures, and the contributions sections 20.2 through 20.5 (Göller) and sections 47.3 through 47.3.6.2 (Müller). My special gratitude goes to their truly collaborative spirit.

Contributions to this book have been delivered by (in alphabetic order): By Carlos Agon, and Gérard Assayag (both IRCAM) with their precious Lambda-calculus-oriented presentation of the object-oriented programming principles in the composition software OpenMusic described in chapter 52, Moreno Andreatta (IRCAM) with an elucidating discourse on the American Set Theory in section 11.5.2 and section 16.3, Jan Beran (Universität Konstanz) with his contribution to the compositional strategies in his original composition [103] in section 11.5.1.1, as well as with his inspiring work on statistics as reported in chapters 43 and 44, Chantal Buteau (Universität and ETH Zürich) with her detailed review of chapter 22, Roberto Ferretti (ETH Zürich) with his progressive contributions to the algebraic geometry of inverse performance theory in sections 39.8 and 47.2, Anja Fleischer (Technische Universität Berlin) with her short but critical preliminaries in chapter 23, Harald Fripertinger (Universität Graz) with his 'killer' formulas concerning enumeration of finite local and global compositions in sections 11.4, 16.2.2 and appendix C.3.6, Jörg Garbers (Technische Universität Berlin) with his portation of the RUBATO® application to Mac OS X, as documented in the screenshots in chapters 40, 41, Werner Hemmert (Infineon) with a very up-to-date presentation of room acoustics in section A.1.1.1 and auditory physiology in appendix B.1 (we would have loved to include more of his knowledge), Michael Leyton (DIMACS, Rutgers University) with a formidable cover figure entitled "Dark Theory", a beautiful subtitle to this book, as well as with innumerable discussions around time and its reduction to symmetries as presented in chapter 48, Emilio Lluis Puebla (UNAM, Mexico City)

[2] A legal matter file is contained in the book's CD-ROM, see page li.

with his unique and engaged promotion and dissipation of mathematical music theory on the American continent, especially also in the preparation and critical review of this book, Mariana Montiel Hernandez (UNAM, Mexico City) with her critical review of the theory of circular forms and denotators in section 6.5 and appendix G.2.2.1, Thomas Noll (Technische Universität Berlin) with his substantial contributions to the functorial theory of compositions, and for his revolutionary rebuilding of Riemann's harmony and its relations to counterpoint, Joachim Stange-Elbe (Universität Osnabrück) with a very clear and innovative description of his outstanding RUBATO® performance of Bach's contrapunctus III in the *Art of Fugue* in sections 42.2 through 42.4.3, Hans Straub with his adventurous extensions of classical cadence theory in section 26.2.2 and his classification of four-element motives in appendix O.4, and, last but not least, Oliver Zahorka (Out Media Design), my former collaborator and chief programmer of the NeXT RUBATO® application, which has contributed so much to the success of the Zürich school of performance theory. To all of them, I owe my deepest gratitude and recognition for their sweat and tears.

My sincere acknowledgments go to Alexander Grothendieck, whose encouraging letters and, no doubt, awe inspiring revolution in mathematical thinking has given me so much in isolated phases of this enterprise. My acknowledgments also go to my engaged mentor Peter Stucki, director of the MultiMedia Laboratory of the Department of Information Technology at the University of Zurich; without his support, this book would have seen its birthday years later, if ever. My thanks also go to my brother Silvio, who once again (he did it already for my first book [670]) supported the final review efforts by an ideal environment in his villa in Vulpera. My thanks also go to the unbureaucratic management of the book's production by Birkhäuser's lector Thomas Hempfling and the very patient copy editor Edwin Beschler. All these beautiful supports would have failed without my wife Christina's infinite understanding and vital environment—if this book is a trace of humanity, it is also, and strongly, hers.

Vulpera, June 2002

Guerino Mazzola

Volume III Contents

Part XV Gesture Philosophy for Music

Part XVII Concept Architectures and Software for Gesture Theory

Part XIX Gestures in Music and Performance Theory, and in Ethnomusicology

Book Set Contents

Part I Introduction and Orientation

Part II Navigation on Concept Spaces

Part III Local Theory

Part IV Global Theory

Part XV Gesture Philosophy for Music

Part XVII Concept Architectures and Software for Gesture Theory

Part XIX Gestures in Music and Performance Theory, and in Ethnomusicology

Leitfaden

Leitfaden I & II

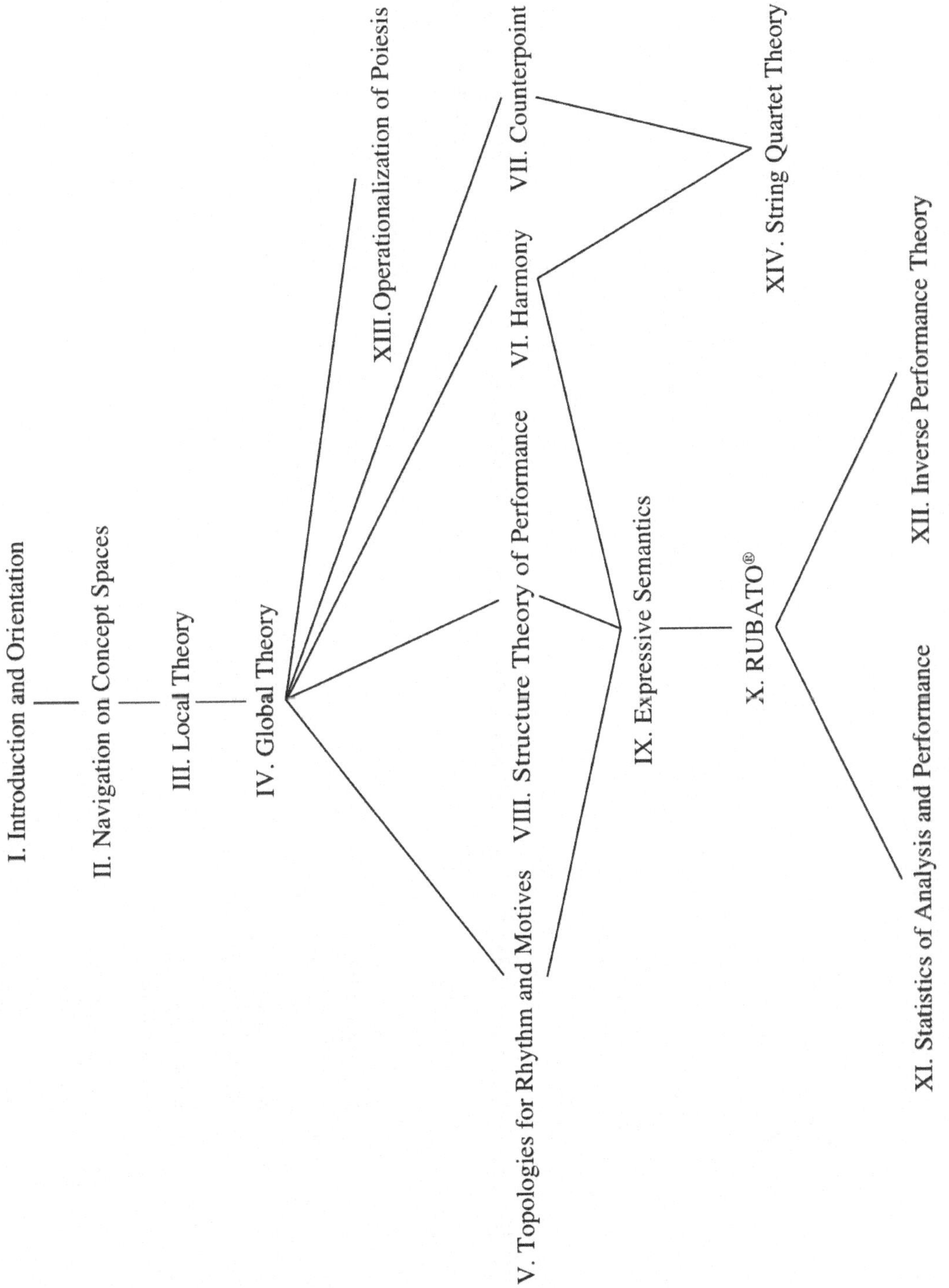

Leitfaden III

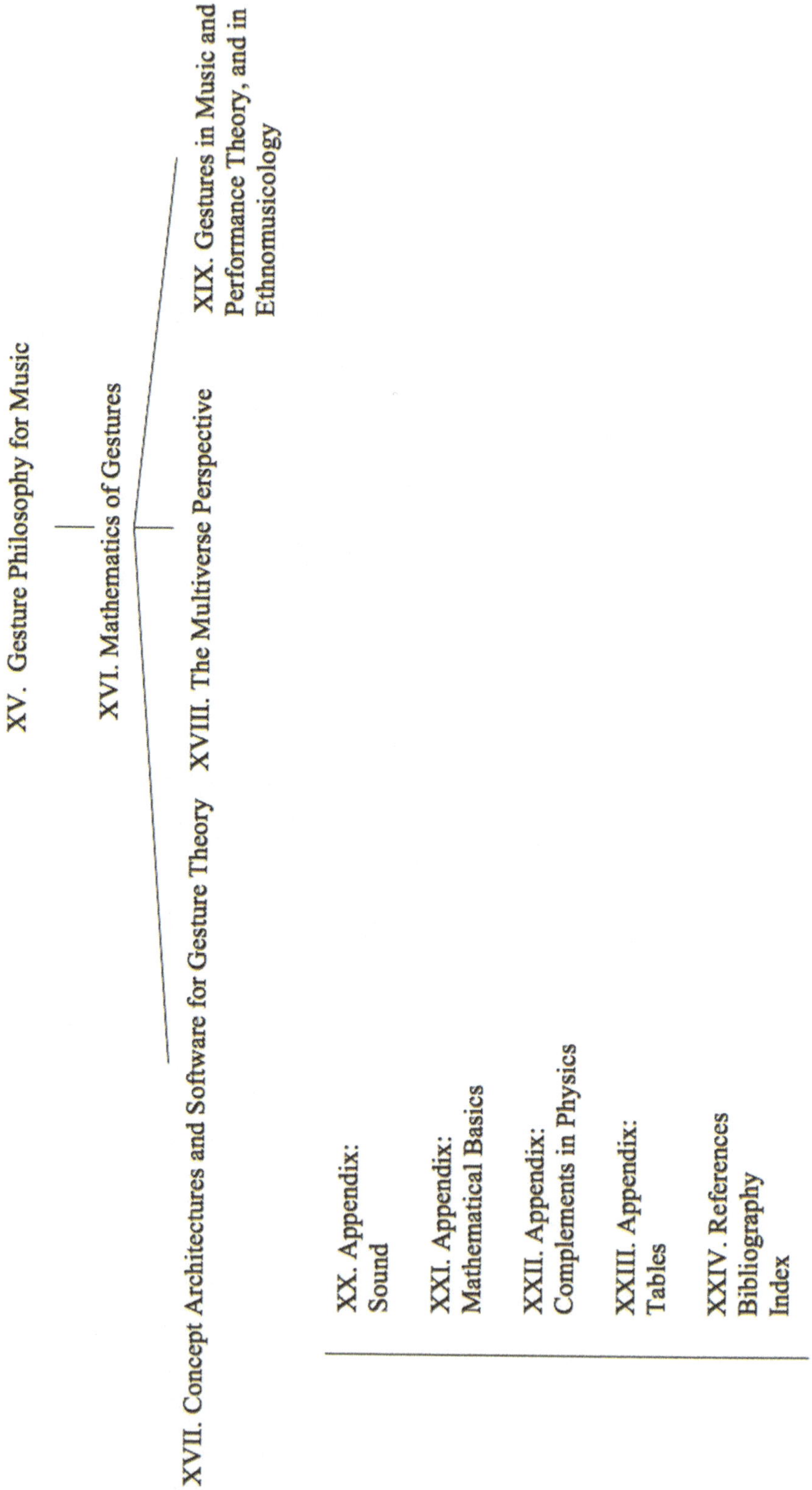

ToM_CD

- ToM_CD
  - ToposOfMusic.pdf
  - legal
    - GPL.pdf
    - README.txt
  - software
    - presto
      - documentation
        - hbdisk
        - README.txt
      - examples
      - music
        - czerny_chopin
        - mystery_child
        - synthesis
          - audio_files
          - presto_files
      - programs
        - prestino.prg
        - presto
        - README.txt
      - sources
        - README.txt
        - source1
        - source2
    - rubato
      - macrubato
        - examples
        - program
          - rubato
          - README.txt
        - source
      - nextrubato
        - documentation
        - source
      - performances
        - contrapunctus_III
          - audio_files
          - stemmata
        - kuriose_geschichte
        - traeumerei
          - alptraeumerei
          - takefive

Part XV

Gesture Philosophy for Music

56

The Topos of Gestures

Summary. This is the third[1] part of *The Topos of Music* and deals with gestures. We summarize the trajectory gestures took from the first edition of *The Topos of Music* to the present second edition.

– Σ –

As already mentioned in the preface to the second edition of this book, work on the first edition in 2002 had ended when its author discovered (in the context of a talk he gave on the 18th of May at IRCAM about his improvisational technique and its relationship to mathematical music theory) that gestures were as essential if not more prominent than abstract formulas for his improvisational practice.

At first this was a considerable shock in view of the imminent publication of *The Topos of Music*. But it soon turned out to define the very power of Mazzola's research in future years. The third part of this book in its second edition, fifteen years after the first edition, is the present state of the art, an art that is far from completed, to be clear.

The role of gesture theory can be understood from its ontological status. We refer to Section 57.1 for a more detailed discussion of musical ontology with gestures. Ontology of music answers the questions "Where?" (realities), "Why?" (semiotics), and "How?" (communication). This was discussed in Chapter 2.

With gestures, we add the answer to the question: "how does music *come into* being?" It is in fact not true that music is just a collection of facts (done things), but it is strongly focused on the action of making, be it the performance of a score, the improvisation in jazz or the free setting in contemporary creative musicking.

The topos of gestures is not thought of as a topos in Grothendieck's understanding, although the study of gestures involves a number of mathematical topoi. The title, above all, focuses on the second understanding of "topos", namely the Kantian conceptual localization[2] of music in its gestural unfolding. It is a substantial extension of the concept of music from facts to gestures, from its static ontology to an ontology of the making. In this sense, we are far from a terminal theory of gestures, and this is good: otherwise the ontology of making would be contradicting itself.

[1] The first part is theory, while the second is performance.

[2] See our catchword on page ix.

G. Mazzola et al., *The Topos of Music III: Gestures*, Computational Music Science,
https://doi.org/10.1007/978-3-319-64481-3_1

57

Gesture Philosophy: Phenomenology, Ontology, and Semiotics

57.1 A Short Recapitulation of Musical Ontology

Summary. This short section recapitulates the global architecture of the ontology of music.

– Σ –

Ontology of music includes three dimensions: realities, semiotics, and communiction. It also includes the extension of ontology to the fourth dimension of embodiment. We call this extension "oniontology" for reasons that will become evident soon.

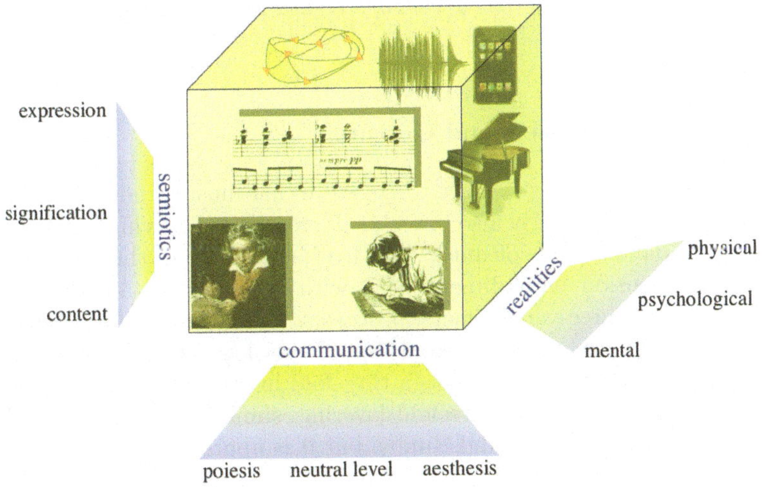

Fig. 57.1. The three-dimensional cube of musical ontology.

57.1.1 Ontology: Where, Why, and How

Ontology is the science of being. We are therefore discussing the ways of being which are shared by music. See Chapter 2 for a detailed discussion. As shown in Figure 57.1, we view musical being as spanned by three 'dimensions', i.e., fundamental ways of being. The first one is the dimension of realities. Music has a threefold articulated reality: physics, psychology, and mentality. Mentality means that music has a symbolic reality that it shares with mathematics. This answers the question of "where" music exists.

G. Mazzola et al., *The Topos of Music III: Gestures*, Computational Music Science,
https://doi.org/10.1007/978-3-319-64481-3_2

The second dimension, semiotics, specifies that musical being is also one of meaningful expression. Music is also an expressive entity. This answers the question of "why" music is so important: it creates meaningful expressions, the signs which point to contents.

The third dimension, communication, stresses the fact that music exists also as a shared being between a sender (usually the composer or musician), the message (typically the composition), and the receiver (the audience). Musical communication answers the question of "how" music exists.

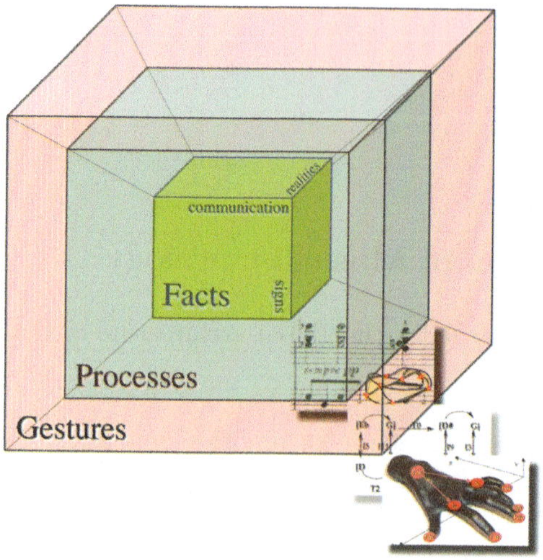

Fig. 57.2. The hypercube of musical oniontology.

57.1.2 Oniontology: Facts, Processes, and Gestures

Beyond the three dimensions of ontology, we have to be aware that music is not only a being that is built from facts, finished results; no, music is strongly also processual, creative, and living in the very making of sounds. Musical performance is a typical essence of music that lives, especially in the realm of improvisation, while being created. The fourth dimension, embodiment, deals with this aspect, it answers the question "how *come into* being?" It is articulated in three values: facts, processes, and gestures. This fourth dimension gives the cube of the three ontological dimensions a threefold aspect: ontology of facts, of processes, and of gestures. This four-dimensional display can be visualized as a threefold imbrication of the ontological cube, and this, as shown in Figure 57.2, turns out to be a threefold layering, similar to an onion's. This is the reason why we coined this structure "oniontology"—sounds funny, but it is appropriate terminology.

57.2 Jean-Claude Schmitt's Historiographic and Philosophical Treatise "La raison des gestes dans l'Occident médiéval"

"La raison des gestes dans l'Occident médiéval" was published in 1990 [946]. It is the most complete and important contribution to the history of the concept, philosophy, and social and religious roles of gestures during the early centuries of our modern Western culture. It starts with a summary of the ancient Greek and Roman traditions: Plato, Aristotle, Cicero, and Quintilian, and then draws a trajectory through the Middle Ages, starting with the early writings of Martianus Capella (between 410 and 470), culminating in the sophisticated and detailed writings of Paris-based theologist Hugues de Saint-Victor (1096-1141), and terminating with a detailed discussion of the transition of the Christian culture during the twelfth and thirteenth centuries to an "intellectual Renaissance" where new technologies, the new medical paradigm of surgery, the first universities, and the rediscovery of ancient traditions generate new perspectives on the phenomenon of gestures.

Before we discuss the most evolved definition of a gesture by Hugues de Saint-Victor, let us summarize the relation of gestures to the Christian religion and church in the Middle Ages as it is described and analyzed by Schmitt. As gestures are always related to the human body, their role is strongly related to the position of the body in the early Christian tradition. This position has two contradictory faces. On the one hand, our body is the "prison of the soul" and also occupies the role of a cause of sin and mean animalistic behavior. In this face, the body's gestures must be limited and controlled by the high moral and holy principles of the Christian soul that seeks its salvation in God. This perspective explains the repression of mimes, histrions and jugglers, and also of feminine body movements or more generally of any erotically colored gestural utterances. On the other hand, the central essence of Christian religion is described by the Eucharist, the transformation of bread and wine into Christ's body and blood. Here, the body signifies incarnation of a divine existence, and therefore it receives a totally different role, not sinful and animalistic, but realizing the embodiment of a divine revelation. In this second face, gestures, above all the priests' precisely canonized gestures when delivering the holy communion, but also in the ecstasy of mystical Christians, such as stigmatist Francis of Assisi, are understood as connectors that transcend verbal expressivity. According to Schmitt, what is common to these two faces is that gestures are (1) *expressive* in the sense that the relate the internal human moods or emotions or thoughts to the external reality of visible bodies, (2) *communicative*, i.e., they transfer contents as signs (Hugues de Saint-Victor calls them *signa*), and (3) *technical* in the sense that they ask us to make something.

It is remarkable that many modern approaches to a theory of gestures germinated in the Middle Ages (if not in Greek Antiquity), for example the association of gesture with language (the basis of Adam Kendon's and David McNeil's theories; see Section 57.7), the idea of music being more than sound, including gestures in its dancing expressivity, or more generally the theories of non-verbal communication.

Apart from the fact that Schmitt recognizes the Middle Ages as a "civilization of gestures", a fact that positions the project of this book in a long cultural tradition, it is interesting to discuss the concept of a gesture that has been defined in Hugues de Saint-Victor's *De institutione novitiorum*, written in Paris around 1140. In Chapter XI, he starts with a definition of a gesture:

Gestus est motus et figuratio membrorum corporis, ad omnem agenda et habendi modum.

Gesture is the movement and figuration of the body's limbs with an aim, but also according to the measure and modality proper to the achievement of all action and attitude.[1] That gesture is the human body's movement (motus) is in harmony with traditional, ancient musically and rhetorically motivated conceptualizations. But the specification of the movement as also the body's *figuratio* is new and specific to Hugues de Saint-Victor's setup. Schmitt interprets figuration as as the creation of a configuration of the body's members which express the soul's movements. We cannot follow this interpretation; it seems too strongly directed towards a semantic view (gestures are expressing internal contents of mood, etc.). We would prefer to simply see the configuration of the body's limbs in this characterization, meaning that a gesture is more than a general movement; it is the movement of an articulated whole, a combination of parts (the limbs) that relate to each other in a complex anatomic architecture. The second couple of properties: action and attitude, which are attributed to *modus*, translated as measure and modality by Katsman, are more difficult to understand. Katsman and also Schmitt use longer circumscriptions of *modus* in its relation to action and attitude. The Latin wording is very short and elegant, being reduced to the preposition *ad*, meaning that *modus* is *for*, *specified by* action and attitude. This means that the body's movement and (con)figuration are directed to a modality, which is specified by action and attitude.

57.2.1 Comments

We don't see the semantic valuation as explicitly as Schmitt does. It is a movement and figuration of the body with a specification, not a general movement, but carrying a modality. This modality is not a priori

[1] Following the translation by Roman Katsman [524]; see also `http://sites.utoronto.ca/tsq/12/katsman12.shtml`.

meaningful; it is an action and attitude, not more and not less. One could object that Hugues de Saint-Victor's comments about the nature of a gesture as a sign would suggest a semiotic specification in his definition. This seems plausible in the context of Hugues de Saint-Victor's Christian background: A gesture must be meaningful to comply with the general principle of meaningful life *sub specie aeternitatis divinae*; this is evidenced in Schmitt's concluding chapter, *L'Efficacité symbolique*: Everything is meaningful, if not magic, and incessantly observed by God. We would prefer not to load this definition too much with a semiotic scheme. This needn't mean that gestures in Hugues de Saint-Victor's understanding are a presemiotic concept, but it could mean that gestures not only are carriers of given contents but could be involved in the creation of contents, that they could be semantic generators. We shall come back to this question when discussing the modern French philosophy of gestures in Chapter 58.

57.3 Vilém Flusser's *Gesten: Versuch einer Phänomenologie*

57.3.1 A Short Introduction to Flusser's Essay

Flusser's German essay [328] was first published in 1991. The author focuses on a phenomenological approach, i.e., following his own words in the first chapter, a non-historical philosophical perspective, where freedom is not bound to linear time, but gets off ground without any preconception and describes or analyzes gestures in their singular specific shapes. These are the chapters: 1. Geste und Gestimmtheit (Gesture and Coherence) 2. Jenseits der Maschinen (Beyond Machines) 3. Die Geste des Schreibens (The Gesture of Writing) 4. Die Geste des Sprechens (The Gesture of Speaking) 5. Die Geste des Machens (The Gesture of Making) 6. Die Geste des Liebens (The Gesture of Loving) 7. Die Geste des Zerstörens (The Gesture of Destroying) 8. Die Geste des Malens (The Gesture of Painting) 9. Die Geste des Fotographierens (The Gesture of Photographing) 10. Die Geste des Filmens (The Gesture of Filming) 11. Die Geste des Maskenwendens (The Gesture of Mask Turn-over). The first chapter is an introduction to the general topic of gesture phenomenology. Here, Flusser gives a definition of the gesture concept, which is a thoroughly semiotic one: gestures have meaning. We will discuss this in the next section. He however stresses that there is no general theory of what he calls "interpretation of gestures", which means that there is no theory that would offer a scientific approach to what gestures mean. The keyword "Stimmigkeit" ("coherence") means a symbolic transformation of "Stimmung" ("mood"), an artistic transfiguration, using gestures, of the emotional atmosphere. The author is aware that this is a circular statement: Understanding gestures means using gestures to explain their transformational power of creating/giving meaning.

The second chapter is an attempt to position the gestural phenomenon within a triple characterization of how we work in this world: in the author's words, *ontology* describes the world how it is, *deontology* looks at how the world should be, and *methodology* deals with the ways we act to actually change the world. The remaining chapters discuss (but not exhaustively) a spectrum of gestural phenomena that the author considers important.

57.3.2 The Semiotic Neurosis

Flusser's phenomenology of gestures is one big conjuration of the concept of a gesture. His definition is essentially negative: Freedom, that part that cannot be satisfied in the causal determination of the human body's intentional movements and its associated tools. He searches desperately for the meaning of that movement. For, what is meaningless cannot be understood. This is the semiotic trap which has been avoided by Châtelet, Alunni and their French associates.

My mathematical music theory is also pre- or a-semiotic. The musical realization of symmetries, groups, gestures, etc. is not a preliminary form of meaning. Remember Hanslick [438]: "The content of music are sounding moved forms." What would be the meaning of a dodecahedron, a sphere, a fractal geometry? This is it: Nothing a priori, as much as a pirouette of a dancer.

The semiotic neurosis is virulent. I (Mazzola) had described that already in my novel fragment "Das Geschlecht des Klaviers" ("The Gender of the Piano"). And in that artistic Harakiri jazz performance where content and expression were provocatively exchanged.

Meaning is always a *alquid pro aliquo*, a replacement action. This should not happen ad infinitum (Hegel's bad infinity where $1 \mapsto \ldots n \mapsto n+1 \mapsto n+2 \ldots$). To avoid this pathology (if one does not want to end up in meaningless dead ends), the only solution is to introduce circularity: The system of signs is autoreferential; it has, as a directed graphical structure, cycles. Using this strategy, a semiotic can work perfectly, eventually coming back to itself. The question of the function of circularity however remains; to state it in a provocative style: What is the meaning of circularity, apart from the fact that it eliminates bad infinity?

Let us first focus on circularity as a special case of a reference, and let us say "pointer" instead of "reference", since this concept describes better the activity of pointing that generates the referential relation. In circularity, the pointer action remains valid. But the target of a pointer is the same as the starting instance: $x \mapsto x$. It is action or movement, without moving anything; nothing happens to x. It is a conceptual pirouette, turns around itself. The autoreference, the self-pointer unmasks exactly the concept of a pointer since it is, in the circularity, a pointing without any effect. Is a gesture this pointing without considering any visible effect on its object?

Let us have a look at Frege's interpretation of the relation of circular pointing. Here it is nothing but the ordered pair (x, x). For Frege, the reference is reduced to the empty relation of a formal juxtaposition x to x. This is what already Châtelet had recognized. But in our case, the Fregean emptiness is also a conceptual one: One ends up in the not-understanding if the pointing movement is not embedded in the mathematical formalism. The pirouette of thought is a thought of the pirouette. The thought of a pirouette is a pirouette of the thought, same altogether.

Once the related components (*relata*) are removed, what remains is the plain pointing movement, the elementary finger pointing described by Tommaso Campanella! And this one is a "pure" gesture, the pointer qua pointing, without de Saussure's *signifiant* and *signifié*.

This is why a gesture is presemiotic: It is a condition for the very concept of a sign. The pointing action is that part of a sign which is preconceived as a gesture. Signs are compound concepts.

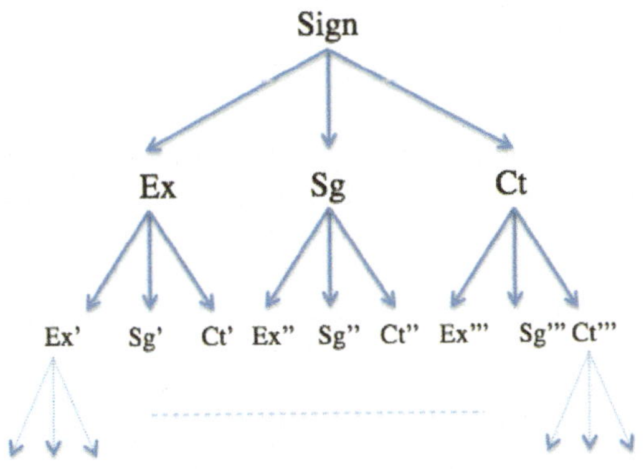

Fig. 57.3. The threefold ramified Hjelmslev sign gesture.

What is the role of Hjelmslev's construction here? His idea is the following: Instead of explaining the three parts of a sign, expression Ex, signification Sg, and content Ct, Hjelmslev conceives all three parts as pointers to the same concept; they are signs by themselves, i.e., $Ex \mapsto (Ex', Sg', Ct')$, and so on. It is a threefold ramified gesture; see Figure 57.3.

Hjelmslev's idea is pure gesturality. He replaces de Saussure's inhomogeneous conceptuality by the homogeneous threefold pointing structure. This is utterly elegant, but it simultaneously positions gesturality at the first position. *His conceptual substance of a sign is a threefold ramified gesture.*

Conclusion: The semiotic concept of a sign needs the preconception of a gesture, in Hjelmslev's homogeneous approach even the simple gesture defined by a threefold ramification.

Remark 1. It is coherent with the French school (specifically with regard to Alunni) that gestures are generators of meaning since the gesture concept is fundamentally responsible for the sign concept (not only for examples of signs!). Gestures in first place enable the structure of signification, of the pointer to meaning.

This is a giant step: Conceptual mathematics above all needs the mathematical theory of gestures. I believe that Grothendieck took exactly this giant step: Moving away from 20th century mathematics of structures (after the 19th century mathematics of objects) to the 21st century mathematics of concepts. Motives, scheme, etc. are just excellent examples of this conceptual research.

57.4 Michel Guérin's *philosophie des gestes*

57.4.1 The Essay's Structure

This is a typical French non-semiotic approach; however it is similar in its phenomenological style to Flusser's essay. Guérin discusses four gesture types which he considers to be elementary, i.e., the simplest ones and a complete set of such elements. They are: Faire (make), don(ner) (donate, gift), écrire (write), danser (dance). In a second part of the essay, Guérin summarizes these four aspects and observes (p.79) that in occidental thinking, the thought of a gesture has never occurred; he also refers to Bergson for this.

57.4.2 Gestural Ontology and Four Elementary Gestures

The author does not give a definition of a gesture, but exhibits a number of characteristics of that concept, e.g., (p. 13) as a relation between body and mind, (p. 32) the gesture informs by its deformation; it desires a remote form and caresses a near one; (p. 35) when matter is reduced to formulas, the gesture has no countenance (*prise*) anymore; (p. 45) the gesture's essence is its contact; (p. 72) the gesture of dance is an expression that expresses nothing; (p. 76) dance touches the sacred; in its pure form it has no intention; (p. 80) gesture is ambiguous between history and project; (p. 82) the sense of gesture is found in the dialogical relation of mode and function; (p. 84) the work (l'oeuvre) is the perfect circle of an exploited gesture (I would also refer to Mallarmé's *le livre* here); (p. 95) at the beginning is the gesture (auto-motion); (p. 115) the first art is dancing the gesture; (p. 129) thinking is the action of the presence, the presence of the action.

In the concluding second part (added later to the first part) he reconsiders the "quadrature" (the four elementary gesture types) and adds the ontology of presence in the dancing thought (following Nietzsche), as well as its mirror nature (the same but turned around), not really well explicated; I know better mirror analyses in my own work. He then adds the "Figure" concept following Rilke, in that we think in figures, the forms of gestures. Gesture has finality without end (*finalité sans fin*), and gesture is finiteness of the circular loop foot-hand. It generates space from this circularity, it generates every thought, the announcement, the angel's appearance before any content is transferred. On page 108, a mathematical allusion is made: *le geste est son proper "mathème", il est coextensive à son apprendre; il s'enseigne lui-même.*

Although, as a substitute for a definition, these characteristics are important, it is not clear whether they are generic or only valid for their specific gestural types. The layer of reality of these gestures is not specified. Is writing a physical gesture? Or a mixed reality? Dancing seems physical, but it is evidently also highly symbolic, etc.

In particular, the last example: the gesture of thought, is that a symbolic gesture? Then, in what presence would it happen? Here the question of its reality/realities is particularly delicate: What is the relation of a mental gesture's reality to the reality of thoughts in general? This is very important since gestures of thought are considered as being basic.

Also the blurring of biological, anthropological, and symbolic levels of "realities"/perspectives is confusing, in particular when they are (ab)used to generate evidence.

The discourse about the gesture of writing is badly written. Also full of those very French word games of alliteration type, such as *La lettre tue parce qu'elle se taît.* Especially in view of the fact that all those negative statements about the separating role of writing are *written.*

It is further not clear why the author specifies the four types as gestures. While this is clear for dancing, it is not clear for donation. Not every action is a gesture. We are not informed why he qualifies them as gestures, even if one agrees intuitively. This is the typical philosophical style: abstract and imprecise.

Many of the above characteristics are mysterious, for example, the above *il s'enseigne lui-même*, what does that mean? How does such a self-instruction work? Is there a subject for that? Etc.

Finally, there is absolutely no structure theory. Everything is utterly fuzzy.

57.5 Flusser and Guérin: Some Consequences

From the two contributions we draw these conclusions:

- The ontology of gestures is a proper ontological topic of philosophy.
- Gestures are dialogical, live in presence, are circular, elastic, are presemiotic, and are as such already differentiated (being gestural can be ramified into different types).

The big question would be this one: If this differentiation is not a semiotic one, i.e., "writing" is not the meaning of such a gesture, what else can it be?

We are asked to develop an anatomy of gesturality, and to do so, we need a structure theory. Evidently, we can describe the structure of natural gestures, such as writing, using the mathematical theory of gestures. And this for the entire quadrature. But for the gesture of thinking, I am not so sure, except if the general mathematical theory of gestures (over topological categories) would be taken as a candidate of a general theory of gestural thinking. This would then be Guérin's *mathème.*

Of course, the category-theoretical theory of gestures would not be a philosophical first movement since it is embedded in mathematical concept architectures. But we don't ask for such a philosophically foundational theory. We have to be very careful here in the interpretation of the musical ontology. It is not claimed that one of its dimensions, or even dimensional values, can be reduced to others. Gestures can also be communicative, and specific in their realities. It is only the semiotic dimension that would not be specific in gestural perspectives. Which does not mean that this dimension is absent altogether, but it can and will be near to vanishing.

Philosophy confirms the necessity of an ontological dimension of embodiment. It also, in principle, confirms the necessity of a structural theory of gestures, and let us call this the anatomy. But this anatomy will not presuppose semiotics as an essential conceptual ingredient, only—if at all—as a technical tool. Mathematics would be such a tool.

We could call this a height structure, as opposed to a depth structure. We should liberate ourselves from the tyranny of the vertical: things might be fundamental on top, not down there. In circular conceptual architectures, every locus without which other loci fall apart would be called "essential". Are gestures essential? Their *mathème* is, there is no valid *theory* of gestures (not only a philosophy) without a mathematical structure theory. The claim of a *mathème* is empty if it isn't made explicit as a mathematical building. This necessity is what we have to explicate.

Evidently, many philosophers and, helas, also music theorists and computer scientists, maintain that a definition of "gesture" and a corresponding structure theory are superfluous. However, physics teaches us that this is erroneous. The physical concept of time as a real (or complex) coordinate has a trivial ontology. But its behavior under transformations is characteristic for the comprehension of its ontology. Ontology is above all reified in the concept's behavior, not in its static Kantian "Ding an sich". This is Yoneda: What time is as such is irrelevant. It is how it behaves (as a functor). Being as a "so-sein" (suchness), not only "da-sein" (being here, existing). This is also essential for gestures. Abstract existence is irrelevant and may be replaced by suchness. Better: Existence is the locus in the topos and this is the isomorphism class of the functor, i.e., Yoneda is also a theorem about the replacement of existence by suchness. In fact, all characteristics are verbs, suchness (answering *How is the gesture behaving?*), never existence (except perhaps presence, that prae-esse, being in front of you). The later contributions by Châtelet are not present here, but they will be in complete coherence with the present ones.

Ontology is no longer fundamental now as a first existence, but is as a first suchness. How gestures are manifested, their suchness, this is where they are (exist) and what they are. Existence survives as suchness.

Remark 2. The philosophical part should not try to use the fundamental role of gestures as an ontological first thing, but only (1) stress its independence from other ontological dimensions, (2) elaborate their specific attributes, and (3) explicate the program to work these attributes out, a thing that may fail in parts.

57.6 A Program

Here is our proposed program:

1. Gestures are presemiotic. This has been described above.
2. They are non-essential. Their essence is contact. They are dialogic (see there!).
3. Gestures are dialogic. They are neither subjective nor objective, but grounded in the second person (you); see Benveniste for this. They are also formally this irreducible movement $I \to X$.
4. They are auto-motion. This follows from the dialogical explanation above.
5. They are elastic. This is covered by the hypergesture concept which generalizes homotopy.
6. They are an interface between body and mind. This is covered by our discussion of cognitive aspects, in particular the action-body, mirror neurons, Yoneda-on-the-body, etc.
7. They include a *mathème*, a self-instructing mathematical structure. This is the mathematical theory of gestures.
8. They are *in presence*. They might be covered by the flow theory developed in the free jazz discourse.
9. Circularity of gestures. This is a delicate topic, to be discussed below.

57.6.1 Circularity

Here is a model of how gestures could involve circularity on the conceptual level. Circularity is not covered by *finalité sans fin* and similar attributes. They are merely word games.

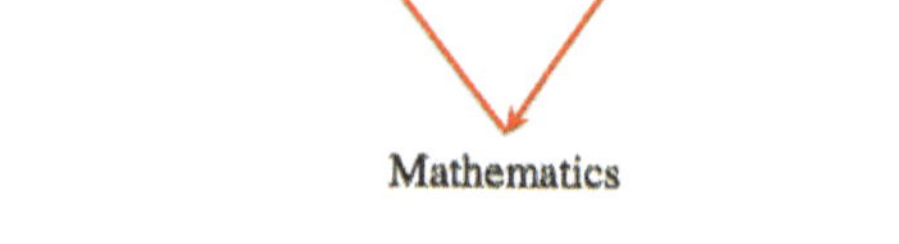

Fig. 57.4. The circularity of gestures and their mathematical theory.

We had seen that the presemiotic character of gestures is evidenced in Hjelmslev's idea of purely gestural description of the sign structure; see Figure 57.3, where the threefold non-terminating gestural skeleton defines the structure of a sign. If we understand the mathematical theory of gestures as an outcome of the semiotic system of mathematics, semiotics is induced by the concept of a gesture, but induces the mathematical structure theory of gestures (their *mathème*); see Figure 57.4. But the latter is the precise framework to yield a valid definition of a gesture. This is a type of essential circularity. None of its points can be omitted without a loss of other points.

When arguing about gestures, one may start in an adequate point of the circular cycle and then proceed to the next point without claiming that any of these points be more fundamental than others. It is an operational discourse, not an essentialistic one.

57.7 The Semiotic Gesture Concept of Adam Kendon and David McNeill

In this section we refer to an important Anglo-Saxon tradition of gesture theory. We discuss the two well-known theories of Adam Kendon and David McNeill; see Figure 57.5. We focus on the semiotic aspect of these theories; psychology as such is not our focus here.

Fig. 57.5. Adam Kendon (left) and David McNeill.

We first refer to *Adam Kendon's* book "Gesture—Visible Action as Utterance" [530]. He first discusses possible definitions of a gesture, then recapitulates historical contributions from Classical Antiquity, then from the nineteenth century, and finally from the twentieth century. Next he discusses classification issues and gesture units and phrases. He then thematizes semantic issues. Discussions of pointing and other more specific gesture types follow. The book—which is a rich source of examples about speech-related gestures—terminates with a discussion of gestures without speech (when they replace words), gestures and sign language, and a summary of the state of the art. Similarly to David McNeill's approach, Kendon's is a theory of language (in the broader sense, including non-verbal systems; we come back to this in the following discussion of McNeill's work). He stresses the "very intimate way in which gesture is integrated with speech." And he concludes: "For a truly inclusive view of human language, gesture must be taken into account."

We want to focus on Kendon's attempts to define gesture, rather than summarize the book, since his approach to this concept is, together with McNeill's theory, a very special understanding of the phenomenon of gestures.

The second chapter, "Visible action as gesture," is devoted to the problem of defining gestures. The author gets off the ground with suggesting that "'Gesture' is a term for visible action when it is used as an utterance or as a part of an utterance." He then asks what 'utterance' is and "how actions in this domain are recognized as playing a part in it." His first clarification refers to Ervin Goffman's definition [375], namely that utterance is "any ensemble of action that counts for others as an attempt by the actor to 'give' information of some sort." This generality is however not what Kendon will use; it is also too general since anything can be interpreted as information. Such a definition would be abstract nonsense. Kendon then cites the *Oxford English Dictionary* (2nd edition 1989) where 'gesture' is defined as "a movement of the body, or any part of it, that is expressive of thought or feeling." After an analysis of the intentionality in utterances, he concludes: "'Gesture' we suggest, then, is a label for actions that have the features of manifest deliberate expressiveness." The critical concept is 'expressiveness', but Kendon makes this clear in his comment following that definition: "The more a movement shares these features (manifest deliberate expressiveness, G.M.), the more likely it is to be given privileged status in the attention of another and to be seen as part of the individual's effort to convey meaning." So expressiveness is about conveying meaning. This is an unequivocally semiotic statement: Gestures are expressing meaning, they are signs (although complex ones) that communicate meaning. The information in Goffman's definition is made precise as semantic information. In other words: gestures are signs in the sense of semiotics. And they are always related to the body's actions, no more abstract concept of a gesture is addressed. This will be confirmed by the second gesture theorist of this school.

Our reference to *David McNeill* is due to his book "Gesture & Thought" [741]. In the introduction, he defines the general scope of his theory: "Now [the emphasis] is how gestures fuel thought and speech.(...)

Gesture, language, and thought are seen as different cognitive, and ultimately biological levels. (...) Gestures are active participants in speaking and thinking." This makes clear that McNeill is restricting his gesture theory to gestures that are in different ways coexisting with language; this is also the reason he uses the concept of "language" in two ways: first as the linguistic phenomenon, and then in a larger sense—we might say: in a semiotic sense—as a system where the linguistic part of language is joined by the gestural part of language. Let us denote the language concept that includes gestures by "Language". Referring to neuroscientist Antonio Damasio, McNeill states that "language is inseparable from imagery", and that the imagery is covered by gestures. The connection between these two components is existential, and also, language evolution could not have happened without gestures: "To treat gestures in isolation from speech makes no more sense than to read a book by only looking at the 'g's." In this sense, McNeill takes over from Kendon what he calls the "Kendon continuum", namely a sequence of roles that gestures can play in relation to the linguistic language part; see Figure 57.6. To the extreme left, we have gesticulation, where gestures are completely secondary with respect to language, and to the extreme right, we have the case of sign language, where gestures are given the structure of a full-fledged linguistic semiotics.

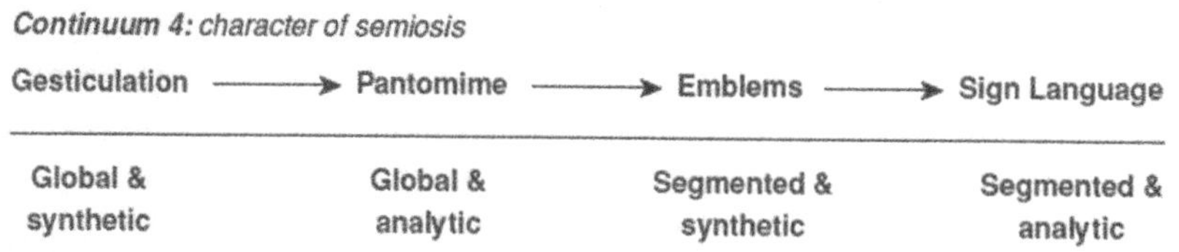

Fig. 57.6. The semiotic aspect of Kendon's continuum.

What is essential in this approach is that gestures in McNeill's theory always "carry meaning", they are in a general semiotic sense signs. Accordingly, this language enables synonymous signs: a gesture might have a word that points to the same meaning, McNeill calls such words "lexical affiliates". Another semiotic property of gestures is what Karl-Erik McCullogh calls "semiotic components". These are meaning-bearing parts of gestures, of the hand's orientation for example. More generally, gestures here are also given an anatomy: They unfold in time, and define groups of gestural movements that McNeill calls "phrases", namely gestural units that have a complete inner structure: the sequence of phases: from *preparation* to *prestroke* to *stroke* to *stroke hold* to *poststroke* and to *retraction.* Gestures are also given two viewpoints: the third-person *observer* viewpoint, where a gesture presents an entity in a narration, and the first-person *character* viewpoint, which represents the speaker who is "inside the gesture space".

McNeill stresses a different structure of the semiosis in his theory. He claims that the Saussurean semiotics is a static one, which needs a complementary view of a "dynamic" semiosis, i.e., one where meaning is not a fact, but a process, following the ideas of Russian psychologist Lev Semyonovich Vygotsky [1100]. For Vygotsky, language and thought are two overlapping instances, and neither of them contains the other. This contribution is enriched by the insights of Maurice Merleau-Ponty that language is not the thought's dress, but its body. The semiosis for McNeill is given an ontology of an existential action, with a reference to Martin Heidegger's understanding of language as a "house of being", not a detached formal representation of an object.

The concrete implication of these ideas culminates in what McNeill calls the "Growth Point" (GP). From McNeill's home page we read: "A growth point (GP) is a dynamic unit of online verbal thinking. It is sui generis—not the same as units of synchronic linguistic description (words, morphemes, etc.). The GP combines unlike modes of cognition imagery and linguistic categorial content.(...) It does not exist at all times, and comes into being at some specific moment; the formation of a growth point is this moment, theoretically, and it is made visible in the onset of the gesture." The GP formalizes the interaction of image (gesture) and linguistic code.

57.7.1 Comments

The theories of Kendon and McNeill are above all theories of gestures that are intimately related to language. Moreover, gestures are conceived as bodily utterances; gestures in another context, such as melodic gestures in music or dance gestures of ice skaters, are not addressed. Moreover, the expressivity of gestures as they are defined in this school is a semiotic one: gestures express meaning, they are signs in a semiotic system of a generalized language (including gestures as well as linguistic signs).

57.8 Juhani Pallasmaa and André Chastel: The Thinking Hand in Architecture and the Arts

Juhani Pallasmaa's beautiful book "The Thinking Hand" [814] deals with "the essence of the hand and its seminal role in the evolution of human skills, intelligence and conceptual capacities." And the rationale for this program is that "the hand has its own intentionality, knowledge and skills." Pallasmaa, a famous Finnish architect and writer, evidently focuses on topics that pertain to the culture of architecture, specifically (and these are the chapter titles): the mysterious hand, the working hand, eye-hand-mind fusion, the drawing hand, embodied thinking, body self and mind, emotion and imagination, and, finally, theory and life.

Our interest in this book, and also the following "Le geste dans l'art" by the famous French art historian André Chastel, lies in their focus on the hand as a central gestural instance. This does not mean that symbolic intellectual human skill must stem from the hand, as some anthropologists (such as neurologist Frank R. Wilson [1135]) argue, or as Anaxagoras claimed that humans are intelligent because they have hands. But it is evident that the hand and the brain are deeply interconnected, in the sense that the hand is neurologically and physiologically distributed all over the nervous and muscular systems. Pallasmaa makes an attempt to define the hand and concludes from these facts that "the hand is fundamentally beyond definability." The problem of such a definition seems to be less the anatomy or physiology than the understanding of the hand's functionality, and the latter is obviously related to the gestures hands are performing. We shall see in the next Section 57.9 that hand gestures with their complex ambiguity of sense production are prominent in literary contexts. The potentiality of such gestures is also the key to the idea of the hand gestures being "the tongue and general language of Human Nature, which, without teaching, man in all regions of the habitable world do at the sight most easily understand," as the English physician John Bulwer writes in 1644 [158]. We may anticipate here that it will be shown that gestures in the mathematical sense of the word (to be defined in Section 61.5) in fact can give rise to symbolic objects, such as abstract mathematical groups, the key word being the *fundamental group* of a topological space; see Section 78.2.10.2.

The non-definability of the hand's dynamic potential is not only a given fact, it is also the consequence of the common lack of language power in the discourse about gestures. The very concept of a gesture is extremely demanding, and this may also be one reason why there is no gesture theory in music: it is simply beyond the language power of scientists educated in the humanities to be able to even describe a gesture, and in particular a hand's gesture with its multiple finger configuration, and its complex spatio-temporal unfolding. That the hand's gestures are so central in Pallasmaa's approach is also important in view of the theory of embodiment which he sets forth. It is evident from his arguments that the hand's gestures (and not only its anatomy and physiology) must play the role of a link between the body and the mind. This is a crucial statement for the ontology of gestures. They could offer the missing link in the cartesian divide, a link that is also still missing in the embodied AI approaches, which believe in emergent properties of (even cheap) embodied design.

Chastel's book [188] is a collection of three published papers: "L'art du geste à la Renaissance", "Sémantique de l'index", and "Signum harpocraticum". They all deal with hand gestures in painting. The general program behind his exposé is what he calls "Prolégomènes à une critique de la *gesturalité* dans l'art." His analyses focus around Renaissance paintings, but also deal with Greek vases and reach to 20th century art. The most basic gesture he exhibits is the pointing gesture (index), that gesture which Tommaso Campanella [945] had called the only unconditionally understandable and basic human gesture.[2]

Chastel, fully in congruence with Schmitt's treatise [946] which we have discussed in Section 57.2, refers to Cicero and Quintilian and then recognizes the character of those classical gestures as tamed phenomena: "Les gestes que retient le peintre, ce sont naturellement les gestes stéréotypés, répétitifs, tous construits et codifiés dans le vécu." More precisely: "Il n'est pas absurde de considérer ces formes comme les éléments d'une sorte d'*ars memoriae* religieux."

From this situation of tamed gesture, Chastel discovers a strong disgression into a totally different gestural ontology which was initiated in Leon Battista Alberti's theoretical writing, *De pictura* (1435), where he writes: "Coi movement dell membra mostran movement dell'animo." Chastel comments on it saying that it is a program of painting where the painted figures no longer represent a condition or a quality, but *a being*. And Leonardo da Vinci adds: "Lo bono pittore ha da dipingere due cose principali, cioè l'homo e il concetto della mente sua. Il primo è facile, il secondo difficile perché s'ha a figurare con gesti e movimenti delle membra." And Chastel summarizes that: "La naiveté sémantique du geste, sur laquelle nous avons édifié notre exposé, était en somme compromise par la théory des *moti*." It is the theory of movements, these characteristics of *untamed* gestures, which gives Chastel the argument for a radical change of gestures with Alberti and da Vinci. He concludes that: "Le geste de l'index était si remarkable, si chargé de sense pour Léonard qu'il est devenu avec lui une sorte de geste pur." The reference to the extremely complex and powerful configuration of 130 pointing gestures in Leonardo's *Last Supper* illustrates this theory. Summarizing, the hand's gestures—above all the pointing gesture—are for Chastel and Pallasmaa germinal phenomena that initiate an autonomous ontology of gestures beyond their plain semiotic role as nonverbal signs.

57.9 Émile Benveniste and Marie-Dominique Popelard/Anthony Wall: Gestures as a Dialogical Category

Émile Benveniste in [100] sets up a refined analysis of the Saussurean sign structure that connects the signifier (*signifiant*) to its signified (*signifié*) by means of the signification process (*signification*). Benveniste's analysis of signification stresses the role of the subjects involved in the process when pronouns are involved. The classical *trias* of first, second, and third person, *je/tu/il* (I/you/he), reveals a fundamentally different situation when the deictic nature of these pronouns is at stake. He separates the first two personal pronouns *je/tu* from the third one considering the mechanisms that enable their full meaning. Whereas the first two persons involve a substantial subjective involvement, the third person, which he qualifies as a "non-person," are independent of subjective dimensions. It is in fact well known that the pronouns *je/tu* are strong shifters; their full meaning is only achieved by the integration of the underlying subjects which perform enunciations including those pronouns.

This shift from a formal signification process to one that essentially involves the subject(s) of enunciation is not only a formal enrichment of Saussure's abstract scheme, it also creates an ontological dimension in that the existence and presence of subjects underlying the pronouns *je/tu* becomes a *conditio sine qua non* for the possibility of meaningful fulfillment of these pronouns' *signifiés*. In the spirit of dialogical linguistics, as forwarded by Michail Bakhtin, [70] the first two persons pertain to a relational understanding of linguistic utterances. It is stated that *je/tu* implicitly or explicitly always presuppose the presence of a co-enounciators, meaning that the ontology realized by these pronouns always includes the presence of the Other that is addressed when using *je/tu*. The Saussurean abstract scheme thereby morphs to a speech act where the deictic completion enforces a co-existence of the Other. This is not a direct necessity of gesturality, but in the action that such a co-existence includes (addressing your words to another person), it opens up a dimension that transcends the objective facticity of the third person.

[2] It is remarkable that recent research by Michael Tomasello and collaborators in neuropsychology [1062, 1063] has demonstrated that pointing is a basic and exclusive human ability that children learn at the age of one year, and that this gesture is a root of human language development.

Benveniste's approach is exemplified and discussed in Popelard's and Wall's short, and easy-to-read, book "Des faits et gestes" [857]. Their discourse traces a dialogue about gesture and fact between a philosopher (Popelard) and a literary scientist (Wall). Their understanding of gestures does not include a precise definition,[3] but a phenomenological analysis, mainly focusing on gestures as complements of language, and this in the spirit of Kendon (whom they cite) and McNeill. They attribute to gestures a fundamental role: "Ce sont les gestes qui se servent de nous." It is here a typically French approach to gestures: They are not just a variant of language (in its more general concept), but genuine communicative phenomena.

They discuss gestures working out two main examples: Michelangelo's *Creation of Adam* and Stendhal's *Le rouge et le noir.* In both instances, they focus on the hand's gesture, in Michelangelo's case between God and Adam, and in Stendhal's case between Madame de Rhênal and her lover Julien. Their interaction is a dialogical one that is based on a mutual communication with its inherent ambiguity in the sense that these gestures don't communicate given contents, but create a joined sense uniquely by their bidirectional exchange. Their gestures don't share the classical Saussurean *signifé,* but create a sense that is built upon the second person's co-presence: "Aucun geste se fait tout seul." And: "Les faiseurs de gestes sont des co-gestionnaires." They create a "nous", a we-community[4] that gives their gestures their sense. This is a remarkable shift from Saussure's *signifé* to a sense that transcends the semiotic setup.

[3] On page 10, they however state that "un geste engage un corps qu'il met en mouvement: en temps et en espace, un corps se trouve animé. Comment? En signifiant quoi? Faire un geste dirait quelque chose. La question est de savoir quoi."

[4] It might be a bit unprecise to identify the second person singular pronoun with the first person plural pronoun here. Benveniste would not have accepted this identification. Rather he would prefer understanding that the second person, "you", is a shared identity of intimacy without the ontology of a "public" community of the plural.

58

The French Presemiotic Approach

Summary. The French school of diagrammatic philosophers was inspired by Gilles Deleuze's comments on Bacon [258] and then elaborated upon and deepened by gesture theorists and philosophers, such as Gilles Châtelet [190] and Charles Alunni [24]. This important French approach to gestures reveals a delicate aspect of embodiment in that gestures are conceived as being presemiotic. Gestures—except when 'tamed' by social codes—are not signs in a semiotic environment.

– Σ –

The French school of diagrammatic philosophers was inspired by Gilles Deleuze's comments on Bacon [258] and then elaborated upon and deepened by gesture theorists and philosophers, such as Gilles Châtelet [190] and Charles Alunni [24]. It is remarkable here that we encounter thinkers who are approaching embodiment not directly in a body-centered discourse, but instead focus on gestures as dynamic structures that act upon and through the body in a physical sense, but also more abstractly. We shall come back to this important nuance of embodiment when discussing the problem of the subject that is generated upon the body's basic existence. The French diagrammaticians typically focused on gestures qua dynamic layer of embodiment, an approach that had already been inaugurated by the French mathematician Jean Cavaillès, who had claimed that [181, p. 178] «Comprendre, c'est attraper le geste et pouvoir continuer.» ("Understanding is catching the gesture and being able to continue.") This characteristic French dancing thought (also shaped in Pierre Boulez's reflection on gesture in music [141]) was in fact stated with respect to mathematical theories, and as such it was one of the very first principles of gestural embodiment in mathematics, an idea now quite fashionable through the work of Lakoff and Núñez [570], but also anticipated in Châtelet's observation [190] that the Fregean concept of a function f in mathematics is a dramatic (and questionable) abstraction that replaces the moving gesture from argument x to its functional value $f(x)$ by a kind of disembodied 'teleportation', where the evidence of the functional relation is wrapped and hidden if not destroyed.

This important French approach to gestures[1] reveals a delicate aspect of embodiment in that gestures are conceived as being presemiotic. Gestures—except when 'tamed' by social codes—are not signs in a semiotic environment. They are not a realization of Ferdinand de Saussure's famous signification process from the expressive signifiant to the content of signifié [933]. Châtelet (loc. cit.) is very clear in this point: «Le concept de geste nous semble crucial pour approcher le mouvement d'abstraction amplifiante des mathmatiques. (. . .) Un diagramme peut immobiliser un geste, le mettre au repos, bien avant qu'il ne se blottisse dans un signe, et c'est pourquoi les géomètres ou les cosmologistes contemporains aiment les diagrammes et leurs pouvoirs d'évocation préemptoire.» ("The concept of a gesture seems crucial to approach the amplifying movememt of abstraction in mathematics. (...) A diagram can immobilize a gesture, put it to rest long before it is hidden in a sign; this is why geometers and contemporary cosmologists love diagrams and their power of preemptive evocation.") A gesture can be immoblilized by a diagram (which in this French theory is a kind of disembodied gesture) before it becomes a sign. And Alunni confirms this creative presemiotic role of gestures: «Ce n'est pas la règle qui gouverne l'action diagrammatique, mais l'action qui fait émerger

[1] There are other French approaches to gesture theory, such as Geneviève Calbris, for example.

G. Mazzola et al., *The Topos of Music III: Gestures*, Computational Music Science,
https://doi.org/10.1007/978-3-319-64481-3_3

la règle.» ("It is not the rule which governs diagrammatic action but it is action which causes the rule to emerge.") This is a theory of gestures that diverges from that of the Anglo-Saxon school[2] centered around Adam Kendon and David McNeill [741], where (loc. cit., p. 58) it is stated that "the gesture is created by the speaker as a materialization of meaning."

58.1 Maurice Merleau-Ponty

Summary. The French philosopher Maurice Merleau-Ponty embarked in 1945 on a new positioning of human language in his influential *Phénoménologie de la perception* [751]. The pivotal point of his approach consists in a revision of the relation between language and thought.

– Σ –

In some sense, the preceding section was a preliminary discourse on the French theory of gesture, which is radically different from the semiotically loaded approaches of Kendon and others. Historically, reverberating Valéry's visionary anticipation of the mental dimension of hand gestures (see the head of this chapter), the French philosopher Maurice Merleau-Ponty embarked in 1945 on a new positioning of human language in his influential *Phénoménologie de la perception* [751]. The pivotal point of his approach consists in a revision of the relation between language and thought [751, p. 211]: "The word and speech must somehow cease to be a way of designating things or thoughts, and become the presence of that thought in the phenomenal world, and, moreover, not its clothing but its token or its body." This entails a deep change of the localization of contents (loc. cit.): "We find here, beneath the conceptual meaning of the words, an existential meaning which is not only rendered by them, but which inhabits them, and is inseparable from them. (...)

This power of expression is well known in the arts, for example in music. The musical meaning of a sonata is inseparable from the sounds which are its vehicle: before we have heard it no analysis enables us to anticipate it; once the performance is over, we shall, in our intellectual analysis of the music, be unable to do anything but carry ourselves back to the moment of experiencing it. During the performance, the notes are not only 'signs' of the sonata, but it is there through them, it enters into them." Merleau-Ponty then gives this embodiment a more concrete shape: "The spoken word is a genuine gesture, and it contains its meaning in the same way as the gesture contains it. This is what makes communication possible." The meaning addressed here is however not the naive semantic layer of a sign. Merleau-Ponty specifies: "The spoken word is a gesture, and its meaning, a world." This dramatic restatement of "meaning" destroys all those formally semiotic perspectives, meaning a world is a dispersive action blurring all delimitations of clear-cut semantic units. We shall come back to this point while discussing Châtelet's philosophy later in Section 58.4.

58.2 Francis Bacon and Gilles Deleuze

Summary. The philosopher Gilles Deleuze's 1981 work discusses *Francis Bacon: La logique de la sensation* [258], dealing with the painter's sensational confessions made during David Sylvester's *Interview with Francis Bacon: The Brutality of Fact* [1029]. In these reflections, Deleuze introduces what is now known as the *pensée diagrammatique*, the French school of gestural philosophy.

– Σ –

A strong proclivity for embodied approaches to the arts was initiated by those painters who displayed an innovative understanding of what the action of painting means. It is known that Jackson Pollock stressed gestural action as opposed to illustration. And in a remarkable series of interviews [1029], Francis Bacon made clear that his diagrammatic gestures were more important to his creative work than mentally planned strategies: "The marks are made, and you survey the thing like you would a sort of graph. And you see

[2] There are other Anglo-Saxon approaches to gesture, for example, by Jürgen Streeck, Eve Sweester, and Katharine Young.

within this graph the possibilities of all types of fact being planted. (...) In the way I work I don't in fact know very often what the paint will do, and it does many things which are very much better than I could make it do. Is that an accident? (...) What has never yet been analyzed is why this particular way of painting is more poignant than illustration. I suppose because it has a life completely of its own. (...) So the artist may be able to open up or rather, should I say, unlock the valves of feeling."

These insights imply a shift from the abstract semiotics of expression-signification-content to a more embodied understanding of speech and its "score" in codified language. Such a trend was then strengthened by the philosopher Gilles Deleuze's 1981 work that discusses *Francis Bacon: La logique de la sensation* [258], dealing with the painter's sensational confessions made during David Sylvester's *Interview with Francis Bacon: The Brutality of Fact* [1029]. In these reflections, Deleuze introduces what is now known as the *pensée diagrammatique*, the French school of gestural philosophy. Deleuze takes from Bacon's statements the word "graph" and translates it to "diagramme" and even "geste".

The crucial statement made by Francis Bacon about his working experience and approach is this: "The marks are made, and you survey the thing like you would a sort of graph. And you see within this graph the possibilities of all types of fact being planted. (...) In the way I work I don't in fact know very often what the paint will do, and it does many things which are very much better than I could make it do. Is that an accident? (...) What has never yet been analyzed is why this particular way of painting is more poignant than illustration. I suppose because it has a life completely of its own. (...) So the artist may be able to open up or rather, should I say, unlock the valves of feeling." Bacon's graphing action does not illustrate a fact, but innervates the work of art such that it becomes a living organism of poignant tension.

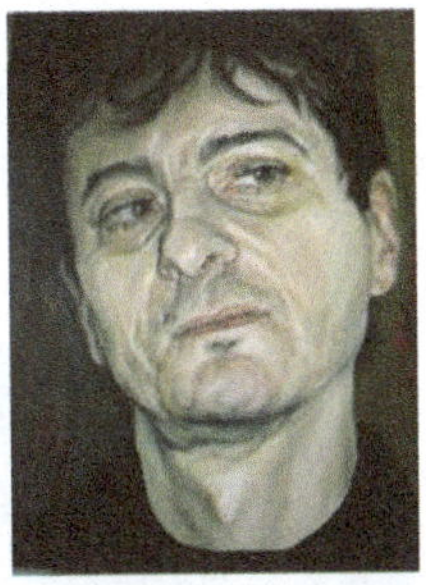

Fig. 58.1. Charles Alunni (Oil painting by Dominique Renson).

58.3 Jean Cavaillès and Charles Alunni

Summary. For the mathematician Jean Cavaillès (Figure 58.2), gestures are not just carriers for the transportation of content, but elastic bodies that must be caught like balls in a game, and which require continuation in order to keep the game going. In a far-reaching consequence, the philosopher Charles Alunni (Figure 58.1) has accepted not only the presemiotic status of gestures, but also their creative power [91]: "It is not the rule that governs the diagrammatic action, but the action that causes the rule to emerge."

– Σ –

In the French school, a gesture is a presemiotic concept that need not convey meaning; it is understood as being capable of generating content, but this is not essential. This latter approach enables a much subtler deployment of gesture in collaborative environments, since it is vital to collaboration to create mutual understanding without presupposing shared knowledge. Jean Cavaillès summarizes the French philosophy of gestures in a concise way that entails heavy consequences for the entire conceptualization of the art of collaboration [181]: "Understanding is catching the gesture and being capable of continuing." So gestures are not just carriers for the transportation of content, but elastic bodies that must be caught like balls in a game, and which require continuation in order to keep the game going. Deleuze completes Merleau-Ponty's approach in that he now emphasizes the logical and therefore mental side of the gestural innervation of artistic creativity, and also the distributed identity suggested by Bacon's approach: The artist's ego is now spread over the entire painting, it reverberates with the painter's action points and opens the locked "valves of feeling".

Fig. 58.2. Jean Cavaillès, mathematician and pioneer of the French philosophy of gestures.

This revelation had an incredible impact on the French philosophers, and it also reassembled forerunners and future leaders of the diagrammatic school. The most notable, following the doyen Valéry, was the mathematician and philosopher Jean Cavaillès (1903-1944), who in his rejection of "uniformity" was shot by the Nazis.

His definition of "understanding" as being a gestural information exchange gives gestures a decisive role for the transmission of thoughts in that they act not as external carriers of thoughts, they are the thoughts themselves. That which is understood is reified in the gestural continuation, not in the thoughtful implosion. In a far-reaching consequence, the philosopher Charles Alunni has not only accepted the presemiotic status of gestures, but also their creative power [91]: "It is not the rule that governs the diagrammatic action, but the action that causes the rule to emerge." Gestures not only make communication, but they "make sense", make what will later be followed on the level of facticity. The rule, the law, that which separates the truth from the false. Alunni's step looks tiny, but it is the reversal of the old laws. Gestures are, in Alunni's approach, the law-makers, not the ornaments of established truths.

58.4 Gilles Châtelet

Summary. The most difficult and radical of the French diagrammaticians is mathematician and philosopher Gilles Châtelet (1945-1999), see Figure 58.3. We shall devote this section to his revolutionary insight.

– Σ –

Fig. 58.3. Gilles Châtelet (Gravure by Jean-Claude Darras).

His writings are therefore difficult, they merge the cold and precise abstraction of mathematics and the hot conceptual magma of philosophy. Perhaps this mixture was the right one to enable him to set forth a theory of gestures that would, in the long run, have a significant and lasting impact upon the inner life of the sciences. Châtelet's radically presemiotic gesture theory is described in his book *Figuring Space* [190] (the French title *Les enjeux du mobile* would be better translated as *The Stake of the Mobile*). For Châtelet, gestures are definitively not identical with diagrams, they are wild vibrations and can be disembodied by diagrams and thereby are transformed from the ontology of the making to that of diagrammatic processes [190, p.9/10]: "A diagram can immobilize a gesture, put it to rest long before it hides itself within a sign, and this is why the contemporary geometers or cosmologists love diagrams and their power of preemptive evocation."

Let us now take a glance at Châtelet's somewhat arcane characterization of gestures [190]:

The gesture is not substantial: it gains amplitude by determining itself. Its sovereignty is equal to its penetration and that is why we refer to the gesture's 'accuracy': the precision of the strike is proof of the reverberation of its skill. The gesture inaugurates a family of gestures, whereas the rule only enunciates 'instructions', a protocol for decomposing the action into endless repeatable acts. The gesture possesses a historical exemplariness: if one can speak of an accumulation of knowledge over the course of successive generations, one should speak of gestures inaugurating dynasties of problems.

Similarly to Cecil Taylor's statement about rhythm-sound being found in the amplitude of each time unit, the gesture is not substantial, but pulsates in its own penetration. It has no 'material kernel', all is in its mobility, it is essentially self-referential. The rule is not self-referential but atomizes actions into 'instructional units', which have no sense except to be repeated ad infinitum. The historical unfolding of a gesture is a dynasty of problems, not the sedimentation of facticity. Gestures only survive in their pulsating movement, they absorb results and heat them up into germinating abysses. To date, a history of gestures has not been written—perhaps such an undertaking will be as difficult and problematic as the history of free jazz beyond the banal collection of factual residues.

The gesture is not a simple spatial displacement: it decides, deliberates and suggests a new modality of 'moving oneself'—Hugues de Saint-Victor defined the gesture as 'the motion and figuration of the members of the body according to the measure and modalities of all action and attitude'. The gesture refers to a disciplined distribution of mobility before any transfer takes place: one is infused with the gesture before knowing it.

The definition of Saint-Victor describes the anatomy of a gesture. It is a distributed mobile figuration, not a moving point. The potentiality of the figuration to be deformed is already an expression of the gesture's mobility.

The gesture is elastic: it can crouch on itself, leap beyond itself and reverberate, whereas the function gives only the form of the transit from one external term to another external term, the act exhausts itself in its result. The gesture is therefore involved with the implicit pole of the relation.

This is the core difference with mathematical functions. Châtelet is probably the first to have observed the illusion of arrows for functions. Compared to a function, a gesture is a living animal, which has no mutually external parts. The question arises of course about how the elasticity and living character can be conceived beyond the metaphorical imagery. But it is evident, that we need a new conceptualization here, and that Châtelet is still far from having built such a renewal. His language is as elastic as his gestures.

The gesture envelops before grasping and sketches its unfolding long before denoting or exemplifying: already domesticated gestures are the ones that serve as references.

This is again a clear negation of any semiotic casting of gestures. Only domesticated gestures can be 'slaves' of meaning and carry it around in denotation and exemplification. The referential arrow of pointers that characterize semiotic signification are not yet activated, gestures do not hit and prick their targets, but envelop and sketch. They are ontological sketches of processes and facts.

A gesture awakens other gestures: it is able to store all the allusions' provocative virtualities, without debasing them into abbreviation.

Gestures do not abbreviate or put to sleep, but penetrate without becoming tired. They are constantly transcending facticities and melting them to virtualities, similarly to heating up wax in order to reshape it. They might be called the fires of thought, or, in Taylor's terms (from *Unit Structures*): naked fire gestures.

In 1991, Cecil Taylor recorded the video *Burning Poles* [1039] with William Parker on flute and bass, André Martinez on drums and percussion, and Tony Oxley on drums. On the 10:20 minute piece *Poles*, Taylor appears as magic cantor and dancer; he floats about his open grand piano, singing and reciting cryptic incantations while scratching and plucking the internal strings of the piano, or hitting the strings with a soft mallet. The music and performance is a percussive shaping of the body of time in a tense trajectory of gestures, and is an excellent illustration of Châtelet's circumscription of "gesture". Taylor's performance radically differs from the music as construed in *Unit Structures*—there are no structures to be broken, no starting and terminal points of functions, no empty time spaces where time shards are imbricated. Taylor invokes a big, smoothly flowing gestural organism that feeds its identity by its incessant unfolding and caressing of time and space. The piano's strings are not merely objects for sound production, but extensions of Taylor's aesthetic of creating and sculpting innervated space-time. Although some of the lyrics of Taylor's *Sprechgesang* may convey normal English words, their extramusical meaning is completely irrelevant; as any kind of localized or preconceived meaning is irrelevant, the gestures are pulsating in a germinal ontology of pure making. No concrete result is achieved, no objective message is conveyed, no truth is established, and when it is over, nothing remains except the desire to review the performance, to delve again into that dis-objected making.

The underlying question to all these gestural approaches is what it means to think in terms of gestures. Or, coming back to Adorno's and Wieland's program (see also Section 60.2), what it means to think of music in its performance. Adorno's answer is very clear: "Performing music [is] making music." And he means "bringing into existence" by "making". Music is only in its very making, exactly what Merleau-Ponty's statement, which we discussed earlier 58.1, means. This entails that thinking of music must happen in its making, an insight that is in fact widely accepted. The German music psychologist Helga de la Motte-Haber writes [234, Vol. X]: "Musical thinking is fundamentally a thinking within music. Where difficult interpretations have to be discussed, not only musicians, but also authors such as August Halm or Hermann

Kretschmar refer to illustrative examples. (...) Analyses of music are often difficult to read, because the structure of linguistic sentences is limited in its power of adaptation to musical facts. You may operate with predicates and objects, but subjects are problematic. (...) Therefore unnatural passive constructions are added in order to eliminate situations where the subject just be hidden."

The problem is however not the subject as such, but its embedding in the mental space of music. This space comprises all three levels of embodiment: facts, processes, and gestures. The subject is not relevant on the factual level, a standard situation for common scientific knowledge. One may write huge treatises on music on that level. That type of insight is important and established, but the thinking of music beyond facticity is less codified. In fact, it tends to escape codification when approaching the inner regions of embodiment. This is what "thinking within music" means: thinking beyond facticity.

It is remarkable that Châtelet is aware of this way of thinking, when he reflects upon the famous paradigm of "thought experiments" in physics. Here is where he brings in the necessity of gestures for scientific thinking [190]:

One could even say that the radical thought experiment is an experiment where Nature and the Understanding switch places. Galileo sometimes puts himself in Nature's place, which, in its simplicity, could not have chosen to 'move at an inconceivable speed an immense number of very large bodies, to produce a result for which the moderate movement of a single body turning around its own centre would suffice'. Einstein was in the habit of saying that it was necessary to put oneself in God's place to understand Nature. There is nothing surprising therefore in these teleportations where Einstein takes himself for a photon and positions himself on the horizon of velocities (...), in these immersions where Archimedes, in his bathtub, imagines that his body is nothing but a gourd of water. Thus, to understand the photon, it is necessary to become a photon and, to understand floating, it is necessary to turn oneself into a wineskin! The thought experiment taken to its conclusion is a diagrammatic experiment in which it becomes clear that a diagram is for itself its own experiment. The gestures that it captures and particularly those that it arouses are no longer directed towards things, but take their place in a line of diagrams.

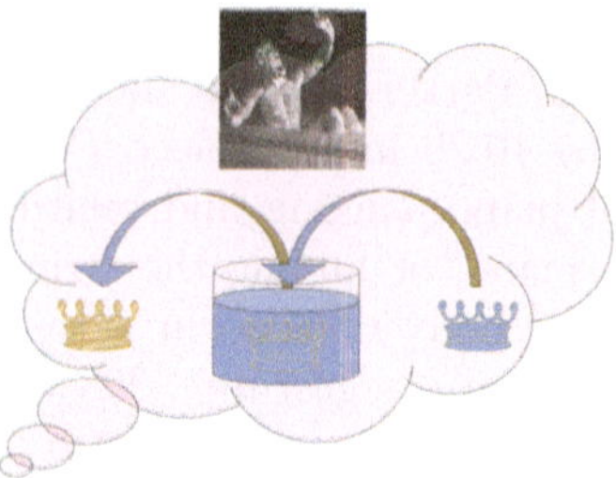

Fig. 58.4. Archimedes in the bathtub with his thought experiment.

To understand the gestural virtuality of thought experiments, let us consider Archimedes' bathtub situation (Figure 58.4). The king's crown must be checked on the specific weight of its material: Is it gold or is it a faked composition? Archimedes imagines the crown being immersed in water. Then the crown is removed from water, but only virtually, in order to keep the water shape as is. This action is followed by a virtual replacement of the hollow volume by water in the shape of the crown. Then physics is reset and since that crown-shaped water will not move, the weight of that water volume is balanced by the force of the surrounding water. Again, this imaginary construction is applied to the crown being re-immersed in its original position (everything is impossible in the real world!), and we conclude that the force of the surrounding water acted upon the real crown equals the weight of that water volume (with opposite signs). Therefore, the bathtub experiment is a sequence of virtual gestures on virtual objects, which are, in Châtelet's reading, just parts of the gestural configuration.

This text is not only remarkable as such, but can be used for a thought experiment. We take the text and virtually remove all physical instances, replacing them by adequate musical ones. And here is the virtual text:

One could even say that the radical thought experiment is an experiment where Music and the Understanding switch places. Murail sometimes puts himself in Music's place, which, in its simplicity, could not have chosen to 'move at an inconceivable tempo an immense number of very large sounds, to produce a result for which the moderate movement of a single sound turning around its own centre would suffice'. Bach was in the habit of saying that it was necessary to put oneself in God's place to understand Music. There is nothing surprising therefore in these teleportations where Murail takes himself for a sound and positions himself on the horizon of tempi (...), in these immersions where Beethoven, at his piano, imagines that his body is nothing but an envelope of waves. Thus, to understand the sound, it is necessary to become a sound and, to understand waves, it is necessary to turn oneself into a vibration! The thought experiment taken to its conclusion is a diagrammatic experiment in which it becomes clear that a diagram is for itself its own experiment. The gestures that it captures and particularly those that it arouses are no longer directed towards things, but take their place in a line of diagrams.

This thought experiment makes evident that de la Motte-Haber's "thinking in music" is akin to making an experiment of thought, and that this making is a gestural category. Châtelet's thinking in physics offers a model for thinking in music, and that model is based upon a gestural action in a virtual space-time. It is a way of thinking in the making. This is not to say that such activity does *not* entail facts! Archimedes' insight creates a fact, and the king's valuation of that artisan who faked the crown in sentencing him to death is anything but virtual. Let us terminate this philosophical perspective by Marcel Marceau's wonderful saying:

To mime the wind, one becomes a tempest.
To mime a fish, you throw yourself into the sea.

59

Cognitive Science

Summary. This chapter deals with gestural arguments in cognitive science. We first discuss the idea of a "science of embodiment", then discuss the neurological vicinity of speech and manual gesture centers (Broca area), also referring to Merleau-Ponty's linguistic philosophy and to the 3D Mental Rotation experiments by Shepard and Cooper. In this context, Donald's gestural anthropology is mentioned, together with Valéry's philosophy of dance and Taylor's critique of disembodied music. The chapter terminates with a fascinating discussion of musical gesture from the perspective of disability studies. Its focus is on two hand-impaired jazz pianists, Horace Parlan and Oscar Peterson.

– Σ –

59.1 Embodiment

Summary. What is the substance, the essence, of the concept of the body that persists in medicine, yoga, psychology, performance, and philosophy?

– Σ –

The question relating to the semiotic state of a gesture not only separates this French school from the mentioned Anglo-Saxon school, it also reveals a deep problem concerning the semiotic understanding of embodiment. What does it mean to be embodied? What is that body whose substance reifies gestures? Such questions have been dealt with by linguists, of course, since Maurice Merleau-Ponty, the figurehead of embodied linguistics, exposed the role of words as gesturally active bodies, and not as clothes and carriers of our thoughts and emotions; see Section 58.1. In a very lucid exposition [1088], the Italian semiotician Patrizia Violi has discussed the semiotic embodiment problematic. Her discourse rightly points out the polysemy of the word "body", which is evident: the word means very different things in medicine, yoga, psychology, performance, and philosophy. But the underlying question is what Violi puts into evidence: What is the substance, the essence, of this concept that persists in all these approaches? "In order to develop a fully embodied theory of semiosis we certainly need a bringing together of body and subject, and to do this we must develop an approach to subjectivity which is quite different from the transcendental Ego that is implicit in the classical structuralist framework."

It is remarkable how the Saussurean disembodied abstraction in his sign concept (signifiant/signifié) resembles the Fregean abstraction in his disembodied concept of a mathematical function (see Section 57.3.2). Violi confronts the linguistic category of a subject with the ontological one of the self as a being (also referring to Émile Benveniste's investigations on enunciation theory [100]) and observes that "the theory of enunciation removes the issue of embodiment altogether." And she concludes that we are confronted with the paradoxical fact that "on the one hand there is a theory of embodiment without subject, on the other a theory of the subject without a body." We learn that Violi opens a question which by far transcends the conceptual/semiotic analysis of "body" or "embodiment", but points to the problem of how substantial these

G. Mazzola et al., *The Topos of Music III: Gestures*, Computational Music Science,
https://doi.org/10.1007/978-3-319-64481-3_4

concepts could be in the very making of thoughts. The linguistic category she alludes to is a gate to the problem of why embodiment is such a basic category of any type of articulation, verbal and beyond. Or, to re-state in Merleau-Ponty's spirit: If the words are the thoughts' body, what can we draw from such a body's anatomy when we think?

59.1.1 Embodiment Science

In light of the many and diverse fields from the full range of disciplines, in which embodiment plays an ever increasing role, we find numerous arguments in favor of the creation of a human body/embodiment science. We emphasize that we are hardly the first to have considered such a multidisciplinary endeavor. Two such enterprises come from France and the United States:

The first is the initiative of Frenchman Bernard Andrieu[1] entitled *Le site du corps*, it can be found online [35]. Andrieu describes his approach as follows: «Pour étudier comment le corps a été étudié et interprété par la science et la philosophie, notre recherche est organisée autour de deux grands axes, développés en parallèle depuis 1986: la désincarnation scientifique du corps et la description du sujet incarné par les philosophies du corps. Notre postulat est de lier l'épistémologie et l'ontologie du corps en démontrant comment la modélisation du corps présuppose, en le laissant le plus souvent dans l'implicite, une conception ontologique du corps.» ("In order to study how the body has been investigated and integrated by science and philosophy, our research is organized around two great axes, developed simultaneously since 1986: the scientific disembodiment of the body and the description of the embodied subject by philosophies of the body. Our postulate is to connect epistemology and ontology of the body by showing how the body's modeling presupposes, keeping it however mostly implicit, an ontological concept of the body.") The initiative is very well organized, including a number of books,[2] research, international connections, and the French internet network. In particular, it lists all doctoral theses in France in the years 1971-2007 with the word "corps" in their titles. The site reveals a great awareness of the multidisciplinary nature of the subject and also offers a historical perspective of the body. Although the site is centered around French contributions and perspectives, we consider this initiative a vital step towards an institutionalization of a science of the human body. Andrieu's initiative also comprises the journal "Corps" [36].

Another enterprise is more philosophical, though not restricted to philosophical discourse. It was created and directed by Richard Shusterman,[3] who has institutionalized his theoretical and practical approaches in the *Center for Body, Mind, and Culture* at Florida Atlantic University [974]. Like Andrieu, Shusterman has created an incredibly rich resource for scholarly research, which also includes information on mind-body practices such as Feldenkrais, Alexander Technique, and Confucian and Zen approaches, which are beyond the purview of our discussion, but which are relevant to understanding Shusterman's comprehensive approach and his understanding of philosophy as a quest for good life. His seminal paper *Somaethetics: A Disciplinary Proposal*, published [975] in 1999, displayed a model of what could be regarded as a philosophy of the human body and embodiment. In his critical review of Alexander Baumgarten's Aesthetica, Shusterman proposes a discipline of "somaesthetics," defined as "the critical, meliorative study of the experience and use of one's body as a locus of sensory-aesthetic appreciation (aesthesis) and creative self-fashioning." In remarkable agreement with Andrieu's approach, Schusterman states that "Beyond the essential epistemological, ethical, and sociopolitical issues, the body plays a crucial role in ontology." He adds that "analytical philosophy examines the body as a criterion for personal identity and as the ontological ground (through its central nervous system) for explaining mental states." On this ground, Shusterman then defines three fundamental dimensions of somaesthetics:

1. *Analytical somaethetics*, describing "the basic nature of bodily perceptions and practices and also of their function in our knowledge and construction of reality."

[1] Professor at the Faculty of Sport/UHP of Nancy University, France.

[2] Among others, one on a philosophy of the body, the skin, suntan, etc.

[3] Dorothy F. Schmidt Eminent Scholar in the Humanities, Professor of Philosophy and English and Director of the Center for Body, Mind, and Culture at Florida Atlantic University. See `http://www.fau.edu/humanitieschair`.

2. *Pragmatic somaethetics*, "having a distinctly normative, prescriptive character—by proposing specific methods of somatic improvement and engaging in their comparative critique."
3. *Practical somaesthetics*, "being concerned not with saying but with doing, this practical dimension is the most neglected by academic body philosophers, whose commitment to the discursive logos typically ends in textualizing the body."

He views the disciplinary proposal in a "double meaning: as a branch of learning or instruction and as a corporeal form of training and exercise." He strongly argues for a wider conception of philosophy, including "the ancient idea of philosophy as an embodied practice, a way of life."

Given the above tour d'horizon as well as Andrieu's and Shusterman's strong arguments for a science of the human body and embodiment, we can now attempt to describe the state of the art of the scientific discipline of embodiment. We focus on the analytical rather than the practical dimension, and on the global picture of such a scientific proposal. We do not yet problematize the concepts of "body" or "embodiment" in the manner of Violi, but will postpone this topic until Section 59.1.1.1. We embrace the artistic insights exposed in our tour d'horizon as a strong argument for further investigating the architecture of a science that could contribute to the unification of methodologies and findings regarding the role of the body in the construction of higher cognition, intelligence, knowledge, and culture. Recent insights from the hard sciences, neurobiology and embodied artificial intelligence draw a picture of a new type of homunculus. Whereas the classical homunculus was a topographic image of the interface between the body's limbs and their sensorio-motor innervations, we now display an integrated map of action and perception. This unified action-perception topography of the body is what we would call the action-homunculus. It is the integral of all the dimensions that define our physiological existence as a space-time of action and perception. It is the global interface with extensional reality of the human existence. Although we are far from understanding its detailed mechanisms, we can conjecture that this layer is the bodily basis for whatever we may experience physiologically as living beings.

This is one extremal set of data, from which we have to draw a line to the other extremal set of human existence: cognition, emotions, and higher human dimensions such as language, non-verbal semiotic systems (traffic, fashion, alimentation, social interaction, emotions, sexual and erotic codes, and so on), logic, mathematics, and the arts. Let us call this layer the *cognitive stratum*.

It is one of the big implicit axioms of all scientific and artistic approaches to embodiment that the layer of the action-homunculus (or any corresponding concept they would associate with the body's basic reality) is the first level of a cognitive leading to the cognitive stratum. We use the term cognitive embryology because it is a second axiom of this philosophy of embodiment that the cognitive stratum is the completion of a complex organic evolution and not something which is added at once and stemming from a metaphysical platonic sky or from some divine fertilization. This is a strong assumption that contradicts the Leibnitz monadic principle: there are no two worlds, our thoughts are created from the action-homunculus and not imported from another reality.

59.1.1.1 The Cognitive Layer

This setup is not a fact, but a basic approach to an attempt to understand cognition and intelligence in a very precise evolutionary architecture. The *cognitive layer* is the grownup shape of the action-homunculus. Although this approach is far from complete, we must ask whether such a strong fundamental assumption or principle can be the basis of a scientific (multi)discipline. The model of a scientific discipline par excellence is modern physics. It is an extremely successful science with a complex theory and an overwhelming arsenal of applications in other sciences and in practice. Physics may alter its shape as theory and experiments evolve, but it is also built upon a very abstract and invariant principle, namely that nature is governed by a small set of laws. At present, the architecture of these laws is dramatically simple: Nature is completely described by three types of forces: electro-weak force, strong force and gravitation. And all efforts in the Big Science of elementary particle physics are united by the axiom that these forces should be expressions of one basic force, an endeavor coined ToE: Theory of everything. Nothing confirms such a far-out principle, but physicists agree that this is the big challenge.

In view of this exemplary model of a hard science it is reasonable to make an analogous effort in the understanding of the human condition, namely that the cognitive stratum is the endpoint of a cognitive embryology based upon the action-homunculus.

What is the critical point in such a scientific enterprise? The big question is this: How is it possible to describe and understand the transition from the action-homunculus to the cognitive layer? In embodied Artificial Intelligence, the problem is manifest: All efforts on the level of robotics and other cheap or expensive embodiment machines are in a desperate search for emergent properties. There is no general methodology that would suggest how higher cognitive performance can emerge from low-level actions. The same is true when dealing with neurophysiological approaches to action-perception: We know quite a lot about the mechanisms of action-perception cycles, but there is a critical missing link from the action-homunculus to the cognitive stratum. We suggest that a science of the body and of embodiment should declare as its principle that we must find this missing link in the same spirit as that of modern physics' search of the unified force in the ToE.

Therefore, we propose a model of such a science that connects the action-homunculus to the cognitive stratum by a layer which may be called the action-to-cognition layer, in short A2C. We propose an A2C which is built upon the theory of gestures. It is not, however, a defining feature of this science but a proposal—motivated from gesture theory in music—of how such a layer may be conceived.

An approach to dealing with the missing link between the action-homunculus and the cognitive stratum is deduced from musical gesture theory. Its motivation stems from the well-known fact that in music, thoughts are known to be embodied in the performer's gestural actions. In this general shape, the idea is, however, metaphorical and abstract. Can we hope to make this more precise? Let us first recapitulate what is given: We have to search for a connecting structure between action-homunculus and the cognitive stratum. Now, the action-homunculus is already wellprepared to deal with gestures: actions are no longer a static body shape, actions can be taken as elementary structures to enable gestures. The action-homunculus is a body in action.

Understanding the musical theory of gestures from a more philosophical point of view may give a hint of how this problem can be attacked. In the concept of a gesture, we have two material ingredients: the skeleton and the body (see Section 61.5). The skeletal ingredients act upon the body: the vertexes and arrows are embodied by points and continuous curves on the body. The latter may be thought of as being given from the action-homunculus as a space, where action takes place. When accessed from the layer of gestures, this body is not directly grasped; instead we only experience it through the configurations of continuous curves defined by the map g on the skeleton. The ontology of the body is not directly accessible, we are forced to act via gestures upon this body X. The body X only exists as an embodiment of the possible skeleta. This means that X is an abstract entity and that its ontology for us humans is only visible when we let our gestures act upon this body X. We would like to call the totality of gestural actions (all the maps g with variable skeleta $\mathcal{D}$) the body's embodiment. We cannot know X, but only how it "feels" like acting on X. So is X forever hidden to our insight? Not really, and this is one of the most beautiful results from modern mathematics, category theory, to be precise. The result is known as Yoneda's Lemma. Expressed in common language it is known as the Yoneda philosophy (see Section 9.3), which states that a mathematical object is completely known if we know how it looks when observed from all objects of the same type. For example, a topological space is completely known if we know all the continuous maps from any other topological space. Now, in our case the set of all gestures targeting one and the same body is precisely the situation described by Yoneda's philosophy. This means that once we know all gestures on X, the system $C(X)$ is completely known, and then the points of X, being the constant curves on X, are also known. In other words: Yes, the system of all gestures on X gives a lot of information about X.

Our interpretation of these facts is that the embodiment of the body by the system of all gestures creates a machinery which we may control and which allows us to instantiate X when experienced through our gestures. The ontology of X is not given directly, but mediated through our gestural dynamics. And, if we use the recursive machinery of hypergestures, the passage to the cognitive stratum will occur in a series of hypergestural intermetidate layers, unfolding from simple gestures on X to highly hypergestural ones in hypergesture spaces. This means that we have good reason to conjecture that the missing A2C link between the action-homunculus and the cognitive stratum might be realized by the category of gestures acting upon the action-homunculus' body. Violi's subject may then be interpreted as being the identification of the body when the system of gestures operates in the creation of the ontology as mediated by Yoneda's Lemma.

59.2 Neuroscience

Summary. We discuss some neuroscientific arguments about the hypergestural reality in cognition, in particular mirror neurons.

– Σ –

Hypergestures are gestures of gestures; see Section 61.6 for a precise definition. Hypergestures are attractive, but are problematic with respect to many situations in the perception of families of shapes. Robert S. Hatten [446] has observed that we often encounter discrete sequences of events, gestures and shapes as opposed to continuous spectra. This seems to create a problem when applying the hypergesture concept to such discrete situations.

Suppose that we have the basic situation of two gestural configurations, for example two drum gestures or two spatial gestures of a pianist's hand, as shown in Figure 59.1. Such configurations are not naturally connected by a continuous family of intermediate states. If we want to build a hypergestural curve connecting

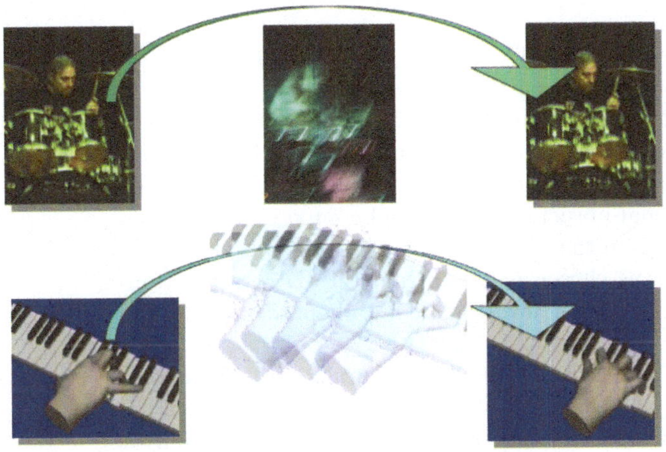

Fig. 59.1. Two situations of discrete families of gestures, which are thought to be deformed hypergesturally into each other by a virtual continuous deformation.

them, we would need to invent a non-existent auxiliary structure. Is this legitimate? Is there any reason beyond the inner logic of the hypergestural system to do so? Or is this just a theoretical artifact to connect disconnected items?

Merleau-Ponty's philosophical claims are supported by recent results from cognitive science and neuroscience. Maurizio Gentilucci and Michael C. Corballis [362] have proposed a theory of gradual transition from manual gesture to speech: "It is supposed that speech itself a gestural system rather than an acoustic system, an idea captured by the motor theory of speech perception and articulatory phonology. Studies of primate premotor cortex, and, in particular, of the so-called "mirror system" suggest a double hand/mouth command system that may have evolved initially in the context of ingestion, and later formed a platform for combined manual and vocal communication. In humans, speech is typically accompanied by manual gesture, speech production itself is influenced by executing or observing hand movements, and manual actions also play an important role in the development of speech, from the babbling stage onwards." The premotor cortex

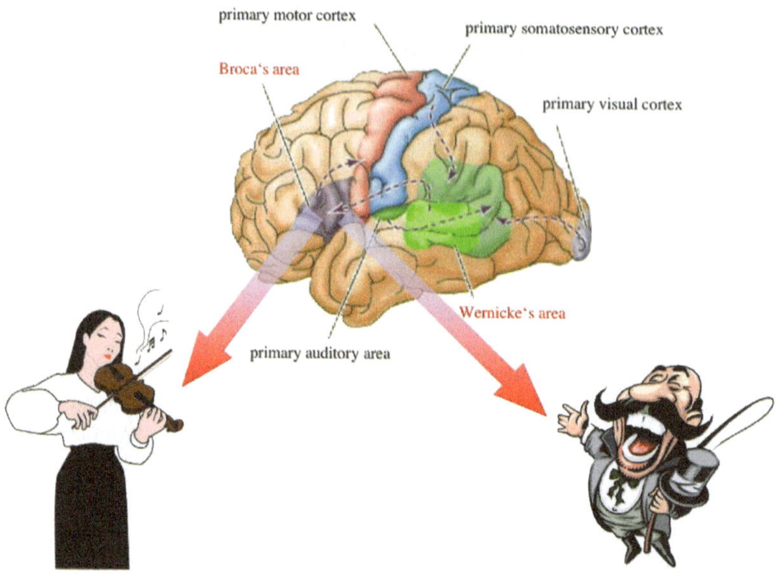

Fig. 59.2. The left hemisphere of the human brain, showing Broca's area and its significance for fine-grained and fast motor control in musicians and the production of speech.

lies above Broca's area, which is responsible for the speech production, as opposed to Wernicke's area, which takes care of speech perception; see Figure 59.2.

Merleau-Ponty's theories are confirmed by the topological vicinity in Broca's area of language and gesture as motor-related human competences. Recent results investigating the relation between sight reading competences of professional musicians and spatial competences have revealed a strong correlation in the same Broca area [986]. More precisely, the spatial competence was chosen to be the mental 3D rotation competence of spatial objects as first investigated by Shepard and Cooper [959].

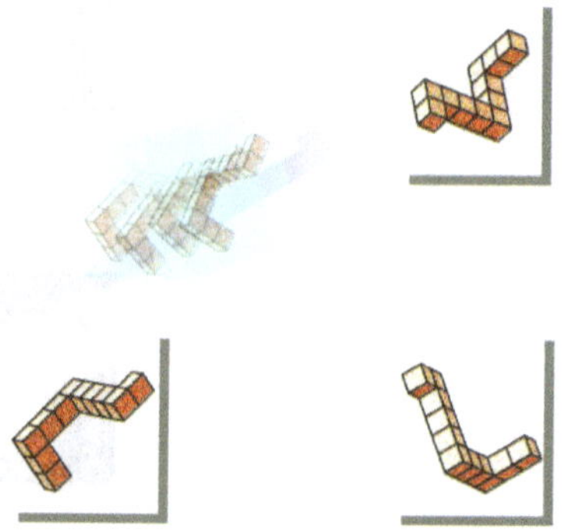

Fig. 59.3. 3D mental rotation is virtual gestural activity.

This competence means the velocity and accuracy of being able to decide whether two 3D objects are a rotational displacement (rotation plus translation) of each other or a rotational displacement of the mirror image of each other. It was shown in the Cooper-Shepard experiments that the comparison of such spatial objects was effectively performed by their mental, i.e., virtual, rotation (3DMR = 3D Mental Rotation). This indicates that the comparison of two spatial objects is realized by a motor action: moving them around in a mental space. This is very near to Châtelet's thought experiments. We compare shapes by a very common gestural action: taking the objects in our hands and moving them around.

The result of those investigations comparing musical skills and 3DMR are remarkable: It turns out that professional musicians have a significantly better spatial competence than control groups. And that the neuronal gray matter is more abundant in the relevant Broca area. In other words: musical activity (playing an instrument while sight reading the score) and mental rotational skills are managed by the same locality of the brain. The experiment's result is also true in the other direction: Persons suffering from amusia are significantly weaker in their ability to mentally rotate objects [116]. In a nutshell, it has been revealed that musical imagination and 3DMR are both driven by a profiled gestural motor activity.

Coming back to our problem of associating two gestures, we are now motivated by the findings of cognitive neuroscience to model such an activity by a hypergestural curve in a mental space which connects the respective gestures. Such a hypergestural continuation between discrete shapes is not artificial, but cognitively motivated. We in fact perform hypergestures when comparing objects and identifying their commonalities. This can again be restated with a more philosophical wording: Understanding is throwing around gestures through hypergestural trajectories and catching them in order to continue. Cavaillès is confirmed in an astonishingly modern sense!

Summarizing, we have found a geometric definition of a gesture, which explodes in its power of accumulated applications in a hypergestural construct. This 'heating up' of the gestural concept not only enables a rich repertory of concrete shapes in space-time, but also captures the paradigmatic action of comparing shapes or gestures that are only given as discrete points, but are cognitively correlated in a continuous way, germinating from motor actions as required in music, speech, and spatial competences. Broca's area, which hosts these neural processes, seems to accumulate the hypergestural potency of understanding in the making.

In neuroscience there are a number of discoveries and associated theories relating the action-body to higher cognitive functions. Let us refer to the most prominent one, associated with the discovery of mirror neurons by Giacomo Rizzolatti and collaborators in 1992. These neurons were discovered in the ventral premotor cortex of the monkey [910]: "The fundamental characteristic of these neurons was that they discharged both when the monkey performed a certain motor act (e.g., grasping an object) and when it observed another individual (monkey or human) performing that or a similar motor act." Similar findings were made for humans, and also for acoustical, not only visual, input. These results have been interpreted as evidence that human cognition is a closed circuit process: The passive perception is mirrored in the action layer, and there is growing evidence that, vice versa, action also induces sensorial neurons (visual cortex) to fire [285]. It has induced a series of models of human cognition and intelligence, individually, socially, and culturally. And it is astonishing that Cavaillès' above statement that understanding means catching the gesture and continuing, is nothing else but the mirror neuron mechanism in neuroscientific terms: Perception activates the neurons for action, you continue the action when understanding. This philosophy is akin to other neuroscientific approaches, such as Wolfgang Prinz's theory [864] of common coding of perception, cognition, and action, stating that these three human performances share one and the same code, a theory that had been suggested by the American psychologist William James and neuroscientist Roger Sperry [997]. The theory of mirror neurons, which can be interpreted as an embodied version of ideas that were also virulent in the above French gesture theory, has entailed a number of models for different areas of science. Let us mention two of them. Neurologist Vilayanur S. Ramachandran argues [877] that mirror neurons are the agents for the human capacity for learning and imitation, and that they will play the role for psychology that DNA did for biology: "Mirror neurons will provide a unifying framework and help explain a host of mental abilities that have hitherto remained mysterious and inaccessible to experiments."

59.2.1 Embodied AI

This research makes clear that a precise mathematical conceptualization of gestures and their embodied aspects is feasible. We have published two papers [720, 723] on this theory and shall come back to it in more detail below, when discussing the missing link between embodiment and cognition. The point in this conceptualization is that we shall be able to shed light on Violi's question concerning subject and embodiment, but also on the problem of how gestures and embodiment are related, a question that remained unanswered in the above French approach to gestural embodiment. Violi's question is also critical in the ongoing embodied AI research, as typically represented by the breath-taking investigations with robots and biological simulators conducted by Rolf Pfeifer and collaborators [846]. They quite radically state that "embodiment is an enabler for cognition or thinking: in other words, it is a prerequisite for any kind of intelligence. So, the body is not something troublesome that is simply there to carry the brain around, but it is necessary for cognition." The problem here is that the body can be simulated by electro-mechanical machines (sometimes even by what they call "cheap design," simple robots that do things just because their mechanical configuration enables it, such as walking down the street), but there is no evidence or even experimental proof that the gestural dynamics of these machines can transcend the level of a banal vaudeville show. In Violi's words: What must be added to transform the body into a subject? And vice versa: In what way is the subject acting upon its body? After the failure of the brain-oriented symbolic AI and the nerve-oriented connectionist AI, we have now arrived at the upside-down variant that starts at the feet and aims at reaching the brain's performance through the body's intelligence, so to speak. The critical question arises from the problem of understanding the added value of action, of gesture, of what makes the dynamic body become a subject without any postulate of divine instantiation of a soul. In Pfeifer's statement, the body was recognized as being necessary for intelligence. In a trivial sense, this is true, since disembodied clouds of intelligence have never been observed to date. But the underlying thesis is that it is also sufficient that intelligence is an emergent property of complex dynamic body configurations.

59.3 Anthropology

Summary. The Australian anthropologist Merlin Donald maintains that mirror neurons offer a basic mechanism for a cultural mimesis thesis.

– Σ –

The Australian anthropologist Merlin Donald maintains [273] that mirror neurons offer a basic mechanism for a cultural mimesis thesis: "Most importantly, the fundamentals of articulatory gesture, from which all languages are built, were put in place when mimetic capacity emerged."

In the vein of Donald's anthropological perspective, Italian performance artist Romina De Novellis considers the body's nonverbal communication as an embryonic form of 'high culture'. Therefore, to reach an integral vision of these phenomena and their origins, the investigation of infancy is essential. De Novellis's inquiry refers to a process similar to Darwinian evolution [244] that is structured in firmly linked stages of development. Her anthropological investigations refer to a 20th century perspective on the body, called *Les mutations du regard* by Alain Corbin, Jean-Jacques Courtine, and Georges Vigarello in the third volume of their comprehensive analytical work *Histoire du corps* [217]. These mutations affect not just the idea of the body as a means of communication but also as a projection of the mental and imaginative worlds, two elements important to Saussure in his semiotic analysis of human existence.

Curiously, the origins of the popular tradition of the physical body are recovered in the trance state, where the body becomes a communicative container of language. In the 1950s, Diego Carpitella and Ernesto De Martino [658], during a series of trips around southern Italy, recovered many testimonies of women who represented the performative behaviors of the attarantate tradition in the area of Salento, where folk rituals and festivities give us a quite primordial representation of the body in trance, expressing a psychological journey of emotions. When dealing with trance and bodily manifestations of emotions, it becomes necessary to understand the body as a representation, as a performance of its proper psychological dynamics.

While the evolutionary emergence of the mimetic capacity is a anthropological argument, the problem we referred to in the review of Violi's discussion remains unanswered: How can higher cognition emerge from the embodied action scheme in humans? How can we describe the evolution of language, logic, mathematics, music, and other arts from the action-perception cycles and mirror neurons?

59.4 Dance

Summary. At present, dance as a way of thinking in the embodied making of space-time has become the basic method for understanding embodied movement as thought. We discuss Cecil Taylor's, Paul Valéry's, and Rudolf Laban's ideas on dance.

– Σ –

Our introduction of dance to music theory goes back to the presentation of Mazzola regarding his formal tools in shaping free improvisation at a conference at IRCAM in 2002 [709], as already described in the second edition's preface. He learned following a dramatic insight that free improvisation is not about abstract algebraic tools, those theories which he had developed in this book's first edition [714], but about gestures that are embodied in the dancing hands of the acting pianist. This opened his approach to the vital shaping force of the body in the art of music, an approach supported by the insights of Cecil Taylor, the monstre sacré of improvised piano music, stating [996] that: "I try to imitate on the piano the leaps in space a dancer makes." Taylor quite radically opposes a representative type of Western music culture: "David Tudor is supposed to be the great pianist of the modern Western music because he's so detached. You're damned right he's detached. He's so detached he ain't even there. Like, he would never get emotionally involved in it; and dig, that's the word, they don't want to get involved with music. It's a theory, it's a mental exercise in which the body is there as an attribute to complement that exercise. The body is in no way supposed to get involved in it."

All of a sudden, the body became a core creator, not only the carrier of detached spiritual entities. And the famous mime Marcel Marceau echoed Taylor in saying: «Dans le vide de l'espace quelqu'un dessine, crée à travers son corps l'infini du temps. Les mains bavardent, le buste s'exalte, le regard s'illumine et la scène se remplit petit à petit. » ("In the emptiness of space somebody draws, creates with his body the infinity of time. The hands chatter, the bust heats up, the view is illuminated and the scene is filled bit by bit.") The body was recognized for not only inhabiting a given space-time, but even creating it. It was the visionary French poet and philosopher Paul Valéry who summarized these artistic perspectives on embodiment with his famous inscription about the fundamental role of the hand in human cognition and creativity[4] on Paris' Palais Chaillot:

Dans ces murs voués aux merveilles
J'accueille et garde les ouvrages
De la main prodigieuse de l'artiste
égale et rivale de sa pensée
L'une n'est rien sans l'autre.
(In these walls devoted to the marvels
I receive and keep the works
of the artist's prodigious hand
equal and rival of his thought
one is nothing without the other.)

It is not surprising that Valéry wrote an essay on the philosophy of dance [1074], in which he concludes not with a scholarly description of dance, but by suggesting we start dancing our thoughts instead of thinking about dance. This image of a thorough consciousness of the primacy of embodiment in the arts is completed by the fact that in dance and its theories, embodiment has been a strong and important approach, as borne out by Jaques-Dalcroze's eurhythmics [743] and by Rudolf Laban's work that for the first time succeeded in creating a subtle dance notation. But even more, Laban's geometric language was able to make evident the role of the human body in the very definition of space.[5]

At present, dance as a way of thinking in the embodied making of space-time has become the basic method for understanding embodied movement as thought [869]. However, it remains problematic to determine what precisely is the thought that dance is expressing or, more radically, what precisely is a "thought"?

59.5 Disabled Gestures Versus Gestures Disabled: Parlan's Versus Peterson's Pianism

Summary. This essay discusses musical gesture from the perspective of disability studies. Its focus is on two hand-impaired jazz pianists, Horace Parlan and Oscar Peterson. Its theme is the distinction between "disabled gestures" (Parlan) and "gestures disabled" (Peterson), with its principal methodology the analysis of recorded performances. Parlan, right side hemiplegic, as a result of childhood polio, always played the piano as a person with a disability. Thus, he played the instrument with a unique approach to gesture and

[4] The magic of the hand has recently been described in a marvelous book by architecture theorist Juhani Pallasmaa [814]; see also Section 57.

[5] See also the treatise [460], dealing with Laban's subtle contribution to the notation of dance, where the movement is never formally described, but only pointed at and left to the dancer's existential embodiment.

texture, the latter defined here as "the polyphony of gestures." An analogy is made to juggling, because, with significant right hand limitations, unable to produce bass, harmony, and melody simultaneously, Parlan must constantly switch between these elements. His intriguing gestural-textural idiom thus emerges from an obviously uniquely configured mind-body: "disabled gestures." Peterson, disabled late from a stroke in 1993, continued to perform as an essentially one-handed (right hand) pianist, with very limited left hand contributions, consisting of slow moving bass lines and, rarely, dyads. His post-stroke playing is most often "less," a textural subset of his prior, virtuosic, Tatum-esque style, and also of gestural-textural norms for jazz piano in tonal idioms. His playing is perceived as having missing elements; thus, "gestures disabled." The essay concludes with a contemplation of the potential application of the disabled gestures/gestures disabled binary, in particular a consideration of its use with instruments besides piano, for which it is excellently suited. There is a brief consideration of the binary's suitability to guitar, in reference to jazz musician Django Reinhardt and blues artist CeDell Davis.

– Σ –

59.5.1 Performative Gestures: Disabled Jazz Pianists

The musical traditions that interest me most originate in American vernacular traditions. They include rock, jazz, and blues. I have argued previously [623, 624, 625] that these, which value wide interpretive latitudes in performance—including improvisation, composing/arranging one's own part, and openness to broad variance in rhythm/phrasing—tend to provide greater opportunities to and allow acceptance by musicians whose bodies do not enable them to meet those normative requirements of instrumental or vocal technique that pertain most notably to the "standard repertoire" of Western classical music. Non-normativity does not, however, necessarily preclude virtuosity, and may even be the source of unique gestural styling.

The artist to whom I've given most attention in my research is jazz pianist Horace Parlan [623, 624, 625]. Hemiplegic, largely paralyzed on his right side, as a result of childhood polio, Parlan played mostly with his left hand, which supplies bass notes, chords, and faster moving melodies, obviously by necessity alternating these functions most/much of the time, though he is quite adept at including chords with his left hand melodies. Oscar Peterson [626], a better-known pianist, became disabled late in life, as a result of a stroke and lost most of the use of his left hand, which he used only for slow moving bass lines, chording and playing melodies with the right, sometimes simultaneously.

I distinguish between these pianists by characterizing Parlan's playing as comprised of "disabled gestures" and Peterson's post-stroke playing as composed of "gestures disabled." The distinction has much to do with the different trajectories of the artists' careers. Parlan was disabled quite young and, from the beginning of his studies, always played the piano with physical limitations. With his right hand nearly immobile, Parlan was never able to play the standard classical repertoire or any conventional two-handed textures and was thus required to develop a highly individualistic approach, characterized by a fast and facile left hand and arm and a slow right locked into a limited number of dyad formations that nearly always play successions of parallel intervals. Peterson, by contrast, was known for his exceptionally virtuosic technique throughout his career, even receiving acclaim from many critics after his stroke for having maintained much of that prowess [626], praise I would argue combined wishful thinking, understandable support for an esteemed artist and, to the degree that it was true, referred only to the pianist's unaffected right hand and arm.

This essay is principally concerned with Parlan's "disabled gestures," which are more distinctive and interesting than Peterson's "gestures disabled." The latter derive from a more conventional approach to the instrument, that is, Peterson's own pre-stroke technique. I have treated Peterson's post-stroke career and playing at great length elsewhere [626] and offer a consideration of his work here primarily in the interest of a contrast to Parlan's, as both an individual artist and an exemplar of a particular relationship between disability and gesture.

59.5.2 Horace Parlan: Disabled Gestures

Unlike disabled classical pianists who never use their impaired hand at all, Horace Parlan[6] used his immobile but large right hand in a manner akin to the mallets of a xylophone or marimba, striking the keys with two fingers at a time in dyads; typically, depending on the fingers used parallel octaves or thirds, less often fourths. He could not use his thumb, which significantly limited the speed of his right hand, typically used for melody by jazz pianists. This made right hand legato possible for Parlan only with the use of the damper pedal, which he employs extensively. Parlan's functional limitations have never held him back from a distinguished career as a sideman or leader. Notably, he has often performed in the most challenging, exposed contexts for a pianist, as a soloist and, famously, in duets, with tenor saxophonist Archie Shepp.

Like any jazz pianist, Parlan played differently according to context, in particular depending on whether or not he has the support of bass and, to a lesser extent, drums. But his disability also—inevitably—caused him to produce different gestures and textures than those produced by pianists with two fully able hands. Before observing the distinctiveness of Parlan's playing, we must consider how an audience perceives a pianist's gestures.

We should assume a degree of synesthesia on the part of most of the audience. Note that I have been avoiding the term "listeners," because I propose that even those who cannot see the pianist for whatever reason (such as those listening to an audio recording), unless they have never had sufficient vision (or, perhaps rarely, because they have simply never seen a pianist), are imagining a visual element, whether it is some semblance of an actual pianist, a sequence of contours, or some combination thereof. It is also likely that the music is received to some degree tactilely.

59.5.3 Parlan with Bass (and Drums)

The best resource for viewing Parlan's playing with rhythm section, in particular the tonal support of bass, is the Archie Shepp Band video, *The Geneva Concert* [960], recorded in concert at the New Morning Jazz Blues Festival (portions of the concert are easily accessed online on YouTube and other websites). Understandably, most of the shots of Parlan playing are taken during his solos, with occasional shots of comping, that is, his mostly chordal accompaniments to instrumental solos, by saxophonist Shepp and bassist Wayne Dockery, and to Shepp's vocals. It is illuminating to see how he plays, with an exceptional left hand and a right hand whose fingers are essentially immobilized but used in a manner somewhat analogous to mallet percussion, such as vibraphone, marimba, and xylophone. At the same time, it is important to recognize that there is no simple, precise correspondence between the gestures as seen and as heard. The distribution of musical material between the hands, and thus also the polyphony of gestures, is not always audibly obvious, arguably less so than in the playing of able-bodied jazz pianists, particularly since Parlan will often transfer melody from his right hand—with which he could only play parallel dyads and at no more than a moderate tempo—to his exceptionally nimble left, without shifting register, or divide melody between the hands.

Although Parlan obviously played technically and texturally quite differently from other jazz pianists, due to his disability, this can easily go unnoticed, even among highly skilled and knowledgeable listeners. I discovered this at the 2010 University of Guelph Jazz Festival Symposium, at which, as keynote speaker, I lectured about Parlan and his hand impairment. After my talk, I met several jazz scholars who had known and admired Parlan's playing for decades, but who were unaware of his limitations. (Because Parlan has lived in Denmark and mostly worked in Europe since 1973, he is mostly known in North America through his audio recordings.) Even more striking are those critics who misunderstand his hand impairment in ways that are belied fairly easily by listening alone. Thom Jurek [514] characterizes Parlan's fourth and fifth fingers as "useless," while Stephen Thomas Erlewine [304] writes that

> If it weren't for the inventive chord voicings and percussive right-hand attack, it would be impossible to tell that he was missing two fingers on his right hand, since his playing is remarkably agile and fluid.

[6] See `https://www.theguardian.com/tv-and-radio/2015/jul/11/horace-parlan-tribute-david-hepworth`.

Yet another anonymous writer describes Parlan as "an excellent modern jazz pianist, with a strong left hand," without reference to his disability [271].

The extent to which Parlan played right hand octaves, easy to hear and even easier to see, belies the observations of the two critics regarding Parlan's useless or missing fingers. That he could not use his right thumb at all, but used all his immobile fingers, invites percussive playing, but this is not a characteristic of Parlan's right hand only, nor did he at all eschew legato, though, as previously noted, unable to use his (typically all-important) right thumb, he often relied on the damper pedal. While his hand impairment obviously affected Parlan's chord voicings, as his characteristic right hand intervals are octaves or thirds, he also often chorded with the left hand alone. At no time in any video of Parlan's playing have I observed him playing anything but dyads with his right hand: mostly octaves and thirds, but also sevenths and, rarely, fourths. If he was able to play one note at a time or more than two notes simultaneously with his right hand, he chose not to. I suspect, though, that this enormous consistency of right hand texture is a forced choice, though one that Parlan used to great and unique advantage.

Once one is aware of Parlan's disability and knows what to listen for, the differences in his gesture/texture from that of his pianist colleagues become more readily apparent, even when he was supported by a rhythm section, our concern at this time. These differences are, not surprisingly, even more obvious when he was not supported by bass and drums, either playing when solo or in his renowned duet recordings with Shepp (to which we will turn later). Although, out of necessity—but surely also in part owing to an outstanding harmonic imagination—he voiced chords differently from able-bodied pianists, and often only with his left hand, the differences are not such that even the most astute listener is likely to attribute them to his hand impairment without access to video. Although his technical proclivities largely emerged from his impairment, an able-bodied pianist could certainly play as he does, although, to the best of my knowledge, none do. Further, the idiosyncrasies of his playing are often manifested in things he *does not* do, and listeners are, of course, far more likely to focus on what a performer *does* than the infinite number of roads not taken.

59.5.4 Parlan with Rhythm Section

As a soloist supported by bass and (usually) drums, Parlan gestured quite differently from other pianists. Because his pitch/rhythmic language, though personal and imaginative, grounded him in the post-bop idiom of his collaborators, the uniqueness of his pianistic textures and contours only appear as remarkable as they are once one is aware of his impairment and knows what to listen for and, when possible, to watch.

As can be witnessed in this quartet performance with Archie Shepp, among the configurations of harmony and melody he employs—like most post-bop pianists supported by a rhythm section, Parlan left bass lines to his bassist—Parlan often played melody with his able left hand, while comping simultaneously, if sparsely, with the same hand and less often with his right. Left hand melody with right hand comping, spare though the chords may be, is, of course, essentially the inversion of the typical configuration of melody and harmony.

Parlan's left hand melodies often extend into a higher range than that hand typically goes and exhibit a greater proclivity for continuous ascent—that is, toward and sometimes beyond the body—than his right hand melodies. Thus, although there is a certain physical relaxation characteristic of the hand's motion toward the body, its melodic ascent translates in sound into heightened tension, the opposite affect from the natural tendency toward tonal descent in right hand melodies as it moves inward.

Other features of Parlan's soloing include slower-moving lyrical right hand melody in parallel intervals, at times joined by the left with chords, all in rhythmic unison; antiphony at the phrase level between left and right; and, perhaps the most visually striking, melody divided between the hands, often in oblique motion, with a moving left against a static right. All these gestural templates were strongly idiomatic to the pianist's impaired right hand. Any of the performances from *The Geneva Concert* [960] offers significant opportunities to see as well as listen to Parlan's idiosyncratic technique, but "Arrival," "Steam," and "Sophisticated Lady" particularly stand out. The first two are especially notable as they are uptempos, something not found at all in Parlan's recorded solo oeuvre and only occasionally in his recorded duets with Shepp. It appears that tempo was not a limitation at all for Parlan when there was a rhythm section to support him. The inclusion of bass seems more important than drums, both because of its obvious ability to provide and because, elsewhere, in Parlan's own DVD documentary, *Horace Parlan by Horace Parlan* [819], he chose to perform accompanied only by bass.

59.5.5 Parlan as Soloist

The only video I am aware of that is available commercially or on social media that features Parlan unaccompanied (there are no videos available of the many Shepp/Parlan duets) is his arrangement of the traditional spiritual "Deep River," from the video documentary *Horace Parlan by Horace Parlan* [819]. (On all other performances in this film, Danish bassist Jimmi Pederson accompanies Parlan.) The principal difference between Parlan's solo playing and his work with rhythm section is, much like it is for any jazz pianist in a similar tonal idiom, the necessity of providing bass notes. The main technical distinction between Parlan and able-bodied pianists is, of course, the larger responsibility of his left hand, which must also supply harmony and any fast-moving melody. These three functions cannot all be simultaneous, though, and the bass line, which is uniformly sparse in these performances—mostly comprised of isolated if prominent pitches—but no less important for being so, alternates with either chords or melody. Left hand chords accompanied Parlan's right hand melody, which is characteristically slow to moderate in speed and in parallel octaves, made legato via the damper pedal. Left hand melodies in this no-rhythm-section context were even more often long ascending than when Parlan played with a rhythm section (even when it is bass only, as in this video). These ascents sound and feel as if they are a natural continuation of the leap up from bass notes. This is particularly notable toward the end of "Deep River." Further, the trajectory of these melodies is often completely ascending, that is, without any deviating descending motion, however brief, such that they have the gestural sense of arpeggios and thus a quasi-harmonic feeling, regardless of whether their interval content readily suggests chords or not. Thus, a sense of harmony emerges from a gesture that suggests it—that is, continuous ascent—rather than from intervallic content.

(Later in the documentary, during Parlan's performance of his composition "Arrival"—also heard in *The Geneva Concert*—we get a brief glimpse of his left foot working the damper pedal. This is doubtless a consequence of his impairment and it is possible that, because it changes his orientation to the keyboard, it has an influence on his playing, leading him to favor the upper register.)

In addition to "Deep River," on the abovementioned documentary, Parlan has recorded three solo albums: *Musically Yours* (1979, re-released 2010) [821], *The Maestro* (1982; re-released 1995) [818], and *Voyage of Rediscovery* (1999; re-released as *Horace Parlan in Copenhagen*, 2008) [820]. The gestural/textural characteristics described above in the pianist's performance of "Deep River" are all in evidence throughout these recordings. After one has seen Parlan's playing, the distribution of musical material between the hands can be determined in these audio-only recordings with some confidence. These three albums comprise a large database of solo performances that reveals Parlan's tendencies in a manner that cannot be determined from a single piece, albeit the one available on video. The relationship to his hand impairment is clear.

Taken as a body of work, the most striking aspects of Parlan's solo playing are the complete lack of uptempos, the large portion of the time, perhaps even a plurality, spent playing left hand melody alone (accompanied by the pedal resonances of previously played notes in all registers), and the sparseness and exposure of bass notes. Although Parlan could play melody with his left hand as fast as necessary for any tempo, as evidenced in Archie Shepp's Geneva concert, the limitations of his right hand appear to prevent him from accompanying the left appropriately to enable quick tempi. It is possible that there was another reason for the pianist's having chosen only slow and medium tempos, but their absolute consistency over three albums (plus "Deep River") strongly suggests that fast solo playing was not an option.

Of course, the lack of fast tempi in Parlan's recorded solo *oeuvre* is striking only when that repertoire is considered as a whole, not in individual works. Much more distinguishing, if not unique, is the great preponderance of solo, left hand melody in every performance. The pianist's solo output, like much of his work with rhythm section, is largely a one-handed affair. Since Parlan could not possibly provide all functions typically delivered by solo jazz pianists in tonal idioms—bass, harmony, and melody—he often alternated sparse bass notes with abundant, virtuosic, left hand melody, occasionally buttressed by chords.

In addition, he less frequently comped with the right hand. As previously noted, this left hand melody-in-the-midrange/right hand chords-above texture is a distinguishing characteristic of Parlan's playing that also contributes to his unusual chord voicings. As previously mentioned, at times the right hand joins in melody, often repeating a dyad in alternation with moving notes in the left, creating a rapid-fire oblique motion. At other times, when the left hand is not soloing (supported only by damper pedal resonances), more typical textures are employed, such as left hand bass and chords (in alternation), supporting right hand melody in parallel intervals, or chordal passages employing both hands.

Finally, the role of the bass notes requires an accounting. Parlan was a post-bop, rather than a stride player, who only occasionally visited the lower register. Typically, only one bass note, or occasionally a bass dyad, is played at a time, in alternation with another left hand function, harmony or melody. These alternations can be quite rapid. What is striking, though, is that the right hand, as best I can tell, never strikes notes simultaneously with the left hand when the latter is in the bass register. This may simply be a style feature of choice, or it may also be related to Parlan's right side impairment that affects both his hand and foot.

The result of the forced choices that emerge from Parlan's impaired right hand and extraordinarily nimble left are more transparent in Parlan's solo playing than when he is supported by a rhythm section. Far from making him a lesser player than his peers, his inventiveness with texture and gesture in light of his technical limitations are a source of great fascination. That he could not "do everything at once" makes for a rapid alternation of textures and registers, somewhat evocative of the Baroque *stile brise*, which may be most familiar in lute music and Johann Sebastian Bach's unaccompanied works for violin and cello. The most unusual texture, though one Parlan uses a great deal, is the solo left hand melody. Its alternation with occasional bass notes give the music a stop-time quality, made even more interesting by the rhythmic unpredictability of the timing of the bass and the asymmetry of the left hand phrases of varying lengths.

There is certain strangeness about this nearly monophonic piano texture that the late disability theorist Tobin Siebers [968] argues is intrinsic to Modernist aesthetics. Given his extraordinary tonal, especially harmonic, imagination, it is not unreasonable to place Parlan in the company of such great post-bop solo players as Bill Evans and Thelonious Monk. Comparisons to the latter are especially apt, given that Parlan shared Monk's affinity for sparse textures and lack of interest in overt displays of technique. In that context, it is noteworth that Parlan recorded two solo performances of Monk's classic "Ruby, My Dear," on both *Musically Yours* [821] and *The Maestro* [818].

59.5.6 Parlan's Duets with Archie Shepp

There are no videos[7] of the Parlan and Shepp duo available commercially or on social media. The roles Parlan assigned to his hands in these performances must therefore be conjectured from a copious body of audio recordings, consisting of five albums, both studio and in concert. The textures and gestures employed seem in essence the same as those in his solo recordings. Notably different from Parlan's solo albums, though, are there occasional uptempo pieces with a proclivity for steady tempos, with the notable exception of the album *Goin' Home* (1979, re-released 2010) [961], a collection of spirituals and similar works, often performed in a rubato style, evocative of preaching. As an accompanist, Parlan is distinguished for his eschewal of walking bass—not surprising given that his bass notes are always sounded alone—and a nearly consistent avoidance of steady enunciation of the beat, with the exception of a couple of pieces, "Backwater Blues" (from *Trouble in Mind*, 1980, re-released 2010) [962] and "Mama Rose" (from *En Concert: First Set*, 1987) [963], that employ ostinatos. That the pulse is felt so strongly without either artist constantly marking it—Parlan in particular shapes time without consistently playing on the beat, while Shepp is often given to Coltrane-like "sheets of sound" that go "outside"—turns the lack of rhythm section into an advantage, with a kind of seemingly effortless swing. I am tempted to suggest that the lightness of texture of this duo mirrors and magnifies Parlan's own playing, in which the tonal tasks of bass, harmony, and melody are out of necessity constantly juggled.

[7] But see `https://www.juno.co.uk/products/archie-shepp-horace-parlan-trouble-in-mind/552212-01/`.

Parlan's solos in these duo performances are similar in gesture and texture to those on his solo recordings. His support of Shepp occasionally employs a stride texture, and at times utilizes ostinatos, but more often he only comps, with either left hand or both hands, in much the same manner as he plays with rhythm section in the Geneva concert. His comping is heavily syncopated, leveraging the phenomenon that his—and Shepp's—playing around the beat with great momentum results in the beat being heard and, perhaps even more, felt, whether it is actually enunciated or not.

Of particular interest is "Deep River," from the album *En Concert: Second Set* [964], which Parlan also plays solo in his documentary, as discussed above. Parlan's proclivity for long, ascending left hand lines is at its most marked, both under Shepp's tenor saxophone performance of the melodic "head" and in the solo that follows, as well as in the pianist's own solo. The impression of melody-as-harmony in such passages is reinforced with skillful pedaling. With a duration of 11:54, the performance is nearly twice as long as Parlan's video solo (6:31). The two performances of Duke Ellington's "Sophisticated Lady," from *Reunion* [965] and *Second Set* [964], are also standouts. Parlan plays mostly chords here, but topped with such a strong melody that the impression is more of counterpoint than homophony. *Reunion* also includes "Cousin Flo," in which, unique among these recorded duets, Parlan plays only a left hand basso ostinato much of the time, transposing it according to blues chord changes.

59.5.7 Disabled Gestures

That Parlan's hand impairment and the unique gestures and textures it generates in his playing have been largely overlooked—or perhaps "over-listened"—surely owes in part to the limited, even nonexistent, opportunities those of us in the Western Hemisphere have had to see him play. But, more important, they also owe to habits of listening that are so focused on tonality and those aspects of rhythm most closely related to tonality. Parlan was an artist with an extraordinary tonal, especially harmonic, imagination and, although it is beyond the purview of this essay, some of his invention appears to be a function of the need to voice chords differently, owing to his right hand impairment. This hand is best at reaching large dyads, such as sevenths. And since Parlan often played melody with chordal punctuations in his left hand while comping with his right, chord voicings are often interesting.

More important though, is what I previously called "juggling." Parlan had to do so much of the work of a tonal jazz pianist with his left hand—bass, harmony, and melody—that he had to alternate these functions rapidly. This creates a shaping of time based on shifts of register and function. Perhaps most notable are the many times when all that is freshly sounded is left hand melody, accompanied by recently pedaled notes in one or both hands. These resonances are not drones—most of these pieces have steadily moving harmonies—but they are un-damped resonances that accumulate into chords in a manner evocative of hammer dulcimers such as the Persian *santur*, Indian *santoor*, and Chinese *yangqin*, giving Parlan's often monophonic playing an Eastern quality that I have not heard elsewhere. That there is an expectation in piano playing of a more consistently polyphonic texture only adds to the special quality of Parlan's artistry.

Of course, the character of Parlan's playing is not only a function of a left hand equal to the demands of multitasking, but also of the impaired right hand whose limitations impose greater work on the left. Parlan seems not to play simultaneously with his right hand while he is striking bass notes. He only uses his right hand for harmonic dyads, slow moving melodies in parallel octaves, and shared melodies with the left, in which the right often plays repeated notes in alternation with the left's melodic motion. Thus, Parlan's right hand is limited in its independence, its textural capabilities, and its speed, all of which contribute—positively—to a sparseness like no other.

This may be most readily appreciated in his duets with Archie Shepp, where bass notes are relatively infrequent and Parlan's comping is usually rhythmically quite varied and highly syncopated, though it is firmly wedded to the beat, in contrast to Shepp's frequent "sheets of sound." That both players are characteristically intense—Shepp with a technique, sound, and style honed in free jazz, Parlan with extraordinary harmonic invention and richness, even within impairment-imposed textural limitations—is tempered and given a sense of flight when unencumbered by the bottom and beat of a rhythm section. The steady pulse is instead kept, perhaps even performed, by listeners, who perform a calculus upon the duo's rhythmic inventions, especially Parlan's sense of swing.

Beyond the specifics of Parlan's music per se, his playing within such forthright limitations provides an important lesson in the genesis of gesture and texture. Parlan's playing in an improvisational idiom reminds us as few others can that we all think and perform with our bodies. It is only the atypicality of Parlan's body—that is, its right side and especially the right hand—that makes the corporeality of performance so obvious. While another pianist might be able to emulate Parlan, Parlan has only ever been a pianist with a very specific limiting impairment, to whom a particular textual and gestural imagination is absolutely idiomatic and frames his musical imagination.

This relationship of corporeality to invention is true of every body, of course, but the presence of standard technique and repertoire as defined in Western classical music—whose hegemonic influence is so strong within certain sectors that it can safely be referred to simply as "music"—has deflected attention from the physicality of music making. Canonic works are composed for "piano," not the particular physical idiosyncrasies of any single "pianist." This even applies to those works composed for a single hand—almost always the left—since the technique is normalized for the one able hand, while the disabled hand is exiled from performance.

I find in Parlan's unique approach to the piano based on his impairment "disabled gestures"—a positive concept—because they are the natural outcome of a lifetime of making the musical best of significant physical limitations. In this context, it makes sense to think of the piano and the human body as interacting instruments. Although neither a guitar nor a saxophone has the polyphonic capabilities of the piano, I suspect that few among us think of them as "less" than the piano, in part because they both are readily capable of many things, such as access to considerable timbral variety, that are challenging at the keyboard and not in the repertoire of every skilled pianist. Likewise, because he thought and played with the body he had (as we all do, if less noticeably), Parlan had easier access to what is gestural and texturally special about his playing than an able-bodied pianist whose imagination is differently wired corporeally.

59.5.8 Gestures Disabled: Oscar Peterson

Fig. 59.4. Oscar Peterson. Photo Reuters/Jean-Bernard Sieber-ARC, reproduced with permission.

Those "gestures disabled" that are exemplified in Oscar Peterson's playing from 1994 to his death are something quite different from the "disabled gestures" described above in the music of Horace Parlan.

Peterson was a very different pianist with a very different impairment history. As quoted in Block [121], a self-proclaimed "traditionalist", long known for his formidable technique and aggressive playing, Peterson had a major stroke in 1993. After a period of rehabilitation, he resumed performing in 1994 and performed until his death in 2007.

The stroke seriously affected his left side. Post-stroke recordings—ten CDs, one DVD, and a radio transcription of National Public Radio's *Marian McPartland's Piano Jazz* [742]—all indicate that Peterson could only play single notes and, though rarely, dyads with his left hand, and only slowly, apparently using arm, rather than finger weight. Much of the time in recordings [742, 832, 833, 834, 835, 836, 837, 838, 839, 840, 841, 842], he appears to be playing with his right hand only, an assessment whose reliability owes to substantial registral spreads between the bass line and upper voices when the left hand is used.

While critical assessment of Peterson's post-stroke playing runs the gamut from enthusiasm to vicious condemnation, critics, liner note authors, and the pianist himself (especially in his conversation with McPartland) are unanimous in that Peterson played "less" after acquiring his disability. "Less" in this context is not meant aesthetically; rather, the artist was incapable of delivering much texture with his left hand. And, although Peterson's right hand seemed unaffected and he remained capable of playing extremely fast with it, as in his performance with McPartland, many critics, particularly Stephen Holden [477] and the pianist himself (as quoted in Moon) [762], felt that his playing in this period was more restrained and lyrical. Holden in particular felt that, as I have observed with Parlan, Peterson used his forced choices to musical advantage.

While I am inclined to agree with Holden that, at least in some ways, Peterson became a more thoughtful musician as a result of this major physical challenge, I am less concerned with that aesthetic judgment than with drawing a distinction between Parlan's "gestures disabled" and Peterson's "disabled gestures." As we have seen and heard, Parlan, a disabled pianist his entire life, out of necessity developed an approach to gesture and texture that differs radically from that of any of his colleagues, most notably in the preponderance of monophony (accompanied only by previously pedaled notes) and the rapid alternation of registers and tonal functions that he cannot fulfill simultaneously.

Peterson, by contrast, had had a long career as an able-bodied pianist known for his dazzling technique prior to acquiring his disability late in life. While there was a new lyrical bent in his playing, even at times eschewing technical virtuosity altogether [742], there is never in him the sense of gestural/textural innovation that emerges from Parlan's lifetime of thinking and playing with an impaired body. Peterson either plays as much of what he would have played pre-stroke or he plays simpler—but not unprecedented—textures that one might encounter in another pianist.

It is not only because Peterson's disability was acquired late that his playing is—by his own admission—"less," rather than gesturally-texturally creative. It is also because the nature of Peterson's impairment was different from Parlan's or that of any of the famous disabled classical pianists. It is a simple question of which hand was impaired and to what degree. Like Parlan, all the famous classical pianists with hand injuries—Paul Wittgenstein, Leon Fleisher, and Gary Graffman (I exclude composer Robert Schumann from this list as a non-performer)—lost right hand function, or, in Wittgenstein's case, the right arm. The best-known one-handed classical repertoire such as the Ravel and Prokofiev concertos and the Brahms arrangement of J. S. Bach's unaccompanied violin *Chaconne*—is for the left hand. This owes only in part to the existence and commissions of famous injured virtuosi. Some of these works exist for other reasons, for example, as etudes. Finally, it is widely acknowledged that the left hand is better suited to one-handed pianism.

Pianist-composer Hans Brofeldt [155] attributes this advantage to the left thumb being more apt for melody than the right hand fifth ("pinky") finger and to the left arm's greater capacity for upward leaps, an auspicious feature of Parlan's playing, as he quickly shifts registers and thus also functions, due of course, to the left thumb's advantageous position over the right's. But I would also add Brofeldt's characterization that the left side has more ready access to the critical bass range of the instrument. Of course, Parlan made significant use of his impaired hand, far more than Peterson does, because both of the latter's more significant disability and that disability's very late acquisition. (Brofeldt is also adamant that *any* hand injury renders that hand utterly useless for a pianist, a classical musician's perspective fortunately not shared by either Parlan or Peterson.)

As Holden observed [477], Peterson was late in life essentially a one-handed pianist, far more than Parlan. There are no dazzling shifts of function in Peterson's playing. The left hand only plays slow bass lines. The right plays melody and comps, sometimes during solos doing both at the same time. Like Parlan, Peterson had large hands that facilitate this. From a gestural-textural perspective, Peterson is most interesting playing unaccompanied, which, unlike Parlan, can be heard only very rarely, in two recorded performances, both of his own compositions: "Love Ballade" (on McPartland's 1997 radio program) [742] and "When Summer Comes," on *The Very Tall Band: Live at the Blue Note* (1999) [837]. The former includes little or no improvisation, the latter a bit more, with more jazz feeling but not a great deal of virtuosity: both are essentially Romantic in style. Peterson tells McPartland that "Love Ballade" is influenced by Chopin; thus the use of "ballade" rather than "ballad". In both pieces, the right hand employs Romantic piano textures, most notably arpeggiation beneath melody. Peterson composed more after his stroke, which Robert Sandall [271] associates with the pianist's reduced technical capacities and Richard Palmer [815] proposes was a means of continued creativity during Peterson's initial post-stroke inability to play. Like Parlan's, Peterson's solo performances avoid uptempos. As further evidence of the pianist's limitations, "When Summer Comes" is by far the shortest of the three unaccompanied solos by the stars of The Very Tall Band, the others by vibraphonist Milt Jackson and bassist Ray Brown. (When McPartland plays a solo during their radio program, she tells Peterson it is to give him a medically needed break) [742].

The above discussion illustrates Peterson's "disabled gestures." In his impressively frank conversation with McPartland, he professes his significant limitations and aptly characterizes the new limitations on his playing as "still me." "Me" is often, though not always, characterized by relentlessly fast right hand melodies, at times supported by chords in that same hand, but less support from the left, only in the form of simpler bass lines and (as best I can tell) no chords whatever. Some textural elements are either conspicuously missing or reduced in density from what they once were in Peterson's playing and remain the gestural-textural norm in post-Swing Era tonal jazz piano. These omissions—of anticipated normative texture, not necessarily of aesthetic value—are heard—and likely otherwise synaesthetically perceived—as empty time/space. The support of such aggressive guitarists as Lorne Lofsky and Ulf Wakenius during essentially one-handed solos exemplifies these "gaps" in musical texture, reminding the audience of the missing or reduced left hand contribution, an absence felt more than when, for example, Parlan (who never plays with guitar comping or other chordal support) plays left hand melody alone over bass (and drums). Over a lifetime of performance with an impaired hand, Parlan, by contrast, mastered the "disabled gestures" of keeping all musical elements in play through rapid alternation and register transfer; his disability resulted not in an aesthetic deficit, but in a display of extraordinary prowess and invention, and a particularly creative shaping of time by means of shifting gestures and textures.

Post-stroke, Peterson's playing in various ways manifested capacities diminished from the pianist's legendary Tatum-esque technique. With a rhythm section (or McPartland's second piano), he played reduced textural elements, such as comping with only the right hand and performing much-simplified, slow bass lines. But the conceptions of gesture and texture remain the same as prior to his stroke. Whether he was more dependent on his accompanists than before is a matter of debate and Peterson at times seems to have changed his mind (or, I prefer not to think, contradicted himself) [626].

59.5.9 Conclusion

Perhaps the distinction between disabled gestures and gestures disabled might best be illustrated by harkening back to my experience presenting Horace Parlan's music to an audience that thought it knew him and was surprised to learn of his impairment. Were I to have presented on Oscar Peterson (and assuming his stroke was not known to the audience), the likelihood that "something was missing" would, I believe, be far greater. Even if Parlan's audience were to have noticed his unique approach to gesture and texture—which might be overshadowed by his tonal creativity, I doubt that many would speculate that it arose from disability. Other pianists, like Thelonious Monk and Cecil Taylor, are also geniuses of gesture and texture, but their creativity does/did not emerge from physical limitation. "Disabled gestures," then, are roughly analogous to, for example, "blues gestures" or "Italianate gestures;" an element of style, not of defect.

Post-stroke, Peterson, on the other hand (pun not so much intended as inevitable), plays in a manner in which the "holes" in the otherwise normative texture are obvious. It would not be challenging to ascertain that the diminished texture results from diminished physical capacity. The lack of (much) left hand is made more obvious when, as was typical in this period (and much before Peterson as well), guitar supplies much of the harmony, albeit in a very different manner than piano. "Gestures disabled," then, exhibit a loss, perhaps not aesthetic—a matter of critical debate [626]—but surely a deficiency in texture that deviates from a norm indicated by style and those musical elements that remain.

It remains to be seen the degree to which the distinction between disabled gestures and gestures disabled is operative for instruments beyond piano, for which it is extremely apt. I have, in my earlier work, alluded to that distinction in a discussion of the legendary hand-impaired guitarist Django Reinhardt [624]. It is even more applicable to the underappreciated Delta blues master, CeDell Davis [656], both of whose hands are impaired by childhood polio. (Owing to a series of strokes, Davis no longer plays at all, but still sings.) In both guitarists' work, impairment is an impetus for creativity, that is, for disabled gestures, for style rather than defect. The application of these categories is a project that has only begun.

59.6 Aristotle, Blind Lemon Jefferson, and Vilayanur S. Ramachandran Walk into a Bar: Blues, Blindness, *Politics*, and Mirror Neurons

59.6.1 Introduction

I propose elsewhere in this volume that the perception of musical gesture is inherently synaesthetic, not only aural, but visual and often tactile, as well. Surely the case for visual perception of musical gesture is supported by references to "high" and low" pitch, "horizontal and vertical" tonal relationships, and musical "form" and "architecture," "tone color," and, of particular relevance here, the "blues." Some visual analogies such as these are culturally and temporally contingent. "High" and "low" pitch had opposite meanings for ancient Greek and medieval theorists. "Horizontal" and "vertical," understood to be unfolding ("melodic") and simultaneous ("harmonic") relationships, are so construed because of their display in standard Western staff notation. The value of visual imagery so evoked is thus, to say the least, problematic, as it has no clear and unambiguous relationship to perception.

My focus here, though, is in the blues, a musical idiom synaesthetically named. As a scholar of disability studies in music, I find it most ironic that it has been said that the archetypal bluesman (rarely a woman), the person best qualified to deliver this sound-as-color imagery, is blind. (In an incidence of triple synesthesia—aural/visual/tactile—it is often said that the "blues is a feeling," a clause that produced 43K Google hits in a search on 25 July 2015.) Not only have there been numerous blind blues musicians, such as Blind Lemon Jefferson, Blind Blake, and Blind Willie McTell, there have been a few sighted musicians (even, at one time, Bob Dylan, alias "Blind Boy Grunt") who have appropriated blind identity [1], doubtless believing, like neurologist Oliver Sacks [925], the "Blind" is "almost an honorific."

59.6.2 Division by (Almost) Zero: Many Blind Bluesmen but Few Blind Blues

It is a great curiosity that there are virtually no blues songs about blindness. The term "blues" applies not only to a musical idiom, but also to an emotional reference that the *Oxford Dictionary of American English* (online) defines as "feelings of melancholy, sadness, or depression." The three blues songs identified by Taylor and Hughes [1038] as concerning blindness, all by blind artists—"I Must Be Blind, I Cannot See" (Blind Roger Hays), "Lord I Wish I Could See" (Blind Gary Davis), and "Stone Blind Blues" (Sleepy John Estes) all characterize that disability in emotionally "blue" terms, not surprising giving the hardships sightlessness must have imposed on African-American men in the mid-20th century. Given its impact on livelihood, education, socialization, and other activities, blindness must have been a central and disadvantaging factor of identity in these and other blind blues musicians' lives; its virtual absence as a theme of blues songs is a question that must be explored.

Blues scholars Yuval Taylor and Hugh Barker [1038], who compiled the very short list of blues songs about blindness, offer two possible explanations for the dearth of such compositions. The first is that the artists chose not to call attention to their disabilities. This is at least in part belied by the number of them—apparently a majority during the pre-electric period of the blues [120]—who used "Blind" in their stage names. The second reason Taylor and Hughes propose [1038], that the audience for the blues was and remains overwhelmingly sighted and thus not interested in songs about blindness, is compelling, owing to the rationale, which we shall shortly see, behind artistic renderings of negative emotion.

In "Blues and Catharsis," philosopher Roopen Majithia [1009] references Aristotle on music in observing that the artistic performance of negative emotions is a means by which such feelings are purged. It makes sense then that the subject matter of blues songs appeals more broadly to the experiences of listeners than would a song concerning a particular disability they have not experienced (and which perhaps they find too frightening to contemplate). As Majithia states (p. 90):

> [T]he blues has always been about universal themes: love and loss; trying to find strength in the face of misfortune; metaphoric, often comic, vignettes of sex; and so on—themes that increasing occurred in the parlance of urban culture in ways that resonated with the experiences of people, cutting across race and gender.

If Majithia is to be believed—as he is by me—audiences for the blues (far) beyond the African-American audiences for whom and from whom the music originated are able to "see" themselves in (African-American) culture-specific situations, so long as they are able to generalize these to more universal conditions that resonate with their own experience. (A serious blues enthusiast and activist, Majithia, who grew up in India, may himself be Exhibit A for the universality of his beloved music.) Majithia sees this universalization cutting across not only racial/cultural, but also gender lines. What he does not reference—and what Taylor sees as likely impregnable—is that barrier which comes between the experiences of the disabled and the able-bodied. In the case of the blues, whose archetypal blind artists are at once attributed sage-like qualities perceived as emerging from their particular suffering, it seems that that suffering is magnified by its silence, that is, its exclusion from the subject matter of blind artists' repertoires.

59.6.3 Seeing Blind Blues: Gesture, Flow, Circuitry, and Amplification

That this topic has landed in a book concerning gesture theory owes to a kind of synaesthetic performance, a way of visualizing Aristotle's theory of music and emotion, as put forth in his *Politics*, localized to the blues by Majithia, made specific to blind bluesmen by me, and (as we will soon see) illuminated by the discoveries of Vilayanur S. Ramachandran [877]. The cathartic performances referenced (and to some degree specified through a discussion of modes and emotions) by Aristotle, and of which Majithia regards the blues as exemplars, form a kind of complete circuit that is easy to envision. Majithia, who particularly values the blues in live, intimate performance, sees a link between catharsis and community that evokes the image of a circuit, in which audience and artist feed upon each other's emotionally purgative energies. As long as the flow is based on the two terminals'—audience and artists'—perception of commonalities of experience/emotional response or "feeling," the flow between terminals is unproblematic.

That circuit, however, is problematized by the complications that arise from one of its terminals—the blind artist—being perceived as "other," as paranormal—gifted with special *insight* while lacking *sight*—by an audience that understands itself as normative, as is the tendency of communities grounded in common interests (in this case, the blues). That blind musicians possess special insight is a belief not only apparently held by the blues community, for whom the sightless minstrel is archetypal, but also in Japanese [663] and Ukrainian [544] traditions. Particularly apropos here, given the importance of Aristotelian philosophy upon Majithia's analysis of the blues, is the honored position of the blind minstrel in ancient Greek civilization, best known today through Homer, though certainly more as a literary than a musical figure.

Although Majithia mentions many blues artists in his essay, none of them are blind. I have no reason to believe that this is intentional. His clear preference is for electric blues, in which blind artists are, perhaps surprisingly, not uncommon. But they far less likely to be called "Blind" [120], which likely explains their lesser collective notoriety than in acoustic blues.

As a musician myself, one who amplifies his instruments as needed, I imagine the (mostly) sighted audience's and blind artists' circuit as a feedback loop—as it would be even with sighted artists—but one whose flow of energy requires transducers or whose flow of sound requires preamplifiers, so different are the charges emitting from the two generators (performers and listeners). That feedback by nature is both a type of amplification quite analogous to the energy buildup that is the successful communication between audience and performers as well as a kind of mirror relationship—though one in which the reflections grow louder, not smaller—is apropos.

I qualify this characterization by noting that it is less so that audience and (blind) artists are different than that they are perceived as different. In all likelihood, it is the audience that marks the difference between sighted humanity and blind humanity more powerfully. The audience has always had the capacity not only to recognize certain blues artists as blind, but to foreground disability as an aspect of their identity—even to essentialize it—by naming so many of these artists "Blind," while, for all practical purposes, denying those artists the right to express that important facet of their identities in song. This is an exemplar of the social challenges that have likely always and in all places been part and parcel of life with a disability.

A discussion that has migrated from Aristotle and politics to imagined renderings of that politics as circuitry is itself a circuit that must be completed with a reference to neuroscientist Ramachandran [877]. Ramachandran's extension of Giacomo Rizzolatti's discovery of the behavior of mirror neurons—simply put, that the same cell fires in a human (or other primate) when perceiving an activity as in the being performing it—seems compellingly foreshadowed by Aristotle's observations on shared catharsis between musicians and audience, rendered with eloquent specificity in Majathia's descriptions of live performance of the blues in intimate settings. My concern, of course, is that there is something troubling about the metaphorical mirror when the performer is blind and his (almost always a man) blindness is at once valorized and denied, the latter by being essentially banned from his performance repertoire.

59.6.4 Epilogue: Puns as Gestures

I cannot exit this stage without an exegesis of the opening clause of the title of this essay. I propose that the associative value—the meaning, as opposed to their sound—of words is received as a polyphony of percussion, likely the snare drum rimshots the jokes that make us laugh against our better natures.

As most Americans (and perhaps others) will know, the title of this essay is a reference to the catch phrase "a priest, a minister, and a rabbi walked into a bar" that begins many jokes. The line is a rarity among its kind in that it is funny in and of itself, regardless of what follows, owing to its improbability, both for its consortium of clergy and their role as communal social drinkers. In their place, I have inserted three intellectual giants in their respective fields, whose juxtaposition across time, space, field, and nationality is significantly more improbable than the imbibing clergy who inspired their encounter in my title. My own experience of the code switching in this title mirrors the synaesthesia I ascribed to the blues early in this essay, in which blues music (aural) becomes color (visual) becomes "a feeling" (tactile). Here the associative value of the title (meaning/thought) evokes the percussion associated with standup comedy (aural), which becomes tactile (because the code switching is complex enough to require intellectual labor that is *felt* as well as *thought*).

It is often said that a joke that needs to be explained is not funny. In this case, I do not think (and hope) the joke in my title requires an explanation to be funny. I do think humor, as gesture across the senses, is a subject worthy of contemplation. Most importantly, though, the juxtaposition of Greek philosophy, African-American music, disability studies, and contemporary neuroscience makes not just for a funny title, but also is a serious matter of interdisciplinary illumination, one that, paradoxically if metaphorically, even sheds light upon blindness.

Check for updates

60

Models from Music

In these walls devoted to the marvels
I receive and keep the works
of the prodigious hand of the artist
equal and rival of his thought
one is nothing without the other.
Paul Valéry, inscription at the Palais Chaillot, Paris

Summary. We discuss contributions from music theory, performance, and technology to gestural modeling.

– Σ –

Music in theory and performance is a field where thinking and embodied making, the mindful gesture, have been a traditional topic of attempts at theorization. It is a classical saying that music must be thought in the making, in performance. These insights are shared by Theodor Wiesengrund Adorno [8], Roger Sessions [972], Alexandra Pierce [848], Renate Wieland [1068], and Manfred Clynes [206], and we may just cite one representative statement by Adorno: "Notation wants music to be forgotten, in order to fix it and to cast it into identical reproduction, namely the objectivation of the gesture, which for all music of barbarian cultures martyrs the eardrum of the listener. The eternization of music through notation contains a deadly moment: what it captures becomes irrevocable.(...) Musical notation is an expression of the Christianization of music.(...) It is about eternity: it kills music as a natural phenomenon in order to conserve it—once it is broken—as a spiritual entity: The survival of music in its persistence presupposes the killing of its here and now, and achieves within the notation the ban from its mimetic representation." Despite these harsh insights, Robert S. Hatten rightly complains that: "Given the importance of gesture to interpretation, why do we not have a comprehensive theory of gesture in music?" This reflects the dichotomy between performance practice, where embodiment and gesturality are well-known perspectives, and theoretical understanding, which might also be difficult because music theory has been a very rigid, algebraically shaped formalism that has had no chance to deal with topological considerations needed for gestural and embodied analysis.

This dichotomy appears condensed in David Lewin's celebrated book on musical transformation theory [605], where he asks: "If I am at s and wish to get to t, what characteristic gesture should I perform in order to arrive there?" He thinks about a dancer, about the embodied musical subject, and the metaphor is quite strong. However, Lewin's transformational theory is all but gestural; it is a very conservative formalism of abstract mathematical functions in the spirit of Emmy Noether of (at that time) modern algebra in the late 1920s, and there is no topology, no continuity or even homotopy, at all. Gesture research in music has however taken place in the field of computer-aided performance and in our own computerized implementation of musical performance gesture theory for the pianist's hand.

G. Mazzola et al., *The Topos of Music III: Gestures*, Computational Music Science,
https://doi.org/10.1007/978-3-319-64481-3_5

60.1 Wolfgang Graeser

Summary. Graeser's tragic biography and work: from symmetry theory to gestures.

– Σ –

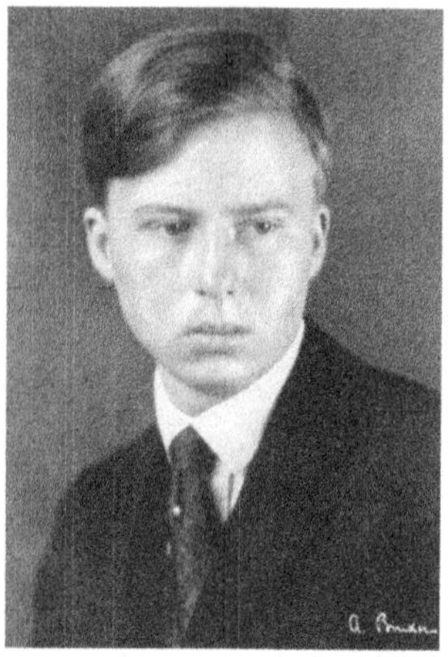

Fig. 60.1. Wolfgang Graeser. Zentralbibliothek Zürich, Mus NL 162 (Nachlass Wolfgang Graeser).

Perhaps the most radical and tragic first gesturally oriented scientific approach to music is that of Wolfgang Graeser (1906-1928), a German mathematician and music theorist, who had studied mathematics, physics, music, and oriental languages since the tender age of seventeen in Berlin and Zurich. He became famous for his symmetry-oriented analysis of Bach's *Art of Fugue* [387]. But a dramatic change in his understanding of music took place when he saw how dancers were rehearsing with Bach's Goldberg Variations. This led to his understanding that all of the musical essence was expressed in the dancers' bodily movement. His last writing was consequently an essay on embodiment [388], wherein he concluded: "Now we comprehend the body uncaged and without veiling insinuations." This explosion in his understanding was more than an intellectual insight, it effectively opened to him a completely new view of the essence of art. But it was also too much of a revolution in his understanding of human expression. At age twenty-two he committed suicide, overwhelmed by his deep and by that time lonely insights.

60.2 Adorno, Wieland, Sessions, Clynes

Summary. This section discusses four contributions to gesturality from prominent music performance professionals and theorists. Their position is quite radically opposed to the European tradition of score-based music.

– Σ –

In the domain of music, gestures were first thematized in musical performance theory. This seems quite natural, since our hands and limbs are in many ways the physical medium between a written score and the sound of a musical piece. It is the philosopher and musicologist Theodor Wiesengrund Adorno, who made the first argument for a gestural understanding of music [8] in 1946: "Correspondingly the task of the interpreter would be to consider the notes until they are transformed into original manuscripts under the insistent eye of the observer; however not as images of the author's emotion—they are also such, but only accidentally—but as the seismographic curves, which the body has left to the music in its gestural vibrations." Adorno argues for what Mazzola had called "the score as a repertory of frozen gestures" in [718]. He does not argue for the emotional message of gestures. Rather, he argues for their nature as "vibrating" bodily utterances. At first sight, this may look overly materialistic and far from the symbolic meaning of musical creation, but we shall see in a moment that Adorno insinuates a spiritual component in the gestural dynamics. This perspective is in fact supported by the very history of score notation. Originally, scores encoded the gestural hints in the graphemes of Medieval neumes. These graphemes then successively morphed to the present notation, which has abstracted neumatic threads to discrete point symbols; see also Section 86.2.3.

Adorno's student, Renate Wieland, and her fellow scholar Jürgen Uhde [1068] make the teacher's approach more explicit and apply it the their system of piano performance. Wieland: "Originally affects were actions, related to an exterior object, along the process of internalization they were detached from their object, but they are still determined by the coordinates of space. (...) There is therefore something like gestural (space) coordinates." She makes clear that gestures are abstractions from concrete actions, however they remain geometric entities in some more generic space. Wieland also argues that the emotional connotation in music originally is e-motion, out-movement, and so the gestural transmutation is not an artificial construct, but the restatement of the original phenomenon.

60.2.1 Theodor Wiesengrund Adorno

Theodor Wiesengrund Adorno has done deep analyses of performance, in particular with respect to their subcutaneous gestural implications as published in his posthumous work *Zu einer Theorie der musikalischen Reproduktion* [8]. It is interesting to see how Adorno gets off the ground with his gestural discourse on the same basis as Sessions and Clynes, namely a radical critique of the score-based reduction of music (translated from [8, p. 227/8]):

> *Notation wants music to be forgotten, in order to fix it and to cast it into identical reproduction, namely the objectivation of the gesture, which for all music of barbarian cultures martyrs the eardrum of the listener. The eternization of music through notation contains a deadly moment: what it captures becomes irrevocable.*
>
> (...)
>
> *Spatialization (through notation) means total control. This is the utopic contradiction in the reproduction of music: to re-create by total control what had been irrevocably lost.*
>
> (...)
>
> *All making music is a recherche du temps perdu.*

And later on (translated from [8, p.235]):

> *Musical notation is an expression of the Christianization of music.*
>
> (...)
>
> *It is about eternity: it kills music as a natural phenomenon in order to conserve (or "embalm" G.M.) it—once it is broken—as a spiritual entity: The survival of music in its persistence presupposes the killing of its here and now, and achieves within the notation the ban (or "detachment" G.M.) from its mimetic*[1] *representation.*

To begin with, Adorno, Sessions, and Clynes agree upon the fact that music notation, and its score, abolishes music, which is fixed and cast into a format for identical reproduction. It does so in objectifying the gesture and thereby martyring the eardrum, an act of barbarian culture. It is remarkable that musical notation is related to barbarian culture. The eternization of music in the notation's casting is killing music; it retains a dead body, not the living music. This eternity of dead—in fact, embalmed—bodies appears as a Christian ritual of sacred denaturation. The procedure of notation kills the music's here and now; its expressivity is annihilated, banned forever. The notational process kills through spatialization, which means total control. Time does not fly by anymore, a note is a point in a dead space of eternity. Adorno views this as being the great contradiction of notation in that it claims total control for a reproduction of what has been irrevocably lost. It is a *temps perdu*, and making music is doomed to a *recherche du temps perdu*.

Adorno then makes important comments on what he views as being the gestural substance of music (translated from [8, p.244/5]):

> *As each face and each gesture, each play of features, is mediated by the I, so the musical moments are the very arena of mimic in music. What must be read and decoded within music are its mimic innervations.*
>
> (...)
>
> *However, a pathetic or cautious or expiring location does not signify pathos, caution, or expiration as a spiritual thing, but maps the corresponding expressive categories into the musical configuration, and those who want to perform them correctly have to find those encapsulated gestures in order to mimic them.*

[1] For Adorno, "mimesis" means "expression of expression," and this is precisely our context: The expression as content is expressed via rhetorical shaping.

(...)

Finding through reading: the decoding work by the interpreter; the very concept of musical performance is the path into the empire of mimic characters.

(...)

The spatialization of gestures, that impulse of neumatic notation is at the same time the negation of the gestural element.

(...)

By the visual fixation, where the musical gesture is positioned into a simultaneous relation to its equals, it ceases to be a gesture, it becomes an object, a mental thing.

Here Adorno refers to the mimetic category in his theory. It is the category "expression of expression." So it is about the expression of emotions, for example, not about emotions, and it is about the musical image of these expressions. Therefore, we have to read those mimic innervations of gestural expressions in music. Musical performance deals with the explication of those hidden innervations, with the action of displaying them in the making, here and now. And it now becomes clear that the neumatic notation creates static photographs of those gestures, which negate them by this spatial fixation. The spatial trace of a gesture is its negation, freezing it as a spatial object.

We should, however, briefly digress on the very concept of a space here, since it is not what a geometer or a physicist would call a space. In physics, a space is a geometric entity that can have different interpretations, so space-time is (locally speaking) a four-dimensional real vector space, and the mathematical structure of time is not different from that of the three space coordinates. Of course, the Lorentz metric distinguishes time in the metrical structure of space-time, but it is still a metrical space. In performance theory of music, time has a radically different role. The four-dimensional space of onset, pitch, loudness, and duration for piano music, which is used in score notation, does not have the ontology of musical time. Under no circumstances would the onset or duration coordinates be accepted as representing the time that takes place in performance. This *differentia specifica* in the performative time concept is related to gestures, not to geometric representation. For Adorno, gesture has an existential character; it cannot be objectivized; it only exists in the moment of the making; it is mediated by the I, which cannot be cast in a dictionary—the I is the non-lexical, the shifter, par excellence. However, it is not part of the subject, it is not subjective as opposed to being objective (the score objects are so). I is only mediated by the I, it seems to lie between subject and object; therefore, the utterance of a gesture is neither object nor subject.

Adorno continues (translated from [8, p.269]):

The true reproduction is the mimicry of a non-existent original.

(...)

But this mimicry of the non-existent original is at the same time nothing else but the X-ray photography of the text.

(...)

Its challenge is to make evident all relations, transitions, contrasts, tension and relaxation fields, and whatever there is that builds the construction, all of that being hidden under the mensural notation and the sensorial surface of sounds.

The true reproduction is not a reference to an object out there; the original is non-existent, and it is not the I, which would be an existent entity. It is something mysterious since there is an X-ray procedure, but it does not show something hidden in the dead object of the score. It is as if that mystery would be brought to existence by the very X-ray procedure. The innervation must be made, not only discovered and pointed to.

Adorno's concept of a gesture is as difficult as it is radically different from what can be described in terms of traditional subject-object duality.

Let us see what Adorno concludes from all these subtle reflections (translated from [8, p.269,270,271]):

What happens in true performance is the articulation of the sensorial appearance that reaches into the most hidden details, wherein the totality of the construction, the gesture of the work, reveals its mimical execution.

(...)

The concept of clarity defines the degree of an analytical performance: everything that exists as relations within the mensural text must become clear, but this concept cannot be understood in a primitive way, i.e. as a clarity of every single relation, but as a hierarchy of clarity and blurredness in the sense of the clarity of the overall structure, the mimic gesture.

And he summarizes this entire perspective on gestural performance (translated from [8, p.247]):

Correspondingly the task of the interpreter would be to consider the notes until they are transformed into original manuscripts under the insistent eye of the observer; however not as images of the author's emotion—they are also such, but only accidentally—but as the seismographic curves, which the body has left to the music in its gestural vibrations.

60.2.2 Renate Wieland

As a student of Adorno, piano pedagogue Renate Wieland (Figure 60.2), in collaboration with her colleague Jürgen Uhde, has developed a theory of piano performance that is based upon Adorno's gestural philosophy.

Fig. 60.2. Renate Wieland.

The remarkable feature of this work is that she succeeds in
(1) giving her approach a clear-cut separation from emotional dramaturgy and
(2) reshaping gesture theory in an explicit geometric language.

She makes these two points very clear in her text (translated from [1068, p.169]):

Musical gestures are perceived in the free conducting movement, in the playing movement and sublimated in the spiritual mimesis of pure imagination. Whatever the level, such experiments are always within space. Originally, affects were actions, related to an exterior object, along the process of interiorization they were detached from their object, but they are still determined by the coordinates of space.

(...)

Language reminds us everywhere of the connection of affect and movement and of the way gestures behave in space. It speaks about hautiness, elevation and inclination, about greatness of mind, pettiness, about respectful and forward, etc.

(...) *There is therefore something like gestural coordinates; they can help ask how the gestural impulse out of the inner is projected into space, how it wants to expand, which direction is dominant: Is its energy vertically or horizontally active? Does it rather propagate ahead or backward? Upward or downward? To the right or to the left? Are forces acting more concentrically or excentrically? Does the gesture rather point "inward," as we read in Schumann's work, or "outward"? Which amplitude does the expression choose? Does it live in all spatial dimensions, and with what proportion and intensity?*

She reminds us of the etymology of the word "emotion": *ex movere*, to move from inside out. She also makes clear that the original setup is now internalized, but that it remains a spatial concept. She then gives examples of etymological shifts, which are parallel to this internalization process: Words now mean abstract things, but when we go to the kernel of a meaning, it is related to a spatial action. So the mimetic action in Adorno's sense is the expression of that spatially conceived gesture in the realm of musical space. She adds the following excellent illustration of a gestural mimesis in music (translated from [1068, p.169]):

Models of contrast between extreme vertical and horizontal gestures are found in Beethoven's Bagatelle op.126,2.
(...)
Aggressively starting initial gestures are answered by flat, conciliating gestures, where the extremes are polarized to the outermost in the course of the piece. In this way, asking again and again, gesture becomes plastic in the end. But it only succeeds insofar as it constitutes a unity, is emanated from one inner central impulse.
(...)
Gestures are the utmost delicate; where their unity is disturbed, their expression immediately vanishes.

It is again in Adorno's and Session's spirit that she views gestures as being extremely unstable in their existentiality: Nothing is easier than to disturb and vaporize a gesture. It is by this fact that Gilles Châtelet, one of the fathers of French gesture theory, has characterized gestures as being the smile of existence [190].

Wieland finally transcends her approach in a seemingly breathtaking intensification, which reads as follows (translated from [1068, p.190]):

The touch of sound is the target of the comprising gesture; the touch is so-to-speak the gesture within the gesture, and like the gesture at large, it equally relates to the coordinates of space.
(...)
The eros of the pianist's touch is not limited to the direct contact with the key, the inner surface of the entire hand pre-senses the sound, etc., etc.

She introduces what one could call the reverberance of a gesture, namely the gesture within a gesture, meaning that a gesture can incorporate other gestures, can become a gesture of gestures. We shall see later in Section 61.6, relating to our own research, that this concept is very powerful for the theory of gestures in that it enables complex imbrications of gestures, so-called hypergestures, for the construction of movements of movements of movements..., an idea that is crucial in the dynamics of musical utterances.

60.2.3 Roger Sessions

The dramatically intense but still underestimated role of gestures in performance has been described in a beautifully clear way by American composer and music critic Roger Sessions in his book *Questions About Music* [972, Chapter III]:

It is fairly obvious, I suppose, that our total awareness of movement—which in essence signifies our awareness of time as a process*—demands sustained attention, which is limited to the duration of the specific act of movement in question; it holds us captive, as it were, for the duration. We are aware of a beginning and an end. In respect to space on the other hand, the words "beginning" and "end" have an essentially metaphorical meaning; they represent boundaries or limits that remain even after we have become aware of them, as does all that lies between. Our attention is our own to husband and deploy as we wish. We can withdraw it and absent ourselves merely by averting or closing our eyes, and return whenever and for as long as we wish.*

What I am saying is that we experience music as a pattern of movement, as a gesture; and that a gesture gradually loses its meaning for us insofar as we become aware of having witnessed it, in its total identity, before. If it is to retain this meaning in its full force, it must be on each occasion reinvested with fresh energy. Otherwise we experience it, to an increasing degree, as static; its impact, as movement, diminishes, and in the end we cease to experience it as movement at all. Its essentially static nature has imposed itself on our awareness.

This is why I am convinced that the performer is an essential element in the whole musical picture. It is why I came to realize that my earlier dreams—that composers might learn to freeze their own performance, in wax or otherwise (tape recorders had not been invented at that time)—were, to put it bluntly, quite ill-directed. They were ill-directed, above all, for the reasons I have been outlining; a gesture needs constant renewal if it is to retain its force on subsequent repetitions. Composers above all should know this, especially if they have developed the practice of taking part in performances of their own work. Each performance is a new one, and the work is always studied and approached anew, even by the composer. The same, it should be obvious, is true of professional performers. I would go even much further and point out that there is no such thing as a "definitive" performance of any work whatever. This is true even of performances by the composer himself, in spite of the fact that recordings of his performances of his own work should be made and preserved, for a number of quite obvious reasons.

Fig. 60.3. Roger Sessions.

Session's discussion of movement as a processing of time leads him to acknowledge that this dynamical action is a gesture—not only in the making, but also in the music's perception. So he gives the argument for a messaging of gestures, and by means of gestures, which is our topic in this chapter. It is remarkable that he then recognizes that a gesture cannot preserve its meaning except in its energetic refreshment on each occasion of performance. This is very similar to the French theory of gestures; see Chapter 58 and [721, Chapter 7.2], which stresses the impossibility to tame living gestures.

He moreover recognizes the performer's essential role in the "whole music picture" and also reminds composers, himself included, that their work of musical creation is not accomplished until it is performed. This does not mean that a composer must intervene in the performance of his/her works. Some are dead and simply cannot do this anymore. No, it means that the completion of a musical work cannot be achieved before its performance has occured. In this sense, performance is strongly what semioticians call a *deictic* part of the musical sign system: Musical signs reach their full meaning only and essentially through their pragmatic instantiation.

This second insight is strongly related to the gestural aspect since gestures are not lexicographic, they are shifters, as Sessions stresses with his "French" view on gestures. We are not astonished that Manfred Clynes refers to Session's writings in his critique of score-based music.

60.2.4 Manfred Clynes

The Australian pianist and theorist Manfred Clynes conceived expressivity as a shaping of performance in pulses, those embodiments of *essentic forms*, via specific deformations of duration and loudness; see [205, Section 13.3]. He claimed that such pulses were characteristic of the emotional expressivity of composers such as Beethoven, Mozart, etc. Clynes' pulses are not only emotional categories, but also, and perhaps more significantly, curves of gestural utterances. Clynes accordingly constructed and patented a machine, the *sentograph*, providing us with an interface to grasp such gestural movements. Following Clynes' ideas, Hungarian composer Tamas Ungvary has constructed a sentograph that can be used by improvising composers in order to play/create music by gestural input [1069]. Ungvary replaces the usual encoding of sound events at discrete points in a parameter space by an intrinsically gestural input that is given by variable pressure and angle on a joystick (Figure 60.4). Despite the fascinating perspective on musical creation, the

Fig. 60.4. Tamas Ungvary playing the sentograph. The joystick is accessed with the right middle finger.

gestural input remains very abstract insofar as no significant movement of the fingers is possible. The musician has to stay in contact with that fixed piece of metal and cannot move freely in space. This restriction heavily limits the natural human need for movements when gestures have to be created from the living body. Perhaps a more natural encoding of the input parameters would improve the expressive power of this interesting machine.

60.3 Johan Sundberg and Neil P. McAngus Todd

Summary. These two authors discuss gestural aspects in computational performance theory and cognitive modeling.

– Σ –

On a more down-to-earth level, gesture has been studied by Johan Sundberg and collaborators. In a paper entitled "Is the Musical Ritard an Allusion to Physical Motion" [560], Sunberg and Ulf Kronman have studied final ritard as a phenomenon akin to physical ritard. The model conjectures that a tempo decrease at the end of a musical piece would be related to a quadratic function, which appears for mechanical ritard with a constant force. So we suppose that we are given a constant force F, and its action on a given mass m, which generates a constant deceleration $a = F/m$ according to Newton's second law. Given an initial velocity v, the velocity after t seconds is $v - a.t$. Hence the distance $s(t)$ traveled after t seconds is $s(t) = \int_0^t v - a.\tau d\tau = t.v - a/2.t^2$. If the final velocity at time t_0 is 0, we have $t_0.a = v$, whence $s(t_0) = (v/a).v - a/2.(v/a)^2 = v^2/2a$. Therefore velocity at time t is $v(t) = v.\sqrt{1 - s(t)/s(t_0)}$. Supposing that this physical situation relates to the musical one by a constant c, i.e., $s(t) = c.E(t)$, E being the symbolic onset, we get $T(t) = T(t_0).\sqrt{1 - E(t)/E(t_0)}$. This implies

$$T(E) = T(E_0).\sqrt{1 - E/E_0},$$

namely the tempo at onset E being the above function of the tempo $T(E_0)$ at the beginning E_0 of the ritard, the onset E and the beginning onset E_0. The experimental situation is shown in Figure 60.5. The parabolic tempo curve relates to the phase I in the left graph. Phase II is interpreted as a linear tempo decrease.

Besides the poor fit of the measured tempo with the mathematical curve, the question arises why such a mechanical function should hold. What is the musical analog to mass, what is the force analog to a constant mechanical force? We do not see any musical structure entailing such a mechanical model. It is interesting

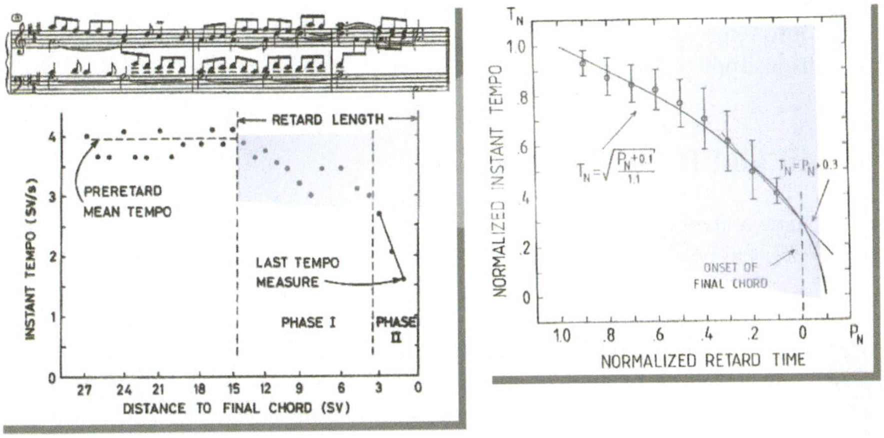

Fig. 60.5. The parabolic tempo curve (right figure) relates to the phase I in the left graphic. Phase II is interpreted as a linear tempo decrease.

that the ritard phase II relates to a quite sophisticated harmonic and melodic musical process, which is not taken into account.

Another mechanical model of agogics has been proposed by Neil P. McAngus Todd in [1060]. He rightly observes that the final retard is only a very special agogical situation and therefore models his tempo curves according to a superposition of accelerando/ritardando units that are defined by a triangular sink potential V. Accordingly, tempo is defined as a velocity v, and the total energy of the system, $E = \frac{1}{2}mv^2 + V$—supposed to be constant (why so?)—gives the velocity formula $v = \sqrt{2(E-V)/m}$. Todd further supposes that there is an intensity variable I for loudness, with a relation $I = K.v^2$ that is common to many physical systems. This yields the relation $I = 2K(E-V)/m$ and sums up to an aggregated formula $I = \sum_l 2K(E-V_l)/m_l$ if the grouping of the piece is taken into account. The idea is that there is a physical energy and intensity parameter system that controls the "surface" of the tempo (= velocity) via classical energy formulas. The background structure is an energetic one, i.e., the tempo curve and loudness are expressions of mechanical dynamics. The author comments on his method as follows [1060, p.3549]:

> *The model of musical dynamics presented in this paper was based on two basic principles. First, that musical expression has its origins in simple motor actions and that the performance and perception of tempo/musical dynamics is based on an internal sense of motion. Second, that this internal movement is organized in a hierarchical manner corresponding to how the grouping of phrase structure is organized in the performer's memory.*

The author also suggests a physiological correlate of this models:

> *...it may be the case that expressive sounds can induce a percept of self-motion in the listener and that the internal sense of motion referred to above may have its origin in the central vestibular system. Thus, according to this theory, the reason why expression based on the equation of elementary mechanics sounds natural is that the vestibular system evolved to deal with precisely these kinds of motions.*

Todd refers to the insights of neurophysiologists that the vestibular system is also sensitive to vibrational phenomena. The musical expressivity is therefore understood as an effect of transformed neurophysiological motion.

The drawback of this approach is that finer musical structures are not involved in the structuring of the energy that shapes tempo/intensity. And even if that could be done, there is an essential kernel of this shaping method that should be based upon paradigms of motion. These paradigms do not however appear clearly in the above approach. More precisely: The complex motion dynamics of the vestibular system cannot easily be mapped onto the structures of performative expressivity. What is the operator that transforms whatever structures of motion into expression parameters? If music were isomorphic to motion, no such isomorphism could be recognized from Todd's approach.

60.4 David Lewin and Robert S. Hatten

Summary. Here we have a strong argument for a gesture theory in music and performance theory, however without a strictly theoretical conceptualization.

– Σ –

Fig. 60.6. David Lewin.

Coming from a seemingly opposite position, namely music theory, the American music theorist David Lewin (Figure 60.6) introduced in 1987 the gestural perspective in his seminal book *Generalized Musical Intervals* [605]. Well, nearly, since the theory and the textual representation are more complex. Lewin's book describes what is now called "transformational theory", later adapted by his student Henry Klumpenhouwer to become K-nets [599]. Such a network replaces an "amorphous" set of tone objects by a diagram, where the tone objects are placed at the diagram's vertices, while the diagram's arrows designate (affine) transformations mapping tone objects into each other; see [34] for a modern interpretation of this theory in terms of category theory. The strictly scientific setup of transformational theory is not really gestural. Lewin argues against what they call the "cartesian thinking", which observes musical objects as *res extensae*.

Opposed to this passive attitude, Lewin suggests that transformations between musical points (such as pitch classes, for example) are the new path to pursue. In [605, p.159], we read: "If I am at s and wish to get to t, what characteristic gesture should I perform in order to arrive there?" Now, this language sounds very gestural, but is dependent upon different mathematical principles. Let us clarify this subtle mathematical point, which may escape the non-professional. Lewin's theory uses classical transformations and then, in Klumpenhouwer's networks, diagrams of transformations.

We have shown in [719] that Klumpenhouwer's and Lewin's transformational networks are typically points of projective limits of diagrams of affine transformations in musical standard spaces (pitch class spaces, for example). This is a giant step ahead, since projective (and inductive) limits are related to processes, namely the underlying diagrams. Diagrams are systems of transformations between a set of spaces and they relate points in those spaces by determined transformations, see Figure 60.7. But they are not identical with the point systems generated by the so-called limit construction. An intuitive, and incidentally mathematically correct, way of characterizing diagrams is as generalized equations, whereas the objects from the limits are solutions of such equations. So the diagrams play the role of industrial plants, producing facts (*factum*, what is made), namely Klumpenhouwer's K-nets. So the Lewinian digression from cartesian facticity (or extensionality) is the step to processes, but not to gestures.—We have to discuss this difference more precisely in order to understand the missing processes and gestures. In a diagram of transformations, these arrows, which encode transformations, are intuitive graphemes. They are used everywhere in mathematics to denote functions, transformations, or homomorphisms. In category theory, such arrows are called morphisms, and their meaning is absolutely abstract. But already in the classical language as framed by set theory, arrows denote functions f in the sense of Gottlob Frege.

What is such a function? It is (together with its domain X and codomain Y) a set f of ordered pairs (x, y), where the second component y is denoted by $f(x)$. So these two components have nothing more in common than their being part of a set f of ordered pairs. There is no interior relation beyond this association. Coming back to the arrow notation $f : X \to Y$ for such a function, the arrow has absolutely no relation to the interior of its shaft. One could as well write $f : X \Diamond Y$. This is a dramatic fact: Arrows suggest a movement, but this is merely illusory. Nothing moves. This has been observed by the French philosopher and mathematician Gilles Châtelet [190]. For example, if we take the matricial representation of a rotation R in three-space, the matrix has no relation to the rotational movement. Matrices are functional objects and do not imply any continuous displacement of points whatsoever. The only relation of matrices to real movements is created by the calculation of so-called eigenvectors, which may eventually help define a rotational axis, and then an angle or rotation, and finally (!) give the option to realize this rotation by successively increasing the rotational angle from zero to the actual value.

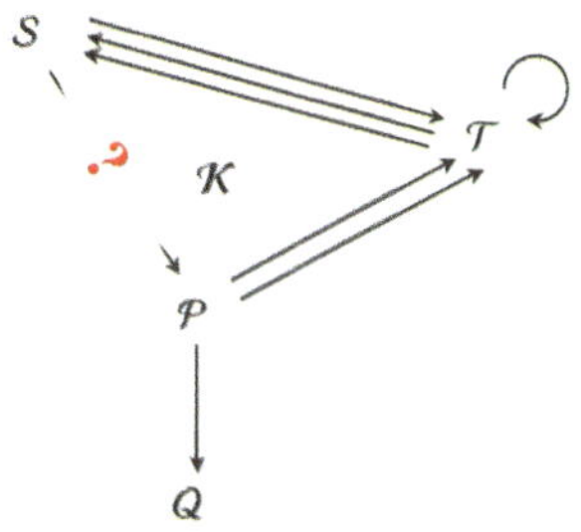

Fig. 60.7. A diagram $\mathcal{K}$ of transformations between spaces $\mathcal{S}, \mathcal{T}, \mathcal{P}$, and $\mathcal{Q}$.

It should be clear by now that the arrow notation for functions is intuitively associated with movements, but does not correspond to any movement at all. Coming back to Lewin's transformational theory, this means that his language that refers to moving from one point to another, and *a fortiori* his suggestion of a gesture relating point s to point t, are different from the reality of his mathematical formalism. He speaks about gestures, but writes about processes. In short: his theory is processual, half way between facts and gestures in terms of the axis of embodiment. It would be very interesting to investigate Lewin's text with that subtext of gestural thinking in mind, since he repeatedly uses this metaphor in a speaking way. With regard to his question about the movement of s to t, he adds [605, p.159]: "This attitude is by and large the attitude of someone *inside* the music, as idealized dancer and/or singer. No external observer (analyst, listener) is needed."

Fig. 60.8. Robert S. Hatten.

This is a remarkable statement, which leads to the question set forth by music theorist Robert S. Hatten in his book with the now explicitly gesture-related title *Interpreting Musical Gestures, Topics, and Tropes* [446]: "Given the importance of gesture to interpretation, why do we not have a comprehensive theory of gesture in music?" For Hatten gesturality became a core topic when he learned that performance of classical piano music, Mozart, Beethoven, Schubert, is strongly determined by gestural attitudes. This is best exemplified when comparing Glenn Gould's interpretation of Beethoven's op. 57, *Appassionata*, to Vladimir Horowitz's version. Gould's performance completely lacks gesturality. His so-called "analytical' reading is the opposite of what Adorno had recommended, and amounts to Beethoven minus gestures, a substantial negation given the strongly gestural nature of Beethoven's music. Hatten confirms this in theory, as does Gould by his contrafactual experiment.

Hatten's definition of a gesture reads as follows: "Gesture is most generally defined as communicative (whether intended or not), expressive, energetic shaping through time (including characteristic features of musicality such as beat, rhythm, timing of exchanges, contour, intensity), regardless of medium (channel) or sensory-motor source (intermodal or cross-modal)." He distinguishes his understanding of gestures from the school of Adam Kendon and David McNeill in that (1) semantic aspects are not characteristic and (2) he stresses "energetic shaping through time", an interesting wording, since the main subject is "shaping", an action, not shaping of something, but pure action. The making in itself becomes a central feature, not the resulting facts generated by the making! And he adds, in remarkable congruence with Wieland's abstract geometry of gestures, that: "at a higher, more symbolic cognitive level, the representation of gesture may be considered amodal, in that it is not restricted to any particular modality."

This shift away from the conservative semiotic perspective on gestures can also be observed in psychology. In Susan Goldin-Meadow's book *Hearing gesture: How our hands help us think* [377], a title reminiscent of Paul Valéry's phrase at the head of this chapter, she investigates the role of gestures in the development of a child's ability to reason mathematically. She writes: "Advances in mathematical reasoning are very likely to come first in gesture—and they do. (...) Do new ideas always come first in gesture, regardless of domain?"

Coming back to the context of musical gestures, the question of semiotics of gestures arises when we display the overall image of traditional Western musical performance; see Figure 60.9. This process starts from the score, which is a text of more or less analyzed symbols. The score symbols are then "thawed" and unfold in gestures, which interact with the interface of an instrument and thusly induce sounding events. (The reversed process of freezing gestures is concretely taking place in a MIDI recording session.) The meaning of music is thereby guaranteed by the dominant role of the score. The entire process is only produced in order to rhetorically communicate the given meaning that was recognized in the score's symbolic code. This canon is

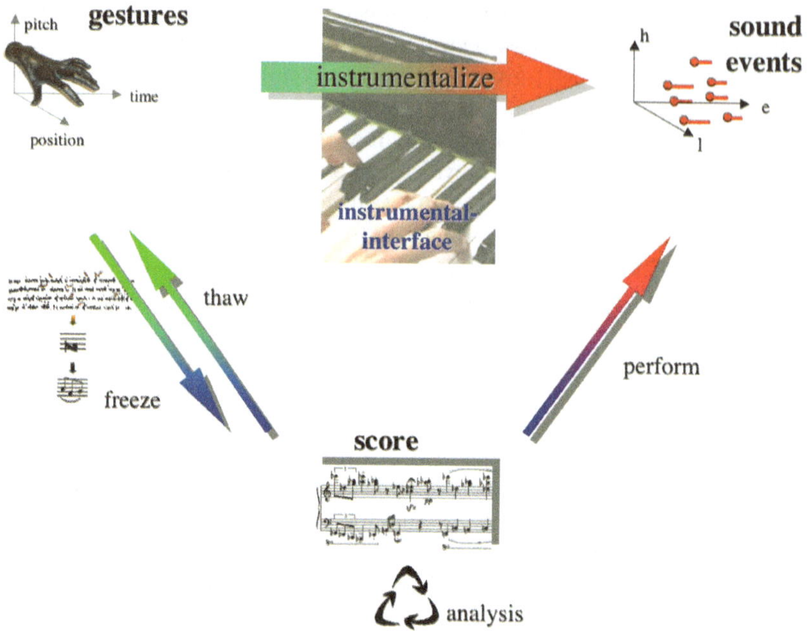

Fig. 60.9. The triangle of Western musical performance.

broken up once the lower vertex of the triangle, the score, is not present. It is not a general truth that music must be produced starting from a given abstract score text. It could as well start from the gestural utterance and its interaction with the instrumental interface that acts like a dance floor for a gestural dance. In other words, the semiotic approach to gestures in music is traditionally related to the score-driven production of music.

But there are many musics that are not score-driven, and free jazz is one of them, while (standard) jazz is not; it is framed in the scores of song forms, lead sheets, and similar ready-mades. It was a dramatic insight in Mazzola's own development as a pianist and as a theorist in mathematical music theory for him to learn that his free jazz playing had nothing to do with score-driven logic and meaning (see [718] for that affair).

Once making music has been freed from the interpretational task of classical Western performance, the question of meaning becomes secondary, making music is no longer just an expression of given meaning. Semantics is no longer the core business, and, as Ornette Coleman states: it is no longer a question of playing the background of other things, such as meaning, symbols, calculations, everything but music.

To summarize, gestures have been recognized as essential to music. The layer of semiotic musical functions has been recognized as being unimportant, or only important in relation to gesture. The background of score-driven rhetoric has been abolished by free jazz and other gesture-oriented music, liberating the musician to freely dance on the instrumental interface. A wonderful example of free jazz without any reference to score or associated semiotics is the double LP *Mu* [195] by Don Cherry, on pocket trumpet, piano, bells, flute, percussion, and voice, and Ed Blackwell, on drums, percussion and bells. The music here is completely free of global strategies. Nobody tells Don and Ed where to go. They just throw gestures at each other and play a game of free gestural dialogs without pre-meditated meaning and significance, creating new sounds that facilitate a gestural dance by maintaining a sonic equilibrium. Although there is a strong reference to something like folk sound, the music is not following any specific ethnic tradition, it is just free playing. Blackwell's percussion is not bound by strict rhythmic frames, and often utilizes the more loose concept of "phase shifting" heard in many varieties of African drumming. Often, he abruptly stops patterns and lets the empty space of time go by, and then takes up another germ of time without any reference, without any obligation to mean anything. The freshness of this music is exactly rooted in its independence from given semantics. It "don't mean a thing," but it has got so much swing.

60.5 Marcelo Wanderley and Claude Cadoz, Rolf Inge Godøy and Marc Leman

Summary. This section describes more technologically oriented conceptualizations of gestures. Refer to [168] and [371].

– Σ –

In [168], Claude Cadoz and Marcelo Wanderley give a summary of the gesture concept and its specific structure for the performing musician. Their discourse is based on a list of definitions of a gesture that can be found in dictionaries and gesture research resources. All these definitions are stated in common language and share the following two characteristics:

- Gestures are defined as human body movements that
- carry information or, in other words, are expressive.

This implies that all these definitions interpret gestures as a special type of signs, semiotic instances that express some content or meaning. Carrying information is however arcane: how is information present in a gesture, and where in the movement's anatomy is this information located? Also, the more contemporary concept of information is not defined. And what information is considered relevant?

Cadoz and Wanderley however take a typically French turn and propose a definition that is not semiotic: In [168, p.73], we read: "We consider that the word gesture (or the French equivalent geste) necessarily makes reference to a human being and to its body behavior—whether they be useful or not, significant or meaningless, expressive or inexpressive, conscious or not, intentional or automatic/reflex, completely controlled or not, applied or not to a physical object, effective or ineffective, or suggested."

The main point of their paper is—following their own words [168, p.74]—to discuss human-human and/or human-machine communication through gestures in a musical context. When moving to the concept of a musical gesture, they recognize that theirs may be quite different from the general concept. In fact, one of the musical gesture definitions states that "the notion of a musical gesture that at the time it occurs involves no actual human movement but merely *refers* to it is quite common."

Although they refrain from presenting a universally valid definition of the gesture concept, they admit [168, p.74] that "in essence, the direct or indirect reference to human physical behavior tends to be the common denominator to all the notions."

After the general exposition of the gestural concepts, the authors proceed to a more detailed typology of musical gestures. They refer to Christophe Ramstein's [878] methodology for analyzing instrumental gestures [168, p.74]:

- a *phenomenological* approach, i.e., a descriptive analysis;
- a *functional* approach, referring to the possible functions a gesture may perform in a specific situation;
- an *intrinsic* approach (from the musician's point of view), it is based on the conditions of gesture production by the performer.

In the phenomenological approach, they propose the classification that uses gestural primitives. Following Insook Choi [196], *gestural primitives* are "fundamental human movements that relate the human subject to dynamic responses in an environment." Choi proposes three types of gestural primitives, both device- and signal-independent:

- Trajectory-based primitives: e.g. changes of orientation;
- Force-based primitives: e.g. gradient movements;
- Pattern-based primitives: e.g. quasi-periodic movements.

In the functional approach,[2] following Cadoz's proposal [169], they identify three different functions (complementary and dependent on each other):

- material action, modification and transformation of the environment—the *ergotic* function;
- perception of the environment—the *epistemic* function;
- communication of information towards the environment—the *semiotic* function.

In view of the French generalization to non-semiotic gestures, it is remarkable, if not quite strange, that in the following discussion of an instrumental gesture the authors state [168, p.79]: "Instrumental gesture is considered as a "communication modality" complementary to empty-handed gestures. They are therefore singular in that they possess, *à la fois,* all three characteristics of the gestural channel: ergotic, epistemic and semiotic."

They then focus on gesture typology with this rationale [168, p.82]: "The importance of gesture typologies is then not to completely describe acoustic musical instruments but to provide general guidelines for the design of gestural input devices, mostly regarding the presence of different types of feedback related to different gestures." In the their typology of instrumental gestures, they distinguish between three types:

- *Excitation* gesture; it can be instantaneous (percussive or picking) or continuous.
- *Modification* gesture; it can be parametric (continuous variation of a parameter, such as vibrato) or structural (when the modification is related to categorical differences, such as the insertion/removal of an extra part, e.g. a mute in the case of the trumpet, or a register in an organ).
- *Selection* gesture (a choice among multiple similar elements in an instrument).

The system is then illustrated by a number of case studies, involving different instruments, such as cello, clarinet, and bagpipe.

Godøy's and Leman's book [371] is an edited volume that recollects a number of contributions around the topic of musical gestures, meaning gestures that arise while producing or perceiving music. The contributors include well-known researchers in that field, such as Marcelo Wanderley, Frédéric Bevilacqua, Roberto Bresin, Antonio Camurri, and Albrecht Schneider, for example. The initiative for this book came from a European research project, COST87—ConGAS—Gesture Controlled Audio Systems, running from 2003 to 2007. The book is conceived as a representation of a highly interdisciplinary collaboration, however without aiming at a final discussion of the book's topic.

The book is divided into three parts. Part I, " Gestures in Music", introduces definitions, examples, and a history of gestures in music. Part II, "Gestural Signification", provides a theoretical framework for the formation of signification in gesture in music. Part III, "Gesture Generation and Control", concerns the processing and control of gesture in music.

Answering the question "Why Study Musical Gestures" (title of Chapter 1), it is stated that "we believe that musical experience is inseparable from the sensation of movement, and hence, that studying theses gestures, what we call *musical gestures*, ought to be a high priority task in music research."

In Chapter 2, besides categorizing gestures as poietic or aesthetic utterances, it is stated that the core role of gestures is to provide a "bridge between movement and meaning", it "surpasses the cartesian divide between physics and mind." This is a clear conceptualization of "gesture" as a semiotic entity: it has meaning, although its production is different from meaning that we deal with in linguistic systems: gestures "demonstrate", while languages "say" or "denote". And accordingly, Part II deals with the semantic aspect of this gesture concept.

The definitions of "gesture" are taken from three perspectives: communication, control, and metaphor. Communication happens when gestures are vehicles for meaning in social interaction, control when gestures are elements of computational and interactive systems, and metaphor when gestures work as concepts that project physical movements on cultural topics. Summarizing, they define "musical gesture as an action pattern that produces music, is encoded in music, or is made in response to music." They distinguish between four types of functional aspects of musical gestures: sound-producing, communicative, sound-facilitating, and sound-accompanying, and they represent these four characteristics quantitatively in the 2D plane with two

[2] The intrinsic approach is not explicitly discussed in this paper.

axes, one spanned by the couple "sound producing/sound-accompanying" , the other by the couple "sound-facilitating/communicative". This representation is used to position musicians as opposed to dancers at their musical gesture values; see Figure 60.10.

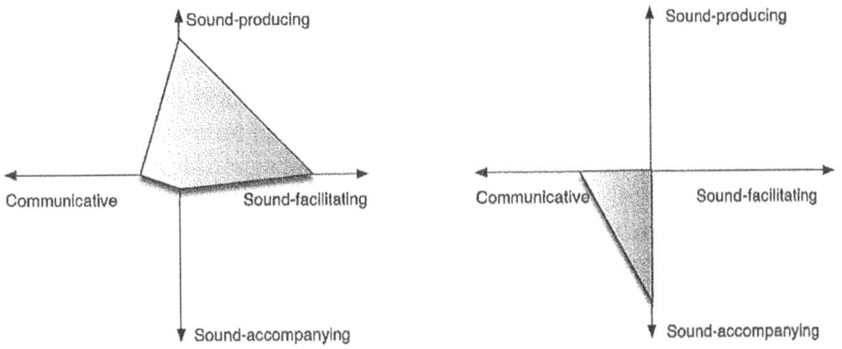

Fig. 60.10. Four types of functional aspects of musical gestures. Left: Musicians, right: dancers.

Chapter 2 concludes stating that "Up until now, there is no single unequivocal definition of gesture, although most authors seem to agree that gestures involve both body movement and meaning." Chapter 3 discusses gestures in performance for a number of classical and electronic musical instruments. Chapter 4 is a historical overview and, after an interesting survey, starting with Aristoteles and Platon, concludes: "The motional and gestural qualities of music have been known since antiquity, and have been developed in various musical styles and genres." However, the reference to Jean-Claude Schmitt's important book about medieval gesture theory is missing, and also the reference to the French gesture philosophy, which is unfortunately completely absent (except for Maurice Merleau-Ponty) in this book.

The gestural production of embodied meaning is addressed in Chapter 6 of Part II. Meaning results from a chain of transformations (synaestetic, kinesthetic, cenaesthetic) from sonic features to cognitive strata. The third, "canaestetic transformation can be seen as a precondition for a fully symbolized type of meaning formation." This confirms the initial principle that "gestures can be understood as close to body movements and close to meaning." "Gesture can be considered as a hierarchically structured action pattern to which we can have mental access."

The concept of "gesture" is differentiated according to the three personal dimensions of "I", "You", "He/She/It". The first person shows a gestural ontology that relates to the concepts of flow, presence, and cause-effect. The second person shows that gestures are more than isolated phenomena, they pertain to a dialogical dimension (also stressed by Émile Benveniste's theory of pronouns; see also Sections 57.9 and 59.1), and this is also confirmed by neuroscience (as made evident by the mirror neuron phenomenon to which this text does not refer). The third person perspective deals with the objectively measurable traces of gestures.

The chapter terminates with the "question to what extent music *is* gesture." It is concluded that music contains gesture but music is also (auto)referential: its internal gestures refer to one-another. As a final conclusion, gestures are characterized as *multi-modal* (audio, motor, etc.), *multi-level* (space-time hierarchies), and *monistic* (bridging the cartesian divide) phenomena. The last Section 7 of this part deals with basic bio-kinetics in activation and signification.

The third part of that book deals with more technical topics: gesture generation and control, discussing gesture and timbre or the conductor's gestures and their mapping to sound synthesis.

Part XVI

Mathematics of Gestures

Check for updates

61

Fundamental Concepts and Associated Categories

Summary. This chapter introduces the definition, some basic propositions and first examples regarding the mathematical concept of a gesture for topological spaces. It also includes a short discussion of the topos-theoretic logic that is implied by the topos of directed graphs.

– Σ –

61.1 Introduction

This chapter presents a programmatic category-oriented framework for the description of the gestural relations between musical and mathematical activities, a first publication was given in [719]. This relation may be described in terms of adjointness between functors, which extend the functorial setup discussed in the first edition [714] of this book. Thus, on a meta-level, the relations between musical and mathematical activities are investigated from a mathematical point of view.

Far from being isomorphic, music and mathematics seem to involve some common structures which can be related by one of the most powerful concepts of category theory: the notion of adjoint functors. This construction, proposed by Daniel Kan in the 1950s as a technical device for the study of the combinatorial properties in homotopy theory [518], turns out to be the most adequate tool to link three main categories: the equations or formulas (category of spectroids), the diagram schemes (category of directed graphs) and the gestures (category of diagrams of curves in topological spaces). In a schematic way, we may view mathematics and music as adjoint functors between the categories of formulas and gestures:

$$\text{formulas} \underset{\text{mathematics}}{\overset{\text{music}}{\rightleftarrows}} \text{gestures}$$

The category of directed graphs, which has been recently proposed as a foundational concept in mathematics for both classical and categorical set-theory [280], seems to provide a musically interesting mediating structure between the two other categories, on which music and mathematics act in adjoint positions. By means of diagrams, mathematics turns gestures into formulas. In fact, a diagram is a system of transformational arrows. On such a system you may follow different paths starting and ending at the same two points. These paths can be viewed as gestural movements. If two such paths commute, i.e., they yield the same composed transformation, then we have exactly what is called a formula or equation: Two expressions yield the same result. Quite generally speaking, formulas are commutativity relations between gestural paths. Conversely, musical activity "unfreezes" formulas into gestures that can be described as the unfolding of formulas in space-time.

With such a conceptual framework we want to include embodied performance into the formalization of musical structures. In [720] we argued that the categorical presentation of Klumpenhouwer networks[1] as

[1] See also Section 65.2 for a mathematical description of Klumpenhouwer networks.

G. Mazzola et al., *The Topos of Music III: Gestures*, Computational Music Science,
https://doi.org/10.1007/978-3-319-64481-3_6

elements in limits of diagrams of spaces and transformations has some important operational consequences. A parallel argumentation applies to the usage of the category of directed graphs of curves in topological spaces as a theoretical framework of gesture theory. From a purely theoretical aspect, "gestures of gestures" (or hypergestures) as well as "natural gestures" are canonically defined, as we will see by discussing the case of the gesture of a finger of a piano player's hand and its hypergestural generalizations. As in the case of the development of category and topos theory, as discussed by Mac Lane in [636], the notion of gesture as suggested here offers a good illustration of "collision" between algebraic and topological methods.

But there is another intriguing aspect of this new categorical setup for musical gestures which deals with the philosophical ramification of category theory. Category theory is more than a useful universal language, eventually providing the theoretical setting for the foundations of mathematics. When applied to a complex human activity such as music, category theory offers the conceptual framework generating a new theoretical perspective of the relations between the philosophy of music and the philosophy of mathematics, in fact, by shedding new light on the understanding of the genesis and ontology of musical and mathematical activities.[2]

The case of gesture theory suggests that we can naturally transfer Mac Lane's conception of mathematics as "an elaborate tightly connected network of formal systems, axiom systems, rules, and connections" [638, p.417] to music. The adjoint functors that we establish between the formal category of formulas and the functional universe of gestures suggest that musical activity could be also conceived as arising from a formal network based on some dynamic concepts that evolve according to their function. This framework has some very interesting philosophical consequences, especially when trying to update the debate on the relation between the structural conception of mathematics and the structuralist approach to music.[3] We suggest that mathematical structuralism could be taken as a philosophical position for the music-theoretical activity once it is accepted that mathematical music theory is about music conceived as a structured system. As rightly observed by Elaine Landry and Jean-Pierre Marquis in an interesting attempt at putting category theory into a historical, foundational and philosophical context [575], "the problem with standard structural approaches is that they cleave to the residual Fregean assumption that there is one unique context that provides us with the pre-conditions for the actual existence of 'structures' or for the possible existence of types of structured systems" [575, p.21]. And as the categorical framework suggests that "mathematical concepts have to be thought of in a context that can be varied in a systematic fashion" [575, p.21], the categories of formulas, diagrams, and gestures in music suggest that the functorial adjointness that we shall present provides a general framework for the study of gestures in a given musical context.

Unfolded from the scant category of digraphs, the linear categories associated with digraphs (spectroids [351]) and the categories of gestures split the structural content of the morphism concept: Whereas the algebraic context of spectroids (which also includes the transformational approach to music theory) leaves the morphism concept in its abstract setup inherited from the classical Fregean approach to functions, i.e., the totally abstract relation between argument and functional value, the category of gestures is built upon morphisms which are derived from continuous curves, such that the movement from argument to value is mediated along the entire curve following the curve parameter. *The gesture is a morphism, where the linkage is a real movement and not only a symbolic arrow without bridging substance.* The arrow is a symbol of category theory which suggests a bridge between domain and codomain and thereby points to a metaphor

[2] Essentially, this is due to Alexander Grothendieck's reinvention of the point concept in algebraic geometry. He redefines a point as being a morphism $f : X \to Y$ in the category of schemes, and conversely, using the Yoneda Lemma, a morphism $f : X \to Y$ in any category can be viewed as a point in the presheaf associated with Y. This means that the original Euclidean point concept (*punctus est cuius pars nulla est*) is replaced by the elementary concept of a morphism, which determines the ontology of a category. This point of view suggests that such morphisms, which are commonly understood as "arrows", induce a dynamical aspect; a morphism is the movement of an arrow. In this spirit, mathematical activities are presently being debated as gestural movements along such arrows, instead of abstract manipulation of symbols [24]. But the arrow-oriented approach to mathematics also enables a description of basic musical concepts as activities in terms of morphisms. David Lewin's transformational theory [605], Thomas Noll's harmonic morphology [802] and our categorical approach to performance theory [714, Chapter 35] are examples thereof.

[3] See [31] for a detailed discussion on the emergence and the rise of the notion of mathematical structure in music from an algebraic perspective.

overloaded by embodiment. However, according to Jean Cavaillès [91] "Comprendre est attraper le geste et pouvoir continuer." This means that human evidence and operational competence are intimately tied to the embodied movement, and this is a gesture, not the abstract arrow.

We therefore argue that *the gestural movement should be considered as being the missing link between abstract formulaic intelligence and plain bodily gesticulation.* And it offers a generative force in the creation of significance and regularity of rule-based systems. This refers to a substantial insight of Charles Alunni [24]: "Ce n'est pas la règle qui gouverne l'action diagrammatique, mais l'action qui fait émerger la règle."

61.2 Towards a Musical String Theory

It is well known that the formal description, classification, and analysis of music in terms of local or global systems of note sets in adequate parameter spaces does not grasp the full reality of music. In fact, beyond this music-theoretical reality, music must be performed, i.e., the formally parametrized note configurations must be mapped into physically meaningful spaces, see [243] for a musicological account of performance theory; see Part IX and Part X for a more mathematical and computational overview. Although performance theory may yield sophisticated sounds from a "mechanical" score, it does not embody the musician's instrumental activity. When we play music, we make the performed sounds. This is more than a set of sound events, more than a CD recording may ever trace on its acoustic level. The important role of embodiment of sounds is rightly testified by the strong need for concerts, where performance is not only heard, but also experienced from the musicians' bodies in movement. Understanding of music is strongly enhanced if not enabled by means of its presentation in moving bodies, or, to put it more concisely: in musical gestures.

It may be argued that gestures are, like performances of shaped sounds, rhetorical means to convey a meaning which in principle is faithfully represented in the score's content. This is however erroneous for two reasons. To begin with, score signs are not unambiguously loaded with meaning. The creation of signification of musical signs shares a deictic nature. Only the user (in particular: the performing artist) can complete the partial meaning of musical score signs. It can be shown that there is an infinity of such completions, be it on the symbolic level 13.4.1, or be it on the level of agogics, articulation, or dynamics 46. The second—really dramatic—reason is that there are many types of music which are not subjected to the scheme "score to performance," at least not if score means a structural or processual scheme controlling sound production. Much jazz music, for example, is defined through its bodily realization rather than as a projection of score information. Cecil Taylor's piano music is a stunning example of the primacy of bodily gestures (what Volker Spicker calls the "Abstraktmotiv" [998]) over structures, a fact often pointed out by Taylor himself. And for Beethoven, as also observed by Robert S. Hatten [446], ruling out gestural shaping misses the musical contents, as beautifully illustrated by Glenn Gould's notorious "contrafactual" recording of op. 57 *Appassionata*, where, for example, the cascade in bars 14-15 of the opening movement is not played as a gestural cascade, but as a static structure of arpeggiated VII-chords, which destroys completely the inherent movement, as performed by Vladimir Horowitz, say.

Despite the intuitive understanding of what a gesture is, including body movement and semantics, a precise conceptualization looks less easy. We agree with Jean-Claude Schmitt [946] that the medieval definition of a gesture, as given by Hugues de Saint-Victor, remains one of the most adequate, at least when referring to the concrete human body: "Gestus est motus et figuratio membrorum corporis, ad omnem agendi et habendi modum." Gesture is the movement and figuration of the body's limbs with an aim, but also according to the measure and modality proper to the achievement of all action and attitude. Most important is that it is an articulated figuration, a composition of parts (limbs), and that it includes a movement of that figuration in the space-time of the given body. Moreover, it serves for any (omnem) mode of action and attitude, so it has a purpose or target, but it does not, automatically, point to a semantic level, it only reaches the mode of an activity/habit.[4] Often, in the definition of a gesture, the sign character (signifier–signification–signified) is included, for example in Adam Kendon's and David McNeill's approach [530, 741];

[4] This is also confirmed by the comment of Jean-Claude Schmitt [946] in his comment on Saint-Victor's definition: "Le geste enveloppe avant de saisir et esquisse son déploiement bien avant de dénoter ou d'exemplifier; ce sont les gestes déjà domestiqués qui font référence."

see also Section 57.7. Together with the French diagrammatic philosophers, such as Charles Alunni [24], we do not share this perspective and would rather say that a gesture is a presemiotic entity, although it may be a component of a sign. But this is not mandatory. For example, in dance or in music, gestures are often bare of meaning. They stand for themselves, and this in particular as esthetic entities. We should also add that our reference to Saint-Victor does not mean that we shall limit ourselves to this definition. It is only a point of departure, and we shall transcend its strict reference to the human body and its limbs and consider much more abstract or metaphoric configurations in the following mathematical theory of gestures, including gestural objects relating to graphics and sounds. But what is important for our understanding of the concept is that in Saint-Victor's definition, the *composed parts* on the one hand and the *presemiotic nature* of a gesture on the other are recognized.

The shift from abstract algebra and transformational paradigms to gestures is not a purely formal one, since the basic object given by a gesture is not a "point" within an algebraically shaped space, which may be connected to other points by functional correspondence. A gesture, by its parametrized curve character, has interiorized the transformational approach: its endpoints are intrinsically related by the curve parameter. This is in contrast to the functional correspondence, where an arrow effectively has nothing which exists in its shaft, it is a purely graphical symbol. No additional transformational input is needed. This is an enrichment which has been realized for similar constructions in the physical string theory of elementary particles. In this language, particles are also curves of an ample inner nature, when compared to the classical point-like elementary particle models. For example, in string theory, the electric charge of an electron may be constructed geometrically as a winding number of the electron loop around a supplementary compact dimension, i.e., by an instance of the fundamental group of the compact dimension. In this spirit, the gestural approach is an enrichment of musical object categories, which enables a refinement of the conceptual anatomy and at the same time a rapprochement to the human reality of making music. We could call it a *musical string theory*, but see Chapter 75.

61.3 Initial Investigations: Diagrams of Curves

Fig. 61.1. The technical description of hand and elbow movements in Marek's treatise on piano playing shows gestures in the form of complex curve configurations.

The consultation of detailed technical treatises on virtuoso piano playing reveals a high consciousness of the gestural side of this art. For example, the classical book *Lehre des Klavierspiels* [651] by Ceslav Marek, a student of Theodor Leschetitzki, a legendary student of Franz Liszt, teaches that the movements of the hand and the elbow are types of gestural curve configurations; see Figure 61.1.

In his PhD thesis [771], Stefan Müller has implemented this approach in software to obtain a computer-generated virtual piano performance, transforming the data of a standard score into movements of a computer-graphical hand model. This means that we had to define a system of curves in space-time, which describe the movements of a hand that plays a given score. The problem is extremely complex, since one has to deal with (1) the complex geometry of the human hand, (2) the dynamic conditions of Newton's law, which describes the possible trajectories of the fingers' and carpus' masses as a function of available

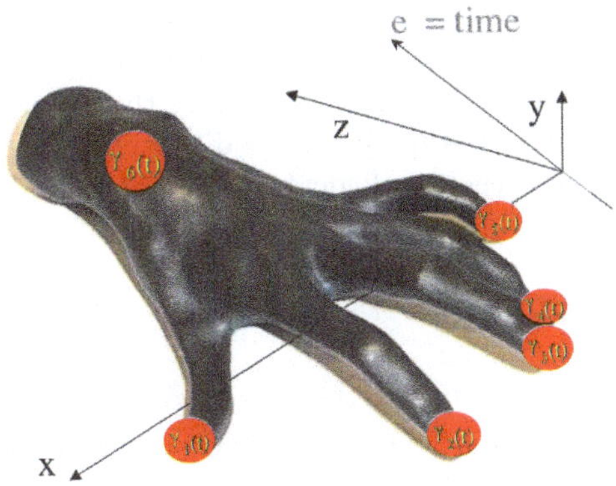

Fig. 61.2. The hand model used for a computer-graphic simulation of a pianist's performance. This hand is modeled after Chopin's.

forces, and (3) the question concerning optimal solutions to such system conditions. We modeled the hand by parameterized space-time curves $\gamma_i(t)$ of the five tips and the carpus; see Figure 61.2. The curve parameter t is an abstract parameter, which cannot be identical with the time value for the following reason. In fact, scores have successive notes of different pitch *for the same finger* that must be connected by a continuous curve while time remains unchanged! This symbolic gesture must be taken as a default setup, which is later deformed to comply with physical and geometric conditions. See Chapter 78 for the theory of performance based upon symbolic and physical gestures.

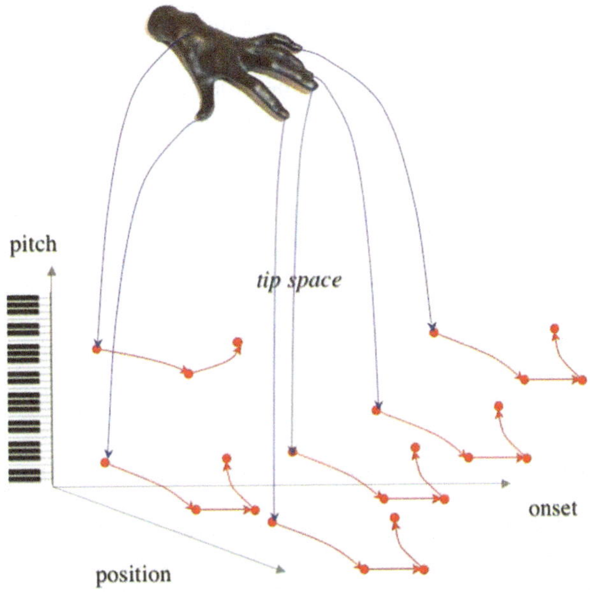

Fig. 61.3. The six space-time curves of one hand of a pianist as modeled by Stefan Müller in [771].

An animation of such a performance was shown at the ICMC 2003 [772], the performed piece being the right hand of Carl Czerny's small exercise op. 500. The overall picture of such a configuration of space-time curves of fingers and carpus is shown in Figure 61.3.

The lesson drawn from this preliminary study about gestures is that they are not single curves, but complex configurations of curves (in our example, each finger shows a concatenation of three curves, one for moving down to the key, then one while resting on that key, and a third for moving away from the key). And the space of such curves may show an arbitrarily complex topology (for the fingers, one is confronted with a manifold in a high-dimensional real space, which is defined by the geometric and physical constraints); see Section 78.2.3 for a more concise representation of the hand's geometry. We shall now step over to a general definition of a gesture, which comprises the topological as well as the configurational aspects of the concept.

61.4 Modeling a Pianist's Hand

One of the most evident gestural expressions is the body movement of a performing musician. Think of the gestural power of famous pianists. It is logical that one should therefore attempt to model gestures of musicians. In collaboration with my PhD student Stefan Müller, I embarked in the modeling of the pianist's hand [772] on the level of computer graphics. The idea was not only to model the hand's movements, but also to implement a software that could transform the abstract symbols of a score into hand movements that were adequate for the rendering of the score on a piano keyboard.

The project had three components:

1. Modeling the hand with its spatio-temporal trajectory in the movement.
2. Transforming abstract score symbols of notes (what we call deep-frozen gestures, since they historically stem from neumatic abstraction, neumes being gestural signs) into *symbolic gestures*, i.e., curves in a space related to the piano keyboard geometry.
3. Deforming symbolic hand gestures into physically valid spatio-temporal curves of the pianist's hand.

61.4.1 The Hand's Model

This task was accomplished with a simplified representation of the hand by six curves $\gamma_i(t)$ in physical space-time with space axes x, y, z to denote the momentous position of the hand, and e, the physical time of that position. The curve parameter t is not the physical time, it is just an abstract curve parameter (Figure 61.2). The curves $\gamma_1, \ldots \gamma_5$ represent thumb, index, middle, ring, and little finger, respectively, while γ_6 represents the carpus.

These curves are subjected to geometric constraints G resulting from their connectivity as parts of the hand's geometry. We refer you to [772] for more details. And the curves are subjected to mechanical constraints M, which means that if we think of the ith finger's mass m_i, and if the pianist is capable of exerting a maximal force of K_i upon that mass, then Newton's second law imposes the inequality

$$m_i|d^2\gamma_i^{space}/de^2(t)| \leqslant K_i$$

at any curve parameter value t, where γ^{space} is the three-dimensional spatial part of the curve.

61.4.2 Transforming Abstract Note Symbols into Symbolic Gestures

Refer to Figure 61.4 for the following discussion. In traditional performance theory, we look at the transformation $\wp_{score}$ of score symbols into sound events. This is shown in the bottom row of the rectangular diagram of Figure 61.4. In the gestural extension of this disembodied process, we have to create the sonic result via gestural actions. The sounds are just the result of physical gesture curves interacting with the keyboard; these curves are shown in the right top corner of the diagram.

In order to generate these physical curves, one first has to unfreeze the note symbols and to transform them into gestural symbols. This unfreezing process is shown in the left top-bottom half of the diagram. This does not create physical gesture curves, but only symbolic gesture curves, which are faithful representations of the note symbols. This process resembles the MIDI interpretation of notes insofar as the commands associated with notes are abstract movements: In MIDI, a note is defined by an ON command, which means,

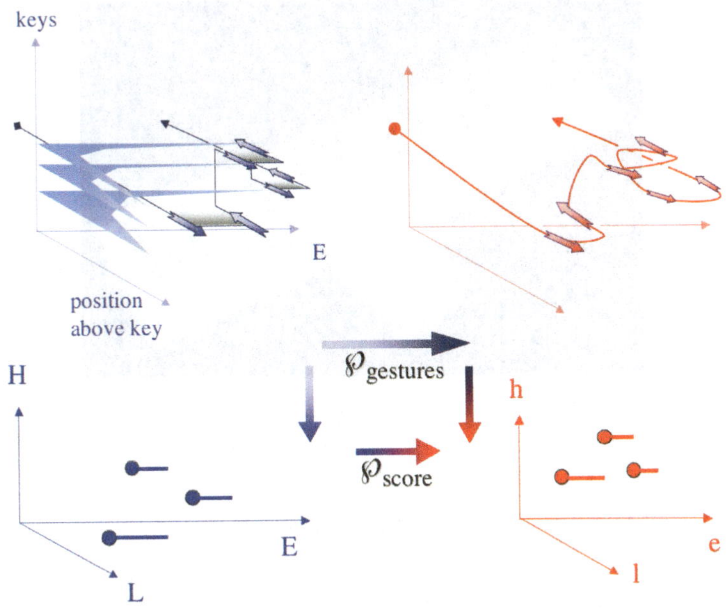

Fig. 61.4. The four levels of performance: symbolic score representation (left bottom), performed sound events (right bottom), symbolic gesture curves (left top), physical gestures (right top).

go down to that pitch at a determined moment, and the MIDI velocity used to move down to the key defines loudness. Then the finger remains in that position until the OFF command tells it to move up again, etc.

This representation defines a very abstract gesture, but it is this that tells the fingers in a qualitative way how to move. This movement is shown in the top left corner. We see the *symbolic gesture* associated with a sequence of three notes in the bottom left corner. The finger moves down over a first key, then remains there and after its duration moves up, changes the key coordinate, goes down to the second key with a second velocity, remains there for its duration, moves up, shifts to a third key position, moves down with the third velocity, remains for its duration, and finally moves up. All these phases are connected in a continuous curve, which has angles, i.e., it is not differentiable, and whose movement is orthogonal to the time axis E when moving down at a determined velocity.

61.4.3 From Symbolic Hand Gestures to Physical Gestures

The third step toward gestural performance is the horizontal transformation on top of the diagram in Figure 61.4. The given symbolic gesture does the right thing, but it does not move within the geometric and physical constraints. These constraints define a subspace of the space of all continuous curves, in fact a manifold $X(G, M)$ in terms of global geometry. We are given the symbolic gesture curve from the top left data and now have to create a physically valid deformation thereof, i.e., one that fulfills the geometric and mechanical constraints G, M. This is a very delicate operation. Essentially, it boils down to looking at the symbolic gesture $\gamma_{\mathrm{Symbolic}}(t)$ and then searching for one $\gamma_{\mathrm{Physical}}(t)$ that is as near as possible to $\gamma_{\mathrm{Symbolic}}(t)$ and lives in $X(G, M)$. The delicate point is that it is often not possible to cope with all conditions for a perfect performance, since, for example, physics does not allow for infinite velocities. So when the finger has to play two different keys in immediate succession without a pause, the duration of the first note must be shortened in order to jump from the first to the second key. Such difficulties must be met by defining distance between curves in such a way that musical constraints are given a high weight. For example, the prescribed key coordinates cannot be changed, while durations may be changed, but only a little. It may then happen that there is no solution to a given score input and its associated symbolic curve. This must be possible as a function of the anatomic and physical constraints given by human conditions. We have implemented this process and have performed a simple Czerny exercise in a movie that is illustrated in Figure 61.5.

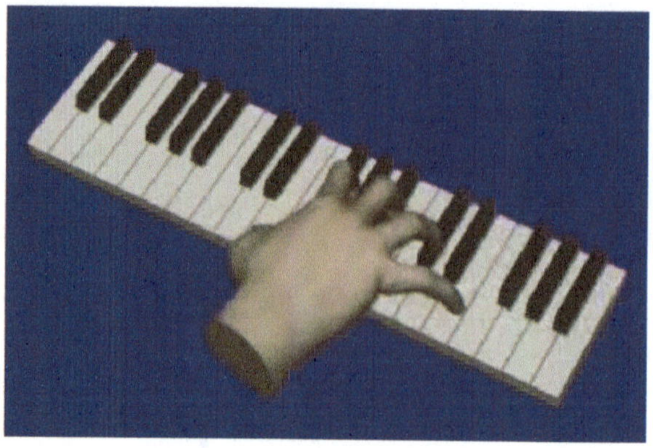

Fig. 61.5. A Czerny exercise played by the computer-graphical model of a pianist's hand according to the gestural "unfreezing" process described in the text.

61.5 The Mathematical Definition of Gestures

In the following definition of a gesture, we shall rely on Saint-Victor's definition and implement the (con)figuration of a gesture by the articulation of diagrams. We shall then describe the movement in the parametrization of curves representing the figuration, and we shall formalize the body's space-time by a topological space, where the movement takes place. The semantics of gestures will not be our concern here; this must be dealt with *after* a thorough investigation of the formal mathematics of gestures.

Let us first review the category **Digraph** of directed graphs, in short: digraphs. This is a basic category for algebra as well as for topology. In the naive setup, its objects are functions $\Gamma : A \to V^2$ from a set $A = A_\Gamma$ of *arrows* to the cartesian square $V^2 = V \times V$ of the set $V = V_\Gamma$ of *vertices*. The first projection $t = pr_1 \circ \Gamma$ is called the *tail* function, the second $t = pr_2 \circ \Gamma$ is called the *head* function of the digraph. For an arrow a, the vertices $t(a)$ and $h(a)$ are called its head and tail, respectively, and denoted by $t(a) \xrightarrow{a} h(a)$. A morphism $f : \Gamma \to \Delta$ of digraphs is a couple $f = (u, v)$ of functions $u : A_\Gamma \to A_\Delta, v : V_\Gamma \to V_\Delta$ such that $v^2 \circ \Gamma = \Delta \circ u$.

It can be shown that **Digraph** is a topos, see Appendix Section J.1.3. In particular, it has a a final object, $1 = t \circlearrowleft T$, which is embedded by the true morphism $T : 1 \to \Omega$ into a subobject classifier

$\Omega =$ 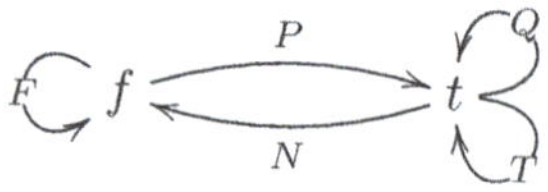

Every evaluation $\Gamma @ \Omega$ describes the set of subdigraphs of Γ, together with its canonical Heyting logic. This fact may be used to introduce gestural logic; see Section 61.13.

We need a special subcategory of digraphs, the *spatial digraphs*. Such a digraph is associated with a topological space X and denoted by $\overrightarrow{X}$. By definition, the arrow set is $A_{\overrightarrow{X}} = I@X$, the set of continuous curves $c : I = [0, 1] \to X$ in X, while the vertex set is $V_{\overrightarrow{X}} = X$, $h(c) = c(1)$, and $t(c) = c(0)$. A spatial morphism is a digraph morphism $\overrightarrow{f} : \overrightarrow{X} \to \overrightarrow{Y}$ canonically induced by a continuous map $f : X \to Y$. The subcategory of spatial digraphs and morphisms is denoted by *SpaceDigraph*. A spatial digraph is more than a digraph: it is also a topological digraph in the following sense. The set $A_{\overrightarrow{X}} = I@X$ of arrows of $\overrightarrow{X}$ is a topological space by the compact-open topology, and the head and tail maps $h, t : I@X \to X$ are continuous. Moreover, for a continuous map $f : X \to Y$, the arrow map $I@f : I@X \to I@Y$ is continuous.

Given a digraph Δ and a topological space X, a Δ-*gesture in* X is a digraph morphism $\delta : \Delta \to \overrightarrow{X}$, i.e., a realization of the abstract vertices and arrows within a topological space, as shown in Figure 61.6. It is essential here to distinguish the curve parameter of a gesture from the time parameter, which intervenes in a number of common gestures. For example, when drawing a curve on a sheet of paper, this gesture, $\delta : \Delta \to \overrightarrow{X}$, would have the arrow digraph $\Delta = \uparrow$ with $\uparrow = \bullet \to \bullet$ and the space $X = \mathbb{R}^2 \times \mathbb{R}$, whose first two coordinates denote the points of the paper surface, whereas the third denotes the physical time when a point at given parameter value is drawn on the paper sheet. We discuss the time parameter in the following Example 59.

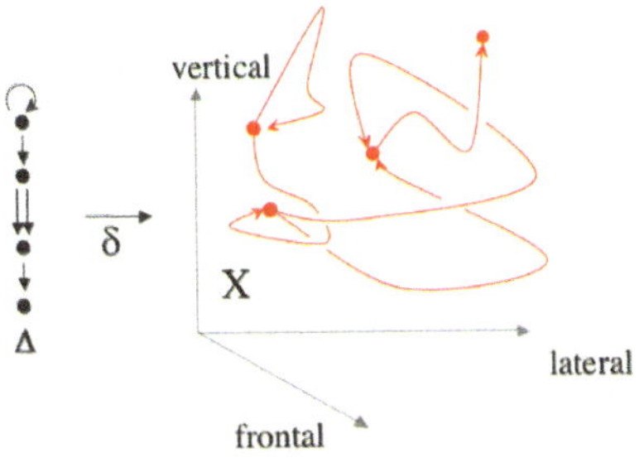

Fig. 61.6. A gesture in the ordinary 3-space.

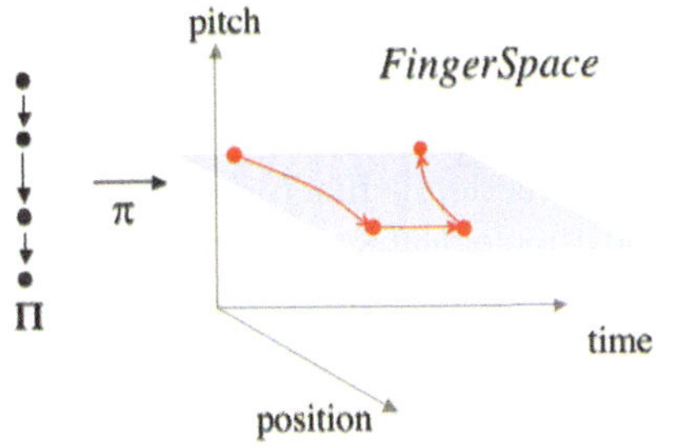

Fig. 61.7. An elementary finger gesture of a pianist's hand.

Given two gestures, $\delta : \Delta \to \overrightarrow{X}, \gamma : \Gamma \to \overrightarrow{Y}$, a morphism $f : \delta \to \gamma$ is a digraph morphism $f : \Delta \to \Gamma$ such that there exists a spatial morphism $\overrightarrow{h} : \overrightarrow{X} \to \overrightarrow{Y}$ which commutes with f, i.e., $\overrightarrow{h} \circ \delta = \gamma \circ f$. This defines the category *Gesture* of gestures, and we have a projection $p : Gesture \to$ **Digraph**, which sends the gesture $\delta : \Delta \to \overrightarrow{X}$ to the underlying digraph Δ. So it is essentially a forgetful functor: it cancels out the spatial interpretation of the given digraph.

Example 59 Let us give here an elementary example of the gesture of a finger of a piano player's hand. The gesture represents the movement of a finger going down to a key, staying in that position for the duration of a tone, and then moving back upwards to be ready for the next movement. This gesture has three segments, which we formally relate to a gesture defined on the digraph, $\Pi = \bullet \xrightarrow{d} \bullet \xrightarrow{h} \bullet \xrightarrow{u} \bullet$. The space of this gesture is three-dimensional, i.e., $FingerSpace = \mathbb{R}^3$, where the first coordinate is the height above the keyboard (position, positive direction for approaching the piano), the second is the time, and the third parametrizes the pitch (the coordinate of the key on the piano's keyboard). Then a finger gesture $\pi : \Pi \to \overrightarrow{FingerSpace}$ is a diagram of curves as shown in Figure 61.7.

Example 60 A more involved example is provided by the gestural space of an entire hand, as investigated in [772] in the context of the simulation of a pianist's playing from a given score, under constraints from the hand's geometry as well as the physical dynamics from Newton's second law (limits of forces avaliable from the pianist's performance). Figure 61.2 shows the hand position in time. It is given by six points $\gamma_i(t)$ in the 4-space $\mathbb{R}^4$ with three space coordinates (pitch, vertical position above keyboard, and horizontal frontal position from the key face line) and one time coordinate. The fingertips are represented by the indexes $i = 1, 2, \ldots 5$, while the carp is represented by $i = 6$. Observe that the time coordinate is not the curve coordinate t! A gesture's curve coordinate is an abstract parametrization of the curve in a given space, not the material time coordinate, which may also be absent, as shown in the gestures from Figures 61.9 and 61.10.

61.6 Hypergestures

If gestures of a certain kind are themselves conceived as points in a space, one may study gestures within such a gesture-space. These may be called hypergestures. In order to define them, we need to know how to turn the set of gestures $\Delta@\overrightarrow{X}$ into a topological space. Now, the special case $\Delta = \uparrow$ means that we have the topological space $\uparrow @\overrightarrow{X} \xrightarrow{\sim} I@X$ of continuous curves $c : I \to X$ (with the above mentioned compact-open topology). The general case follows from the observation that Δ is the colimit of the following diagram

$\mathcal{D}$ of digraphs: We take one arrow digraph $\uparrow_a = \uparrow$ for each arrow $a \in A_\Delta$ and one bullet digraph $\bullet_x = \bullet$ for each vertex $x \in V_\Delta$. We take as morphisms the tail or head injections $\bullet_x \to \uparrow_a$ whenever $x = t(a)$ or $x = h(a)$. Then evidently, $\Delta \xrightarrow{\sim} \mathrm{colim}\mathcal{D}$. Therefore, the set of gestures $\Delta @\overrightarrow{X}$ is bijective to the limit $\lim \mathcal{D}@\overrightarrow{X}$ of a diagram of topological spaces $\uparrow @\overrightarrow{X} \xrightarrow{\sim} I@X$. The topology of $\Delta @\overrightarrow{X}$ is defined as the limit topology of this diagram, the space with this topology is denoted by $\Delta\overrightarrow{@}X$. In the case of a metric d defining the topology of X, it is well known that the compact-open topology on $I@X$ coincides with the topology defined by the metric $\overrightarrow{d}(\gamma, \gamma') = \sup_t d(\gamma(t), \gamma'(t))$. And then, the topology of $\Delta\overrightarrow{@}X$ is defined by the metric $\Delta\overrightarrow{@}d(\gamma, \gamma') = \sum_{a \in A_\Delta} \overrightarrow{d}(\gamma_a, \gamma'_a)$.

This construction is functorial in both arguments: If $h : \Gamma \to \Delta, f : X \to Y$ is a couple of morphisms of digraphs and topological spaces, respectively, then the canonical map $(f, h) = \overrightarrow{f} \circ ? \circ h : \Delta\overrightarrow{@}X \to \Gamma\overrightarrow{@}Y$ is continuous and functorial in h, f. For example, if $h : p \to \Delta$ is the embedding of a single vertex p, the map $(Id_X, h) : \Delta\overrightarrow{@}X \to p\overrightarrow{@}X$ is the restriction of gestures to the point p. If we select a particular gesture ξ in $p\overrightarrow{@}X$, i.e., a point $\xi \in X$, then the fiber $(Id_X, h)^{-1}(\xi)$ is the set of gestures sending p to ξ. If in particular $\Delta = 1$, this fiber is the set of loops in ξ.

By the above, one may now repeat the gesture construction and consider the topological (and especially, the metric) space of hypergestures $\Gamma\overrightarrow{@}\Delta\overrightarrow{@}X$, hyperhypergestures $\Lambda\overrightarrow{@}\Gamma\overrightarrow{@}\Delta\overrightarrow{@}X$, etc. Notice that in particular, the space $\uparrow \overrightarrow{@} \uparrow \overrightarrow{@}X$ is the topological space $I^2@X$ of homotopies in X, i.e., hypergestures generalize homotopies between continuous curves. Here is a useful proposition concerning the order in which hypergestures are constructed:

Proposition 61 (First Escher Theorem[5]) *If Γ, Δ are digraphs and X is a topological space, then we have a canonical homeomorphism*

$$\Gamma\overrightarrow{@}\Delta\overrightarrow{@}X \xrightarrow{\sim} \Delta\overrightarrow{@}\Gamma\overrightarrow{@}X.$$

This results from the above fact that a digraph is the colimit of its arrows, glued together on their head and tail points. On the one hand this entails that the hypergesture space $\Gamma\overrightarrow{@}\Delta\overrightarrow{@}X$ is the limit of the spaces with arrows $\uparrow \overrightarrow{@}\Delta\overrightarrow{@}X$, which are in turn identified with the spaces of curves $I@\Delta\overrightarrow{@}X$, but these are the limit of spaces $I@I@X$ over the arrow system of Δ. On the other hand, one may also first start with the limit over the arrow system of Δ and then pass to the limit over the arrow system of Γ. Thus, the two limit systems can be applied in any order, and this means that the two topological spaces in question are isomorphic by universal properties of colimits and limits. One may also use the fact that double limits exchange.

Corollary 23 *The action*

$$\overrightarrow{@} : \mathbf{Digraph} \times \mathbf{Top} \to \mathbf{Top} : (\Gamma, X) \mapsto \Gamma\overrightarrow{@}X$$

canonically extends to an action (denoted by the same symbol)

$$\overrightarrow{@} : [\mathbf{Digraph}] \times \mathbf{Top} \to \mathbf{Top} : (W, X) \mapsto W\overrightarrow{@}X$$

of the free commutative monoid **[Digraph]**, *i.e., the monoid of commutative words $W = \Gamma_1\Gamma_2 \ldots \Gamma_k$ over the alphabet* **Digraph** *of digraphs (the objects only). It is defined [6] inductively by $\Gamma_1\Gamma_2 \ldots \Gamma_k\overrightarrow{@}X = \Gamma_1\overrightarrow{@}(\Gamma_2 \ldots \Gamma_k\overrightarrow{@}X)$ and $\emptyset\overrightarrow{@}X = X$.*

Example 61 Referring to the finger gestures described in Example 59, we want to discuss an example of piano finger hypergesture. Before so doing, let us mention a quite intriguing statement of Renate Wieland and Jürgen Uhde in [1067]:

> Die Klangberührung ist das Ziel der zusammenfassenden Geste, der Anschlag ist sozusagen die Geste in der Geste.[7]

[5] The reference to Maurits Cornelis Escher is given because in his art, flipping roles are an important artistic tool.
[6] To be precise, this action is defined up to homeomorphisms.
[7] The touch of sound is the target of the embracing gesture, the keystroke is so to speak the gesture within the gesture.

Although the precise meaning of this statement is somewhat cryptic, it could be interpreted as arguing that a gesture may be thought of as being built from other gestures. Recall that a hypergesture is built from a system of homotopies of the curves that compose its vertex gestures. Now, let us take a hypergesture $\eta :\uparrow\to \overrightarrow{FingerSpace}$. This means that we have a continuous curve of finger gestures $\eta(t)$, $t \in I$, from the initial finger gesture $\eta(0)$ to the final finger gesture $\eta(1)$.

Musically, this means that in the given *FingerSpace*, two gestures can be related to each other by thinking of the final gesture as a result of an infinite series of intermediate gestures. So they may be connected cognitively by just deforming the original gesture to obtain the final one. This seems to be obvious in our example, since no obstruction to such a deformation is indicated. The next example shows that in a less trivial space, viz. the torus, such deformations are not always possible and from a cognitive point of view, this would lead us to conceive radically different strategies in the management of gestures. In fact, the existence of hypergestures relates to the fundamental group of the underlying topological space.

Fig. 61.8. There is no hypergesture from η to ν (left), while these two gestures are isomorphic (right).

Example 62 Take the (two-dimensional) torus $X = \mathbb{T}^2$ and the final digraph 1; see Figure 61.8. Then, if $\eta, \nu : 1 \to \overrightarrow{\mathbb{T}^2}$ are the horizontal equatorial circle curve at the origin 0 of $\mathbb{T}^2$ and the vertical meridian curve at the origin, respectively, there is no hypergesture of type $\uparrow$ from η to ν, whereas the diagonal reflection on $\mathbb{T}$ yields the morphism $Id_1 : \eta \to \nu$.

This makes clear the difference between hypergestures and morphisms: Hypergestures realize the "arrows" between vertex gestures as curves, whereas morphisms realize them by transformations between the vertex curves.

61.6.1 Spatial Hypergestures

Intuitively, gestures allude to a movement in space, where the curve parameter plays the role of time. This is however misleading and restrictive. Let us explain it by the representation of complex bodies as spatial hypergestures. If we start with the space of loops in 3D, i.e., the gesture space $X = 1\overrightarrow{@}\mathbb{R}^3$, a gesture of diagram Γ in X is a hypergesture $\beta \in B = \Gamma\overrightarrow{@}1\overrightarrow{@}\mathbb{R}^3$; see Figure 61.9 for an example.

With spatial hypergestures, one can essentially do all the computer graphics constructions, such as spline surfaces of Bézier type [710]. Such a surface, when defined by a grid of $(n+1) \times (m+1)$ points, appears as a hypergesture $\uparrow^n \overrightarrow{@} \uparrow^m \overrightarrow{@}\mathbb{R}^3$, where $\uparrow^n$ is the digraph consisting of $n+1$ vertices, and having one arrow from vertex i to vertex $i+1$ for all $i = 0, 1, 2, \ldots n-1$. Then, using this type of purely spatial hypergestures β, which we temporarily call body, we may model its movement in time as a gesture $\mu \in\uparrow \overrightarrow{@}B\times\mathbb{R}$, having values in the space of pairs (β, τ) of bodies and times. This allows for the description of realistic body movements (viz. animated graphics) for dance or sports, for example. In fact, $\uparrow \overrightarrow{@}B \times \mathbb{R} \xrightarrow{\sim}\uparrow \overrightarrow{@}B\times \uparrow \overrightarrow{@}\mathbb{R}$, i.e., such a movement is a pair of a $\uparrow$-hypergesture of bodies in $\uparrow \overrightarrow{@}\Gamma\overrightarrow{@}1\overrightarrow{@}\mathbb{R}^3$ and a time gesture in $\uparrow \overrightarrow{@}\mathbb{R}$. Figure 61.10 shows four stages of a deformation of a body of type $\Sigma\overrightarrow{@}1\overrightarrow{@}\mathbb{R}^3$, where Σ is a digraph with four vertices z, w_1, w_2, w_3 and three arrows, $a_i : z \to w_i, i = 1, 2, 3$.

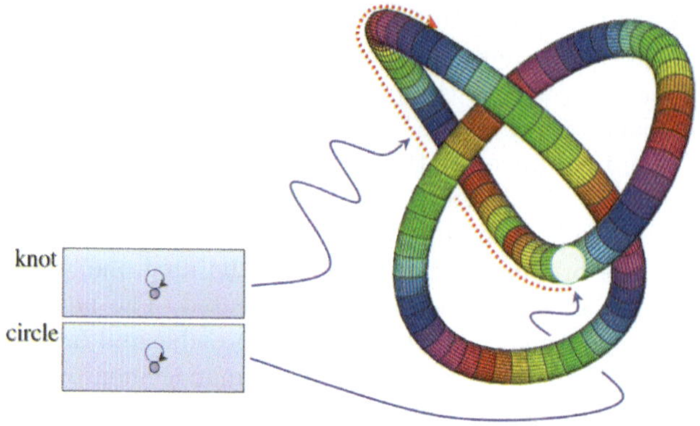

Fig. 61.9. A knot represents a complex hypergesture built from loop gestures.

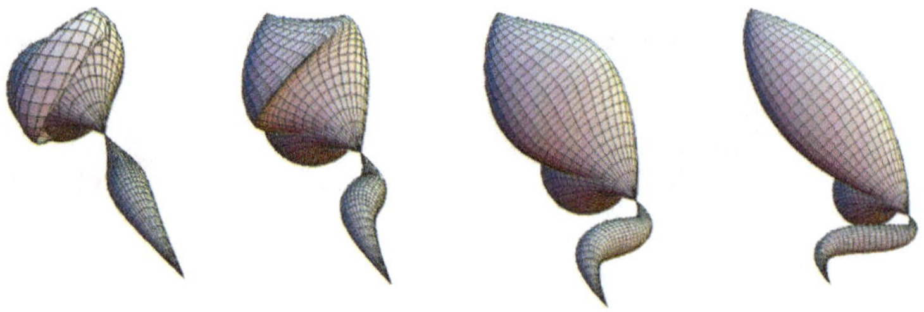

Fig. 61.10. Four stages (from left to right) of a "dancing" hypergesture of type $\uparrow$ with values in spatial hypergestures of type $\Sigma\overrightarrow{@}1\overrightarrow{@}\mathbb{R}^3$.

$\Sigma =$

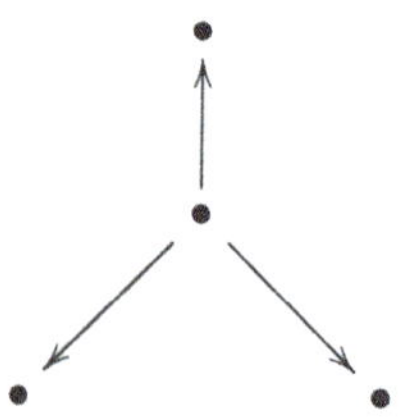

In this example, we may reinterpret the hypergesture in $\uparrow \overrightarrow{@}\Sigma\overrightarrow{@}1\overrightarrow{@}\mathbb{R}^3 =\uparrow \Sigma 1\overrightarrow{@}\mathbb{R}^3$ by use of Proposition 61 and its Corollary 1: While our gesture is a curve of Σ-shaped hypergestures of loops, we may use the isomorphism $\uparrow \Sigma 1\overrightarrow{@}\mathbb{R}^3 \xrightarrow{\sim} 1 \uparrow \Sigma\overrightarrow{@}\mathbb{R}^3$ to view the hypergesture in question as being a loop of hypergestures which are curves of Σ-shaped gestures in X.

61.7 Categorically Natural Gestures

We now proceed to the construction of a gesture for each digraph, and this in a so-called natural way, i.e., such that the category of digraphs is related to the associated gestures in a structurally compatible way, which mathematicians call "natural" (see [637] for a background). We have spatialization functor *Space* : **Digraph** $\to$ *SpaceDigraph* as follows: We first take the colimit $|\Delta|$ of the following diagram of topological spaces (see Figure 61.11): For every arrow of Δ, we take one copy of the unit line $I = [0, 1]$, and for each vertex one copy of the singleton space $\{*\}$. Then we take the maps from the singleton spaces to the line copies $* \mapsto 1$ or $* \mapsto 0$ for each coincidence of arrow heads or tails and arrows.

The colimit topology is this: The *skeletal space* $|\Delta|$ is all the copies of the unit line I being glued together in their common vertices, and the open sets in $|\Delta|$ are the sets intersecting in an open set for each line I. This space is in fact a metric space, i.e. we view it as being given the induced metric from $\mathbb{R}^{V_\Delta} \oplus \mathbb{C}^{A_\Delta}$ on the

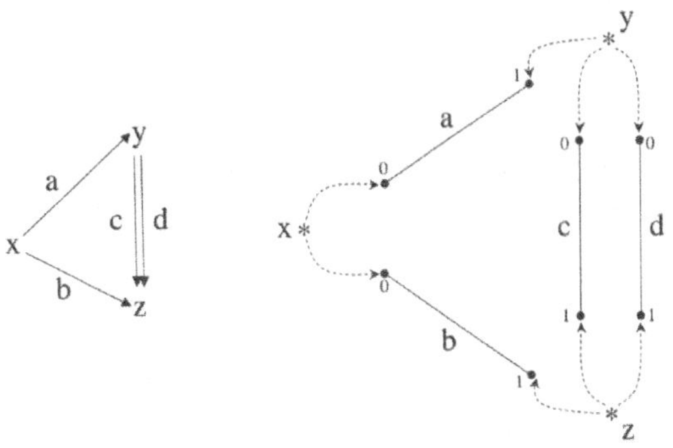

Fig. 61.11. To the right, the diagram for the colimit topology derived from the digraph to the left.

subspace consisting of basis vectors e_v indexed by vertices $v \in V_\Delta$. The arrow $a \in A_\Delta$ is realized by a cartesian product of the line $(1-s)e_t + se_h, s \in I$ for an arrow $t \xrightarrow{a} h$, times the unit circle $S_1 = \exp(2\pi i s), s \in I$, in $\mathbb{C}$ indexed by the arrows. This means that we have a twisted line $((1-s)e_t + se_h, \exp(2\pi i s)), s \in I$ in the direct sum $c_a(t) = \mathbb{R}e_t \oplus \mathbb{R}e_h \oplus \mathbb{C}e_a \subset \mathbb{R}^{V_\Delta} \oplus \mathbb{C}^{A_\Delta}$. We write $\vec{\Delta} = \overrightarrow{|\Delta|} = I@|\Delta|$. The spirals serve two needs: (1) They disambiguate arrows which share heads and tails and (2) they make loops that look like circles, and consequently spirals for "non-loops" are the most natural way to "draw" them.

Moreover, we have natural transformation $g : \mathrm{Id}_{\mathbf{Digraph}} \to \vec{?}$, which means that there is a gesture $g(\Delta) : \Delta \to \vec{\Delta}$ for each digraph, which varies functorially. The gesture maps the vertices v of Δ to the points $g(\Delta)(v) = e_v$ of $|\Delta|$, and the arrows $t \xrightarrow{a} h$ to the curves $g(\Delta)(a) = c_a$ in $|\Delta|$ (to ease intuitive understanding, we use the representation in the metric space $\mathbb{R}^{V_\Delta} \oplus \mathbb{C}^{A_\Delta}$, but in reality we refer to the corresponding structures in the defining colimit). The gesture $g(\Delta) : \Delta \to \vec{\Delta}$ has the following universal property:

Proposition 62 *For any morphism $f : \Delta \to \Gamma$ of digraphs and any gesture $\gamma : \Gamma \to \vec{X}$, there is a unique continuous map $|f| : |\Delta| \to X$ such that its associated digraph morphism $\vec{f} = \overrightarrow{|f|} : \vec{\Delta} \to \vec{X}$ induces a morphism $g(f) : g(\Delta) \to \gamma$ of gestures.*

This follows by standard arguments from the universal property of the colimit topology on $|\Delta|$. In fact, the identity maps on the unit intervals defining the colimit $|\Delta|$ must be mapped into the curves defining the gesture γ, and this uniquely determines the morphism $|f|$ by the universal property of colimits.

Corollary 24 *The maps $\Delta \mapsto \vec{\Delta}$ and $f \mapsto \vec{f}$ define a functor $\vec{?}$:* **Digraph** *→ SpaceDigraph and the gesture $g(\Delta) : \Delta \to \vec{\Delta}$ defines a natural transformation $g : Id_{\mathbf{Digraph}} \to \vec{?}$, also called the* natural gesture associated with the digraph Δ.

Corollary 25 *The gesture $g(\Delta) : \Delta \to \vec{\Delta}$ defines a functor g :* **Digraph** *→ Gesture which is left adjoint to the projection p : Gesture →* **Digraph**, *in symbols $g \dashv p$. This means that*

$$\mathbf{Digraph}(\Delta, p(\gamma)) \xrightarrow{\sim} \mathit{Gesture}(g(\Delta), \gamma)$$

is a bijection, which is functorial in both arguments Δ, γ.

61.8 Connecting to Algebraic Topology: Hypergestures Generalize Homotopy

Algebraic topology is a fundamental field in mathematics that induced the most spectacular change in mathematical thinking in the 20th century. As the name suggests, this field connects the category $\mathcal{T}$ of topological spaces and continuous maps with algebraic categories, such as the category $\mathcal{G}$ of groups and group homomorphisms. One of the most successful processes in algebraic topology is the transition from a topological space X to an associated group, namely the *fundamental group* $\pi_1(X)$ of X (subjected to some technical conditions, which me may neglect in this introduction). The method that enables such a construction is most remarkable if not paradoxical. The fundamental group is built on continuous curves $c : I \to X$, that is, those objects which give rise to gestures in X. One would intuitively expect that such curves are already fuzzy objects, and not quite what could give rise to precise abstract algebraic structures such as groups. But the opposite is true: Continuous curves are too precise objects. What is needed in algebraic topology are classes of curves that emerge from making curves even less precise! The technical concept for this operation is called homotopy. Two curves c_0, c_1 in X are called homotopic if they can be deformed into each other in a continuous way. Technically, this means that there is a continuous map $C : I^2 \to X$ such that $c_0(t) = C(0,t)$ and $c_1(t) = C(1,t)$ for all $t \in I$. For the curves of two gestures of same skeleton, such a homotopy is shown in Figure 61.12.

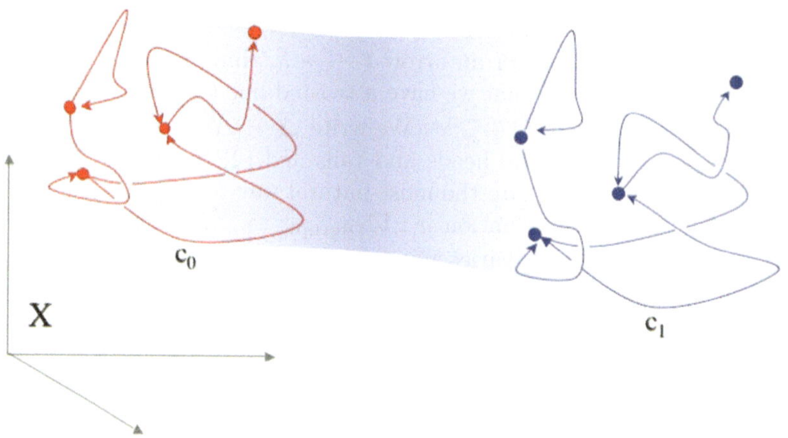

Fig. 61.12. In this homotopy, the left gesture is deformed to the right one by a continuous family of intermediate gestures. The gesture's curves (e.g. c_0 and c_1) are homotopic to each other.

The reader will immediately recognize that such a homotopy of gestures is a special case of a hypergesture with skeleton $\uparrow = \bullet \to \bullet$. In other words, *homotopy is a special case of a hypergesture.* Now, the relation of homotopy is an equivalence relation, and one may therefore consider equivalence classes of homotopic curves (or gestures), the so-called homotopy classes. This means that we do not look at curves, but at fuzzy drawings of such curves: It does not matter where we draw precisely our lines, a small deviation will not change the homotopy class!

The miracle or paradox is that under such a smearing of curves, symbolic structures become real. The precise construction runs as follows. We consider the set $I@X$ of all curves in X. We do not consider any homotopic curve pairs, but only pairs c_0, c_1 such that they coincide at their endpoints, i.e., homotopies such that $c_0(0) = c_1(0)$ and $c_0(1) = c_1(1)$. We denote such a "pointed" homotopy class of curve c by $[c]$. We may now concatenate curves: If curve $e, f : I \to X$ are such that $e(0) = f(1)$, then we build the new curve $e \circ f$ defined by $(e \circ f)(t) = e(2t)$ for $0 \leqslant t \leqslant 0.5$ and $(e \circ f)(t) = f(2t-1)$ for $0.5 \leqslant t \leqslant 1$; see Figure 61.13.

This concatenation is not associative, i.e., if we have three curves e, f, g with common contact points, then $e \circ (f \circ g) \neq (e \circ f) \circ g$, but if we define the concatenation of homotopy classes by $[e] \circ [f] = [e \circ f]$, then, whenever all concatenations are defined, we have $[e] \circ ([f] \circ [g]) = ([e] \circ [f]) \circ [g]$, which we write $[e] \circ [f] \circ [g]$. This means that the most basic algebraic law of associativity is enabled by the fuzzy construction of (pointed) homotopy classes. The pointed homotopy classes of curves in X together with the composition of classes defined by their concatenation are called the *fundamental groupoid of* X, denoted by $\Pi_1(X)$, where as the substructure of those curve classes with a point $x \in X$ as common start- and endpoint is

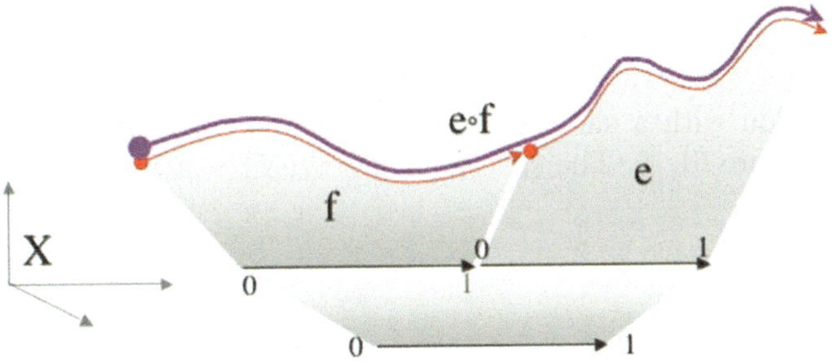

Fig. 61.13. Composition of curves with common endpoints.

called the *fundamental group of* X and is denoted by $\pi_1(X,x)$. Recall that the fundamental group is really a construction that relates to hypergestures of skeletons $\uparrow$, which connect loop gestures of skeleton **1**.

The structure $\pi_1(X,x)$ is in fact a mathematical group, with the class of the constant curve in x as neutral element and the inverse of a curve class $[c]$ being defined by $[c]^{-1} = [c']$ with $c'(t) = c(1-t)$. If X is path-wise connected (i.e., any two points can be joined as endpoints of a continuous curve), then $\pi_1(X,x)$ is independent of x, and we denote it by $\pi_1(X)$. Which groups do we encounter by such a construction? The most important group of integers $\mathbb{Z}$ is the fundamental group of $X = S^1$, the circle. It is also the group of the 3D space, where a column is taken away, as shown in Figure 61.14. This is a very important space for dancers; they may move in curves around the columns, but if such a curve goes around the column, it is not homotopic to the constant curve, the column is a real obstruction in this space and is responsible for the emergence of $\mathbb{Z}$. The standard loop gesture for $n \in \mathbb{Z}$ is the function $c_n(t) = e^{i2\pi nt}$, if we view the circle as a subset of the plane of complex numbers.

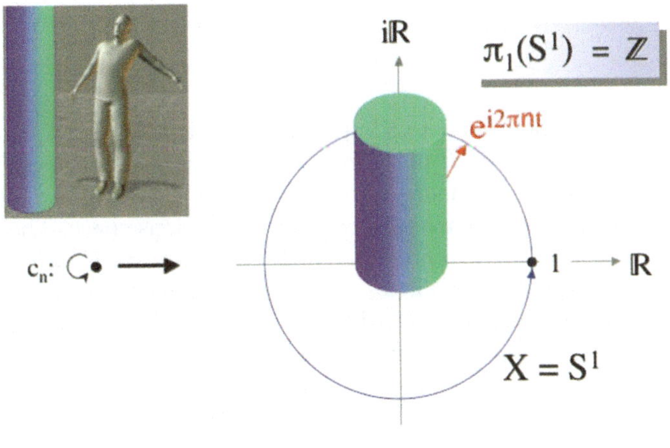

Fig. 61.14. The fundamental group of the circle is the same as that of a 3D space with a column. The number $n \in \mathbb{Z}$ is represented by the loop gesture $c_n(t) = e^{i2\pi nt}$.

Besides this basic group, one also reaches all finite cyclic groups $\mathbb{Z}_n$, for example the important group $\mathbb{Z}_{12}$ of pitch classes. The following is a remarkable result in algebraic topology,; see also [970][993, p.147]:

Theorem 37 *For any group G there exists a topological space X and a point $x \in X$ such that $G \xrightarrow{\sim} \pi_1(X,x)$.*

This shows that all groups, those fundamental structures of mathematics, arise from loop gestures and derived hypergestures.[8] We shall discuss the gestural realization of such algebraic structures in more detail in the next section.

We conclude this section with a short presentation of the construction of modular groups $\mathbb{Z}_n$ from fundamental groups. See Figure 61.15. The spaces having $\pi_1(X) = \mathbb{Z}_n$ are the lens spaces $L_{n,1}$. These are

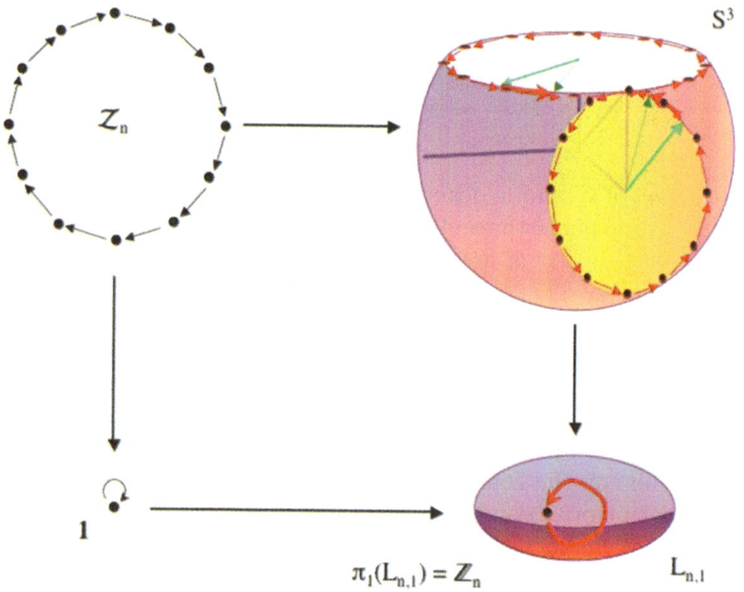

Fig. 61.15. The fundamental groups of lens spaces $L_{n,1}$ are the modular groups $\mathbb{Z}_n$. The generator of $\mathbb{Z}_n$ is represented by the movement on the n-hour clock by one unit, as represented on the upper level as a gesture on the circular skeleton $\mathcal{Z}_n$.

quotient spaces of the three-dimensional sphere S^3 under an action of the group $\mathbb{Z}_n$ as follows: One represents points of S^3 as pairs (x, y) of points $x, y \in S^2$ of the 2-sphere S^2, lying on tangent circles cut out by a common radius as shown in 61.15. We have the canonical action of $\mathbb{Z}_n$ on such pairs by rotating each component, x and y, by the angle $2\pi/n$. The quotient space $L_{n,1} = S^3/\mathbb{Z}_n$ is the lens space. The unit $1 \in \pi_1(L_{n,1}) = \mathbb{Z}_n$ of the fundamental group of the lens space is represented on S^3 by the gesture with skeleton $\mathcal{Z}_n$ and by sending every arrow to the curve segment on the two circles on S^3, which connects two successive hours on the n-hour clock. This is an interesting fact because not only do we have a fundamental group representation of $\mathbb{Z}_n$, but also one which has a very natural gestural interpretation as an elementary movement on an n-hour clock.

61.9 Gestoids

Given a topological space X, consider the category H_X of homotopy classes of curves. Its objects are the elements of X, while the morphism set $H_X(x, y)$ is the set of homotopy classes of curves starting in x and terminating in y. The composition of homotopy classes is the homotopy class of the composed curves. Clearly, this is a groupoid, the inverse of a curve class $[\gamma]$ being the class $[\gamma^*]$ of the inverted curve $\gamma^*(t) = \gamma(1-t)$. In particular, the group $H_X(x, x)$ is the fundamental group $\pi_1(x, X)$ of X in x. The category H_X is called the fundamental groupoid of X. If $\delta : \Delta \to \vec{X}$ is a gesture, the groupoid generated by the arrows and point

[8] It is also well known [569] that the fundamental theorem of algebra can be proved using arguments from homotopy theory.

of δ via the canonical morphism $\Delta \xrightarrow{\delta} \vec{X} \to H_X$ is denoted by H_δ and called *the fundamental groupoid of* δ.

Example 63 If $\Delta = 1$ is the final digraph, and if $\delta : 1 \to \vec{X}$ is a loop $\delta(T)$ in x, then H_δ is the subgroup of the fundamental group $\pi_1(x, X)$ generated by the homotopy class of $\delta(T)$.

We now linearize[9] the fundamental groupoid over a commutative ring R, i.e., the sets $H_X(x, y)$ are taken as a basis over R, and the composition is defined by bilinear extension from the given basis composition. We call this category $R\text{-}H_X$ the *(R-)gestoid of* X. For $R = \mathbb{R}$, the gestoid is given the structure of a linear topological category over $\mathbb{R}$ as follows. We first define a symmetric bilinear form B on every morphism set $\mathbb{R}\text{-}H_X(x, y)$. If p, q are two curve classes, their square distance with respect to B must be $d(p, q)^2 = B(p - q, p - q) = B(p, p) + B(q, q) - 2B(p, q)$, and if we set $B(p, p) = 1$ for all basis vectors p, then this yields the defining values $B(p, q) = 1 - \frac{1}{2}d(p, q)^2$. For every finite subset $F \subset H_X(x, y)$, there is an injection $i_F : \mathbb{R}\langle F\rangle \to \mathbb{R}\text{-}H_X(x, y)$. We take the canonical pseudometric on each space $\mathbb{R}\langle F\rangle$ and then the finest topology on $\mathbb{R}\text{-}H_X(x, y)$ such that all injections $i_F : \mathbb{R}\langle F\rangle \to \mathbb{R}\text{-}H_X(x, y)$ are continuous, i.e., the colimit topology for the diagram $\mathcal{D}_X$ of all finite metric spaces $\mathbb{R}\langle F\rangle$, together with the injections $\mathbb{R}\langle F'\rangle \to \mathbb{R}\langle F\rangle$ for the inclusions $F' \subset F$. Then it is standard to verify that the bilinear composition $\mathbb{R}\text{-}H_X(y, z) \times \mathbb{R}\text{-}H_X(x, y) \to \mathbb{R}\text{-}H_X(x, z)$ is continuous.[10] This means that the linearized category $\mathbb{R}\text{-}H_X$ is topological, i.e., composition, addition, and scalar multiplication of morphism are continuous. The linearized image of the fundamental groupoid H_δ of a gesture δ is called the *(R-)fundamental gestoid of* δ and denoted by $R\mathcal{G}_\delta$. Intuitively, this is the portion of the gestoid of X which is covered by arrows from the given gesture. So the gestoid is a linearized groupoid.

Example 64 In the above Example 63, the fundamental gestoid $\mathbb{R}\mathcal{G}_\delta$ over the reals is the group algebra $\mathbb{R}H_\delta$.

From this construction, the category *R-Gestoid* of R-gestoids is the following. The objects, called gestoids, are the R-linear categories G, i.e., we have bilinear composition, addition, and scalar multiplication of morphisms on the R-modules $x@y$ of morphisms. A morphism $q : G \to H$ is a linear functor, i.e., all maps $q(x, y) : x@y \to q(x)@q(y)$ are R-linear. Moreover, every endomorphism R-algebra is a group algebra, i.e., there is a group G_x such that $x@x \xrightarrow{\sim} RG_x$, and for any two objects x, y, if the morphism set is not empty, then $x \xrightarrow{\sim} y$. So we may select one group G_c per connected component c of the gestoid to describe the group algebras $x@x \xrightarrow{\sim} RG_c$ in that component.

In this way we have associated a gestoid in *R-Gestoid* to a topological space X, and then also to a Δ-gesture $\delta : \Delta \to \vec{X}$ in X. Let us check that this assignment completes to functors on the respective categories. To begin with, if $f : X \to Y$ is continuous, then it maps homotopic curves to homotopic curves, and we deduce a functor $H_f : H_X \to H_Y$, and then by linear extension a linear functor $R\text{-}H_f : R\text{-}H_X \to R\text{-}H_Y$, i.e., a morphism of R-gestoids.

Finally, if $f : \delta \to \gamma$ is a morphism of gestures, then the associated morphism of spatial digraphs is not uniquely determined, but the induced functors on the R-gestoids of the given gestures are well defined. So we have the required functor

$$\mathcal{G} : \textit{Gesture} \to \textit{Groupoid} \to \textit{R-Gestoid}$$

which connects the topological level of gestures via the level of groupoids from algebraic topology to its linearized version of gestoids.

[9] Given a category $\mathcal{C}$ and a commutative ring R, its linearization $R\mathcal{C}$ is the category whose objects coincide with the objects of $\mathcal{C}$, while for two objects X, Y, the morphism set $X@_{R\mathcal{C}}Y$ is the free R-module generated by the set $X@_{\mathcal{C}}Y$ of $\mathcal{C}$-morphisms from X to Y, and the composition of morphisms in $R\mathcal{C}$ is the bilinear extension of the composition in $\mathcal{C}$.

The linearized category $R\mathcal{C}$ is an *R-linear* category, which by definition means that all morphism sets $X@Y$ are R-modules, and that the composition of morphisms is R-bilinear.

[10] Observe that the cartesian product of the colimits $\mathbb{R}\text{-}H_X(y, z), \mathbb{R}\text{-}H_X(x, y)$ identifies to the colimit of the cartesian products of the subspaces generated by finite sets of morphisms from the fundamental groupoid.

61.9.1 The Fundamental Group, Klumpenhouwer Networks, and Fourier Representation

One interesting aspect in this construction is that the R-gestoid of a gesture is based upon groupoids, and these are essentially groups (the automorphism groups in the objects) plus isomorphisms which induce group isomorphisms via conjugation. But this is akin to the background structure of group diagrams used for local networks, and in particular for K-nets. So one might ask whether it is possible to generate gestures which give rise to groupoids relating in a canonical way to given local networks.

This is a central topic in the overall strategy of gestural constructions, since we would like to relate gestures to the abstract algebra of networks or other algebraic concepts, which at first sight have nothing to do with gestures. This concern is intimately related to the fundamental problem of New AI: How is it possible to (re)build symbolic thinking from instances of embodiment. How is it in particular possible to rebuild abstract algebra from gestures? In our context, we have made a step in this direction by use of the powerful tool of fundamental groupoids from algebraic topology. But it is still not a concrete result insofar as the role of this construction is not made explicit or applied to specific problems or constructs from mathematical music theory.

We therefore want to investigate the possible groupoids that intervene in the gestoid construction. Such a reconstruction would entail that, intuitively speaking, we are able to remodel abstract algebraic processes in terms of gestural dynamics. We believe that, in fact, understanding abstract algebra is strongly enhanced (if not enabled) if it uses the gestural embodient, this is what Cavaillès [91] seems to suggest (see the citation in our introduction). For the music, this would mean that we could envisage the question of how to "perform" abstract algebraic structures. This is a deep question, since making music is intimately related to the expression of thoughts. So we would like to be able to express the algebraic insights, as revealed by use of K-nets, or symmetry groups, for example, in terms of musical gestures. To put it more strikingly: Is it possible to play the music of thoughts?

Now, every group, and in particular (for our music-theoretical interests) every finitely generated abelian group, is the fundamental group of a topological space. The latter follows from the fact that such a group is a finite product of cyclic groups, that the fundamental group of a product of topological spaces is the product of the fundamental groups of the factors, and that a finite cyclic group $\mathbb{Z}_n$ is the fundamental group of the lense space $L_{n,1}$, which is the quotient $S^3/\mathbb{Z}_n$ of the 3-sphere $S^3 \subset \mathbb{C}^2$ modulo the group action $k.(z,w) = (u^k z, u^k w), u = e^{2\pi i/n}$ [569, Example 7.15], whereas $\mathbb{Z}$ is the fundamental group of the circle S^1 (or of SO(2), etc.). Therefore, in particular, the classical pitch class group $\mathbb{Z}_{12}$ is in fact a fundamental group, namely that of the cartesian product $L_{3,1} \times L_{4,1}$. Musically speaking this means that

Fact 19 *All groups, in particular the finitely generated abelian groups, such as the pitch class groups $\mathbb{Z}_{12}$, defining K-networks, may be realized via fundamental groups on the level of topological spaces.*

This is a possibility which is most relevant for the problem of effectively playing such networks. But let us explain the situation. We have just learnt that important abstract groups are realized as fundamental groups of topological spaces. This means that curves in topological spaces may represent elements of such abstract groups, whereas the concatenation of such curves realizes abstract group operations. Since the generators of our groups are finite in number, this opens the question whether we may find gestures such that their arrows are associated with the generators of our groups, and whether it is possible to give an interpretation of the involved topological spaces, in particular, the lense spaces, in terms of spaces of more or less complex bodily gestures. This is an open question. But the lense spaces are objects of low dimension and may probably intervene for gestures of conductors' or musicians' limbs. This is plausible from our previous work on a pianist's hand as described in Example 60.

For music theory, the best situation would be the following:

Conjecture 1 *For every K-net, there is a gestoid of the same digraph which contains a K-net isomorphic to the given K-net.*

Let us make this conjecture more precise to give a suggestion of how a K-net could be interpreted within a gestoid. To this end, recall that an $\mathbb{R}$-gestoid has a group algebra $x@x = \mathbb{R}\pi_1(x, X)$ for each of its objects x, and that for any path $f : x \to y$, we have a conjugation isomorphism of $\mathbb{R}$-algebras $Int_f : x@x \xrightarrow{\sim} y@y : z \mapsto f \circ z \circ f^{-1}$. So we are in the category $Alg_{\mathbb{R}}$ of $\mathbb{R}$-algebras, and the given gesture $\gamma : \Gamma \to \vec{X}$ defines a diagram D_γ of $\mathbb{R}$-algebras, the algebra $\mathbb{R}\pi_1(x, X)$ at point x, and algebra isomorphisms. To give a K-net means to select an element of $\lim D_\gamma$. In terms of the category $Alg_{\mathbb{R}}$, this means evaluating the diagram of representable functors $@\mathbb{R}\pi_1(x, X)$ at the address $\mathbb{R}\mathbb{Z}$, since $\mathbb{R}\mathbb{Z}@B \xrightarrow{\sim} B^*$, where B^* is the group of multiplicatively invertible elements of B. If we have a space X with $\pi_1(x, X) \xrightarrow{\sim} \mathbb{Z}_{12}$, then an element $z_x \in \mathbb{Z}_{12}$ corresponds to a point in $\mathbb{R}\mathbb{Z}@B \xrightarrow{\sim} B^*$ for $B = \mathbb{R}\mathbb{Z}_{12}$. The combination of an algebra isomorphism Int_f with a multiplication by a group element simulates the affine morphisms used in K-net theory. Therefore this structure yields a model for K-nets in finitely generated abelian groups.

The intriguing point in this presentation is that the musical interpretation of elements $z \in \mathbb{Z}_{12}$ as pitch classes seems to be somewhat mysterious when reinterpreting $\mathbb{Z}_{12} \xrightarrow{\sim} \pi_1(x, X)$. Why should a multiple loop be associated with a pitch? There is however a very natural interpretation of this mystery in terms of Fourier theory. If a periodic time function $x(t)$ with frequency 1 is represented by a Fourier series $x(t) = \sum_n \gamma_n e^{i2\pi nt}$ with finitely many non-vanishing coefficients γ_n, the functions $\epsilon_n(t) = e^{i2\pi nt}$ are linearly independent, and $\epsilon_n = \epsilon_1^n$ for $n \in \mathbb{Z}$. Therefore the function x is in fact an element of the group algebra[11] $\mathbb{C}\mathbb{Z}$, if we identify ϵ_n with $n \in \mathbb{Z}$. It is now easy to reinterpret the elements of $\mathbb{Z}$ in a natural way as being elements of a fundamental group.

In fact, consider the circle S^1, which we identify with the unitary group $U \subset \mathbb{C}$ of complex numbers of length 1. Then a gesture $\delta : 1 \to \vec{S^1}$ is a loop in S^1. In particular, we have the gesture $\varepsilon : 1 \to \epsilon_1$ associated with the loop $\epsilon_1 : I \to S^1$ in $1 \in U$. This yields the fundamental component of the Fourier representation, and it is a generator of the fundamental group $\pi_1(1, S^1) \xrightarrow{\sim} \mathbb{Z}$. So the Fourier representation corresponds to the formal sum $\sum_n \gamma_n \varepsilon^n$ in the *complex* gestoid $\mathbb{C}\mathcal{G}_\varepsilon$, i.e., the linearization of the fundamental groupoid over the complex numbers. In short:

Fact 20 *The Fourier representation is the linear combination in the complex gestoid $\mathbb{C}\mathcal{G}_\varepsilon$ of gesture classes, which are powers of the fundamental loop gesture in* 1.

The converse, i.e., the reinterpretation of elements in more general gestoids in terms of time functions, is a challenging problem. In particular, we are asked to *reinterpret the loops in the fundamental group $\mathbb{Z}_m$ of a lense space $L_{m,1}$ in terms of acoustically meaningful functions.*

61.10 Gabriel's Spectroids and Natural Formulas

From the category **Digraph** of digraphs, one may also derive algebraic instead of topological structures. On one side, this is motivated by the well-known dichotomy of topology versus algebra as guiding paradigms in mathematics. On the other side, as we shall see in Section 61.10.1, an algebraic perspective on digraphs is musically motivated by the theory of local musical networks.

Our approach refers to Gabriel's construction of spectroids[12] from digraphs [351, 352], which runs as follows. Given a commutative artinian ring k, the category k-*Spectroid* of k-spectroids has the k-spectroids as objects: They are k-linear categories S, where any two different objects $x \neq y$ are non-isomorphic, where the endomorphism algebras $x@x$ of all objects x are local (i.e., the non-invertible endomorphisms of x are a (two-sided) ideal $Rad(x) \subset x@x$), and such that the morphism spaces $x@y$ have finite length over k. We shall henceforth omit the finiteness condition except when explicitly stated. The morphisms between k-spectroids

[11] A *group algebra* is a monoid algebra, where the monoid is a group.

[12] The term spectroid is akin to spectrum. This is intentional since, by Gabriel's thesis [351], one has this fact: For B a k-algebra over an artinian commutative ring k, consider the category $\mathbf{Mod}_B$ of right B-modules of finite k-length, and take the full subcategory $Sp(\mathbf{Mod}_B)$ of representatives of all indecomposable injective modules. Then if every right ideal of B is two-sided, there is a bijection between the objects of $Sp(\mathbf{Mod}_B)$ and the prime spectrum $Spec(B)$, i.e., the set of two-sided prime ideals of B (meaing that any inclusion $aIb \subset I$ for such an ideal I, and $a, b \in B$ implies $a \in I$ or $b \in I$).

are the *local*[13] k-linear functors $f : S \to T$, i.e. those carrying the radical $Rad(S)$ into $Rad(T)$. In the sequel, we fix $k = \mathbb{R}$ as the base ring of spectroids and omit this specification.

The analog construction of gestures is an algebraic interpretation of digraphs instead of a topological one. We interpret vertices as objects and arrows as morphisms in spectroids. More precisely, a digraph is called *radical* iff it is the digraph $Rad(S)$ of a spectroid S having the noninvertible morphisms as its arrows and the domain and codomain maps $d, c : Rad(S) \rightrightarrows S$ as tails $d = t$ and heads $c = h$ of arrows. The category *RadicalDigraph* of radical digraphs has the radical digraphs $Rad(S)$ as their objects and the graph morphisms $r : Rad(S) \to Rad(T)$ induced by morphisms $s : S \to T$ of the underlying spectroids, we then write $r = Rad(s)$. This category is the algebraic analog to the category *SpaceDigraph* of spatial digraphs. A morphism of digraphs $\phi : \Delta \to Rad(S)$ is called a *Δ-formula in S*.

Example 65 Given a digraph Δ, we have the $\mathbb{R}$-category $\mathbb{R}\Delta$, which is the linearized path category[14] $\mathbb{R}\Delta = \mathbb{R}Path(\Delta)$. Consider the digraph $\Delta = Y \circlearrowright t \circlearrowleft X$ and the quotient category S obtained from $\mathbb{R}\Delta$ by division through the ideal generated by the relations $X^2 - YX, Y^2 - XY, XY + YX, X^3$. This means that we have a three-dimensional radical $Rad(S) = \mathbb{R}x + \mathbb{R}y + \mathbb{R}xy$, with the relations $x^2 = yx = -y^2$, $xy = -yx$, and $x^3 = 0$. Now, if we are given a digraph Γ, a formula $\psi : \Gamma \to Rad(S)$ is the assignment of any vector $\psi(a) \in Rad(S)$ to an arrow $a \in A_\Gamma$, since there is only one vertex in $Rad(S)$. Then the formula is completely described by labeling the arrows of Γ with their images and writing down the relations in S. For example, if $\Gamma = 1$, and if the ψ-image of T is $\psi(T) = xy + 3y^2$, then we may write the formula as

$$\psi = \; xy{+}3y^2 \circlearrowright t \; , \; (xy + 3y^2)^2 = 0.$$

Given two formulas $\phi : \Delta \to Rad(S), \psi : \Gamma \to Rad(T)$, a morphism $f : \phi \to \psi$ of formulas is a morphism $f : \Delta \to \Gamma$ of digraphs such that there is a morphism $Rad(s) : Rad(S) \to Rad(T)$ of radical digraphs which commutes with f, i.e., $\psi \circ f = Rad(s) \circ \phi$. The category of formulas is denoted by *Formula*, and we have a canonical projection $q : Formula \to \mathbf{Digraph}$ by forgetting about the radical codomain.

For a digraph morphism $f : \Delta \to \Gamma$, we have the associated functor $Path(f) : Path(\Delta) \to Path(\Gamma)$ and its linearized extension $\mathbb{R}f : \mathbb{R}\Delta \to \mathbb{R}\Gamma$. This defines a functor $\mathbb{R}? : \mathbf{Digraph} \to Spectroid$, and then, by restriction to the radicals, a functor $Rad : \mathbf{Digraph} \to RadicalDigraph : \Delta \mapsto Rad(\Delta) = Rad(\mathbb{R}\Delta)$. For every digraph Δ, we have a canonical formula $r(\Delta) : \Delta \to Rad(\Delta)$ given by the identity on the arrows and vertices.

Proposition 63 *For any morphism $f : \Delta \to \Gamma$ of digraphs and any formula $\psi : \Gamma \to Rad(S)$, there is a unique functor $\mathbb{R}f : \mathbb{R}\Delta \to S$ such that its associated digraph morphism $Rad(f) : Rad(\Delta) \to Rad(S)$ induces a morphism $r(f) : r(\Delta) \to \psi$ of formulas.*

Corollary 26 *The maps $\Delta \mapsto Rad(\Delta)$ and $f \mapsto Rad(f)$ define a functor $Rad : \mathbf{Digraph} \to RadicalDigraph$ and the formula $r(\Delta) : \Delta \to Rad(\Delta)$ defines a natural transformation $r : Id_{\mathbf{Digraph}} \to Rad$, also called the* natural formula associated with the digraph Δ.

Corollary 27 *The formula $r(\Delta) : \Delta \to Rad(\Delta)$ defines a functor $r : \mathbf{Digraph} \to Formula$ which is left adjoint to the projection $q : Formula \to \mathbf{Digraph}$, in symbols $r \dashv q$. This means that*

$$\mathbf{Digraph}(\Delta, q(\psi)) \xrightarrow{\sim} Formula(r(\Delta), \psi))$$

is a bijection, which is functorial in both arguments Δ, ψ.

[13] By definition the non-invertible morphisms of S.

[14] The path category of a digraph Δ has the vertex setV_Δ as the object set and the paths $x \to x' \to \ldots y$ as morphisms from vertex x to vertex y. The composition of morphisms is the composition of paths, while the lazy path of length 0 in x is the identity in x.

61.10.1 Solutions of Representations of Natural Formulas by Local Networks

If we are given a formula, it is natural to ask for its solutions. We do not elaborate on this subject here, but since it is an essential technique for the construction of (local) networks,[15] it is necessary to indicate the relation to these concepts, which are important in music theory [719, 599]. A spectroid is deduced from the situation of a category $\mathbf{Mod}_k$ of k-modules over a commutative ring k. Then the morphism sets $M@N$ are k-modules and the composition is bilinear. One selects a complete set of representatives of isomorphism classes of indecomposable injective modules and considers the full subcategory S defined by these objects. Therefore a formula $\psi : \Gamma \to Rad(S)$ defines a diagram of morphisms in S, which is a special case of a diagram $\psi : \Gamma \to \mathbf{Mod}_k$ as considered in network theory. We therefore call ψ a *generalized formula*. Now, this is equivalent to giving a k-linear functor $k\psi : k\Gamma \to \mathbf{Mod}_k$ defined on the spectroid $k\Gamma$. But if we take the natural formula $r(\Gamma) : \Gamma \to Rad(\Gamma)$, $k\psi$ defines an obvious morphism $f(\psi) : r(\Gamma) \to \psi$. And if we are given relations R among the arrows associated with ψ, this means that we may factorize $f(\psi)$ through the quotient category $k\Gamma/R$ and instead give a morphism $f(\psi)/R : r(\Gamma)/R \to \psi$. Therefore, the generalized formulas in $\mathbf{Mod}_k$ are "representations" of natural formulas or quotients thereof.

To deal with solutions, we need points in the modules, which are the vertex images of the generalized formula. To this end, one selects a module A (an address module in the theory of denotators; see Chapter 6) and then evaluates the representable functors $@\psi(x)$ of the vertex modules $\psi(x)$ at A, whence a diagram $A@\psi$ of sets $A@\psi(x)$, which are related by the evaluated morphisms $A@\psi(a)$ for arrow a of Γ. A solution of the generalized formula ψ at address A is, by definition, an element s of the limit $A@\lim\psi$ of this set diagram. This is precisely what we called a *local network* in the theory of networks [720]. Summarizing:

Fact 21 *Local networks are the solutions of generalized formulas at selected addresses A.*

61.11 The Tangent Category

In algebraic geometry, the Zariski tangent space $T_{X,x}$ of an $\mathbb{R}$-rational point of an $\mathbb{R}$-scheme is the $\mathbb{R}$-linear dual $(m/m^2)^*$ of the quotient m/m^2 of the maximal ideal m of the local ring $\mathcal{O}_{X,x}$; see also Appendix Section F.4.1. In an $\mathbb{R}$-spectroid S, every endomorphism algebra $x@x$ is local with maximal ideal $m_x = Rad(x)$. So we may associate the following linear category with tangent spaces: Consider the quotient category $S/Rad(S)^2$ and take its linear subcategory T_S which for $x \neq y$ has $T_S(x,y) = Rad(S)(x,y)/Rad(S)^2(x,y)$, while $T_S(x,x) = \mathbb{R} \oplus Rad(x)/Rad^2(x)$. So the endomorphisms in $T_S(x,x)$ are identified with the formal sums $\mu + \epsilon.t, \mu \in \mathbb{R}, t \in Rad(x)/Rad^2(x)$, which are added component-wise and multiplied under the infinitesimal condition $\epsilon^2 = 0$ via $(\mu + \epsilon.t) \circ (\nu + \epsilon.s) = \mu\nu + \epsilon.(\nu t + \mu s)$. If for $x \neq y$, $f \in T_S(x,y)$, then we formally write $\epsilon.f$ instead of f, and then $\epsilon.f \circ (\mu + \epsilon.t) = \epsilon.\mu f$. This means that we view the tangent category as built from "tangent vectors", as indicated by the ϵ-coefficient.

If $f : S \to U$ is a morphism of spectroids, then, being local, it factorizes to $f_R : S/Rad^2(S) \to U/Rad^2(U)$ and then evidently also takes the tangent subcategories into one another, yielding a functor $T_f : T_S \to T_U$. This defines a functor on the category of spectroids.

Example 66 If $S = \mathbb{R}1$ with the final digraph $1 = t\circlearrowleft$, then T_S looks as follows: It has just one object x and its endomorphisms $T_S(x,x)$ are the algebra $\mathbb{R}[\epsilon]$ of dual numbers $\mu + \epsilon.t, \mu, t \in \mathbb{R}$. Such a number represents the tangent of length t at the point μ of the affine line $\mathbb{A}$ over the reals.

Then, if we are given a formula $\psi : \Gamma \to Rad(S)$, we consider the morphisms in T_S defined by ψ as follows: for $x \neq y$ and an arrow $x \xrightarrow{a} y$, we take $\psi^a = \epsilon.(\psi(a) \mod Rad^2)$, and for a loop $x \xrightarrow{a} x$,

[15] Recall that a local network is described by a diagram of modules M_i and affine module homomorphisms, together with an element $m_i \in M_i$ for each vertex module M_i, such that any two such elements are mapped to each other by the given homomorphisms. In other words, a local network is an element of the limit of the given diagram. See [720] for more details on this construction.

we take $\psi^a = 1 + \epsilon.(\psi(a) \mod Rad^2)$. Therefore, in the tangent category T_S, we have the exponential rule for endomorphisms on x, i.e., for two loops $x \xrightarrow{a,b} x$, we have $\psi^{a+b} = \psi^a \circ \psi^b$, while for $x \neq y$, loops $x \xrightarrow{a} x, y \xrightarrow{c} y$, and an arrow $x \xrightarrow{b} y$, we have $\psi^b \circ \psi^a = \psi^b = \psi^c \circ \psi^b$. Denote then by T_ψ the subcategory of T_S generated by the arrows $\psi^a, a \in A_\Gamma$ and their inverses—if they exist—and call this the *tangent category of the formula* ψ. The inverses are of course precisely the automorphisms $(\psi^a)^{-1} = 1+\epsilon.(-\psi(a) \mod Rad^2)$, which we may also think of being generated by new "inverse arrows" $-a$ on Γ, much as it is done in the Bass-Serre theory of graphs of groups [970]. So we also write $\psi^{-a} = 1 + \epsilon.(-\psi(a) \mod Rad^2)$ for these inverses. Observe that for a formula $\psi : \Gamma \to Rad(S)$ of a finite digraph Γ, the tangent category T_ψ has only finitely many morphisms between different objects and the endomorphism monoids are finitely generated commutative groups.

In analogy to the fundamental group construction on the topological branch of our discussion, we may now linearize the tangent category T_ψ of a formula. We denote this by $\mathcal{F}_\psi$ and call it *the radical formoid* of the formula ψ. In the above-mentioned situation of a finite digraph of the formula ψ, the radical formoid has finite-dimensional morphism vector spaces $x@y$ for different objects x, y, whereas the endomorphism spaces are group algebras $x@x \xrightarrow{\sim} \mathbb{R}T_\psi(x,x)$ over the finitely generated commutative groups $T_\psi(x,x)$.

Given two formulas $\psi : \Gamma \to Rad(S)$, $\phi : \Delta \to Rad(U)$ and a morphism $f : \psi \to \phi$ of formulas, we have an auxiliary morphism $s : S \to U$ of spectroids such that $Rad(s)$ commutes with $f : \Gamma \to \Delta$. So for any arrow $a \in A_\Gamma$, we have $s(\psi(a)) = \phi(f(a))$. Moreover, $s(1_x) = 1_{s(x)}$. Therefore for the residual categories modulo Rad^2, we have the residual functor $s/Rad^2 : S/Rad(S)^2 \to U/Rad(U)^2$, which on loops $x \xrightarrow{a} x$ in x acts by $s/Rad^2(1_x + \psi(a) \mod Rad^2) = 1_{s(x)} + s(\psi(a)) \mod Rad^2 = 1_{s(x)} + s(\phi(f(a)) \mod Rad^2$, and on arrows $x \xrightarrow{a} y$ for $x \neq y$ yields $s/Rad^2(\psi(a) \mod Rad^2) = \phi(f(a)) \mod Rad^2$. This shows that we have a functor $\mathcal{F}_f : \mathcal{F}_\psi \to \mathcal{F}_\phi$ associated with f, and this assignment is obviously functorial. Therefore we have a functor $\mathcal{F} : Formula \to \mathbb{R}$-*Formoid*, the latter category being, by definition, the category of linear categories generated from subcategories of residual categories mod Rad^2 of spectroids.

Example 67 In the above Example 66, if we take the natural formula $r(1) : 1 \to Rad(\mathbb{R}1)$, sending the true arrow T to the radical residue ϵ, we get $T_{r(1)} \xrightarrow{\sim} \mathbb{Z}$, the integer z corresponding to the power $1 + \epsilon.z$ in $\mathbb{R}[\epsilon]$.

We are now ready to observe the overall path from the final digraph $1 = t \circlearrowright T$ when the gestural and formulaic functors are applied. On the one hand, the topological construction of the fundamental gestoid $\mathcal{G}_{g(1)}$ is the category $\mathbb{R}\mathbb{Z}$, the group algebra over $\mathbb{R}$ of the group of integers, and the latter comes in via the generator of the fundamental group of the circle S^1. On the other hand, the radical formoid $\mathcal{F}_{r(1)}$ is also the group algebra $\mathbb{R}\mathbb{Z}$, where $\mathbb{Z}$ now comes in as the group generated by the "tangent vector" $1+\epsilon$, corresponding to the residue ϵ of the generating loop T of the linearized path algebra $\mathbb{R}1 \xrightarrow{\sim} \mathbb{R}[T]$ associated with the final digraph 1. So both constructions yield isomorphic categories

$$\mathcal{G}_{g(1)} \xrightarrow{\sim} \mathcal{F}_{r(1)}$$

when taking the gestural or formulaic branch. Of course, this identification of the results from the two paths is not possible for general digraphs, since the gestoid is built from a groupoid, whereas the formoid stems from a spectroid, where different objects are never isomorphic. So for general digraphs Γ we have $\mathcal{G}_{g(\Gamma)} \not\xrightarrow{\sim} \mathcal{F}_{r(\Gamma)}$.

This can be made more precise on the endomorphism algebras. The fundamental gestoid $\mathcal{G}_{g(\Gamma)}$ of a digraph Γ is easily calculated: It is well known [590, Theorem 10.7] that the fundamental group of a graph is free, the number of generators being given from the number of edges added to a spanning tree. So normally, this is a non-abelian group, whereas the formoid $\mathcal{F}_{r(\Gamma)}$ has abelian groups defining the endomorphism algebras.

Example 68 Let us discuss the example of the classifying digraph Ω shown in Section 61.5. Here, the gestoid $\mathcal{G}_{g(\Omega)}$ is as follows: The natural gesture of Ω has the topological space

$|\Omega| =$ 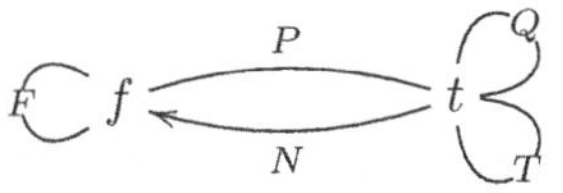

whose fundamental group is the free group $\pi_1(|\Omega|) = \langle T, Q, R, S \rangle$ generated by the four loops[16] $T, Q, R = NP$, and $S = NFP$. The groupoid of the space is then uniquely described by the two objects f, t, their groups $f@f \xrightarrow{\sim} t@t \xrightarrow{\sim} \pi_1(|\Omega|)$, the isomorphism $f@f \xrightarrow{\sim} t@t$ being given by conjugation $N?P$, whereas the morphisms between the two objects are $t@f = Nt@t \xrightarrow{\sim} t@t, f@t = Pt@t \xrightarrow{\sim} f@f$. The groupoid is then the two-object linear category with the group algebras $\mathbb{R}\langle T, Q, R, S \rangle$ as endomorphism algebras.

On the other hand, the category $\mathbb{R}\Omega$ of Ω has two objects and the linear combinations of paths with given start and end as morphisms, where we then have to divide by the paths of length $\geqslant 2$. This division yields the endomorphism algebra $f@f \xrightarrow{\sim} \mathbb{R}[\epsilon]$ and $t@t \xrightarrow{\sim} \mathbb{R}[\epsilon_1, \epsilon_2]$ (two infinitesimals ϵ_1, ϵ_2 with $\epsilon_1^2 = \epsilon_2^2 = \epsilon_1\epsilon_2 = \epsilon_2\epsilon_1 = 0$). Then the image $1 + \epsilon_F$ of the loop F generates $\mathbb{Z}$ and we have the group algebra $\mathbb{R}\mathbb{Z}$ as endomorphism algebra of $\mathcal{F}_{r(\Omega)}$ in f. In t, we have the group algebra $\mathbb{R}\mathbb{Z}^2 \xrightarrow{\sim} \mathbb{R}\mathbb{Z} \otimes_{\mathbb{R}} \mathbb{R}\mathbb{Z}$ as endomorphism algebra of $\mathcal{F}_{r(\Omega)}$.

61.12 The Diamond Conjecture

The preceding constructions are by no means in a complete theoretical shape. For example, the formulaic branch does not rely on the real numbers, any commutative field k or artinian commutative ring would equally provide a reasonable setup. For example, the formoids over the fields of positive characteristic p yield group algebras $\mathcal{F}_{r(1)} \xrightarrow{\sim} k\mathbb{Z}_p$ over the cyclic group $\mathbb{Z}_p$, and this algebra would be needed to reflect the appearance of fundamental groups $\mathbb{Z}_p$ for lense spaces, as mentioned above. Moreover, the reduction modulo Rad^2 is not mandatory. More general reductions modulo $Rad^n, n > 2$, would also yield interesting information. For example, we could take care of longer paths in categories $k\Gamma$ than just those of length 1.

In order to shape the hypothetical ideal architecture, which should encompass and harmonize the two branches, we present the subsequent diamond diagram. It is topped by a hypothetical category X, which should unite the algebraic and the topological branches. The hypothetical functors G, F, u, v, x, y are shown in dotted arrows. It is further conjectured that for the functors G and F, we have the adjunctions $\mathcal{F} \dashv G$ and $\mathcal{G} \dashv F$. Finally, it is conjectured that the ascending/descending arrows can be completed to a commutative diagram by two ascending/descending arrows, $x, y/u, v$.

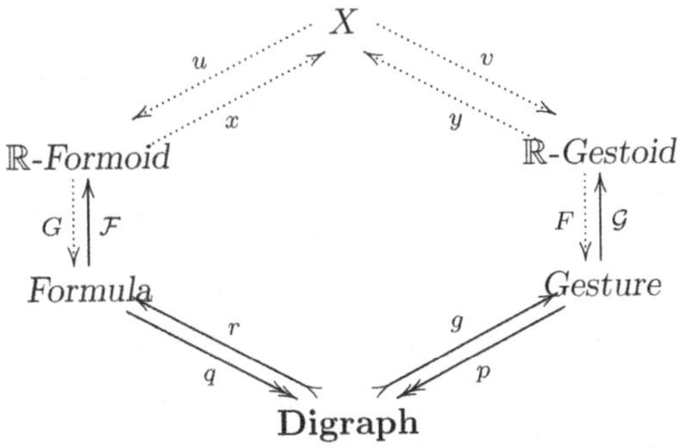

In other words, starting from a common basis of elementary mathematical structures, i.e., the digraphs, we have a double unfolding of musically relevant structures: gestures and formulas. Our hypothesis is that there exists a "universe" X, whose ontology englobes these two branches in a natural way and at the same time expresses a unified comprehension of music.

[16] This follows immediately from the Seifert-Van Kampen theorem [590, Chapter 10]. In fact, $|\Omega|$ is the colimit of a loop graph L_1 and the colimit L_2 of three copies of the loop graph. L_2 is obtained by gluing together the loop graphs in a nondegenerate point, so their fundamental group is free over three generators. And the gluing of L_1 with L_2 is in one nondegenerate point f, so the total fundamental group is free over the three generators of $\pi_1(L_2)$ and the generator of $\pi_1(L_1)$, whence the claim.

61.13 Topos Logic for Gestures

This section is quite sketchy, but nonetheless important for future research since it introduces a logical technique for gestures, formulas and their related gestoids and formoids. Since **Digraph** is a topos, every digraph Δ has a canonical logic, i.e., a Heyting algebra on the sub-digraph sets $\mathrm{Sub}(\Delta) = \Delta@\Omega$, for the subobject classifier Ω. The logical operations on $\mathrm{Sub}(\Delta)$ are as follows. If Γ, Σ are two sub-digraphs of Δ, then we have $\Gamma \wedge \Sigma = \Gamma \cap \Sigma$ and $\Gamma \vee \Sigma = \Gamma \cup \Sigma$. The implication $\Gamma \implies \Sigma$ is a bit more involved: We have $V_{\Gamma \Longrightarrow \Sigma} = (V_\Delta - V_\Gamma) \cup V_\Sigma$. The arrows are the following: They include $\Gamma \cap \Sigma$, all arrows on vertices of the intersection $V_\Gamma \cap V_\Sigma$ which are not in Γ, all Δ-arrows on vertices in $V_\Delta - V_\Gamma$, and all Δ-arrows between vertices in $V_\Gamma \cap V_\Sigma$ and vertices in $V_\Delta - V_\Gamma$.

This Heyting logic is identified with the contravariant functor $@\Omega$: **Digraph** $\to$ *Heyting* with values in the category *Heyting* of Heyting algebras. The functor also applies to gestures and formulas by applying the logical operations to the respective domains. More precisely, if $\gamma : \Gamma \to \vec{X}$ is a gesture, then the set of subgestures $Sub(\gamma)$ of γ inherits the Heyting structure on its domain Γ. Mutatis mutandis, the analog Heyting structure is realized for the set $Sub(\psi)$ of a formula ψ.

Example 69 In mathematical music theory, local networks are also used for the description of prominent triadic chords; see [719, 714] for details. In particular, the major, minor, and diminished triadic degrees of the major scale are described by the orbits $\langle f\rangle(x) = \{x, f(x), f^2(x), f^3(x), \ldots\}$ of single affine transformations $f = T^t g$ with $f(x) = t + gx$, acting on a particular pitch class p. Here is a list of such local networks for the all triadic degrees of the C-major scale $C_{maj} = \{0, 2, 4, 5, 7, 9, 11\}$ (identifying c with $0 \in \mathbb{Z}_{12}$, $c_\sharp = d_\flat$ with $1 \in \mathbb{Z}_{12}$, etc.). We have

$$
\begin{array}{rll}
I = \{0,4,7\} = \langle T^7 3\rangle(0), & & 0 \longrightarrow 7 \rightleftarrows 4\\
III = \{4,7,11\} = \langle T^7 3\rangle(11), & & 11 \longrightarrow 4 \rightleftarrows 7\\
II = \{2,5,9\} = \langle T^{11} 3\rangle(9), & & 9 \longrightarrow 2 \rightleftarrows 5\\
IV = \{5,9,0\} = \langle T^9 3\rangle(5), & & 5 \longrightarrow 0 \rightleftarrows 9\\
VI = \{9,0,4\} = \langle T^9 3\rangle(4), & & 4 \longrightarrow 9 \rightleftarrows 0\\
V = \{7,11,2\} = \langle T^5 3\rangle(7), & & 7 \longrightarrow 2 \rightleftarrows 11\\
VII = \{11,2,5\} = \langle T^1 2\rangle(2), & & 2 \longrightarrow 5 \longrightarrow 11 \circlearrowleft
\end{array}
$$

with the two intervals $7 \rightleftarrows 4$ in I, III and $0 \rightleftarrows 9$ in IV, VI being generated by the same transformation each, i.e., $T^7 3$ for the first interval and $T^9 3$ for the second. This defines a digraph Γ_C with the vertex set $V_{\Gamma_C} = C_{maj}$ and the unique arrows $i \to j$ associated with the above transformations; see Figure 61.16. Taking the gesture space $Fingers = \Pi\vec{@}FingerSpace$ as topological target space[17], we fix a hypergesture

$$\gamma_C : \Gamma_C \to \overrightarrow{Fingers}.$$

and then look for the logic on $Sub(\gamma_C)$. We denote by γ_X the restriction of γ_C to a subdigraph X of Γ_C. We also identify the subdigraph associated with a degree X with X. Then the logical operations can be performed on the seven degrees. For example, one has $\gamma_V \implies \gamma_I = \gamma_{I \cup IV \cup VI}$, or else $\gamma_{IV} \implies \gamma_V = \gamma_{III \cup V}$.

Whereas the Heyting-algebra structure of $Sub(\gamma)$ is by construction isomorphic to the Heyting algebra $Sub(\Gamma)$ of its domain $\Gamma = p(\gamma)$, the system of gestoids $\mathcal{G}_{\gamma'} \subset \mathcal{G}_\gamma$ for the subgestures $\gamma' \subset \gamma$ within $Sub(\mathcal{G}_\gamma)$ may look completely different depending on the topology of the space X present in $\gamma : \Gamma \to \vec{X}$ and of the map γ as such. In particular, the characterization of gestoids such as $\mathcal{G}_{\gamma' \wedge \gamma''}$, $\mathcal{G}_{\gamma' \vee \gamma''}$, or $\mathcal{G}_{\gamma' \Longrightarrow \gamma''}$ for subgestures γ', γ'' of γ must be investigated within $Sub(\mathcal{G}_\gamma)$, and, mutatis mutandis, for logical constructions on the branch of formoids.

[17] Of course, one could enrich the context by blowing up the finger space to include the real shape of a finger, but this is not important here.

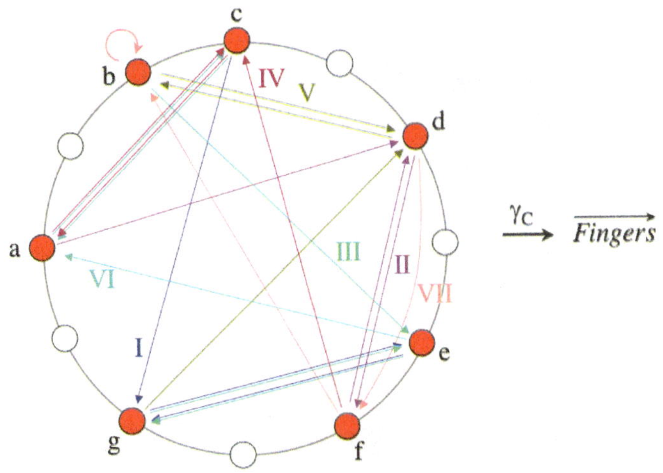

Fig. 61.16. The C-major scale can be covered by a net defined by the seven transformations defining the degrees. Each degree is represented by its digraph, which is a subdigraph of the digraph Γ_C of the scale.

When applying logic to hypergestures γ in spaces $W\vec{@}X$, where $W = \Gamma_1\Gamma_2\ldots\Gamma_k$, Proposition 1 enables logical operations of "inner" digraphs $\Gamma_i, i > 1$, since one may shift those digraphs to the leftmost position and get the commutative word $W = \Gamma_i\Gamma_2\ldots\widehat{\Gamma_i}\ldots\Gamma_k$. Then one may apply logical operations to the digraph Γ_i for the subgestures $Sub(\gamma')$ of the gesture γ' corresponding to γ after the permutation of digraph letters in W, and finally switch back to the corresponding gesture in the original space $W\vec{@}X$.

61.14 The Escher Theorem for Hypergestures

Summary. We discuss some musical applications of the Escher Theorem.

– Σ –

61.14.1 Hypergestures and the Escher Theorem for Fux Counterpoint

The Escher Theorem (see Proposition 61 in Section 61.6) is the following, seemingly inoffensive, statement, which essentially states that iterated hypergestures can be built in any order of the involved directed graphs:

Theorem 38 *Let $\Gamma_1, \Gamma_2, \ldots \Gamma_n$ be n digraphs, X a topological space, and π a permutation of the set $\{1, 2, \ldots n\}$. Then there is a canonical homeomorphism (i.e., a bijection which conserves all topological structures)*

$$\Gamma_1\vec{@}\Gamma_2\vec{@}\ldots\Gamma_n\vec{@}X \xrightarrow{\sim} \Gamma_{\pi(1)}\vec{@}\Gamma_{\pi(2)}\vec{@}\ldots\Gamma_{\pi(n)}\vec{@}X.$$

This does not mean that the hypergestures in the space to the left are the same as those in the space to the right, but there is a one-to-one correspondence among these hypergestures which is perfectly compatible with all topological relations among neighborhoods.

A musical example will make this more lucid. The first species of Fux theory of counterpoint considers two voices, *punctus contra punctum*. There are two melodic lines, the *cantus firmus* (c.f.), and the *discantus*, as shown at the bottom of Figure 61.17. There are two readings of the term "punctus contra punctum": According to the common 'vertical' understanding, the discant 'point' (the upper tone at a given onset in the example) is set against the cantus firmus 'point' (the lower tone at a given onset in the example). This must always be a consonant interval (prime, minor third, major third, fifth, minor sixth, major sixth). The more adequate, but less known, interpretation is the 'horizontal' one: The 'point' is a defined interval at a

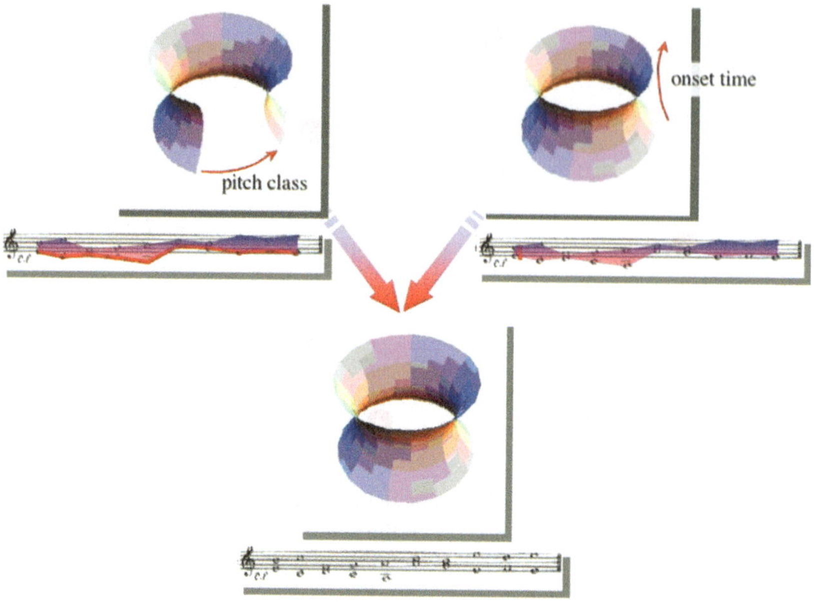

Fig. 61.17. Bottom: The first species Fux counterpoint exhibits two melody lines: the cantus firmus (lower melody, c.f.) and the discantus, upper melody.

given onset time, whereas the 'counter-point' to this is the subsequent interval [924]. This makes more sense since the compositional tension is not vertical, but horizontal. Whatever the interpretation, the result is this double melodic configuration, whose vertical or horizontal genealogy is however no longer retraceable, i.e., the listener or reader of a given contrapuntal composition cannot tell from that data how it was constructed, vertically or horizontally, The result is neutral.

Fig. 61.18. Maurits Cornelis Escher's *Belvedere* as an illustration of the bifurcation of neutral image regions into incompatible perspectives. ©2017 The M.C. Escher Company-The Netherlands. All rights reserved www.mcescher.com.

According to the discussion of continuous extensions of discrete sequences of gestures, if we allow the horizontal (in time direction) continuous extension of the discrete sequence of a two-voice counterpoint melody and simultaneously the vertical (in pitch class direction) continuous extension of the two discrete melodies, then the entire configuration looks like a tubular shape, as shown in Figure 61.17, above the bottom score. The upper left reading is then the common interpretation: connecting two melodies by a curve of melodies in the pitch class circle domain. The upper right reading is the more essential one, showing the time-oriented curve of intervals in pitch class space. In terms of hypergestures, the left one is a loop of lines, whereas the right one is a line of loops. The Escher Theorem lets these two hypergestures correspond to each other. This elementary example shows that the interpretation of a complex shape may give rise to very different readings in terms of skeletons and bodies of hypergestures.

This generates a dramatic hot spot in the interpretational activity, which drives the free understanding of musical shapes. I have to stress the adjective "free" here, since in a standard interpretation only one view is cultivated. It is never a line of loops, but must be strictly read as a loop of lines. Standardization is a strong poison against creativity. Its etymology from "stand hard" is speaking: the softness of interpretation is given no chance. This hot spot is based upon the $n!$ permutations of skeletal digraph sequences that generate the hypergesture space $\Gamma_{\pi(1)} \overrightarrow{@} \Gamma_{\pi(2)} \overrightarrow{@} \ldots \Gamma_{\pi(n)} \overrightarrow{@} X$.

These permutational variants are ways of breaking the delicate equilibrium of an uninterpreted shape. Each permutation generates a completely different hypergesture. One could view these departures from the neutral cusp towards a variety of interpretational sinks as a bifurcation process: The neutral trajectory of perception explodes into $n!$ interpretations, perspectives, or ways of handling one and the same neutral datum. Such a bifurcational process is best illustrated by a work of Maurits Cornelis Escher (see Figure 61.18), where neutral regions of the graphics (imagine a horizontal strip between the upper and lower levels of the architecture) are split into different, mutually exclusive, perspectives. This is the reason, why we coin the above theorem "Escher Theorem". We view this variety of switching into different interpretations, we have chosen the header statement from Saint Augustine: modulation as a switching action between different musical perspectives.

61.14.2 Rebecca Lazier's Vanish: Lawvere, Escher, Schoenberg

The category of (hyper)gestures is significant beyond the reciprocal interplay of free jazz musicians; it is fundamental for a gestural understanding of music. We want to sketch such an aspect here, because our concern is with the sophisticated active way of listening to music, as thematized by a number of musicians, such as Albert Ayler, or scholars, such as Helga de la Motte-Haber.

Listening to and understanding intelligent music is by no means a passive activity; a fortiori, a quality performance of free jazz requires an optimum of activity in order to be able to throw back the gestures one has received. This type of active listening, of course, is not exclusive to free jazz, and in this section we present a short study another type of music that is rich in fluctuation and permutation, viz. Schoenberg's *String Trio*, op. 45, a twelve-tone composition. The Trio is rich in intense gestural textures, mirroring Schoenberg's intense state during the period of its 1946 composition; Schoenberg had just suffered a nearly fatal heart attack.

Rebecca Lazier, inspired by the gestural offerings of the Trio as well as by the Trio's overall beauty, created a choreographed arrangement of the piece for seven dancers, titled *Vanish*, which premiered in New York City in 2002. Lazier describes her program: "In order to create parallels between Schoenberg's musical evolution and the evolution of the moving body and dance composition, I was required to create broader choreographic constructs that consisted of many voices and many experiments."

Her experiment specifically punctuates the gestural strength of the piece, and in turn facilitates a gestural way of listening. Lazier innervates the music with a network of hypergestural projections that are manifested in the gestures of the dancers, elevating the piece to a type of dynamic dancing hypergesture. This may justify the funny title "Making Schoenberg Dance" of her paper in [587]. To return to the subject of "active listening," this dance interpretation intimately intertwines a visual counterpart with the process of invested listening, providing another vital perspective on Schoenberg's music and gestures in general.

When Mazzola presented Lazier's choreography to his students, they were quite surprised about being confronted with a dancing Schoenberg perspective. Isn't this music anything but dance? Does a visual counterpart detract from the strength of the music? Isn't it an abstract, highly intellectual affair? We understand that dodecaphonic threads, the realizations of variants of the composition's *grundgestalt*, are gestures that are thrown into the composition without being explicitly connected with each other *within* the piece's notated structures.

Schoenberg himself was aware of the "dancing" character of such oppositional musical threads in space-time, and in the Trio, the architecture of imperfection is loaded with Michael Cherlin's time shards that float in a space of atemporal emptiness, of broken references and non sequiturs. Such structural ruins cannot be built upon a fixed framework; they are only aesthetically justified as strong gestures in the sense of Châtelet: they gain their amplitude by determining themselves; their self-referential sovereignty equals their penetration. Otherwise, we were lost in the listening adventure of such islands of beauty.

Fig. 61.19. Choreographer Rebecca Lazier.

Lazier symmetrically parallels the five parts of the Trio: Teil 1, Episode 1, Teil 2, Episode 2, Teil 3: Teil 1 is a constructive exposition of dodecaphonic material, based upon seven 'dodecaphonic series for dancers': Each dancer invents a series of movement units according to the sequence

near-death/transience/flight/hover/shiver/...
... slide/propel/crash/soar/fling/tender/explode

of words. These series are exposed in Teil 1 and reflect a well-formed counterpoint of dancers and movements. Teil 3 is a symmetric movement and reprises Teil 1, but also vanishes into oblivion at the end. The middle part, Teil 2, is dominated by what Lazier calls "catching phrase." Here is the description [587]:

> Each dancer was asked to run across the stage space, 'catch' something of any shape or size and then continue running off. The word 'catching' served as the inspiration for the dancers to create their own movements without visual information from the choreographer, hence, the movements were infused with individual perceptions, memories, and psychology.

Lazier does not follow the surface of the musical sound, but the vibrations of the twelve-tone rows, and she does not comment on the structure of the composition by some mirroring or mickey-mousing of musical events, she lets her dancers catch the musical gestures and demonstrate their understanding thereof in their performance of Teil 2. The dancers catch Schoenberg's gestures of rows and the free floating shards of time.

It is also remarkable that the collective and collaborative hypergesture realized by the dancers escapes—has to escape—the choreographer's control. This is her statement: "Movement itself or the syntax was placed beyond my willful control: either in the hands of chance, or in the bodies of the dancers." The dance itself made a revealing impression to my (Mazzola's) students: They felt that the overarching hypergestures were very clearly perceived and that the elasticity of the bodies contributed a lot to the more resonant understanding of the compact musical events and processes. The gestural understanding created what could be called a logic of bodies, and brought out certain aspects of the piece that were shrouded in pure acoustical space. The term "logic of bodies" is not a simple metaphor or even an abuse of the word "logic" in order to give the "primordial" reality of bodies a higher 'intellectual' status. The secret behind this wording is that gestures bear a natural logic in the most formal sense of mathematics. In fact, the category of directed graphs, the skeleta of gestures—which generalize the human skeleta–are a topos, i.e., a mathematical category full of logical and geometric properties. The most interesting fact for us is that every directed graph bears the structure of a Heyting algebra, which is the core structure of intuitionistic logic[18] that was thoroughly investigated by the school of Charles Ehresmann and William Lawvere [376].

[18] In intuitionistic logic, the negation of the negation of a statement is not equal the original statement, but somewhat more true. Also the law of the excluded third is invalid. See also Appendix Section G.5.

This logic is therefore a natural enrichment of the naked skeleton. Its parts behave like truth values that are distributed over the body: Each bone, the chest, the limbs, the pelvis, all these parts are truth values of the skeleton. We shall not delve further into these aspects, but it is essential to know that they are there, ready to be dealt with in a more sophisticated analysis of gestures. It would be important and profitable to enrich Lazier's choreography by a logical analysis of her dancers' body configurations. Lazier's *Vanish* therefore appears as a highly inspiring contribution to the synopsis of Lawvere's, Escher's and Schoenberg's iridescent perspectives, which are somewhat more substantially interrelated than Gödel's, Escher's, and Bach's.

62

Categories of Gestures over Topological Categories

Summary. We generalize the topological approach to gestures, and culminate in the construction of a gesture bicategory, which enriches the classical Yoneda embedding and could be a valid candidate for the conjectured space X in the diamond conjecture [720]; see also Section 61.12. We discuss first applications thereof for topological groups, and then more concretely gestures in modulation processes in Beethoven's Hammerklavier sonata. The latter offers a first concretization of answers to Lewin's big question from [605] concerning characteristic gestures. This research is a first step towards a replacement of Fregean functional abstraction by gestural dynamics.

– Σ –

In Chapter 61, we presented a mathematical model for gestures in music. In that model, a gesture γ is built from two components: a combinatorial "skeleton" represented by a digraph Γ, and a "body", represented by a configuration of continuous curves $\gamma(a) : I \to X$ on the real unit interval I with values in a topological space X, one for each arrow a of the skeleton, and connected according to the digraph's vertex configuration. Given two gestures δ, γ, a morphism $f : \delta \to \gamma$ is a digraph morphism $f : \Delta \to \Gamma$ between the skeleta Δ, Γ of δ, γ, respectively, which "extends" to a morphism of the respective bodies by a continuous map defined on the respective topological spaces. See [719] or Section 61.5 for the formal setup. This defines the category *Gesture* of gestures, which shares the two crucial properties:

- The set of gestures with skeleton Γ and with body in the topological space X is canonically provided with a topology deduced from the compact-open topology on the set $I@X$ of continuous maps from I to X; this topological space is denoted by $\Gamma\overrightarrow{@}X$. We therefore are capable of defining gestures of gestures, namely gestures with values in a topological space $\Gamma\overrightarrow{@}X$. Such gestures are called hypergestures.
- The hypergesture construction entails spaces of iterated hypergestures in the sense that for a sequence $\Gamma_1, \Gamma_2, \ldots \Gamma_n$ of skeleta and a topological space X, we have the space $\Gamma_1\overrightarrow{@}\Gamma_2\overrightarrow{@}\ldots\Gamma_n\overrightarrow{@}X$ of n-fold hypergestures over X. We then have the theorem that this iterated construction yields homeomorphic topological spaces if we permute the order of these skeleta; see Proposition 61 (First Escher Theorem) and Corollary 1. This result is of primordial significance in the creative gestural interaction in free jazz, see [721] for a detailed discussion.

Despite these promising first results, gesture theory is still "adolescent": Here are some questions, which we have encountered after a first critical analysis of the state of the art:

1. In the definition of a gesture, no allusions to transformations are made. We only deal with continuous curve systems. However, many examples from practice are more specific, they also involve transformations generating such curves. The classical and trivial example is a shift from a note x to a note y in a parameter space X, such as $X = \mathbb{R}^m$, to fix the ideas. This shift can be seen as a curve $c : I \to \mathbb{R}^m$ with $c(t) = T^{t(y-x)}(x)$, where $T^d : \mathbb{R}^m \to \mathbb{R}^m$ is the shift operation by d. This defines a special curve, not just any continuous data.

G. Mazzola et al., *The Topos of Music III: Gestures*, Computational Music Science,
https://doi.org/10.1007/978-3-319-64481-3_7

2. A very important point, also related to the previous one, refers to that famous gesture-theoretical question posed by David Lewin [605, p.159]: "If I am at s and wish to get to t, what characteristic gesture should I perform in order to arrive there?" A first observation must be made with respect to the general theoretical background of the question. Lewin poses it in a thoroughly transformational context. His book deals with transformations, with very classical affine functions on musical parameter spaces (mainly pitch class spaces). It is however not the only occasion where he opens up a gestural connotation of his transformational text. In many spots, he uses the word "gesture" and its paradigm, such as dancing and other motional and emotional metaphors. Lewin's gestural subtext is manifestly more than intuitive rhetorics, he means gesture and not just a fancy description of transformational configurations. This is a deep conflict in Lewin's musical thinking: He unfolds a valid transformational theory, but the subtext of gestures is not reflected in this theory. It remains a "dream of continuity while sleeping in the hard 'cartesian' bed of abstract algebra".
   In view of this observations, the question reveals its full power: How could we merge the transformational reality with the dream of continuity? The immediate mathematical response is: "by continuous transformations!" But the question is not solved with this immediate reply, since it is not clear what Lewin means by "characteristic". What is a characteristic gesture in contrast to any gesture? What is the character that has to be grasped?
3. The domain of continuous curves, I, is not only a topological space, but intrinsically has its topology derived from the linear ordering among real numbers. This information was not exploited in the previous theory. In other words: What is the reflection of this ordering relation within the topological space X that embodies a gesture? Is there any rationale to introduce "directions" in X?
4. Relating more specifically to the existent theory of Chapter 61, we see this general picture: The theory unfolds in two branches from the basic category **Digraph** of digraphs. The topological branch unfolds gestures as embodiments of digraphs within topological spaces, whereas the algebraic branch realizes digraphs as special linear categories, namely spectroids, enabeling formulaic structures typically related to commutative diagrams defined by algebraic relations. These ramifications are radically different interpretations of digraphs as basic constructors of mathematical theories. It was conjectured in [720] that one may construct a universal category X above the two categories of the two branches, *Gestoid* and *Formoid*, which would enable one to embed them as special cases of a comprising big structure.
   There are several indications that such a category might exist. The first one is the possibility to rebuild algebraic structures from gestures, more precisely: to rebuild groups through homotopy theory. It is in fact well known that any group is isomorphic to a fundamental group of a topological space [993], and we have more specifically given examples of such spaces for finitely generated abelian groups in [720], including some musical interpretation; see also Section 78.2.10.2. Intuitively speaking, since hypergestures generalize homotopy classes, one may say that *every group is realized by a group of hypergestures of loop gestures.*
   To date, the reverse direction looks less promising. No reasonable way is known to step over to gestures from abstract categories. A universal space as conjectured in the diamond conjecture should deal with this problem. Philosophically speaking, it is the problem of reconstructing gestural instances from general abstract categories.
5. A seemingly different, but in fact very relevant question arises from the deeper understanding of Yoneda's Lemma [637]. The lemma states (among other things) that in any category $\mathcal{C}$, the canonical functor $@ : X \mapsto @X$ sending an object X to its presheaf $@X$ is fully faithful, which means that the morphisms $f : X \to Y$ in $\mathcal{C}$ are in one-to-one correspondence with the morphisms (natural transformations) $F : @X \to @Y$. This sounds abstract, but it means that we may look at abstract morphisms $f : X \to Y$ in terms of ordinary "old-fashioned" Fregean functions on point sets $A@F : A@X \to A@Y$ for each given argument A. This technique is a big help for reconstructing intuitive human manipulation of mathematical objects when dealing with abstract categories. It however does not help us reconstruct the motion, which is intuitively happening, when moving from an argument x of a function f to the value $f(x)$. This deficiency is exactly what lies behind Lewin's question: The transformations are Fregean

functions and do not automatically involve any kind of motion as suggested by gestural utterances. So the question would be whether there is a way to embody Fregean functions within the realm of gestures.

The plan of this chapter is this. We first generalize the idea of a gesture from a purely topological setup to a functorial one, namely the setup of *topological categories*, i.e., categories internal to the category **Top** of topological spaces and continuous maps, replacing the unit interval I by a topological category (see Appendix Section J.4), the simplex category ∇, and continuous functions by continuous functors. This will be used in the second step, where we construct gestures from morphisms in abstract categories. This step is a decisive one towards the incorporation of abstract categories in the framework of gesture theory. It culminates in the construction of a bicategory of gestures for any category and leads to a first answer to the diamond conjecture. In a third step we apply these constructions to the important case of the category canonically associated with a topological group. We also discuss technical tools for overcoming the core problem of the mirror operation, which does not as such offer a gestural interpretation. In the fourth step, we discuss two modulations in Beethoven's Hammerklavier sonata op. 106 in order to apply the gestural approach for a deeper understanding of these modulations. We shall discuss Lewin's question about characteristic gestures. This last discussion reveals the intrinsically dramatic character of gestural interpretations of given scores.

62.1 Gestures over Topological Categories

In this section we set up the framework for a gesture theory that is based upon categories instead of plain topological spaces. In our setup, a category $\mathcal{C}$ is thought of as being a collection of morphisms, together with two maps $d, c : \mathcal{C} \to \mathcal{C}$ (d for "domain", c for "codomain"), and we write $f : d(f) \to c(f)$ to make these maps evident. In what follows, we shall start from a given *topological category* K (see Appendix Section J.4). This means that the collection of morphisms K is a topological space, and that domain and codomain, as well as the composition of morphisms (on the morphism sets with the relative topologies), are continuous.

Here are two basic examples of such categories: (1) The *simplex category* ∇ associated with the unit interval I: Its morphism set is $\nabla = \{(x,y)|x,y \in I \text{ and } x \leqslant y\}$, $d(x,y) = (x,x)$, $c(x,y) = (y,y)$, the composition of morphisms is obvious, and the topology on ∇ is the relative topology inherited from the usual product topology on $I \times I \subset \mathbb{R} \times \mathbb{R}$. (2) The *graph category* associated with any topological space X: Its morphism set is $X \times X$, equipped with the product topology, while we set $d(x,y) = (x,x), c(x,y) = (y,y)$, and again, the composition of morphisms is the obvious one. If no confusion is likely, we denote the graph category of X by X. Clearly, a graph category is a topological groupoid. In particular, the simplex category ∇ is just the subcategory of the graph category I on the pairs $(x,y), x \leqslant y$.

If K, L are two topological categories, a *topological functor* $F : K \to L$ is a functor, which is also continuous as a map between morphism sets. This defines the category **TopCat** (in fact a 2-category, see [131, Proposition 8.1.4]) of topological categories. In order to distinguish the set of topological functors $F : K \to L$ from the larger set $K@L = \mathbf{Cat}(K, L)$ of all possible functors, we write $K©L$ for $\mathbf{TopCat}(K, L)$. If X, Y are topological spaces, then the map which associates with a continuous map $f : X \to Y$ the synonymous continuous functor is fully faithful, so the category of topological spaces is a full subcategory of the category of continuous categories. Therefore we shall henceforth tacitly identify the category **Top** of topological spaces and continous maps with the associated subcategory of topological categories and continuous functors embedded in **TopCat** via the graph category associated with a topological space.

With this in mind, if K is a topological category, the set of continuous curves with values in K is by definition the set $\nabla©K$. Evidently, if K is a topological space, then $\nabla©K \xrightarrow{\sim} I@K$, where $I@K$ is the set $\mathbf{Top}(I, K)$ of continuous I-parametrized curves $c : I \to K$ in the topological space K, the bijection being induced by the restriction of a functor $F : \nabla \to K$ to the canonical diagonal embedding $I \to \nabla$ of the objects in ∇. This set $\nabla©K$ is the object set of a category also denoted by $\nabla©K$ if we take as morphisms between two curves $f, g : \nabla \to K$ the continuous natural transformations $\nu : f \to g$, which means that the defining maps $\nu : I \to K$ are continuous and satisfy the defining commutative squares for natural transformations. We do however want it to become a topological category, and this works as follows: We take the morphism set as being composed by the triples (f, g, ν) as above. The topology is defined by the following construction. The set of objects of $\nabla©K$ is given the compact-open topology induced by the topologies of ∇ and K, the subset of continuous natural transformations $\nu : I \to K$ within $I@K$ is given the topology induced by the compact-open topology on $I@K$. The triples are viewed as points in the product topology on $\nabla©K \times \nabla©K \times I@K$. Clearly, this is a topological category. Also observe that in the case of a topological space K, the compact-open topology of $I@K$ coincides with the topology induced by the isomorphism $\nabla©K \stackrel{\sim}{\to} I@K$ and the compact-open topology on $\nabla©K$.

Example 70 The set $\nabla©K$ can also be enforced for a not a priori topological category K as follows. Take any set $C \subset \nabla@K$ of functors $F : \nabla \to K$ into an abstract category (suppose K small, if set theory matters) and then select the finest topology on K such that all functors of C become continuous. For this construction one writes $\nabla©_C K$ to indicate that K is made a topological category via C, and that this is the set of all continuous curves with respect to this topology.

62.1.1 The Categorical Digraph of a Topological Category

In order to obtain gestures in topological categories, we need to mimic the construction of a spatial digraph [720], see also Section 61.5. To this end, we consider the two continuous tail and head functors $t, h : \nabla©K \to K$, which are defined as follows. If $\nu : f \to g$ is a natural transformation between $f, g : \nabla \to K$, then $t(\nu) = \nu(0) : f(0) \to g(0)$, and $h(\nu) = \nu(1) : f(1) \to g(1)$. So the tail and head maps are not only set maps but functors. Call this diagram of topological categories and continuous functors the *categorical digraph*[1] $\overrightarrow{K}$ *of* K. If we forget about the category and just retain the objects of this configuration, we call it the *(underlying) spatial digraph of* K. In particular, if Γ is a digraph, the set of morphisms $\Gamma@\overrightarrow{K}$ is the set of digraph morphisms into the underlying spatial digraph of K. In other words, such a morphism assigns an object of K to every vertex of Γ and a continuous curve (topological functor) $\nabla \to K$ to every arrow of Γ, with matching sources and targets. We call then, by definition, a *gesture with skeleton* Γ *and body in* K a morphism of digraphs $g : \Gamma \to \overrightarrow{K}$.

If the topological category is a topological groupoid, then we have an easy proposition which guarantees that one may reverse all arrows, in other words: the categorical digraphs of topological groupoids are self-dual.

Proposition 1. *Let K be a topological groupoid. Then we have a duality automorphism $?^* : \overrightarrow{K} \stackrel{\sim}{\to} \overrightarrow{K}^*$ onto the dual digraph $\overrightarrow{K}^*$ (tail and head functors exchanged), which maps a curve $g : \nabla \to K$ to its inverse curve $g^* : \nabla \to K$ defined by $g^*(x, y) = g(1 - y, 1 - x)^{-1}$.*

Therefore, for a topological groupoid K, the set $\Gamma@\overrightarrow{K}$ is in bijection with its dual set $\Gamma^*@\overrightarrow{K}^*$, and then with the set $\Gamma^*@\overrightarrow{K}$ associated by the duality $?^*$. Call the gesture $g* : \Gamma^* \to \overrightarrow{K}$ associated by this bijection with a given gesture $g : \Gamma \to \overrightarrow{K}$ the *dual gesture*. Intuitively it reverses the arrows of the skeleton and the morphisms of the body's curves.

62.1.2 Gestures with Body in a Topological Category

We have constructed the set $\Gamma@\overrightarrow{K}$ of gestures with skeleton Γ and body in a topological category K. In the previous theory described in Section 61.6, this set was enriched to yield a topological space in order to enable the iterative construction of hypergestures. In our present setup, we have to construct a topological category out of the above set. To do so, recall that the special case $\Gamma = \uparrow$ (one arrow between two different vertices) means that we have the topological category $\uparrow @\overrightarrow{K} \stackrel{\sim}{\to} \nabla©K$ of continuous curves $c : \nabla \to K$ (with the above mentioned compact-open topology).

The general case follows from the observation that Γ is the colimit of the following diagram $\mathcal{D}$ of digraphs: We take one arrow digraph $\uparrow_a = \uparrow$ for each arrow $a \in A_\Gamma$ (A_Γ is the set of arrows of Γ) and one

[1] We have chosen this wording as an analogy with the spatial digraph, where the topological *space* is now replaced by the topological *category*. Although this is a diagram of topological categories, and not just of sets, we believe that the intuitive wording is not confusing.

bullet digraph $\bullet_x = \bullet$ for each vertex $x \in V_\Gamma$ (V_Γ is the set of vertices of Γ). We take as morphisms the tail or head injections $\bullet_x \to \uparrow_a$ whenever $x = t(a)$ or $x = h(a)$. Then evidently, $\Gamma \xrightarrow{\sim} \mathrm{colim}\mathcal{D}$. Therefore, the set of gestures $\Gamma @\vec{K}$ is bijective with the limit $\lim \mathcal{D}@\vec{K}$ of a diagram of *the objects of* topological categories $\uparrow @\vec{K} \xrightarrow{\sim} \nabla © K$ (for the digraph's arrows) and $\bullet @\vec{K} \xrightarrow{\sim} K$ (for the digraph's vertices). But the maps between these objects of categories stem in fact from functors (those from the categorical graph $\vec{K}$). Therefore the limit can be taken as one of a diagram of topological categories. This yields a category, whose topology is defined as the limit topology of this diagram. This topological category is denoted by $\Gamma \vec{@} K$. In this category, a morphism is the limit of natural transformations between continuous curves and morphisms between objects of K, the latter representing the end points of the continuous curves.

Example 71 If the topological category K is a topological space, we recover the topological category $\Gamma \vec{@} K$ associated with the topological space $\Gamma \vec{@} K$ in the previous theory of Section 61.6.

The construction of the topological category $\Gamma \vec{@} K$ automatically enables the machinery of hypergestures known from the previous topological space setup. And again, we have the Escher Theorem for topological categories of hypergestures:

Proposition 2. (Escher Theorem for Topological Categories) *If Γ, Δ are digraphs and K is a topological category, then we have a canonical isomorphism of topological categories,*

$$\Gamma \vec{@} \Delta \vec{@} K \xrightarrow{\sim} \Delta \vec{@} \Gamma \vec{@} K.$$

Corollary 1. *The action*

$$\vec{@} : \mathbf{Digraph} \times \mathbf{TopCat} \to \mathbf{TopCat} : (\Gamma, K) \mapsto \Gamma \vec{@} K$$

canonically extends to an action (denoted by the same symbol)

$$\vec{@} : [\mathbf{Digraph}] \times \mathbf{TopCat} \to \mathbf{TopCat} : (W, K) \mapsto W \vec{@} K$$

of the free commutative monoid [**Digraph**], *i.e., the monoid of commutative words $W = \Gamma_1\Gamma_2 \ldots \Gamma_k$ over the alphabet* **Digraph** *of digraphs (the objects only). It is defined inductively by $\Gamma_1\Gamma_2 \ldots \Gamma_k \vec{@} K = \Gamma_1 \vec{@} (\Gamma_2 \ldots \Gamma_k \vec{@} K)$ and*[2] *$\varnothing \vec{@} K = K$.*

With this hypergestural construction, we define the category of gestures with body in K, now also including the morphisms between their skeleta. It is denoted by *Gesture*(K). Its objects are the objects of $\Gamma \vec{@} K$ for any digraph Γ. Given two such gestures $g : \Gamma \to \vec{K}, h : \Delta \to \vec{K}$, a *morphism* $a : g \to h$ is a pair $a = (t, \nu)$, consisting of a digraph morphism $t : \Gamma \to \Delta$, and a morphism $\nu : g \to h \circ t$ in $\Gamma \vec{@} K$, which we also write as a diagram, but with the natural transformation being denoted by a double arrow in order to prevent a wrong intuition about a commutative square:

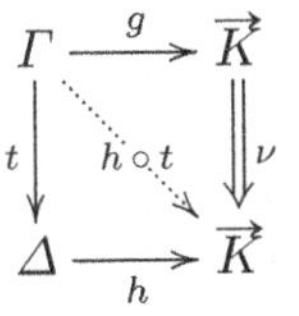

If we are given a second morphism $b : h \to k$, $b = (s, \mu)$, with codomain $k : \Sigma \to \vec{K}$, then the composition $b \circ a : g \to k$ is defined by $b \circ a = (s \circ t, \mu|t \circ \nu)$, where $\mu|t$ means that the natural transformation μ from h to $k \circ s$ is "restricted" by the digraph morphism t.

The category *Gesture*(K) therefore contains two types of subcategories: On the one hand the (comma category) topos $G(K) = \mathbf{Digraph}/\vec{K} \subset$ *Gesture*(K) of gestures with body in K, the morphism being the digraph morphisms of gesture skeleta commuting with the domain and codomain gestures. On the other

[2] $\varnothing$ denotes the empty word.

hand, we have, for each skeleton Γ, the topological category $\Gamma \vec{@} K \subset Gesture(K)$. Finally, for each digraph morphism $t : \Gamma \to \Delta$, we have a canonical continuous restriction functor $|t : \Delta \vec{@} K \to \Gamma \vec{@} K$. Here is the overall picture:

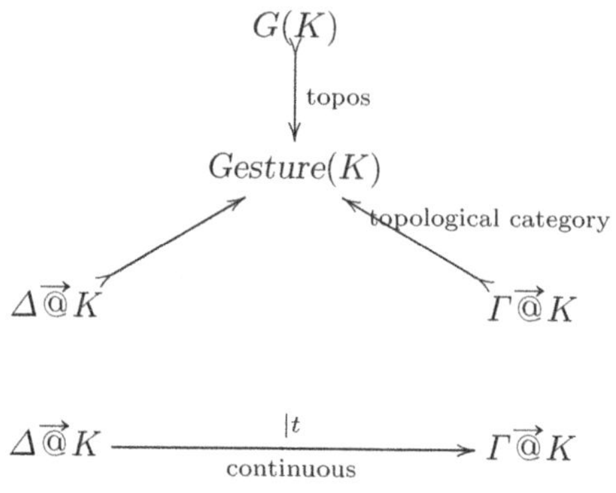

62.1.3 Varying the Underlying Topological Category

For a continuous functor $F : K \to L$ between topological categories, we have a canonical morphism of categorical digraphs $\vec{K} \to \vec{L}$, which sends vertices to vertices, namely by the given functor $F : K \to L$, and sends curves $f : \nabla \to K$ to curves $F \circ f$, whereas continuous natural transformations $\nu : f \to g$ are sent to the continuous natural transformations $F \circ \nu : F \circ f \to F \circ g$. Call this morphism a *spatial (categorical) digraph morphism* and denote it by $\vec{F}$. This morphism canonically induces a functor $Gesture(F) : Gesture(K) \to Gesture(L)$, which is compatible with the above subcategories as shown by the following commutative diagram:

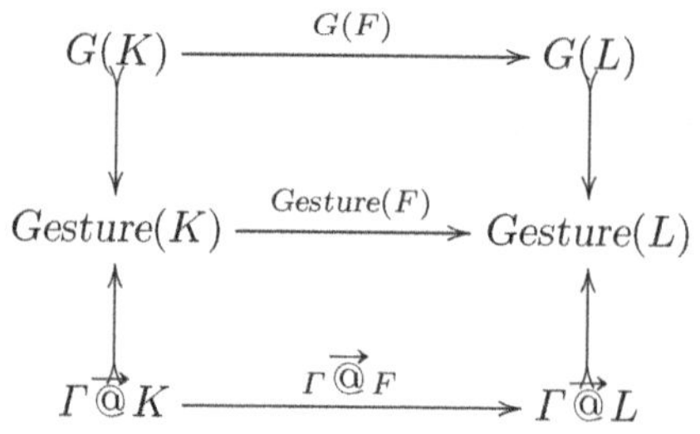

While the functor $\Gamma \vec{@} F$ is continuous, the functor $G(F)$ has a number of well-known properties of functors between topoi [639, Ch. IV.7]. The first of these properties is that $G(F)$ is right adjoint to the base change functor $\times \vec{F} : G(L) \to G(K)$ which associates with a gesture $g : \Gamma \to \vec{L}$ the fibre product gesture $g \times \vec{F} : \Gamma \times_{\vec{L}} \vec{K} \to \vec{K}$. Furthermore, the base change $\times \vec{F}$ is a logical functor (i.e., it preserves all topos-theoretical constructs, such as sub-object classifiers, finite limits and colimits, and exponentials, and has also a right adjoint). Paired with its right adjoint $\times \vec{F}_*$, the base-change functor defines a geometric morphism $G(K) \to G(L)$ [639, Ch. VII.1]. We shall come back to these facts later in Section 62.2.4, when discussing the gestural part of Yoneda's Lemma.

62.2 From Morphisms to Gestures

To conceive of a general method for generating gestures from morphisms $f : X \to Y$ in abstract categories, we start with a heuristic consideration. Suppose that we are working in a musical parameter space $\mathbb{R}^2$, which we endow with the structure of the Gaussian plane of complex numbers. Take a rotation $e^{i\theta} : x \mapsto x.e^{i\theta}$ on $\mathbb{R}^2$. In linear algebra, this morphism $f = e^{i\theta}$ is an encapsulated function, which has no relation to a gesture, but acts by Fregean "teleportation" on x. We are stressing this fact since in contradiction to the algebraic reality, our intuition of a rotation by angle θ is different in that we imagine a continuous rotational movement of x around the space origin until it reaches the final position $x.e^{i\theta}$. This process is visualized by the trace of x while rotating, i.e., by a continuous curve $c_x : I \to \mathbb{R}^2 : t \mapsto x.e^{i\theta t}$ on a circle of radius $|x|$. Each intermediate position $x.e^{i\theta t}$ corresponds to a factorization $f = e^{i\theta(1-t)} \circ e^{i\theta t} = f_{1-t} \circ f_t$ of f. In other words, the curve $c : I \to \mathrm{GL}_2(\mathbb{R})$ is a curve of factorizations of the given morphism f. This restatement of the gesture c in terms of factorizations means that c is viewed as being an "infinite" factorization insofar as the factors are parametrized by the curve parameter $t \in I$.

This enables us to rethink the basic elements of a gestural interpretation of morphisms in abstract categories. To this end, we fix a morphism $f : X \to Y$ in a category $\mathcal{C}$. The *category* $[f]$ *of factorizations of* f is defined as follows. Its morphism are the triples (u, g, v) of morphism $u : X \to W, g : W \to Z, v : Z \to Y$ such that $v \circ g \circ u = f$. The domain map is $d(u, g, v) = (u, Id_W, v \circ g)$, while the codomain map is $c(u, g, v) = (g \circ u, Id_Z, v)$. Suppose we have two morphisms $(u, g, v), (r, h, s)$ such that $c(u, g, v) = d(r, h, s)$, $h : Z \to Q$, then their composition is the morphism $(u, h \circ g, s)$, as shown in the following commutative diagram:

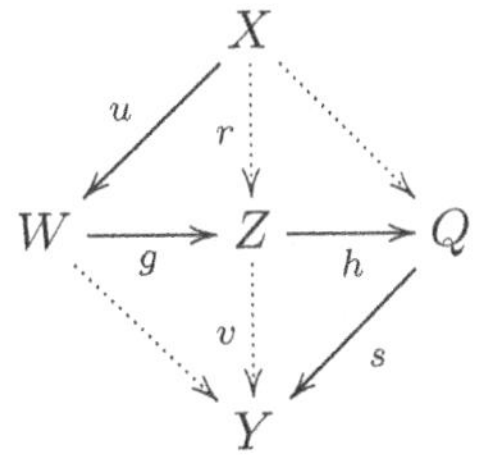

This construction entails a number of evident facts: To begin with, the category $[f]$ has the initial object (Id_X, Id_X, f) and the final object (f, Id_Y, Id_Y). Moreover, if $k : Y \to E$ and $l : A \to X$ are morphisms, then there are two functors $[k\circ] : [f] \to [k \circ f]$ and $[\circ l] : [f] \to [f \circ l]$, respectively, sending (u, g, v) to $(u, g, k \circ v)$ and to $(u \circ l, g, v)$, respectively (keeping the above notations). If $\mathcal{C}$ is a topological category, then so is $[f]$, if it is viewed as a subset of $\mathcal{C}^3$. Also, the two functors $k\circ, \circ l$ are continuous.

For any two objects X, Y in $\mathcal{C}$ we now build the disjoint sum $[X, Y] = \coprod_{f \in X@Y}[f]$ of the factorization categories $[f]$ (including the coproduct of topologies on the $[f]$). Therefore $\nabla@[X, Y] = \coprod_{f \in X@Y} \nabla@[f]$, and, if we endow $[X, Y]$ with the coproduct topology, also $\nabla©[X, Y] = \coprod_{f \in X@Y} \nabla©[f]$. The above construction of functors from morphisms also works in this coproduct situation, and also mutatis mutandis for topologies on these categories, i.e., conserving the above notations, we have two continuous functors $[k\circ], [X, Y] \to [X, E]$ and $[\circ l] : [X, Y] \to [A, Y]$, and their associated curve functors $\nabla©[k\circ] : \nabla©[X, Y] \to \nabla©[X, E]$ and $\nabla©[\circ l] : \nabla©[X, Y] \to \nabla©[A, Y]$.

Example 72 If $f = Id_X$, then $[f]$ is the category of sections and retractions of X, since its objects are the triples (u, Id, v) such that $v \circ u = Id_X$.

Example 73 The category $\mathcal{C}$ is defined by a topological group G, i.e., as a category, has one single object and the group elements as morphisms, then $[f] \xrightarrow{\sim} G$, where G is the graph category of the topological space G. More explicitly, the morphisms of $[f]$ are the triples (u, g, v) of elements of G such that $v \circ g \circ u = f$. Since any two of them are free and determine the third, we take the morphisms as being the pairs $(d, c) \in G \times G$, where we have $u = d, g = c \circ u^{-1}, v = f \circ u^{-1} \circ g^{-1}$. The topology is the product topology of $G \times G$.

For example, if $\mathcal{C}$ is defined by the cartesian product group $G = \mathbb{R}^n \times \overrightarrow{GL}_n(\mathbb{R})$ of the additive group $\mathbb{R}^n$ and the general affine group $\overrightarrow{GL}_n(\mathbb{R})$, $[f] \xrightarrow{\sim} \mathbb{R}^n \times \overrightarrow{GL}_n(\mathbb{R})$, the topological space category of pairs (x, g) of points x in $\mathbb{R}^n$ and affine transformations $g : \mathbb{R}^n \xrightarrow{\sim} \mathbb{R}^n$. We then have a continuous (group action) functor $\epsilon : \mathbb{R}^n \times \overrightarrow{GL}_n(\mathbb{R}) \to \mathbb{R}^n : (x, g) \mapsto g(x)$ into the topological category $\mathbb{R}^n$ deduced from the group $\mathbb{R}^n$. Therefore, from a gesture $g : \Gamma \to \mathbb{R}^n \times \overrightarrow{GL}_n(\mathbb{R})$, we deduce a gesture $\epsilon \circ g : \Gamma \to \mathbb{R}^n$. The latter is a gesture whose curves are just continuous curves in real n-space, but they are not arbitrary, since they are induced by curves of points and linear transformations. This very special case reveals the power of our construction of factor categories: They include the concept of gestures of transformations of points, and not only abstract topological gestures. But they are much more powerful since a gesture in $\mathbb{R}^n \times \overrightarrow{GL}_n(\mathbb{R})$ might well specify curves that are more general than just curves of transformations, but let us make this more precise.

As in our initial example in $\mathbb{R}^2$ of a rotational curve $c(t) = (x, e^{i\theta t})$, we may vary the transformation and fix the point x, but we may as well just take an arbitrary continuous curve $d(t) = (x(t), Id_{\mathbb{R}^2})$ in $\mathbb{R}^2$ and let the transformation remain the identity. More generally, we may vary both, the point and the curve, and consider a curve $e(t) = (x(t), g(t))$ in $\mathbb{R}^2 \times \overrightarrow{GL}_2(\mathbb{R})$. This opens the concept of a gesture, whose curves are characteristic in that they may pertain either to transformational constructs, to purely topological rationales, or to both. Such a setup works for any (topological) group action on a given module, such as, for example, the musically relevant action of the general affine group $\overrightarrow{GL}(\mathbb{Z}_{12})$ on the pitch class group $\mathbb{Z}_{12}$ (with the discrete topology, for example).

62.2.1 Diagrams as Gestures

Example 72 suggests that one should take a closer look at the category of factorizations for module categories, since sections and retractions define direct summands in the abelian categories. To this end, we first construct certain standard gestures. To begin with, let $g : W \to Z$ be any morphism in a category $\mathcal{C}$. Then there is a functor $\searrow (g) : \nabla \to \mathcal{C}$ with

$$\searrow (g)(x,y) = \begin{cases} Id_W & \text{if } x = y = 0, \\ g & \text{if } 0 = x < y, \\ Id_Z & \text{if } 0 < x. \end{cases} \tag{62.1}$$

This construction method enables the construction of gestures from diagrams in categories as follows. Suppose that a category K is small. Then take the topology on K such that all functors $c : \nabla \to K$ are curves, i.e., we take $\nabla © _{\nabla @ K} K$. Consider K as a digraph with the two maps $d, c : Mor(K) \to Ob(K)$ from the morphism set $Mor(K)$ to the object set $Ob(K)$. Then we have the following morphism of digraphs $\searrow : K \to \overrightarrow{K}$ which sends a morphism $f : X \to Y$ to the curve $\searrow (f)$ with tail X and head Y. Therefore, if we have any diagram $\delta : \Delta \to K$ in the category K, we may compose it with $\searrow$ and obtain a gesture $\searrow \circ \delta : \Delta \to \overrightarrow{K}$, which we denote by $\overrightarrow{\delta}$ and call the *discrete gesture associated with the diagram* δ. This evidently extends to a *discrete gesture functor* $\overrightarrow{?} : \Delta @ K \to \Delta @ \overrightarrow{K}$ from the category of diagrams and natural transformations to the category of gestures of these spaces.

62.2.2 Gestures in Factorization Categories

In our context, a morphism (u, g, v) in a factorization category $[f]$ with

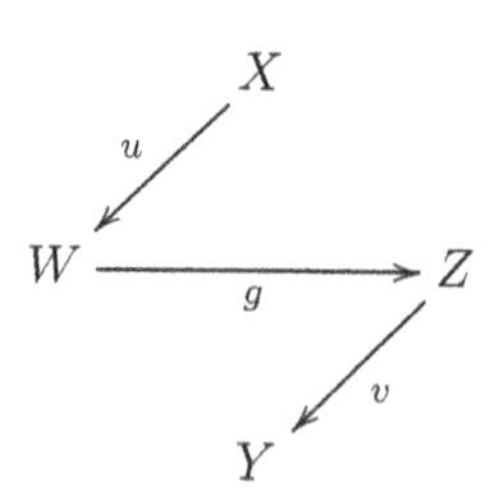

yields a curve $\searrow (u, g, v) : \nabla \to [f]$. This construction can be iterated in the sense that for any sequence $g_. = (g_i : W_i \to W_{i+1})_{i=0,1,\dots m-1}$ of length m of morphisms in $\mathcal{C}$, there is a functor $\searrow (g_.) : \nabla \to \mathcal{C}$ where the restriction $\searrow (g_.)|\nabla_{[i/m,(i+1)/m]} \to \mathcal{C}$ to the full subcategory $\nabla_{[i/m,(i+1)/m]} = \{(x,y)|i/m \leqslant x \leqslant y \leqslant (i+1)/m\}$ of ∇ is the above one-step construction for $g_i : W_i \to W_{i+1}$. This entails that we may also consider curves $\searrow (u, g_., v)$ associated with the chain $(u, g_., v)$ of morphisms in $[f]$:

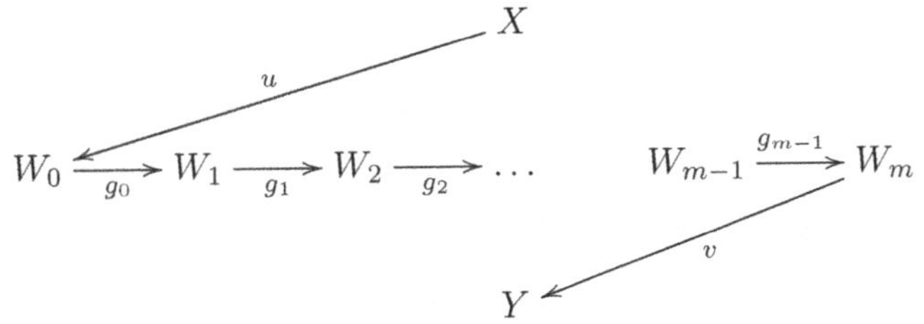

Given two such curves $\searrow (u, g., v), \searrow (u, h., v)$ of the same length m, the second one involving the morphisms $h_i : Z_i \to Z_{i+1}$ and $u' : X \to Z_0, v' : Z_m \to Y$, a morphism $\nu_. : \searrow (u, g., v) \to \searrow (u, h., v)$ is a natural transformation consisting of a chain of morphisms $\nu_. = (\nu_0 : W_0 \to Z_0, \ldots \nu_m : W_m \to Z_m)$ such that we have this commutative diagram:

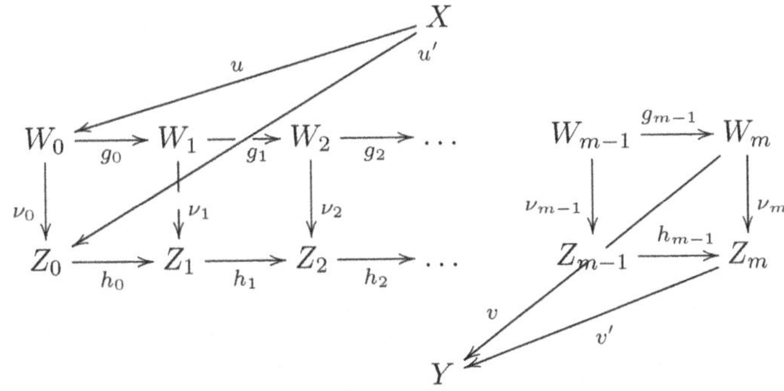

62.2.3 Extensions from Homological Algebra Are Gestures

Now, if we consider the special case where $\mathcal{C} = {}_R\mathbf{Mod}$, the category of left R-modules and linear homomorphisms over a commutative ring R, then we may take the factorization category $[0]$ of the zero homomorphism $0 : 0 \to 0$ on the zero module. We may further consider two exact sequences, one $g_.$ of modules $W.$, and one $h_.$ of modules $Z.$, to generate curves, which we should call *exact curves*. Then the morphism $\nu.$ is just a morphism between exact sequences, which means that the category of exact sequences is a canonical subcategory of the category of curves in $[0]$. In particular, if we look at such short exact sequences (length 2), and we restrict ourselvesto morphisms between sequences of common initial module W and terminal module Z, we obtain the groupoid of exact sequences, and the isomorphism classes define the classical set $Ext_R(Z, W)$ of congruence classes of extension of Z by W [635]. Consequently, we have this fact:

Fact 22 *The categories of factorization are a natural extension of structures from homological algebra encountered, for example, in the construction of* $Ext^n_R(Z, W)$.

62.2.4 The Bicategory of Gestures

Suppose that for two morphisms $f : X \to Y, g : Y \to Z$ in a category $\mathcal{C}$ we are given topologies on the factorization categories $[f], [g]$ such that the two functors $[g\circ] : [f] \to [g \circ f], [\circ f] : [g] \to [g \circ f]$ are continuous (e.g. if $\mathcal{C}$ is topological). Write $G[f]$ for the gesture topos $G([f])$. For such morphisms, we denote the categorical digraphs $\overrightarrow{[f]}$ by $\overrightarrow{f}$, and the corresponding morphisms of categorical digraphs, such as the ones derived from $g\circ$ and $\circ f$, are denoted by $\overrightarrow{g\circ}$ and $\overrightarrow{\circ f}$. We therefore have two morphisms of categorical digraphs $\overrightarrow{g\circ} : \overrightarrow{f} \to \overrightarrow{g \circ f}$ and $\overrightarrow{\circ f} : \overrightarrow{g} \to \overrightarrow{g \circ f}$, which induce the two canonical functors between topoi, $[g\circ] : G[f] \to G[g \circ f]$ and $[\circ f] : G[g] \to G[g \circ f]$. Taking coproducts on the skeleta, this induces a functor between topoi,

$$b_* : G[g] \times G[f] \to G[g \circ f],$$

which maps a pair $\gamma : \Gamma \to \overrightarrow{g}, \phi : \Phi \to \overrightarrow{f}$ of gestures to the gesture

$$\overrightarrow{\circ f}\gamma \sqcup \overrightarrow{g\circ}\phi : \Gamma \sqcup \Phi \to \overrightarrow{g \circ f}.$$

On the other hand, the two base changes $\overrightarrow{g\circ}$ and $\overrightarrow{\circ f}$ induce[3] base change functors $\times\overrightarrow{g\circ} : G[g \circ f] \to G[f]$ and $\times\overrightarrow{\circ f} : G[g \circ f] \to G[g]$, which we combine to get the base change functor

$$b^* : G[g \circ f] \to G[g] \times G[f].$$

By use of routine topos-theoretical arguments, we have this result (recall from [639] the definition of a geometric morphism of topoi):

Theorem 1. *Given the above conditions and notations,*

(*i*) *the base-change functor b^* is a logical functor, i.e., it conserves all topos-theoretical structures, subobject classifier, finite limits, colimits, and exponentials.*
(*ii*) *The coproduct functor b_* is left adjoint to b^*, and*
(*iii*) *there is functor $a : G[g] \times G[f] \to G[g \circ f]$, which is right adjoint to b^* such that the pair (a, b^*) is a geometric morphism of topoi.*
(*iv*) *If f or g is the identity, then b_* is isomorphic to the identical functor. If $h : Z \to W$ is a third morphism, also sharing the above properties of f, g, then the functor b_* is associative up to isomorphisms.*

Adding up all the factorization categories relating to morphisms $f : X \to Y$, we define the coproduct category $X☞Y = \coprod_{f \in X@Y} G[f]$. If we are given a second morphism $g : Y \to Z$ with the above conditions still holding, then we have a functor deduced from the above functor b_*, notated with capital letters:

$$B_* : Y☞Z \times X☞Y \to X☞Z \tag{62.2}$$

It is associative up to isomorphisms and has the identity gesture $\emptyset \to \overrightarrow{Id_X}$ for each object X. If these constructions work for all objects and morphisms (e.g. if $\mathcal{C}$ is topological), then the composition functors (62.2) define a bicategory [637], the *gesture bicategory of* $\mathcal{C}$ denoted by $\mathcal{C}^{☞}$. This is nearly a 2-category, except that composition is only associative up to isomorphisms. This being so, the "morphic" half of Yoneda's Lemma would consist in characterizing the functors (62.2)—or else the geometric functors between the topoi $G[g] \times G[f]$ and $G[g \circ f]$—which stem from composing morphisms in the original category $\mathcal{C}$. This would enable us to think of morphisms as being represented by gestures and to calculate all of the category's operations on the level of gestures. Given that the classical "objective" Yoneda Lemma already takes care of the reconstruction of point sets from abstract objects by the transition from $\mathcal{C}$ to $\mathcal{C}^{@}$, this hypothetical "morphic" Yoneda Lemma would give us back the full gestural intuition on the level of $(\mathcal{C}^{@})^{☞}$ while working in abstract categories.

62.2.5 Entering the Diamond Space

In view of the preceding results, we have set up a concept space, as made explicit in the gesture bicategory construction $\mathcal{C}^{☞}$, which embraces the topological gesture theory of our former work [719] as well as the diagram theory backing the network approach from Lewin's and Klumpenhouwer's transformational theory, but also basic constructions from homological algebra, such as congruence morphisms between extensions in abelian categories. The philosophy of this approach is that the concept of a categorical gesture, although completely in the vein of gestural reflections fostered by musical requirements, is flexible enough to include the extremal cases of "discrete" and properly "continuous" gestures as well. The relation between these two cases being that continuous gestures are a kind of limit of factorization when the factors are becoming more and more "fine grained" until they are parametrized by continuously varying real parameters; see Figure 62.1 for an intuitive image of factorization granularity. We therefore argue that this space construction is a good candidate for our conjectural space X as described in the Diamond Conjecture [719, Section 9], see also Section 61.12. At this stage, we do not however yet state that this space X has been found since a number of tests have to be performed in order to learn about power vs. deficiencies of the present approach.

[3] Recall that the product in a comma category is the fiber product in the original category.

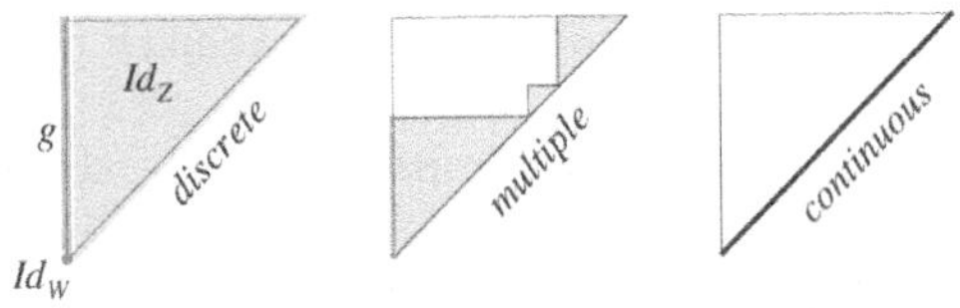

Fig. 62.1. Factorization on ∇ from one "discrete" step $\searrow$ (g) through a finite series of (discrete) factors to the "limit" of a continuous curve of factors.

62.3 Diagrams in Topological Groups for Gestures

In this section, we want to make explicit the transition from the type of diagrams used in transformational theory to gestures. Of course, we are not dealing here with the above embedding of diagram categories into gesture categories, but want to transform discrete gesture curves into continuous curves that enable an infinity of intermediate stages between the starting and the ending position of diagrammatic arrows.

To this end we first discuss the gestures with values in the factor category discussed above, namely starting from the topological group $G = \overrightarrow{GL}_n(\mathbb{R})$, so that $[f] \xrightarrow{\sim} \overrightarrow{GL}_n(\mathbb{R})$, the topological space category of affine transformations $g : \mathbb{R}^n \xrightarrow{\sim} \mathbb{R}^n$, whose morphisms are parametrized by pairs $(d, c) \in G \times G$ of group elements, the transition morphism $g : d \to c$ representing the transformation $g = c \circ d^{-1}$. We then know that the data of a curve $\delta : \nabla \to G \times G$ is equivalent to its diagonal restriction to the objects $I \rightarrowtail \nabla$, i.e., to a continuous map $\delta_I : I \to G$, where $G \rightarrowtail G \times G$ identifies G with the diagonal in $G \times G$ by the diagonal embedding (much as I is in ∇). Since I is connected, the image of δ_I must be either in the connected component $G^+ = \overrightarrow{GL}_n^+(\mathbb{R})$ or in the complementary connected component $G^- = \overrightarrow{GL}_n^-(\mathbb{R})$ of G, where $\overrightarrow{GL}_n^+(\mathbb{R})$ ($\overrightarrow{GL}_n^-(\mathbb{R})$) is the group (coset) of affine transformations with linear part in the subgroup $\mathrm{GL}_n^+(\mathbb{R}) \subset \mathrm{GL}_n(\mathbb{R})$ (in the coset $\mathrm{GL}_n^-(\mathbb{R}) \subset \mathrm{GL}_n(\mathbb{R})$) of transformations with positive (negative) determinant. Therefore any gesture with body in $G \times G$ with connected skeleton must have all its object curves either in G^+ or in G^-.

Therefore connectedness of I implies that we cannot connect transformations of different determinant signatures, e.g., the identity Id_G for $G = \overrightarrow{GL}_2(\mathbb{R})$ and the mirror transformation $m = \begin{pmatrix} -1 & 0 \\ 0 & 1 \end{pmatrix}$. This is a major problem for the continuous gestualization of discrete gestures. In fact, if we take the morphism $Id_G \to m$, there is no continuous curve starting in Id_G and ending in m. Why should this be required? If we had such a curve, $\alpha_I : I \to G$, with $\alpha_I(0) = Id_G, \alpha_I(1) = m$, we could use it to generate a continuous curve of points $\alpha_I.x(t) = \alpha_I(t)(x) \in \mathbb{R}^2$ by evaluation of the curve at a given initial point $x \in \mathbb{R}^2$ and parameter $t \in I$ as explained in Example 73.

In order to understand the specific problem which appears with mirroring, let us look at the generators of $G = \overrightarrow{GL}_2(\mathbb{R})$ and their musical meaning (see Section 8.3 for a detailed discussion). They are (1) translations $T^{(1,0)}$ by one unit in horizontal direction, (2) all positive dilations of the first coordinate, (3) the above mirror m, (4) the horizontal transvection $t = \begin{pmatrix} 1 & 0 \\ 1 & 1 \end{pmatrix}$, and the 180° rotation $R = -Id$. For all these generating transformations g_i, except for the mirror m, there is a continuous curve $\delta_i : I \to G^+$ starting at Id_G and ending at g_i. This means that all transformations $g \in G$, which can be written as products of these generators without m, have a continuous curve $\delta : I \to G^+$ such that $\delta(0) = Id_G, \delta(1) = g$. Therefore, evaluating at a point $x \in \mathbb{R}^2$ yields curves $\delta.x(t) = \delta(t)(x) \in \mathbb{R}^2$ that are induced by curves of transformations. This was also used in the component *BigBang Rubette* for composition in the Rubato Composer software environment [729]; see also Chapter 69. But the case of m does not work as is; in other words, mirroring is a non-gestural operation. In order to pass to the mirror of an object, one has to traverse the singular state of a flattened object in the mirror. The change of determinant sign is the hard point, so we are not in a state of overcoming this problem within the given space. We do not want to delve into the deep and metaphorically loaded topic of the mirror, but it is clear that the mystery of the mirror transformation must relate to the fact that there is no gesture, no continuous transition from the original to the mirror image. Vampires have no reflection in mirrors, and superstition is abundant with mirrors. Some "imaginary process" must be happening when we switch to the mirror world.

There is a well-known intuitive solution of the mirror problem, which you may find whenever you ask a person to describe what movement is the reflection of a plane figure at a line in the plane: He would immediately make that movement the one of leafing a book's page. Leafing turns the original figure to its mirrored version. The point is that instead of mirroring x to $-x$, it lifts it into a new dimension and rotates the point in this dimension until it comes down to $-x$. This procedure is more accurately described by complexification of real vector spaces. In the one-dimensional case $\mathbb{R}$, the mirroring $m(x) = -x$ is embedded in the Gaussian complex number plane $\mathbb{C} \xrightarrow{\sim} \mathbb{R}^2$. Here, we have the rotation defined by multiplying x by the complex unitary number $e^{i\theta}$. The image $x.e^{i\theta}$ is the vector x looking in direction of $e^{i\theta}$. If we consider the curve $\gamma(t) = e^{i\pi t} \in \mathrm{GL}_2(\mathbb{R}), t \in I$, then the evaluated curve $\gamma.x(t) = \gamma(t)(x) = x.e^{i\pi t}$ rotates x in a half circle to $-x$. This means that the mysterious mirroring has been demystified by an inoffensive gestural curve through complex numbers. And halfway on that curve we have its imaginary position $\gamma.x(1/2) = i.x$, the purely imaginary position of the curve, where its real projection vanishes.

This means that complex numbers solve the problem of the real singularity by lifting the mirror movement orthogonally to the real axis in an imaginary realm. It might be that one reads our description as a mystification of complex numbers, but the resolution of the negation $x \mapsto -x$ by a rotation in a new dimension is no overinterpretation of complex numbers. A strong argument for this "gestural" reinterpretation of negation is in fact provided by the proof of the fundamental theorem of algebra using fundamental groups in the Gaussian plane, see [569], for example. The fundamental theorem of algebra is the most important single thorem of algebra whose proof can be based upon the thoroughly gestural toolbox of algebraic topology.

This being so, if we are given any transformation $h \in \overrightarrow{GL}_n(\mathbb{R})$, then we may complexify it, which means that we write $h = T^s \circ \eta, s \in \mathbb{R}^n, \eta \in \mathrm{GL}_n(\mathbb{R})$ and then tensorize it with the complex number $e^{i\theta} \in \mathrm{GL}_2(\mathbb{R})$ as above and obtain the transformation $h \otimes e^{i\theta} = T^s \circ (\eta \otimes e^{i\theta}) : \mathbb{R}^n \otimes \mathbb{C} \to \mathbb{R}^n \otimes \mathbb{C} : x \otimes y \mapsto s \otimes 1 + \eta(x) \otimes y.e^{i\theta}$. The determinant of a tensor product $\eta \otimes \kappa$ of linear maps $\eta \in \mathrm{GL}_u(\mathbb{R}), \kappa \in \mathrm{GL}_v(\mathbb{R})$ being $det(\eta)^v det(\kappa)^u$, we have $det(h \otimes e^{i\theta}) = det(\eta)^2 det(e^{i\theta})^n = det(\eta)^2 (1)^n > 0$. This means that complexification of any transformation h with the rotation $e^{i\theta}$ in $\mathbb{C}$ turns it into a transformation $h \otimes e^{i\theta} \in \overrightarrow{GL}^+_{2n}(\mathbb{R})$. In particular, if $\gamma : I \to \overrightarrow{GL}^+_n(\mathbb{R})$ is a positive curve, then we obtain a curve $\gamma \otimes e^{i\pi ?} : I \to \overrightarrow{GL}^+_{2n}(\mathbb{R}) : t \mapsto \gamma(t) \otimes e^{i\pi t}$ such that its value starts at $t = 0$ with $\gamma(0) \otimes 1$ and ends at $t = 1$ with $\gamma(1) \otimes -1 = -1\gamma(1) \otimes 1$. If n is odd, this yields a negative determinant transformation $-1\gamma(1)$.

For even n, this does not work directly, but one may then select a direct decomposition $\mathbb{R}^n = V \oplus W$ with odd dimension $dim(W)$. Then we take again $\mathbb{R}^n \otimes \mathbb{C} = V \otimes \mathbb{C} \oplus W \otimes \mathbb{C}$, but this time apply the complex rotation only to the second summand $W \otimes \mathbb{C}$. Denote this restricted rotation by $e^{i\pi ?}|W$. Then if $\gamma : I \to \overrightarrow{GL}^+_n(\mathbb{R})$ is a positive curve (positive determinants), we obtain a curve $\gamma \otimes e^{i\pi ?}|W : I \to \overrightarrow{GL}^+_{2n}(\mathbb{R}) : t \mapsto \gamma(t) \otimes e^{i\pi t}|W$ such that its value starts at $t = 0$ with $\gamma(0) \otimes 1$ and ends at $t = 1$ with $\gamma(1) \otimes -1|W = -1|W\gamma(1) \otimes 1$, where the latter means that -1 is only applied to the subspace W coordinates, and thus yields a negative determinant transformation. This second case is not as invariant as the first one for odd n since one has to select a direct decomposition $\mathbb{R}^n = V \oplus W$, but the variety of choices offers a strong tool for turning general curves of transformations with positive determinants into gestural curves of opposite determinant signature. To terminate these constructions on the curve level, we add the reversed construction, which takes the positive curve $\gamma : I \to \overrightarrow{GL}^+_n(\mathbb{R})$, but tensorizes it with the reversed complex rotation, i.e., $\gamma \otimes e^{i\pi(1-?)}$ in the odd-dimensional case, and $\gamma \otimes e^{i\pi(1-?)}|W$ in the even-dimensional case.

This construction is easily generalized to the level of general gestures. To do so, we have to display the signatures of all the vertices of the gesture's skeleton Γ, since we want to define curves that change the determinant sign via the preceding complexification technique. More precisely, we suppose that we are given a signature for each vertex, which means that we have a digraph morphism $\sigma : \Gamma \to Sig$, where Sig is the complete digraph with two vertices $\oplus, \ominus$ (positive, negative). The fiber of $\ominus$ is the set of vertices with negative determinant, the complementary fiber of $\oplus$ is the set of vertices with positive determinant. This means that the arrows mapping to $\ominus \xrightarrow{-+} \oplus$ are those from a transformation with negative determinant to a transformation with positive determinant, etc. The complexification tools are only needed for the $+-$ and $-+$ arrows. The $+-$ case refers to the maps $\gamma \otimes e^{i\pi?}$ or $\gamma \otimes e^{i\pi?}|W$ according to the dimension being odd or even, whereas the $-+$ case refers to the maps $\gamma \otimes e^{i\pi(1-?)}$ or $\gamma \otimes e^{i\pi(1-?)}|W$. The case $++$ takes just the positive curves as is, tensored with the identity on $\mathbb{C}$, while the case $--$ takes a negative determinant copy of the positive curve. It is not relevant which negative copy we take, as long as the bijection $\mathrm{GL}_n^+(\mathbb{R}) \sim \mathrm{GL}_n^-(\mathbb{R})$ of negation is fixed once for all. The essential point is that all these complexified transformations start and end on transformations that leave the real subspace $\mathbb{R}^n \otimes 1$ invariant. The intermediate transformations however map $\mathbb{R}^n \otimes 1$ to a more general subspace of $\mathbb{R}^n \otimes \mathbb{C}$ as shown in Figure 62.2:

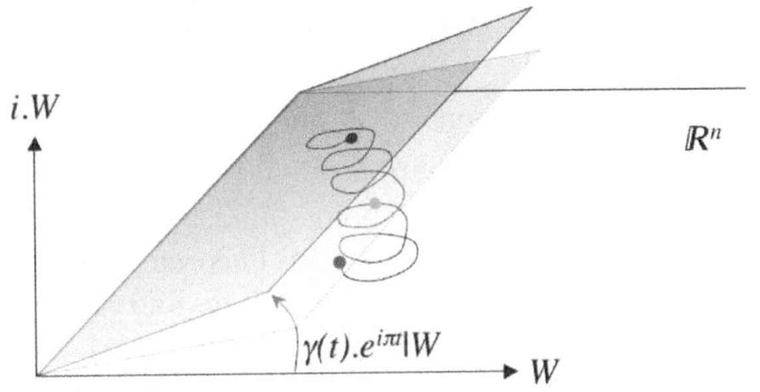

Fig. 62.2. The "leafing" transformation of a point in the original real space into complexified space as a function of the complex factor, eventually producing a change of the determinant's signature.

With this general method in mind, we now have to deal with the transformation of a discrete curve corresponding to a transformation (u, g, v) in $[f]$. In view of the above complexification method, we may concentrate on the case of $u = Id, det(v) > 0$. Our plan is to construct a continuous curve of transformations $\gamma : I \to \overrightarrow{GL}_n^+(\mathbb{R})$ from Id to v. The shifting part being trivial, we may focus on $v \in \mathrm{GL}_n^+(\mathbb{R})$. From matrix theory it is well known that $\mathrm{GL}_n(\mathbb{R})$ is generated by subgroups isomorphic to $\mathrm{GL}_2(\mathbb{R})$. So we can write v as a product of transformations v_i affecting only two coordinates. This reduces the problem to $n = 2$. We also have $v = \begin{pmatrix} det(v) & 0 \\ 0 & 1 \end{pmatrix}.s, s \in \mathrm{SL}_2(\mathbb{R})$. But $\mathrm{SL}_2(\mathbb{R})$ is generated by transvections $u(b), b \in \mathbb{R}$, and the 180° rotation w [578, XI, §2]:

$$u(b) = \begin{pmatrix} 1 & b \\ 0 & 1 \end{pmatrix}, w = \begin{pmatrix} 0 & 1 \\ -1 & 0 \end{pmatrix}$$

which both are endpoints at $t = 1$ of countinuous curves $\delta(t) = u(tb), \rho(t) = e^{i\pi t}$, the latter being the rotation as defined by the Gaussian plane. Therefore the curve

$$v(t) = \begin{pmatrix} (1-t) + t.det(v) & 0 \\ 0 & 1 \end{pmatrix}.\delta(t).\rho(t)$$

does the job in two dimensions. In short: $\mathrm{GL}_n^+(\mathbb{R})$ is arcwise connected, and we have just given a constructive proof thereof involving standard generators. In short, given any $v \in \mathrm{GL}_n^+(\mathbb{R})$, we find a continuous curve $v(?) : I \to \mathrm{GL}_n^+(\mathbb{R})$ with $v(0) = Id, v(1) = v$. Together with the signature changing tools discussed above (using the signature morphism σ), we have this theorem:

Theorem 2. *If $\delta : \Gamma \to \overrightarrow{GL}_n(\mathbb{R})$ is a diagram of (non-singular) affine transformations on $\mathbb{R}^n$, then there is a gesture $\delta \otimes \mathbb{C} : \Gamma \to [Id_{\mathbb{R}^n \otimes \mathbb{C}}]$, whose morphisms for the extremal ∇-morphisms $(0,1)$ of the curves $\delta \otimes \mathbb{C}(a)$*

of arrows a in Γ are the transformations $(Id_{\mathbb{R}^n\otimes\mathbb{C}}, \delta(a))$. *The morphisms between curve parameters s and t, $s \leqslant t$, are the factorization quotients*

$$\delta\otimes\mathbb{C}(a)(t)\circ(\delta\otimes\mathbb{C}(a)(s))^{-1}.$$

This is not a deep theorem, but it enables the extension of discrete gestures associated with diagrams in real n-space gestures to continuous gestures that may have to traverse complexification but start and end at the given real transformations. This is a general theorem, which just guarantees the said extensibility and stresses the fundamental role of complex numbers. But when thinking about Lewin's question concerning the characteristic gestures, although this might be the generic, it is not necessarily the characteristic one in n-space. We shall discuss this issue in Section 62.4.3.

Coming back to the evaluation map ϵ from Example 73, we also have an evaluation functor $\epsilon : \mathbb{R}^n\otimes\mathbb{C}\times\overrightarrow{GL}_n(\mathbb{R})\otimes\mathbb{C}\to\mathbb{R}^n\otimes\mathbb{C}$ on topological categories. When we take a gesture $\gamma:\Gamma\to\mathbb{R}^n\otimes\mathbb{C}\times\overrightarrow{GL}_n(\mathbb{R})\otimes\mathbb{C}$, ϵ yields a gesture of points in $\mathbb{R}^n\otimes\mathbb{C}$, and if the initial and terminal values of the curves of transformations of γ leave $\mathbb{R}^n$ invariant, the gestures in $\mathbb{R}^n\otimes\mathbb{C}$ also start and end in real points.

Example 74 A prototypical example would consist of a network of points in $\mathbb{R}^n$ connected by affine transformations, i.e., a diagram of points and transformations $\delta:\Delta\to\mathbb{R}^n$, such that for an arrow $x\xrightarrow{a}y$ in Δ, we have a non-singular affine transformation $\delta(a)$ of points $\delta(x),\delta(y)\in\mathbb{R}^n$ with $\delta(a)(\delta(x))=\delta(y)$. Focusing on the transformations, one therefore has a diagram as in Theorem 2, which has all its curves starting at the identity and ending on the different $\delta(a)$ for arrows a of δ. So on transformations, the diagram is a star-shaped one with the identity as center and radiating to each of the $\delta(a)$. However, on the different starting points $\delta(x)$, the star is uncoupled in order to be able to transform all the $\delta(x)$ in the particular curves that traverse complex spaces when determinant signs are changed to between 1 and the signature of $det(\delta(a))$. This means that we are given pairs $(\delta(x),\delta\otimes\mathbb{C}(a))$ with variable transformations (i.e., curves) and fixed points that are transformed according to the transformation curves by $t\mapsto\delta\otimes\mathbb{C}(a)(t)(\delta(x))$. The straightforward generalization is to define non-constant curves in the points, too, i.e., $t\mapsto\delta\otimes\mathbb{C}(a)(t)(\delta(x)(t))$, which comprises the two extremal cases of the purely transformational curves and the purely "topological" point curves with constant transformation, usually the identity.

Example 75 If S^1 denotes the unit circle group, which we may view as an extension of the pitch class group, consider the topological group $G = T^{S^1}\rtimes\mathbb{Z}_2$ generated by translations $T^t, t\in S^1$, and the reflection $-Id$ (the generator of $\mathbb{Z}_2$). This group has two connected components, namely $G^+ = T^{S^1}$ and $G^- = T^{S^1}.-Id$. As for the general affine group, this group defines a category with one object, and we also have $[Id]\xrightarrow{\sim}G$, the topological space category with morphisms $(d,c)\in G\times G$ representing the quotient $g = c\circ d^{-1}$. As for the general affine group, the morphisms $Id\to T^t$ can easily be extended to continuous curves, while for $Id\to T^t.-Id$ this is not possible for the same reason as before. We may however resolve the conflict by again adding a dimension and embedding S^1 in the sphere S^2 as one of its meridian circles. What was previously done by the rotation in $\mathbb{C}$ now works by the half circle rotation around the polar axis or another axis through the embedded S^1.

62.4 Modulations in Beethoven's "Hammerklavier" Sonata op.106/Allegro: A Gestural Interpretation

The following section is not the first occasion where gestural aspects of Beethoven's compositions have been discussed; see Robert S. Hatten's study [446], or Jürgen Uhde's and Renate Wieland's books [1068, 1067], for example. Our discussion however differs from earlier investigations in these two points:

- To begin with, we are applying the previous categorical theory of gestures and do not stick to the more metaphorical and intuitive usage of the term "gesture" in previous studies. Of course, this is also a risky enterprise since many statements, which may be acceptable or valid on those more intuitive levels of conceptualization, might become questionable, dubious or even untenable when made mathematically explicit.

- Second, we focus on very delicate modulatory processes in the Allegro movement of Beethoven's op.106, see Section 28.2 for a detailed discussion. These are known to have quite non-standard appearances, partly breaking with standard combinations, such as the modulation $B_\flat$-major $\rightarrow$ G-major to the secondary theme in the exposition, where the modulation to the dominant is expected by standard sonata theory [659], and partly because some so-called "catastrophe" modulations deviate from standard modulation processes altogether.

We shall not reconsider a basic discussion of these modulatory processes, but rely on the established methods and results as exposed in [714, ch.28.2]. However, the gestural aspects of these modulations will open considerations of dynamical nature that do not rely solely on these methods and results. We therefore hope that the following discussion is also useful for a basic discourse on gestures in modulatory processes.

Why is this a desirable topic? The argument is that a purely structural analysis of modulatory processes (among others) may fail to capture the energetic understanding, the dramatic tension of the musical deployment. We are not claiming here that gestural analysis is comprising all such aspects, but it seems worthwhile to approach those energetic and dramatic tensions by gestural dynamics since the theory of gestures is an ideal mediator between static structures and energetic processes without having to recur to psychological, narrative, or other extra-musical categories. Our hope is nevertheless that gestural considerations might *eventually* converge to a fairly complete understanding of what is called the "dramatic content" of absolute music, such as Beethoven's late works.

One word about the *intrinsic usage of gestures* in the following analysis, as opposed to transformational structures. Is it really necessary to work with gestures? Couldn't one as well restate most if not all those gestural reflections in terms of transformational (hyper)networks? It is true that some of the following gestures (e.g. the gestures α_i and α'_j shown in Figure 62.5) seem to be "overdressed" versions of transformational networks. There are (at least) three arguments against such a suspicion:

- Gestures are completely different objects from networks. Intuitively speaking, networks only deal with start and end points of gestures. Also, hypergestures are generalized homotopies, which networks are not. Many of the following hypergestures are intrinsically continuous constructions, which require different, namely topological, technical tools than transformational networks, which are essentially built upon affine algebraic transformations. This is particularly dramatic in the context of the complexification gestures, which move *as* curves out of the real spaces into complex superspaces.
- Lewin's own unsolved dilemma is that he imagined continuous movements (his dancers!), but worked with agebraic (Fregean) transformations. What we offering here (and in the paper [720] with Moreno Andreatta) is nothing less than the one-to-one construction of Lewin's dreams in terms of precise mathematics.
- The very language of gestures opens a style of thoughts and a paradigm of understanding that the transformational paradigm would not have offered. In mathematics, the modern conceptual linear algebra opened so many new ideas that would never have been conceived of in terms of old-fashioned matrix calculus. Of course, once you have the idea, it is possible to translate it back into the old language, but this restatement is only possible *ex post.* Or, to put it into Lewinian dance language: How can you understand the dancer's touch point configuration with the dance floor if you are not told how he or she is connected by his or her real movements?

62.4.1 Recapitulation of the Results from Section 28.2

The modulation architecture in the Allegro movement of op. 106 is derived from a model of tonal modulation that uses inversional and transpositional symmetries on pitch class sets as "modulators", i.e., as operators, which transform pitch class sets (in 12-temperament) into each other. The tonalities in this model are the triadic interpretations $X^{(3)}$, i.e. coverings by the seven standard triadic degree chords $I_X, II_X, \ldots VII_X$ of the twelve diatonic scales $X = C, D_\flat, D, E_\flat, E, F, \ldots B_\flat, B$. According to a fundamental hypothesis on this composition, a hypothesis that is derived from classical analyses by Erwin Ratz and Jürgen Uhde (see Section 28.2), the system of possible modulators is the automorphism group $Aut(\mathfrak{M}_0)$ of the diminished seventh chord $\mathfrak{M}_0 = \{d_\flat, e, g, b_\flat\}$. This group partitions the set of tonalities into two orbits, the eight element orbit $W = Aut(\mathfrak{M}_0).B_\flat^{(3)}$ (the "world") consisting of the tonalities of scale set $\{B_\flat, D_\flat, E, G, A, C, E_\flat, G_\flat\}$, and the four element orbit $W^* = Aut(\mathfrak{M}_0).D^{(3)}$ (the "antiworld") consisting of the tonalities of scale set $\{D, F, A_\flat, B\}$.

This implies that modulations according to the given model are only admitted within W or within W^*, since no modulators are available in order to switch between these worlds. It turns out that what Uhde has coined catastrophe modulations are exactly those when Beethoven switches between world W and antiworld W^*. And here, the "responsible" diminished seventh chord appears with a nearly obsessive density, annihilating any melodic or tonal framework. All other modulations, within the world or within the antiworld, obey strictly the modulation rules provided by our model. Moreover, the modulators in these cases not only act as hidden symmetries but are also visible as symmetries between note groups that are within the modulating score segment, see Section 28.2.

Although the above results are describing the abstract modulatory structures and also the modulator symmetries in a strikingly precise way, which by far exceeds the predictive power of general modulation theories, the dramatic character of these modulations is not represented. In fact, much more is happening here than a verifiable instantiation of the model's abstract characteristics. The richness of the modulation dynamics has an impact that cannot be comprised by transformational diagrams connecting groups of notes. And this precisely, since diagrams incorporate no real movement, because their arrows are just as "cartesian" as plain set theory. We have discussed this topic in detail in the theoretical part of this chapter and in previous work [719]. This is the reason why we propose drawing a gestural picture of these modulation processes, which transcends the results as described in Section 28.2. So the following discussion is not about the previous analysis and its model, but about the added value that gestural reflections can contribute to the understanding of the note-wise embodiment of the composer's ideas following his famous statement[4]: "Was der Geist sinnlich von der Musik empfindet, das ist die Verkörperung geistiger Erkenntnis."

62.4.2 The Modulation $B_\flat$-major $\rightsquigarrow$ G-major Between Measure 31 and Measure 44

The first modulation, $B_\flat$-major $\rightsquigarrow$ G-major, in the transition (measures 39-46) to the second subject could in principle be performed by use of a "pedal modulation" [948]. We do however not encounter this modulation, but 'merely' a sequence of $VII_{G\text{-major}}$-degrees whose top notes are shifted by minor thirds from each other, i.e., exactly the situation of the pivot VII and the third translation, as predicted by the modulation with restricted modulators (Section 28.2.2).

This compact description from Section 28.2.4 however does not grasp the elaborate note process around that abstract fact of the $VII_{G\text{-major}}$ degree. This process consists of four groups, (A) measures 31 to 34, (B) measures 35-36, (C) measures 37-38, (D) measures 39-44. We do not discuss the concluding figure in measures 45-46, where the modulation is already terminated, and refer to [718, 9.2.1] for that matter. The entire process is typical for many of the modulations in this movement: It seems as if there were obstructions to a fast and easy modulation, which have to be surmounted. In the present case, the fanfare of part (B) is repeated in the subsequent part (C), but the second appearance first neutralizes $I_{B_\flat\text{-major}}$ to the simple note d on the third beat of measure 37, which is the third of $B_\flat$-major and the fifth of G-major. The next chord then replaces the $e_\flat$ from the original fanfare by $f_\sharp$ and creates $V_{G\text{-major}} = IV_{A\text{-major}} = I_{D\text{-major}}$. The movement $e_\flat \mapsto f_\sharp$ is a minor third (in terms of chromatic pitches and pitch classes, the present model is not based upon tonal alterations). This short formula is ambiguous in terms of which symmetry might have caused it. We have two candidates: $f_\sharp = T^3(e_\flat) = T^9.-1(e_\flat)$, transposition or inversion. The general modulation model with unrestricted modulators would yield the inversion as modulator, in fact, $T^9.-1.B_\flat^{(3)} = G^{(3)}$. The fanfare of part (C) could therefore also result from the inversional modulator acting upon $e_\flat$. But $T^9.-1 \notin Aut(\mathfrak{M}_0)$. Therefore only T^3 can transform $B_\flat^{(3)}$ into $G^{(3)}$. But this is not clear in part (C). A modulation process

[4] What the spirit perceives through the senses from music, is the embodiment of spiritual insight.

Fig. 62.3. The modulation process $B_\flat$-major $\rightsquigarrow$ G-major between measure 31 and measure 45.

has happened, but neither is it evident which symmetry was applied, nor is it terminated since the cadence part in the target tonality is not achieved in part (C). This state of ambiguity is expressed by the fermata in measure 38. It is a moment of hesitation of uncertainty: What happened, where are we? Could we really go on in G-major and step over directly to the last quarter of measure 46? Playing this shortened version sounds like not having digested the process, like stepping into a new tonality in a haphazard way without having made clear how we left the old one.

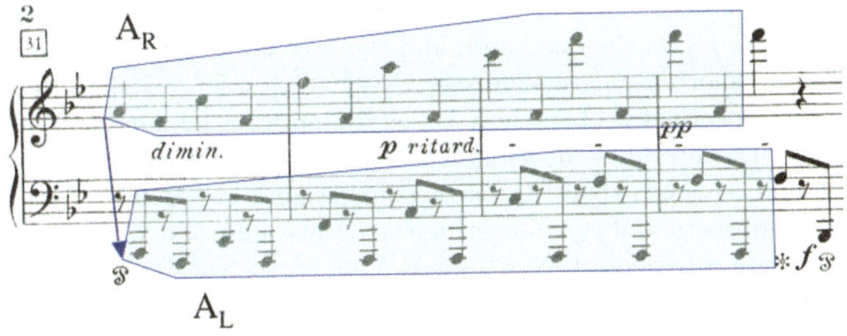

Fig. 62.4. The echo hypergesture preceding the modulation $B_\flat$-major $\rightsquigarrow$ G-major.

Of course, the plain appearance of degree $VII_{G\text{-major}} \subset V^7_{G\text{-major}}$ makes the target clear and cadences the new tonality. But again, this would also be true if we made that brute connection to the last quarter of measure 46, since that one initiates an arpeggiated $VII_{G\text{-major}}$. The point is that the stopping movement terminated by the fermata in measure 38 was not only defined there, but started much earlier in part (A). Harmonically, this part is a repeated arpeggio of $V_{B_\flat\text{-major}}$, terminating on the descending fifth step $f \mapsto b_\flat$ at the end of measure 34. This is by no means remarkable. But the shape of the arpeggio is! To begin with, part (A) splits into two subsets A_R and A_L, whose onset and pitch relate by a downward shifting,

$A_L = T^{(1/8,-12)}.A_R$, which creates a deeper and weaker (eighth notes) echo A_L of the right hand part A_R (time in 1/8, pitch in semitone units). We shall henceforth focus on the parameters onset and pitch and position these two parameters in $\mathbb{R}^2$, onset for the first, pitch for the second coordinate. Supposing for the moment that A_R and A_L are gestures, this echo turns out to be the endpoint of a hypergesture curve $A(t) = T^{t(1/8,-12)}.A_R, t \in I$ that is the evaluation of the curve $T^{t(1/8,-12)}, t \in I$, of transformations at A_R, see Figure 62.4.

This descending echo is the outer shape of a movement that becomes already visible in the initial gesture framed by A_R. How is this gesture constructed? Refer to Figure 62.5 for the following discussion. To begin with, we have seven small descending interval gestures $\alpha_1, \alpha_2, \alpha_3, \alpha'_1, \alpha'_2, \alpha'_3, \alpha_4 \in \uparrow \overrightarrow{@}\mathbb{R}^2$, which are induced by descending translation curves T_i along the vectors $v_1 = (1/4,-4), v_2 = (1/4,-7), v_3 = (1/4,-12), v'_1 = (1/4,-16), v'_2 = (1/4,-19), v'_3 = v_4 = (1/4,-24)$ to the same periodically repeated f, the fifth of the given tonality. Since we have this note as a fixed reference point, we view the translations as being the dual curves $T_i = S_i^*$ to the translation curves S_i associated with $-v_i$, i.e., $S_i(0) = Id, S_i(1) = T^{-v_i}$. All T_i are then evaluated at the seven instances of f, the evaluation starting at the higher interval note and ending at f.

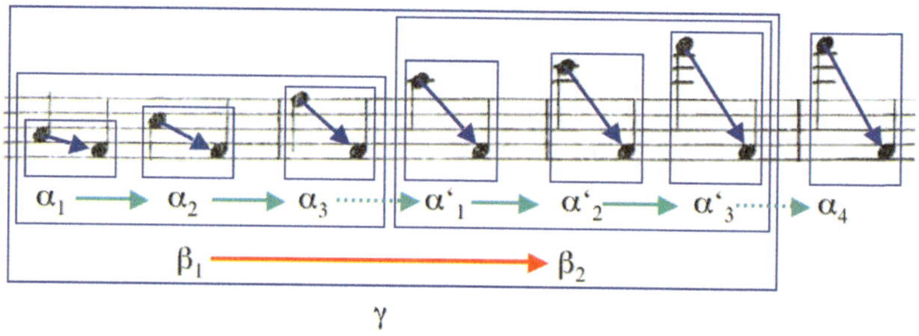

Fig. 62.5. The right hand hypergesture underlying A_R.

Recall from Section 61.6.1 that we denote by $\uparrow^n$ the digraph consisting of $n+1$ vertices, and having one arrow from vertex i to vertex $i+1$ for all $i = 0, 1, 2, \ldots n-1$. Then we have a hypergesture $\beta \in \uparrow^6 \overrightarrow{@} \uparrow \overrightarrow{@}\mathbb{R}^2$ that deforms α_1 into α_2, etc., and α'_3 into α_4. We leave it to the reader as an exercise to describe these deformations explicitly in terms of homotopies of translation curves over the category of affine transformations on $\mathbb{R}^2$. The hypergesture's β projections $p_1, p_2 : \uparrow^6 \overrightarrow{@} \uparrow \overrightarrow{@}\mathbb{R}^2 \to \uparrow^6 \overrightarrow{@}\mathbb{R}^2$ via the head and tail maps on $\uparrow$ yield the periodic gesture of successively shifted f's on the one hand, and the ascending upper voice gesture on the other. But there is more: We look at the shorter hypergestures $\beta_1, \beta_2 \in \uparrow^2 \overrightarrow{@} \uparrow \overrightarrow{@}\mathbb{R}^2$, restrictions of β to the three vertices $\alpha_1, \alpha_2, \alpha_3$ and $\alpha'_1, \alpha'_2, \alpha'_3$, respectively. Then they are endpoints of a hypergesture $\gamma \in \uparrow \overrightarrow{@} \uparrow^2 \overrightarrow{@} \uparrow \overrightarrow{@}\mathbb{R}^2$. It projects to the f hypergesture and shifts the subgesture of β_1 built from the three f's in time (by 3/4) to the corresponding subgesture of β_2. The subgesture of β_1 connecting a, c, f is shifted by an octave and time (by 3/4) to the corresponding subgesture of β_2 connecting the octave shift of a, c, f. Putting everything together, including the echo, we obtain a hypergesture $\gamma^+ \in \uparrow \overrightarrow{@} \uparrow \overrightarrow{@} \uparrow^2 \overrightarrow{@} \uparrow \overrightarrow{@}\mathbb{R}^2$. Observe that such hypergesture constructions go beyond hypernetwork constructions since they are intrinsically topological.

Fig. 62.6. The beginning of part (D) shows an ascending twofold octave echo gesture on the fifth d.

So we are facing a hypergesture γ^+ that is traced on the "dance floor" of the score by part (A), and which is a strong and multilayered expression of the descending movement towards the fifth (for the dominant) of $B_\flat$-major. Everything is stopping at this point, also supported by the "ritardando" and "diminuendo" performance directions. The process's period, expressed in the fanfare part (B), opens the transformed fanfare in part (C) as discussed above. But the mentioned transformational ambiguity needs to be resolved, and this is achieved in part (D) by use of a counter-gesture corresponding to the hypergesture of part (A). How is this one structured, and how does it respond to the halting hypergesture of part (A)? The first two measures, 39 and 40, show a quadruple appearance of degree $V_{G\text{-major}}$ as initiated in (C). After the fermata, a new movement is initiated, as shown in Figure 62.6. This is similar to the octave descending echo gesture in part (A). However, now it is ascending in two octave steps $X_1 \to X_2 \to X_3$ (we refrain from a more precise gesture description, since this is straightforward here). This opening gesture towards the new tonality indicates that the moment of halting and cadential termination is over: the movement is now reversed, ascending towards new horizons, new skies.

It is not surprising that the following configuration of gestures appearing in part (D) is a complete mirror of the configuration encountered in part (A). The hypergestural anatomy is shown in Figure 62.7. Let us

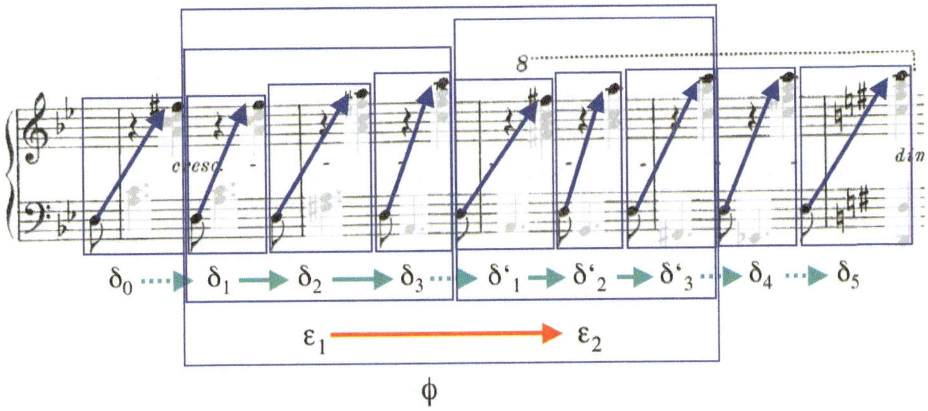

Fig. 62.7. The hypergestural configuration in part (D) mirrors the one from part (A).

describe what this "mirroring" looks like in detail. To begin with, we have a strong similarity of hypergestural configurations. We first have nine small ascending interval gestures $\delta_0, \delta_1, \delta_2, \delta_3, \delta'_1, \delta'_2, \delta'_3, \delta_4, \delta_5 \in \uparrow \overrightarrow{@}\mathbb{R}^2$, which again are anchored in the periodically repeated fifth d of the given tonality G-major and add up like β in (A) previously to a hypergesture $\epsilon \in \uparrow^8 \overrightarrow{@} \uparrow \overrightarrow{@}\mathbb{R}^2$. The three gestures $\delta_1, \delta_2, \delta_3$ and $\delta'_1, \delta'_2, \delta'_3$ define hypergestures $\epsilon_1, \epsilon_2 \in \uparrow^2 \overrightarrow{@} \uparrow \overrightarrow{@}\mathbb{R}^2$, much as β_1, β_2 did in the previous situation. And these two hypergestures also are connected to a big hypergesture ϕ, similarly to the above hypergesture γ. And again, the gestures δ_i, δ'_i in ϕ are anchored on the fifth and have the upper voice showing a characteristic chord. While in part (A) this was the arpeggiated $V_{B_\flat\text{-major}}$, here it is the arpeggiated modulation pivot $VII_{G\text{-major}}$ (which is the upper triad in the score's $V^7_{G\text{-major}}$). Moreover, as already mentioned previously in this section, the pivotal $VII_{G\text{-major}}$ appears arpeggiated in steps of two ascending minor thirds, which again stresses the third transposition symmetry against the inversion that would have modulated this configuration without restricted modulators.

62.4.3 Lewin's Characteristic Gestures Identified?

The situation is full of similarities above the level of the elementary interval gestures in (A) and in (D). But these are not similar. They are the carriers of what had been called "mirroring" above. What does this look like? Let us take the paradigmatic example of the two gestures α_1 in (A) and δ_1 in (D). We had associated gesture α_1 with the descending onset-time shift transformation $t_A = T^{(1/4,-4)}$, and δ_1 can be associated with the ascending onset-time shift transformation $t_B = T^{(1/4,28)}$. The straightforward curve of transformations is then $T_1 = S_1^*$ for α_1 and $T_B(t) = T^{t.t_B}, t \in I$, for δ_1. Evaluating these transformational gestures (curves) at the anchor points $f, d \in \mathbb{R}^2$ yields two curves $T_1.f, T_B.d \in \uparrow \vec{@}\mathbb{R}^2$. Although these curves of transformations and of evaluated points are mathematically straightforward, they are not evident per se. Why should the curve connecting start and end be a linear trajectory? One could also select $T'_B(t) = T^{t^2.t_B}$, for example. The trace would be the same. This simple example shows that the gestures on the score's dance floor are multiple. Only the points of contact are unchanged. The difference on the gestural level pertains to the interpretative effort (the "aesthesic" position in Valéry's wording), not to the work's neutral level, see Section 2.2.2 or [704]. On the gestural interpretation level, there is a manifold of solutions beyond the generic one.

A priori, there are essentially two levels where a solution may be sought out: the evaluated points level, i.e., the curves in $\uparrow \vec{@}\mathbb{R}^2$, or the transformational level, i.e., the curves of transformations in $\uparrow \vec{@}[Id_{\overrightarrow{GL}(\mathbb{R}^2)}]$. (We put aside the "mixed" solutions with variable points and transformations in $\uparrow \vec{@}[Id_{\mathbb{R}^2 \times \overrightarrow{GL}(\mathbb{R}^2)}]$ as already described in Example 74.) The difference of these approaches lies in their semantic power, which is expressed in the mathematical manifold which they describe. The simple curve level $\uparrow \vec{@}\mathbb{R}^2$ offers a big topological space, but no a priori semantics. Any curve would do, be it induced by a physical hand movement rationale as developed in [772], or by any poetic phantasy of spatial curves. In contrast, the transformational level $\uparrow \vec{@}[Id_{\overrightarrow{GL}(\mathbb{R}^2)}]$ defines a repertory, which is more expressive as it refers to curves of transformations, such as rotational curves, or curves of transvections, which, for example, may be loaded by musical meaning. This situation is much the same as for transformational theory, where two determined notes are connected by an affine transformation out of a set of transformations, which is essentially the stabilizer subgroup of the point of departure, and where the selection of an element of that stabilizer expresses a semantic choice—except that the manifold of curves is much larger than it is for that theory.

Let us look at the expressive richness on the hypergesture level, which defines curve gestures between gesture α_1 and gesture δ_1. One first interpretative action is the earlier defined dual gesture construction $T_1 = S_1^*$ for α_1. It exchanges the start and the end of the gesture and so doing means that the perspective is taken from the higher note $a = T_1(0)$ towards the anchor note $f = T_1(1)$, but the underlying logic stems from the dual gesture S_1, evaluated to the dual curve α_1^* that views a as the endpoint of a movement starting in f. Duality is interpreted as going backwards, coming back to the root f, although we are moving forwards in terms of the onset parameter. This is the first part of the mirroring operations. It is not a gesture, but a reinterpretation of the given gesture's curve parameter. Next, we want to compare α_1^* to δ_1. They are both ascending gestures, and they do so from the same scale degree, the fifth (call it dominant, if that matters) of the given tonality. The first looks backwards, the second forwards in time. Looking into the past and then into the future is a simple dramatic logic of the gestural trajectory. *We view this duality argument as a first example of thinking about Lewin's "characteristic gesture" in the sense that the gestural operation is a characteristic for the expression of a specific musico-logical thought.*

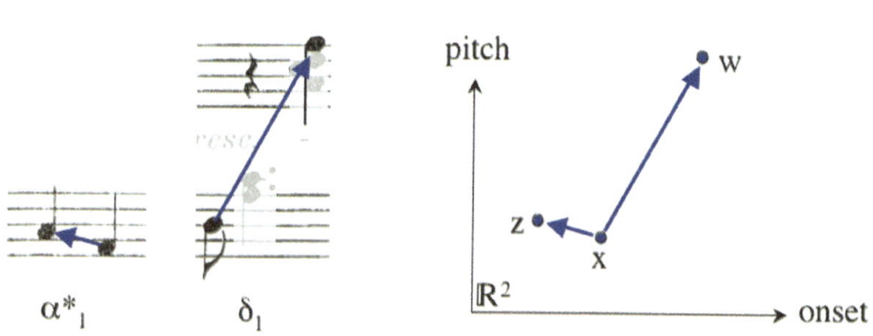

Fig. 62.8. The elementary gestures and their abstract representation for the mirroring operations.

There are several options to connect α_1^* to δ_1 in a hypergestural curve in order to construct a further differentiation of Lewin's "characteristic" movements. To ease the formal setup, let us think about both gestures starting from the same point $x \in \mathbb{R}^2$ and ending in z for α_1 and in w for δ_1. So this prototypical representation has $\alpha_1^*(t) = T^{t(z-x)}.x$ and $\delta_1(t) = T^{t(w-x)}.x$, see also Figure 62.8.

On the level of transformations, we have the prototypical gestures $T_\alpha(t) = T^{t(z-x)}, t \in I$, and $T_\delta(t) = T^{t(w-x)}, t \in I$, representing α_1^* and δ_1. In order to connect them by a curve (in fact a homotopy), one may set $T_\alpha(s)(t) = T^{(Q(s)(z-x))t}\phi(s)$ where $\phi : I \to \mathrm{GL}_2(\mathbb{R})$ is a loop starting and ending at Id, and where $Q(s) = T^{s(w-z)}$. The left part of Figure 62.9 shows this construction when evaluated at x and with trivial

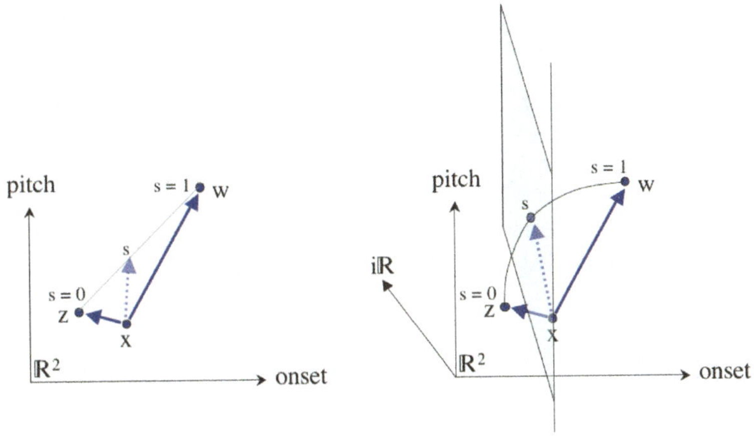

Fig. 62.9. The elementary gestures and their abstract representation for two mirroring operations.

ϕ. One may also refer to the complexification technique described above in Section 62.3 and set

$$Q(s) = \begin{pmatrix} e^{-i\pi s}((1+s) - s\Delta_1) & 0 \\ 0 & ((1-s) + s\Delta_2) \end{pmatrix}$$

with $\Delta_i = (w_i - x_i)/(z_i - x_i), i = 1, 2$, where $Q(s)$ rotates the onset part by $e^{-i\pi s}$ (clockwise) and produces an imaginary onset for $s \neq 0, 1$; see Figure 62.9, right part. It is fundamentally different from the first solution in that it *means* mirroring time instead of just pointing to continuously changing interval directions in the first choice. The difference in these gestures is that they express in characteristic ways the mirroring operation from the backwards oriented interval in (A) to the forwards oriented one in (D). These operations represent totally different musical thoughts. We would adhere more to the second choice in the sense that it is dramatic and coincides with the fermata: For a short moment, time becomes imaginary (here also imaginary in terms of complex numbers), and when we have transgressed that "higher sphere of pure imagination", we are heading for a new tonal region.

62.4.4 Modulation $E_\flat$-major $\rightsquigarrow$ D-major/B-minor from W to W^*

This modulation is a catastrophe in the sense of Uhde since it leads to the antiworld W^*. As we may recognize already from the score in Figure 62.10, measures 189-197 are of a dramatic shape. Any elaborate motivic or harmonic effort is postponed in favor of a pertinent rhythmical presentation of diminished seventh chords. An approach to modulation via the inversion between e and f (provided by the modulation Theorem 30 in Section 27.1.4), $U_{e/f} = T^9.-1$, $VI_{E_\flat\text{-major}} \mapsto I_{D\text{-major}} = U_{e/f}.VI_{E_\flat\text{-major}}$ (measures 189-192), fails; the resolution of all alteration signs indicates the exit from tonal space. We hear the "generator" of the catastrophe, the diminished seventh chord as such. The situation before the modulation is similar to that in the previous modulation, where we also moved down and stayed on the fifth until the tonic was reached at the end of measure 34. In measures 187-188, the dominant degree $V_{E_\flat\text{-major}}$ appears four times, each time

Fig. 62.10. The modulation process $E_\flat$-major $\rightsquigarrow$ D-major/B-minor between measure 189 and measure 197.

initiating a downward four-note motif $b_\flat, a_\flat, g, f$ that finally reaches the tonic and degree $I_{E_\flat\text{-major}}$ at the beginning of measure 189. In this measure a two-measure-periodic rhythmic duration sequence of multiples of eights, namely 1/8, 3/8, 1/8, 3/8, 1/8, 3/8, 1/8, 2/8 (followed by a 1/8 rest in the left hand), which concatenates twice the rhythm of the first four notes of the fanfare, is established and repeated without exception five times until measure 198, where the $I_{D\text{-major}}$ is reached. The rhythmical energy then breaks down to an even shorter rhythm, namely the very beginning rhythm 1/8, 3/8 of the fanfare.

The beginning of this rhythmical pattern also parallels the ambiguous situation in measures 37-38. There, it was shown that the transition from $e_\flat$ to $f_\sharp$ was ambiguous, being either an inversion under $T^9.-1$ or a transposition T^3. Since the admitted symmetries exclude the inversion, the transposition was left and also showed its pivotal seventh degree in the following measures. But now, the situation is significantly different. The same inversion $T^9.-1$ does in fact transform $E_\flat$-major to D-major (or B-minor for that matter). But this is forbidden as above. Moreover—and this is different here—there is no other symmetry in the group of modulators that would do the job. We are confronting two tonalities in different worlds, $E_\flat$-major $\in W$ and D-major $\in W^*$. As can be expected from our previous modulation, the transformation $e_\flat \to f_\sharp$ also appears in the transition from the third of $VI_{E_\flat\text{-major}}$ to the third of $I_{D_{\text{-major}}}$ in the action $T^9.-1(VI_{E_\flat\text{-major}}) = I_{D_{\text{-major}}}$ on the neigboring degrees in measures 190-191. The impossibility of applying this modulator coincides with the above rhythmic pattern. This time there is no modulator. And as opposed to the fanfare that works in our previous case, it cannot be completed, i.e., it is blocked in its salient initial stage.

62.4.5 The Fanfare

In order to discuss the gestural interpretation of this process, we have to investigate the fanfare in more detail. We shall lead this discussion in a less technical style regarding the gestural formalisms, because technicalities can easily be filled out by the attentive reader, and because the point here is less the technical than the semantic level enabled by our gestural toolbox.

As is evident from the previous description, we have to focus on the rhythmic structure of this gesturality. To do so, we look at the time coordinates of the fanfare: onset and duration, see Figure 62.11. They show a bipartite gestural anatomy. We recognize two groups of gestures: the two ascending arrows to the left, and the two horizontal arrows to the right. These groups correspond to the pitch-ascending part and the pitch-descending part in the fanfare.

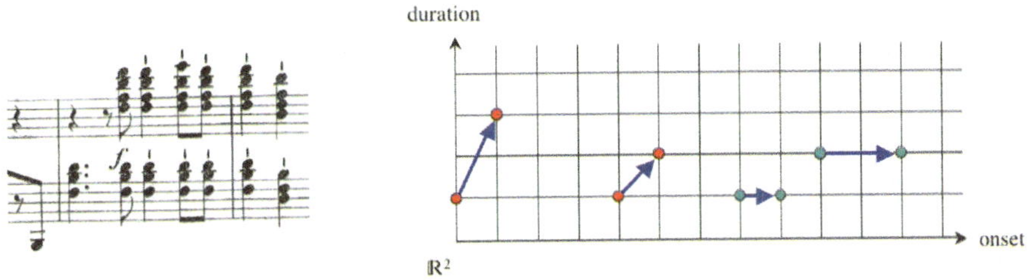

Fig. 62.11. The representation of the fanfare on the time-related plane of onset and duration.

So we view these structures as a gestural construction, which starts with the first ascending curve and deforms to the second ascending curve. This defines a hypergesture in the rhythm-space of onset and duration. In our understanding, the ascending character means that we address a downbeat, a halting energy. This elementary gesture (first arrow) is deformed to a second appearance (second ascending arrow). This deformation is shown as hypergesture ρ in the left lower part of Figure 62.12. This interpretation is ontologically non-trivial since it creates a continuous transition from the initial note to the second longer one, which amounts to imagining an entire curve of intermediate notes that succeed each other in infinitely near onset times and durations. This enrichment in fact fills out the empty time-space that is not denoted on the score by what in our musical imagination takes place while the first note is being heard/played. The hypergesture connecting the first and second arrow gestures is the connection of this first rhythmic step to the second in the same way, but conceptually and in the perceptive/performative level at a higher stage of imaginative coherence.

The first hypergesture ρ is followed by a second hypergesture σ, which deforms an arrow connecting two eighth notes to the arrow between two quarter notes. This time, the deformation of these arrows is not the hypergesture connecting a repeated halting movement, but expresses the halting movement of a regular succession of notes of same duration. It is not the repetition of a halting movement, but the halting of a repetitive movement: the roles of repetition and halting are exchanged. In order to connect these two hypergestures ρ and σ, we give two gestural transitions: First, the initial hypergesture ρ is rotated (in a rotative curve gesture) into the intermediate hypergesture ρ' (upper left corner in Figure 62.12). Then, ρ' is deformed into σ, but also replaced by its dual ρ'^* (not shown explicitly in the figure, since it just reverses the homotopy direction). This guarantees that the hypergesture moves from a lower duration to a higher one as required in σ. So we have a hypergesture of skeleton $\uparrow^2$ connecting ρ and σ. The other variant avoids duality, but uses a diagonal mirror operation to flip ρ into ρ'' in the upper right corner of Figure 62.12. Of course, this requires a complexification of the real 2-space, which we do not draw to keep the visualization simple. Again, from the intermediate ρ'', we deform down to σ by a similar transformation curve as that for the preceding case between ρ' and σ. And again, we have a hypergesture of skeleton $\uparrow^2$, this time generating σ from ρ via ρ''.

Again, we have different characteristics, which are addressed in these two paths from ρ to σ: The first keeps orientation (by a rotation in the onset-duration plane), but has to reverse curve time in the duality switch, while the second reverses onset and duration (through an imaginary rotation in complex numbers). According to the above statement that the second σ is not the repetition of a halting movement (which ρ was), but the halting of a repetitive movement. The exchange of onset and duration in the imaginary mirror path seems more to the point of this rhythmical construction. The dialectic pairing of ρ and σ in this interpretation resolves the repeated attempt to halt time in ρ by its hypergestural deformation to a completed

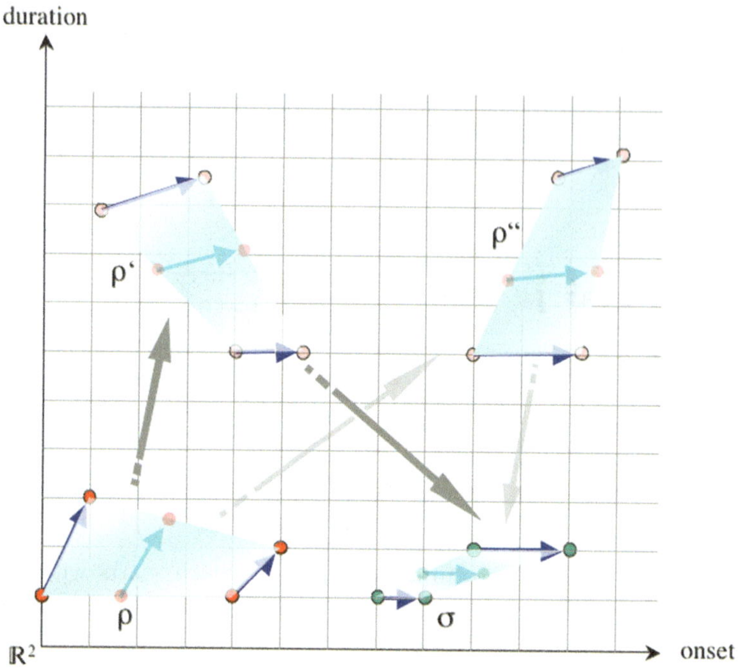

Fig. 62.12. The first hypergesture ρ is followed by a second hypergesture σ, which deforms an arrow connecting two eighth notes to the arrow between two quarter notes. We show the hypergestural connection between these two hypergestures.

repetition, and then into its halting slow-down. From this point of view, the repeated presentation of the fanfare's first part ρ in the catastrophe modulation perpetuates that internally already prototyped repeated halting and thereby expresses in an unfolding of the "idea in a nutshell" the failure to release the tension and to modulate in a well-structured way into the antiworld. The final reduction to the initial arrow of ρ in the ritardando measures 199-200 completes this failure and brings the energies to their exhaustion, the gesture dissolves.

62.5 Conclusion for the Categorical Gesture Approach

In this chapter, we have constructed a categorical framework of gestures, generalizing the topological approach from Chapter 61, and culminating in the construction of a gesture bicategory, which enriches the classical Yoneda embedding. This framework could be a valid candidate for the conjectured space X in the diamond conjecture (see Section 61.12). Future research will have to investigate typical algebraic categories of modules or topoi above module categories, or word monoids (in particular regarding scale theory [274], which deals a lot with factorization!), which are representative of algebraic music theories. We have discussed first applications of this framework for topological groups, and then more concretely gestures in modulation processes in Beethoven's Hammerklavier sonata. The latter have offered a first concretization of answers to Lewin's big question from [605] concerning characteristic gestures. The characters in our setup have been provided by well-chosen transformational gestures and their semantic interpretation in terms of dramatic instances. Despite these concrete examples, the present research has not solved the morphic half of Yoneda's idea, namely the fully gestural reconstruction of arrows in abstract categories, but it is a first step towards a replacement of Fregean functional abstraction by gestural dynamics. The fact that core constructions in homological algebra, such as extensions, are naturally incorporated in this approach is a sign of having embarked in the right direction.

62.6 Functorial Gestures: General Addresses

To this point, mathematical gesture theory has been developed under the tacit assumption that the classical addressed approach in functorial mathematical music theory, based upon the category $\mathbf{Mod}^@$ of modules, would not apply to gestures. However, such a restriction is superfluous, and we will develop here the functorial extension of the theory of local gestures in topological categories. We shall later include this extension into the global gesture theory, too, see Section 66.8.

In the categorial context, a continuous curve is a continuous functor $g : \nabla \to X$ with values in a topological category X. Its "points" are the values at the objects $x \in \nabla$. To generalize this situation to more functorially conceived points, we take the domain $\nabla \times A$ for a general address, i.e., a topological category $A \in \mathbf{TopCat}$. An A-addressed curve in X is a continuous functor $g : \nabla \times A \to X$. Evidently, if $f : B \to A$ is an address change in **TopCat**, we get a new curve $g.f := g \times (Id_\nabla \times f)$ from g and f. Also, if $x \in I$ is an object of ∇, we have the injection $f : 1 \to \nabla : 0 \mapsto x$ and the associated address change curve $g|x : A \to I \times A \to X : a \mapsto g(x, a)$; we shall need this latter construction especially for $g|0$ and $g|1$.

As in the zero-address case, we have a topological category (with object set) denoted by $\nabla©_A X := (\nabla \times A)©X$, and continuous natural transformations $\nu : g \to h$ of curves $g, h : \nabla \times A \to X$ as morphisms. The definition of the topology on this set is completely analogous to the construction in the zero-addressed case, see Section 62.1. Clearly, the address change $f : B \to A$ now induces a continuous functor $©f : \nabla©_A X \to \nabla©_B X$. Also, if $x \in I$ is an object (a point) of ∇, we have a continuous functor $©|x : \nabla©_A X \to A©X$ into the topological category $A©X$ with the compact-open topology. In particular, we have the tail and head functors $t_A = ©|0, h_A = ©|1 : \nabla©_A X \to A©X$, which define the *$A$-addressed categorical digraph $A@\overrightarrow{X}$ of X*. The digraph of the object maps of $A@\overrightarrow{X}$ is called the *spatial A-addressed digraph of X*. Again, similarly to the zero-addressed case, an *A-addressed gestures with skeleton digraph Γ and body X* is a digraph morphism $g : \Gamma \to A@\overrightarrow{X}$ into (the spatial digraph of) $A@\overrightarrow{X}$. The set of these gestures is denoted by $\Gamma@_A\overrightarrow{X}$.

We have the following functorial maps: Given an address change $f : B \to A$, a continuous functor of topological categories $m : X \to Y$, and a digraph morphism $t : \Delta \to \Gamma$, we have

$$\begin{aligned}
\Gamma@_f X &: \Gamma@_A\overrightarrow{X} \to \Gamma@_B\overrightarrow{X},\\
\Gamma@_A m &: \Gamma@_A\overrightarrow{X} \to \Gamma@_A\overrightarrow{Y},\\
t@_A X &: \Gamma@_A\overrightarrow{X} \to \Delta@_A\overrightarrow{X}.
\end{aligned}$$

To turn $\Gamma@_A\overrightarrow{X}$ into a topological category, denoted as earlier by $\Gamma\overrightarrow{@}_A X$, observe that Γ is the colimit of the diagram $\mathcal{D}$ of digraphs described in Section 62.1.2. Then we have $\Gamma@_A\overrightarrow{X} \xrightarrow{\sim} \lim_\mathcal{D} A@\overrightarrow{X}$ since $\uparrow @_A\overrightarrow{X} \xrightarrow{\sim} A@\overrightarrow{X}$, which turns $\Gamma@_A\overrightarrow{X}$ into a topological category since **TopCat** is finitely complete.

One expects the Escher Theorem 61 to be true for general addresses, too. We have this extension, which is valid for locally compact Hausdorff addresses:

Proposition 3. (Functorial Escher Theorem) *If Γ, Δ are digraphs, X is a topological category, and A, B are two locally compact Hausdorff topological categories, then we have a canonical isomorphism of topological categories,*

$$\Gamma\overrightarrow{@}_A\Delta\overrightarrow{@}_B X \xrightarrow{\sim} \Delta\overrightarrow{@}_B\Gamma\overrightarrow{@}_A X.$$

Modulo taking limits, the proof boils down to the special case where $\Gamma = \Delta = \uparrow$. Then we have to prove that there is a (canonical) isomorphism $\uparrow \overrightarrow{@}_A \uparrow \overrightarrow{@}_B X \xrightarrow{\sim} \uparrow \overrightarrow{@}_B \uparrow \overrightarrow{@}_A X$ of topological categories. Because of $\uparrow @_A\overrightarrow{X} \xrightarrow{\sim} A@\overrightarrow{X}$, this means that $A@\overrightarrow{B@\overrightarrow{X}} \xrightarrow{\sim} B@\overrightarrow{A@\overrightarrow{X}}$, i.e.,

$$(\nabla \times A)©((\nabla \times B)©X) \xrightarrow{\sim} (\nabla \times B)©((\nabla \times A)©X).$$

This isomorphism follows from the following series of isomorphisms that are all due either to the universal property of the compact-open topology (Mathematical Appendix, Section J.4.1.2, Theorem on Exponential Correspondence) or to the isomorphism $X \times Y \xrightarrow{\sim} Y \times X$ of cartesian products. We start with the right expression above.

$$(\nabla \times B)©((\nabla \times A)©X) \xrightarrow{\sim} (\nabla \times B)©(\nabla©(A©X)) \xrightarrow{\sim}$$
$$\nabla©(B©(\nabla©(A©X))) \xrightarrow{\sim} \nabla©(B \times \nabla©(A©X)) \xrightarrow{\sim}$$
$$\nabla©(\nabla \times B©(A©X)) \xrightarrow{\sim} \nabla©(\nabla©(B©(A©X))) \xrightarrow{\sim}$$
$$\nabla©(\nabla©((B \times A)©X)) \xrightarrow{\sim} \nabla©(\nabla©((A \times B)©X)) \xrightarrow{\sim}$$
$$(\nabla \times \nabla)©((A \times B)©X) \xrightarrow{\sim} (\nabla \times \nabla)©((A \times B)©X) \text{ permute the two copies of } \nabla.$$

The last expression is the result we get from the same procedure, but starting from the left expression above, and we are done.

We shall give musical applications of this functorial formalism in Section 78.4.1.

62.7 Yoneda's Lemma for Gestures

The classical Yoneda Lemma (Appendix Section G.2) deals with the Yoneda functor $Y : \mathcal{C} \to \mathcal{C}^{@} : X \mapsto @X$, and in its general shape states that for any presheaf (contraviariant functor) $F \in \mathcal{C}^{@}$ and object X in $\mathcal{C}$, we have a bijection

$$Nat(@X, F) \xrightarrow{\sim} X@F.$$

In our present situation, we also have presheaves, but they are different from the classical ones. For a topological category X, we have the presheaf (denoted by the classical symbol as no confusion is likely) $@X$: **Digraph** × **TopCat** → **TopCat** : $(\Sigma, A) \mapsto \Sigma\overrightarrow{@}_A X$. It is not representable in the classical sense, but nevertheless we have a representational situation here. Denote by $\overrightarrow{\mathbf{TopCat}}^{©}$ the category of presheaves F : **Digraph** × **TopCat** → **TopCat**; its objects are also called *gestural presheaves*. Observe that the natural transformations in this category need to refer to morphisms in **TopCat**; we then also denote by $Nat_{TC}(F, G)$ the set of morphisms $f : F \to G$ in $\overrightarrow{\mathbf{TopCat}}^{©}$.

Definition 111 *A gestural presheaf* F : **Digraph** × **TopCat** → **TopCat** *is said to be* gesturally representable *iff it is isomorphic in* $\overrightarrow{\mathbf{TopCat}}^{©}$ *to* $@X$ *for a topological category* X. *We then also say that* F *is* represented by the gesture space X.

In what follows, we shall prove a Yoneda Lemma that identifies morphisms in $\overrightarrow{\mathbf{TopCat}}^{©}$, i.e., natural transformations $f : @X \to F$ of gestural presheaves with an evaluation of determined functors at X. To this end, we consider the category **TC-Digraph** of TC-digraphs that are internal to $\mathbf{TopCat}^{©}$, the category of presheaves on **TopCat** that have values in **TopCat** (we also call them "continuous presheaves"). A TC-digraph D is given by natural head and tail morphisms $D = \eta, \tau : C \rightrightarrows P$ between (continuous) presheaves $C, P \in \mathbf{TopCat}^{©}$, C being called the *presheaf of curves*, while P is called the *presheaf of points*. A morphism $\vec{f} = (f^{\nabla}, f^{\cdot}) : D_1 \to D_2$ for $D_1 = \eta_1, \tau_1 : C_1 \rightrightarrows P_1, D_2 = \eta_2, \tau_2 : C_2 \rightrightarrows P_2$ is the usual pair of morphisms $f^{\nabla} : C_1 \to C_2, f^{\cdot} : P_1 \to P_2$ that commutes with head and tail morphisms. Every gestural presheaf F gives rise to such a TC-digraph $F^{\nabla} \rightrightarrows F^{\cdot}$ that evaluates to $A@F^{\nabla} = (\uparrow, A)@F, A@F^{\cdot} = (., A)@F$, the evaluation at the line digraph $\uparrow$ and the singleton digraph ., while the head and tail morphisms are defined by the two injections $. \rightrightarrows \uparrow$. This digraph is denoted by $\vec{F}$. In particular, if $F = @X$, we have $A@X^{\nabla} = (\uparrow, A)@X = A@X^{\nabla}, A@X^{\cdot} = (., A)@X = A@X^{\cdot}$, where $X^{\nabla} := \nabla@X, X^{\cdot} := X$. This digraph is denoted by $\vec{X}$. With this digraph formalism, we have a Yoneda Lemma for gestural presheaves F that are limits of curve functors, i.e. $(\Sigma, A)@F \xrightarrow{\sim} \lim_{\mathcal{D}}(F^{\nabla}, F^{\cdot})$, where $\mathcal{D}$ is the usual diagram of digraphs whose colimit is Σ. Call such functors *limiting functors*. For example, all the gesturally representable functors are limiting.

Theorem 39 (Yoneda Lemma for Functorial Gestures) *For a topological category* X *and a limiting gestural presheaf* F, *we have a bijection*

$$Nat_{TC}(@X, F) \xrightarrow{\sim} \mathbf{TC\text{-}Digraph}(\vec{X}, \vec{F}).$$

If $\overrightarrow{f} = (f^\nabla, f^\cdot)$ is a morphism $\overrightarrow{X} \to \overrightarrow{F}$, the classical Yoneda Lemma states that this is equivalent to having a pair $f^\nabla \in X^\nabla @F^\nabla, f^\cdot \in X@F^\cdot$ such that the two images $\eta f^\nabla, \tau f^\nabla : X^\nabla \rightrightarrows F^\cdot$ defined by the functors $\eta, \tau : F^\nabla \rightrightarrows F^\cdot$ and the two two images $f^\cdot h, f^\cdot t : X^\nabla \rightrightarrows F^\cdot$ defined by head and tail morphisms $X^\nabla \rightrightarrows X$ coincide[5]*, respectively:*

$$\eta f^\nabla = f^\cdot h,$$
$$\tau f^\nabla = f^\cdot t.$$

The proof runs as follows. For a natural transformation $f : @X \to F$, the evaluation $(\Sigma, A)@f : \Sigma\overrightarrow{@}_A @X \to (\Sigma, A)@F$ for general digraphs Σ and topological categories A commutes with its evaluations $(\uparrow, A)@f : \uparrow \overrightarrow{@}_A X \to (\uparrow, A)@F$ $(., A)@f : .\overrightarrow{@}_A X \to (., A)@F$ at the arrow $\uparrow$ and the point digraph ., which means that, in view of the limiting character of these functors, the arrow and point evaluation functors determine one-to-one the original morphism f. But these two evaluations mean the commutativity (with the left and right vertical arrows, respectively) of this functor diagram:

$$\begin{array}{ccc} A@X^\nabla & \xrightarrow{A@f^\nabla} & A@F^\nabla \\ h \downarrow\downarrow t & & \eta \downarrow\downarrow \tau \\ A@X^\cdot & \xrightarrow[A@f^\cdot]{} & A@F^\cdot \end{array}$$

But this is equivalent to the Yoneda evaluation at A of the digraph morphism $\overrightarrow{f} : \overrightarrow{X} \to \overrightarrow{F}$, and we are done.

This lemma has a deep impact on the gestural understanding of artistic utterance. While the functor F is not defined by gestures, nor has its values in gestural structures, the functors $@X$ are gestural by their very construction. The natural transformations $f : @X \to F$ define gestural perspectives on F, our understanding of F in terms of gestural functors. One could call the entire big functor $Nat_{TC}(@?, F)$ the *functorial gestural aesthetics of* F. It tells us how much we can know about F in terms of gestural constructions. This is an important tool to discuss musical constructions that are *not*, a priori, gestural in nature. Such a situation may occur typically in electronic music, but also in classical constructions of scores that are not derived from gestural aspects. In terms of the classical Yoneda Lemma, we could consider the category **TC-Digraph** and the Yoneda functor Y : **TC-Digraph** $\to$ **TC-Digraph**$^©$, and the above big functor would mean restricting the functorial domains to the subcategory $@TC \subset$ **TC-Digraph** of gesturally representable functors $\overrightarrow{X}$, and asking whether the functor Y : **TC-Digraph** $\to$ **TC-Digraph**$^©$ $\to @TC^©$ is still fully faithful.

62.8 Examples from Music

In the following three examples, we shall illustrate this situation, namely for constructions of sound waves, spectral compositions, and MIDI-related ON-OFF processes.

In all three examples, we shall use non-representable functors of the same nature: powers functors, which are well-known to be non-representable as **TopCat** is not a topos.

62.8.1 Collections of Acoustical Waves

The first example considers the topological space $Z = \mathcal{C}^n([0,1], \mathbb{R})$ of n times differentiable functions on the unit interval, describing sound events in a defined time interval $[0, 1]$. The functors are $A@F_Z^\nabla = 2^{A@Z^\nabla}$, whose elements are sets g of morphisms $g_i : A \to Z^\nabla$, or, equivalently, morphisms $g_i : \nabla \times A \to Z$, the latter being interpreted as curves with values in Z that are parametrized by values in A, i.e., for each $a \in A$, we have a curve $g_{i,a} : \nabla \to Z$, which is equivalent to a curve $[0, 1] \to Z$ as we are dealing with topological

[5] This can be restated as a diagram limit condition.

spaces here. Technologically, this means that we are given a processual setup that creates a curve of sound events for each parameter choice $a \in A$. Such a situation is standard when working with Max MSP software, for example.

Next, we also need the functor $F^\cdot$, which we define by $A@F^\cdot_Z = 2^{A@Z}$. Its values are the sets of A-parametrized sound events. The head and tail transformations are the evident maps $\eta, \tau : A@F^\nabla_Z \to A@F^\cdot_Z$ that send curves to their head and tail values. These sets are given the indiscrete topology.

It is a bit tricky to find morphisms $f : @X \to F_Z$. One way of doing so is to think about the above condition $\eta f^\nabla = f^\cdot h$, $\tau f^\nabla = f^\cdot t$. We have to select two sets, $f^\nabla \subset X^\nabla @Z^\nabla, f^\cdot \subset X@Z$, such that these conditions hold, meaning that for every $g^\nabla \in f^\nabla$ there is a $g^\cdot \in f^\cdot$ such that $\eta g^\nabla = g^\cdot h$, and a $g^\cdot_* \in f^\cdot$ with $\tau g^\nabla = g^\cdot_* t$, and vice versa. For example, $\eta g^\nabla = g^\cdot h : X^\nabla \to F_Z$ means that for every curve $\kappa : \nabla \to X$, we have $g^\nabla(\kappa)(1) = g^\cdot(\kappa(1))$. These are quite involved conditions. But it is easy to find solutions. Take any set $f^\cdot \subset X@Z$. Then define f^∇ as follows. For every $g^\cdot \in f^\cdot$, define $g^{\cdot,\nabla}(\kappa, m) = g^\cdot(\kappa(m))$ for any curve $\kappa : \nabla \to X$ and morphism (!) $m \in \nabla$. The set f^∇ is built from these morphisms $g^{\cdot,\nabla}$, and it is evident that this is a solution to our problem. It is also clear that this solution holds for any topological space Z.

62.8.2 Collections of Spectral Music Data

The second example uses the same architecture as the previous one, but Z is now a different space that is related to spectral composition methods. Instead of sets of parametrized sound events in $C^n([0,1], \mathbb{R})$, we now define Fourier coefficients that are time-dependent. More precisely, we define sound events as functions of time $x \in \mathbb{R}$ via the Fourier expressions

$$w(x) = \sum_n^{\pm\infty} c_n(x) e^{i2\pi n\nu(x)x},$$

where every member of the function sequence $c_n, \nu : \mathbb{R} \to \mathbb{C}$ has compact support, we have real values for ν, and the sequence produces a convergent sum at all times. Call Z_{spec} the topological space of these sequences with one of the usual topologies (in fact defined by scalar products). Then our digraph of presheaves $F^\nabla_{Z_{spec}} \rightrightarrows F^\cdot_{Z_{spec}}$ represents sets of parametrized time-dependent Fourier coefficients used in classical spectral compositions.

62.8.3 MIDI-Type ON-OFF Transformations

For our third example, we take $A@F^\cdot_{EHL} = 2^{A@\mathbb{R}^{EHL}}$ and $A@F^\nabla_{EHLDGC} = 2^{A@\mathbb{R}^{EHLDGC}}$, where $\mathbb{R}^{EHL}$ is the three-dimensional real vector space of note events with onset (E), pitch (H), and loudness (L), whereas $\mathbb{R}^{EHLDGC}$ is the six-dimensional real vector space of note events with onset (E), pitch (H), loudness (L), duration (D), glissando (G), and crescendo (C). Again, the powersets are given the indiscrete topology. Here, the head and tail functors are defined by the typical operators from the MIDI ON and OFF functions, namely η (for ON) is defined by the first projection $p_{EHL} : \mathbb{R}^{EHLDGC} \to \mathbb{R}^{EHL} : (x, y, z, u, v, w) \mapsto (x, y, z)$, while τ (for OFF) is defined by the alteration function $\alpha : \mathbb{R}^{EHLDGC} \to \mathbb{R}^{EHL} : (x, y, z, u, v, w) \mapsto (x+u, y+v, z+w)$. In this situation, we can construct a morphism as follows. We again start with a set $f^\cdot \subset X@\mathbb{R}^{EHL}$. The set $f^\nabla \subset X@\mathbb{R}^{EHLDGC}$ consists of these morphisms: For every $l : X \to \mathbb{R}^{EHL}$ in $f^\cdot$, we take its two composed morphisms $l_0 : X^\nabla \to \mathbb{R}^{EHL} : \kappa \mapsto l(\kappa(0))$ and $l_1 : X^\nabla \to \mathbb{R}^{EHL} : \kappa \mapsto l(\kappa(1))$. Then we define members of $f^\nabla \subset X@\mathbb{R}^{EHLDGC}$ by taking for each $l : X \to \mathbb{R}^{EHL}$ the function $l^\nabla : X^\nabla \to \mathbb{R}^{EHLDGC} : \kappa \mapsto (l_1(\kappa), l_0(\kappa) - l_1(\kappa))$. It is immediate that this defines a solution.

63

Singular Homology of Hypergestures

Summary. In this chapter we interpret the basic cubic chain spaces of singular homology in terms of hypergestures in a topological space over a series of copies of the arrow digraph $\uparrow$. This interpretation allows for a generalized homological setup. The generalization is (1) to topological categories instead of topological spaces, and (2) to any sequence of digraph $(\Gamma_n)_{n\in\mathbb{Z}}$ instead of the constant series of $\uparrow$. We then define the corresponding chain complexes, and prove the core boundary operator equation $\partial^2 = 0$, enabling the associated homology modules over a commutative ring R. We discuss some geometric examples and a musical one, interpreting contrapuntal rules in terms of singular homology.

– Σ –

63.1 An Introductory Example

Let us give an introductory example of the Escher Theorem, which has musical relevance, before we embark on the homological theme. Let G be a topological group, X a topological space, and $G \times X \to X$ a continuous group action. Denote by ${}^G X$ the topological category whose objects are the elements of X, and whose morphisms $g : x \to y$ are the triples $(x, y, g) \in X^2 \times G$ such that $y = gx$, the topology of ${}^G X$ being induced from the product topology on $X^2 \times G$. If the topologies are all indiscrete, a continuous curve $F : \nabla \to {}^G X$ is just a functor.

A classical example for such a topological category ${}^G X$ from transformational music theory is the canonical action of the general affine group $G = \overrightarrow{GL}(\mathbb{Z}_{12}) = T^{\mathbb{Z}_{12}} \rtimes \mathbb{Z}^{\times}$ on the pitch class set $X = \mathbb{Z}_{12}$, together with the indiscrete topology. Recall that in Section 62.2.1, we have constructed special curves, so-called *discrete* gestures, $\searrow(g) : \nabla \to {}^G X$ for every morphism $g : x \to y$ as follows:

$$\searrow(g)(s,t) = \begin{cases} Id_x & \text{if } s = t = 0, \\ g & \text{if } 0 = s < t, \\ Id_y & \text{if } 0 < s. \end{cases}$$

We may therefore identify morphism of ${}^G X$ with such discrete gestures. Composing such discrete gestures, we may more generally define discrete gestures $\searrow(g_1, g_2, \ldots g_n)$ for any sequence $g_1, g_2, \ldots g_n$ of morphisms which can be composed to $g_1 \circ g_2 \circ \ldots g_n$, see Section 62.2.1 or [723, 3.1] for details. We have $\searrow(g_1, g_2, \ldots g_n) \in \uparrow \overrightarrow{@} {}^G X$.

In our musical example, we may interpret a consonant interval as being a discrete gesture

$$\searrow(g) \in \uparrow \overrightarrow{@} {}^{\overrightarrow{GL}(\mathbb{Z}_{12})}\mathbb{Z}_{12}$$

associated with a morphism $g : c \to d$ from a cantus firmus pitch class c to discantus pitch class d, where $g = T^k$ is the translation by a Fux consonance $k = d - c \in K = \{0, 3, 4, 7, 8, 9\}$. If $\searrow(g_1), \searrow(g_2), \ldots \searrow(g_n)$

G. Mazzola et al., *The Topos of Music III: Gestures*, Computational Music Science,
https://doi.org/10.1007/978-3-319-64481-3_8

is a sequence of n such discrete interval gestures, stemming from morphisms $g_i : c_i \to d_i, i = 1, \dots n$, we have the morphisms $f_i : \searrow (g_i) \to \searrow (g_{i+1}), i = 1, \dots n-1$, defined by the natural transformations on curve arguments $0, 1$ via $f_i(0) = T^{c_{i+1}-c_i}, f(1) = T^{d_{i+1}-d_i}, i = 1, \dots n-1$. This defines a hypergestural curve

$$\searrow (f_1, f_2, \dots f_n) \in {\uparrow\uparrow} \,\overrightarrow{@}\, \overrightarrow{GL}^{(\mathbb{Z}_{12})}\mathbb{Z}_{12}$$

which formally represents a first species contrapuntal sequence of n consonant intervals. Applying the Escher Theorem for the exchange permutation of the first and the second digraph $\uparrow$, the sequence $\searrow (f_1, f_2, \dots f_n)$ corresponds to the discrete curve from the cantus firmus curve $\searrow (c_{1,2}, c_{2,3}, \dots c_{n-1,n})$ defined by the translational morphisms $c_{i,i+1} : c_i \to c_{i+1}$ to the discantus curve $\searrow (d_{1,2}, d_{2,3}, \dots d_{n-1,n})$ defined by the translational morphisms $d_{i,i+1} : d_i \to d_{i+1}$ for $i = 1, 2, \dots n-1$. These two interpretations of a contrapuntal sequence correspond to the two musicological interpretations of the word "contra": The sequence $\searrow (f_1, f_2, \dots f_n)$ is the original meaning, i.e. that the opposition is between successive intervals, whereas the other interpretation, the discantus curve being the "opposition" to the cantus firmus line, is well known, but historically not adequate; see [924] for details. The Escher Theorem mediates between these two interpretations.

The present homological approach is deduced from the following observations. To begin with, singular homology is based on continuous functions on standard objects, either n-dimensional simplexes or n-dimensional cubes [635]. It is well known that both, the simplicial and the cubical homology, yield the same homology groups. Our approach is based on cubic homology; see [999] for a good presentation of cubic homology. This one considers continuous functions $s : I^n \to X$ on the n-dimensional cubes, n-fold cartesian products $I^n = I \times I \times \dots I$ of the real unit interval I, with values in a topological space X. These functions are called *singular n-cubes.* Let us look at some singular cubes on the torus surface $X = \mathbb{T}^2$, see Figure 63.1. A singular 0-cube is a map σ_0 from the singleton I^0 to $\mathbb{T}^2$, a singular 1-cube is a continuous line map $\sigma_1 : I \to \mathbb{T}^2$, a 2-cube is a continuous square surface map $\sigma_2 : I^2 \to \mathbb{T}^2$.

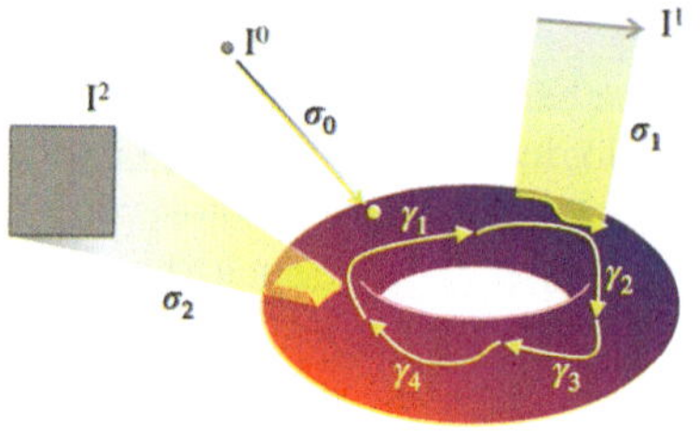

Fig. 63.1. Singular cubes on the torus. A 0-cube is a point, a 1-cube is a continuous line map, a 2-cube is a continuous square surface map. And four 1-cubes circumscribing the torus hole.

In homology, for a given (unitary) commutative ring R, one builds the modules $C_n(R, X)$ of n-chains, i.e. the formal R-linear combination of singular n-cubes. The "yoga" of homology is that one can map n-chains to $n-1$-chains by the R-linear boundary homomorphism $\partial_n : C_n(R, X) \to C_{n-1}(R, X)$ such that $\partial_n \circ \partial_{n+1} = 0$. This means that $B_{n+1} = Im(\partial_{n+1}) \subset Z_n = Ker(\partial_n)$. The quotients $H_n(R, X) = Z_n/B_{n+1}$ are called *nth homology modules.* One of their significations is that their dimension may measure holes in the topological space X. Let us make a simple example for the torus. If we look at a singular 1-cube σ_1, its boundary $\partial_1\sigma_1$ is an alternate sum of its faces (the two curve endpoints): $\partial_1\sigma_1 = \sigma_1|_{x=1} - \sigma_1|_{x=0}$, where x is the curve parameter on I. If we look at a singular 2-cube σ_2, its boundary $\partial_2\sigma_2$ is an alternate sum of its four faces $\sigma_2|_{x=0}, \sigma_2|_{x=1}, \sigma_2|_{y=0}, \sigma_2|_{y=1}$, which are the restrictions of σ_2 to its four "face" subsets defined for the conditions $x = 0, x = 1, y = 0, y = 1$ on the I^2's coordinates x and y. These four contiguous singular 1-cubes define a 1-chain, namely the alternate sign sum $\sigma_2|_{x=1} - \sigma_2|_{x=0} + \sigma_2|_{y=0} - \sigma_2|_{y=1}$. But it is clear that if we concatenate four contiguous singular 1-cubes $\gamma_1, \gamma_2, \gamma_3, \gamma_4$ as shown in Figure 61.8, their sum cannot stem from the boundary of a singular 2-cube, since the torus has a hole that prevents the existence of such a surface on $\mathbb{T}^2$. This means that the hole circumscribed by the chain $\gamma_1 + \gamma_2 + \gamma_3 + \gamma_4$ in Z_1 defines a non-zero element of $H_1(R, \mathbb{T}^2)$.

The connection between homology and gesture theory is as follows. We have

$$\mathbf{Top}(I^n, X) \xrightarrow{\sim} \mathbf{Top}(I, \mathbf{Top}(I, \ldots \mathbf{Top}(I, X)\ldots)).$$

Then, because $\mathbf{Top}(I, X) \xrightarrow{\sim} \uparrow @\vec{X}$, we may identify $\mathbf{Top}(I^n, X)$ with the space $\uparrow @ \uparrow @ \ldots \uparrow \vec{@} X$ of n-fold hypergestures in X for the sequence $\uparrow, \ldots \uparrow$ of n copies of the one-arrow digraph $\uparrow$. In other words, the singular n-cubes in X are very special hypergestures.

This means that the basic objects of (cubic) homology are special hypergestures. Therefore it is straightforward to consider the task of defining a homology theory for hypergestures, and we shall do so for the categorical hypergestures. *This means that we shall introduce homology based upon a sequence $\Gamma_1, \Gamma_2, \ldots \Gamma_n$ of arbitrary digraphs instead of the classical sequence of one-arrow digraphs $\uparrow$.* Recall from Section 62.2.3 that homological extensions are also special concepts derived from gestural constructions related to categories of factorizations. So the present approach is a further example of the conceptual homological power of mathematical gesture theory.

63.2 Chain Modules for Singular Hypergestural Homology

As it is standard in homology theory, we shall consider a commutative ring R as the basic coefficient domain for modules in homology. To begin with, we need the modules generated by singular cubes. If $\Gamma_1, \Gamma_2, \ldots \Gamma_n$ is a sequence of n digraphs, and if K is a topological category, a *gestural n-cube* is, by definition, a hypergesture $s \in \Gamma_n \Gamma_{n-1} \ldots \Gamma_1 \vec{@} K$. We denote by $R\Gamma_n \Gamma_{n-1} \ldots \Gamma_1 \vec{@} K$ the free R-module over the basis $\Gamma_n \Gamma_{n-1} \ldots \Gamma_1 \vec{@} K$. The elements of this module are called *gestural n-chains*. The 0-chains are the elements of RK, the free R-module over the objects of K.

Since we shall consider n-chains for sequences of not necessarily identical digraphs, we need to specify the definition of such modules for variable digraphs. The idea is that to define boundary homomorphism ∂ between any such chain modules, we need a sufficiently (but not too) general context where boundary homomorphisms can be defined. One could define our theory for any infinite sequence $\Gamma_{.} = \Gamma_1, \Gamma_2, \ldots$ of digraphs. In the present context, we want to choose a type of digraph sequence that provides us with a finiteness condition, namely that our infinite sequence $\Gamma_{.}$ should use only a finite number of digraphs. Let us denote them by $\Gamma_0, \Gamma_1, \ldots \Gamma_{d-1}$. Then the sequence $\Gamma_{.}$ may be encoded as a d-adic number $0.c_1, c_2, \ldots c_i, c_{i+1}, \ldots$, where each entry $0 \leqslant c_i < d$ refers to one of the digraphs of our selection $\Gamma_0, \Gamma_1, \ldots \Gamma_{d-1}$. The total information, the sequence of digraphs and the basic topological category K, will be encoded by the d-adic "number" $K.c. = K.c_1, c_2, \ldots c_i, c_{i+1}, \ldots$. For example, if we take the classical sequence in cubic singular homology in a topological space X of constant digraphs $\uparrow$, encoded by $\varnothing = \Gamma_0, \uparrow = \Gamma_1$, we have the encoding $X.1111\ldots$; or for a finite theory, $X.1111\ldots11000000\ldots$. We could also take $\varnothing = \Gamma_0, \uparrow = \Gamma_1, t \circlearrowleft T = \Gamma_2$, the last digraph being the final digraph $\mathbf{1}$ with loop T in the category of digraphs. The code $S^1.210000\ldots$ would then encode the hypergestures in $\uparrow \mathbf{1} \vec{@} S^1$ in the unit circle S^1, the space $\varnothing \uparrow \mathbf{1} \vec{@} S^1$ being a singleton.

Given these conventions and a sequence $K.c.$, we can now consider the n-chain modules. The module C_n of all n-chains is the direct sum of all modules defined by selecting any partial sequence of n (not necessarily contiguous) digraphs in the digraph sequence. This means that we have to look at the n-length hypergestures defined by a partial sequence in the d-adic "number", $c_{j_1}, c_{j_2}, \ldots c_{j_n}$, with $j_1 < j_2 < \ldots j_n$, on $\Gamma_{c_{j_n}}, \ldots \Gamma_{c_{j_1}} \vec{@} K$. We define $C_n = C_n(K.c.)$ to be the direct sum of all these length n hypergesture R-modules $R\Gamma_{c_{j_n}}, \ldots \Gamma_{c_{j_1}} \vec{@} K$, with the restriction that for any such n-length sequence, we take it only once if it occurs for different index sequences. For example, in the classical case of cubic singular homology $X.1111\ldots$, we have $C_n(K.1111\ldots) = R \uparrow\uparrow \ldots \uparrow \vec{@} K$, the R-module generated by the hypergestures on n copies of $\uparrow$ with values in K. We have built the sequence of n-chain modules, starting with $C_0(K.c.) = RK$. The next step is to define the boundary homomorphims $\partial_n : C_n \to C_{n-1}$. As usual, we set $C_n = 0$ for negative n, and therefore automatically $\partial_n = 0$ for $n \leqslant 0$.

63.3 The Boundary Homomorphism

The classical homology boundary operator $\partial_n = \partial$ on n-chains $c = \sum_i r_i g_i$ is the R-linear extension of its action on the basis, the gestural n-cubes:

$$\partial(\sum_i r_i g_i) = \sum_i r_i \partial g_i.$$

So let us select a length-n sequence $\Gamma_0, \Gamma_1, \ldots \Gamma_i, \ldots \Gamma_{n-1}$ of digraphs and look at the boundary operator on a single gestural n-cube $g \in \Gamma_0 \Gamma_1 \ldots \Gamma_i \ldots \Gamma_{n-1} \overrightarrow{@} K$. The classical formula has two components: the face operator $g \mapsto g_{i,\alpha}$, and then the alternating combination of faces:

$$\partial g = \sum_{i=0}^{n-1} \sum_{\alpha=0,1} (-1)^{i+\alpha} g_{i,\alpha}.$$

The face operator creates a chain in C_{n-1}, and this works by means of evaluation of the original singular cube on $n-1$-dimensional subcubes of g, its faces. The classical situation makes two choices to define such faces. First, we select a dimension $i = 0, 1, \ldots n-1$ in the cube I^n. This coordinate space I has the two extremal values, $\alpha = 0, \alpha = 1$. The faces in this dimension are the restrictions of the singular cube g (which is a function!) to the subsets $I^n|\alpha = 0, I^n|\alpha = 1$, which are functions on the remaining $n-1$ coordinates, in fact singular $n-1$ cubes; we denote them by $g_{i,\alpha}$.

This classical setup can be generalized to gestures as follows. First, observe that in the gestural understanding of the unit interval, 0 is just the tail argument, and 1 is the head argument of the continuous curve given on the digraph $\uparrow$. In other words, if we have a singular cube $g : I \to X$, this corresponds to the gesture $g^* : \uparrow \to \overrightarrow{X}$. The evaluation at tail t and head h of $\uparrow$ then corresponds to the restriction of g^* to the two vertices of $\uparrow$, yielding two points $g^*(t), g^*(h)$ of X, and the classical formula suggests that we take the head point with positive, and the tail point with negative sign α, yielding $g^*(h) - g^*(t)$.

This situation is suitable in the simple case of one single arrow, but for general digraphs, we need a formula that takes care of all possible arrows (if any). The idea is this: if we look at what happens in the above evaluation at head and tail, we recognize that the head value is the result of calculating the gesture on the digraph that results from (1) omitting the tail vertex t and (2) taking the digraph resulting from all that is left after removing the arrows that are connected to t. In other words, the general procedure for a digraph Γ is to (1) select an arrow $a \in A_\Gamma$ in the arrow set A_Γ of Γ, then (2) take the tail t_a of a, and then(3) restrict to the digraph $\Gamma|a^-$ obtained from removing t_a and all arrows connected to t_a. The analogue procedure for the head of a would yield the restricted digraph $\Gamma|a^+$, resulting from removing the head h_a as well as all arrows connected to this head. So, for our elementary situation, we have $\uparrow|a^- = \{h\}, \uparrow|a^+ = \{t\}$.

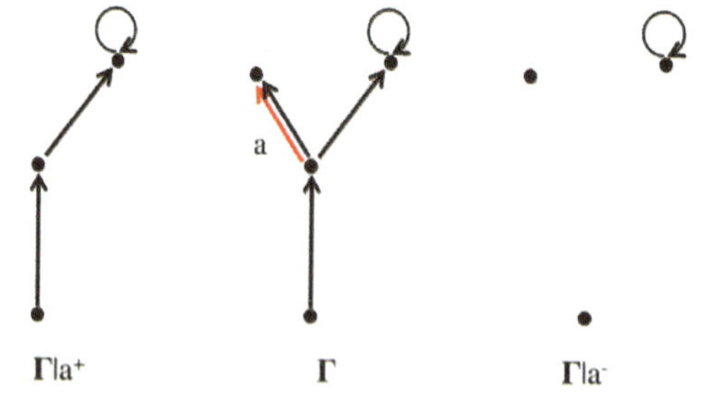

Fig. 63.2. The two reductions of the skeleton digraph of a gesture.

This construction will work if it manages to exhaust all arrows. More precisely, we may define the face of a gesture $g \in \Gamma \overrightarrow{@} K$ on a discrete digraph Γ as the sum $\sum_{v \in V_\Gamma} g(v)$ of the values of g on the digraph's vertices. The general case for these face operations can be set recursively as follows: We restrict the situation to the digraphs $\Gamma|a^-$ and $\Gamma|a^+$, see Figure 63.2.

In the simple situation of $\Gamma = \uparrow$, we have only points left. In the general case, there will be arrows left in both, $\Gamma|a^-$ and $\Gamma|a^+$ after omitting the tail and head of arrow a. But we now have less arrows and vertices. Therefore we can apply the recursive definition of face construction. This boils down to this definition:

Definition 112 *Let $g : \Gamma \to \overrightarrow{K}$ be a gesture. Then its face $g^\square$ is defined as follows, with values in the chain module RK:*

1. *If $A_\Gamma = \emptyset$ (discrete skeleton), then we set*
$$g^\square = \sum_{v \in V_\Gamma} g(v).$$
2. *The general case is defined recursively on the number of arrows in A_Γ:*
$$g^\square = \sum_{a \in A_\Gamma} (g|a^-)^\square - (g|a^+)^\square.$$
3. *The face homomorphism is canonically extended by linearization to the face homomorphism $R\Gamma \overrightarrow{@} K \to RK$ on chains $x = \sum_g c_g g$ by*
$$x^\square = \sum_g c_g g^\square.$$

So far, our procedure for gestures was restricted to the digraph that is the first (left) in the sequence of digraphs defining hyergestures. This corresponds to taking the faces with respect to the first coordinate of the hypercube I^n. In the definition of the boundary operator ∂, we take faces with respect to all the other coordinates. In order to deal with this more general coordinate selection, we use the Escher Theorem. It will allow us to transform the ith coordinate space to the first one and to perform face operators there. More precisely, if $g \in \Gamma_0\Gamma_1 \ldots \Gamma_{n-1} \overrightarrow{@} K$, and if $0 \leqslant i < n$, we consider the permutation π_i that maps the ith index to the first and leaves the others unaltered, resulting in a sequence $\Gamma_i\Gamma_0 \ldots \Gamma_{i-1}\Gamma_{i+1} \ldots \Gamma_{n-1} \overrightarrow{@} K$, and then the Escher Theorem yields a hypergesture $g_i \in \Gamma_i\Gamma_0 \ldots \Gamma_{i-1}\Gamma_{i+1} \ldots \Gamma_{n-1} \overrightarrow{@} K$, corresponding to g. We also extend by linearization this map $g \mapsto g_i$ to the chain modules generated by these gesture spaces. To g_i we apply the face operator, resulting in a hypergesture $g_i^\square \in \Gamma_0 \ldots \Gamma_{i-1}\Gamma_{i+1} \ldots \Gamma_{n-1} \overrightarrow{@} K$, the *ith face of g*. With this construction, the boundary operator is defined by
$$\partial_n g = \sum_{i=0}^{n-1} (-1)^i g_i^\square,$$
yielding a chain in C_{n-1}, and we finally extend this operator linearly to C_n. Observe that the power $(-1)^i$ is the signature of π_i. It is immediate from the above linearization procedures for Escher correspondences and faces that for any chain $x \in C_n$, we have
$$\partial_n x = \sum_{i=0}^{n-1} (-1)^i x_i^\square.$$

It remains to be shown that this boundary operator verifies the crucial equation $\partial^2 = 0$, which then enables us to define the homology modules:

Proposition 64 *With the above notations, the composition $\partial_{n-1} \circ \partial_n$ is the zero homomorphism $C_n \to C_{n-2}$.*

Proof. It suffices to prove $\partial^2 = 0$ on hypergestures $g \in \Gamma_0\Gamma_1 \ldots \Gamma_{n-1} \overrightarrow{@} K$. We step through all important points of the proof and leave details to the reader. If all digraphs are discrete, we have $\Gamma_0\Gamma_1 \ldots \Gamma_{n-1} \overrightarrow{@} K \xrightarrow{\sim} K^{\Gamma_0 \times \Gamma_1 \times \ldots \Gamma_{n-1}}$, and the vanishing of ∂^2 follows from the proof idea in the classical case of cubic homology, namely that for $i \leqslant j$, $g_i(x)_j(y) = g_{j+1}(y)_i(x)$, and they cancel out since the signs of the -1 coefficients are different. The same argument works for the case where all digraphs Γ_i are isomorphic to $\uparrow$ or discrete.

For the general case, we have

$$\begin{aligned}
\partial_{n-1}(\partial_n g) &= \sum_{i=0}^{n-1}(-1)^i \partial_{n-1}(g_i^{\square}) \\
&= \sum_{i=0}^{n-1}(-1)^i \sum_{a}^{A_{\Gamma_i}} \partial_{n-1}((g_i|a^-)^{\square}) - \partial_{n-1}((g_i|a^+)^{\square}) \\
&= \sum_{i=0}^{n-1}\sum_{j=0}^{n-2}\sum_{(a,b)}^{A_{\Gamma_i}\times A_{\Gamma_j}} \\
&\quad (-1)^{i+j}(((g_i|a^-)_j^{\square}|b^-)^{\square} - ((g_i|a^-)_j^{\square}|b^+)^{\square} - ((g_i|a^+)_j^{\square}|b^-)^{\square} + ((g_i|a^+)_j^{\square}|b^+)^{\square}).
\end{aligned}$$

Similar to the classical case we need to show that for any sign combination $\alpha, \beta \in \{+,-\}$, and $i \leqslant j$, we have

$$((g_i|a^\alpha)_j^{\square}|b^\beta)^{\square} = ((g_{j+1}|b^\beta)_i^{\square}|a^\alpha)^{\square}.$$

The left side is

$$((g_i|a^\alpha)_j^{\square}|b^\beta)^{\square} = ((\sum_{c}^{A_{a^\alpha}}(g_i|a^\alpha|c^-)^{\square} - (g_i|a^\alpha|c^+)^{\square})_j|b^\beta)^{\square}.$$

Since Escher permutation in j, the face operator, and restriction $|b^\beta$, are linear, this yields

$$((g_i|a^\alpha)_j^{\square}|b^\beta)^{\square} = \sum_{c}^{A_{a^\alpha}}(((g_i|a^\alpha|c^-)^{\square})_j|b^\beta)^{\square} - (((g_i|a^\alpha|c^+)^{\square})_j|b^\beta)^{\square}.$$

But evidently, the double restriction $|a^\alpha|c^\pm$ commutes, therefore $g_i|a^\alpha|c^\pm = g_i|c^\pm|a^\alpha$ and also $g_i|c^\pm = h_i$ for an Escher transformed h_i of an h defined on $c^\pm$ instead of Γ_i in the ith coordinate. Therefore, recursion on the size of Γ_i yields

$$((h_i|a^\alpha)_j^{\square}|b^\beta)^{\square} = ((h_{j+1}|b^\beta)_i^{\square}|a^\alpha)^{\square}$$

for all these restricted hypergestures h. Since these restrictions cover all of a^α, this implies $((g_i|a^\alpha)_j^{\square}|b^\beta)^{\square} = ((g_{j+1}|b^\beta)_i^{\square}|a^\alpha)^{\square}$, and we are done since these components appear with opposite sign.

Since by this result, $Im(\partial_{n+1}) \subset Ker(\partial_n)$, we can define homology modules:

Definition 113 *With the above notations, for a sequence $K.c.$ of digraphs $c.$ and topological category K, we have a chain complex*

$$\ldots \longrightarrow C_n(K.c.) \xrightarrow{\partial_n} C_{n-1}(K.c.) \longrightarrow \ldots C_0(K.c.) \longrightarrow 0$$

and we may define hypergestural homology modules

$$H_n(K.c.) = Ker(\partial_n)/Im(\partial_{n+1})$$

for all n.

The following result about functoriality of homology is straightforward:

Proposition 65 *If $f : K \to L$ is a topological functor, and if we are given a sequence $\Gamma.$ of digraphs, encoded as number $c.$ as above, then the canonical morphism of chain complexes $C(f : C(K.c.) \to C(L.c.)$ induces a canonical sequence, functorial in f, of homology module homomorphisms $H_n(f) : H_n(K.c.) \to H_n(L.c.)$.*

This functoriality implies that each sequence $\Gamma.$ of digraphs generates a sequence of invariants on topological categories K that comprise the classical homology modules associated with the constant sequence $K.\bar{1}$.

Example 76 Let us first look at a simple example involving different digraphs, namely the case mentioned above, $K.210000\ldots = K.21\bar{0}$, encoding the hypergestures in $\uparrow \mathbf{1}\vec{@}K$ in K. We have this chain complex:

$$0 \longrightarrow R \uparrow \mathbf{1}\vec{@}K \xrightarrow{\partial_2} R \uparrow \vec{@}K \oplus R\mathbf{1}\vec{@}K \xrightarrow{\partial_1} RK \longrightarrow 0$$

If $g \in \uparrow \vec{@}K$, we have $\partial g = g(h) - g(t)$, which in a musical situation expresses the formal difference between end state of gesture g and its initial state.

If $g \in \mathbf{1}\vec{@}K$, then $\partial g = 0$. Defining $D_1 : R \uparrow \vec{@}K \to RK$ by linear extension to the free module by $g \mapsto \Delta(g) = g(h) - g(t)$, we have $Im(\partial_1) = Im(D_1)$ and therefore

$$H_0 = RK/Im(D_1).$$

If R is a field, and if K is a topological space, $dim(H_0)$ is well known to be the number of pathwise components of K.

Further, $Ker(\partial_1) = Ker(D_1) \oplus R\mathbf{1}\vec{@}K$. Let us calculate $Im(\partial_2)$. For $g \in \uparrow \mathbf{1}\vec{@}K$, we have

$$\partial_2 g = g_0^{\square} - g_1^{\square}$$

with $g_1^{\square} = 0$ since $a^{\pm} = \emptyset$, and $g_0^{\square} = g(t) - g(h) \in \mathbf{1}\vec{@}K$. As above, we now define $D_2 : R \uparrow \mathbf{1}\vec{@}K \to R\mathbf{1}\vec{@}K$ via $D_2(g) = g(h) - g(t)$, and we have $Im(\partial_2) = Im(D_2) \subset R\mathbf{1}\vec{@}K$. Therefore

$$H_1 = Ker(D_1) \oplus R\mathbf{1}\vec{@}K/Im(D_2),$$

where the right factor is generated by homotopy classes of loops, and the left one is generated by loops. Finally, we have $H_2 = Ker(D_2)$, the space of loops of loops.

For a first topological category, take $K = S^1$, the topological space of the unit circle. The factor $R\mathbf{1}\vec{@}K/Im(D_2)$ is the fundamental group R-algebra of homotopy classes of loops, since all classes are represented by loops at a fixed point. Their group composition is also well defined: $H_1(S^1.21\bar{0}) = Ker(D_1) \oplus R\pi_1(S^1)$.

For a second topological category, we consider the topological category defined by a topological group $K = G$, which is quite the contrary to the category defined by a topological space: we have a single object, the group's identity $Id \in G$, whereas the morphisms $x : Id \to Id$ are the group elements, and their composition $x \circ y$ is the group operation. A curve $g : \nabla \to G$ is a continuous function g with $g(y, z) \circ g(x, y) = g(x, z)$ for any $0 \leqslant x \leqslant y \leqslant z \leqslant 1$. For example, if $G = \mathrm{SL}_2(\mathbb{R})$, then

$$\text{the shearing } s(x, y) = \begin{pmatrix} 1 & y - x \\ 0 & 1 \end{pmatrix} \text{ and the dilation } d(x, y) = \begin{pmatrix} e^{y-x} & 0 \\ 0 & e^{x-y} \end{pmatrix}$$

are examples of such curves.

To calculate H_0, observe that G has a single point. Furthermore $Im(\partial_1) = Im(D_1) = 0$, since all objects of ∇ go to the identity for a curve $\nabla \to G$. Therefore $H_0 \xrightarrow{\sim} R$.

To get H_1, let us look at a curve $F : \nabla \to \mathbf{1}\vec{@}G$. It is defined by a natural transformation $\mu(x, y) : F(x) \to F(y)$ with $\mu(y, z) \circ \mu(x, y) = \mu(x, z)$ for any morphisms $x \leqslant y, y \leqslant z$ in ∇, i.e. we have a commutative diagram of group elements for any morphism $s \leqslant t$, continuous in x, y, s, t, as follows:

$$\begin{array}{ccc} F(x)(s) & \xrightarrow{\mu(x,y)(s)} & F(y)(s) \\ {\scriptstyle F(x)(s,t)}\downarrow & & \downarrow{\scriptstyle F(y)(s,t)} \\ F(x)(t) & \xrightarrow[\mu(x,y)(t)]{} & F(y)(t) \end{array}$$

where all objects of the square are the identity of G. This implies

$$\mu(x,y)(t) = F(x)(s,t)^{-1} \circ \mu(x,y)(s) \circ F(y)(s,t).$$

Taking $s = 0$, the natural transformation $\mu(x,y) : F(x) \to F(y)$ is defined once we have $\mu(x,y)(0) : F(x)(0) \to F(y)(0)$ such that $\mu(y,z)(0) \circ \mu(x,y)(0) = \mu(x,z)(0)$. But we may set this initial value to Id. Suppose that we are given two curves $F_0, F_1 : \nabla \to G$. We may define an intermediate curve $F(x) = F_1$ for every $x > 0$. Evidently, then the natural transformation $\mu(x,1) = Id$ connects $F(x)$ to F_1 for all $x > 0$. To connect F_0 to $F(x) = F_1$, we take the initial transformation $\mu(0,x) = Id$ and then the following special case of the above formula

$$\mu(0,x)(s) = F_0(0,s)^{-1} \circ Id \circ F(x)(0,s) = F_0(0,s)^{-1} \circ F_1(0,s).$$

It is easy to verify that this rule defines a continuous curve from F_0 to F_1. This means that any two curves F_0, F_1 are endpoints $F_0 = F(0), F_1 = F(1)$ of a curve $F : \nabla \to \mathbf{1}\overrightarrow{@}G$, and this means that $Im(D_2)$ is generated by all differences of any two gestures in $\mathbf{1}\overrightarrow{@}G$. Moreover, since all gestures $g \in \uparrow \overrightarrow{@}G$ have $D_1(g) = 0$, $Ker(D_1) = \uparrow \overrightarrow{@}G$. Therefore $Ker(\partial_1) = C_1(K.c.) = \uparrow \overrightarrow{@}G \oplus R\mathbf{1}\overrightarrow{@}G$. Therefore

$$H_1 \stackrel{\sim}{\to} R \uparrow \overrightarrow{@}G \oplus R,$$

where the second factor represents the quotient of $R\mathbf{1}\overrightarrow{@}G$ modulo the module of gesture differences. The second homology group H_2 is described as in the general case above.

Example 77 This example relates to the category ${}^{G}X$ derived from a group action $G \times X \to X$ as described above. We suppose here that its topology is indiscrete, as in the case of the musical category ${}^{\overrightarrow{GL}(\mathbb{Z}_{12})}\mathbb{Z}_{12}$.

For the classical cubic homology we have this result:

Theorem 40 *The first homology module of an indiscrete topological category ${}^{G}X$ is the free R-module*

$$H_1({}^{G}X) = R^{G\backslash X},$$

where $G\backslash X$ is the set of orbits of this action.

Its proof is quite involved and resides on a series of lemmata; we leave it as an exercise to the reader.

64

Stokes' Theorem for Hypergestures

Summary. As singular homology is strongly related to de Rham cohomology, in particular by Stokes' classical theorem, it is natural to ask for such a theorem in our context of hypergestures. But there is a deeper reason for such a project, namely the idea that music theory of hypergestures could provide us with models of energy exchange in gestural interaction. In such a (still hypothetical) theory, Stokes' theorem would play a crucial role regarding questions of energy conservation (integral invariants).

– Σ –

64.1 The Need for Stokes' Theorem for Hypergestures

Stokes' classical theorem states

$$\int_C d\omega = \int_{\partial C} \omega,$$

where C is a compact oriented k-dimensional manifold with boundary and ω is a $k-1$-form on C. The operator $d\omega$ is the exterior derivative of ω, and ∂C is the boundary of C, see Appendix Section J.8. It is well known that this formula is valid for slightly more general situations, namely, where the boundary is not a manifold, but has singularities such as "corners" and the like, see [617, 11.4].

Stokes' theorem is of primordial importance in many fields of physics, e.g. in mechanics (integral invariants, see [2]) or in electrodynamics (relating differential and integral forms of Maxwell's equations [497]). The reason why we are interested in such a theorem for mathematical music theory is twofold: On the one hand, we have initiated a homological study of hypergestural structures [727] (see Chapter 63) which has also provided us with applications to counterpoint theory [16] (see Chapter 79). As singular homology is strongly related to de Rham cohomology, in particular by Stokes' theorem, it is natural to ask for such a theorem in our context of hypergestures. But there is a deeper reason for such a project, namely the idea that music theory of hypergestures could provide us with models of energy exchange in gestural interaction. In such a (still hypothetical) theory, Stokes' theorem would play a crucial role regarding questions of energy conservation (integral invariants).

64.2 Almost Regular Manifolds, Differential Forms, and Integration for Hypergestures

We first need to specify the basic concepts that contribute to the Stokes statement. We are aware of the somewhat sloppy style in this quite standard part of the chapter; the reader is kindly asked to fill out the standard technical details.

G. Mazzola et al., *The Topos of Music III: Gestures*, Computational Music Science,
https://doi.org/10.1007/978-3-319-64481-3_9

64.2.1 Locally Almost Regular Manifolds

In the present context we need hypergestures in manifolds since we are dealing with differentiable structures. We however need quite general manifolds in the sense of what are called "almost regular manifolds" in [617] or even more singular manifolds, where the boundaries have corners. To understand our requirement we look at typical manifolds in the context of hypergestures. In [719], we have introduced a standard topological space $|\Sigma|$ associated with a digraph Σ, see Section 61.7. It is the colimit of the digraph's arrow set A_Σ, the gluing operation being performed on the digraph vertices set V_Σ. This topological space specifies one line chart $|a| \xrightarrow{\sim} I = [0,1]$ per arrow a and a point chart $|x|$ for each isolated vertex x. The specification of this atlas is mandatory since we don't want to glue two consecutive arrows $x \xrightarrow{a} y \xrightarrow{b} z$ to one line. The differentiability in the connecting vertex y is suspended. Or it may also happen that three or more arrows share a vertex, and then the differentiability in such a vertex would not make sense. We call *skeletal space* the manifold $|\Sigma|$ associated with skeleton Σ.

The best conceptual approach to this situation is to embed such a manifold in a differentiable manifold M as a subset whose charts are manifolds with boundary isomorphic to the unit interval I or to a zero-dimensional point manifold 0. We next need cartesian products of such manifolds when hypergestures are discussed. This means that we have to consider products of type $|\Sigma_1| \times |\Sigma_1| \times \ldots |\Sigma_n|$. These manifolds are living in cartesian products of their carrier manifolds $M_1, M_2, \ldots M_n$, and the typical boundary of a product $|\Sigma_1| \times |\Sigma_2|$ is $\partial(|\Sigma_1| \times |\Sigma_2|) = \partial|\Sigma_1| \times |\Sigma_2| \cup |\Sigma_1| \times \partial|\Sigma_2|$, see Figure 64.1 for an example.

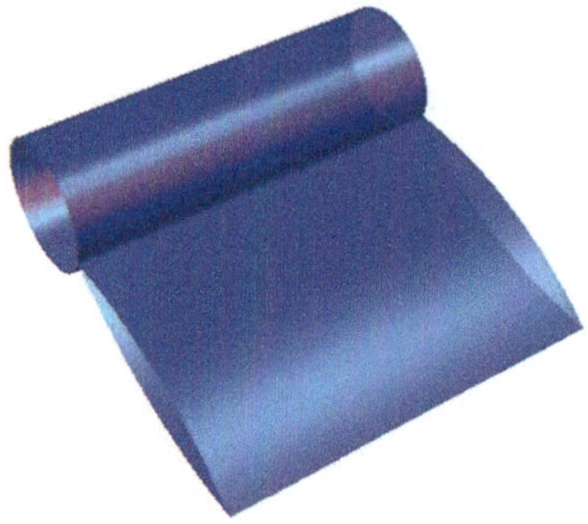

Fig. 64.1. A skeletal space.

But observe that due to singular points in digraphs, such products can be inhomogeneous in their dimension. A product may be a disjoint union of submanifolds of different dimensions.

To get a reasonable category of such manifolds, we consider differentiable morphism $L \to M$ of the carrier manifolds L, M of $\mathcal{L}, \mathcal{M}$, respectively, that restrict to atlas-compatible maps $f : \mathcal{L}^I \to \mathcal{M}^J$, where I, J designate the atlases of $\mathcal{L}, \mathcal{M}$, respectively. Atlas-compatibility means that, as in mathematical music theory of global compositions, we are also given a map $g : I \to J$ such that f sends I-chart $\mathcal{L}_i$ to J-chart $\mathcal{M}_{g(i)}$. We denote this *category of locally almost regular manifolds* by $LARM$. Such a manifold need not have a determined dimension, but may have several dimensions according to connected components and charts. In what follows, we shall call dimension $dim(\mathcal{L})$ of an almost regular manifold $\mathcal{L}$ the maximal dimension of such components. The submanifold of $\mathcal{L}$ of a determined dimension k will be denoted by $\mathcal{L}^k$.

The most important application of $LARM$ for the Stokes theory lies in a reinterpretation of hypergestures. Suppose we are given a hypergesture $c \in \Sigma_1\Sigma_2 \ldots \Sigma_n \overrightarrow{@}\mathcal{L}$ over n skeleta $\Sigma_1, \Sigma_2, \ldots \Sigma_n$ with values in a locally almost regular manifold $\mathcal{L}$. By the very definition of hypergestures, and by the adjointness property of the manifold $|\Sigma|$ associated with skeleton Σ (Proposition 62), as well as the adjointness of the cartesian product and repeated function spaces (also known as currying in computer science), $\Sigma_1\Sigma_2 \ldots \Sigma_n \overrightarrow{@}\mathcal{L} \xrightarrow{\sim} |\Sigma_1| \times |\Sigma_2| \times \ldots |\Sigma_n| © \mathcal{L}$, the set of continuous functions from the cartesian product of the skeletal manifolds to $\mathcal{L}$. Within this function set, we exhibit the differentiable morphisms and denote their set by $|\Sigma_1|\times|\Sigma_2|\times\ldots|\Sigma_n|@\mathcal{L}$. The morphisms in the latter (more precisely: their corresponding hypergestures) are called *differentiable hypergestures*; the set of these hypergestures is also denoted by $\Sigma_1\Sigma_2\ldots\Sigma_n@\mathcal{L}$. In the context of the Stokes theorem, we need differentiable singular n-cubes. Their generalization to hypergestures are differentiable gestural n-cubes, namely the elements of $\Sigma_1\Sigma_2\ldots\Sigma_n@\mathcal{L}$. The free module $\mathbb{R}\Sigma_1\Sigma_2\ldots\Sigma_n@\mathcal{L}$ of $\mathbb{R}$-linear combinations of differentiable gestural n-cubes (the module basis) defines the *(differentiable) n-chains over $\Sigma_1, \Sigma_2, \ldots \Sigma_n$* with values in $\mathcal{L}$.

64.2.2 Differential Forms

On a locally almost regular manifold $\mathcal{L}$ (we omit the atlas if possible to ease notation), differential forms can be considered in the sense that they are defined on each chart as usual. If such a chart $\mathcal{L}_i$ has dimension n, the differential forms of dimension $k \leqslant n$ define non-trivial real vector spaces $\bigwedge^k \mathcal{L}_{i,x}$ at each point x of $\mathcal{L}_i$. A differential k-form ω on $\mathcal{L}$ is a differentiable section in each chart $\bigwedge^k \mathcal{L}_i$. Since our manifolds are of different dimensions locally, we will have to deal with forms that don't have the same dimension everywhere, they are not homogeneous. We therefore consider the direct sum $\bigwedge^{\oplus k}\mathcal{L} = \bigoplus_{l\leqslant k}\bigwedge^l \mathcal{L}$. If we take a differential form $\omega \in \bigwedge^{\oplus k}\mathcal{L}$, its l-component will be denoted by ω_l. As in the classical case, for a morphism $f : \mathcal{L} \to \mathcal{M}$ of locally almost regular manifolds, one has the canonical inverse image $f^*\omega \in \bigwedge^k \mathcal{L}$ for $\omega \in \bigwedge^k \mathcal{M}$.

In the classical case, one has the exterior derivative operator $d : \bigwedge^k \mathcal{L} \to \bigwedge^{k+1}\mathcal{L}$ with $d^2 = 0$. For the non-homogeneous case mentioned above, we need a derivative operator $d_\oplus$ defined by $d_\oplus\omega = (\omega_0, d\omega_0, d\omega_1, d\omega_2, \ldots)$ for $\omega = (\omega_0, \omega_1, \omega_2, \ldots)$. For this operator, we have $d^2_\oplus\omega = (\omega_0, d\omega_0, 0, \ldots)$. And as in the classical case, the operators d and $d_\oplus$ commute with inverse images.

64.2.3 Integration

Modulo linear extensions to n-chains, we need to define $\int_c \omega$ for a gestural n-cube $c \in \Sigma_1\Sigma_2\ldots\Sigma_n@\mathcal{L}$. As usual, the formula is defined to mean $\int_{|\Sigma_1|\times|\Sigma_2|\times\ldots|\Sigma_n|} c^*\omega$, which amounts to restricting ourselves to the special case $\mathcal{L} = |\Sigma_1| \times |\Sigma_2| \times \ldots |\Sigma_n|$. We shall define the integral by recursion on the hypergestural parameters and recalling the Fubini theorem for iterated integration [999, Theorem 3-1]. Let $(\lambda, t) \in T|\Sigma_1|_\lambda$, the tangent space at $\lambda \in |\Sigma_1|$. This argument defines a form $c^*\omega_{\lambda,t} \in \bigwedge^{\oplus(n-1)}|\Sigma_2| \times \ldots |\Sigma_n|$, and we may suppose by recursion that $I(\lambda, t) = \int_{|\Sigma_2|\times\ldots|\Sigma_n|} c^*\omega_{\lambda,t}$ is defined, which yields an element of $\bigwedge^{\oplus 1}|\Sigma_1|$. So we are left with the definition of the integral for $n = 0, 1$. If $n = 0$, $c \in \mathcal{L}$, and $\omega \in \mathcal{F}(\mathcal{L})$ is a function. Then we set $\int_c \omega = \omega(c)$. In dimension $n = 1$, there are three cases for Σ_1:

1. If $A_{\Sigma_1} = \emptyset$, then set $\int_c \omega = \sum_{i\in V_{\Sigma_1}} \omega_0(c(i)) = \sum_{i\in V_{\Sigma_1}} \int_{c(i)} \omega_0$.
2. Recall from [727, Section 3] that for an arrow a of Σ_1, a^- denotes the subskeleton of Σ_1 after taking away the tail $t(a)$ and all arrows connected to $t(a)$, and a^+ denotes the subskeleton of Σ_1 after taking away the head $h(a)$ and all arrows connected to $h(a)$. In this second case, we suppose that there is at least one arrow a, but both A_{a^-} and A_{a^+} are empty. This means that, besides isolated vertices, there are either a number of loops on a single vertex or a number of arrows between two distinct points. This is the classical one-dimensional situation for integration on the unit interval. So we define $\int_c \omega = \sum_{a\in A_{\Sigma_1}} \int_a \omega_1 + \int_{\text{isolated vertices}} \omega$, where $\int_a \omega_1$ is the evident classical integration.
3. In the third case, there is an arrow a such that $A_{a^-} \cup A_{a^+} \neq \emptyset$. We then set the recursive formula $\int_c \omega = \sum_{a\in A_{\Sigma_1}} (\int_{c|a^-} \omega - \int_{c|a^+} \omega)$, a formula that reminds us of the definition of the face operator ?$^\square$ given in [727, Definition 3.1].

64.3 Stokes' Theorem

For the proof of Stokes' theorem for hypergestures, we need a technical lemma. It refers to the Escher theorem operation on chains $c \in \Sigma_1\Sigma_2 \ldots \Sigma_n @\mathcal{L}$ which generates a chain $c_j \in \Sigma_j\Sigma_1\Sigma_2 \ldots \widehat{\Sigma_j} \ldots \Sigma_n @\mathcal{L}$.

Lemma 1. *If $c \in \Sigma_1\Sigma_2 \ldots \Sigma_n @\mathcal{L}$ is a differentiable n-cube, $1 \leqslant j \leqslant n$, $a \in A_{\Sigma_j}$, and $\lambda \in |\Sigma_1|$, then we have*

$$(c_j|a^{\pm})^{\square}(\lambda) = (c(\lambda)_j|a^{\pm})^{\square},$$

and therefore also

$$(c_j)^{\square}(\lambda) = (c(\lambda)_j)^{\square}.$$

The lemma follows from the observation that (1) the face operator yields the same linear combination on both sides since it acts on the same $\Sigma_j|a^{\pm}$, and (2) the evaluation at λ is taken on the same face operator result.

Theorem 41 (Stokes' Theorem for Hypergestures) *Let $c \in \mathbb{R}\Sigma_1\Sigma_2 \ldots \Sigma_k @\mathcal{L}$ be a k-chain in a k-dimensional locally almost regular manifold $\mathcal{L}$, and let $f \in \bigwedge^{k-1}\mathcal{L}$. Then*

$$\int_c d^{\oplus} f = \int_{\partial c} f.$$

Proof. We can of course restrict to gestural k-cubes. For $k = 1$, f is a function on $\mathcal{L}$ and $c \in \Sigma @\mathcal{L}$. Let first $A_{\Sigma} = \varnothing$. Then $\int_{\partial c} f = \sum_{i \in V_{\Sigma}} f(c(i))$, whereas $\int_c d^{\oplus} f = \sum_{i \in V_{\Sigma}} (d^{\oplus} f)_0(c(i)) = \sum_{i \in V_{\Sigma}} f(c(i))$ yields the same. For the second case, $A_{a^-} \cup A_{a^+} = \varnothing$, but since arrows exist, we may focus on the subskeleton bearing those arrows, the discrete part having been already dealt with. Here,

$$\begin{aligned}
\int_c d^{\oplus} f &= \sum_{a \in A_{\Sigma}} \int_a df \\
&= \sum_{a \in A_{\Sigma}} \int_{\partial a} f \\
&= \sum_{a \in A_{\Sigma}} f(c(h(a))) - f(c(t(a))) \\
&= \int_{\partial c} f,
\end{aligned}$$

this is the classical case. For the third case, $A_{a^-} \cup A_{a^+} \neq \varnothing$, we have

$$\begin{aligned}
\int_c d^{\oplus} f &= \sum_{a \in A_{\Sigma}} \int_{c|a^-} d^{\oplus} f - \int_{c|a^+} d^{\oplus} f \\
&= \sum_{a \in A_{\Sigma}} \int_{\partial(c|a^-)} f - \int_{\partial(c|a^+)} f \\
&= \sum_{a \in A_{\Sigma}} \int_{(c|a^-)^{\square}} f - \int_{(c|a^+)^{\square}} f \\
&= \int_{\partial c} f
\end{aligned}$$

by recursion and since ∂ and $?^{\square}$ coincide in dimension one.

The case of higher dimensions runs as follows:

$$
\begin{aligned}
\int_c d^{\oplus} f &= \int_{|\Sigma_1|} \int_{c(\lambda)} d^{\oplus} f \ (\lambda \in |\Sigma_1|) \\
&= \int_{|\Sigma_1|} \int_{\partial c(\lambda)} f \ \text{(recursion)} \\
&= \int_{|\Sigma_1|} \sum_j (-1)^j \int_{(c(\lambda)_j)^{\square}} f \\
&= \sum_j (-1)^j \int_{|\Sigma_1|} \int_{(c(\lambda)_j)^{\square}} f \\
&= \sum_j (-1)^j \int_{|\Sigma_1|} \int_{(c_j)^{\square}(\lambda)} f \ \text{(Lemma 1)} \\
&= \sum_j (-1)^j \int_{(c_j)^{\square}} f \\
&= \int_{\partial c} f.
\end{aligned}
$$

This concludes the proof of Stokes' theorem.

65

Local Facts, Processes, and Gestures

Summary. In this chapter we describe the mathematical framework for the three fundamental layers of musical ontology: facts, processes, and gestures. The layer of facts is described by the theory of local and global compositions, a major topic in American Set Theory [765] and in the European school developed by the author and his collaborators [682]. The second layer is captured by the American Transformational Theory [605, 538] and, again in Europe, by the author's theory of categorical limits (and colimits) as embedded in topos theory [714]. The third layer has been the author's main concern in the last ten years [720, 723, 727], also paralleled by American research such as Robert S. Hatten's work [446].

– Σ –

All of this is also strongly motivated by software developments in the Rubato Composer environment [739, 730], especially related to the universal music data format of denotators for the topos of presheaves over modules. Since the present software development in collaboration with Florian Thalmann [730] has progressively focused on a seamless integration of factual, processual, and gestural aspects, a 'global' music theory requires a more and more unified view of musical ontology. Therefore we want to present this "theory of everything" in its mathematical shape, including some interesting new examples and results regarding global networks, especially a global "Zarlino network" motivated by neo-Riemann theory, and gestures over locally compact points that occur in mathematical performance theory, as well as functors relating singular hom of hypergestures and hypernetworks.

65.1 Categories of Local Compositions

On the level of facts we deal with sets of musical objects, such as chords, motives, etc. It is the level of mathematical music theory which is related to American Set Theory [765] and the early work of the Zurich school [670, 682]. We have formalized this level in [714] with categories of local compositions. The minimal workable category for this type of theory is the category $\mathbf{Mod}^{@}$ of presheaves over modules. Let us shortly recall the conceptual framework developed in Section 7.4. The usual objects in this music theory are triples (K, A, F), where $K \subset A@F$, A is a module, and F is a presheaf. We usually write $K \subset A@F$ and call this an *objective local composition in F with address A*. In this chapter we don't deal with more general functorial local compositions, namely subfunctors $K \subset @A \times F$. We henceforth omit the specification "objective" for all theories of this chapter. If $K \subset A@F$, $L \subset B@G$, then a morphism $K \to L$ is a pair (f, α) where $\alpha : A \to B$ is an affine module morphism and $f : K \to L\alpha$ is a set map, the set $L\alpha$ being the image of L under the map $\alpha@G : B@G \to A@G$, such that there exists a natural transformation $h : F \to G$ with $f = A@h|K$. We write $f : K \to L$ if no confusion is likely. The category of (objective) local compositions is denoted by *ObLoc*. The subcategory of objective local *modular* compositions *ComLoc* is the full subcategory of *ObLoc* whose local compositions K have their functor F being represented by a module M over a commutative ring R, i.e. $K \subset A@M$. For a zero-addressed modular local composition $K \subset 0@M$, its module RK is the

G. Mazzola et al., *The Topos of Music III: Gestures*, Computational Music Science,
https://doi.org/10.1007/978-3-319-64481-3_10

submodule of M generated by all differences $k - k_0$ for a fixed $k_0 \in K$. The map $K \mapsto RK$ extends to a functor from zero-addressed modular local compositions to modules.

Typical modular local compositions are chords Ch of pitch classes, i.e. zero-addressed local compositions $Ch \subset 0@\mathbb{Z}_{12}$, while Thomas Noll's self-addressed chords [802] are $\mathbb{Z}_{12}$-addressed local compositions $Sh \subset \mathbb{Z}_{12}@\mathbb{Z}_{12}$. A set S of dodecaphonic series would be a local $\mathbb{Z}^{11}$-addressed composition $S \subset \mathbb{Z}^{11}@\mathbb{Z}_{12}$. The retrograde operation on dodecaphonic series is represented by the address change $R : \mathbb{Z}^{11} \to \mathbb{Z}^{11}$ sending the canonical affine basis vector e_i to e_{11-i} for $i = 0, 1, \ldots 11$ ($e_0 = 0$). David Lewin's *time spans* are elements of $\mathbb{R}@\mathbb{R}$, i.e. self-addressed onsets, while his interval operation is an address change, but see page 103 for details, including those about intervallic invariance for time spans. Non-modular local compositions are often generated from power set constructions. For example, the non-representable functor $2^{\mathbb{R}\oplus\mathbb{Z}}$, evaluating at address A to $A@2^{\mathbb{R}\oplus\mathbb{Z}} = 2^{A@\mathbb{R}\oplus\mathbb{Z}}$, can be used to describe sets of A-addressed melodic motives with real onsets and integer pitches.

Local modular compositions have been classified (i.e. their isomorphism classes have been calculated) for the most important modules. Complete class lists for zero-addressed and self-addressed chords of pitch classes as well as 2-, 3-, and 4-element zero-adressed melodic motives are given in Chapter 11. We come back to the more general classification theorem for global modular compositions, as demonstrated in Chapter 15, in Section 66.2.

65.2 Categories of Local Networks

Networks are the mathematical shapes associated with processual or transformational approaches to music theory—and to other branches of the sciences, such as artificial and natural neural networks and object-oriented programming. For example, Karl Pribram's theory of dendritic neural processing [862] could be understood using the formalism of global networks. Instead of sets, one now uses (projective) limits of diagrams, i.e. the musical objects now represent diagrams of transformations between a list of elements. This approach has been forwarded by David Lewin [605] and Henry Klumpenhouwer [538] and was anticipated by the author regarding circle chords in [670] (see Figure 65.1) and elaborated in Noll's dissertation on transformational harmony [802]. It is remarkable to see that local networks arise when we deal with nerves of global compositions (see Section 66.1.1), and especially when we deal with module complexes used in the classification theorem of global compositions (see Section 66.2).

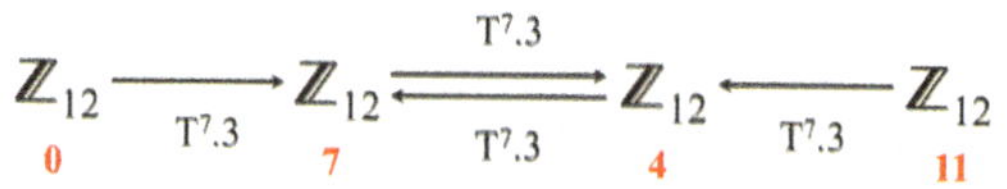

Fig. 65.1. A major seventh chord is generate by two pitch classes 0, 11 and a single transformation $f = T^7.3$. It unites two circle chords, the major and the minor chord, generated by f, starting from 0 or from 11, respectively.

Limit objects have been introduced in mathematical music theory with the development of the universal denotator format in the Rubato project since 1992 [690, 689] and systematically used in [714]. Here is the general framework to deal with such structures. Networks have two components: a digraph Γ and a category $\mathcal{C}$. A *local network* is a digraph morphism $g : \Gamma \to \mathcal{C}$ of the underlying digraphs. Observe that even if Γ is discrete, i.e. has no arrows, the map g indexes the codomain objects of $\mathcal{C}$. A local network is always an indexed family of objects and morphisms, not just a set. A local network may identify different vertices or arrows. If no vertices or arrows are identified, g is called *faithful*. Faithful local networks without arrows are just bijections of the vertex sets to subsets of category objects. But they are more than these object sets, they "order" them by the digraph's vertices. Local networks can be transformed according to two operations. If $g : \Gamma \to \mathcal{C}$ is a local network, if $t : \Delta \to \Gamma$ is a digraph morphism, and if $f : \mathcal{C} \to \mathcal{D}$, then the composition $f \circ g \circ t : \Delta \to \mathcal{D}$ is a new local network. For example, if g is a faithful local network on a discrete digraph Γ, then the automorphism group $Aut(\Gamma)$ of Γ is just the permutation group of its vertices, and the orbit $g \circ Aut(\Gamma)$ can be identified with the underlying image set of g. This technique has been used for the classfication of local compositions in [670].

Morphisms between local networks are defined as follows: If $g : \Gamma \to \mathcal{C}$ and $h : \Delta \to \mathcal{D}$ are local networks, a morphism $f : g \to h$ is a triple $f = (\gamma : \Gamma \to \Delta, u : \mathcal{C} \to \mathcal{D}, n : h \circ \gamma \to u \circ g)$, consisting of a digraph morphism γ, a functor u, and a natural transformation n. This defines the (big) category *LocNet* of local networks, and we denote by *LocNet*$(\mathcal{C})$ the subcategory of networks over $\mathcal{C}$, with morphisms $f = (\gamma, Id_{\mathcal{C}}, n)$, and by *LocNet*$(\Gamma, \mathcal{C})$ its subcategory of local networks $n : \Gamma \to \mathcal{C}$, with morphisms $f = (Id_\Gamma, Id_{\mathcal{C}}, n)$. For

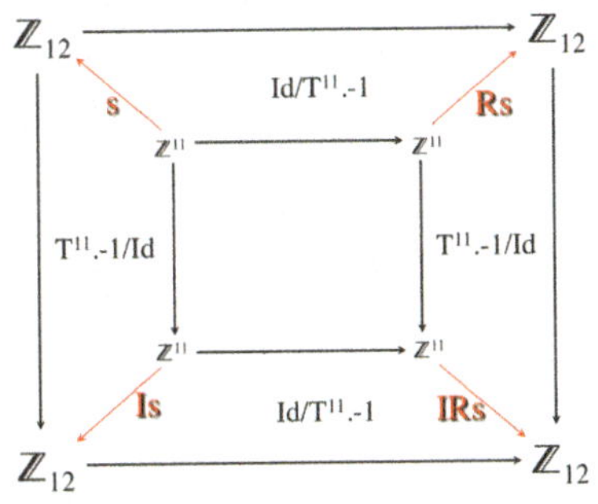

Fig. 65.2. A dodecaphonic network with address $\mathbb{Z}^{11}$.

music, a most prominent example of a category of networks is the category *LocNet*$(\int_{\mathcal{C}})$ over the *category* $\int_{\mathcal{C}}$ *of points over a category* $\mathcal{C}$. This is defined as follows. Its objects (the "points") are the A-addressed points x of presheaves F over $\mathcal{C}$, i.e. $x \in A@F$, or equivalently with Yoneda: $x : @A \to F$. The morphisms $f : x \to y$ from point $x : @A \to F$ to point $y : @B \to G$ are pairs $f = (\alpha : A \to B, h : F \to G)$ of morphisms in $\mathcal{C}$ and in $\mathcal{C}^@$, respectively, such that $h \circ x = y \circ @\alpha$; we notate these morphisms by h/α. Figure 65.2 shows such a network with $A = \mathbb{Z}^{11}, F = @\mathbb{Z}_{12}$, and four points s, Rs, IRs, Is, representing a dodecaphonic series s and its standard transforms by retrograde R and inversion I. The retrograde is a morphism α on the address $\mathbb{Z}^{11}$, whereas the inversion is a morphism h on the presheaf $@\mathbb{Z}_{12}$.

For many examples, it is even reasonable to restrict to the subcategory *FlatLocNet*$(\mathcal{C})$ of *flat local networks over* $\mathcal{C}$. Its objects are local networks in *LocNet*$(\int_{\mathcal{C}})$ which have their points all at one and the same address A and presheaf F, which means that the natural transformations associated with arrows of the digraph are all endomorphisms of F, i.e. we work in the category $\int_{End(F)}$ (with variable F). *Flat morphisms* between such flat local networks are defined as follows: If $g : \Gamma \to \int_{End(F)}, h : \Delta \to \int_{End(G)}$, g on address A, h on address B, then the digraph transformation $t : \Gamma \to \Delta$ and address change $\alpha : A \to B$ induce a natural transformation $n(\phi) : g \to h \circ t$, $\phi \in Nat(F, G)$, which is defined by the following morphism family of points at vertex $i \in V_\Gamma$:

$$\begin{array}{ccc} @A & \xrightarrow{g_i} & F \\ \alpha \downarrow & & \downarrow \phi \\ @B & \xrightarrow{h(t(i))} & G \end{array}$$

Such a morphism is denoted by $\phi/^t\alpha$ or ϕ/α if $t = Id_\Gamma$. For example, if we go back to Figure 65.2, we can interpret it as a flat morphism from the left vertical network g on $F = @\mathbb{Z}_{12}$, with one arrow $T^{11}. - 1/Id$ connecting s to Is, to the right vertical network h, also on F, with one arrow $T^{11}. - 1/Id$ connecting Rs to IRs. The morphism $\phi/\alpha : h \to g$ is defined by $\phi/\alpha = Id/T^{11}. - 1$. The classical K-nets [537] are all zero-addressed and flat local networks, often on $F = @\mathbb{Z}_{12}$ or on $2^{@\mathbb{Z}_{12}}$. However, their morphisms are not flat in general, but see [720] for a category-theoretical discussion of K-nets.

65.3 Categories of Local Gestures

If local networks refined local compositions by introducing functions that connect their elements, local gestures now refine local networks in that these functions are replaced by continuous curves from arguments to function values. This refinement is not only a topological task, but also takes care of the categorical context of local networks.

Here is the formal setup in a short recapitulation, but refer to Chapter 62 for a detailed discussion. Continuous curves are topological functors $f : \nabla \to K$ from the simplex category ∇ to a topological category K. A curve is a morphism in the category **TopCat** of topological categories. To distinguish topological functors from category-theoretical ones, we write $K \copyright L$ for $\mathbf{TopCat}(K, L)$. The set $\nabla \copyright K$ of curves in K becomes a category with $\nabla \copyright K$ as its object set, while a morphism $\nu : f \to g$ between curves f and g is a continuous natural transformation, meaning that $\nu : I \to K$ is continuous. $\nabla \copyright K$ becomes a topological category as follows: The morphisms are viewed as triples $(f, g, \nu) \in \nabla \copyright K \times \nabla \copyright K \times I \copyright K$, each space given the compact-open topology. This construction generalizes the purely topological situation where K is a topological space and curves are continuous maps $I \to K$, see 62.1 or [723].

The *spatial digraph* $\overrightarrow{K}$ of a topological category K is the digraph with $V_{\overrightarrow{K}} = K, A_{\overrightarrow{K}} = \nabla \copyright K$, and $t(g) = g(0), h(g) = g(1)$. Then a *(local) gesture* is a digraph morphism $g : \Gamma \to \overrightarrow{K}$, the digraph Γ is called the gesture's *skeleton*, and the category K its *body*. The set $\Gamma @ \overrightarrow{K}$ becomes a topological category by the fact that Γ is the colimit of a diagram $\mathcal{D}$ of arrows and points, so that $\Gamma @ \overrightarrow{K} \xrightarrow{\sim} \lim \mathcal{D}\overrightarrow{K}$, but for the one-arrow digraph $\uparrow$, we have $\uparrow @ \overrightarrow{K} \xrightarrow{\sim} \nabla \copyright K$, a topological category. Therefore the limit $\Gamma @ \overrightarrow{K}$ is a topological category, which we denote by $\Gamma \overrightarrow{@} K$.

Morphisms between local gestures are defined as follows. Let $g : \Gamma \to \overrightarrow{K}, h : \Delta \to \overrightarrow{L}$ be two gestures. Then a morphism $f : g \to h$ is a triple $f = (t : \Gamma \to \Delta, n : K \to L, \nu : \overrightarrow{n} \circ g \to h \circ t)$, where t is a digraph morphism, n a topological functor, and ν a natural transformation. Composition of morphisms is evident. This defines the category *LocGesture* of local gestures.

65.3.1 Local Gestures on Topological Categories of Points

An important example of a topological category that plays the role of the a gestural body is given by the category of points over a topological category $\mathcal{C}$ as follows. For a topological category $\mathcal{C}$, define the *continuous*[1] *presheaf* category $\mathcal{C}^\tau = \{F \in \mathcal{C}^@, F = \text{continuous}\}$, the morphism set $Nat(F, G)$ being given by the topology generated by the condition that all maps $A@ : Nat(F, G) \to (A@F)@(A@G)$ are continuous, where we take the compact-open topology on a set $S@T$ of morphisms between topological spaces S and T.

We want to show that the Yoneda map $Yoneda : \mathcal{C} \to \mathcal{C}^\tau$ is defined and continuous. To begin with, we show that for an object A of $\mathcal{C}$, the functor $Yoneda(A) = @A$ is continuous. Let $g : C \to B$ be a morphism in $\mathcal{C}$. Then the map $g@A : B@A \to C@A$ is continuous. In fact, the composition map $c : B@A \times C@B \to C@A$ is continuous by definition. The embedding $(?, g) : B@A \to B@A \times C@B : x \mapsto (x, g)$ is continuous, and $g@A = c \circ (?, g)$, whence the claim. The functor $Yoneda$ is also continuous, i.e. we show that the morphism maps $\xi : A@B \to (@A)@(@B)$ are continuous. By definition, we have to show that the maps $X@\xi : A@B \to (X@A)@(X@B)$ are continuous for all objects X of $\mathcal{C}$. Let $K \subset X@A$ be compact and $U \subset X@B$ be open. Denote by $[K, U]$ the set of all continuous functions $f : X@A \to X@B$ such that $f(K) \subset U$, a member of the subbasis of the compact-open topology on this function space. Show that $X@\xi^{-1}([K, U])$ is open. Let $p \in X@\xi^{-1}([K, U])$, which means that under the composition map $* : A@B \times X@A \to X@B$, $\{p\} \times K$ maps into U. Since K is compact, there is a neighborhood S of p and a neighborhood T of K such that $*(S \times T) \subset U$. This means that $S \subset X@\xi^{-1}([K, U])$ and we are done. Next we show that the Yoneda map $Y : Nat(@A, G) \to A@G : h \mapsto A@h(Id_A)$ is continuous. This is the composition of two continuous maps: the evaluation $A@ : Nat(@A, G) \to (A@A)@(A@G)$ and the identity evaluation $(A@A)@(A@G) \to A@G : f \mapsto f(Id_A)$. The latter is continuous since the inverse image of an open set $U \subset A@G$ is the subbase open set $[\{Id_A\}, U]$. Finally, we show that the inverse

[1] Continuous meaning that these presheaves map into the category **Top** of topological spaces.

$Y' : A@G \to Nat(@A, G)$, which maps $x \in A@G$ to the following natural transformation at object X: $X@Y'(x) : X@A \to X@G : g \mapsto G(g)(x)$, is continuous if G maps to locally compact Hausdorff topological spaces. We call such a presheaf *locally compact Hausdorff.* We have to show that for every X, the map $X@Y' : A@G \to (X@A)@(X@G)$ is continuous. Let $K \subset X@A$ be compact and $U \subset X@G$ open. Let us show that $X@Y'^{-1}([K, U])$ is open. The map $X@Y'(x)$ has this factorization

$$X@Y'(x) : X@A \xrightarrow{G} (A@G)@(X@G) \xrightarrow{?(x)} X@G$$

into two continuous factors. The first is independent of x and maps K to the compact set $L = G(K)$. Suppose that $x \in X@Y'^{-1}([K, U])$ This means that for all $l \in L$, $l(x) \in U$. Since l is continuous, there is an compact neighborhood $W_l(x)$ of x with $l(W_l(x)) \subset U$, meaning that $l \in [W_l(x), U]$. Since L is compact, there are finitely many $l_1, \ldots l_k$ such that $L \subset \bigcup_i [W_{l_i}(x), U]$. Therefore $L \subset [W(x), U]$ if $W(x)$ is a compact neighborhood of x contained in the finite intersection neighborhood $\bigcap_i [W_{l_i}(x), U]$. Therefore, for every $y \in W(x)$, $l(y) \in U$ for all $l \in L$, i.e. $W(x) \subset X@Y'^{-1}([K, U])$, QED.

This shows that under the local compactness condition on G, we have a Yoneda homeomorphism $Y : Nat(@A, G) \xrightarrow{\sim} A@G$. Let us therefore restrict to the subcategory **LCTop** of locally compact topological Hausdorff categories, i.e. the morphism sets are all locally compact Hausdorff and the target category of presheaves is **LCTop**, which means that $\mathcal{C}$ is locally compact Hausdorff and the Yoneda embedding is $Yoneda : \mathcal{C} \to \mathcal{C}^{lc\tau}$ with $\mathcal{C}^{lc\tau} = \{$functors $f : \mathcal{C}^{opp} \to$ **LCTop**$\}$, the category of locally compact Hausdorff presheaves. We may therefore topologize the set of morphisms $x@y$ between points $x : @A \to F, y : @B \to G$ as follows: We take the product topology on the cartesian product $Nat(@A, F) \times Nat(@B, G) \times A@B \times Nat(F, G)$, and given its subset $x@y$ the induced topology. With this topology, the composition of morphisms $y@z \times x@y \to x@z$ is continuous. In fact, the composition of morphism sets $B@C \times A@B \to A@C$ is continuous by definition on $\mathcal{C}$. Let us see that the composition $Nat(G, H) \times Nat(F, G) \to Nat(F, H)$ is also continuous. It is evidently sufficient to show this on the evaluation of all these presheaves at an object X, i.e.

$$(X@G)@(X@H) \times (X@F)@(X@G) \to (X@F)@(X@H)$$

is continuous. But this is true because our presheaves are locally compact. Therefore the category $\int_{\mathcal{C}}^{lc}$ of points over a locally compact category $\mathcal{C}$ and with values in locally compact presheaves is a topological category. Local gestures in this category are called *gestures of locally compact points over $\mathcal{C}$.*

And here is an example about gestures of locally compact points in musical performance theory. Refer to Parts VIII through XII and [725] for the mathematical theory of performance. We start with the small locally compact topological category **Frame**$_n$ of n-dimensional frames. The objects are *frames*, i.e. cubes $c = [a_1, b_1] \times \ldots [a_n, b_n] \subset \mathbb{R}^n$, $a_i \leqslant b_i$ for all $i = 1, \ldots n$. They are the frames where a piece of music has its notes in an n-dimensional parametrization. The morphisms are the inclusions of frames. They will be used to generate restrictions of performance vector fields. The locally compact topology of this category is given by the Euclidean metric d on $\mathbb{R}^{2n}$, where the frames are represented as points. The functor $F :$ **Frame**$_n^{opp} \to$ **Top** which we consider here is the functor of $\mathcal{C}^1$ vector fields, i.e. if c is an n-dimensional frame, $F(c) = \{v : c \to Tc | v$ is a $\mathcal{C}^1$ vector field$\}$, Tc being the tangent bundle of c. Since c is compact, the metric $(v, w) = max_c(d(v(x), w(x))$ for the Euclidean distance d on vectors in $\mathbb{R}^n$ defines a locally compact topological space. In performance theory, such fields are called performance fields, the most prominent example being the tempo and tuning curves, both one-dimensional performance fields. In performance theory one considers operators on performance vector fields, which alter performance fields according to given rationales, typically stemming from musical analysis. Here we just suppose that such an operator is a differentiable automorphism on F, i.e., an automorphism of the tangent bundle $F(c)$ which is natural in c, i.e., commutes with restrictions to smaller frames. This is a common situation in performance theory. Let us understand what a morphism between two points $v_1 : @c_1 \to F, v_2 : @c_2 \to F$ is. It is described by a subframe inclusion $\alpha : c_1 \subset c_2$ and an operator $\phi : F \xrightarrow{\sim} F$ such that the diagram

$$\begin{array}{ccc} @c_1 & \xrightarrow{v_1} & F \\ \alpha\downarrow & & \downarrow\phi \\ @c_2 & \xrightarrow{v_2} & F \end{array}$$

commutes. But this means that the restriction $v_2 \circ \alpha$ of v_2 to the subframe c_1 is the transformed version $\phi \circ v_1$ of the field on the smaller frame. In other words, $\phi^{-1} \circ v_2 \circ \alpha = v_1$, the restricted field is the transformed version of the restricted larger field under the inverse of the operator ϕ. This is exactly what one needs in performance theory to step from a "mother performance" to its "daughter performance". Now a curve in this topological category of points is a curve of such restrictions and operators, i.e. a continuous transition from the mother performance to the daughter performance. In performance theory, the successive refinement of performances works according to the performance stemma. This is an "inheritance" tree digraph Γ starting at a primary performance and ramifying to daughter performances for parts of the original piece. This means that for each vertex x of the stemmatic tree, we have a frame c_x and a performance field v_x. If this vertex ramifies to a number of daughter performance vertices $x_1, \ldots x_k$ this means that we have a number of subframes $c_{x_i} \subset c_x$, usually disjoint from each other, such as the two subframes separating a piano performance of the right from the performance of the left hand. For each of these subframes we have a curve of performance operators $\phi_i(t)$ and of restrictions $\alpha_i(t)$, $t \in I$, which defines the transition to the daughter performance field for each i as described above.

65.4 Connecting Functors

Local compositions, networks, and gestures are connected by a number of functors. To make the parallelism of gesture and network constructions more visible graphically, we write $\Gamma \circlearrowright C$ for $LocNet(\Gamma, C)$.

- **Gestures to Networks**: Let $g : \Gamma \to \overrightarrow{K}$ be a local gesture. We have a functor **form** : $\overrightarrow{K} \to K$ ("**form**" for "formalize", a terminology relating to forms and denotators in the Rubato Composer environment [730]) sending a curve $c : \nabla \to K$ to the morphism $c(0,1) : c(0,0) \to c(11)$ and a vertex (an object!) in $\overrightarrow{K}$ to itself. If $\nu : g \to h$ is a morphism in $\overrightarrow{K}$, its restriction to the endpoints of curves defines a morphism **form** $\circ\, \nu$ of associated local networks. This creates a functor **form** : $\Gamma \overrightarrow{@K} \to \Gamma \circlearrowright K$.
- **Networks to Compositions**: For a functor from networks to compositions, we have to restrict to flat local networks. Then we have a functor **fact** : $FlatLocNet \to ObLoc$ assigning to a flat local network $g : \Gamma \to \int_{End(F)}$ at address A the local composition $\{g(i) | i \in V_\Gamma\} \subset A@F$. If $h : \Delta \to \int_{End(G)}$ is a second flat local network, and if $\phi/^t\alpha : g \to h$ is a flat morphism, then we obtain a morphism of local compositions **fact**$(\phi/^t\alpha)$: **fact**$(g) \to$ **fact**(h) as follows: The element $g(i), i \in V_\Gamma$, is mapped to the element $h(t(i))\alpha$. We have to show that this map is in fact a morphism of local compositions. Since the morphism $\phi/^t\alpha$ is flat, we have $h(t(i))\alpha = A@\phi(g(i))$. Therefore the map does not depend on i, but only on the element $g(i)$, it is a well-defined set map. And the map is induced by the natural transformation ϕ on the underlying functors.
- **Compositions to Networks**: If we are given a local composition $K \subset A@F$, it defines a discrete network $n(K)$ on the discrete digraph K with the identity $K \to K$ as network map. If a morphism $f/\alpha : K \subset A@F \to L \subset B@G$ is given, there is a map $f' : K \to L$ that induces f, i.e. $\alpha \circ f' = f$ since $\alpha : L \to L\alpha$ is surjective. Evidently, the flat network morphism $(f', \alpha, \phi) : n(K) \to n(L)$ projects to f/α under the above functor **fact**. But the map (f', α, ϕ) is not a functorial assignment.
- **Networks to Gestures**: This transition is more delicate. We first of all need a topological category K to be able to define gestures there. There is a trivial solution for non-topological categories: Just give them the indiscrete topology. Then by the discrete gesture construction in Section 62.2.1, for a morphism $g : W \to Z$ in a category C there is a functor $\searrow(g) : \nabla \to C$. This is automatically a curve for the indiscrete topology on C. This construction yields a functor $\Gamma \circlearrowright C \to \Gamma \overrightarrow{@C}$, right-inverse to the functor **form** if the category C is given the indiscrete topology. No solution of this transition is known for general topological categories. However, for local networks of points in $\mathbb{R}^n$ (variable n), the method defined by

the Bruhat decomposition of linear transformations as described in [730, 2.3] yields gestures lying above given local networks for the projection functor **form**, however not in any functorial way. The Bruhat method is very important for implementations of gestural user-interfaces for musical composition.

65.5 Hypernetworks and Hypergestures

The fact that the spaces $\Gamma \overrightarrow{@} K$ of local gestures with skeleton Γ and body K are new topological categories and the parallel fact that spaces of local networks $\Gamma \circlearrowright \mathcal{C}$ in categories $\mathcal{C}$ also are new categories enable the iteration of such constructions. We may consider the topological local gesture category $\Delta \overrightarrow{@} \Gamma \overrightarrow{@} K$ of local gestures with skeleton Δ and body $\Gamma \overrightarrow{@} K$ as well as the category $\Delta \circlearrowright \Gamma \circlearrowright \mathcal{C}$ of local Δ-networks with values in the category $\Gamma \circlearrowright \mathcal{C}$. Local gestures for such an iteration are called *hypergestures*, local networks for such an iteration are called *hypernetworks*. It is immediate that hypernetworks and hypergestures are functorial in the basic categories, i.e. for a sequence of digraphs $\Gamma_0, \Gamma_1, \ldots \Gamma_{n-1}$, if $f : K \to L$ is a topological functor of topological categories, we have a canonical topological functor $\Gamma_0 \overrightarrow{@} \Gamma_1 \overrightarrow{@} \ldots \Gamma_{n-1} \overrightarrow{@} K \to \Gamma_0 \overrightarrow{@} \Gamma_1 \overrightarrow{@} \ldots \Gamma_{n-1} \overrightarrow{@} L$, and if $f : \mathcal{C} \to \mathcal{D}$ is a functor, then we have a canonical functor $\Gamma_0 \circlearrowright \Gamma_1 \circlearrowright \ldots \Gamma_{n-1} \circlearrowright \mathcal{C} \to \Gamma_0 \circlearrowright \Gamma_1 \circlearrowright \ldots \Gamma_{n-1} \circlearrowright \mathcal{D}$.

Moreover, since we have the **form** functor $\overrightarrow{K} \to K$, we have the corresponding functor from hypergestures to hypernetworks,

$$\overrightarrow{@}2\circlearrowright : \Gamma_0 \overrightarrow{@} \Gamma_1 \overrightarrow{@} \ldots \Gamma_{n-1} \overrightarrow{@} K \to \Gamma_0 \circlearrowright \Gamma_1 \circlearrowright \ldots \Gamma_{n-1} \circlearrowright K,$$

which, for topological functors $f : K \to L$, yields the commutative diagram

$$\begin{array}{ccc} \Gamma_0 \overrightarrow{@} \Gamma_1 \overrightarrow{@} \ldots \Gamma_{n-1} \overrightarrow{@} K & \xrightarrow{\overrightarrow{@}2\circlearrowright} & \Gamma_0 \circlearrowright \Gamma_1 \circlearrowright \ldots \Gamma_{n-1} \circlearrowright K \\ {\scriptstyle \overrightarrow{@}_f}\downarrow & & \downarrow{\scriptstyle \circlearrowright f} \\ \Gamma_0 \overrightarrow{@} \Gamma_1 \overrightarrow{@} \ldots \Gamma_{n-1} \overrightarrow{@} L & \xrightarrow{\overrightarrow{@}2\circlearrowright} & \Gamma_0 \circlearrowright \Gamma_1 \circlearrowright \ldots \Gamma_{n-1} \circlearrowright L \end{array}$$

65.5.1 Escher Theorems

Escher theorems have been proved for topological and categorical hypergestures in [720, 723], see Sections 61.14, 62.1.2. Their proof can be based upon the fact that double limits commute. This proof carries over to hypernetworks.

Theorem 3. (Escher Theorem) *For a sequence of digraphs $\Gamma_0, \Gamma_1, \ldots \Gamma_{n-1}$, a topological category K, a category $\mathcal{C}$, and a permutation $\pi \in S_n$, we have canonical isomorphism of topological categories or categories, respectively:*

$$\Gamma_0 \overrightarrow{@} \Gamma_1 \overrightarrow{@} \ldots \Gamma_{n-1} \overrightarrow{@} K \xrightarrow{\sim} \Gamma_{\pi(0)} \overrightarrow{@} \Gamma_{\pi(1)} \overrightarrow{@} \ldots \Gamma_{\pi(n-1)} \overrightarrow{@} K,$$
$$\Gamma_0 \circlearrowright \Gamma_1 \circlearrowright \ldots \Gamma_{n-1} \circlearrowright \mathcal{C} \xrightarrow{\sim} \Gamma_{\pi(0)} \circlearrowright \Gamma_{\pi(1)} \circlearrowright \ldots \Gamma_{\pi(n-1)} \circlearrowright \mathcal{C}.$$

65.6 Singular Homology of Hypernetworks and Hypergestures

Singular homology for topological spaces K has been generalized in [727] to hypergestures since singular cubic n-chains, i.e., continuous maps $c : I^n \to K \in \mathbf{Top}(I^n, K)$ in a topological space K, can be viewed as iterated 1-chains $I \to \mathbf{Top}(I^{n-1}, K)$, i.e., they are n-fold hypergestures in $\uparrow \overrightarrow{@} \uparrow \overrightarrow{@} \ldots \uparrow \overrightarrow{@} K$. Generalizing this fact, we have looked at n-fold hypergestures $c \in \Gamma_0 \overrightarrow{@} \Gamma_1 \overrightarrow{@} \ldots \Gamma_{n-1} \overrightarrow{@} K$ for topological categories K instead of n-chains, see Chapter 63. Homology is then defined by a procedure that is independent of the special topological conditions, see Chapter 63. We first introduce a face operator $c^{\square} \in K$ for a gesture

$c \in \Gamma \overrightarrow{@} K$. For a given commutative ring R, we define the module $C_k(R, K, \Gamma_*)$ of k-chains as the free R-module generated by all k-fold hypergestures of partial and different k-length sequences of skeleta from the original sequence $\Gamma_* = \Gamma_0, \Gamma_1, \ldots \Gamma_{n-1}$. We then define a boundary operator using the Escher theorem as follows. Let $c \in \Gamma_0 \overrightarrow{@} \Gamma_1 \overrightarrow{@} \ldots \Gamma_{n-1} \overrightarrow{@} K$ and call $c_i \in \Gamma_i \overrightarrow{@} \Gamma_0 \overrightarrow{@} \ldots \Gamma_{i-1} \Gamma_{i+1} \ldots \Gamma_{n-1} \overrightarrow{@} K$ the corresponding hypergesture given by the Escher theorem. Then we set $\partial_n c = \sum_{i=0}^{n-1} (-1)^i c_i^{\square}$. This map is extended linearly to the module $C_n(R, \Gamma_., K)$. The important fact is that $\partial_{n-1} \circ \partial_n = 0$, i.e. $Z_{n-1} = Ker(\partial_{n-1}) \subset B_n = Im(\partial_n)$. The quotient module $H_{n-1}^{gest}(R, K, \Gamma_*) = Z_{n-1}/B_n$ is called the $n-1st$ *gestural homology module* of the data R, K, and Γ_*.

This procedure is totally independent of the specific nature of chains. This means that we can also define homology with the above procedures *mutatis mutandis* for hypernetworks instead of hypergestures. The only difference is that we now start from any category $\mathcal{C}$ instead of a topological category K, and that we take hypernetworks instead of hypergestures. This enables the definition of *network homology modules* $H_{n-1}^{net}(R, \mathcal{C}, \Gamma_*)$. It is immediate that if we are given a functor $f : K \to \mathcal{C}$, then the above hypergesture-to-hypernetwork maps induce homology module homomorphisms

$$H_k(f) : H_k^{gest}(R, K, \Gamma_*) \to H_k^{net}(R, \mathcal{C}, \Gamma_*)$$

for all indices k in our sequence. Let us summarize this discourse by saying that homology theory is the classical technical tool to connect hypergestures or hypernetworks of different levels. It has been a key to the solution of the Weil and Fermat conjectures, and this is why we expect them to give us essential insights to gestural and transformational music theory.

66

Global Categories

Summary. We discuss global categories of compositions, processes, and gestures.

– Σ –

66.1 Categories of Global Compositions

Let us first recall the category of objective global compositions introduced in Chapter 13. We will not deal with non-objective global compositions in this chapter. Global compositions arise from the music-theoretical interpretation of sets of musical objects X by coverings with specific subsets $Y \subset X$. For example, a major scale $X \subset \mathbb{Z}_{12}$ may be covered by the set $C = \{I_X, II_X, \ldots VI_X, VII_X\}$ of seven degree triads. Or a musical analysis of a composition, given as a set $X \subset N$ of notes in a note space N, may be interpreted analytically with a covering $C = \{C_1, \ldots C_n\}$ consisting of subsets $C_i \subset X$. The covering then defines an atlas of charts which intersect in a specific way and define a global structure, similarly to a geographic atlas or a differentiable manifold or an algebraic variety, which we call a(n objective) *global composition*. Here is the definition:

Definition 114 *A(n objective) global composition consists of these components:*

1. *a set G, the* support, *and a finite covering I of G by non-empty sets,*
2. *an address module A,*
3. *a family $(K_t \subset A@F_t)_T$ of A-addressed local compositions,*
4. *a surjection $I_? : T \to I$,*
5. *a bijection $\phi_t : K_t \xrightarrow{\sim} I_t$ for each $t \in T$,*
6. *for each couple $s, t \in T$ such that $I_s \cap I_t \neq \emptyset$, the induced bijection*

$$\phi_{s,t} : \phi_s^{-1}(I_s \cap I_t) \xrightarrow{\sim} \phi_t^{-1}(I_s \cap I_t)$$

is an isomorphism

$$\phi_{s,t}/1 : (\phi_s^{-1}(I_s \cap I_t), A@F_s) \xrightarrow{\sim} (\phi_t^{-1}(I_s \cap I_t), A@F_t)$$

of local compositions.

The data 3.-5. are called an *A-addressed atlas* of the global composition which is often denoted by G^I. The bijections ϕ_t are called the *charts* of the atlas. Two atlases are *equivalent* if their disjoint union is also an atlas for G^I. The global composition is defined by an equivalence class of atlases.

If G^I and H^J are two global compositions, a morphism $f : G^I \to H^J$ is a triple of maps $f = (f_1 : G \to H, f_2 : I \to J, \alpha : A \to B)$ with $f_1(i) \subset f_2(i)$ for each $i \in I$, such that for each chart $K_t \subset A@F_t$ for I_t in G^I, and $L_s \subset B@G_s$ for $I_s = f_2(I_t)$ in H^J, the map $f_1|K_t : K_t \to L_s$ is a morphism $f_1|K_t/\alpha$ of local compositions. This defines the category *ObGl* of (objective) global compositions. It contains the full subcategory of *ObLoc*, the latter being identified with the global compositions having a singleton covering.

G. Mazzola et al., *The Topos of Music III: Gestures*, Computational Music Science,
https://doi.org/10.1007/978-3-319-64481-3_11

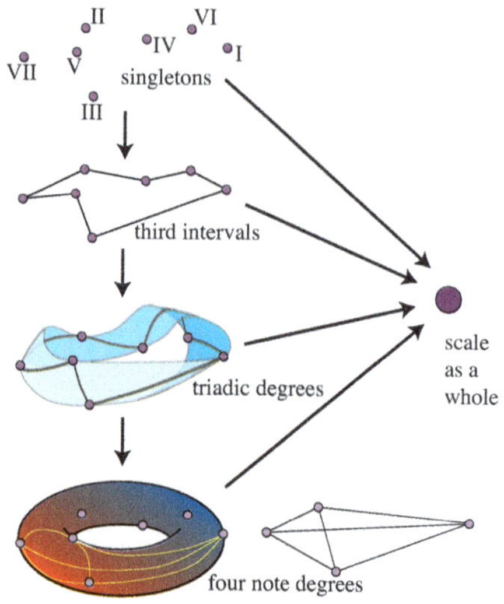

Fig. 66.1. The morphisms between nerves of different coverings (singletons, thirds, triads, tetrades) of the diatonic scale.

66.1.1 Simplicial Methods

Before we discuss some important examples, we introduce a powerful geometric method for understanding global compositions: the simplicial functor. It will also be useful *mutatis mutandis* for global networks and global gestures. We will not explicate those aspects here since they are straightforward once the case of global compositions is understood. This method is based on the *nerve* functor $N : ObGlob \rightarrow \mathbf{Cat}$ mapping a global composition G^I to the simplicial complex $N(G^I)$ of its covering I. In combinatorial topology, this is also called the *covering's nerve*. It is a small category, whose objects are finite, non-empty sets $\sigma \subset I$ such that their intersection $\cap\sigma$ is not empty. An *n-simplex* is a simplex having $n+1$ elements. Its morphisms are the inclusions $\sigma \subset \tau$. Nerves define a functor since (with the above notation for a morphism, $f : G^I \rightarrow H^J$) $f_2|\sigma : \sigma \rightarrow f_2(\sigma)$ defines a natural transformation between associated nerves. Nerves are visualized by the so-called *geometric realization*, or *geometric nerve*, which shows 0-simplices as points, 1-simplices as lines connecting two 0-simplex points, 2-simplices as triangular surfaces, 3-simplices as tetrahedra volumes, and so on for higher dimensional geometric shapes. It turns out that frequent geometric constructions in neo-Riemann theory are special cases of this simplicial method, rather than so-called duality constructions.

Figure 66.1 shows the geometric nerve morphisms between different coverings of the diatonic scale: From one-element to thirds to the classical triadic to the tetradic covering, we have morphisms of global compositions induced by the identity on the supporting scale and embeddings of singletons in intervals, intervals in triads, and triads in tetrades. One sees that the famous harmonic band the nerve of the triadic covering, a Möbius strip, is embedded in the full torus nerve for the tetradic covering.

Moreover, a simplex σ induces a local composition on its intersection, which we also denote by $\cap\sigma$. This defines the contravariant functor, the *simplicial weight* $\cap N(G^I) : N(G^I)^{opp} \rightarrow ObLoc$ of local subcompositions of G^I induced on the simplex intersections. The simplicial weight is a powerful technique for the description and classification of global compositions. We may attach the isomorphism class to each local composition of the simplicial weight of G^I and then obtain the *class nerve* $CN(G^I)$. In some cases, this is a full set of invariants for G^I, i.e. its isomorphism class is determined by its class nerve. Figure 66.2 shows the intervallic class nerve of a three-element motif in $\mathbb{Z}_{12}^2$. Such three-element motives are in fact classified by their intervallic class nerves, while four-element motives are not classified by the class nerve of their three-element submotif coverings, see Section 14.5 and 17.1 for examples and counterexamples. The class nerve of the triadic Möbius strip of the diatonic scale is shown in Figure 66.3.

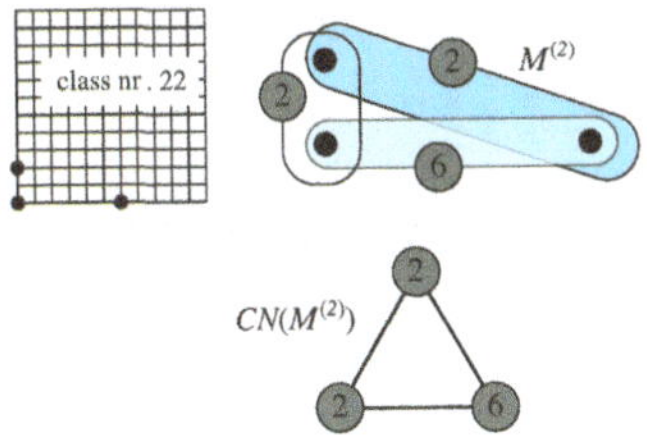

Fig. 66.2. The class nerve $CN(M^{(2)})$ for the intervallic motif interpretation of motif class nr. 22 from list O.3. The class numbers 2, 2, 6 are from the list O.1.

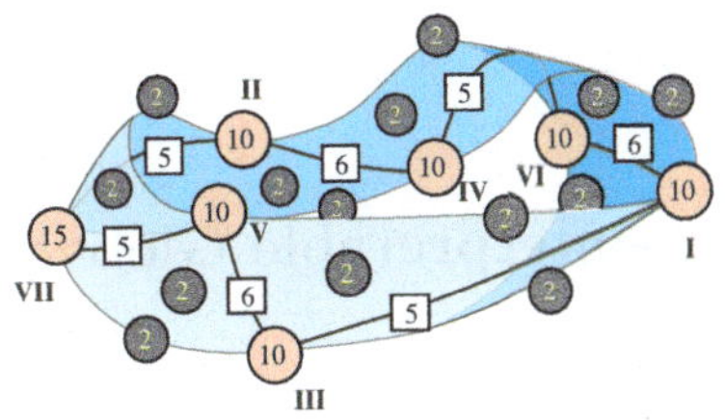

Fig. 66.3. The class nerve of the harmonic strip; the numbers are the class numbers of chords, intervals, and notes in the classification list N.1.

In view of the often high-dimensionality of nerves of compositions it is recommended to restrict the geometric perspective to low dimensions, and this is the standard approach in neo-Riemann theory. The general procedure is this: One takes the simplicial weight functor $\cap N(G^I) : N(G^I)^{opp} \to ObLoc$, selects a subcategory **Sel** $\subset ObLoc$ and considers the fiber of $\cap N(G^I)|$**Sel** over **Sel**. In neo-Riemann theory, one considers typically the covering (3) of the local composition $\mathbb{Z}_{12}$ of pitch classes by major and minor triads. This defines a global composition $\mathbb{Z}_{12}^{(3)}$ which has up to five-dimensional simplices (a pitch class can sit in six different triads). One therefore only looks at simplices σ having at least two elements in $\cap\sigma$. The fiber over his subcategory defines a subcategory of the nerve, in fact Jack Douthett's and Richard Steinbach's "Chickenwire" category [276] with only zero- and one-dimensional simplices. Like many approaches to global theory, neo-Riemannian geometry constructs global compositions starting from a local composition G and defines global compositions G^I from a covering I of G by interesting subcompositions. This evidently defines a global composition whose support is G, and whose charts are the local compositions from I, the gluing isomorphisms being all identities. We call such a global composition an *interpretation of* G. A global composition which is isomorphic an interpretation is called *interpretable*. It is one of the main results of classification theory to have classified many types of global compositions and to have given criteria for interpretability.

66.2 Classification of Global Compositions

Here is the most general classification theorem for global compositions, proved for zero-addressed global compositions in [670, p.52 ff.] and then for general addresses in Section 15.3.2, Theorem 18.

Theorem 42 (Geometric Classification of Global Compositions)
Let A be a locally free module of finite rank over the commutative ring R. Consider the A-addressed global compositions G^I with the following properties (*):

- *the modules $R.G_i$ that are generated by the charts G_i, i.e., by all differences $x - y, x, y \in G_i$ within the chart spaces, are locally free of finite rank;*
- *the modules of affine functions* $\mathbf{Aff}(G_i)$ *are projective.*

Then there exists a subscheme J of a projective R-scheme of finite type whose points $\omega : Spec(S) \to J$ are in one-to-one correspondence with the isomorphism classes of global modular compositions at address $S \otimes_R A$ sharing the properties (*).

Special cases of this result are the complete lists of isomorphism classes for local compositions $K \subset \mathbb{Z}_{12}$ (zero-addressed chords), zero-addressed motives of 1,2,3, and 4 elements $M \subset \mathbb{Z}_{12}^2$, and self-addressed chords in $\mathbb{Z}_N$ or in the dual number space $\mathbb{Z}_N[\varepsilon]$; see Appendices N.1 and O. The proof of this theorem resides on the construction of the *resolution* $\Delta(G^I) \to G^I$ of a global composition, a free global composition over $N(G^I)$, and on the classification of orbits of modules of affine functions on the resolution under the action of the automorphism group of this resolution. This technique vastly generalizes the idea of investigating local compositions as orbits under the permutation group of their discrete indexing networks as described earlier in Section 65.2.

66.3 Non-interpretable Global Compositions

The existence of non-interpretable global compositions is a dramatic step in the understanding of music, a detailed discussion of such a situation is given in Chapter 17. The reason is that the combination of local charts on a global composition can generate musical connectivity that is not representable in a single composition. But it has deeper consequences, namely that we can also create non-interpretable global networks and gestures using the classification of global compositions. We come back to this context when discussing global networks and gestures.

Here are two examples of non-interpretable global compositions, both zero-addressed.

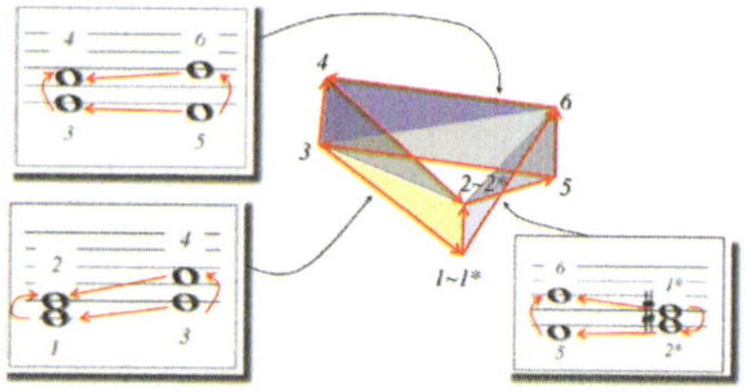

Fig. 66.4. A global Möbius-type network. It is not interpretable because its associated global composition shown in Figure 66.5 is not interpretable.

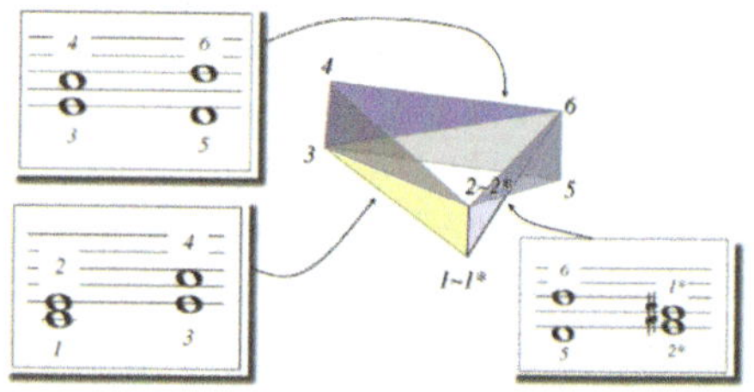

Fig. 66.5. The global, non-interpretable compositions associated with the global network in Figure 66.4.

The first example is very simple to construct. We take a support set $G = \{x_0, x_1, x_2, x_2'\}$ of four points and cover it by two charts $G_1 : \{x_0, x_1, x_2\} \xrightarrow{\sim} \mathbb{Z}_3 : x_i \mapsto i$ and $G_2 : \{x_0, x_1, x_2'\} \xrightarrow{\sim} \mathbb{Z}_3 : x_0 \mapsto 0, x_1 \mapsto 1, x_2' \mapsto 2$. The charts are glued together by the identity on the two-element intersection $\{x_0, x_1\}$. This global composition is not interpretable since in any $\mathbb{Z}_3$-vector space, the three points of the charts would be collinear, and therefore only three points would result instead of four.

The second example is more refined since it does not simply double points by gluing special subsets; refer to Figure 66.5. This global composition has a support of six points $G = \{x_1, x_2, x_3, x_4, x_5, x_6\}$. It has three charts $G_i, i = 1, 2, 3$, defined as follows: All charts are in zero-addressed local compositions in $\mathbb{R}^2$, namely $G_1 : \{x_1, x_2, x_3, x_4\} \xrightarrow{\sim} \{y_1 = (0,0), y_2 = (0,1), y_3 = (1,0), y_4 = (1,1)\} \subset \mathbb{R}^2 : x_i \mapsto y_i$, $G_2 : \{x_1, x_2, x_5, x_6\} \xrightarrow{\sim} \{y_1^* = (1,1), y_2^* = (1,0), y_5 = (0,0), y_6 = (0,1)\} \subset \mathbb{R}^2 : x_i \mapsto y_i^*, i = 1, 2, x_i \mapsto y_i, i = 5, 6$, and $G_3 : \{x_3, x_4, x_5, x_6\} \xrightarrow{\sim} \{y_3 = (0,0), y_4 = (0,1), y_5 = (1,0), y_6 = (1,1)\} \subset \mathbb{R}^2 : x_i \mapsto y_i$. To decide whether this is an interpretable global composition or not, we apply a criterion from Section 15.2.2. Let f be an affine function on G. Then evidently, writing $f_{ij} = f(x_i) - f(x_j)$, we have $f_{21} = f_{43}$, $f_{21} = f_{56}$, $f_{56} = f_{34} = -f_{43}$. Therefore $f_{43} = -f_{43} = 0$, and $f_{21} = f_{56} = f_{34} = 0$, no affine function can separate points x_2 from x_1, x_5 from x_6, and x_3 from x_4. According to Proposition 16 in Section 15.2.2, G is not interpretable.

66.4 Categories of Global Networks

Global networks as interpretations of local networks related to K-nets have been considered implicitly by Robert Peck [829] in neo-Riemannian theory. Similarly to global compositions we may define global networks as follows:

Definition 1. *A global network consists of these components:*

1. *a digraph Γ, the* support, *and a finite covering I of Γ by non-empty sub-digraphs,*

2. an address module A,
3. a family $(k_t : \Delta_t \to \int_{End(F_t)})_T$ of A-addressed flat and faithful local networks,
4. a surjection $I_? : T \to I$,
5. an isomorphism $\phi_t : \Delta_t \xrightarrow{\sim} I_t$ for each $t \in T$,
6. for each couple $s, t \in T$ such that $I_s \cap I_t \neq \varnothing$, the induced isomorphism

$$\phi_{s,t} : \phi_s^{-1}(I_s \cap I_t) \xrightarrow{\sim} \phi_t^{-1}(I_s \cap I_t)$$

extends to a flat isomorphism of flat local networks

$$\phi_{s,t}/1 : (\phi_s^{-1}(I_s \cap I_t) \to \int_{End(F_s)}) \xrightarrow{\sim} (\phi_t^{-1}(I_s \cap I_t) \to \int_{End(F_t)}).$$

The data 3.-5. are called an *A-addressed atlas* of the global network, which is often denoted by Γ^I. The bijections ϕ_t are called the *charts* of the atlas. Two atlases are *equivalent* if their disjoint union is also an atlas for Γ^I. The global composition is defined by an equivalence class of atlases.

If Γ^I and Σ^J are two global networks, a morphism $f : \Gamma^I \to \Sigma^J$ is a triple of maps $f = (f_1 : \Gamma \to \Sigma, f_2 : I \to J, \alpha : A \to B)$ with $f_1(i) \subset f_2(i)$ for each $i \in I$, such that for each chart $\Delta_t \to \int_{End(F_t)}$ for I_t in Γ^I, and $\Lambda_s \to \int_{End(G_s)}$ for $I_s = f_2(I_t)$ in Σ^J, the map $f_1|\Delta_t : \Delta_t \to \Lambda_s$ induces a flat morphism $\psi_t/^{f_1|\Delta_t}\alpha$ of flat local networks. This defines the category *GlobNet* of global networks. It contains the full subcategory of *FlatLocNet*, the latter being identified with the global networks having a singleton covering.

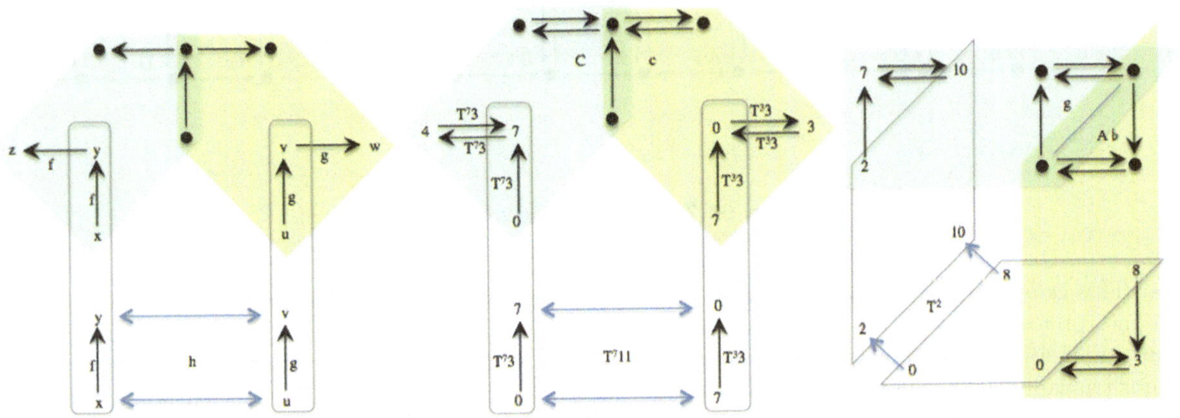

Fig. 66.6. The gluing lemma (left) and the two gluing types in the neo-Riemann global network (middle and right).

66.4.1 Non-interpretable Global Networks

To ease notation, we shall denote zero-addressed flat local networks on digraphs Γ with values in modules M by $f : \Gamma \to M$ instead of $f : \Gamma \to \int_{End(@M)}$. Let us first prove a simple criterium with splitting circle maps. Consider the following simple zero-addressed global network construction; refer to Figure 66.6, left network. The global network is supported by the digraph Γ on the top of Figure 66.6, and we are given two charts of three points and two arrows each, intersecting at the middle vertical arrow. The left chart $G_1 : \Gamma_1 \to \mathbb{Z}_{12}$ is a diagram $x \xrightarrow{f} y \xrightarrow{f} z$, the right chart $G_2 : \Gamma_2 \to \mathbb{Z}_{12}$ is a diagram $u \xrightarrow{g} v \xrightarrow{g} w$. The charts are intersecting on the subdiagrams $x \xrightarrow{f} y$ and $u \xrightarrow{g} v$, which are connected by an isomorphism of local networks h. Call this global network G.

Lemma 2. *If* $(\{x, y\}, \mathbb{Z}_{12})$ *is a generating local composition, the global composition* G *is not interpretable by a faithful local network on any* $\mathbb{Z}_{12}$*-module* M.

In fact, suppose that there is a local network $L : \Gamma \to M$ whose interpretation $L^{\{\Gamma_1, \Gamma_2\}}$ is isomorphic to G. Let x', y', z' be the points in M corresponding to the points x, y, z of chart G_1. Then there is an isomorphism of local compositions $f : (\{x, y, z\}, \mathbb{Z}_{12}) \xrightarrow{\sim} (\{x', y', z'\}, M)$, and hhence an isomorphism of the underlying modules $f_0 : \mathbb{Z}_{12}\{x, y, z\} \xrightarrow{\sim} \mathbb{Z}_{12}\{x', y', z'\}$. Since $\{x, y\}$ is generating, $\mathbb{Z}_{12}\{x, y, z\} = \mathbb{Z}_{12}$, and its image $\mathbb{Z}_{12}\{x', y', z'\}$ is isomorphic to $\mathbb{Z}_{12}$. But as $\mathbb{Z}_{12}$ is self-injective, this image is a direct summand of M: $M = \mathbb{Z}_{12}\{x', y', z'\} \oplus M' \xrightarrow{\sim} \mathbb{Z}_{12} \oplus M'$. Observe that h is in fact a module isomorphism since $(\{x, y\}, \mathbb{Z}_{12})$ is generating. The analogous statement is therefore true for the second chart G_2 since $(\{u, v\}, \mathbb{Z}_{12})$ is also generating via the *isomorphism* h. Since u, v map to x', y', too, $\mathbb{Z}_{12}\{u, v, w\}$ maps to the same module as $\mathbb{Z}_{12}\{x, y, z\}$ does. Therefore both charts map to the common coset $x' + \mathbb{Z}_{12}\{x', y', z'\}$ in M. Transposing on M by $T^{-x'}$, we may suppose that the images of both charts lie in $\mathbb{Z}_{12}\{x', y', z'\}$, which means that we may suppose wlog that $M = \mathbb{Z}_{12}$. But then the transformation f on the first chart must coincide with the transformation g on the second chart in the local network in M. This implies that $z' = f(y') = g(y') = w'$, where w' is the third point on the second chart. But then there are only three points on this network, it is not faithful, and we are done.

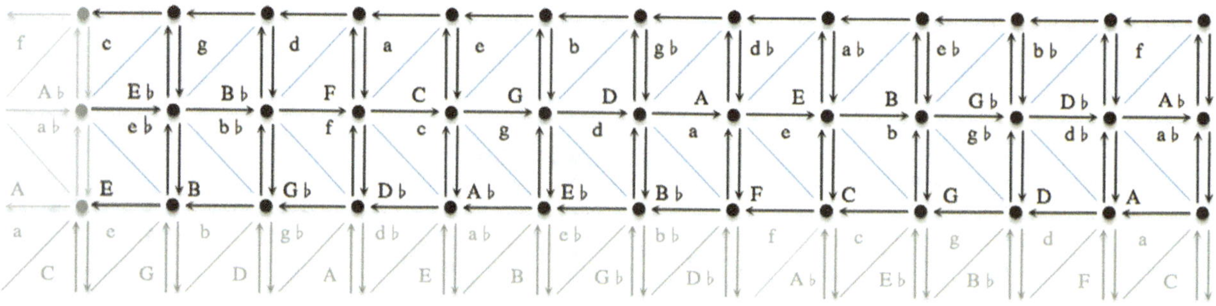

Fig. 66.7. The global Zarlino network.

This criterium applies to show that a global network related to neo-Riemannian approaches is in fact not interpretable. The network is shown in Figure 66.7. It is motivated by the desire (1) to use the fact that major and minor triads are circular chords as shown in Figure 65.1, and (2) to connect major and minor triads under the classical Zarlino inversion that connects major to minor triads exchanging tonic and fifths and exchanging major third with minor third. We represent all major and minor triads as circle chord networks. All gluing data are identities except the two types shown in the middle and right graphics on Figure 66.6. Since the middle gluing data are present in each major/minor couple of the Zarlino network, the network cannot be interpretable. It is a torus-shaped grid with shifted horizontal gluing connections, and every triadic chart appears twice, once in a fifth tower, and once in a fourth tower.

A second criterium for non-interpretability of global networks is the existence of a global functor **globfact** : *GlobNet* → *ObGlob*, which maps interpretable global networks to interpretable global compositions. We leave it to the reader to define this functor. It is essentially the extension of the **fact** functor for local theories described in the second item of Section 65.4, just gluing together charts of local objects. An example of this criterium is the global network shown in Figure 66.4. It projects to the non-interpretable global composition shown in Figure 66.5 and discussed in Section 66.3. Therefore it is also not interpretable.

The classification of global networks is not settled, and it is far from evident. We are afraid that it is as difficult as the classification of the representation of digraphs in linear algebra. The latter is a *wild* problem in the sense that one could classify all module categories if that were possible. Therefore the factualization functor **globfact** : *GlobNet* → *ObGlob* is essential in that the classification of the codomain category helps us understand the isomorphism classes of global networks.

66.5 Categories of Global Gestures

Here is the definition of a global gesture, similar to the definition of global networks or compositions:

Definition 115 *A global gesture consists of these components:*

1. *a digraph Γ, the* support, *and a finite covering I of Γ by non-empty subdigraphs,*
2. *a family $(k_t : I_t \to \overrightarrow{K_t})_I$ of faithful*[1] *local gestures,*
3. *for each couple $s, t \in I$ such that $I_s \cap I_t \neq \emptyset$, the identity $Id_{s,t}$ of digraphs $I_s \cap I_t$ extends to an isomorphism $f_{s,t} = (Id_{s,t}, n_{s,t}, \nu_{s,t})$ of local gestures,*

$$f_{s,t} : k_s|I_s \cap I_t \xrightarrow{\sim} k_t|I_s \cap I_t.$$

The data in 2. and 3. are called an *atlas* of the global gesture, which is often denoted by Γ^I. The local gestures k_t are called the *charts* of the atlas. Two atlases are *equivalent* if their disjoint union is also an atlas for Γ^I. The global gesture is defined by an equivalence class of atlases.

If Γ^I and Σ^J are two global gestures, a morphism $f : \Gamma^I \to \Sigma^J$ is a pair of maps $f = (f_1 : \Gamma \to \Sigma, f_2 : I \to J)$ with $f_1(i) \subset f_2(i)$ for each $i \in I$, such that for each chart $k_t : I_t \to \overrightarrow{K_t}$ in Γ^I, and $l_t : I_s \to \overrightarrow{L_s}$ for $I_s = f_2(I_t)$ in Σ^J, the map $f_1|k_t : I_t \to I_s$ induces a local gesture morphism $(f_1|I_t, h_t, n_t) : k_t \to l_s$. This defines the category *GlGesture* of global gestures. It contains the full subcategory of *LocGesture*, the latter being identified with the global gestures having a singleton covering.

The functorial transition from global networks to global compositions cannot be copied without problems to the transition from global gestures to global networks. The point is that the local transition functor **form** : $\Gamma \overrightarrow{@} K \to \Gamma \circlearrowright K$ does not conserve faithfulness in general since many gestures g from x to y can give the same morphism $g(0,1)$ from x to y. And the natural transformations for gestures over locally compact points also do not yield flat local networks and morphisms. In this generality, gestures are definitely richer than networks.

66.6 Globalizing Topological Categories: Categorical Manifolds

Intuitively, a categorical manifold is a manifold whose charts are topological categories. We first have to recall from Appendix J.4.2 that a (small) category is a directed graph $\Gamma : \mathbf{C_1} \to \mathbf{C_0^2}$, with an arrow set called morphism set $\mathbf{C_1}$, and a vertex set called object set $\mathbf{C_0}$, together with two maps $e : \mathbf{C_0} \to \mathbf{C_1}$, $m : \mathbf{C_1} \times_{\mathbf{C_0}} \mathbf{C_1} \to \mathbf{C_1}$ for identification of objects and identity morphisms and composition of morphisms. A topological category is the same, except that all objects and maps are in the category **Top** of topological spaces and continuous maps. A *topological digraph* is a digraph $\Gamma : A_\Gamma \to V_\Gamma^2$ that is internal to **Top**, i.e., the objects and maps are in **Top**. Therefore, a topological category is a topological digraph that is enriched by the two continuous maps e and m, together with the category-theoretical axioms.

Definition 116 *A* categorical manifold *is a finite covering I of a topological digraph Σ by non-empty open sub-digraphs $\Sigma_\iota \subset \Sigma$ (meaning that their arrow and vertex sets are open in Σ's arrow and vertex sets) which are all endowed with e_ι, m_ι and define topological categories; they are called the manifold's* charts. *Moreover, the non-empty intersections $\Sigma_\iota \cap \Sigma_\kappa$ are also supposed to be topological categories that are induced by the super-categories $\Sigma_\iota, \Sigma_\kappa$. Such a categorical manifold is denoted by Σ^I.*

If Σ^I, Δ^J are two categorical manifolds, a morphism $f : \Sigma^I \to \Delta^J$ is a pair of morphisms $f = (\phi, F)$, $\phi : \Sigma \to \Delta$ being a continuous digraph map, and $F : I \to J$ a set map such that $\phi(\Sigma_\iota) \subset \Delta_{F(\iota)}$ for all $\iota \in I$, and the map $\phi|\Sigma_\iota : \Sigma_\iota \to \Delta_{F(\iota)}$ is a morphism of topological categories.

This defines the category **CatMan** *of categorical manifolds.*

[1] Meaning that the digraph morphism k_t is mono.

Example 78 Every topological category is a categorical manifold with a single chart. Every covering of a topological category by a set of open non-empty subcategories defines a categorical manifold, since the non-empty intersections of any two subcategories are subcategories. We call categorical manifolds that are isomorphic to such objects *interpretable categorical manifolds*, in analogy to the situation for global compositions. We don't know whether there are non-interpretable categorical manifolds. The conjecture is "yes".

We can visualize this construction on the topological category ∇. Take an open interval covering $J = \{[0, b_1[,]a_2, b_2[, \ldots]a_n, 1]\}$ of its object set $[0, 1]$. Figure 66.8 shows the categorical manifold ∇^J.

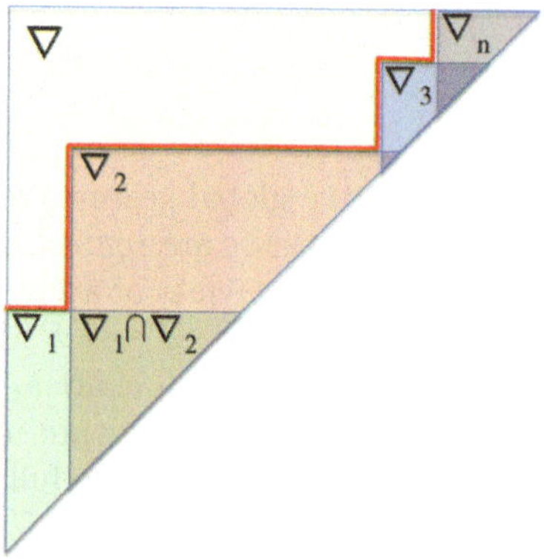

Fig. 66.8. The categorical manifold obtained from an open interval covering of the object set of ∇.

Example 79 If Σ^I is a covering of a digraph Σ with the discrete topology, we may consider for every subdigraph Σ_ι of this covering its path category $Path(\Sigma_\iota)$. Adding the paths of these subdigraphs to Σ defines a larger graph, which we denote by $Path(\Sigma^I)$ and call the *path manifold of* Σ^I. We have the intersection formula $Path(\Sigma_\iota) \cap Path(\Sigma_\kappa) = Path(\Sigma_\iota) \cap \Sigma_\kappa)$. $Path(\Sigma^I)$ is a categorical manifold with the discrete topology. Check that this is not a special case of Example 78.

Example 80 See Figure 66.9 for this example. Take two copies ∇_1, ∇_2 of the basic topological category ∇. Select two open intervals $]a_1, b_1[,]a_2, b_2[\subset I$, and consider the full open subcategories $\nabla(a_1, b_1) \subset \nabla_1$, $\nabla(a_2, b_2) \subset \nabla_2$ defined on the objects from $]a_1, b_1[,]a_2, b_2[$, respectively. The evident linear bijection $q :]a_1, b_1[\overset{\sim}{\to}]a_2, b_2[$ defines an isomorphism of categories $q : \nabla(a_1, b_1) \overset{\sim}{\to} \nabla(a_2, b_2)$. Now glue ∇_1, ∇_2 together along the subcategories $\nabla(a_1, b_1), \nabla(a_2, b_2)$ via q. This colimit Γ^I of topological digraphs defines a categorical manifold that has two charts, $I = \{\nabla_1, \nabla_2\}$, whose intersection is the glued part $q : \nabla(a_1, b_1) \overset{\sim}{\to} \nabla(a_2, b_2)$. One sees that even if two morphisms, one in ∇_1, the other in ∇_2, are such that their domains and codomains could define a composed morphism, this will not be possible in general in Γ^I; this manifold is not a category anymore. Check that this is a special case of Example 78.

Example 81 A practical musical example results from the hand's space-time $M_{\mathbb{C}} = W \times \mathbb{C}$ (see Section 78.2.3 for W). This topological space (qua topological category) may be covered by open subsets that reflect regions describing where the hand may move within the total space, and where it may not be positioned. Such constraints can occur for particular hand anatomy or for requirements that result from the composition's structure. The resulting categorical manifold is of type defined in Example 78.

To get an idea of the hand's space-time $M_{\mathbb{C}}$ and its possible categorical manifolds, we can think of the keyboard-related space. In principle, every point can be reached by a pianist's hands. However, physiological and physical limitations impose some space-time restrictions.

To study the space-configurations in the keyboard-related space in detail, we can cover it with open subsets instead of regarding it as a whole, following the physical limits of physiology (muscle forces, etc.):

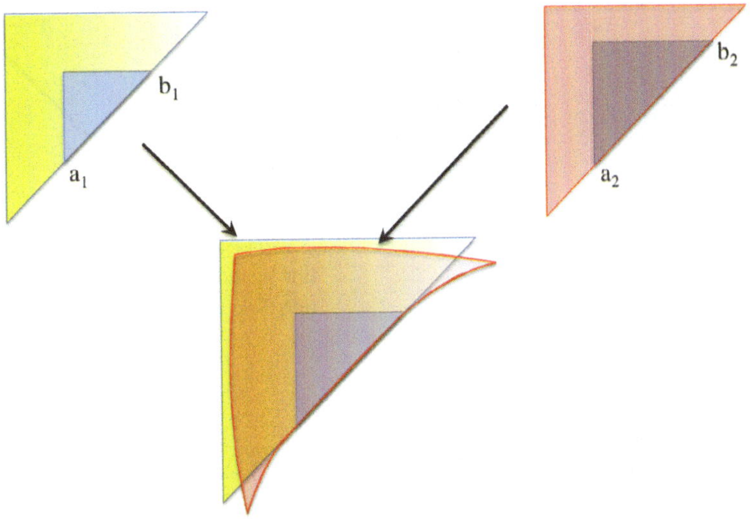

Fig. 66.9. The categorical manifold obtained from gluing two copies of ∇ along ∇-shaped subcategories.

one open space-time set for every fingertip and finger, taking into account the possible articulations; one open set for the motion of each hand's palm. Which point of keyboard space can be reached, and how fast, depends on the degree of freedom of articulations, with constraints due to flexibility and speediness. For example, the same fingertip cannot touch two keys at the opposite extremes of the keyboard if sufficient time is missing to perform such a movement. This defines a region in space-time. While writing a symbolic score, the composer has to verify the realizability of such gestures to avoid any Procrustes effect.

The following proposition is immediate from the fact that the categories of topological digraphs and of topological categories are finitely complete (check the existence of fiber products and of the terminal object).

Proposition 66 *The category* **CatMan** *is finitely complete.*

This being so, we may define gestures with values in categorical manifolds. Take a digraph Γ and a categorical manifold Σ^I. Then a gesture $g : \Gamma \to \Sigma^I$ is defined as follows. To begin with, let $\Gamma = \uparrow$ (see Section J.1.1). Then for the basic topological category ∇, which is a special categorical manifold, we have the morphism set $\nabla @ \Sigma^I$ in **CatMan**. This set is covered by the subsets $\nabla @ \Sigma_\iota$ for the charts Σ_ι of I. These subsets are already known to define topological categories. And their intersections are the sets $\nabla @ (\Sigma_\iota \cap \Sigma_\kappa)$, which are topological categories for the same reason. Therefore $\nabla @ \Sigma^I$ is given the structure of a categorical manifold that we also denote by $\overrightarrow{\Sigma^I}$. Since every digraph Γ is the colimit of the diagram $\mathcal{D}$ of its arrows and vertices, we may define $\Gamma @ \Sigma^I$ as $\lim \mathcal{D}$, the limit (that exists by Proposition 66) of its arrow values of type $\nabla_\delta @ \Sigma^I$, ∇_δ the ∇ for arrow $\uparrow_\delta$,

$$\Gamma \overrightarrow{@} \Sigma^I = \lim_\delta \nabla_\delta @ \Sigma^I = \lim_\delta \overrightarrow{\Sigma^I}.$$

Definition 117 *With the above notations, a gesture $g : \Gamma \to \Sigma^I$ is by definition an element of the categorical manifold $\Gamma \overrightarrow{@} \Sigma^I$.*

Figure 66.10 shows a gesture in a topological manifold.

In particular, we may now build hypergestures over categorical manifolds, a necessary prerequisite for an Escher Theorem over categorical manifolds. Here is this theorem, whose proof is immediate as it follows the same ideas used for the proof for topological categories.

Theorem 4. (*Escher Theorem*) *For a sequence of digraphs $\Gamma_0, \Gamma_1, \ldots \Gamma_{n-1}$, a categorical manifold Σ^I, and a permutation $\pi \in S_n$, we have canonical isomorphism of categorical manifolds:*

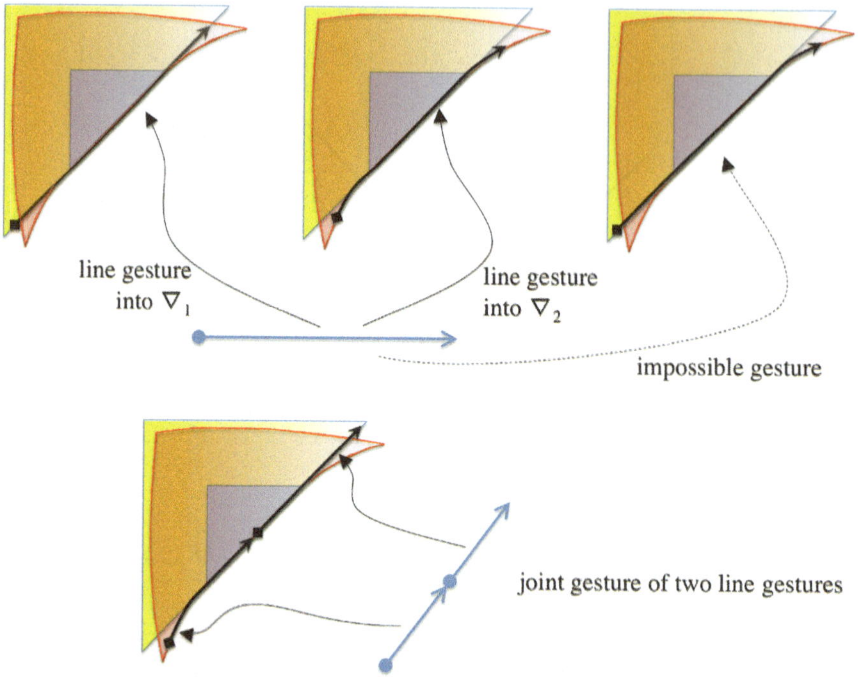

Fig. 66.10. Top line: Left and middle show two line gestures (defined on the line digraph ↑), while the right one is not allowed since it does not map into a chart. Bottom line: Here we see a gesture that is defined on the chain digraph of length 2, and this one is allowed as its connecting point maps into the intersection category.

$$\Gamma_0 \overrightarrow{@} \Gamma_1 \overrightarrow{@} \ldots \Gamma_{n-1} \overrightarrow{@} \Sigma^I \xrightarrow{\sim} \Gamma_{\pi(0)} \overrightarrow{@} \Gamma_{\pi(1)} \overrightarrow{@} \ldots \Gamma_{\pi(n-1)} \overrightarrow{@} \Sigma^I.$$

66.7 Globalizing Skeleta

The next step toward globalization of gestures deals with global skeleta. Suppose that we are given a global digraph Γ^G, i.e., a covering G of digraph Γ by non-empty subgraphs G_ι. The category *GlobalDigraph* of global digraphs has as morphisms morphisms of digraphs, together with maps of the coverings that respect images of covering subgraphs. In the situation of global gestures, as defined in Definition 115, we are given such a global digraph, and we suppose that every chart G_ι is embedded in the categorical manifold Σ^I. In our situation of a global gesture with values in Σ^I, this means that we are given local gestures $g_{\iota,\kappa} : G_\iota \cap G_\kappa \rightarrowtail \overrightarrow{\Sigma^I}, g_{\kappa,\iota} : G_\iota \cap G_\kappa \rightarrowtail \overrightarrow{\Sigma^I}$ that can be transformed into each other by an automorphism $f_{\iota,\kappa} : \Sigma^I \xrightarrow{\sim} \Sigma^I$ of the underlying categorical manifold[2] Σ^I. A global gesture is given by its local charts $g_\iota : G_\iota \rightarrowtail \overrightarrow{\Sigma^I}$. This means that the global gesture is an element $g = (g_\iota)_\iota \in \prod_\iota G_\iota \overrightarrow{@} \Sigma^I$, together with the above automorphism conditions, which define a determined subset of $\prod_\iota G_\iota \overrightarrow{@} \Sigma^I$. Since this cartesian product of categorical manifolds is a categorical manifold, the only question is whether the automorphism conditions define a submanifold. We know that $\prod_\iota G_\iota \overrightarrow{@} \Sigma^I$ is covered by categories that intersect in categories. The question is: are the intersections of these categories with the set of global gestures also categories? The answer is yes, since the automorphism conditions have nothing to do with the structure of natural transformations which define morphisms in such manifold categories. This means that the set of global gestures $g : \Gamma^G \to \Sigma^I$ is a categorical manifold, which we denote by the known symbol $\Gamma^G \overrightarrow{@} \Sigma^I$. Observe that if Σ^I is a topological manifold, then so is $\Gamma^G \overrightarrow{@} \Sigma^I$.

[2] In this context, to simplify the formalism, we only deal with local gesture chart isomorphisms where the natural transformations $n_{s,t}$ in Definition 115 are identities.

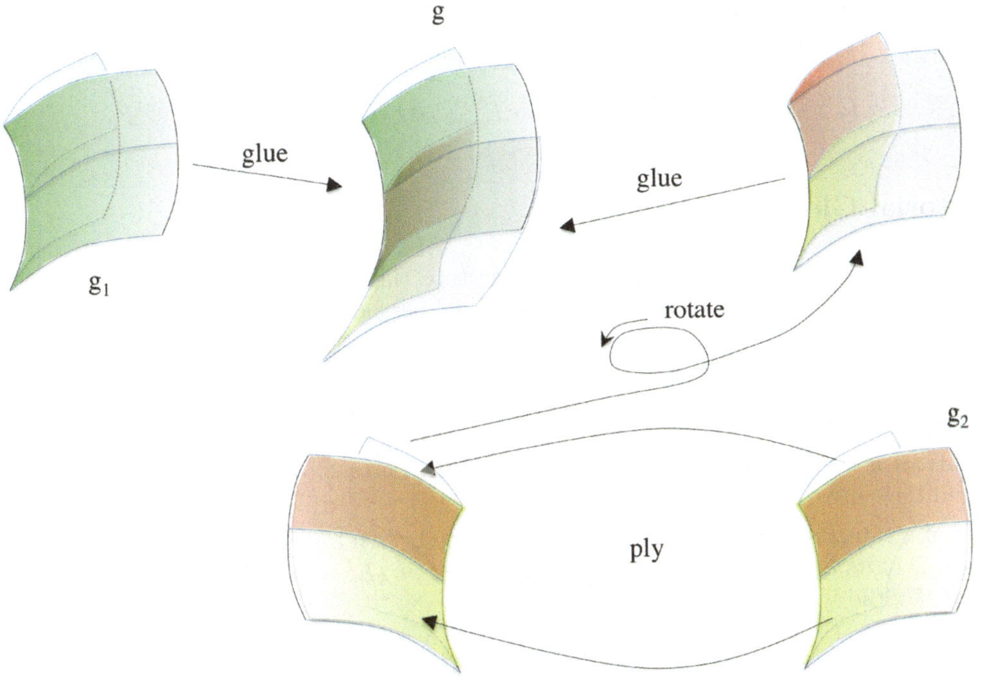

Fig. 66.11. The two hypergestures $g_1, g_2 \in \uparrow^2 \overrightarrow{@}\Xi\overrightarrow{@}RT$ are glued together to a global gesture g on $(\uparrow^3)^J$, $J = \{\uparrow^2, \uparrow^2\}$, along their intersection on $\uparrow$ and exchanging the two arrows of the bifurcated digraph Ξ (two arrows starting from the same vertex) of the gestures in space-time RT.

Example 82 Global skeleta reflect the classical situation of a global composition (see Chapter 13) that is a patchwork of local subcompositions (charts). Since we are now dealing with gestures, the patchwork is not a set-theoretic one, but deals with skeleta. Figure 66.12 shows a simple example. The composition's gestural setup is defined on the skeleton $[3]_L \sqcup_{[2]} [3]_R$, and this skeleton is seen as a union of subdigraphs $J1$ and $J2$. Their intersection $J1 \cap J2$ is the line digraph [1] from point 1 to point 2.

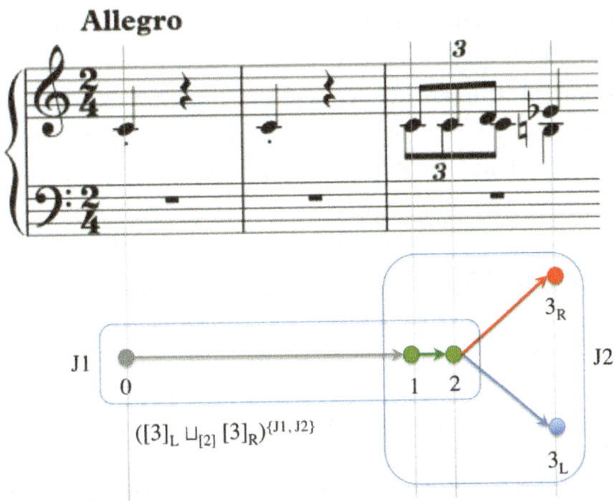

Fig. 66.12. A global digraph for a simple composition. See Example 82 for details.

The final result here would be a global Escher Theorem, which we state as a conjecture without proof:

Theorem 5. (*Global Escher Theorem*) *For a sequence of global digraphs* $\Gamma_0^{G_0}, \Gamma_1^{G_1}, \ldots \Gamma_{n-1}^{G_{n-1}}$, *a categorical manifold* Σ^I, *and a permutation* $\pi \in S_n$, *we have canonical isomorphism of categorical manifolds:*

$$\Gamma_0^{G_0} \overrightarrow{@} \Gamma_1^{G_1} \overrightarrow{@} \ldots \Gamma_{n-1}^{G_{n-1}} \overrightarrow{@} \Sigma^I \xrightarrow{\sim} \Gamma_{\pi(0)}^{G_{\pi(0)}} \overrightarrow{@} \Gamma_{\pi(1)}^{G_{\pi(1)}} \overrightarrow{@} \ldots \Gamma_{\pi(n-1)}^{G_{\pi(n-1)}} \overrightarrow{@} \Sigma^I.$$

66.8 Functorial Global Gestures

As described in Section 62.6, we may also unfold gesture theory in a functorial setup, i.e., for general addresses. In this section, we want to include the functorial perspective into the global concept architecture.

We again have to start with gestures with values in a categorical manifold, but this time the gestures are A-addressed for a topological category A.

Take a digraph Γ and a categorical manifold Σ^I; then an *A-addressed gesture* $g : \Gamma \rightarrow_A \Sigma^I$ is defined as follows. To begin with, let $\Gamma = \uparrow$. Then for the topological category ∇ we have the morphism set $\nabla @_A \Sigma^I$ in **CatMan**. This set is covered by the subsets $\nabla @_A \Sigma_\iota$ for the charts Σ_ι of I. These subsets define topological categories, and their intersections are the topological categories $\nabla @_A (\Sigma_\iota \cap \Sigma_\kappa)$. Therefore $\nabla @_A \Sigma^I$ is given the structure of a categorical manifold that we also denote by $A @ \overrightarrow{\Sigma^I}$. Since every digraph Γ is the colimit of the diagram $\mathcal{D}$ of its arrows and vertices, we may define $\Gamma \overrightarrow{@}_A \Sigma^I$ as $\lim \mathcal{D}$, the limit of its arrow values of type $\nabla_\delta @_A \Sigma^I$, ∇_δ the ∇ for arrow $\uparrow_\delta$,

$$\Gamma \overrightarrow{@}_A \Sigma^I = \lim_\delta \nabla_\delta @_A \Sigma^I = \lim_\delta A @ \overrightarrow{\Sigma^I}.$$

Definition 118 *With the above notations, an* A-addressed gesture $g : \Gamma \rightarrow_A \Sigma^I$ *is by definition an element of the categorical manifold* $\Gamma \overrightarrow{@}_A \Sigma^I$.

The last step toward globalization of gestures deals with global skeleta for A-addressed gestures. Suppose that we are again given a global digraph Γ^G (a covering G of digraph Γ by non-empty subgraphs G_ι). For global gestures, we are given local A-addressed gestures $g_{\iota,\kappa} : G_\iota \cap G_\kappa \rightarrowtail_A \overrightarrow{\Sigma^I}, g_{\kappa,\iota} : G_\iota \cap G_\kappa \rightarrowtail_A \overrightarrow{\Sigma^I}$ that can be transformed into each other by an automorphism $f_{\iota,\kappa} : \Sigma^I \xrightarrow{\sim} \Sigma^I$ of the underlying categorical manifold Σ^I.

Again, a global A-addressed gesture is given by its local charts $g_\iota : G_\iota \rightarrowtail_A \overrightarrow{\Sigma^I}$. This means that the global gesture is an element $g = (g_\iota)_\iota \in \prod_\iota G_\iota \overrightarrow{@}_A \Sigma^I$, together with the above automorphism conditions, which define a determined subset of $\prod_\iota G_\iota \overrightarrow{@}_A \Sigma^I$. These automorphism conditions define a submanifold as already discussed above. Therefore the set of global gestures $g : \Gamma^G \rightarrow_A \Sigma^I$ is a categorical manifold, which we denote by the known symbol $\Gamma^G \overrightarrow{@}_A \Sigma^I$.

This construction verifies all the functorial properties that one expects. For an address change $f : B \rightarrow A$, a categorical manifold morphism $m : \Sigma^I \rightarrow \Xi^J$, and a global digraph morphism $t : \Delta^H \rightarrow \Gamma^G$, we have a (functorially) corresponding morphism of categorical manifolds:

$$\begin{aligned} \Gamma^G @_f \Sigma^I &: \Gamma^G \overrightarrow{@}_A \Sigma^I \rightarrow \Gamma^G \overrightarrow{@}_B \Sigma^I \\ \Gamma^G @_A m &: \Gamma^G \overrightarrow{@}_A \Sigma^I \rightarrow \Gamma^G \overrightarrow{@}_A \Xi^J \\ t @_f \Sigma^I &: \Gamma^G \overrightarrow{@}_A \Sigma^I \rightarrow \Delta^H \overrightarrow{@}_A \Sigma^I. \end{aligned}$$

The last step of Escher theorem type is this theorem, whose proof is obvious from the preceding theorem versions:

Theorem 6. (*Global Functorial Escher Theorem*) *For a sequence of global digraphs and addresses*

$$\begin{aligned} &\Gamma_0^{G_0}, \Gamma_1^{G_1}, \ldots \Gamma_{n-1}^{G_{n-1}}, \\ &A_0, A_1, \ldots A_{n-1}, \end{aligned}$$

a categorical manifold Σ^I*, and a permutation* $\pi \in S_n$*, we have canonical isomorphism of categorical manifolds:*

$$\Gamma_0^{G_0} \vec{@}_{A_0} \Gamma_1^{G_1} \vec{@}_{A_1} \ldots \Gamma_{n-1}^{G_{n-1}} \vec{@}_{A_{n-1}} \Sigma^I \xrightarrow{\sim} \Gamma_{\pi(0)}^{G_{\pi(0)}} \vec{@}_{A_{\pi(0)}} \Gamma_{\pi(1)}^{G_{\pi(1)}} \vec{@}_{A_{\pi(1)}} \ldots \Gamma_{\pi(n-1)}^{G_{\pi(n-1)}} \vec{@}_{A_{\pi(n-1)}} \Sigma^I.$$

67

Mathematical Models of Creativity

Summary. We claim that category theory is a mathematical theory, proceeding from the observation of mathematical activities and gestures, and constructing a mathematical theory as a kind of algebra of these gestures. Especially, categoricians observe their own activity, and so category theory is also constructing a mathematical theory of itself, of its own system of gestures. We imagine that this theory can be used to model any activity, by a parallel action with the categorical activity. This *categorical modeling* is what we need for a mathematical holding of mathematical creativity because every activity is in fact somehow an activity of modeling.

– Σ –

67.1 Forewarning: Invention of Gestures in Mathematics

67.1.1 Thinking Exactness, Like a Rolling Mind

The question with respect to any given activity α is: in this area of activity α what is being modeled, and how does it work? Our modeling will be the construction of a relation between the work into α and the work into category theory, a relation from acts in α to acts in category theory. Categorical modeling seems possible when α = mathematics, when α = music.

For us creation and invention in mathematics is analogous to creation and invention in music, the mathematician playing with theories and models resembles the musician playing with instruments and interpretations. We do not argue and expand this view; we let that to our reader only as a conducting wire.

Our final and highest claim will be (for future works): this approach of creativity in mathematics by category theory can be elaborated as a mathematical method.

⋆

With respect to the question of modeling of creativity as a matter of priority we have to lower our attention to Truth, Being, and Phenomenon, and we have to consider the truth conditions of mathematical activity. A working mathematician does not think of truth or reality, but only of exactness. He is constructing proofs, and for that his base is a strong observation of the mathematical activity.

Neither ontology nor phenomenology, but *actology*: we believe to such an extent that mathematics is a thought *and* an act, mainly at this very moment where the opening thought cancels itself as a known thought. Even though the three terms ontology, phenomenology, and actology are in a borromean configuration, each term is what holds the two others parallel to the borromean configuration of "the Being, the World, the Act", or its first cousin: "Truth, Horizon, Exactness". This configuration probably yields an explanation of why creativity in mathematics is possible. But at first, to develop a method of invention, we have to emphasize actology as such and the question of *gestures*.

G. Mazzola et al., *The Topos of Music III: Gestures*, Computational Music Science,
https://doi.org/10.1007/978-3-319-64481-3_12

We believe that acting mathematical thought has to transit through the effective act of doing a calculation. And at the time of this act, logical prescriptions and certainties of truths are fired, being replaced by the *risk of exactness*. Here we consider "calculation" in a very large sense, including arithmetics, algebra, combinatorics, geometrical constructions, recursions, calculus of limits, and even what can be named algebraic argumentation from "mathematical logic". We highlight the fact that "mathematical logic" is an algebro-geometrical interpretation of the old philosophical logic; in some sense it is philosophical logic upside down: the question of the *truth* in language and legitimate words is replaced by the question of *exact* writing and clear visibility of a coincidence.

We emphasize the fact that exactness is related to connectivity and continuity, as in the sentence: "this figure is exact." This is far from the idea of "truth", a notion implying two points, the first one being the idea of a fixed datum; and the second one, as a consequence, the fact—or the order perhaps—that we have to participate in its coming. So exactness is not an order, it is just a geometrical noticing which is saying about a given situation: there are no gaps, no interruptions, the process is continuously working. Hence it is directly related to the mathematical meaning of "exact" within the notion of exact sequence or exact square. And this notion is enough to control and to give sense to mathematical activity.

Creativity in mathematics is possible under this condition, when you play with exactness of calculations as a child innocently plays with rolling stones or drawings on a beach. So you do gestures into writings, forgetting singular interpretations of these gestures, remembering that any previous gesture is due to an oversight: *a priori* you are innocent in any assigned sense, divested of unadorned truth or useful reality, *a posteriori* only do you risk a sense for your gesture. This is the *mathematical pulsation*.

67.1.2 Thought as an Algebra of Gestures

Category theory has several aspects. It is an ordinary mathematical theory; it provides a mathematical return of a particular historical examination of mathematical gestures in the first part of the 20th century. At the turn of mathematical structuralism and universal algebra, at the beginning of algebraic topology, it unifies previous mathematical fields such as groups and lattices. It allows us for example to understand invariants, dualities and completions. And so it is useful in various mathematical areas such as algebraic geometry, computing sciences, physics, as well as for modeling in engineering science. In these fields this tool allows advances without a priori questions on foundations or logic, but rather as a continuation of the algebro-geometric intuition observable with gestures, and offering new gestures.

But also some tools in category theory, e.g. adjunctions, limits, completions, and topoi, have philosophical interpretations, and we can use category theory to express some philosophy such as ontology, philosophy of difference, dialectics, as well as various metaphysics or pragmatics or even logic. Category theory allows a modeling of philosophy, in fact from all aspects of mathematical analysis of mathematical gestures.

Augustus De Morgan knew well that mathematics and logic are different when he said (quoted in [816]): "The two eyes of exact science are mathematics and logic." Today we can emphasize this thought as follows. Unlike the foundational perspective of set theory and logic, category theory provides a downwards analysis of mathematical gestures, mainly of structurings, and it could even be understood as a theory of theories. And category theory seems to be *useful for exposition and development of mathematical creativity*, as *a mathematical guide towards a method of invention* of gestures in mathematics (and reflexively in category theory itself). This point is our objective.

⋆

Apart from this forewarning and the conclusion, this section has four parts, where our ideas are repeated, without being identically reiterated. Here are some of the ideas discussed into these parts.

In the first part (Sections 67.2 and 67.3) we ask for a method of invention, namely a categorical method for almost-free mathematical invention, away from stricto-sensu logical and set theoretical paths, from 'applications', in fact related to the mathematical pulsation as *a mathematical gesture* into writings and among diagrams. We highlight the fact that objects are semblances; we discuss simple objects and structures. We introduce the question of necessity of objects as universes and completions. We hope that this introduces good examples of mathematical gestures and exactness.

In the second part (Section 67.4) we propose a mathematical description of mathematical gestures, we discuss history of mathematics as a tool for invention, and we examine the nature of mathematical productions as a writing of gestures, supported by their own history.

In the third part (Section 67.5) the accent is put on the analysis/synthesis process, and the invention of the mathematical tools for that, namely the invention of the coordination, putting gestures into equations, curves and shapes calculus, converging toward general functional analysis.

In the fourth and last part (Section 67.6), we insist on the stage setting of mathematical works in terms of modeling of gestures as transitive actions, and on the analysis of the living shapes of these stages. Hence a final proposal of general cohomology as ethical directive to close back the method.

⋆

This chapter being only a reflection or an obsessive rumination, the reader will find many examples and related theorems in the referenced publications. Here all our examples can easily justify the categorical approach.

We hope this short overview will help our reader learn the link proposed by our title between pulsation, transit and history, and its pertinence with respect to mathematical creativity or invention of gestures, in order to model shapes or forms (categorically), and to compute invariants and curvatures (cohomologically).

67.2 Method and Objects, Summarily Explained: I—Preamble

67.2.1 Prelude to a Discourse of a Method: "Caminos", "Aletheia", Irreverence

67.2.1.1 Categorical Modeling, Method, Estrangement, Intellectuality

In this writing we would like to contribute to an examination of a new method in mathematics that we name *categorical modeling*, useful as a *method of invention* in many mathematical disciplines, and (why not?) especially in this young branch of mathematics named category theory, whose real purpose is the mathematical examination of mathematical gestures and activities.

But we have not to confuse the mathematical branch "category theory" with the method "categorical modeling". In *category theory*, we *explain* by diagrams, we construct universal diagrammatical explanations standing back mathematical gestures, we develop in itself algebras of diagrams and structures, and again we construct diagrammatical explanations standing back categorical gestures, and so on. Hence $\textsc{Cats} \subset \textsc{Maths}$ i.e., category theory is a part of mathematics, it can be purely developed in itself. In *categorical modeling* we *replace* mathematical situations by categorical situations such as diagrams, inventing a kind of "functor" $\textsc{Maths} \longrightarrow \textsc{Cats}$ ("functor" in its original linguistic sense, as in Carnap's works). It is a branch of applied mathematics, as in fact formal logic is, and it needs more than pure mathematics. For instance, it needs teaching and history of mathematics, going off pure mathematics, whereas in category theory (as in any part of) MATHS, it is possible to work "internally", in the semi-closed algebraic thinking of categorical notions and algebras.

The exact meaning for us of "modeling" is expressed by the French word: *modelage*. So a "model" is not an emergency tank of knowledge for future actions, but at first a peculiar track of an action, to be read. Working in pure mathematics is also a matter of modeling. By the term "modeling" we want to cover all meanings of the word "model": construction in set-theoretical theory of models or in universal algebra and with structures in category theory, as well as modeling in science and technology by numerical models and optimization; and making of a sketch or a mockup, or a design; and computation within a given algebraic organization.

⋆

Estrangement is a key word for anyone who wants a chance to solve a problem: do not try to force the solution, let the calculus expand its own naturalness; let your mind forget the concrete aspect of the question, and move around its form, only proposing new terms; at the beginning a problem has to be nothing else than a pure abstract object, and slowly we have to discover links with similar objects. This point is decisively paradoxical: given a singular datum, strangely we proceed to its *categorization* (in a linguistic, but also in a mathematical sense), i.e., we invent a system of objects *different* from the singular datum, but which are *similar* (sic!), among them and with the singular datum, with possibilities of *modifications* between them, and on this system, we introduce discrimination by properly adapted concepts. So we reach an estrangement (a foreign land) of the initial singular datum, and our new knowledge of the singular datum will be the shape of this foreign land. It works as *negative* or *apophatic theology*, saying that we can never truly define God in positive words, as in cataphatic theology; we can only speak of God by negation, speaking of God only in terms of what He is not. In our domain, the domain of all possible calculus, it is an unsecured loan, but with a chance of getting off ground. So in mathematics we do work in harmony with the proverb: why go straight to the point when it is possible to bypass? Or as expressed in a funny way by Henri Poincaré: "Mathematics consists of proving the most obvious thing in the least obvious way." (quoted in [816]).

This is our method to grasp the renewed shape of the problem. And to do so, we need to accept that the "reality" which was so transparent and obvious becomes opaque [367] and incomprehensible; then we can start to write down a non-conformist calculus expressing the new shape we are trying to imagine. In passing we should note that the *shape* here is analogous to the *form* in [400], [401]: the disposition of the object in front of a background.

Positively, after its initial darkening step, another name for this "estrangement" could be "structuring"! The only thing you have to do is to try some constructions or deconstructions, inside any world that you build or destroy, etc. You have to structure and to observe effects of your functional data on these structures. You structure a geometrical situation when you add new lines to the initial figure, and this has two effects; on the one hand it complicates and darkens the picture, but on the other hand in this obscurity we can clarify some new terms and bring them to the forefront. The same works when we structure an algebraic situation, when we state that in fact a precise group law is involved, and this allows us use of terms in this structure. Clearly structuring is a main resource for categorization and estrangement, a second one being doubt about a chosen structuring, opening a new field for re-structuring, by what we name mathematical pulsation.

And of course we have to apply this 'key word' (estrangement or structuring) to our problem here, which is to determine a creative method.

⋆

Our method will be via universal properties, but it is not at all "the" universal method. Many things are invented *inside* a given closed area of mathematics such as elementary geometry, combinatorics, algebra, etc., without any categorical recourse, even if most often the *a posteriori* explanation in terms of categories would be possible. In many cases to intend to make use of category theory would be exaggerated; for example to compute

$$\lim_{x\to 1-}\prod_{n=1}^{\infty}\left(1+\frac{x^{2n}}{\sqrt{n}-\frac{1}{2}}\right)\left(1-\frac{x^{2n+1}}{\sqrt{n}+\frac{1}{2}}\right)=2,$$

what we have to do is to construct a path into an internal space made of reasonings with inequalities, increases, algebraic operations, numbers, etc. *A priori* we do not need external construction of functors, even if *a posteriori* our space of reasoning can be reformulated as a category, or if *a posteriori* our equality comes as evaluation of an invariant.

⋆

So to do mathematics of any kind, as well as to develop our method as a mathematical trick, we have to write an unknown evolving path in the open unknown space of mathematical entities, according to the deep view of Antonio Machado [634, p.222]:

> *Caminante, son tus huellas el camino, y nada más; caminante, no hay camino, se hace camino al andar. Al andar se hace camino, y al volver la vista atrás se ve la senda que nunca se ha de volver a pisar. Caminante, no hay camino, sino estelas en la mar.*

(Wanderer, it is your tracks which are the road, and nothing else. Wanderer, there is no road, walking makes the road ...).

In fact these open *caminos* exist in the mathematical world, as well as in the world of language, in the world of music, etc. When we invent our thoughts (i.e., what we were saying just a short moment prior to thinking it) we do not yet know the end of our sentence; when we play music the same phenomenon occurs, etc. Nevertheless we suppose some future existence of a foreign land or more accurately a foreign structure into which our invention is thrown.

The invention of a musical, a poetical, or a mathematical object is the invention of such an open camino; from now on, we will speak of an open invention of a path, and for us, these paths are exactly the mathematical creations, performed as if *throwing a "letter" (or a "name") into an unknown future structure.*

⋆

Toward a method—assuming that vast unifying meaning of "modeling"—a first principle is that *any mathematical datum* (number, law, table, figure, geometry, theory, function, representation, etc.) *is an object*; a second principle is that *any gesture* among these objects (to write, to erase, to replace, to modify in any way, etc.) *is an arrow.* A third principle is that *gestures or arrows are objects*, and *objects are "semblances"*, i.e., they are valuable only as pretexts for actions (compositions and factorizations with new arrows). So "Categorical modeling" means a development of general mathematical modeling with the starting view and the support of category theory.

Of course our purpose is somewhat auto-reflexive, as categorical modeling is supposing category theory, and can be applied to itself and to searching in category theory, in such a way to merge with itself.

And therefore we use the locution "DISCOURSE *of* A METHOD" to signify that we are looking for a discourse coming from inside the method (as if the method were speaking to itself), and not for a discourse outside the method on the subject of a method.

Clearly the question of our research is to identify *a* method, and not *the* sole method; exactly as classical algebra was not the sole method for mathematics. Moreover we claim that this method is very good for *a posteriori* analysis of almost all inventions, but we do not claim that it enables many discoveries. Any real new discovery in fact needs new inventiveness in *categorical modeling* in order to be understood by that method. So this method is not complete, and has to be thought and used in an open way, producing its own dialectical development.

What we want to stress is that for René Descartes, to use algebra was a decisive promising gesture in geometry; and similarly with categorical modeling and category theory—in fact a somewhat natural extension of algebra (and this point should be necessarily explained at length)—we cherish the hope of fruitful new adventures in the world of mathematics.

Actually by "a method" we mean a method of navigation in the open space of mathematical creation or truth. The method is an equipment of human understanding for travel across the seas of sciences, as expressed by Francis Bacon [59, p.77], as quoted in [80, p.35]. But for us this is not only a method of discovering (as the scientific method is for Bacon), but also a method of invention. The question is not only to discover the laws of nature with the help of convenient tools to bring them out, but to open the ways of invention, by an observation of the human activity of invention (into the dialectic—via analysis and synthesis—between objects and arrows).

⋆

In previous publications we made other partial examinations, with subjects as models, signs and forms, see for example [424]. Here we have to use these examinations and several of our epistemological studies recently, such as [425], [426], [427], [428]. We use also our earlier works on *actions* and *machines* [405] and *ébauches de structures* (or *draft structures, rough sketches of structures*), [402] considered as generalizations of sketches, and allude to other papers.

We will miss here many decisive explanations of the interpretations of categorical results (in logic, epistemology, and philosophy; in structuralism and model theory, etc.), and we do not consider the question of the *theory of knowledge*, of its relationship to metaphysics, language, etc.; it has to be for future works, with the right historical explanations; here we consider all that as being a little out of our subject, which is related to a method of invention.

We think also that at least implicitly any very *technical progress* results from a *historical reconstruction* of a personal view of some past mathematical data, as a *teaching aid* for our auto-education in mathematics. But we pass to further reflections the examination of this philosophical knot of pedagogy, education, science and history, the knot whose name is simply—for us and from an aesthetic point of view—*mathematics*. Any desire for a method based on mathematical activity has to pay attention to that.

Of course, from a technical point of view, here a lot of results of many categorists—included in our own formation's story—could explicitly or implicitly take an active part in our reflections. We use only three decisive classical notions: adjoint functors or construction of free objects, limits or gluing, Kan extensions. The free objects act as imaginary data, something like stratagems (cunnings or ruses of reason), whereas limits and extensions act as computing machines. A mathematical development consists of a succession of mechanical actions (general substitutions and gluings) among stragems (figures and terms): that is what we named a path. We use the Yoneda Lemma in the perspective of shape theory and representation, sketches and topoi. We consider these tools as variations around the following theme: to provide a diagrammatic presentation of mathematical gestures of analysis and synthesis, estrangement and structuring, in most of mathematical works. But for the moment we do not try to develop some more technical tools necessary for our subject, such as cohomological algebra in its very general state; we content ourselves with some brief indications.

⋆

We would like to adopt the attitude of a *pensive mathematician*, i.e., the attitude of a working mathematician engaged in *mathematical intellectuality* according to that nice notion of François Nicolas, of trying to transmit in common language the mathematical thinking, in answer to any demand of mathematics.[1]

Mathematical intellectuality has three components: *theoretical* organization of mathematical thinking in relation with rationality, *logical* development of mathematical discourse in relation with its performative and demonstrative method, *aesthetic* relation with sensible domains, through beauty and intuition. Any wish for a method of invention has to assume these components.

As Nicolas spelt out [789], mathematically intellectuality of René Guitart is centered on a decision to make nondiscernable the work of thoughts immanent in the mathematical text and the work of the mathematician reading and appropriating this text. So we work according to what he called "Guitart lemma": *every mathematical text is isomorphic to the set of its relations with the world of mathematicians.*

This is one of the reasons we believe that mathematics is a collective invention, in which the part played by each player is not at all definitive, depending precisely on the future invention of new interpretations. For mathematical creation, we depend on our personal knowledge of history, and onf the web of our mathematical interlocutors: this is the breeding ground for our ingenuity.

The point here, for our method, is the transmission of the system of mathematical thinking as being the objective shape of mathematical invention; we believe that category theory is well adapted to this transmission. A strong argument for that is that category theory started as a "mathematical history" of mathematical practices, of inventive mathematical gestures, with an accent on searching for invariants via new structuring, and distancing ourselves from previous knowledge, representations or direct calculus.

⋆

Our method will be at first directed by the will *to estrange from our private intellectuality*; at the mathematical level we have to leave our habitus, even our skills and convenient interpretations, to start again with fresh interpretations. And what is more, estrangement goes through *irreverence* relative to definitions, logic rules and paradoxes. Let us remember that "irreverence is basically the champion of liberty" (Mark Twain) and "the essence of mathematics is liberty" (Georg Cantor): we mean that estrangement, via its basic irreverent behavior, is near the idea promoted by Byers [166] that creating mathematics is supposing

[1] For Nicolas, there is also a comparable *musical intellectuality* [789], [790].

a natural familiarity with a kind of provocation of ambiguity, contradiction and paradox. This is included in the idea of distancing ourselves from well-known reasons and of *mathematical pulsation.*

⋆

In continuing with this introductory section we will observe the institution of algebra as a method of invention, by Descartes—even though naturally he was not the creator or inventor of algebra (or of geometry)—and its significance for Locke with respect to general rational creativity, further than the scope of magnitudes and quantities. Also we will introduce Vico's view of mathematics with respect to 'ingenium" and creativity, a kind of criticism of the idea of a method, and iat the same time a strengthening of the possibility of creativity in mathematics. As he said: "to prove is to create." We will conclude in the style of Badiou's platonism, by his following idea: to do mathematics is a positive reinforcement of our tendency of pure thinking, and so it is nice for pleasure and the good life—in spite of our enslavement by numerical control through averages and statistics.

Perhaps mathematics produces true propositions; but more important is that when it does so, mathematics is *beautiful.* The beauty of mathematics lies on the final production of an unexpected clearing after a long and difficult walk in the dark of rationality (see Alain Badiou [65, p.12]).

Through pedagogical, scientific and historical aspects, we have to emphasize this beauty out of a specific method of invention—namely the *algebra of categories*—that we want to promote.

This is why we consider mathematics as a craft or even an art (motor of the art of rational thinking, according to Descartes) or a production of the *ingenium* according to Vico (but certainly not as pragmatic scientific knowledge), and category theory as an observatory of the beauty of mathematical gestures.

So we are with our backs to the wall: is it reasonable to believe that we can construct a method to produce free beauty? Free beauties of variations on truths? A method of mathematical invention of mathematics?

67.2.1.2 With René Descartes

The discipline today named *algebraic geometry* has its roots in works of Pierre de Fermat and René Descartes. However Descartes specifically promoted *algebraic geometry* as a method—that is to say an algebraic method of analysis and synthesis—for geometrical problems; this sets a concrete example for its "Discours de la méthode" ("Discourse on the Method of Rightly Conducting the Reason, and Seeking Truth in the Sciences"); we can say that his *"Géométrie"* is one of the "Discourses *of* the Method": through algebra, the Method itself is speaking.

For Descartes, algebra-for-geometry was not only a new discipline, but it was more, a mathematical method of description and observation of mathematical invention in geometry, a mathematical process to discover possible ways of invention of ancient and modern geometers. Algebraic computations replace deductions in the style of Aristotle; elementary mathematical gestures are reduced to rewritings and substitutions. Then truth will not proceed from an eternal logical principle external to the mathematician, or even from geometrical evidences, but from actually present gestures of the mathematician, the computations that algebraic expressions encode. Rather than a "foundation" at the beginning of mathematics, algebra is a codification process for mathematical gestures. As a discipline, its problems come from questions about clarification of gestures, included in algebra itself.

Of course when doing such a drastic reduction of geometry to algebra, by introduction of the operations of arithmetics in geometry, we lose something, which is the natural spirit of geometry, and the corresponding gestures. These gestures (to draw a line or a circle, to draw a tangent to a curve, to look at specific triangles and associated points, at conics, and focus, to cut and to glue figures, etc.) are replaced with combinations of algebraic operations (additions, multiplications, substitutions). This cartesian reduction is itself a high-level gesture, which modifies the corresponding internal systems of gestures in geometry and in algebra, and also modifies the sets of simple elements, the nature of compositions and decompositions.

Later, in 1655, John Wallis introduced the plane and the (cartesian) axes—the so called cartesian coordinates [1104]. In fact Jakob and Johann Bernoulli were the first to speak of "cartesian coordinates" [957, p.342]). They achieved a proof of the cartesian translation process, reaching Descartes' double replacement gesture:

$$\text{Conic} \leftrightarrows \text{second degree equation.}$$

Into this transitory mental space that he had introduced between geometry and algebra, René Descartes invented a new object, *Descartes' oval.*

On the one hand the oval is described by an equation (of the fourth degree), as a solution to the anaclastic problem. On the other hand (a branch of) it is constructed as a mixture (gluing or graphical juxtaposition) of the two "gardener constructions" for ellipses and hyperbola [264, vol. VI, p.428]. Hence the mathematician decides to extend his algebraic and his geometrical activities:

$$\text{second degree} \rightarrow \text{fourth degree equation,} \qquad \text{hyperbola and ellipse} \rightarrow \text{ovals.}$$

The oval can be constructed as follows (gardener-type construction, see below Figure 67.1 and the figure of the mechanism).

The three points F, K, G are fixed. A rope $G\underline{C}K\underline{C}E$ of length f is attached at G and E (fixed on FE with $FE = L$), and is stretched passing through $\underline{C}$ (moving on FE), K and $\underline{C}$ again, and E. When the rod FE turns around F, the point $\underline{C}$ moves on the oval.

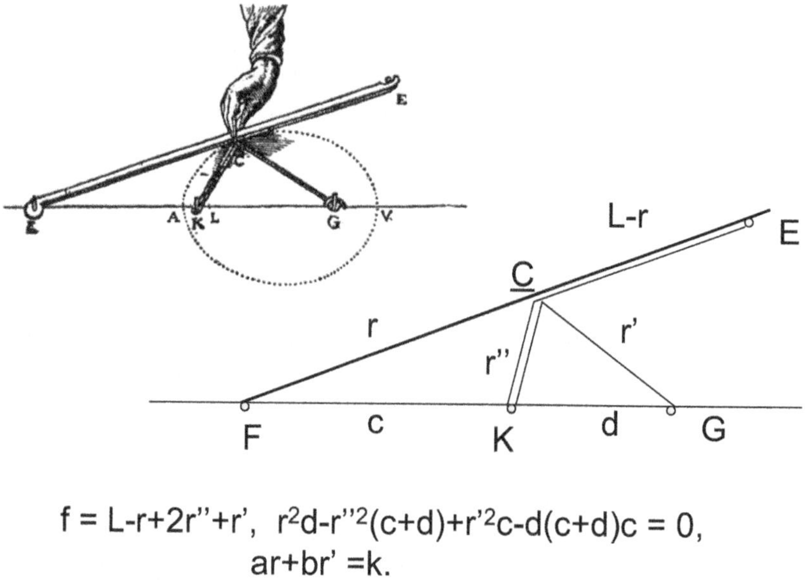

Fig. 67.1. Cartesian oval: from a mechanism to a diagram and a tripolar equation.

The proposal of this oval partakes in the invention of the general ideal of a curve [80]. In the 19th century it will be an important geometrical actor in the algebra of elliptic functions. Today there are a lot of constructions for ovals, as for example these two very easy ones(Figure 67.2):

This "object" is at the beginning of a fruitful story [76, 77]; an example of a historical scenery, see Section 67.4.2.3. Its first effect is to open the closed universe of conics to higher dimensions, into the infinite universe of algebraic curves. This opening is clear from the gardener construction and bipolar equation as well from rectangular equations: If the bipolar equation of a branch $ar + br' = k$ is written, with $r + nr' = q$, with $n = \frac{b}{a}$, $q = \frac{k}{a}$, then the complete rectangular equation (for the two branches) can be written [379, Tome I, p.218], with one focus F at $(0,0)$ and the other G at $(p,0)$, with $FG = p = c + d \geqslant 0$,

$$[(1-n^2)(x^2+y^2) + 2n^2px + q^2 - n^2p^2]^2 = 4q^2(x^2+y^2),$$

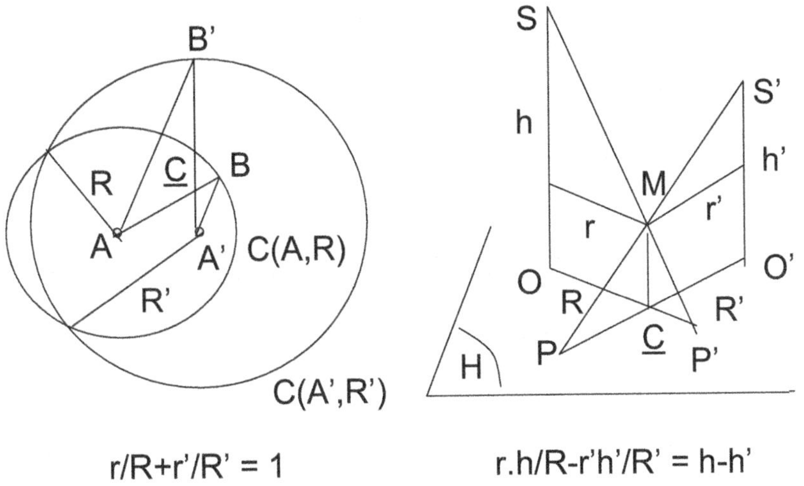

Fig. 67.2. Cartesian oval: more elementary construction and reduction to bipolar equation.

and if $n = \pm 1$, it reduces to the conic $\frac{(x-\frac{p}{2})^2}{\frac{q^2}{4}} - \frac{y^2}{\frac{q^2-p^2}{4}} = 1$.

Looking at the two other constructions, we see the left plan construction uses two circles, $C(A, R)$ and $C(A', R')$. We obtain a point $\underline{C}$ of the oval as the intersection of AB and $A'B'$, with AB' and $A'B$ parallel, $B \in C(A, R)$, $B' \in C(A', R')$. We do not know who the inventor of this construction is. The right construction (given by a perspective in three dimensions) is the proof by De Tranquelléon in 1864 [266] of the result of Quételet and Chasles exhibiting an oval as a projection of the intersection of two right circular cones with parallel axes. We let the reader prove these constructions and deduce two parametrizations of the oval.

These pictures are also a mathematical object, a presentation of this other object which is the oval. In fact we claim that mathematical objects do not really exist: there are only presentations (constructions or parametrizations, etc.) of objects, ready for manipulation and transformation. In other words, an "object" is only a name for an ideal reduction of several equivalent presentations, as a functional data ready for action. This is the case even with drawings of various complete ovals, such as the blue and the red ones below (Figure 67.3), with equations

$$(-3x^2 - 3y^2 + 24x + \lambda - 36)^2 - 4\lambda(x^2 + y^2) = 0,$$

with $\lambda = 1$ for the blue and $\lambda = 26$ for the red. These ovals have the same focus at $O = (0, 0)$ and $O' = (3, 0)$, and their polar equations are $r - 2r' = \pm\sqrt{\lambda}$. They do not have the same shape, the two branches of the blue being convex (in this case the external branch is also called *oviform*), and the external branch of the red nit being convex (in this case the external branch is called *cordiform*). This difference of shape depends on the different values of λ, according to a critical value c with $1 < c < 26$: for any λ the external branch is not convex if $c < \lambda \leqslant 26$, the two branches are convex if $1 \leqslant \lambda \leqslant c$. So the fact that we see as a quality (obviously visible but *a priori* not calculable) may be represented by a quantity c (*a priori* invisible but calculable; in fact c becomes "visible" by the race of its calculations). We let the reader determine the exact value of c. Also we let him pursue his own examination of this family of ovals: is it true or not that for a value k of λ, with $k > 26$, the two branches will contact each other? Could you determine this k? Then is the corresponding oval a known other special curve, with an easy special construction, etc.? Of course at this moment the important fact is that we considered one oval as a member of a family, and this suggested and allowed discovery of new properties.

⋆

So we see that in mathematics Descartes invents a method (the algebraic geometry)—i.e., he does the mathematical gesture of *junction of algebra and geometry* (and in parallel *junction of algebra and mechanical tools*) by his decision to show "how the operations of arithmetic are related to the operations of geometry:

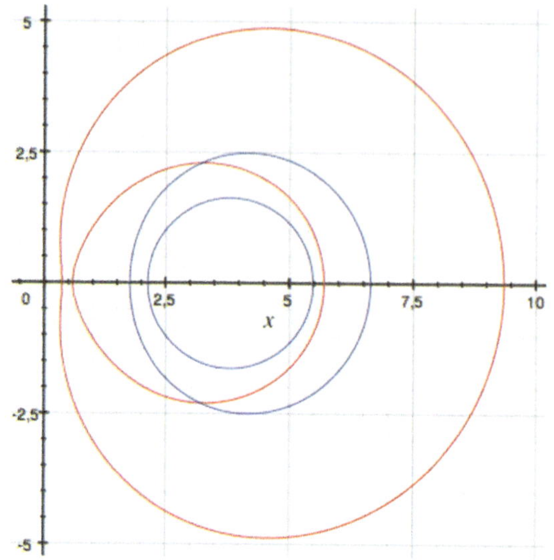

Fig. 67.3. Two ovals in the movement of a one-parameter family.

just as arithmetic consists of only four or five operations [...] so in geometry, to find required lines it is merely necessary to add or subtract other lines [...]"; so at the level of mathematics his method insists on a unifying coherence between two specific processes of analysis and synthesis. And inside this method, on a roundtrip between algebra and geometry, he invents an "object" (the oval), and a tentative classification of its types. In fact this object is a true creation because it opens new ways for activities in the area of algebraic geometry, hence virtually containing future new gestures: studies into the infinity of algebraic curves, via equations and/or via linkages. The pulsation between these two approaches was achieved and abridged by Kempe's theorem [76]. In the 19th century the link with elliptic functions was invented and discovered [77].

All of this constitutes a living historical scenery (in the sense of Section 67.4.1.4), from which one can continue to work, to exert on mathematics. For example, we let our reader work on this:

Exercise 1. Express the rectangular equation of an oval in such a way that its "composition" by gluing of a hyperbola and an ellipse becomes obvious.

With our present experience in mathematics, we can interpret Descartes' work as a mathematical expression of the analysis and synthesis in terms of decomposition and recomposition, substitution and parametrization, into a situation or a category, and also the transformation of the situation. This is as much a method to discover as a method to learn or to teach (see [854]), and to invent a proof.

67.2.1.3 In the School of the Mathematicians, According to John Locke

In the conception of John Locke—an attentive reader of Descartes—a mathematician is concerned with numbers and figures, with space; he works by "varying the idea of space, and thereby making still new compositions, by repeating his own ideas, and joining them as he pleases [...] And so he can multiply figures ad infinitum." [616, 2.13.5. p.151]. Mathematicians have to find proofs, what Locke called *intermediate ideas* between two ideas: they succeed in inventing new configurations, and the mathematical truths so discovered or invented lie on an *a priori* principle: the immutability of relations between immutable things [822, p.113]. At that point, Descartes would say that our mathematical evidences lie on the fact that "there is no malin génie".

Perhaps nowadays we can determine new mathematical concepts, or new mathematical structures; anyway we want to stress this fact: any such new idea is (or has to be) *a step* in a proof; and a proof is a path or a pavement of such ideas, from one idea to another. Algebra is a beautiful mathematical creation as a method for constructions of intermediate ideas, as expressed by Locke [616, IV-17.11, p.679]:

> Because we perceive not intermediate ideas to show conclusions. Our reason is often at a stand because it perceives not those ideas, which could serve to show the certain or probable agreement or disagreement of any other two ideas: and in this some men's faculties far outgo others. Until algebra, that great instrument and instance of human sagacity, was discovered, men with amazement looked on several of the demonstrations of ancient mathematicians, and could scarce forbear to think the finding several of those proofs to be something more than human.

By way of a first portrait of a working mathematician at the end of Descartes' century, we quote from John Locke in 1690 [616, IV-12.7, p.637-638]:

> The true method of advancing knowledge is by considering our abstract ideas [...]. General and certain truths are only founded in the habitudes and relations of abstract ideas [...]. By what steps we are to proceed in these, is to be learned in the schools of the mathematicians, who, from very plain and easy beginnings, by gentle degrees, and a continued chain of reasonings, proceed to the discovery and demonstration of truths that appear at first sight beyond human capacity. [...] the admirable methods they have invented for the singling out and laying in order those intermediate ideas that demonstratively show the equality or inequality of unapplicable quantities [...] but whether something like this, in respect of other ideas, as well as those of magnitude, may not in time be found out, I will not determine.

Also, Locke wrote [616, IV-12.15, p.642-644]:

> Clear and distinct ideas with settled names, and the finding of those intermediate ideas which show their agreement or disagreement, are the ways to enlarge our knowledge. But whether natural philosophy is capable of certainty or not, the ways to enlarge our knowledge, as far as we are capable, seem to me, in short, to be these two:
> The first is to get and settle in our minds determined ideas of those things whereof we have general or specific names; [...] we should put together as many simple ideas as [...] may perfectly determine the species [...]
> The other is the art of finding out those intermediate ideas which may show us the agreement or repugnancy of other ideas which cannot be immediately compared.
> 15. Mathematics an instance of this. That these two [...] are the right methods of improving our knowledge in the ideas of other modes besides those of quantity, the consideration of mathematical knowledge will easily inform us.[...] Further, it is evident that it was not the influence of those maxims which are taken for principles in mathematics that hath led the masters of that science into those wonderful discoveries they have made. [...] I suppose, scarce ever come to know that the square of the hypotenuse in a right-angled triangle is equal to the squares of the two other sides. The knowledge that "the whole is equal to all its parts," and "if you take equals from equals, the remainder will be equal," etc., helped him not, I presume, to this demonstration: and a man may, I think, pore long enough on those axioms without ever seeing one jot the more of mathematical truths. They have been discovered by the thoughts otherwise applied: the mind had other objects, other views before it, far different from those maxims, when it first got the knowledge of such truths in mathematics, which men, well enough acquainted with those received axioms, but ignorant of their method who first made these demonstrations, can never sufficiently admire. And who knows what methods to enlarge our knowledge in other parts of science may hereafter be invented, answering that of algebra in mathematics, which so readily finds out the ideas of quantities to measure others by; whose equality or proportion we could otherwise very hardly, or, perhaps, never come to know?

We gave these rather long quotations because, with respect to our interest in creativity and methods, three important ideas are advanced by Locke, even if he was not at all a mathematician himself.

- The first idea is that of *intermediate ideas*, with which proofs are composed, as long continuous sequences of obvious facts, and possibly out of prescribed calculus.
- The second is that mathematical modeling could be possible out of the domain of quantities, etc., in many or all areas of human knowledge.
- The third is that creativity and invention have to come from other disciplinary domains, by the *thoughts otherwise applied* (here we will speak of *mathematical pulsation*).

Therefore we need a method strong enough to study mathematically the mathematical creativity. We hope for a sort of extended algebra, in general applicable to the observable mathematical activity made of pulsating analysis and synthesis, of an open construction of continuous paths in any virtual space of mathematical manipulations.

67.2.1.4 Methods and Creativity, with Giambattista Vico

Based on our reading of Giovan Battista Vico's book of 1708 [1085] (considered as "the second Discourse on the method" after Descartes'), and the comments by Ennio Floris [327] in 1974, we comment on the will for a method of invention in relation to conceptions of doubt, intuition and dialectic between analysis and synthesis, especially in a mathematical context.

According to Floris, Vico was making a synthesis of humanism and cartesianism. On the first hand, humanism proceeded from a vision through philology and poetry, art, imagination and *ingenium*; with this "ingenium", the human being is a maker, an inventor. On the other hand, with cartesianism, using philosophy and critical thinking and science, the human being is a spirit. Rather than an epistemological and encyclopedic attempt, the work of Vico is a philosophy of the history of culture, as a production of human craft. He wants to bring man back to his own painful self-consciousness, far from the dispersion of his sensations, in the heart of creativity. For Vico this self-consciousness is constructed by culture, and both humanism and cartesianism are necessary there. The point is that, for Vico, creativity proceeds as well from ingenium and making of works as from cartesianism and thinking. For Vico the analytic method of Descartes is the inverse of the process of creation.

Sceptical doubt by sophists meant that it is radically impossible to get certainty about truth of principles or adequacy to things; they are only opinions, as effects of language. Resistance to this *sceptical doubt*, had been the nerve of philosophical quarrel for Socrates, Platon, and Aristoteles. They introduced the *dialectical doubt* into argumentations, to convince opponents against their own opinions. In fact, according to Vico, this was too weak, because external persuasion is captive of likelihood. So with Descartes, doubt became internal in the human being, and the corresponding dialectic resides in his own mind. With this kind of *cartesian doubt*—the first stone for cartesian method—imagination is completely separate from plausibility, and *evidence* (i.e., positive lack of doubt) became a criterion for truth.

Now for Descartes, the method consists of doubting, distinguishing and clarifying the evidence of simple things, as expressed in the *Regulae*. From the point of view of Vico, the difficulty here is that analysis prevails over synthesis; synthesis is only possible on the condition of *a priori* analysis and reduction to simple elements; analysis is an *a posteriori* construction. So we can say that the Descartes' method is an *analytical method*, particularly in the case of mathematics where algebra is an analytical tool for reduction of geometry to simple things such as lines. In some sense—at least in mathematics—to doubt is to analyze, and it is not enough to create.

For Vico, Descartes the mathematician (the geometer) is a solver of enigmas, or a constructor of riddles and puzzles, by a method that now everybody can use, the algebrico-analytical method. And he asked if, by including mathematics in a method, Descartes has not closed the way to new discoveries, separating the thinker from the ingenium (e.g. non-analytical geometry) as a source of invention.

For Vico, mathematics are a production of *ingenium*, using a *synthetical method* and *intuitio* as underlying dynamic, whereas in the analytical method the dynamic is *intuitus*.

Vico said that *to proveis to reduce to the field of thoughts*, and *this reduction is the characteristic of mathematical thinking*, the elements of which are creations of ingenium. The evidence comes from the conscience the mind has that in a deduction process it is creator of the links. So Vico concluded (see [327, 2nd part; No. 97]):

To prove is to create.

⋆

For us a mathematician is an engineer, a solver of puzzles, and this is far from logic and truth: it is a question of exactness. To prove is to exhibit a path or a link between two analyses, as in a proof of the theorem

$$2 + 2 = 2 \times 2,$$

considered as a link between two analyses of the "same" object named "4".

This is reminiscent of Andrée Ehresmann's *Multiplicity Principle* with multifold or *multifaceted* objects in the theory of living systems as proposed in [295] and recently in [296] (see our Section 67.6.2.1). A very simple case of this Multiplicity Principle is, with a, b, c, d some elements:

$$\{a, b\} \cup \{c, d\} \simeq \{a, c\} \cup \{b, d\};$$

this expresses two different and unmatchable cuttings of $\{a, b, c, d\}$, two really different analyses of 4 as sum. In fact this is more subtle than $2 + 2 = 2 \times 2$, which is induced by a matrix (see Section 67.5.4.2); with $2_1, 2_2, 2_3, 2_4 \simeq 2$, we have this instead:

$$2_1 + 2_2 = 2_3 + 2_4.$$

So to exhibit a new path at a higher level is to create, and in this perspective mathematics is a living system.

⋆

We want to make precise what a proof is by giving two examples.

Here, in Figure 67.4, we give an example of a mathematical creation of a path. This example was exposed in a talk to psychoanalysts [429] to show them how a mathematical proof works and *a priori* is not related to logic.

The problem is to find how to transform continuously the state 1 into the state 10, and the series of pictures shows elementary steps on a path toward a solution: it is a proof. There are two stiff metallic rings and two flexible but not elastic ropes (green and black). We see on the sequence of pictures that the movement is possible, and how it is possible by a "collaboration of the two ropes, contrary to our first hope to succeed by a move of one or the other separately." We let the scrupulous reader formalize that series in a finite series of finite instructions. As George Pólya wrote [854, p.121]: "It is an excellent intellectual exercise to try to build a formal proof of what we have in intuition, and to see the intuition of what is formally proved." This suggestion is nothing else than to practice the pulsation between blind formal level and clear intuitive understanding. In our case here, the proof shown by ten pictures is already a decomposition of the gesture into a sequence of more obvious gestures; these obvious gestures are intuitive because we have a direct skill with them, we know how to decompose them again into Reidmeister's moves [885] (or more precisely into what we call *stiff-Reidmeister*'s moves), i.e., how to reduce them to a formal algebraic word. The final reduction will be to drop the supposition that all that is written is in a concrete physical space.

Our reader will compare this proof in geometry with the following in algebra, where we start with a gesture given by a series of elementary gestures, in order to extract the explicit value of γ from the implicit condition $\sqrt{\alpha+\gamma} \pm \sqrt{\beta+\gamma} = w$, where we are supposing $\epsilon = \pm 1$, and $\alpha, \beta, \gamma \geqslant 0$, and $w \neq 0$.

We obtain γ by formulas (9) or (11):

$$\sqrt{\alpha+\gamma} + \epsilon\sqrt{\beta+\gamma} = w \quad (1)$$

$$\alpha + \gamma + \beta + \gamma + 2\epsilon\sqrt{(\alpha+\gamma)(\beta+\gamma)} = w^2 \quad (2)$$

$$w^2 - (\alpha + \gamma + \beta + \gamma) = 2\epsilon\sqrt{(\alpha+\gamma)(\beta+\gamma)} \quad (3)$$

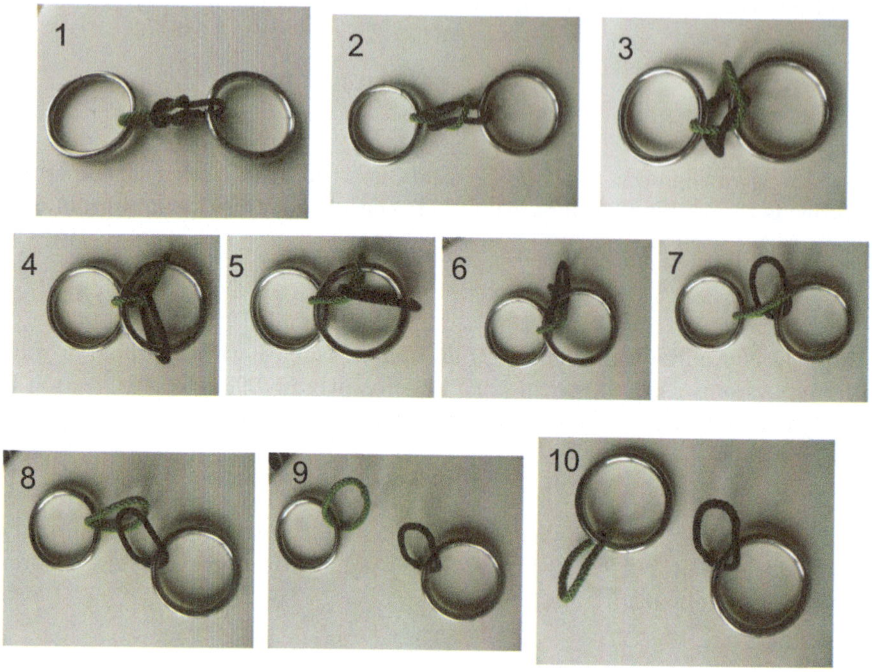

Fig. 67.4. A sequence of elementary gestures as a proof of unknotting a situation.

$$w^4 - 2w^2(\alpha + \gamma + \beta + \gamma) + (\alpha + \gamma + \beta + \gamma)^2 = 4(\alpha + \gamma)(\beta + \gamma) \tag{4}$$

$$w^4 - 2w^2(\alpha + \gamma + \beta + \gamma) + (\alpha + \gamma)^2 + (\beta + \gamma)^2 - 2(\alpha + \gamma)(\beta + \gamma) = 0 \tag{5}$$

$$w^4 - 2w^2(\alpha + \gamma + \beta + \gamma) + ((\alpha + \gamma) - (\beta + \gamma))^2 = 0 \tag{6}$$

$$w^4 - 2w^2(\alpha + \beta + 2\gamma) + (\alpha - \beta)^2 = 0 \tag{7}$$

$$w^4 - 2w^2(\alpha + \beta) + (\alpha - \beta)^2 = 4w^2\gamma \tag{8}$$

$$\gamma = \frac{w^4 - 2w^2(\alpha + \beta) + (\alpha - \beta)^2}{4w^2} \tag{9}$$

$$\gamma = \frac{\left[w^2 - (\alpha + \beta)\right]^2 - (\alpha + \beta)^2 + (\alpha - \beta)^2}{4w^2} \tag{10}$$

$$\gamma = \frac{\left[w^2 - (\alpha + \beta)\right]^2 - 4\alpha\beta}{4w^2} \tag{11}$$

This proof will be a tool to study more properties of ovals.

★

In these two proofs we observe that: we write and we see a finite sequence of modifications in some writing, according to admitted specific laws of transformations, and this is axiomatic (the laws of transformations are of our own decision). We let the reader express in ordinary language the right reason for which each step in these proofs is valuable. This "exercise" *is the proof* for each reader of the proof. Do it!

We produced a ruled transformation of a diagram (e.g. picture 1 or equation 1) to another (Figure 67.4, part 10 or equation (10)), and we are sure of its exactness, and this "proof" is again a diagram. And this proof is now the description of a new authorized gesture: we have created a gesture. Is there a reason to believe that it is truth, or even that "to be truth" makes sense? Exactness is not identical to truth, even if we can try to say that the proposition that a given proof is exact is a true proposition. Casually we evolute from the register of an act of "writing something" toward an act of "speaking of a writing". Under our eyes we have exact constructions of *mathêmata*, we can see them, and we know that they are abbreviations of a sequence of gestures.

If you stay at the level of seeing *exactness* you will get the real pleasure of the mathematician, appreciating the *ingenuity* of the situation (the presence of the *ingenium*), and living a moment of resolution, the surprise that something in the writing had been broken or unknotted, unravelled: *something come loosed*; hence the possibility of mathematical jokes and questions (you have the knack, what is the trick?); but if you change your mind in favor of the observation of *truth* of the fact, that it is exact, you can only get the pleasure of the philosopher receiving knowledge: *something is reached*, you are in charge of that.

⋆

In passing, we would like to show the superiority of diagrammatic writing over language. Let us consider the following mathematical joke W as an object of meditation:

W: We have three types of mathematicians: those with ability to count, and those without.

Is it a truth proposition; is it even a proposition? If it is, what should be its truth value ? But it is easy to represent the meaning of this joke by a mathematical diagram:

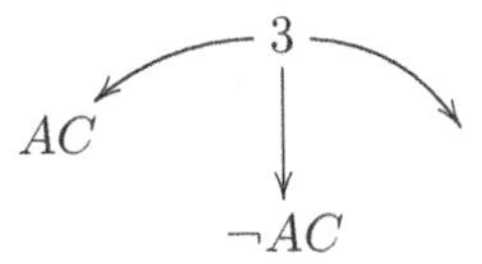

Here the strange construction (the "3" in the sentence, only followed by two items AC and $\neg AC$) is no more thought of as a mistake, but as a voluntary gap: in the diagram it is rendered by the fact that there are three arrows, but the third is not completed by a target (this being more or less forbidden by the fact that the two names of the first two target are complementary).

⋆

A mathematical object which has been constructed in a very complex way, as the field of real numbers $\mathbb{R}$, can become "intuitive" as primary data, and in this sense a simple thing. On the other side, a simple and intuitive thing, such as a point, could be decomposed again, becoming complex data.

We claim that real mathematical invention works as the dialectic between these two directions of thinking and ingenium, intuitus and intuitio, analysis and synthesis. Mathematical productions, which ultimately are proofs (i.e., paths in various heterogenous systems of writings), appear in the living movement of this dialectic of composition/decomposition.

Looking ahead at the next sections, let us say that of course this dialectic can be modeled in terms of adjunctions (adjoint functors), and in terms of duality between limits and colimits in a category.

In this perspective, simplicity of objects is very relative; objects themselves are semblances. If we include mathematics in a new method around this dialectic (as for instance categorical modeling), of course we know that we risk closing the way to new inventions; but also we know that by doing so we are creating a particular piece of mathematics.

67.2.2 Our Posture

67.2.2.1 Towards the True and the Being, Mathematically: On the Road Again

In René Descartes' view, geometry is presented as an application or an illustration of the method. For us it will be different; practices of arithmetic and geometry and modern mathematics will come first, as a preliminary condition of any thinking for a method. In fact, for Descartes it was perhaps as for us, and the method came from his *Regulae* and his previous works on geometry.

A clear useful definition of what art is, what science is, seems to be out of reach to give. It is even difficult to give scientificity critera. Nevertheless in the case of mathematics we propose that it is a science or an art according to whether its knotting with logic is assumed or not, and simultaneously according to whether we consider mathematics as an act or as a knowledge. We do not need to make an effectively strong distinction between art and craft; both are a question of know-how (savoir faire), as opposed to science, which lays on the question to make it known (faire savoir). At first it is as an art (and/or a craft) that mathematics has to be considered regarding the development of its creativity. Let us nevertheless start with the scientific side.

We agree with Alain Badiou's view (as in [62] and [65]), where he says that mathematics is a science going toward Truth, i.e., toward the Being in itself [62, p.384]. He specifies [62, p.395] that higher arithmetics is necessary because it forces us to use pure thinking, as far as we are human subjects oriented toward the truth. He also says that *mathematics is the thinking of multiple in itself* [61, p.121]; hence the real truths will be assertions about organizations of multiplicities. But we consider that the job of a mathematician is not to claim that these assertions are truths, but to construct these organizations: this is what he can create; he is an architect, not a logician.

Our consideration of the "multiple" a priori can start with continuity or fusion as in geometrical spaces, rather than with discrete values and distinct elements as in sets. But a minimal mathematical work is with the pure development of the possibilities of finite combinations among multiplicities of distinct elements, starting from "Nothing" i.e., from the empty set $\varnothing$: arithmetic of sets. But generally a mathematical truth is *a true rational contribution to ontology*: it is an assertion in a given language, claiming the existence (or not) of a construction with given tools of a mathematical object (or arrow) satisfying a given property, expressed in another given language.

With its principle of analysis (or de-construction, or de-structurating) and synthesis (or structuring, or gluing), mathematics is at the heart of any doctrine of a scientific method, such a method being considered as *the doctrine of assertions pretending to be truths* (in the words of Fichte [316]); but truth is only a horizon of mathematical activity, a possibility of qualification of its proofs. Mathematics is not a tool to discover truths (necessarily necessary), but a method (necessarily contingent), and it is a fact that contingently (i.e., according to an open geometrical way, and not in virtue of a closed logical process) this method produces truths.

But in spite of appeals to languages, mathematical truths are not equivalent to logical truths. As "mathematical causality" is not "logical causality" [951]. In fact for us mathematical truths are closer to the initial question of *aletheia* [265] than to the question of logical truths. And especially when it is about creating mathematics, e.g., solving a problem, the point is not to follow blindly logical rules, but the personal engagement by a clear claim for a new road we imagine. Living mathematical truths are constitutively related to the intuition tied to our visual perception. Let us recall this thought of René Thom, namely that we believe in logical rules because we have a geometrical interpretation of implication by the means of inclusions (of extensions of concepts) that we see.

If from A we "deduce" B, a fact represented by "$A \Rightarrow B$" or by "$A \subset B$", etc., or by $A \xrightarrow{f} B$, and if we assume A, then we can say that B holds: but very precisely the "truth" here resides in the deduction itself, i.e., the $\Rightarrow$" or the "$\subset$", etc., or f, and not in B. The true resides in the path, not in the target, and when we see the path we know the truth. A deduction seen as a path is a displacement, a modification, a variation or a change, a gesture, symbolized by arrows; mathematics is the calculus with these arrows and their compositions.

So the conception based on ontology, insisting that a constitutive link between Being and Thinking does exist, or more or less equivalently based on a theological claim about a constitutive dependence between God and Truth [1051], is not a necessary hypothesis to promote mathematical activity. The mathematical creation needs only a "surveillance" of exactness in production of gestures, and so is a thinking.

In the categorical framework, everything (languages, constructions, tools, objects, properties) becomes an object of a category, or, equivalently, a functor. So any mathematical claim becomes an organization of functors, a *diagram*.

Our conception of a method starts with this initial position about mathematics, and the belief that when we are doing mathematics we are creating truths, as diagrams. To do mathematics is *to invent* various exact mathematical truths; sometimes these truths are really mathematically valuable, i.e., are nice and beautiful, and then we think that we *discovered* something of the Truth of the Being. We invent truths and so we discover beauties.

Then the method will *indicate* some style of gestures. These indications are not prescriptions, only opened possibilities. The system of such indications can appear somewhat axiomatic, but it has in fact to be completely opposed to a formalism.

A priori this *mathematical method* (of invention) is not a logical or a rational method, with a rather pragmatic goal, but only *a system of figures and computations*, with an infinity of possible interpretations in retrospect, as for a writing we have a lot of readings: meaning is always an afterthought. And justly our main indication is the care about *pulsation*, the skill to take old meanings out of the writings, and then to vest these writings with new significations.

As is the case in the art of painting, or with a method of playing piano (and then hopefully of producing music...), here we are looking for a method of playing mathematics (and so of producing truths...). We do not pursue any deeper analysis of outbuildings and relations between science and art, beauty and truth, no hypothesis on what is really useful in this somewhat axiomatical writing, at the expense of philosophy: the crucial point is that as a creator (of proofs) a mathematician is not exactly a scientist, or even an engineer, but an artist. If it were necessary to define the philosophical position of a mathematician at work, very often they say that all mathematicians are Platonists; frankly there we prefer to speak of *anti-pragmatism*.

The things that a mathematician creates are writings destined for reading, written in words, letters, figures and numbers, with a flavor of *geometry-and-algebra*, that is to say with a double inclination toward visual interpretations in terms of movements. The mathematician writes movements into the mathematical matter made of mathematical writings. These movements are given as functions among objects and relations. In the categorical perspective they can be formulated as diagrams, or systems of diagrams, or machines (see Section 67.4.2.1).

For intuition, logic comes after geometry, and algebra is a kind of discrete geometry, in one-and-half or in two dimensions. Let us recall a famous sentence by Sophie Germain, that "Algebra is but written geometry and geometry is but figured algebra" [764, no. 1706]. In fact this relates to the ability to see: we see figures, we say formulas, but also conversely. And the nodal point for any mathematical enterprise is there: "voir ce que l'on dit, dire ce que l'on voit" [412, p.162]. This pulsation in some sense is absorbed into the notion of *diagram*, in such a way that diagrammatic (and categorical) thinking subsumes algebra and geometry and their interaction. In fact, any pulsation provides a progress of mathematics, as expressed by Lagrange [764, no 1707]: "So long as algebra and geometry proceeded separately, their progress was slow and their application limited, but when these two sciences were united, they mutually strengthened each other, and marched together at a rapid pace toward perfection."

This is to make up on the road again, as auto-expansion of the mathematical system of writings and representations, as an estrangement with no worry about foundation and origin, and a priori free of logical constraint. Such a freely and rigorously walking makes the road. Our best hope is that categorical modeling is a method to create in this way.

67.2.2.2 *Calculo, Ergo Sum*: *Mathêma* and Doubt

Cogito ergo sum can mean: *I shake some ideas, so I do exist.* We suggest to strengthen that with: *Calculo, ergo sum.* In mathematics, *to calculate* is to work something out in one's head, planning one's move, putting in sight a gesture; so we do some gestures in mathematical writing; each gesture is a calculation, and each gesture is calculated. Creation in mathematics is invention of new calculus *alias* new ways of surveillance of exactness. Under this explanation, "calculo" appears as a special case of "cogito": it is "cogito," mindful of exactness. This "calculo" works as well with arithmetics as with geometry.

In Euclid's *Elements* [306] two aspects of logic are in action: grammar and apagogy.

On the one hand, in mathematics logic is only a grammatical tool to correctly express links between steps in geometrical constructions. Today, it is no longer the true "logic of the ancient philosophers" but already a geometrical (or diagrammatical) process, and its replacement by Descartes by algebraic rules is perfectly rightful (= exact).

On the other hand, logic is mainly a tool for de-monstration when a possibility of monstration is causing controversy. Monstration is this type of proof that shows obviously a thing or a path from one thing to another (what we call in fact a *pure proof*). But in the case where such a path does not exist, a proof of this non-existence could be possible by a reasoning (but not a direct monstration), the famous *reductio ad absurdum* (apagogy). Some people—as Badiou [63, p.249] or Szabó (quoted by Badiou)—say that there is mathematics only when there is *stricto sensu* a *reasoning*, and more strongly when there is explicitly reductio ad absurdum. Badiou [63, 64] does insist on that point: The knot between mathematics and logic was introduced by Parmenide, in his creation of onto-logy, when "he found philosophy on proposing a knot between three concepts: the Being, the Thought, the Non-Being [...] this knot is Borromean [...] and that holds on a *matêma*: the reductio ad absurdum." [63, p.9-10]

In our perspective of effective practice and creation of mathematics, we think this is too much. Yes, it is true that reductio ad absurdum "is moving into the supposition of false" ([63, p.218]); but also direct deduction is moving into the supposition of truth! Hence direct or reverse deduction both *are moving into the supposition of logic* as the question of the alternative of false or true, and the idea of Truth. A deduction there is not only a proof, it is a proof that something is true, according to the determination of the idea of Truth as a fact expressed within language. The function of such a deduction is to state a truth, and this has to be prescriptive. The prescriber is the Being. For the working mathematician, the supposition of True as prescribed by the Being is not necessary; he needs a lesser notion of "truth", the cartesian notion of evidence, and the notion of exactness. Then mathematical logic is only a tool to construct new paths of exactness, using a new mathematical space of writing in which these paths are drawn. Hence mathematical logic is a modeling of ancient logic, in which ultimately truths are modeled by exactness.

In our view, "Truth" is just the last myth, the myth saying that from now mytho-logy is finished, and in place of all poetical fictions we have to believe in Truth, as exemplified by production of *matêmata*. It is possible to product *matêmata*, i.e., absolutely transmissible exact configuration. The myth of the "Truth" says that what is exact is truth, a prescription forever. But "exact" only means *obviously incontestable*, now out of any doubt; but doubt about it has not gone forever.

For those needing an ontology, Being could be the system of all relations (coherences and incoherences) among existing proofs; and this system is living because there is doubt in it. And this is the point: it is from this possibility of doubt about what is exact (an exact computation) that pulsation into interpretations and into writings is possible and can generate invention (new process of calculation).

Now for those needing a Reality, we have to understand how reality is present in Euclid. Euclid is axiomatic, but not in the modern sense according to Hilbert, for instance. In Euclid a concrete interpretation of terms (line, circle, point, triangle, square, etc.) is activated for the imagination of the reader, and simultaneously its pertinence with respect to the truth is suspended. The relation between *oral* speaking and logic is not totally canceled, its effect being represented as the ascendancy of concrete interpretations. This ascendance that comes from orality is clearly not necessary for truth or exactness, but the "feeling of reality" that it gives is a support for the imagination, indispensable for the invention of new exact things. At this point, considering that concrete companion interpretations are contingent necessities for creation, we meet again the question of the pulsation: Pulsation between abstract and concrete, but also pulsation into the multiplicity of possible representations to use as companions in the proof, during the course of an open calculus.

⋆

An example (see [411, Chap.4, Par.31, p.46-47] or [415, p.116]) of a calculation with logical and recursive components is a proof that:

$$(-)^{\mu}: \text{for any sequence } s: \mathbb{N} \to [0,1] \text{ there is an extracted monotone sequence } s^{\mu} = s \circ \phi.$$

Let us make precise that an *extraction* $\phi : \mathbb{N} \to \mathbb{N}$ is defined as a strictly increasing function. To prove that the monotone extraction $(-)^{\mu}$ always exists for a given s, we consider the proposition:

$$E := \text{There is a strictly increasing } \phi : \mathbb{N} \to \mathbb{N}, \text{ such that } s \circ \phi \text{ is monotone.}$$

In fact $E = I \vee D$, with

$$I := \text{There is a strictly increasing } \phi : \mathbb{N} \to \mathbb{N}, \text{ such that } s \circ \phi \text{ is increasing,}$$

$$D := \text{There is a strictly increasing } \phi : \mathbb{N} \to \mathbb{N}, \text{ such that } s \circ \phi \text{ is decreasing.}$$

Hence to prove E it is enough to find a proposition K with $(K \Rightarrow I) \wedge (\neg K \Rightarrow D)$. We take

$$K = \forall A \text{ infinite } \subset \mathbb{N}\ \exists n \in A\ \{p \in A; s(n) \leqslant s(p)\}\text{infinite},$$

$$\neg K = \exists A \text{ infinite } \subset \mathbb{N}\ \forall n \in A\ \{p \in A; s(n) \leqslant s(p)\}\text{finite}.$$

Let us suppose that K is true. If A is infinite in $\mathbb{N}$, we denote by $\sigma(A)$ the first n such that $\{p \in A; s(n) \leqslant s(p)\}$ is infinite, and $\Sigma(A) = \{p \in A; s(\sigma(A)) \leqslant s(p) \wedge p > \sigma(A)\}$. Then we obtain ϕ by recursion: $\phi(0) = \sigma(A)$ and $\phi(n) = \sigma(\Sigma^n(\mathbb{N}))$, and $s \circ \phi$ is increasing.

Now let us suppose that $\neg K$ is true: we have an infinite part A_0 in $\mathbb{N}$ such that for all $n \in A_0$ the set $\{p \in A_0; s(n) \leqslant s(p)\}$ is finite. Consequently, as A is infinite, for any $n \in A_0$, the set $\{p \in A_0; s(n) > s(p) \wedge p > n\}$ is infinite, and we denote by $\sigma(n)$ the first element in this set. Hence we define ϕ by recursion: $\phi(0) = \sigma(0)$ and $\phi(n) = \sigma^{n+1}(0)$, and we conclude that $s \circ \phi$ is strictly decreasing.

This proof can be analyzed as a construction of a path with logical and recursive gestures into a convenient space that it generates, and it is a new gesture that we denote by $(-)^{\mu}$ and name *monotone extraction*. We would like to insist here on the part played by logic (mathematical logic). It is not at all as a foundation and justification of other mathematical gestures, but it is as a super-gesture, a "logical gesture" over the more elementary ones.

We have to be careful with what we call "logical gesture". It does not mean a gesture in this philosophical logic governing by its rules the propagation in the saying of the philosophical truths. It is a gesture in the so-called mathematical logic, in which truth is modeled by exactness, and some handling precautions of writing simulate laws of logic. In fact such a logical method is *transcendent*, because its principle cannot be proved by lower level finite calculations, and it cannot be seen as a construction of a path into the lower level scenary. But of course it requires a decision to introduce a new level to our space of proving, and in this new space, it becomes an official possible path. Clearly it is a great gesture of invention to propose such a transcendent tool for our gesticulations. For instance, the use of *reductio ad absurdum* and of negation, as a mathematical principle for constructing a path of proof, adds a new power to monstration (and de-monstration). As here they can be used to prove that from two objects there are no paths, or to prove that there is a path (but without exhibiting a case).

★

Once this gesture $(-)^{\mu}$ is proved (i.e., is proved to be exact), we can use it for the construction of a new gesture which is the Bolzano-Weierstrass gesture of extraction of a convergent sequence. But we have to be careful: extraction of convergent sequence as well as extraction of monotone sequence exist, but are not at all effective. For instance, we do not know in general if s^{μ} is increasing or decreasing, and we are not even able to give its first value; furthermore s^{μ} is not unique. So s^{μ} is only a virtuality of possibilities, a name for a generic extraction, a letter to pursue new calculations with.

In [415, Par.4, p.118] we use this equivocal operator $(-)^{\mu}$ to prove that there is a map $B : [0,1]^{\mathbb{N}} \to [0,1]$ such that any map $f : [0,1] \to [0,1]$ is continuous if and only if $B \circ f^{\mathbb{N}} = f \circ B$.

Such a B is named a "bout" (a end) operator. Hence continuity becomes an algebraic property *stricto sensu*, and then it is more strictly a question of "calculation". For any s it is legitimate to use with the above property the univocal notation $B(s)$ or s_B (rather than s_{∞}) for a number well defined associated to

s; $B(-)$ is a well defined gesture of calculation, contrary to the operators $(-)^\mu$ or lim (when it exists, we write $\lim s = s_\infty$).

The reader has to ruminate on the distinction between s_B and s_∞. Let us remark that if we interpret a sequence s as a "semi-arrow" (or a jet, as in a water jet, or a flung stone), with a source and a priori without a target, then the operation B consists of completing any semi-arrow into an arrow (this is not reached by the partial operation $(-)_\infty$); and in fact this completion process is equivalent to the topological structure of $[0, 1]$, as it determines the continuity.

⋆

Turning toward itself, by a kind of a orienteering, a method for invention in mathematics has to have a supposition on the internal subject of mathematics, on what it is about. For us it is about the effects of the invisible entities at work on its visible calculations, as in ambiguity in the writings of multiplicities, the definitive gap between finite and infinite, or the orientation of the gesture which traced a line.

Another formulation is: mathematics deals with continuity, as a guarantee for coherence in the succession of its own gestures. For that matter, a mathematical gesture is a path or connection filling a gap by a continuity datum, a comment connecting previous gestures. Hence the importance of calculations with zig-zag, exact sequences or with our *exact squares* [408], all that can replace logical manipulations.

Ultimately we can say that we calculate the shapes of calculations and their connections, trying to construct exactness exactly. This idea has to become integrated into any fruitful method.

67.2.2.3 Semblance of Object: The Dodecahedron

Progressively in the rest of this story our notions of gesture, arrow, diagram, and exactness will become clearer. In the meantime as an aperitif we conclude this section with the example of the following view on Euclid's books [306].

This set of thirteen books named *Euclid Elements* is a very rich gesture, composed of more elementary gestures (definitions, theorems, proofs as algorithms or programs) converging toward the contemplation of the final object named *regular dodecahedron* and denoted by $\mathbb{D}$. It is a composite arrow from hypothesis to the final object; the books give a "proof" (a construction) of this object.

Nowadays with $\phi = \frac{1+\sqrt{5}}{2}$ we analytically specify $\mathbb{D}$ by the exact canonical coordinates of its vertices, which are

$$(0, \pm\frac{1}{\phi}, \pm\phi),\ (\pm\phi, 0, \pm\frac{1}{\phi})\ (\pm\frac{1}{\phi}, \pm\phi, 0),\ (\pm 1, \pm 1, \pm 1),$$

and Leonardo da Vinci drew it as the following "object" (Figure 67.5) to see in exact perspective: After

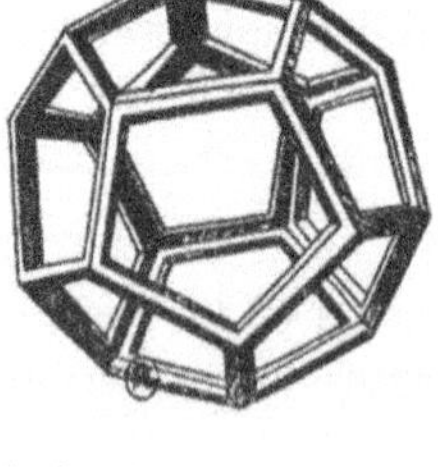

Fig. 67.5. Leonardo drawing of dodecahedron, as an exact presentation in perspective.

Euclid, this "object", and all the geometry as such, will be perceived through new gestures: computations of coordinates, perspectives, dissections, modifications, movements, comparisons, symmetries, group of its isometry, fundamental group, situations into a category of polyhedra or of digraphs, as being itself a category as Polyhedron/$\mathbb{D}$ or Digraph/$\mathbb{D}$. This is another story, in fact a notional living scenery in the sense of Section 67.4.1.4.

For now we highlight the fact that our "object" $\mathbb{D}$ does not exist; we have only several exact presentations thereof: Euclid's description, the systems of coordinates of its 20 vertices, the drawing by Leonardo, etc. The object is a semblance to stimulate gestures to present it, to prove that several exact presentations are exactly equivalent. For us the only true existing datum is the category of comparisons between the various presentations in various situations (= various categories) of this name, $\mathbb{D}$, which is like a cartography of the system of our gestures with $\mathbb{D}$. In this conception, a meaning of a presentation is simply a morphism toward another, as for example in Figure 67.6.

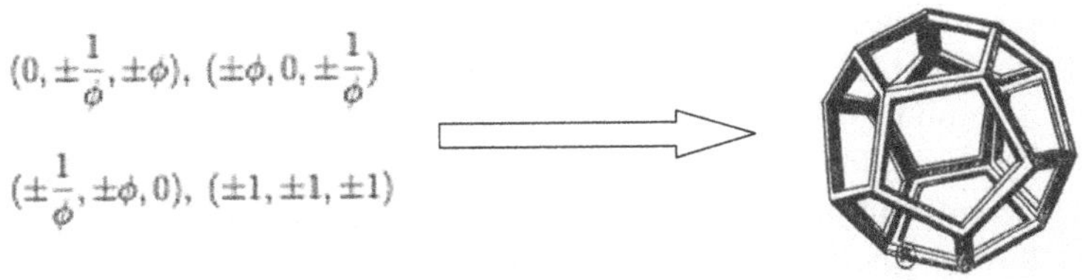

Fig. 67.6. A gesture from a presentation by coordinates to a perspective presentation.

So a complement of our mathematician's saying: "I calculate, therefore I am", is the mathematical object's saying: "you construct and compare presentations of myself, therefore I am a stimulating semblance."

67.3 Method and Objects, Summarily Explained: II—Data

67.3.1 Simple Objects, Structures and Invariants in Mathematics

67.3.1.1 Multiplicity, Ambiguity, Alterity of Objects, Varying Elements of Objects

A mathematical discipline is a closed system of manipulations of mathematical objects and concepts, substitutions according to axioms. A discipline can be a transversal process linking two other disciplines, as for example a theory of representations, a theory of theories, etc. For example we have Euclidean geometry, cartesian algebra, arithmetic, algebraic topology, combinatorial design, functional analysis, etc. A concept is a property of objects in a discipline, or a class of objects in it. For example in the theory of sets, we have the concept or notion of finiteness, in topology the concept of compactness, etc.

A *mathematical object* is any written datum into a given mathematical discipline that a mathematician can write and read, which is able to be interpreted and modified, such as numbers, figures, spaces, groups, etc. The work of a mathematician with his objects is basically to modify them, to construct external movements in the discipline from one object toward another one. But in fact any object is already an abbreviation of an organization of movements at a more internal level; and any external movement between objects can be considered as being itself an object at a higher level.

From the point of view of creativity the very right question about a given object is not if it exists statically and substantially, but what type of mathematical actions it codifies and induces, what its underlying dynamics is. The answer is that an object induces a comparison with others of the same type, or gestures of its modification. In a creative attitude of mind the substantial exhibition has to be replaced by the functional content, the potential system of gestures starting from the object.

A posteriori any "mathematical thing" T can be seen as an object of a category $\mathcal{C}$, or an object of several categories; a multiplicity (of things) is a diagram into a category, or a functor toward a category $T : \cdots \to \mathcal{C}$, whereas a mathematical concept $K : \mathcal{C} \to \ldots$ is a functor on a category $\mathcal{C}$. Hence the evaluation of the fact that T stands to K is given by the composition

$$KT : \cdots \to \mathcal{C} \to \ldots .$$

We consider this disposition as parallel to the analysis of thinking and perception [928]. Our concepts are evolving by analogies, according to our mental constructions with our perceptions in our personal historical scenery, and our perceptions are possible only because we have our concepts. Perceptions and concepts are in permanent continuous interactions, as sensations and meanings. As Emmanuel Sander remarks, the word "sens" in French allows us to tangibly feel this continuity. Sander and Douglas Hofstadter believe that this interaction is carried out by analogy, and our concepts arise and grow from analogy. Our brain asks if the last singular experiment can create a new category useful for understanding the world.

This question of "analogy" is related to the fundamental ability of humans to grasp similarities between phenomena [283], and also to do metaphors. James Geary said: "[The metaphor is] the process of giving the thing a name that belongs to something else. This is the mathematics of metaphor. And fortunately it's very simple. X = Y [Laughter] this formula works wherever a metaphor is present." And in fact an arrow $X \to Y$ acts as a super-metaphor, because it means conflicting notions: similarity and replacement. In some cases we have to read $X \to Y$ as "X is analogous to Y", in other cases as "X will replace Y". So each time we see or use an arrow we enter into a creative ambiguity, because we have to choose between analogy and replacement; in fact we have to choose when we are looking for a determined meaning, but a third way is to wait for that, and to choose after use, at the end of the play. Now that we are aware of these ambiguities, we will use of the word "analogy" to mean "analogy" as well as "metaphor".

By analogy (sic) we consider that mathematical thinking creates by analogy, determining things as objects standing for concepts, and furthermore mathematical thinking creates analogies; this is the heart of mathematical creativity. To stimulate propensity for creation of analogies, we need to have a practice of mathematical pulsation.

Pierre Deligne insists on the fact that often it is fruitful to consider an object as an element among others: *pour comprendre quelque chose, une bonne idée est de l'insérer dans une famille d'objets* [932]. For example, the consideration of symmetry or of Galois' ambiguity proceeds from such a general principle. An important example is also Lamé's approach to a surface as an element of a family of surfaces, namely as a level surface of a coordinate system (see 67.5.1.2). So it is important to look at maps associated to objects, to any multiplicity in which the object can be an element, to families of objects as moduli spaces or fibrations are, or simply to any function $f : E \to I$ interpreted as a family $(E_i)_{i\in I} = (f^{-1}(\{i\})_{i\in I}$ of sets, with therefore $E = \bigcup_{i\in I} E_i$. Given a type $\mathcal{F}$ of objects or figures, we have to observe the set of all the objects of this type on a given support or medium E, that is to say: let us look at a given figure F as an element $F \in \mathcal{F}(E)$, a representative of a concept $\mathcal{F}$ on E.

To consider an object C as an object of a category $\mathcal{C}$, among others of the same nature, is also a sample of the idea which Deligne stresses. Then again an object of a category can be understood by its position among the others in the category; this is the Yoneda Lemma. The deep meaning of this lemma is what we call "évidement" (scooping out): when an object is internally constructed, in a rather empirical and contingent way, with some bricks and glue, then that can be forgotten and replaced by its natural universal and external specification via its position among "its others"; then we know it up to isomorphism, and this is functional enough. By the way the true initially constructed object disappears, and even its necessity disappears. So category theory is a theory of purely functional objects, and an algebra of gestures.

As an example, let us consider the "informal object" $\mathbb{R}$. At first it is "the set of real numbers", given by one precise set theoretical construction. But this fact says almost nothing, only that it is a multiplicity of distinct elements named numbers; the construction implicitly can indicate the purpose of the object. In order to work with $\mathbb{R}$ we have to know two different things. The first is that this object exists in a mathematical world. The second is what precise operations and relations are put in it: algebraic operations, field structure, metric, topology, manifold of various types, linear space, etc. This means that we have to know various categories in which $\mathbb{R}$ is able to live as an "object". Yoneda's Lemma can be interpreted as saying that if we consider $\mathbb{R}$ as an object of a category $\mathcal{C}$, then this datum $(\mathcal{C}, \mathbb{R})$ determines a precise operational context in which we can use $\mathbb{R}$, represented by the localisation $\mathcal{C}/\mathbb{R}$, which is also thinkable as the *shape* of $\mathbb{R}$ within $\mathcal{C}$. So the object $\mathbb{R}$ is ambiguous, its shape being variable with $\mathcal{C}$. Furthermore the mathematical existence of $\mathbb{R}$ is concretely useful through the fact that the functor $F : X \mapsto F(X) = \hom_{\mathcal{C}}(-, \mathbb{R})$ on $\mathcal{C}^{\mathrm{op}}$ is represented by $\mathbb{R}$, i.e., any element x of $F(X)$ is an arrow f toward $\mathbb{R}$, i.e., a formal *family* X parametrized by $\mathbb{R}$,

$X = (f^{-1}(t)_{t\in\mathbb{R}}$, for $t : 1 \to \mathbb{R}$ in $\mathcal{C}$. We say "formal", because f is not really a function or a concrete map, but an arrow in the category $\mathcal{C}$. Hence an arrow as f is in fact the convenient notion of an element of $\mathbb{R}$ relative to $\mathcal{C}$, or a *varying element* of $\mathbb{R}$.

67.3.1.2 The Hexagram of Pascal

In geometry we have, as simple ingredients, points, straight lines, circle, and conics. Then more complex objects (or structures) can be constructed or proved, as for example the "mystic hexagram" of Blaise Pascal. The hexagram is a configuration in the sense of Theodor Reye, and so is an object of a precise category of

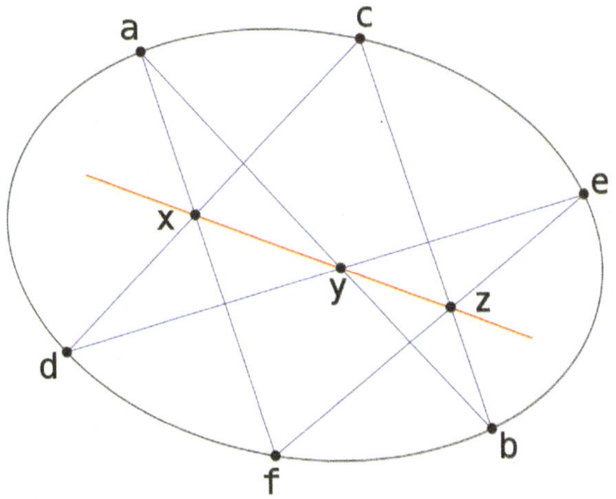

Fig. 67.7. Pascal hexagram for a conic: from a configuration to a construction.

configurations. But this line of thought will not be examined here.

This "mystic hexagram" is shown in Figure 67.7, in which six points a to f are taken on a conic, six lines are drawn, ab, bc, $\ldots ef$, fa, and the intersections

$$x = cd \cap fa, \quad y = ab \cap de, \quad z = bc \cap ef$$

are considered. Then the theorem says:

$$x, y, z \text{ fall into a line.}$$

In this way the hexagram expresses a *property* of any conic. But, using this property, we can transform it into a function, a way to construct a conic from five given points $a, b, c, d, \ldots e$ on it. With $y = ab \cap de$, given any straight line δ (in red on the figure) through y, we consider

$$x = cd \cap \delta, \quad z = bc \cap \delta, \quad f = ax \cap ez.$$

Then the hexagram says that the conic "is" the function $\delta \mapsto f$.

Notice that this is valid as a (not so usual) linear pointwise construction for a circle on which five points are specified. It gives a solution to the problem: to draw points on a circle with five given points with only a rule and a pencil.

And a new question: is it possible to draw points on a circle determined by three given points with only a rule and a pencil ? Of course the three given points plus the two cyclic points at infinity are five points on the circle, and therefore determine linearly all other points. But, if we want a concrete realization with rule and pencil, we have to know how to draw a line from a finite real point to the point I or the point J. Clearly it is not a real line! And what could we think in the case of an hyperbola when we are given the two asymptotes and three points?

67.3.1.3 A Formula of Frye

The first configurations are those that we *see* in elementary figures of geometry, and there the "numbers" are seen as numbers of things of a given type in a given figure. And we count on fingers, it is a gesture. With operations on these numbers we can construct new complex objects, as for example the formula of Roger Frye given in 1988:

$$95800^4 + 217519^4 + 414560^4 = 420841^4.$$

We let the reader discover its meaning. Two very different facts are there: on the one hand we have to do just some tedious verification (by hand or with the help of a computer), a "nice" curious fact; and on the other hand it is exactly a minimal counter-example to a conjecture of Euler, the conjecture saying that the function

$$(\mathbb{N}\backslash\{0\})^3 \ni (x, y, z) \mapsto \sqrt[4]{x^4 + y^4 + z^4} \in \mathbb{R}$$

has no values in $\mathbb{N}$; in fact it has the value 420841. It contrasts with the fact that $(\mathbb{N}\backslash\{0\})^2 \ni (x, y) \mapsto \sqrt[4]{x^4 + y^4} \in \mathbb{R}$ has no values in $\mathbb{N}$, known since Fermat.

67.3.1.4 Finite Configurations: Example of Latin Squares of Euler

After the natural numbers themselves, one of the simplest non-trivial configurations is a sequence of n terms made of the n integers $1, 2, \ldots, n$; such a sequence has a double-sided meaning (hence a pulsation in our intuition, between fixing and move): we can consider it as a fixing or a *disposition* i.e., a linear order on $\{1, 2, \ldots, n\} := [n]$, or as a move or a *permutation*, i.e., a bijection from $\{1, 2, \ldots, n\}$ to itself. Hence a sequence of n terms "is" an object in the category of *order*, or also "is" an object in the category of *presentation of groups*. Nowdays, orders and groups are cases of *categories*. Let us keep in mind that a configuration can be considered as an object in several categories; the alternative here is a living point of pulsation for our actions with it.

A typical mathematical problem with configurations is the following, related to the study of "sequences of sequences" or "tableaux".

Given an integer n, let us construct an $n \times n$ "tableau" with, in each of the n^2-places—namely in the place (i, j) at the intersection of row i and column j—a symbol $k_{i,j}$ taken in a set of n symbols S, in such a way that in each row there are different letters, and in each column too. This datum is named a *Latin square* or a *"squared" permutation* (permutation "carrée") of dimension n, or an $n \times n$-Latin square. For example

the square
$$\begin{matrix} 3 & 2 & 1 & 5 & 4 \\ 5 & 4 & 3 & 2 & 1 \\ 2 & 1 & 5 & 4 & 3 \\ 4 & 3 & 2 & 1 & 5 \\ 1 & 5 & 4 & 3 & 2 \end{matrix}$$
is Latin, and furthermore it has also different symbols in each diagonal.

A Latin square of dimension n is "reduced" if the first row is $(1\ 2 \ldots n)$ and the first column is $(1\ 2 \ldots n)^T$. By a permutation of rows and a permutation of columns any Latin square is transformable in a unique reduced one. The first values of the number R_n of the reduced Latin square are [740]:

$$R_1 = 1,\ R_2 = 1,\ R_3 = 1,\ R_4 = 4,\ R_5 = 56,\ R_6 = 9408,\ R_7 = 16942080, \ldots$$

With R the set of rows and C the set of columns, $\mathrm{Card}\, R = \mathrm{Card}\, C = \mathrm{Card}\, S = n$, and the Latin square is exactly determined by a ternary relation,

$$T \subseteq R \times L \times S = \{(i, j, k);\ i \in R, j \in L, k \in S\},$$

with $\mathrm{Card}\, T = n^2$, which is 3-functional, i.e., such that each coordinate of each element $\theta = (i, j, k) \in T$ is determined by the two others. Hence $k = k_{i,j}$, $i = i_{j,k}$, $j = j_{k,i}$. So a Latin square is naturally a 3-dim object, even if this is against its bending in its presentation as a labelled square, which is a pseudo-2-dim object.

An $n \times n$-*Græco-Latin square* or *Eulerian square* is a pair of two orthogonal Latin squares, i.e., a Latin square $(A_{i,j})$, and another "Latin square," i.e., a squared permutation, but now made of Greek letters $(\alpha_{i,j})$, these two squares being orthogonal, i.e., such that if $(i,j) \neq (i',j')$ then $(A_{i,j}\alpha_{i,j}) \neq (A_{i',j'},\alpha_{i',j'})$. Also we say that these square permutations constitute a set of two *mutually orthogonal Latin squares* (a MOLS). There is a very active branch of mathematics concerned with those kinds of things, *combinatorial design.*

Here is an example of a 5×5-Græco-Latin square, of which the Latin part is reduced, firstly written with letters

$$\begin{array}{ccccc} A\alpha & B\delta & C\beta & D\epsilon & E\gamma \\ B\beta & C\epsilon & D\gamma & E\alpha & A\delta \\ C\gamma & D\alpha & E\delta & A\beta & B\epsilon \\ D\delta & E\beta & A\epsilon & B\gamma & C\alpha \\ E\epsilon & A\gamma & B\alpha & C\delta & D\beta \end{array}$$

and then written with numbers, or using Euler's preferred notation:

$$\begin{array}{ccccc} 1^1 & 2^4 & 3^2 & 4^5 & 5^3 \\ 2^2 & 3^5 & 4^3 & 5^1 & 1^4 \\ 3^3 & 4^1 & 5^4 & 1^2 & 2^5 \\ 4^4 & 5^2 & 1^5 & 2^3 & 3^1 \\ 5^5 & 1^3 & 2^1 & 3^4 & 4^2 \end{array}$$

Of course it is impossible to construct a 2×2-Græco-Latin square. Also it is impossible to construct a 6×6-Græco-Latin square (problem of the 36 officers): in 1782 this was conjectured by Euler [310, Par.140], and in 1900 a proof was published by Tarry [1036], [78]. Also Euler conjectured that there is no solution when $n = 2 \mod 4$; but in 1960, Bose, Shrikhande and Parker [135] [136] proved that for any $n \neq 2, 6$ there is an $n \times n$-Græco-Latin square (see [1011, p.152]).

With the notations of Section 67.5.4.2, an $n \times n$-Eulerian square can be seen as a special map

$$\lambda : [n]^2 \to [n]^2 : (i,j) \mapsto (k^1_{i,j}, k^2_{i,j}),$$

with the properties that λ is a bijection and for any $u \in [n]$ and $e \in \{1,2\}$ the maps

$$p_e.\lambda.R_u : [n] \to [n] : (u,j) \mapsto k^e_{u,j}, \quad p_e.\lambda.C_u : [n] \to [n] : (i,u) \mapsto ke_{i,u},$$

are bijective.

67.3.1.5 Structures or Recreational Mathematics: Same Recourses for Solving

Like hundreds of other objects, the configurations of the three previous objects in Sections 67.3.1.2, 67.3.1.3, 67.3.1.4, are relatively simple "structures", because everybody can see what they are, of what they are made, such as juxtapositions of atoms. It is more difficult to know how they had been invented, and what they mean.

When we stay at an elementary level, mathematical problems and configurations can be viewed as *recreational mathematics.* Another example is the *taquin.*

We consider [405] the "fifteen's puzzle" or "taquin", made of fifteen differently colored squared blocks in a square box, numbered from 1 to 15, with a hole as a sixteenth component, as in the left part of Figure 67.8. The first problem (by Sam Lloyd) is whether by moves where the blocks are not extracted and are only sliding (= moving by successive exchanges of the hole with one of its neighbours) it is possible to reach a state in which the only final modification is that 14 and 15 are exchanged. The situation is naturally described by the datum of a graph—the graph $\mathbb{L}_4 \times \mathbb{L}_4$ with

$$\mathbb{L}_4 = \bullet - \bullet - \bullet - \bullet,$$

Fig. 67.8. Fifteen's puzzle or unnumbered fifteen's puzzle? Two different fibrations to analyze.

and with a fibration $\mathbb{T} \to \mathbb{L}_4 \times \mathbb{L}_4$, whose elements over (i, j)—elements of $\mathbb{T}_{(i,j)}$—are all possible injective configurations $\Delta : \{1, \ldots, 15\} \to \mathbb{L}_4 \times \mathbb{L}_4$ with values $\neq (i, j)$. So the problem is to look at connected components in $\mathbb{T}$. The answer to the problem is "no!" [628], i.e., the two specified configuration are not in the same connected component.

The second problem is the same, about exchanging the fourteenth and fifteenth blocks, but now the problem is about the puzzle on the right, where numbers are no more written; and now the answer is yes! Why? Hint: now, in this problem, the underlying graph is no longer $\mathbb{L}_4 \times \mathbb{L}_4$, but it is $\mathbb{Z}/4\mathbb{Z} \times \mathbb{L}_4 \times \mathbb{L}_4$, with $\mathbb{Z}/4\mathbb{Z} = \{1, q, q^2, q^3\}$, where q represents a global quarter turn of the whole square box.

In solving this problem of the "taquin", in fact we will see that some fibration is in question, and also [628] some problem of invariant calculations is present. Namely it is the question of an invariant, the signature of a permutation associated to the problem. We let our reader read Lucas, and translate his observations into questions about mathematical gestures, and then into category theory problems.

⋆

Systematically mathematicians are continuing the process of successive construction, of *structuration*, according to high level gestures:

Abstraction & Generalization & Specialization.

For example this means that, starting with a datum as a Latin square (or with the datum of a taquin), we abstract by forgetting some specifications, e.g. the fact that it is presented in a square, and keeping in mind other facts, e.g. the fact that we have permutations in the square and the fact that they are acting on Latin squares (or on states of the taquin). We abstract by forgetting: "penser c'est oublier" [1076], and this is an aspect of *mathematical pulsation*, its aspect of estrangement, as expressed in our conclusion (Section 67.7), which means that firstly if by any chance we do some combinations or substitutions, we forget some data within the problem—at the risk wasting time—and possibly recognize the rest as being essential when it is seen in a new context, and at last decide to choose that renewed rest to pursue our job. As in traditional Zen archery this forgetting or oversight is an initial moment of relaxation: see [414] for an explanation of the pulsation in relation to Zen.

With the accent on permutations, and so on elements of groups of permutations, we reach the more abstract structure of a group. The notion of a group itself comes by abstraction from groups of permutations as well as groups of geometrical transformations, by keeping in mind only some properties of the composition law of these permutations or transformations. At this moment we have brought out the part played by the structure of a group into the configuration of a Latin square. But now a group itself is a new type of configuration, which can be studied and classified, and represented as itself. Of course in this general study of groups, it will be permissible and useful to look at special cases, etc.

Groups, lattices, topologies, linear spaces, metric spaces, fields, rings, all these classic structures (the "structures-mères" according to Bourbaki), and also local structures, categories and topoi, arise in a historical process of "structuration" and are basically of the same nature as figures, formulas, configurations, designs or diagrams. However, with a very important precision: an object or a structure, or a type of structure, cannot just be an arbitrary formal free definition; it has to be fruitful, or already it is itself a theorem (a theorem of existence), or it will be quickly clear that it implies interesting facts about previously unsolved questions.

Now, let us give an example. In a particular higher state of mathematics, namely in the mathematical area of algebraic topology, we consider the rather intuitive idea ("intuitive" in this area, of course) that projective modules over commutative rings are like vector bundles on compact spaces. This puzzle has two precise formulations and proofs by Jean-Pierre Serre [969] and then Richard Swan [1028].

We consider that these Serre-Swan theorems are not really different from a puzzle like "taquin", and our effective recourses for solving are the same.

The only difference is that we move at different levels, with different preliminary formation, into another "mathematical history" or personal historical scenery for mathematical activity. In both cases we have to move writings, letters of algebra and figures of geometry, and diagrams; to introduce abbreviations and rules of modifications,; to construct and de-construct; and to structure, by synthesis and analysis. Ultimately any theorem says that two paths of constructions and de-constructions arrive at the same result. In recreational mathematics or in higher mathematics in the construction of the "paths" we have the critical moment when simultaneously we have to invent the "space" in which the path is going now, usually a space of configurations or structures into another space used in a previous step of the path. Our tracks are the road. So creative mathematics are possible as well at the level of recreational mathematics, as in very abstract algebraic topology. The point is the pulsative imagination within analysis-and-synthesis; deeper creation in mathematics is exactly invention of pulsations as methods, that is to say invention of new calculus expressing the tension between analysis and synthesis.

67.3.1.6 Undirectness, Synthetic Thinking and Intuitions

The gestures of folding, bending, cutting, erasing, gluing, etc. are our concrete accesses to any understanding or construction or de-construction of mathematical objects. In algebra it is known through factorizing and expanding expressions, in geometry it is realized in the set of geometrical constructions. But it is of a much more general scope, at any level with any model of any structure, presented by generators and relations. We can do these gestures within the system of elements or within the system of relations, or both.

For example if we read Gaston Tarry [1036] on the problem of the 36 officers, we have to admit that his analysis proceeds by such gestures; in this way he can construct various coordinations of the hypothetical solution object, with various symmetries, modifications, insertions, quotients, group actions, etc., and this allows us to count data of different types, and to conclude when we find impossible values. This is a process of abstraction, combined with structuration, inventions of figures to observe in the given situation, as permutations, paths, or what he calls magic groups (already considered by Euler under the name of "formules directrices"). Also he reduces the analysis by classifying possible Latin squares in a solution up to isotopies or conjugations.

The historical process of structuring by abstraction is parallel to a systematic movement of undirectness in the development of mathematics in the 19th century.

For example, concerning the geometry of curves, from the abridged method of Gabriel Lamé to Etienne Bobillier and Julius Plücker, and to Max Noether, followed by Emanuel Lasker and then Emmy Noether, we start from classical problems on curves through the intersections of given curves, and we reach the invention of Noetherian rings. For the following generations, they think geometry with, the notion of a Noetherian ring at the beginning of their thoughts; they could forget concrete curves and Italian geometry, etc.

The same "synthetic" fact arrives with groups: whatever would be the very good motivations for groups, once assimilated the explanations of geometries in terms of groups according to Klein, the new generations will think directly in terms of groups.

We can perhaps consider that *mathematical intuition*, which is very transversal to the skill for application of logical rules, comes from the habits of working at a given level of abstraction (with numbers, figures, equations, polynomials, spaces, groups, rings, categories, topos, etc.), i.e., in a certain category of objects, which becomes familiar; objects are now very complex configurations which became simple for the mind; it is the skill to move into a given category with such formerly old data. For instance, someone can have or not have the intuition of the object $\mathbb{R}$ (as an object of a category $\mathcal{C}$). Definitely we consider that mathematical activity and mathematical intuition are *absolutely relative*, starting from the datum of a category as an allegedly natural setting of a first organized intuitive world.

67.3.1.7 Categories, Sets, Groups, Lattices, Structures, out of Logical Concern

An idea suggested by several logicians or philosophers [115, 653] is the opposition between sets and categories as a proposal for foundations. More precisely, with Jean-Yves Béziau [115] we can analyze the relations of set theory and category theory with respect to their relation to the question of foundations. Béziau remarks that at least we have three distinct problems: (A): axiomatic foundation of mathematics (to describe a set of axioms from which we can deduce mathematics); (C): conceptual foundation (to describe basic concepts and their links, in order to think mathematics); (L): logical foundation (to show that mathematics are not contradictory). We can add a fourth case: (F): functional foundation (description of a system of mathematical actions which are enough to develop mathematics). Roughly speaking, set theory is good for (A) and (L), and category theory is better for (C) and (F). So the opposition of categories and sets is wrong as an epistemological perspective. Category theory is a method of analysis of the production of mathematical works—and this method itself is mathematical, and a mathematical theory—whereas set theory (and logic) is possibly a ground for all mathematical developments and constructions of structures.

We have to completely dissociate the question of observation, description and control of mathematical gestures, using categories, from the question of foundations and logic, using set theory. The control of the size of universes is interesting for questions on existence based on set theoretical constructions.

Some mathematicians believe that categoricians are mainly interested in foundational questions, and consequently are unable to see that groups are everywhere, that geometrical thoughts innervate good mathematics. In fact the notion of category unified the two geometrical notions of lattice and group.

In [1150, p.37] "category theory" and "structures" are opposed; Bernard Zarca said that the concept of category is a rival of the concept of structure; this is a serious mistake from a historical perspective; it is a wrong interpretation of the discussion in Bourbaki about the question of introduction of categories in Bourbaki's Elements. In fact, the more significative example of category are the categories of fundamental structures (structures-mères), also because each structure-mère "is" a category. So a group is a category, a lattice is a category, and even, indirectly, any object C of a category $\mathcal{C}$ becomes a category, via the consideration of $\mathcal{C}/C$. Furthermore the categorical study of a given concrete category is inspired by practices of Universal algebra with structures (quotients, sub-structures, sums and products, free algebras).

In a decisive manner, by structuring and studying modifications of structures, the categorical point of view participates in the estrangement or undirectness which is necessary foro invention. To work on structures with categorical tools means to use objects, arrows, functors, etc., to represent actions and gestures in a given area of mathematical activity, out of foundational or logical concerns. In this completely relative game, objects and arrows are present at various levels. Arrows are present at high levels, denoting functors between big categories of structures $F : \mathcal{C} \to \mathcal{D}$, as well as at lower levels, inside a given structure C, an object of a category $\mathcal{C}$, representing an equation such as $ax = b$ by an arrow $x : a \to b$.

A mathematical result will become the existence or the nonexistence of a precise object or an arrow or a path or a diagram in a given category of structures, or even only the existence of a category with a given property. So the method of research is to construct categories associated to the problem—with objects some diagrams of structures—to determine the solution as a category with a given condition, and to examine the "formal" universal solution (existing if we put in brackets some essential set theoretic condition of sizes), as for example the *locally free diagram* of the situation [409].

This result on locally free diagrams is available in the context of *mixed sketches* [294], i.e., of categories equipped with specifications of 'virtual' limits and colimits (which are dual processes of analysis of some objects). Hence a mixed sketch is denoted by $\sigma = (\mathcal{C}_\sigma, \mathcal{P}_\sigma, \mathcal{I}_\sigma)$, with $\mathcal{C}_\sigma$ the underlying category, $\mathcal{P}_\sigma$ the specification of limits (projective limits) and $\mathcal{I}_\sigma$ the specification of colimits (inductive limits), and the category of its models or realizations in **Ens** is denoted by $\mathbf{Ens}^\sigma$. If $h : \sigma_2 \to \sigma_1$ is a morphism of sketches, the induced functor between models is $\mathbf{Ens}^h : \mathbf{Ens}^{\sigma_1} \to \mathbf{Ens}^{\sigma_2}$. Now, given $h : \sigma_2 \to \sigma_1$ and $R : \sigma_2 \to \mathbf{Ens}$, a solution to the "problem" represented by (h, R) will be an $\mathbf{Ens}^h$-free structure $F_h(R)$ on R. Such an $F_h(R)$ is as an algebra of h-terms generated by R.

67.3.2 Complete Frameworks, Computations and Representations

67.3.2.1 Do We Need Universes as Complete Global Foundations, or Completions as Locally Achieved Frameworks?

Set theory is not necessary to work in mathematics, but it can be a starting point. An important book in 1970 by Andrée Ehresmann [89] very clearly taught how constructions of universes are possible and useful to aim at the creation of mathematical objects, and especially completions of previously given objects.

In the categorical perspective, as a global starting point we have today the idea of a *topos*, the idea of *fuzzy sets*, or unifying these two, the idea of an *algebraic universe*. See also Section 67.6.1.4. The point is to start with a category equipped with enough operations to construct all the structure objects we could need. The basic operations are: the final object, the kernels, the cartesian product $X \times Y$, and the powerset construction $\mathcal{P}(X)$. Especially there we can construct existing free algebras, functional spaces, completions, etc. The categories of models $\mathbf{Ens}^\sigma$ are examples, and the $\mathbf{Ens}^h$-free structure $F_h(R)$ are also examples.

If X is an object and PX a certain "completion" of X, we think of PX as an algebra of terms on X, with an "inclusion" $X \to PX$, and this is useful as a codomain of maps $Y \to PX$ expressing a relation between X and Y.

Hence these constructions of universes and completions are useful to express mathematical problems according to the scheme:

Is it true that a given map $Y \to PX$ factorizes by a map $Y \to X$?

67.3.2.2 Calculations and Sketches of Gestures

In the common sense, the term "calculation" means concrete computations with integers and Pythagorean tables, or with decimal numbers. So we get concrete values for statistics, estimations and measures or evaluations, and the results are real and actual by the fact of our decision about their interpretation within our experimentations.

In the beginning of 19th century it became common to model physical situations and problems by functions and the optimization of an operator L and evaluation of functional operators. These evaluations are themselves functions from $\mathcal{F}(X, V)$, a functional space, toward a system of numbers W, $L : \mathcal{F}(X, V) \to W$. Given such an L and $f \in \mathcal{F}$, we have the problem of calculating $L(f)$, and then the problem (of optimization) to calculate the best f, for example the one which minimized L. All these are questions of calculation.

In fact, these calculations, with numbers or with functions, are not more concrete than algebraic formal computations in various algebras, groups, etc., or calculations of invariants in any convenient system of values, or even constructions of structures according to specifications. Also, a proof is a calculation, and a geometric construction too; progressively any mathematical gesture could be taken as a case of calculation. Without getting as far as this, we want at least emphasize that "calculations" include "structural calculation", calculation with structures. And we determine a *calculation* as a special gesture with an underlying *program*, producing a *value*, which can be a structure as well as a number.

We have to know that a priori a given calculation is not possible to perform, because we do not have the convenient space into which the ingredients and results can be inscribed. For instance if we work with $\mathbb{Q}$, we cannot calculate $\sqrt{2}$, and we need a completion $P(\mathbb{Q})$, namely $\mathbb{R}$. So before calculating $\sqrt{2}$, we have to 'calculate' $\mathbb{R}$, in such a way that the operation to do, $x \mapsto \sqrt{x}$, becomes a map $\mathbb{Q} \to \mathbb{R}$.

⋆

A categorical model for that idea of "calculation" is possible as the datum of a co-span Π in the category of sketches [410]

$$\Pi :: \delta \xrightarrow{h} \pi \xleftarrow{j} \epsilon,$$

in such a way that given a datum $D : \delta \to \mathbf{Ens}$, the result $\Pi(D)$ of the program Π on it is given as $F_h(D).j$, the evaluation via j of the $\mathbf{Ens}^h$-free structure $F_h(D)$.

All the examples of computations follow this scheme: Ordinary algebraic calculus, execution of programs given by Herbrand schemes, representations of groups, etc.

Of course the computation of $\Pi(D)$ is not always possible, because $F_h(D)$ do not always exist. Nevertheless the locally free diagram theorem asserts that $\Pi(D)$ exist as a Proj-object into $\mathbf{Ens}^\epsilon$. Furthermore, when it is theoretically possible, it can be too complicated, and we have to invent a shorter process, namely for each object X in the underlying category of ϵ a final subcategory of the h-shape of $j(X)$. The calculus of exact square is useful here.

Finally, even if $\Pi(D)$ do exist for any D, this operation is not necessarily "continuous", i.e., exact in the sense of compatibility with limits and colimits; but the analysis of this lack of exactness is possible by cohomological techniques, thanks to the general definition given in [418].

We do consider that a datum such as Π for which the computation $D \mapsto \Pi(D)$ works is a *sketch of gesture* (see the definition of a gesture in Section 61.5), because it provides a functor (a gesture)

$$\mathrm{Mod}(\delta) \longrightarrow \mathrm{Mod}(\epsilon).$$

67.3.2.3 What About Applications, Implements, and Representations?

The art of mathematics becomes a scientific purpose when we are looking for applications in other sciences, including itself, or, when starting from a technical problem, we are trying to implement a solving process.

For instance the Descartes' invention of ovals (Section 67.2.1.2) starts in the realm of physics as a response to an optical problem. Another example is the development of mathematical logic as an implementation of a mathematical process in parallel to the classical logic of speech according to Aristotle. Another example is the theory of statistics as a framework to study populations, to control any democratic or financial process, production of energy, etc. A last example is the quest for an invariant in a mathematical field.

The point is that in all these situations we are looking simultaneously for a double invention of a mathematical function *and* of an interpretation thereof: the function is mathematical, pertaining to the art of exactness; its interpretation is not, and is related to the production of an ideological datum: a law of physics, a law of truth, a repartition function, an invariant. Here the ideological datum is the belief that physics, or truth, or repartition, or mathematics, makes sense. In fact in enriched situations, we are looking for a system of functions *alias* a structure, pretending to model a "concrete" field, such as social life, music, or mathematical activity itself; but there again we are in a tricky situation with the question of interpretation. As expressed in a humorous light in [99] or [437], the difficulty is that on the one hand there are a lot mathematical inventions that can work, and on the other hand decision-makers impose that these inventions work according to their interpretation (or opinion). The mathematical datum is then a pretext for a (con)fusion between ontology (and truth) and hermeneutics (interpretation).

When we are doing mathematics we don't have to feel obliged to believe in science and technology; exactly as Beethoven doesn't have to believe in music to do music. So we do construct functions or structures, exactly, but we do not believe in their interpretations or their truths. Creation in mathematics exists at this point of pulsation between the function that we are making and the interpretation we are not really believing.

Nevertheless, for our method of invention we have to realize that we need interpretations inside our mathematical thinking. This is necessary not only as an external motivation to feed our imagination, but internally as a target or a source for a jet of mathematical gestures. These interpretations can come from physics, sociology, philosophy, and—last but not the least—from history of mathematics (hence in these cases as pseudo-concrete pretext or semblance of problems) but finally these exist within pure mathematical writing itself as what we name *presentations* and *representations*.

★

There are several theories of representations, such as the theory of representation of a given group, or the theory of representation of a lattice, or the representation of a given theory (alias the theory of its models), and more generally the theory of representation of a category $\mathcal{C}$.

Given a group G, we consider the action from the left of G on itself, and that is a maximal representation of G (A dual one is of course by right actions). Such an action is nothing else than a functor $G^{\mathrm{op}} \to \mathbf{Ens}$.

Given a pre-order set $(E, \leqslant)$, or a fortiori a lattice, we look at a representation thereof as being an order map $(E, \leqslant) \to (\{0,1\}, \leqslant)$. The system of these representations is included in the set of functors $(E, \geqslant) \to \{0,1\}$.

Now a group or an ordered set are two special case of a category, with two opposite interpretations of the arrows. In the case of groups, an arrow is an invertible transformation, from one state to another, whereas in the case of an ordered set, the arrow is essentially not invertible.

So, a long time after the junction of algebra and geometry by René Descartes we get the junction of group theory and lattice theory by Alexander Grothendieck, with the idea of a topos. Nowdays, we understand that junction itself is a mathematical gesture.

To begin with, we consider that a representation of a category $\mathcal{C}$ is a functor

$$\mathcal{C}^{\mathrm{op}} \to \mathbf{Ens},$$

and the theory of these representations is the topos $\mathbf{Ens}^{\mathcal{C}^{\mathrm{op}}}$. If we put $P\mathcal{C} = \mathbf{Ens}^{\mathcal{C}^{\mathrm{op}}}$, we can consider that we have a completion of $\mathcal{C}$ in a "universe" Cat. Now the fundamental fact is the property of completion of the "inclusion" $\mathcal{C} \to P\mathcal{C}$, a property known as the Yoneda Lemma. Another possibility of representation of $\mathcal{C}$ is as the category $\mathcal{D}(\mathcal{C})$ of small diagrams in $\mathcal{C}$, of which a completion property is known. We will use this in Section 67.4.2.1 to provide a mathematical description of gestures.

⋆

Now, with a specified theory of representations at hand, we can consider that our effort for creative mathematics will be supported by a two-level pulsation, with crossed effects. On the one hand, a pulsation at the pure writing level itself, in the open space of our writings; and on the other hand a pulsation among interpretations which are helping us pursue the process of imagination. In fact, towards our creation of a gesture, we are moving horizontally and vertically among writings, horizontally without references to interpretations, and vertically inside a given interpretation. Horizontally we take care of exactness or continuity in the space of our writing, and vertically we are concerned with exactness or continuity in the spaces of our interpretations.

Clearly, once the job is finished, we can present all our invented gestures in a straight and uniform way, but this erases the very moment of pulsative invention.

67.4 Creativity in Mathematics: Gestures in Historical Contexts

67.4.1 Creativity: Phenomenology, Psychology and Skills, and Life

67.4.1.1 At the Beginning of Our Creations Are Our Imaginary Gestures

In order to make precise our understanding of the term of "creativity" in general, we start with two quotations.

From Maurice Merleau-Ponty [752, p.82] we quote:

> When we say that each [genuine?] work opens up an horizon of inquiry, we mean that it makes possible what was previously unavailable without it and that it transforms the pictorial enterprise while fulfilling it. Thus two cultural gestures can be identical only under the condition that they are unaware of each other [...] It is therefore essential to art to develop. Art must both change and, as Hegel says, "return to itself" and thus present itself as history.

From Carl Rogers [915] we quote:

> The creative process is defined as the "emergence in action of a novel relational process" from the interaction of a unique organism and its material and circumstantial environment. The creative impulse is "self-actualizing" and arises out of personal need; not all originality is creative, for it must be manifest in the extensional world. Certain inner conditions are prerequisite: lack of rigidity, tolerance of ambiguity, extensional orientation; "an internal locus of evaluation", or a feeling that the creation satisfies and expresses oneself; and the "ability to toy with elements and concepts." For fostering creativity, it is important to make the individual feel worthy "no matter what he does". The teacher must say, "I don't like it", rather than "It is not good". The creative person must have complete freedom to give symbolic expression to his creation. Several specific hypotheses are derived from these assumptions.

⋆

We consider that it is not necessary to suppose that mathematics is not the study of purely imaginary states of things. In fact we think that reality and truth are not given by mathematics, which are rather about multiple realities and complexities, relative truths and ambiguities. To create mathematics is to imagine an exact and clear knowledge of these modalities of knowledge in our mind. As says Barry Mazur, "imagination is held to be a movement" [926], and our mathematical gesture has its source in our personal historical representation of the meaning of old mathematical gestures, alias our imagination.

So we have to reinforce our imagination by reading books presenting mathematics from the point of view of imagination and intuition such as [522, 468, 277]. We will see that what we do when we imagine are gestures: to visualize and to cancel, to categorize and to identify, to systematize and to generalize, to abstract and to specialize, to articulate and to dislocate. These steps are steps in the process of discovering or creating a proof.

⋆

Now, with in our hands these two ideas of gesture and of creative impulse—or imaginative impulse, and whatever we do out of any logical or rational control question, we have to precise what is creativity, as a psychological disposal in the Rogers' style, but also as a technical skill for any craftman capable of the best with gestures—as in the orientation of Merleau-Ponty, and that specifically in mathematics; and conversely how mathematical inventions or discoveries are related to such creativity.

When we "create" some mathematical object, then before anything else our creation "is" a gesture; to be more exact, it is accurately the gesture of creating, the impulse at the root of the movement of creation, as a result of which the so-called created object is produced. The mathematical gesture (to visualize, to cancel, etc.) is a true mathematical datum; it takes place in our memory of mathematics, in our personal notional scenery.

67.4.1.2 Gestures, Diagrams, Computations, Detours, Pulsations

As working mathematicians, our two basic types of gestures are *gluing* and *cutting* inside a system of mathematical writings. These two are not necessarily interpreted in an analytical way as movements, but they can be naturally interpreted so, and they could be noticed by *arrows*. In a synthetic perspective, our gestures are *diagrams*, or designs, or tableaux, or figures, and can be noticed as algebraic data as well as geometric data; we will speak here of *diagrammatical data*. In this way, mathematical problems become questions of enumerations of structures of a given type, and a comparison between types of different diagrams. But the computations could be performed at very different levels; for example we can have to count finite configurations of a given shape, and also may we have to provide information about automorphism groups of very sophisticated and high level structures.

It is interesting to emphasize the distinction between "problems", "open problems", and "open question".

A problem is a question in a given closed system K of computations (numerical, figurative or conceptual), such as the following: Given A and B in K, is it true that $A = B$? Given a property P for data in K, is it true that there is an A such that $P(A)$? How many such A are there? Some problems, such as scholar exercises, are not open; we know the answer. Others are open problems. Then a more open question could be: what is the structure of the system of A in K such that $P(A)$? And a more open question is: for a given A in K such that $P(A)$, is K the best context in which A lives? Then we have to create new closed systems analogous to K.

Creativity in mathematics lies on the skill to move among these levels (problems, open problems, open questions, creations of systems), to do detours, *to open questions.*

A basic point for our method of invention would be precisely this question of the "*detour*", related to what we call the "*mathematical pulsation*". This pulsation itself depends on the possible interpretation of our gesture, as a geometric or an algebraic gesture, or more precisely as a movement into any given axiomatic framework, as well as into another possible one, etc. And, in an open way, these interpretations of our gestures are conditioned by our notional living landscape or scenery. This scenery is an architecture of concepts, and our creative work will add something to this architecture. Our main task would be the target of this evolutive architecture, and how to pursue?

67.4.1.3 Three Pulsations Which Are Internal to Any Mathematical Commitment

When as pupils we started our first trials in algebra, a very strange point was that the same thing can have several names. It is a very fine point in teaching mathematics, and with his student a good teacher has to be clear [413, p.46], [745, p.639]. In fact, more accurately, there is a movement in the course of the computation of the solution of $x-2=0$ in the assignation of meaning of the symbol x. At the beginning, x *is* an unknown, at the end x reaches the value 2, and x *is* 2. Furthermore we can emphasize that known numbers as 2, and unknown as x. They both have to be manipulated in homogenous way; this was invented by François Viète, when he decided to designate by letters known as well as unknown entities.

A second "pulsative" data is the relation of duality between algebra and geometry, as expressed by Sophie Germain: "Algebra is but written geometry, geometry is but figured algebra", as already quoted in Section 67.2.2.1. This pulsation is "solved" by the consideration of diagrams.

A third pulsation is between groups and lattices, a pulsation "solved" by the notion of "category". We like to see a category as the dialectical synthesis of monoïds and orders, and arrows as the dialectical synthesis of the idea of action or transformation and the idea comparison or hierarchy.

67.4.1.4 Creative Mathematics into a Peculiar Notional Living Scenery

Perhaps we have to distinguish between creativity with mathematics and creativity within mathematics as two kinds of creativities, applied and pure; but unless we turn our eyes to automatic applications of mathematics, by an engineer, technician or financial controller, or separate art and craft in mathematical activity, it is difficult to think of *non-creative mathematics*, as it is difficult to imagine that non-creative music exists, and more generally that any non-creative art exists.

We would like to sustain the beautiful posture of Nietzsche when he wrote to his friend Peter Gast: "Das Leben ohne Musik ist einfach ein Irrtum" (Life without music is quite simply a mistake); and also his main idea on mathematics as being an error, but an error necessary for life [427]; and we want to claim ourselves:

> Without creative mathematics thinking is a mistake.

So we have to make clear what we mean by "creative mathematics".

Among several of its comments on Descartes' cogito (the famous "I think, therefore I am"), one of Nietzsche's proposals is: "I live, therefore I think". Again, this can be adventurously commented by: *I compute, so I live*, and *I speak, therefore I think*, etc., transition relations that we can write as formal data: I think $\rightsquigarrow$ I am, I speak $\rightsquigarrow$ I think, etc., or $Z \rightsquigarrow Y$.

Anyone can construct for himself a private constellation of such $Z \rightsquigarrow Y$ organizing his conceptions (or notions, or functions) of actions—that is to say: to be, to think, to live, to speak, to do mathematics, to compute, to construct figures, etc. to sculpt, to paint, to do music, to play piano, to sing, to write, etc., to construct a house, to plant some tomatoes, etc.—as a mental ground space for his own creativity, i.e., the place in which he "expands his life", what we can name his *notional living scenery.*

In [427] we tried to show that the notional living scenary of Friedrich Nietzsche is a kind of evolutive geometric space named "Nietzsche", constituted by his health and his body, his spirit and his thoughts, his will, his life in its entirety, his "style". We think that the situation is not so different for any creative thinker, and especially for a creative mathematician.

From our point of view, initial separations between mental and material, and between art and craft, are not at all pertinent to determine creativity, and a priori creativity is above the domains of its productions. Any creative production has to be simultaneously caught from many different points of this constellation, and according to some $Z \rightsquigarrow Y$ relations. Now, again any given domain is analyzable as a system of $Z' \rightsquigarrow Y'$, with Z' and Y' some objects of this domain. And again these objects (taken as domains) are presentable as new $Z'' \rightsquigarrow Y''$, etc. At last a creative production is nothing else than a proposal of one such $Z \rightsquigarrow Y$ that we call a gesture or a transit. It is creative because through the structure of its private constellation of notions it is connected to the life of the corresponding creator. To create in mathematics is the same as to instill our own mental life into our mathematics, into our personal scenery, transforming it into a web or an architecture of concepts, with new proofs and understandings.

Some works on creativity produced models of a creative space, a space in which creation is expanded, here with the idea of a living scenery, and hereafter with the part played by history, both as a part of what can be named the space of mathematical thinking, we are near the idea of a *concept space* as studied by Margaret A. Boden [123] (as quoted in [32]). She proposed that creativity results from three mental processes: unfamiliar combinations of familiar ideas, navigations, and transformations within a structured concept space. This will be a convenient frame for us, if we do insist on the point that the given structured space is not really given a priori, but is in fact invented and constructed as you go along by navigation and tranformation (cf. the idea of "camino" in Section 67.2.1.1). This "invention" goes by generation of new signs with respect to a given meaningful universe of signs (as proposed in [32, Sec. 2]).

A category-theory model of creativity (invention and novelty) will be very natural in the case of mathematics, and more in the special case of category theory (!), where models of the questions of open paths (caminos) into living sceneries or (structured) concept spaces are realizable via the Yoneda Lemma in terms of creative subcategories, sites, shape theory, and sketches [32].

67.4.1.5 Style and Notional Sceneries in Mathematics as a Natural Language

In contrast to formalists, we consider that mathematics is a *natural language*, the language that mathematicians speak. They "speak" and "write" mathematics within their culture, with their literature and their history. Their creations are, in this language, similar to novels, poems and songs in any other natural language, and to movements in dance. In all these cases, creations are not substantial objects but are functional gestures and *transits*, and above all they are *indirect*; their first value is to be offside.

More precisely, the life in the system of mathematical specialities can be pictured through the present *notional living scenery* used by the *community of mathematicians*, as given by the *MSC* (Mathematical Subject Classification) edited by the AMS [28]. There are 97 columns, from 00xx = General to 97xx = Mathematics education. For example, if a mathematical work is concerned with objects such as fibrations, it is located in the area of "fibered categories", an area spotted by 18D30; 18Dxx is "category with structure", 18xx is "category theory; homological algebra". We can consider that the columns of the MSC are for the mathematician what the musical instruments and pieces of music are for the musician: the space in which they create, creations being movements within this space.

To describe notional sceneries of mathematicians many other approaches are possible, for instance using books on the history of mathematics, on recreational mathematics, on applications of mathematics. All this material determines our culture and so our personal scenery in which we have to enter more, by working.

Also we can use of good introductions to mathematics which provide various styles of division into domains of mathematics. Often the title of the book explains under what point of view the division into domains is realized.

For example let us look at the book of Claude-Paul Bruter [157], "Comprendre les mathématiques". It is a book for teaching, and a presentation of a scenery of mathematics with an anti-formalist approach, against any algebraic virtuosity, claiming to be written from the natural point of view of geometrical deepness. Hence here the order in the presentation of mathematical facts and tools is very specific, and different from the order of MSC, or the order of the Bourbaki's treatise.

The book by Mazzola, Milmeister, and Weissmann [710] is significantly entitled: "Comprehensive Mathematics for Computer Scientists". In this book, an underlying idea is to provide various pure mathematical *tools* for applications, from logic to splines to differential equations to category theory. So the order is different from the order in Bourbaki or in Bruter. Hence another possibility of entrance into mathematics.

Another example is the book by Mac Lane [638], whose title is "Mathematics: Form and Function". In this book, the central idea is a philosophical posture: a twofold background for philosophy of mathematics: there are form and function *of* mathematics, as qualities of the mathematical field, and there are form and function *within* mathematics, as elements in the mathematical business; from our present point of view, this is central to category theory.

Let us stress that in the notional scenery of a given mathematician, we can find data of very different levels, for example information about a philosophical or epistemological possibility for a working mathematician, history of a mathematical object, technical description of a specific algorithm, theoretical scenarios or storylines, theory, counterexamples, and solved and unsolved problems. For example he can know how to solve a second degree equation, how a continuous function induces a morphism of toposes and how the properties of the function is reflected within properties of the morphism, how in the history of mathematics the law for quadratic residues had been introduced and improved, what non-Euclidian geometry is, what the technical signification of the infinite in Cantorian mathematics is, and what its connection with philosophy is. The creativity of a mathematician is dependent on his notional scenery and, in this space, on his *style*. For example the styles of Alexander Grothendieck and of Paul Erdös are different and complementary [1150, p.127].

67.4.1.6 Creativity with Mathematics, in Mathematics: To Prove, to Understand

The two basic elementary gestures in mathematics are: to count with numbers, to draw geometrical figures; manipulations of numbers and figures, with our hands and eyes, paying attention to some writings and readings. We have to word exactly what we see, and to see rigorously what we say.

These manipulations are compositions and decompositions, and the aim is to reach methods for that to prove equality or proportionality between different compositions/decompositions.

Creativity in mathematics is concerned with invention of proofs, in order to discover new methods, if possible new methods of invention, but at least a new interpretation or functionality for an old datum, within mathematics or within any mathematical modeling of a science. Yehuda Rav wrote: "The essence of mathematics resides in inventing methods, tools, strategies and concepts for solving problems which happen to be on the current internal research agenda or suggested by some external application. But conceptual and methodological innovations are inextricably bound to the search for and the discovery of proofs, thereby establishing links between theories, systematizing knowledge, and spurring further developments" [881, p.6]. To create in mathematics is to invent methods for proving and for understanding; to reach these two purposes we have to choose or to invent a "space", to construct a path in this space, and to follow this path: "attraper le geste et savoir continuer" (Cavaillès).

67.4.1.7 Creativity from the Double-Sided Point of View of Categories

Relating to the problem of creativity, the point of view of categories is double-sided. We can construct objects into a given category, or construct functors between two categories.

On the one hand, category theory seems to be concurrent to set theory for a foundational setting. In this case, the things we have to manipulate are sets with structures, or, shortly, *structures*. Hence our activities are within categories of structures (sets, groups, topology, modules, etc.), and the categorical point of view is to describe our activities in these categories in categorical terms, i.e., in terms of universal properties of specific diagrammatical constructions within such a given category. In this case category theory is a mathematical theory, among other, concerned with a given type of structure, the transitivity of arrows seen as functions, and transitive epistemology in any given categorical area.

On the other hand, categories are considered as an homogeneous principle of general explanation of mathematics in terms of functions and forms: the mathematical modeling is reducible to analysis and synthesis through functionality and research of a shape's invariants. In this case, category theory is in fact a mathematical explanation of mathematical activity as constructions of functors between categories. Then category theory is "transverse" to the other mathematical theories (and especially to itself!), concerned with the general transitive process among types of structures.

67.4.2 Determination of Mathematics as a History of Its Gestures

67.4.2.1 Gestures as Transits, Pulsation Among Diagrams, and Machines

Studying the concrete musical gesture of the hand of a pianist in the space $\mathbb{R}^3$, Guerino Mazzola and Moreno Andreatta wisely defined a *gesture* [720, 723, 726] as follows. A gesture from Δ to X is a morphism $\delta : \Delta \to \overrightarrow{X}$ from a digraph (i.e., "oriented graph") $\Delta : A \to V^2$ to the digraph $\overrightarrow{X}$ associated to a space X, where, with $I = [0, 1]$:

$$\overrightarrow{X} : A_{\overrightarrow{X}} = I@X = X^I \to X^2 = V^2_{\overrightarrow{X}},$$

with $A_{\overrightarrow{X}}$ the space of continuous maps from $[0, 1]$ to the space X (with the compact-open topology). The set of gestures from Δ to X is denoted by $\Delta@\overrightarrow{X}$, and this set equipped with a convenient topology is denoted by $\Delta\overrightarrow{@}X$. Then a *hypergesture*—or a gesture of gestures—is a gesture from Γ to $\Delta\overrightarrow{@}X$. To analyze musical composition, a generalization is studied by Guerino Mazzola and Florian Thalmann under the name of *gestural diagram*, which is a continuous diagram in a topological category [730].

⋆

In Mazzola's definitions above, a gesture is a special morphism, with a specific geometric analysis of its composition. Of course a more general categorical definition for any of these morphisms is perfectly possible, as simply is an element in a *functional object* Y^Z, that is to say an arrow $1 \to Y^Z$, or an arrow $Z \to Y$ in a category $\mathcal{C}$ (closed and with terminal object), seen as a trajectory from a typical figure Z toward a backdrop given space Y. In fact such a datum is only the path of a gesture, the gesture itself being an act: stimulating a subject to create such a path. In fact here we can even abandon the geometric feeling expressed in the words "trajectory", "figure", and "backdrop".

For the creative act, in [425] we recall that Gilles Châtelet, with "geometric intuition" as a guide, insists on the movement of thoughts transversal to its own development, what he considers as being on a see-saw machine or on a rocking horse (in French: une *bascule*). Without geometric feeling, this bascule seems to be a case of a *pulsation.*

In [425] Charles Alunni puts the accent on the question of diagrammatical thinking, in a kind of return to Kant and his diagrams. We consider this point as being very good for the examination of creativity, better than the submission to the "geometric intuition". Nevertheless the "good" diagrams could be the diagrams of Peirce, or more restrictively the diagrams of categoricians.

Now, in order to explain what a *mathematical gesture* is, we would like to abandon any geometrical framework, and we have to reduce the geometry to its functional substratum, i.e., to diagrams of arrows. Then any mathematical intervention (geometrical constructions as well as algebraic computations) will be understandable as a game of transits, a mathematical gesture *in a "moving space" of mathematics* (what we have named a notional scenery of a mathematician).

So we prefer to drop the classical geometrical thinking, and the philosophical consideration of Kant's diagrams. We agree with Alunni when he moves towards the views of Bachelard, and the Bachelardian's view of diagrams and mathematical writings, upstream to a hypothetical physical world. For the general analysis of mathematical invention, the image of a moving hand in $\mathbb{R}^3$ or a fixed space X is good, but not enough "abstracted"; we need the idea of a "moving unachieved space" of mathematical writings (diagrams). Our guiding idea will be the idea of a gesture as a general *changing among diagrams*, what before we named a *machine*, and what is interpretable as a system of local actions on the objects and relations of a given situation, and even on concepts about this situation.

⋆

We have to retain two points. The first is that a gesture is given by an arrow, and the second is that the composition of the gesture is given by a concrete analysis of this arrow using a structural decomposition or a matrix (see Section 67.5.4.2).

What is an arrow? Physically it is the graphic symbol "$\longrightarrow$" in the diagram

$$\square \longrightarrow \circ$$

from an object $\square$ toward another object $\circ$ (then a categorical diagram is a system of arrows).

But what is the meaning of an arrow? It is a sign in the sense of Peirce, and a sign is an arrow (see Section 67.5.2), and a sign is a trace of a gesture. We agree with the views of Mazzola and of Alunni that signs are presemiotic: as Mazzola says, they give rise to sign concepts and are not signs a priori; and we can say that an arrow represents the pointing, that is—as also quoted by Mazzola—what Tommaso Campanella considers as the only elementary evidence in human thought.

In mathematics, an arrow $\square \longrightarrow \circ$ represents a transformation or a function associating an element of $\circ$ to any element of $\square$, or an homomorphism between structures. In an abstract category an arrow (a morphism) represents a transit, and even a transformation (precisely, the Yoneda Lemma for a category $\mathcal{C}$ is a transformational representation of $\mathcal{C}$). We use this idea of transit for our transitive epistemology in [426] to understand the analysis of Grothendieck's works by Fernando Zalamea [1149].

But here we can refine the explanation, and say that a transit (according to the Winnicott notion of transitional operations) is something like a semi-arrow, i.e., an "arrow" such as

$$\square \longrightarrow$$

with a source but without any target! This idea is also underlying the reflexions on caminos (Section 80.2) or jets. In Section 67.2.2.2 a model for this idea is given by infinite sequences within a space. In Section 67.5.2.2, with the idea of autograph and autocategory we propose another meaning for an arrow, as a tension between two other arrows. In this case an arrow is going from a source toward a target, but these are no longer objects.

⋆

At first, the only thing that we have to take seriously is the idea of an arrow (or more precisely the idea of the production of an arrow), which we prefer to envisage as a *transit* between objects [426]; then a *mathematical gesture* will be an act of determination of such a transit. And a transit is creative if, according to our quoted words of Rogers, it is an "emergence in action of a novel relational process" $Z \rightsquigarrow X$, or more precisely an arrow $Z \to X$. Eventually the emergence of a creative transit, or even of an ordinary transit, needs a subject, namely a mathematician to provide the creative impulse, and a category as a universe in which the impulse is expanded and constructed, diagrammatically in fact. This last point is completely in accordance with Peirce's viewing of mathematical work as "transformation of diagrams", or if we insist, as diagrammatical transformation of diagrams. So the creative production of gesture is under the condition that our thinking pulsates between analysis and synthesis of diagrams.

There, a gesture is the written trace of an effective transit, written as a *diagrammatic construction of an arrow* in a category. This is related to the construction of a gesture within the category $\mathcal{D}_\Gamma\mathcal{X}$ of small *Γ-diagrams* in a locally small category $\mathcal{X}$, i.e., a functor

$$\Delta \to \mathcal{D}_\Gamma\mathcal{X}$$

that we call a *Γ-machine* from Δ to $\mathcal{X}$, with Δ a locally small category.

This name of "machine" comes from the very special case of Mealy's machines in automata theory, with inputs from a monoid Δ and outputs in a monoid $\mathcal{X}$; here we think of a machine as a *local action* of Δ on $\mathcal{X}$. Here $\mathbb{T}$ is a category whose objects are some categories thought of as "types of shape", $\Gamma : \mathbb{T} \subset \mathrm{Cat}$ is an inclusion functor, a Γ-diagram is a datum $(I; F : \Gamma I \to \mathcal{X})$, and a morphism from such a Γ-diagram to another one $(J; G : \Gamma J \to \mathcal{X})$ is a datum (H, θ), with $H : I \to J$ a functor, and $\theta : F \Rightarrow G\Gamma(H)$ a natural transformation. These machines and categories of diagrams are studied in [403], [405], [406], [407]. A preliminary study is the notion of *ébauche* [402]. It has been proved that $\mathcal{D}_\Gamma\mathcal{X}$ is a Γ-relative lax-co-completion of $\mathcal{X}$ [406, p.403], as, in a very particular case, the powerset $\mathcal{P}(E) = \{S; S \subseteq E\}$ is a lattice co-completion of E (the system of atoms of $\mathcal{P}(E)$); so a machine is a generalization of a binary relation, and so binary relations are gestures. Because of the lax-co-completion property, a machine generates a lax-co-continuous 2-functor, which appears as an analytical description of a gesture:

$$\mathcal{D}_\Gamma\Delta \to \mathcal{D}_\Gamma\mathcal{X}.$$

For the moment we retain that mathematical gestures are things such as machines, or related arrows qua distributors of Jean Bénabou and profunctors of Alexander Grothendieck. In all these cases we can speak of *relations*. In the Bénabou and Grothendieck cases, the analogous of the lax-co-completion $\mathcal{D}_\Gamma\mathcal{X}$ is the co-completion given as $\hat{\mathcal{X}} = \mathrm{Ens}^{\mathcal{X}^{\mathrm{op}}}$, the category of presheaves[2] on $\mathcal{X}$. In fact [407] we have canonical comparisons (with $\mathcal{D} = \mathcal{D}_{\mathrm{Id}_{\mathrm{Cat}}}$)

$$\mathcal{X} \to \mathcal{D}\mathcal{X} \to \mathrm{Cat}^{\mathcal{X}^{\mathrm{op}}} \to \mathrm{Fib}\,\mathcal{X} \to \mathrm{CAT}\,/\mathcal{X},$$

and therefore a more general determination of a gesture from Δ to $\mathcal{X}$ can be as a functor,

$$\Delta \to \mathcal{D}\mathcal{X}, \text{or } \ \Delta \to \mathrm{Fib}\,\mathcal{X}, \text{or } \ \Delta \to \mathrm{CAT}\,/\mathcal{X},$$

or a special 2-functor,

$$\mathcal{D}\Delta \to \mathcal{D}\mathcal{X}, \text{or } \ \mathrm{Fib}\,\Delta \to \mathrm{Fib}\,\mathcal{X}, \text{or } \ \mathrm{CAT}\,/\Delta \to \mathrm{CAT}\,/\mathcal{X}.$$

A special case is examined in Section 67.3.2.2 as a "sketch of gesture".

In this modeling, at this very high level of generality, a gesture is a transformation from one world, the world of things or concepts over Δ, to another world, the world of things or concepts over $\mathcal{X}$; shortly we will say that it is a transit among concepts, a conceptual transit. Here we definitely do not assume some precise control on sizes of objects; hence we can pursue our process, introducing a general "relator" from Δ to $\mathcal{X}$ as a distributor between CAT/Δ and $CAT/\mathcal{X}$ (a "relation" between concepts over Δ and concepts over $\mathcal{X}$), in such a way that gestures are representable relators, a relator being itself a reserve of virtual gestures. As Mazzola says, this was perhaps Grothendieck's hidden approach.

At any rate, a mathematical gesture is not at all an evanescent inmost sentiment, such as an artistic expression of a move, with a pure aesthetic content (as in the French expression: "pour la beauté du geste"). It is the ordinary realistic gesture of a craftsman, and only because of that it is a move in the "art of mathematics", a concrete move of the "hand of the mathematician" in his mental creative space. We believe that this open creative space exists, unachieved and in permanent creation. From a materialistic point of view it is nothing else than the set of all the mathematical writings, and consequently the history of the production of these writings.

So by a *mathematical gesture*, we mean something which is not at all necessarily related to logic and to regular application of logic to the search for solutions of problems: such a logical "method" would be rather stupid with respect to a creative target; it is just the 0-level of gestures, let say the gesture of blind substitution or application of a given rule. The interesting point is the next levels, starting with the gesture to choose a substitution or a rule, even the gesture of choosing a logical framework. Creativity in gesture starts when we start to choose something freely. The idea of a "method" becomes interesting only when we

[2] $\hat{\mathcal{X}}$ is denoted by $\mathcal{X}^{@}$ in the general notation system of this book.

are looking for a *method of invention*, in some way a paradoxical aim: how could we discover from the present state of our current tools any *new* thing? For us, the very nature of mathematics resides in the paradoxical game of its gestures of inventions, in the history of this living game with diagrams.

So two points have to be carefully isolated, logic, and rigor, and we have not to be mistaken about their part. *Logic* could be replaced by mental concentration or internal rigor in the practice inside a specified calculus, to correctly apply substitutions, whereas the true *rigor* in our sense resides in the concentration in order to find new ideas, i.e., to introduce new possibilities of substitutions and representations. A creative process supposes a dialectic between correct substitutions under given ideas and renewing directive ideas.

Furthermore, in order to model creativity mathematically in mathematics, we have two complementary aspects. On the one hand *a posteriori* we can observe a creative process as being after an intriguing event; on the other hand, in an *a priori* perspective, we can observe what was arriving just before the intriguing event, and so before the creating process. At the moment of the event, we have to situate an abrupt change of mode between continuous attention to adequate substitution and a burst for no reason of new directive ideas.

67.4.2.2 To Do and to Apply Mathematics: Mathematical Gestures

To examine creativity in mathematics, it would seem at first that we have to decide what entry within mathematics is to be considered: searching, teaching, or learning. In fact our opinion here is that creativity in mathematics appears in the inseparable triple framework of searching, teaching, and learning, and more according to the sense of mathematics as expressed by its history. This has to be accomplished under the dialectical tension between 'to do mathematics' and 'to apply mathematics', these two being distinct, but each one implying the other.

On the one hand to apply mathematics means we dispose of a known area of abstract mathematics, modeling a given real experimental situation (in physics, mechanics, visual or financial questions, organization of systems, etc.), and that we want to obtain a solution of a problem in the situation via a solution of a mathematical problem in the given modeling, which often is an unknown function to find. To do that we have to solve equations or to construct the solution of a geometric problem, or we even have to invent and develop a new set of mathematical tools, new spaces, or new types of diagrams, and to do mathematics there. So many mathematical areas and objects were created from problems in physics or mechanics, such as mathematically pertinent correlates of observed situations; but also sometime things go in the other direction, as observed by Bachelard: at first we have in our modeling of a situation a significative mathematical datum, and then we are looking for the corresponding real observable thing. In these cases *mathematical gestures* in question are inventions of modelings, followed by resolution of mathematical problems.

On the other hand doing mathematics (e.g. pure mathematics) includes the frequent situation which consists of applying an old piece of mathematics to a new mathematical situation. Descartes says: "chaque verité que je trouuvois eftant vne reigle qui me feruoit aprés a en trouuver d'autres" [264, p.20-21] (as George Pólya quoted in slightly different words: "each problem that I solved became a rule which served afterwards to solve other problems" [853, p.1]).

Some philosophers such as Karl Popper consider that mathematics are tools to describe the world: "Pure mathematics and logic, which permits of proofs, give us no information about the world, but only develop the means of describing it" [859, p.13]. So mathematics can be "applied" to a well determined piece of real situation, to clarify and describe it. But this is not a very convenient way to consider the link between mathematics and the world: it is better to consider that our perception of the world is clear only if at first it proceeds from a mathematical gesture. And in fact mathematics can be applied only to itself: *mathematics develops the means of describing itself*; this is the right setting in which we can understand the possibility of creativity in mathematics.

67.4.2.3 History as Series of Analytico-Synthetical Gestures: Doubt, Obviousness

Obviously, "history" here is absolutely not the name of the academic discipline of the same name (sic); and the same remark is valid for epistemology, philosophy, etc.

According to Hegel, man [the mathematician] is nothing else than the series of his acts; history [of mathematics] is the series of [mathematical] events. The question of the sense is the question of the continuity or structure of this series; the structure of the mathematical series is such that its purpose is to clarify the possibilities of structuring: therefore from the beginning mathematical thinking can be applied to its own development; this is at the center of its history, and it proves that, as Beppo Levi said: "more than anything, mathematics is a way of thinking". We can emphasize that mathematics is the art of thinking that works to erase its own property of being a thought; as a thinking, it works reflexively to resolve itself (se résorber), to reduce its thoughts to its gestures [413, p.101]. So we arrive at the idea that history of mathematics is useful for mathematics exactly if and only if it is the history of the series of these gestures which are the mathematical reductions of intuitive thoughts to finitely generated diagrammatic gestures, and even to the trace of such gestures, by codifications and calculus (calculus being taken in a very large sense), in short, reduction of intuitions to proofs. This history furnishes the real matter for future mathematics, namely the problem of thinking why and how these gestures of reductions themselves are mathematical and mathematically exact. Another formulation, what we call "reduction," is a process of rational writing. A mathematical idea is transformable into the idea of a gesture (within the mathematical world), and we have to produce an exact writing of this gesture in such a way that any mathematician reading this writing should be able to build a replica of the gesture.

If mathematics is an art, adapting our previous quotation of Merleau-Ponty we can say: Mathematics must both change and, as Hegel says, "return to itself" and thus present itself as history". The shape of this history and its auto-move—by application of mathematics to mathematics—is the meaning of mathematics.

In fact application of mathematics to mathematics works by transformation of a solution of a problem into a true mathematical fact, and this fact into a mathematical rule (a solving process), and this rule into a new mathematical object, about which new questions could arise; so, according to these gestures, mathematics enters into itself, and thereby produces its own development, something that we can *a posteriori* understand as its history.

To learn the solution of an exercise, to explain how to understand mathematical things and methods, to search for a solution of an open problem, to prove that a given sentence is true, as well as to construct new objects or new relations, all these levels are completely mixed and inseparable, as far as creativity is the matter at hand. Mathematical activity is the way in which this mixture is accomplished, by writing and reading analytical and synthetical data, doing what we call a *mathematical gesture*. In this view we have to construct the 'true fictional knot' of the history of mathematics.

Hence the history of mathematics that we need is of course a system of internally true fictional narratives, about gestures of reduction alias gestures of invention.

On the one hand, to do so we need a kind of neutralization or *suspension of our current knowledge*, to permit the invented and new thing to enter into our mind. For history in general, this had been expressed in various refined ways by Paul Veyne, Henri-Irénée Marrou, and Carlo Ginsburg; and in the specific case of the history of mathematics, by Évelyne Barbin.

We have to be able to read freshly any old paper, as contemporaries did, forgetting future uses and successes of that given paper, doing the history à la Veyne [1084, p.18]: "s'étonner de ce qui va de soi," as Évelyne Barbin proposed for history of mathematics in [74], a notion that she named "dépaysement". In this way perhaps we will be able to bring out the inventive gesture of the author himself, forgetting the success and errors it obtained in the future.

In the frame of general history, the word of "dépaysement" is also in Marrou [654, p.237], in a rather different but related acception: "Historical knowledge always implies a going off oneself, a "dépaysement", a meeting and a discovery with others" and "history is mainly discovering of a pure alterity." Michel de Certeau, also thought in this direction [186], but not directly on the question of history: "La *xeniteia* est un "dépaysement." Ce mouvement consiste à partir ailleurs, comme Abraham, "sans savoir où." It is a voluntary exile to a foreign land, the monk being invited to leave his past life, without looking back. This is near the idea of estrangement [573] borrowed from Michel de Montaigne, as studied by Carlo Ginsburg [368], and in [574, 369]. As expressed in old French, one has to say "s'estranger."

On the other hand, "history" can be conceived as a tool for invention today of our present new knowledge. Retroactively coming from the past, history becomes an explanation and a justification *a posteriori* of the nature of our knowledge today. This kind of retroactive history looking for precursors is properly "ahistorical", and usually historians consider it as very bad. Nevertheless, working this way we obtain assurances on our subjects of study and our procedures. In the case of mathematics, it is the style of history according to Bourbaki and André Weil, and it is very useful for pursuing our mathematics to encourage our innovations [428]; we get a better understanding of our gestures, without respect for the gestures of our masters. Gestures of masters are read not only in the line of our story of their future.

In history of mathematics, the two sides—estrangement and retroaction—respectively oriented toward alterity (or otherness) and ego, are in "pulsation" with respect to the question of invention, in parallel with the pulsation of doubt and obviousness. At the level of mathematical technical practices we rejoin the notion of *mathematical pulsation.*

67.4.2.4 Rigor and Subjectivity, High Level Gestures

Especially two attitudes are *not* recommended here: systematic pure formalistic attitude and pure pragmatic attitude, in other words, pure formal logic or pure blind application. The point is that when we are in such an attitude we believe that the substance of this attitude is really the substantial nature of mathematics and of mathematical activity. So the naive positivistic logician thinks that mathematics have to be founded on logic, and the naive pragmatic scientist is sure that mathematics are oriented by applications in something that he believes to exist, namely the real world. In fact mathematical activity, and consequently mathematical creativity, resides in the space between these two fires.

The variation of styles—formalism or pragmatic—is fundamental, but this 'opening of mind' itself has not to be absolute. From time to time we have to work inside a given algebraic system, and to discover there new internal or low level gestures, and at other moments we have to leave a given formalism to enter another one, changing the framework: hence we have to accomplish *high level gestures.*

Example 83 For a deliberately elementary example, let us imagine we are computing in the finite field with four elements, namely, modulo 2, with $0, 1$ and the roots α and ω of $X^2 + X + 1 = 0$; and let us suppose we are looking for roots of $P = 0$, with $P = X^3 + X + 1$. Of course we can substitute values $0, 1, \alpha, \omega$ for X, and we have immediately to admit that there is no solution. Here is a first level gesture, namely an *evaluation.* But in a more complicated case, for instance with a ring instead of our field, direct evaluation is not possible, and then our strategy has to be indirect, in a sequence of gestures which seem to take us away from the question. In our example, a first higher level gesture consists of *abstracting* our problem, of considering that we are working in the field of mathematical research named "finite fields theory", represented by the category $\mathcal{C}$ of these fields; and in $\mathcal{C}$ within the object $\mathbb{F}_4$, the finite field with four elements. At this moment we forget little explicit computations with $0, 1, \alpha, \omega$ and the related construction of $\mathbb{F}_4$; we put the accent on the position of this object in $\mathcal{C}$. We continue considering an arrow which is an embedding f_P of $\mathbb{F}_4$ into $\mathbb{F}_{64}$:

$$f_P : \mathbb{F}_4 \to \mathbb{F}_{64}.$$

This gesture is external with respect to $\mathbb{F}_4$ but internal with respect to $\mathcal{C}$. Why to do so? Because $P = X^3 + X + 1$ has three roots in $\mathbb{F}_8$, it has also three roots in an extension of $\mathbb{F}_4$ as $\mathbb{F}_{64}$, because $64 = 2^{2\times 3}$. In fact $\mathbb{F}_{64}$ is the gluing (pushout) of $\mathbb{F}_2 \to \mathbb{F}_4$ and $\mathbb{F}_2 \to \mathbb{F}_8$. After this gesture the initial problem is transformed into the following one: is it true that f_P is invertible? Of course not, because $4 \neq 64$. A posteriori this is not very astonishing; the effective evaluation has been replaced by the effective construction of the pushout. However this pushout is a general construction in finite fields, and the effective computation with $0, 1, \alpha$, and ω is replaced by an instantiation of a general construction.
Later again, for more complicated cases, when it is not easy to decide whether f is invertible, we could do a new high level gesture, considering that $\mathcal{C}$ itself is an object of $\mathbb{C}$, a category of categories, and we can consider as a higher level gesture a functor H from $\mathcal{C}$ to $\mathcal{C}'$, a gesture external to $\mathcal{C}$, and a gesture internal

to $\mathbb{C}$. To conclude that f_P is not invertible it is enough to prove that $H(f_P)$ is not so. For example if $\mathcal{C}'$ is an abelian category $\mathcal{A}$, then f is not bijective as soon as $\operatorname{coker} H(f_P) \neq 0$. Etc.

So we arrive at a *complex gesture* represented by $\operatorname{coker} H(f_P) \in \mathcal{A}$. In this precise gesture, the choice of $\mathcal{A}$, H and f_P are left to the mathematician; it seems to be subjective and free until we obtain a result following this gesture, by an application of a property of $\operatorname{coker} H(f_P)$ to our initial problem.

The initial key of creativity is given by *subjective gestures*, but the confirmation of the mathematical character of a creation proceeds from *objectification of gestures*, i.e., validation of gestures by 'proofs', from the point of view of rigor as a common knowledge for all mathematicians.

So mathematics is the history of the activities of searching, teaching, or learning mathematics, via various level gestures of invention, construction or resolution of mathematical problems, and mathematical creativity lies in this knot: you search to solve problems, you teach to explain what you have found, you learn to get new problems to search and new tools. And what you create is a solution, an explanation, a new problem, a new tool.

67.4.2.5 Problems and Mathematical Pulsation in the Production of Forms

At first glance, a mathematical gesture is destined *to solve problems*. From this point of view we recommend the reading of several important books, such as the books by Hadamard [434] or by Pólya [853, 854], explaining the psychological conditions for invention in mathematics, and how to solve problems. Less known and more elementary are the books of Boirel [129] or Gasquet [360], which insist on the "operational dynamism" of the mathematical activity, on the "mathematical illusion" produced by scholar programs and institutional organizations, which abandon any spirit of invention and creativity.

And precisely creativity is more visible, in a second sight, if we reveal that a mathematical gesture is also destined *to invent problems*. And to invent problems we have *to name*, to introduce some definition: "Almost always a conquest in mathematics does start by a nomination," continuing with Pascal's rule (as quoted in [222, p.102]): "substituer la définition au défini" ("to substitute the full definition to the defined term") [245, p.136, 137]. Some mathematical creativity is there, in this double act of naming and substituting, pulsating between equivocal possibilities of meanings of the name, and various possible effects of its substitutions. This point is similar to an observation by Paul Valéry [1075, Analecta 33, p.188] quoted in [434, p.30]: "It takes two to invent anything. The one makes up combinations; the other one chooses, recognizes what he wishes."

The very point is that constructing a good definition or a nice substitution is not at all trivial; it could take very long to discover and to be precise in such a final simple way. If we cannot do so, we stay at the level of obscure computations and combinatorics, where some fact could arrive by chance... even if we don't know what we wish for exactly.

The aim of mathematical activity is to invent and to solve problems, and, by convenient analytico-synthetic gestures—and these gestures, like any gestures, are equivocal—to reveal a univocal clarification in the *form* of an object or a relation.

This alternative "to solve/to invent" a problem, and consequently a *form* (in the guise of clarification of a problem or of a solution), is in some sense a pre-view of the *mathematical pulsation*, a psychological picture of the attitude needed in order to do what we call *mathematical pulsation*. The pulsation is already perceivable in the ambiguity of the idea of *form*: the word 'form' could mean a configuration, a contour or an outline of an object, but it could also mean the law of construction of this object [245, p.139].

67.4.2.6 History as Imaginary Resource of Necessities for Mathematicians

But don't forget to install your solution or your invention in the realm of the history of older inventions, and with some 'imaginary' questions about various mathematical areas: numbers, figures, finite, infinite, geometric constructions, equations, continuity, general geometry, groups, topology, logics, proof theory, problems in foundations, homology, category theory, etc., etc. This "installation" is a "move to" and a "settle"; hence your last gesture to accomplish your mathematical creation, which we call an *imaginary gesture*.

The way in which you situated your thought and your work among these areas, with many more details among your readings of master as well as of minor works, constitutes your subjective feeling of the history. After preparatory work with logic and chance, sometimes a mathematical invention arrives, but this is always situated by the mathematician within a personal history of mathematics, constituted as a memory of mathematical gestures: definitions, substitutions, algorithms and constructions, writings. In his memory the mathematician believes to have a mathematical familiarity and intuition, and his conscious thought follows this 'intuition', in spite of positive closed logical limits, but with an ultimate aim to invent truth.

Always when he starts his work, a mathematician has the illusion of the existence of such a realm of imaginary thoughts, of such a web; and this is necessary, as a minimal toolbox. But then, on the way to solving a problem, he is able to abandon such pictures for new ones that he invents for the occasion. At such a creative moment, the point is the link with the history, not an objective absolute history (which doesn't exist), but subjective knowledge of the history in his mind. On this path, and only like this, does he know (or believes he knows) what he is doing! Otherwise he is only writing a strange sequence of letters, words and figures.

From our point of view, mathematical creativity is possible on this condition: to be free to invent definitions, objects and relations, and forms in a pulsative status of meanings of these data with respect to the history of mathematics that we have in mind.

In the history of mathematics, from the point of view of a mathematician, we have two components. On the one hand, we have a very personal fiction constructed as a progression by a personal reading of texts of masters, with a choice of what is good to work again: in fact it is a mathematical personal history of the apprenticeship of the given mathematician; in this vein we have the conceptions of Tœplitz and of Weil [428]. On the other hand we have the more 'objective' approach using a historical perspective to get an epistemological "dépaysement" by looking at the meaning for a contemporary [74]; in this case we could observe the finest gestures at the very moment of invention in the text as such.

Let us notice that, for example, the *AMS classification of mathematical subjects* introduced in Section 67.4.1.5 is a kind of fiction or novel about the life and the organization of mathematics today: it participates to the history of mathematics. The same remark works for other "notional sceneries", such as the table of matters of any good general book on mathematics, such as [710] or [638].

67.4.2.7 Fashion, Successes and Errors, Scruples

Examining mathematical inventions, someone could relate them to scientific targets determined by social contexts in which mathematicians are living [1150]. At the very most, this social storytelling could determine choices in fashionable themes of research, and it could explain some contingent conditions according to which the realm of mathematics keeps growing. And then? What about the very moment of creativity? As with artistic creation, social or pragmatic motivations are not really serious. The very moment is the will to create new movements in our mind, to get a new mathematical understanding. To reach this target we need to be free to try new ideas, new combinations, without any guaranty of success. The pulsative attitude starts from this necessity. Concretely, pulsations into combinations of new free ideas are observable in the works of mathematicians in the manner in which they mix methods and intuitions, by observation of their scruples with a very strong worry about exactness.

Rather than in the book of Bernard Zarca [1150] or in any sociological study, we will learn much more about mathematical creativity by reading the book of François Rostand [917], and books of quotations such as [764], [816], [943], where working mathematicians explain their own views on mathematical activity. Among these quotations, and especially in the book of Rostand, we can realize the importance of the question of "*scrupule*", as a counterpart of Descartes' idea of evidence [413]. According to Kant [917, p.111], a scruple is subjective: an argument against a belief, but with only a subjective value. In the scruple we don't know if the obstacle to the belief has an objective foundation. We have to seek the reason for the doubt. The very important result of Rostand shows two things. The first one is that very often mathematicians write or speak about their scruples, mainly about possible errors, but also about the real meaning of some success. The second one is more important: Rostand shows that these scruples are not a kind of extra-commentary about their works, but they are at the very heart of the invention process of their works. They should be read by mathematicians at the same level as the actual text.

For us, this question of scruples (and in parallel the question of evidence) is a constituting factor of mathematical creativity, and especially with regard to the question of mathematical pulsation.

67.4.2.8 Toward Categorical Modeling

These various things having been clarified, we have to get now a deeper understanding of what *mathematical gestures* are, as creations of new objects or relations, and, consequently, what objects and relations are, and forms.

So our proposition will be that any mathematical gesture is an 'implementation' of the very fundamental gesture called *the mathematical pulsation*, and then a convenient mathematical modeling of mathematical activity works through the constructions of objects and relations, as could be exposed by category theory, and as we can observe along all the history of mathematical invention, what we call *categorical modeling.*

67.4.3 Invention in the Art of Mathematics

67.4.3.1 The Truly Creative Mathematician Lives in the Real No-Reality World

The exploration of the so-called "reality" through numerical evaluations, finite tests, statistics, tables of values and maps, doesn't depend really on mathematical thinking. It is just a bad pragmatical aspect of the exploitation of science, when, with no pertinence, mathematics are reduced to numerical computations and evaluations, and exploited for the sake of industrial or financial purposes. In such cases, assuming mathematical computations only works as a magic principle of validation. In this direction, no creativity is possible, because no *beauty* is in question; only a pseudo-pragmatic posture is assumed. From a mathematical point of view, liberal political motivations, around studies on markets and financial supports, and the democratic process such as those of votes and elections, are always ugly. Of course this is not the opinion of all mathematicians.

The true pragmatics starts with the question of the "real world" as a bio-physical unknown entity, and the correlative question of the use there of the mathematical notions.

Different of course are the situations of the theory of numbers, of algebra, of geometry. Digitalization is a consequence of the Pythagorean quest for harmony through numbers and "pixelisation". By such a way nothing is understandable, nothing is of value. From this perspective we get a false conception of mathematics and mathematical creativity, through discourses on real numbers and numerical control of the world. But it is out of such numerical considerations that we would like to think of creativity. In relation with approximations, to construct an effective mathematical process of decisions, serious numerical problems have to be examined after the qualitative problems, with the general treatment of objects and relations.

Grosso modo, we assume that *creativity* in a domain of art depends on the ability to impose a *form*—"to impose a form" is the conception of Nietzsche of what an artist has to do—by performing a system of gestures, by writing and proposing to the senses of the audience some traces that can be read as new and possibly interesting. A useful tool for developing creativity would be any *method of invention.*

Creativity is possible only in a domain of activity which is an art, at least from one point of view; and therefore to speak of creativity in mathematics implies that mathematics is an art. Thereby of course we are absolutely in opposition to the opinion of the famous engineer Sir Alec Issigonis: "All creative people hate mathematics. It's the most uncreative subject you can study." (Quoted in *The Australian* 5 October 1988) [355, p.211].

Therefore we follow the view of Paul Valéry on mathematics, as quoted and praised by Jean Dieudonné [269, p.184-185]: "Mathematics is not the science of quantities; this is twice false: it is not a science and the quantity is not its main subject of investigation. It is an exercise, comparable to dance. [...] mathematics are about searching for properties of a form, and not about any particular problem [...] The eminent intellectual fact is the independence of operations from their contents. [...] When the contents are *created* by the operations themselves, i.e., when some operations are designed, isolated and combined, then we are in mathematics."

We borrow from W.B. Smith a rather strange formulation [355, p.237]: "Mathematics is the universal art apodictic," i.e., the universal art of universal and absolute necessity; for us "universal and absolute necessity" means exactly the laws of the pure combinations of thoughts, and so mathematics is the science of combinations, in the widest sense of the term, including algebraic computations, various calculi, as well as geometric combinations; and ultimately it is also the art of combining these different areas.

When this science becomes an art, i.e., when we are not only solving equations in a given closed framework, but when we invent new frameworks and new opened calculi, then we are in true mathematical creativity. When we do mathematics we are in the process of *calculating* and modifying the possibility of inventing or modifying this process: this science is an art, with a place for creativity.

Mathematics is the *scientific art* of (higher) combinations of areas of (lowest) combinations, that is to say the science of invention of rational methods for the constructions of patterns in the web of our concepts. Godfrey Harold Hardy says that "a mathematician, like a painter or a poet, is a maker of patterns. If his patterns are more permanent than theirs it is because they are made with *ideas*." ([439, p.84], quoted in [355, p.158-159]). Already for Plotinus "geometry is the science of intellectual entities." [851, Fifth Ennead IX.11]

So the mathematician invents new beautiful patterns of 'mathematical ideas', i.e., of mathematical objects, relations or gestures; this is the way in which he creates. If we consider that the musician creates new beautiful patterns of sounds and beats, then, with Sylvester, we can think of mathematics as the "music of reason"(cf. [270], [816]), with patterns of shapes and variations.

We would like to analyze the possibility of the description of a method of invention in the case of mathematics—considered as an art—with *categorical modeling* as a tool. So we are trying to break the following paradox by Bertrand Russell: "It is a paradox in mathematics and physics that we have no good model for teaching of models".

Our starting point will be the belief in the possibility of natural reduction of any mathematical data, ideas, intellectual entities, reasoning, methods, gestures, to two types of entities; *objects* and *relations*, and the modeling of creativity will be a modeling of *invention of objects and relations*, and of sequences of objects and relations, of paths in spaces of objects and relations.

Thanks to category theory, we can confirm, in the case of mathematics, the pertinence of the general method of creativity proposed by Guerino Mazzola in the case of music [726, p.17] as a six-step process: 1. Exhibiting the open question; 2. Identifying the semiotic context; 3. Finding the question's critical sign or concept in the semiotic context; 4. Identifying the concept's walls; 5. Opening the walls and displaying its new perspectives; 6. Evaluating the extended walls.

The Yoneda Lemma says that with $@A = \mathrm{hom}_{\mathcal{C}(-,A)}$ for A an object of $\mathcal{C}$, the Yoneda functor

$$\mathrm{Yon}_{\mathcal{C}} : \mathcal{C} \to \mathbf{Ens}^{\mathcal{C}^{op}} = \mathcal{C}^{@}$$

is full and faithful. As stated in [32], in this context, a *creative subcategory* $\mathcal{A}$ in $\mathcal{C}$ is a subcategory such that the restriction of the Yoneda functor's values to $\mathcal{A}$, given by $A \mapsto @A|_{\mathcal{A}}$, provides

$$\mathrm{Yon}_{\mathcal{C}|_{\mathcal{A}}} : \mathcal{C} \to \mathbf{Ens}^{\mathcal{A}^{op}} = \mathcal{A}^{@},$$

which is still full and faithful.

In this context, a generic model of creativity looks as follows, (1) understand the object A in a category $\mathcal{C}$: (2) this is the category $\mathcal{C}$ where A has been identified, (3) this is A, (4) this is the uncontrolled behavior of $@A$, (5) find an objectively creative subcategory $\mathcal{A}$, (6) calculate the colimit C of a creative diagram, (7) try to understand A via the isomorphism $C \xrightarrow{\sim} A$.

67.4.3.2 Method of Invention: Toward an Art of Functional Modeling

The expression "method of invention" is somewhat paradoxical, but perfectly rational. We borrow this notion from the conceptions of Bacon, Descartes, Mersenne, and Galileo, according to Évelyne Barbin [80]. Of course Bacon is an empiricist, Descartes is dogmatic, and Mersenne is pragmatic; but, as expressed by

Gusdorf [431], quoted in [80, p.11], over the methodological diversity of mental spaces we have a fundamental unity of intention; for these authors the point is to understand the nature *and* to discover and invent in order to progress. At the turn of 1620, science reached a new requirement, and beyond pure speculation; now the question was to control nature, to invent and to master new tools for solving problems: the main tool here would be mathematics, as a method. But, in this revolution, the nature of mathematics itself was modified, and in this transformation more important probably was the invention of the general notion of a curve, in direct relation to the problems of tangents, later followed by the invention of differential calculus. Therefore mathematics explicitly became a method of invention by modeling phenomena and variations. This was achieved with the emergence of the general notion of a function, and the specification of a function by a differential equation (Leibnitz): for a long time, up to today, this method of functions and differential equations remains the main mathematical motor of development and discovery in physics.

And consequently the improvement of this process of functional modeling is the main purpose of pure mathematical research. An ultimate aim of mathematics would be to invent a method of control and development of functional modeling: general notions of transformations, of functions, of calculus with functions, equations and algebras of functions, and then categories of functions or morphisms, etc. This could be seen as an infinite process of structuring systems of objects and their relations.

Then "categorical modeling" would be a good name for the effort of a particular mathematical community toward mathematical creativity with regard to the subject of functional modeling and structuring of mathematical activity as such. The two crucial terms interlaced here in our epistemological dispositive will be the notion of object and the notion of relation, and the question of creativity would be the question of invention of pertinent new objects and new relations.

67.5 On the Mathematical Invention of Coordinations

With respect to our question of creativity and of pulsation in mathematical thinking, the following thought of Valéry is quoted in [128, p.XIII]: *The geometer is going into a strange space, simultaneously pre-existent and arbitrary, necessary and constructed, invented and discovered.*

We believe that the mathematician creates new combinations of ideas and so new configurations in new spaces (i.e., new systems of modifications among some distinctions and identifications); history shows a progressive access to our knowledge of that fact.

Furthermore, it is right that mathematicians propose new configurations, new fibrations and new spaces, and paths in these spaces, and associated invariants in new categories; but at a point in future all that activity has to converge back to solid results in an old traditional area, ending the "stratospheric" travel. A magnificent case of such travel is the achieved history of Wiles' theorem proving Fermat's conjecture.

To understand all of that, at first we have to understand the successive effects of the development of coordinations, alias the game of analysis/synthesis.

In this section we expose a certain notional scenery—in the sense of Section 67.4.1.4—about the emergence of the notion of *coordination* (and subsequently of coordinatization). We follow a historical way, from ancient Greeks to Descartes, Lamé, Ehresmann, and Grothendieck. Many variations would be possible on this question, as it is nothing else than the adventure of the mathematical apprehension of *the dialectic of analysis and synthesis*, through the invention of the modalities of space, form and function. For example, only on the specific question of the notion of a "space" in 20th century, can one read very interesting and advanced books such as [553] or [128].

We refer to a talk given in Thessaloniki in 2009 [423] in a semiotical atmosphere; hence the part played by *arrows* is put in relief. By contrast, fibrations, pro-objects and modern approaches are almost not evoked. In resolutions of problems and mathematical inventions, the notations (or signs) and the scopes (or senses) become intermingled in a kind of pulsation. For teaching it is worthwhile preserving this pulsation and working with it. This pulsation is observed here through the emergence of the idea of coordination in some cases (cartesian rectangular or oblique coordinates, curvilinear coordinates, tripolar coordinates), and toward modern notions of functions and fibrations. So we install a "concrete" historical meaning (filiation) for sites and sketches, fibrations and atlases. For a hermeneutic and a semiotic analysis of this emergence of coordinations, we explain how significations and interpretations could be specified using *arrows*. Then coordinations could be understood as relational systems of coordinates. And thinking in terms of arrows and diagrams in the sense of the theory of categories, we emphasize that nowadays coordinations are nothing else but specifications of projective limits, inductive limits, equational structures, sites (Grothendieck) and sketches (Ehresmann).

67.5.1 Emergence of Coordinations

Hereafter any formula and any figure is a *sign*, and in order to really understand coordinations, a good method would be to determine what signs are used, and, above all, how interactions are working between them.

67.5.1.1 Symptom, Characteristic Equations, Linear Coordinates

In Euclid [306, p.409, Prop. II, 14] it is proved, that given a circle C of diameter BF, if we consider a point H on C and the perpendicular projection E of H on BF, then we get (see Figure 67.9): the square on HE is equal to the rectangle on BE, BF:

$$HE \times HE = BE \times EF.$$

This is not yet an "equation" of the circle C, because it is neither a construction nor an assertion about any

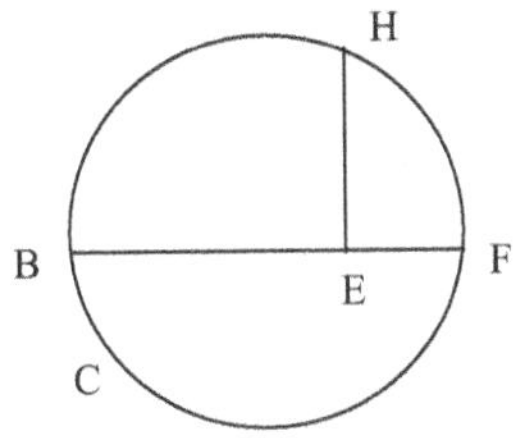

Fig. 67.9. Symptom of a circle, not yet an equation.

arbitrary point in the plane where the circle exists (this plane doesn't exist here as a mathematical object, the only real thing which is considered is the figure of the circle); the information given by our formula is a geometric relation which is defined and true on the circle. Furthermore it works for any circle; it is a *symptom* of any circle. Apollonius [39] used this symptom for a circle to produce symptoms for conics. Nowadays, we metamorphose these symptoms into characteristic equations, namely [389]:

$$y^2 = x(2p + px/a)(\text{hyperbola}), \quad y^2 = 2px(\text{parabola}), \quad y^2 = x(2p - px/a)(\text{ellipse}).$$

Those characteristic equations depend on an effective consideration of the plane, and in this plane on a convenient choice of coordinate axes. They could not be written in Descartes [264], because in fact Descartes did not really introduce the so-called "cartesian axes". He introduced the idea of arithmetization of geometry, by writing arithmetical relations among some lines in the figure. As a symptom, a "cartesian equation" is formed specifically on the curve; but it is not a symptom, because now it is an arithmetical relation which is expressed. The discovery of Descartes is a method to provide such a cartesian equation for arbitrary curves.

The next step, the introduction of the plane and of axes in the plane, and therefore the formulation of characteristic equations, was achieved by Wallis [1104]. We could say that geometry and cartography were unified at this moment. The rectangular coordinates are a mapping, adapted to the basic shape of a square drawn in a plane as a reference. So, considering some given rectangular cartesian axes in a plane, it makes sense to ask for a geometrical characterization of any second degree curves, and to prove (Wallis) that they are exactly conics. Leonard Euler proved that this fact is independent of the chosen rectangular axes. By modification of axes of coordinates, we can explain to what kind of conic corresponds an equation like:

$$ax^2 + 2bxy + cy^2 + ex + fy + g = 0.$$

According to Gino Loria, the first phase of the development of analytic geometry is the period which begins with Descartes and Fermat and ends with Lagrange and Monge. It was at this time that the method of coordinates, outlined by the author of the *Discours de la Méthode* and Fermat, finally became a body of doctrine providing for those who are studying geometry with formulas ready for use, applicable almost automatically, whatever the position of the coordinates axes. Loria thinks that an excellent but not yet perfect textbook on this matter is Biot's book *Essai de géométrie analytique* [118].

After 1802, an important question still to be solved was the transformation of oblique cartesian coordinates. The subject was studied [618] by Carnot, Livet, Français, Hachette, Lamé, Sturm, and Cauchy. Lamé [571] is particularly interested in the subject as a tool for the analysis of crystals, and this is a special case of what he will expand as the general theory of curvilinear coordinates invented by Gabriel Lamé in the 1830s [572]. The rectangular cartesian coordinates were adapted to the shape of a square, and now the oblique coordinates are adapted to the shape of a parallelogram (Figure 67.10).

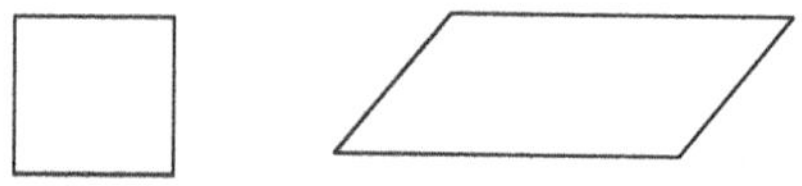

Fig. 67.10. Introducing oblique coordinates, a gesture adapted to crystals.

In an oblique cartesian system of axes, the previous equation represents again an arbitrary conic, because the transformation from any oblique coordinates towards any rectangular coordinates is linear and bijective. But the specification of the type of conic which works in the rectangular case is no more valid directly.

The next natural linear extension of oblique cartesian coordinates (achieved only in the 20th century) is the idea of coordinates in a vector space E with respect to a given linear basis. If a vector space E of finite dimension n is given, a basis in E could be described as the image of the canonical basis of $\mathbb{R}^n$ by a linear isomorphism

$$m : \mathbb{R}^n \to E,$$

and of course this isomorphism is determined by its inverse

$$p : E \to \mathbb{R}^n,$$

which itself is determined by its components $p_i : E \to \mathbb{R}, i = 1...., n$. Then the coordinates of an element x of E are $x_i = p_i(x), i = 1...., n$, and so we write :

$$p(x) = (p_1(x), p_2(x),,, p_n(x)) = (x_1, x_2,, x_n).$$

67.5.1.2 Curvilinear Coordinates as Families of Surfaces or Curves

The idea of curvilinear coordinates probably originates in some Leibnitzian writing on coordination of systems of curves. In the 18th century polar, bipolar, spherical, and cylindrical are used. Polar coordinates were used in Descartes' style, specifically on a given curve, by Bonaventura Cavalieri, Isaac Newton, and Jakob Bernoulli. The use of polar coordinates as a means of fixing any point in the plane and for a systematic study of any curve is proposed by Jakob Hermann in 1729 (see [149, p.76],[213, 462]). Newton is the originator of bipolar coordinates, especially for the study of the 'ellipses of the second order' i.e., the ovals of Descartes. Newton observed [149, p.77] that Descartes handled ovals "in a very prolix manner", without the application of coordinates. If x and y are the distances of a variable point from two fixed poles, their relation for the ovals are

$$a + \frac{e}{d}x - y = 0,$$

and from this equation Newton found the tangent line. Newton noted that if $d = \pm e$, the curve becomes a conic section.

But the true general systematic approach of general curvilinear coordinates is due to Lamé, around 1830 [422]. The motivation of Lamé was the study of physical problems such as the question of the temperature of a given body, and he claimed that the more natural approach is to use a system of coordinates adapted to this body, with a family of level surfaces parallel or orthogonal to the body. Nowdays, we can find a return to these ideas on level surfaces in books such as [811].

The analytical description of curvilinear coordinates is by transformations from rectangular cartesian coordinates.

For example, in the two-dimensional case, if the body is an ellipse, then rather than rectangular or polar coordinates it is better to use an elliptical system of coordinates (Figure 67.11) constituted of confocal ellipses and hyperbolas:

$$\frac{x^2}{a^2\cosh^2\mu} + \frac{y^2}{a^2\sinh^2\mu} = 1, \quad \frac{x^2}{a^2\cos^2\nu} + \frac{y^2}{a^2\sin^2\nu} = 1.$$

A point P of cartesian coordinates (x, y) is located at (μ, ν) such that P belongs to the two corresponding

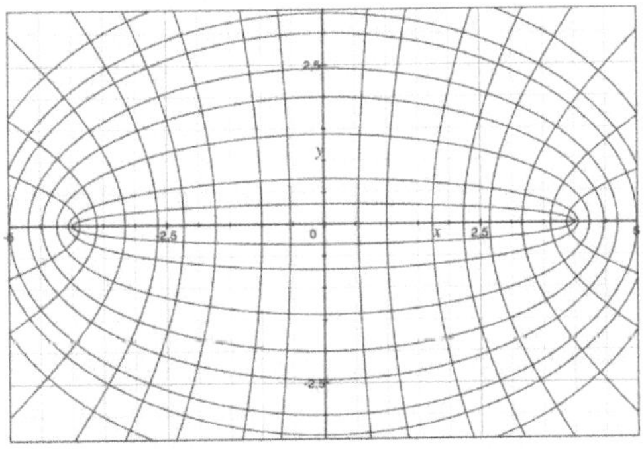

Fig. 67.11. Lamé elliptical coordinates, adapted to the shape of the Earth.

curves. The transformation from (μ, ν) to (x, y) is given by

$$m(\mu, \nu) = (x, y), \quad \text{with} \quad x = a\cosh\mu\cos\nu, \quad y = a\sinh\mu\sin\nu,$$

and conversely the elliptical coordinates are given by

$$p(x, y) = e((x, y)) = (\mu, \nu).$$

67.5.1.3 Tripolar Coordinates, from a Symptom of the Plane

Lazare Carnot [174, p.48] introduced the algebraic relation (a sum of 130 monomial terms) which is satisfied by the ten distances between five points A, B, C, D, and E in the space. This formula is a law of the space; it expresses the proper coordination of the space when the space is endowed with its metric structure. In fact it is for the space something like what the symptom is for a circle; it is *a metric symptom of the space*:

For A, B, C, D fixed, and the fixed quantities $AD = f, AB = g, AC = h, BC = m, CD = n, BD = p$, this symptom expresses, for any point E, the relation which is satisfied by the variable distances $AE = l, BE = q, CE = r, DE = s$. So E is located by such a 4-tuple (tetrapolar coordinates):

$$p(E) = t(E) = (l, q, r, s).$$

These numbers are four, they are linked by one relation, and that is appropriate due to the fact that the space is of dimension three ; they are the coordinates of E with respect to the ground which consists of the space endowed with its metric structure and with the four points A, B, C and D. In 1841, in its first mathematical publication, Arthur Cayley [182] expressed the formulae of Carnot as a determinant. In the case of four points 1, 2, 3, and 4 in the plane, the formula of Cayley is the following symbol (see also Blumenthal [126, p.99]):

$$\det \begin{pmatrix} 0 & 12^2 & 13^2 & 14^2 & 1 \\ 21^2 & 0 & 23^2 & 24^2 & 1 \\ 31^2 & 32^2 & 0 & 34^2 & 1 \\ 41^2 & 42^2 & 43^2 & 0 & 1 \\ 1 & 1 & 1 & 1 & 0 \end{pmatrix} = 0.$$

If we consider in the plane especially three distinct points 1, 2 and 3, with equal distances $12 = 21 = 23 = 32 = 31 = 13$, with value $= 1$ (so 123 is an equilateral triangle), and if a point P is at distances p, q and r of $1, 2$ and 3, P is located by (p, q, r) (tripolar coordinates) agreeing with the symptom; after some computations this becomes:

$$p^4 + q^4 + r^4(p^2q^2 + q^2r^2 + r^2p^2) - (p^2 + q^2 + r^2) + 1 = 0.$$

With $x = p^2, y = q^2, z = r^2$, this symptom could be seen as a representation of the plane as the paraboloid in space with cartesian equation:

$$x^2 + y^2 + z^2(xy + yz + zx)(x + y + z) + 1 = 0.$$

67.5.2 Arrows

Firstly we argue about the fact that any sign "is" an arrow. The games of these arrows provide a coordination of the semiosis, and this is the structural part of the sense. Secondly we make precise that in the case of mathematical discourses, the signs work as diagrams and as abbreviations. The system of signs exhibited above to introduce coordinations could be understood in a very general acceptation in terms of arrows.

67.5.2.1 Semiotics and Hermeneutics

A basic presentation of the problematics of signs and semiotics is given by Umberto Eco [289], but see also [703]; and a clear synthetic presentation of hermeneutics is given by Jean Grondin [393]. We would like to stress the fact that the full sense of a discourse has two complementary and interacting components, semiotics and hermeneutics. On the one hand, any discourse is inscribed in a language, as a phonological production of signs, and there it has a dynamical structure. We know with Jakobson [502, p.78] that a phoneme is a sheer differential sign: the only semiotic content of a phoneme is its difference from other phonemes. So the structure consists of variations and dispositions of phonemes, and it is understandable through coding processes and grammatical analysis.

Ultimately, the question of meaning or signification of a discourse is the problem of articulation and dynamical functioning of a system of representations by signs (linguistic signs). A minimal definition of a sign is in Eco [289, p.45] as an entity which could have a signified object. In Peirce's view, this signified object is the target of an arrow, the body of the arrow itself is the interpreter, and the source of the arrow is the signifier or signifying element

$$\text{signifier} \xrightarrow{\text{interpreter}} \text{object}.$$

In the perspective of Savan [935] or Lizka [614], we have to treat interpretants as translations, or "translatants". Furthermore, each of the three components of a sign could be a sign again. The sense (seen as a global holding of a complex of meanings) is nothing else than the shape of the system, the shape of the structure, as a result of combinations of arrows of signs.

On the other hand, any discourse is dedicated to a promotion of values, and so participates in the open construction of a social culture. The sense is a question of hermeneutics, that is to say a question of interpretation, in the rhetorical tradition; more accurately it is a question of value in the *free commerce of interpretations* [420].

In principle an interpretation pretends to provide an understanding of what truly the discourse would like to say. So it is a question of translation and elucidation of something which is obscure or at least incomplete. In the words of Droysen, Dilthey insists on this point: the goal of hermeneutics is to construct a historical comprehension (Verstehen) rather than a scientific explanation (Erklären); comprehension is delivered as another discourse which draws a kind of arrow in the culture, which is the indication of an orientation from a given point of view. Furthermore, in Gadamer's view, the interpretation depends on a genuine implication of the interpreter; it is a performance, the living gesture of a human being. So, for us, the hermeneutical sense is a kind of arrow,

$$\text{initial discourse} \xrightarrow{\text{interpretation,}} \text{comprehension},$$

the body of which is the subjective act of interpretation, with target the comprehension in discourse given by the interpretation in the realm of "parole", and the source being the initial discourse which was to be interpreted.

On the semiotic side the sense (or meaning) resides in a methodical new combination of some existing arrows, and on the hermeneutic side the sense resides in the subjective elaboration of a single new arrow. Of course we are free to envisage a given culture as a system of cultural signs (including generally some linguistic signs), and so semiotics acts at this level of hermeneutics as well as at the linguistic level. Conversely, a sign considered as an arrow according to the perspective of Savan and Liszka is a kind of elementary interpretation, an element in the hermeneutic realm.

For a given discourse, we have two stakes: its meaning as a shape of an articulation of signifying representations, its sense as a comprehensive interpretation in a culture. Furthermore, the discourse holds these two stakes in a living interaction, through the gesture of speaking (parole). Someone does the act of speaking the discourse, addressing the structure (of a system of phonemes) to someone which is living in a given culture. The question of the full sense of a discourse now is the question of how in the process of speaking ("parole") the structure and the culture interact all together, around the given discourse. So the emergence of interpretation is in the realm of semiosis, and structural semiosis is not separable from games of interpretations. We propose to take care of this observation by thinking in terms of diagrams of arrows.

67.5.2.2 The Case of a Mathematical Discourse

Mutatis mutandis, we can transpose or particularize the previous remark in 67.5.2.1, valid for any discourse, to the special case of a mathematical discourse. There the act of "speaking" is replaced by the act of "doing mathematics", inside the mathematical language. The production of signs is done by mathematical scriptures of computations and figurations. The full sense is performed by confrontation of the given structure of these scriptures (seen also as a system of mathematical representations) with the mathematical history and culture (old theorems, theories and problems), and this provides a mathematical interpretation.

From a semiotical point of view we have in this mathematical game two intertwined levels: at a ground level we have signs for direct scriptural mathematical representations, and at a higher level we could make use of signs for large and complex mathematical interpretations. In this conception, any mathematical data could be understood as an arrow

$$A \xrightarrow{f} B,$$

which could be read

"from the point of view of f, the object A stands for the object B",

as well as

"f is a difference between B and A".

This equivocation on the sense of an arrow (sometimes it means an identification, at other times it means a distinction) is in fact essential to the lively use of arrows in mathematics. We call it the initial pulsation of the arrow. In the sagittal world the basic problem (and tool) is that of creating arrows, by serial or parallel compositions.

An icon of our free organization of analytic thoughts around a given arrow f is the *free autograph* on f (see [424, p.140]), the beginning of which is the picture:

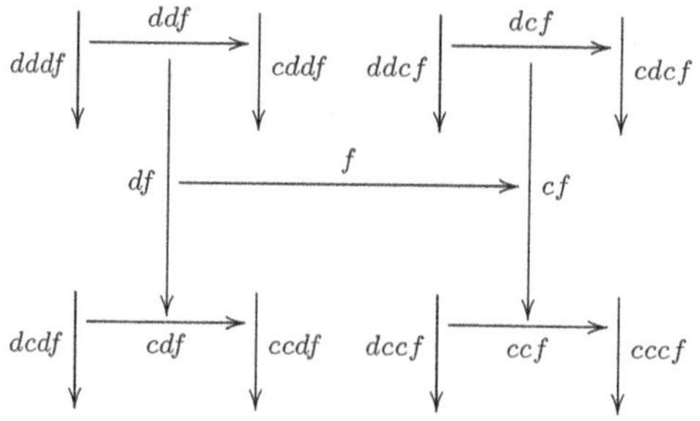

It is easy to relate that picture to the binary development of real numbers, to the construction of Cantor's set: these mathematical objects are elements for "internal coordinations".

67.5.2.3 Coordinations, Diagrams, Abbreviations

Mathematics is neither the science of number and space, nor the science of logic or physics, although such subject-matters are those to which mathematics has been extensively applied, and could be decisive elements in a personal notional scenery.

A better thesis by Cassius Keyser [531] is that mathematical thinking is postulational thinking. For Keyser every successful adventure in postulational thinking eventuates in the establishment of a hypothetical doctrinal function; the function is composed of propositions each of which asserts that some proposition for me is logically implied by a set of other such forms. So, for Keyser, the mathematical questions are those about the world of the logically possible (whereas the scientific questions are those about the actual world). The coordinate enterprises, mathematics and science—both parts of philosophy—,together embrace the whole knowledge-seeking activity of man (by "knowledge" is meant such knowledge as is expressible by propositions). Their combined scope is the two-fold world of the actual and the logical possible, the world of propositional facts and the world of propositional forms, the world whose truth is discoverable by none but empirical thinking, wherein observation is sovereign, and the world whose truth is discoverable by none but postulational thinking, where deduction is sovereign. So, for Keyser, the validity of mathematical propositions is independent of the actual world; the world of existing subject-matters is logically prior to it, and would remain unaffected were it to vanish from being.

Nevertheless this thesis is too much tied to the logical point of view on mathematics, and *we would prefer to think in terms of coordinations rather than in terms of logic.* Here our claim would be:

Mathematics is the art of inventing necessary coordinations in the world of the possible.

Coordinations are nothing else than synthetico-analytical methods for any special science, e.g. for logic, arithmetic, probability, geometry, physics, etc. Conversely, anything in a given specific science which is a pure fact of necessary coordination is a true mathematical point. So in the history of mathematics, the mathematical coordinations have been discovered in some scientific contexts, e.g. in arithmetic, in geometry, in logic, in physics, etc. Arithmetic and geometry are sciences derived from the core of mathematical thinking, when the mathematician constructs an interpretation of his work in terms of listening or sight. Arithmetic and geometry as sciences are already on the side of meaning and sense of mathematics: historically they are two basic ways of interpretation of mathematics. This point applies also to logic and physics.

Much as sciences are domains of applications for mathematics, they are sources of new mathematical ideas as far as they stimulate new practices about coordinations.

Let us observe that, with respect to Keyser's distinction, a true working mathematician is in fact simultaneously a pure mathematician *and* a scientist, because while working, computing, and deducing he is also observing the mathematical entities, and he discovers through these observations. For him the world of mathematical things is the actual world (the one we called *norealworld*—with a positive insight); his observations are assumed as feelings with internal senses inside comprehension. In fact these observations are also actions with a mathematical value, as far as they generate new rigorous coordinations. In some sense the conception of Keyser is near to the conception of Charles Saunders Peirce, who defines mathematics as "the study of hypothetical states of things", and, according to a wording in 1870 of his father Benjamin Peirce: "the science which draws necessary conclusions" [847, pp.227-244].

However Keyser's description is perhaps too much logicist, and it does not meet enough the question of poietics and invention in mathematics. In a recent paper, Daniel G. Campos [172] examines Peirce's propositions in this direction, and he says that Peirce's position is that the creation of mathematical hypotheses is poietic, but it is not merely poietic, and, accordingly, hypothesis-framing is part of mathematical reasoning that involves an element of poiesis but it is not poietic either. So Campos proposed that hypothesis-making in mathematics stands between artistic and scientific poietic creativity with respect to imaginative freedom from logical and actual constraints upon reasoning.

In a letter to Lady Victoria Welby-Gregory, Charles Saunders Peirce wrote: "It has never been in my power to study anything, mathematics, ethics, metaphysics, gravitation, thermodynamics, optics, chemistry, comparative anatomy, astronomy, psychology, phonetics, economics, the history of science, whist, men and women, wine, metrology, except as a study of semiotic." Peirce also treated sign theory as central to his work on logic, as the medium for inquiry and the process of scientific discovery. So from Peirce's perspective, everything in the world is made of signs, of living combinations and productions of signs (the phenomenon of semiosis).

In analyzing mathematical thinking and productions, it will be crucial to exhibit the mechanism of invention of coordinations from the perspective of the semiosis, and to understand construction of coordination through semiotic elaborations. In Peirce's terminology a diagram is a special sign, an icon in fact, exhibiting existing relations among parts of a state of things [193, p.36]; and in Peirce's view *the basic mathematical action is precisely the construction and modification of diagrams.* And we add: constructing or modifying diagrams produces an arrow between diagrams; so the fundamental nature of a mathematical gesture is the production of such an arrow.

There are diagrams with an arithmetical flavor (equational formulas) or with a geometric tendency (geometrical figures), and the "dialectic" between these two aspects is of great importance in the development of mathematics; in some sense the various mathematical results on this point (e.g. principles of duality) express a special mathematical pulsation [412], [419] in the mind of the thinker when he has to choose a sense for interpretation to direct the meaning of his thinking. For instance Cayley's famous diagrams for groups (as well as the diagrammatical forms of Kempe [528]) play exactly this role of a vehicle for the pulsation between the algebraic or arithmetical (tabular) description of a group and its geometric counterpart.

This pulsation between arithmetic and geometry, with other mathematical pulsations, indicates the moment of an inventive gesture in the realm of mathematics. The pulsation is solicited by a diagram that we have to modify, and this is possible because a diagram always bears in itself an *abbreviation* [413, p.162], i.e., an arrow which comes from a ground and is going to a functionality which provides an orientation in mathematical knowledge.

$$\text{old ground} \overset{\text{abbreviation}}{\longrightarrow} \text{new functionality.}$$

To this hermeneutic fact corresponds at the sheer semiotical level the fact that a diagram is always incomplete and therefore open. We speak of pulsation and abbreviation; other authors, such as Châtelet [189] and Alunni [26], speak of a virtual dimension of any diagram as an abstract-machine, prior to any representation. We agree that a diagram is a static picture inciting anyone to a gesture of its modification. The mini-model of that is precisely the initial pulsation of the arrow.

67.5.2.4 The Concrete Map as an Abstract Arrow Abridging a System of Arrows

Today everywhere in mathematics we employ maps or functions

$$f : E \to F : x \mapsto f(x) = y,$$

from a set E to a set F, such a map being the datum of a set E and a set F, and the attribution to any $x \in E$ of an image $f(x) \in F$.

We have to insist on the following point: stricto sensu a map is a gluing of a flow of arrows

$$x \to y,$$

with $x \in E$, $y \in F$, and $y = f(x)$.

So the map $f : E \to F$ is an arrow which is not so "concrete"; it is an abstract abbreviation for an abstract system of arrows (a family of $x \to y$, plus a family of unreached elements). The idea of *the pulsation of the arrow* is made more dynamic when we consider maps, which can be seen as a kind of flow in the semiosis. For example here is pictured the map $q : \{1, 2, 3, 4, 5\} \to \{1', 2', 3', 4', 5'\}$ given by $q(1) = q(2) = 1'$, $q(3) = 2'$, $q(4) = q(5) = 5'$, with two unreached elements $3'$ and $4'$ (Figure 67.12):

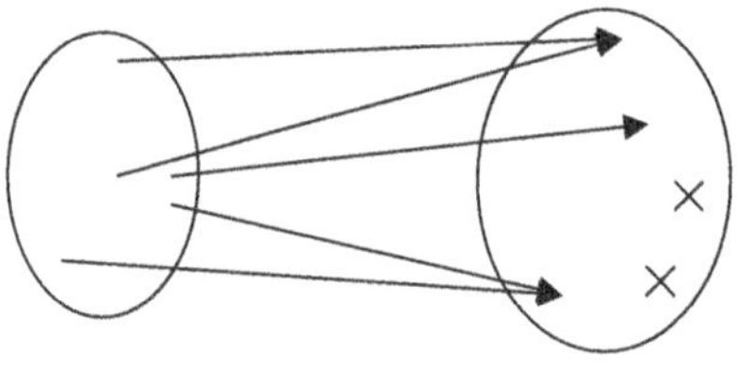

Fig. 67.12. A map as a flow or a family of "formal" arrows.

67.5.2.5 Functional Spaces, Algebras of Functions, Duality

In the history of mathematics, the development leading to the final picture for a function, as given above in the previous Section 67.5.2.4, is related to inventions of curves and coordinations. Coordinations are constructions of maps.

Maps are arrows in the category **Ens** of sets. The general coordination of the semiosis at the mathematical level is expressible in terms of maps, or even in terms of arrows abstractly organized in categories and diagrams. After its unification with algebra, geometry became a question of maps or transformations of spaces. Hence a figure is thinkable as a transformation into a space from an ideal model of this figure, a realization of an idea, as well as the gesture to put the idea into a semantical space.

We do not remember his exact words, but somewhere Borel said that even if the theory of sets would not be so decisive with respect to the problem of foundations for mathematics, at least it allowed the development of functional analysis. (It is impossible to overstate the importance of this fact.) Hence the basic Cantorian construction $\mathcal{P}(E)$ of the powerset of a set E is the fundamental unbounded operation which, with limits and colimits, allows the development of any mathematical construction, and mainly of functional spaces such as $F \subset B^A$, a space of functions from A to B. So we reach an over-level of abstraction, the state in which the elements of a space F are functions among other spaces. Duality consists of the possibility to recover A from F. For example if $A = \mathbb{R}^n$, $B = \mathbb{R}$ and F is the space $L(A, B)$ of linear maps from A to B, then we get a natural isomorphism $A \xrightarrow{\sim} L(F, \mathbb{R})$. The geometry of a space X is now related to algebras of functions on it.

67.5.3 Bodies, Implicit Surfaces, Abstract Relations

67.5.3.1 Relational Coordinations

In 1953-54, in the same vein as Peirce's relational logic, in the framework of the calculus of binary relations à la Peirce-Schreier and Tarski, an abstract analysis of coordination was introduced by Jacques Riguet [894], [895], under the name of *relational system of coordinates*. With the above notations with projections p_i from 67.5.1, let R_i be the equivalence relation (congruence) generated on E by p_i, i.e.,

$$xR_iy \Leftrightarrow p_i(x) = p_i(y).$$

We can forget the external data p_i and just consider the set E as equipped with the family (R_i) of binary relations, and so

$$(E, (R_i))$$

is a relational structure, where, with

$$R^i = \cap_{j \neq i} R_j,$$

we have, for all i:

$$R_i \cap R^i = \Delta_E, \quad R_iR^i = E \times E.$$

Clearly any curvilinear coordinates could be seen as an example of a relational coordination. But this notion allows also the consideration of decomposition of an algebraic structure as a cartesian product of other structures of the same type [187].

In fact, a map is a special case of a binary relation between two sets, and the calculus of compositions and inversions of maps is a part of the general calculus of relatives introduced by Peirce. Nowadays, this calculus is understood as a work in the category of binary relations. So Riguet's presentation is worthwhile by itself, but could also be interpreted in the frame of the involutive category of correspondences.

67.5.3.2 Implicit Surfaces and Spaces

With $0 \in K$ we think of a relation $\Gamma = F^{-1}(0) = \{(x, y) \in A \times B; F(x, y) = 0\}$ from A to B, associated to a function $F : A \times B \to K$, as a coordination relating A to B (F is a *level* function of Γ). That works in the special case of a function $f : A \to B : x \mapsto y = f(x)$, with $F_f(x, y) = 0$ if and only if $y = f(x)$. The set of Γ is an *implicit curve*, and in the case of F_f it is an explicit curve, and so a special case of a parametric curve $I \to A \times B$.

We have the same definitions for three sets A, B, C, with $F : A \times B \times C \to K$, and then $\Gamma = F^{-1}(0)$ is an *implicit surface*, with F as a *level* function. A parametric surface is a function $I \times J \to A \times B \times C$.

Now, given a region $\mathcal{R} \subset A \times B \times C$ containing a body $\mathcal{B}$, we consider three (level) functions $F, G, H : A \times B \times C \to K$—or equivalently one function $L : A \times B \times C \to K^3$—such that for any $p = (x, y, z) \in \mathcal{R}$ there are three unique values $\lambda, \mu, \nu \in K$ such that

$$F(x, y, z) = \lambda, \quad G(x, y, z) = \mu, \quad H(x, y, z) = \nu,$$

and these λ, μ, ν are the curvilinear coordinates of p related to $(F, G, H) = L$. Hence L determines an isomorphism $\mathcal{R} \xrightarrow{\sim} L(\mathcal{R})$ from $\mathcal{R}$ onto its image in K^3. Furthermore if the frontier Σ of $\mathcal{B}$ in $\mathcal{R}$ is a level surface of L, such as $F^{-1}(0)$, then these curvilinear coordinates are said to be *adapted* to $\mathcal{B}$. On Σ we get a double system of curves induced by G and H.

The nice idea of Lamé is multifold, at first, to introduce general curvilinear coordinates, and implicitly general surfaces, as Descartes did with general curves; and then to consider coordinates adapted to a body $\mathcal{B}$ as a natural analytic presentation of this body. Hence a surface dynamically appears as an element in a family of surfaces, even in a triple family of surfaces. And finally a surface, so naturally equipped with a double system of curves, is really a local version of the idea of a Riemann space (as it was said by Élie Cartan), a coherent system of local coordinates (on itself).

67.5.4 Sketches

We go on with the consideration of diagrams in the technical sense in category theory. Of course they are also diagrams according to Peirce, but some very special ones.

67.5.4.1 Coordinations as Categorical Diagrams

Taking seriously into account these ideas that mathematics is the art of invention of necessary coordinations in the world of the possible, and the basic mathematical action is precisely the construction and modification of diagrams, we claim now that we need only categorical diagrams in order to describe the functionality of the mathematical thinking. Specifically we have to show how mathematical coordinations could be realized in the framework of the sagittal icon or more accurately in terms of maps, and as a construction of a categorical diagram. This will be true for any mathematical coordination, but here we will restrict our explanations to the case of the geometric coordinations seen above.

Keeping in mind the general abstract setting on arrows initiated in 67.5.2, we have to strengthen the part played by arrows in coordinations to extract the operational categorical content from the various diagrams shown above (symptoms, equations, graphics). In this way coordination will be directly related to categorical diagrams and the modern idea of a sketch. In this process we emphasize the functionality of coordinates, losing their initial signs.

For instance our initial system of signs for elliptical coordinates (equations and graphics) can now be replaced by a sketch, a matrix-like sketch, specifying components of an arrow p (see Figure 67.13).

67.5.4.2 Projective and Mixed Sketches

The main fundamental elementary tool in category theory is the Yoneda Lemma. We can understand it [417] as a principle allowing us to forget how the objects and arrows were constructed, and to work only with the fact of relative interactions between objects; we speak of a scooping-out of objects and the consideration of the outside as the true substance of which objects are made.

So ultimately any object X in any category $\mathcal{C}$ "is" a system of arrows, namely the category "$\mathcal{C}$ over X", noted $\mathcal{C}/X$, having as objects the arrows of target X in $\mathcal{C}$, and this category $\mathcal{C}/X$ is thought of as the natural *form* or *shape* of X in $\mathcal{C}$. So, any object X is located by the shape of the system of links from others to itself; this is a kind of coordinate system of X tied up to its background $\mathcal{C}$.

The linear isomorphism $m : R^n \to E$ (see, above, our paragraph on linear coordinates) is as a geographical map which allows us to locate points $x \in E$ by using coordinates $p(x) = X$ (or $m(X) = x$). It is a coordination of E, and with this datum E is more structured; so we think also of this datum as a "structuration" of E by a decomposition law.

In fact now, for curvilinear coordinates, we can eliminate the condition that m (or p) is linear; it is sufficient to consider an abstract set E and a bijection $m : R^n \to E$, or its inverse bijection $p : E \to R^n$ or even just a not necessarily bijective map $p : E \to R^n$.

So for the elliptical coordinates we get $p(M) = e(x, y) = (\mu, \nu)$. The datum of p is equivalent (by composition of p with the canonical projections) to $pr_i : R^n \to R, i = 1, ...n$, to the system of maps $p_i = pr_i.p, , i = 1, ...n$, i.e., a cone-shaped diagram with the top E related to the *projective limit* cone with top R^n through a unique arrow $p : E \to R^n$ factorizing the cone (p_i) through (pr_i):

$$(p_i) = (pr_i).p.$$

Furthermore, if the emphasis is kept on the map $m : R^n \to E$, we get a dual approach to curvilinear coordinates, as a parametrization, which could be generalized to a local datum $m : U \to E$, with an open subset U included in R^n. The idea of a manifold now is just the datum of a coherent family (m_j, U_j) of such local parametrizations, and this idea is seizable in the notion of an inductive limit (gluing), and the idea of a mixed sketch, with the factorizations

$$(s_j) = p.(m_j).$$

This is what we can call a *matrix-like sketch*, useful for analysis of any arrow p through

$$p_{i,j} = pr_{i}..p.m_{j}.,$$

and hence any *matrix analysis* $(p_{i,j})$ (Figure 67.13). The true complete sketch is more complicated. After

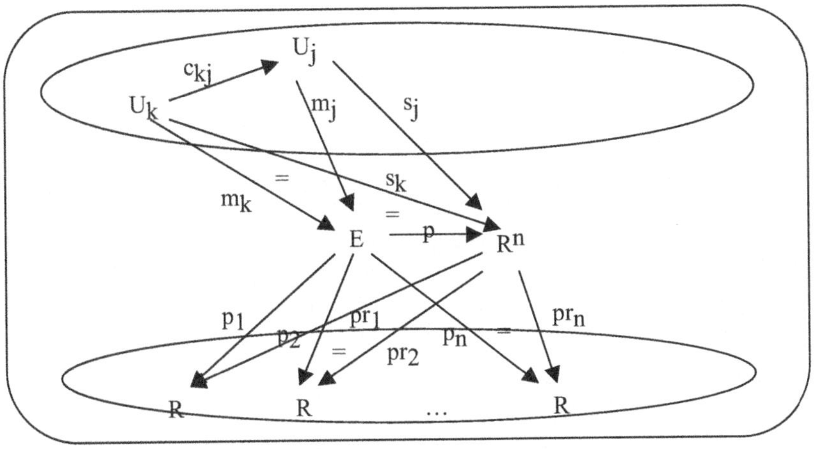

Fig. 67.13. Analysis of a map by colimits and limits, i.e., by a matrix sketch.

the first projective specification, we have to continue, to make precise in detail what each projection p_i is made of, etc. In order to do that we have to add material: basic maps (such as $\cos x$, $\cosh x$,...), compositions of such maps, equations among these compositions, and so on. In fact here we can consider any part of the Handbook of Mathematical Functions [3] as a sketch to be added to specify the concrete values of some function's symbols (arrows) in our sketch (this book is really a coordination of the system of usual functions).

The datum $(p_{i,j}) = (pr_i.p.m_j)$ is an analysis or a *matrix* of p, its analytical coordination in the completed sketch. For example, the 5 × 5-Latin square in Section 67.3.1.4 can be presented as follows.

Another example is the case of the coordination associated to a symptom of the plane through f : $R^3 \to R$ given by $f(x, y, z) = x^2 + y^2 + z^2(xy + yz + zx)(x + y + z) + 1$. A point of the plane is represented by a over-abundant number of coordinates, but these are submitted to a constraint $f(x, y, z) = 0$. So the plane is represented as a kernel p of the map f, which is a projective limit. So we can think of the structure as a sketch again, with a map

$$p : E \to R^3.$$

Those descriptions were possible in fact in the 20th century, when the notion of an arbitrary map and of the category of sets became familiar. As expressed here, the accent is put on the "universal property" of the "projective limit", and this did not make sense before the 1950s. This led in the 1960s to the notion of a sketch defined by Ehresmann between 1966 and 1970 [294], in the framework of the theory of categories. In fact a relational system of coordinates could also be presented as a sketch, with underlying category a category of relations. So as relational systems or as morphisms toward a projective limit, and as realization of a sketch, general curvilinear coordinates on a space, as well as over-abundant coordinations, have now to be understood as diagrammatical structures.

Coordination is an abstract diagrammatic articulation of an object E in the framework of other objects in the same category, an incorporation of E as a source of a sign which is an arrow p. If we compare the part played by this coordination p and equations to the part played by p, we see that in the first case p is external and full (under this name is subsumed a content of figures and equations), and in the second case p is internal and scooped out (it is an element acting in an environment of figures and equations).

67.6 Pulsation in the Living Process of Invention Among Shapes

67.6.1 Productions: Objects and Relations, Problems, Pulsation

67.6.1.1 Historical Transfers of Meanings in the Course of Research

The mathematical invention is performed in the *imaginary* world of mental images of arithmetic, or of geometry, or of algebra, or of various more precise mathematical areas (anyway a world that we do not write down into proofs, but that we talk to ourselves about as a motive), but also at the level of *transfer* between areas. The pulsation acts also at the macro-level of the choice of the area in which the problem has to be examined. So a kind of macro-version of the pulsation is given with the notion of *change of frame* by Régine Douady [275]. In these cases we need knowledge of the history of mathematics, and especially of the different accesses to notions and areas. Also we have to notice the notion of the *use of analogy*, starting from the famous letters from André Weil to his sister in 1940 [1111, p.236]. So the notion of pulsation could be pointed up in the use of mathematical knowlege as a transfer process of data and information between areas established by the historical development. Of course category theory begins with a challenge to treat mathematically such transfers (in this case from topology to algebra), and in such transfers, the analysis of the modification of gluings (naturality of cohomology).

67.6.1.2 The Fundamental Gesture of Pulsation

In Section 67.4.2.2 we claim that creativity in mathematics lies in the inseparable link between searching, teaching, and learning. A strong confirmation of this resides in the nature of the mathematical pulsation [412]. This pulsation is not at all reserved to the higher level of research, for a mathematical elite. It is fully necessary in order to understand any mathematics you are trying to master, and to teach any domain of mathematics. In this direction see analyses [81] and [591].

The mathematician uses entities which are "objects" and their "relations" among them. These entities exist in his mind as indices, but at the very moment of the mathematical act, their existence in a real word is not a pertinent question. Of course he imagines some history about these objects, some significations, and a possible meaning for the mathematical text in which these entities appear; these mental images help him orient his thoughts. These objects and relations are *possible organizational schemes*—or *diagrams*, in the terminology of Peirce—and the mathematical work consists of modifying these diagrams, to compare them by insertions to larger diagrams, or by contractions to smaller ones.

The reader will examine in details two cases.

Example 84 *Resolution of an equation*
A very deep observation by John L. Bell is the following. Along all of its history, mathematics succeeds by changing constant data into variable ones. So in the equation $3x = 6$, the x, at first is unknown, but potentially variable. But when we "discover" that $x = 2$, then no longer is x a variable. The question is: when, in the process of solving the equation, does the nature of x change?

Example 85 *Construction of a geometrical figure*
Nowadays, it is obvious that a geometric figure is drawn in a space; but it was not so obvious in the past, and the underlying space appeared progressively by the fact that something is added to a figure when we work on it: from where or where is now living these new data?

In these two examples the mathematician tries to orient his thinking toward a hypothetical *resolutive state of the modified scheme*. The basic question is: how do we proceed to such modifications? It depends on the knowledge of methods (in the memory of the mathematician, on his knowledge of the history of previous works) of substitutions of signs, and/or of substitutions of significations, or of meanings. We practice analysis in the sense of Condillac, i.e., decomposition and composition of data. At the very elementary level these methods are systems of cutting and gluing of signs. But the very basic question is how he decides about the local orientation of his analysis, of the immediately next gesture in the system of his thoughts? Here is the question of what we name *mathematical pulsation*, depending on the permanent possibility to de-precise data, to render vague data precise. We provide examples of such pulsations: definitions of limits, of tangents, of circles, of complex numbers, etc.

The moral is as follows: to solve problems, i.e., to construct or to specify new objects or relations, or new paths, we need to be relaxed and nevertheless under constraint, as in improvisation: to think deeply in an unknown area, always needing to risk an improvisation and therefore an error. The pulsation is the very moment of effectivity of the ability to improvise a mathematical gesture.

67.6.1.3 Mathematics Invent Effective Transitions Between Possible-Objects

From a philosophical point of view, centered on an epistemology related to the ontological question, we can ask if numbers exist, if circles exist, if groups exist, etc. In fact in mathematical invention they are just tools to regulate the flow of variations as invariants. The mathematical existence of mathematical objects, i.e., the existence from the point of view of the "mathematical life", is a mixed fact, between functionality and imaginary stimulation. If we consider Badiou's view of mathematics as a pure science of possibility of the Being through multiplicities, with a transitional ontology [60], then the theory of category would be an adequate tool to analyze mathematical activity.

Supporting our method at the epistemological level, we would like to propose what could be named a *transitive epistemology*, according to [426], introduced as a comment on [828] and [1149].

If we say that Science is the construction of rational relations among things in order to explain or to predict apparitions of phenomena, then mathematics is the construction of mathematical objects and mathematical relations among these objects in order to exhibit the structure of the rational thinking itself. Mathematics is the art of unveiling the organization of the organizational thinking. Hence the fundamental question of the dependence between physical data and mathematical objects is the question on which the position of Gaston Bachelard is: mathematical objects arrive first, before physical data or the so-called physical objects, or a fortiori before the possibility of a 'real thing', whatever we would like to say by such a term. Because of his knowledge of links between theory of groups and geometry and relativity, Bachelard also thinks that a mathematical object exists at first as a system of relations to assume, as a constraint in mathematical work. For example Charles Alunni [25] explains clearly how for Gaston Bachelard the space of relativity *is* only a system of relations in tensor calculus. And Bachelard wrote [57, p.8]: "la relativité invente l'expérience, elle crée son expérience." In [425] we explain how the positions of Bachelard on objects and relations could be related to the idea of mathematical pulsation.

Today we will say: *any mathematical thing exists as an object in a situation*; any mathematical object exists as an object of a category, i.e., as a possibility of crossing between arrows expressing various potential actions. By functorial modifications, various objectal representations of a given thing are related.

The approach to mathematics from a set theoretical and logical point of view is only one possibility, and a formal presentation of theories is another one. It is important for a creativity theory of mathematical activity to understand that mathematical creation does not exclusively rest on set theory, on logic or on positive formal practice, or on any problematics on foundations. Any creation depends on the revelation of a *new* context, which is seen not as a ground, but as a relative game (see also Chapter 77 on Hesse's *Glass Bead Game*).

67.6.1.4 Diagrams: Sketches and Sites, Topoi and Algebraic Universes

The notion of sketch and of site allows us to produce any regular theory in mathematics as a kind of diagram, and the models of such sketches or sites are again diagrams. All that can be realized internally in any *topos* or *algebraic universe*. The crunch with topos or algebraic universes is to pay attention to the relation between relations and functions, and to the construction of limits. So mathematical activity can be explained as construction of movement in such a "diagrammatic world" to exhibit new objects or new functions. Some other indications on that are given in Section 67.3.2.

67.6.1.5 The Dialectic Resides in Mathematical Acts

The pulsation is an auto-movement of rationality. In each well-defined area of calculus the pulsation works, possibly invisibly; but it becomes visible between the tissue of these well-closed areas. From a philosophical point of view, we would like to explain how this notion allows us to leave a strict analytic philosophy that is concerned with foundations and logic, and to reach dialectical or synthetic thinking of rationality. In this direction companion ideas are in Gaston Bachelard, Gilles Châtelet, and Charles Alunni. Very interesting also are the ideas of Fernando Zalamea [1149]. But we have no place here to develop more explanations (the margin is too small...).

67.6.2 Creativity in the Mathematical World Seen as a Living System of Shapes, in a Categorical Framework

In this section we tackle again the question from [32], with some new precisions of course, but we leave for another occasion the development of higher level techniques; we stay tuned to inescapable basic facts.

67.6.2.1 Living System

A mathematical problem in the previous perspective of shape theory as in Section 67.6.2.3 is the question of production of new data from some specified ingredients, as is the case in any evolutive system. With the help of the theory of sketches (categories with choices of limits and colimits), evolutive systems could be modelized as an MES (*Memory Evolutive System*), according to Andrée Ehresmann and Jean-Paul Vanbremeersch [295]. In fact these MESs are general enough to include as examples many different systems such as real biological organisms or living systems, or the organization of an enterprise.

If we stay at the general level of description of living systems in relation with cohomology, we refer to our paper on Ehresmann-Vanbremeersch theory [421]. Homology or cohomology becomes the step of invariant evaluation of some structural situation, as is the case with curvature, signature, genus, etc.

A new interesting example is the system of globalization of the markets and financial exchanges, as explained in the recent thesis of Baya Mansouri [650]. In this system three basic sets of entities, the States, the Market-Exchange places, and the Enterprises, are given institutions, the existence of each one supposing the existence of both others. Any enterprise creates new products and plans its own regeneration through integrated strategic management; then with these two facts in hand, markets determine dynamical values, and states control good repartitions for their people. A categorical modeling of this will be developed in a forthcoming thesis by Georges Monti [759], essentially in the frame of MES, with the guiding line that today mathematics are about the question of the observation and reading of its own gestures, as argued in this section. So for Monti it is possible to model management and accounting by parallelization of its gestures with gestures in category theory. Pushed by the constraint of the new international and European system for accounting, IFRS, he insists on the historical evolution of mathematics from quantities to qualities to structurating and finally to the description and calculation of gestures; this allows him to suggest a parallel modeling of accounting onto category theory. In this case, but also in many other situations, the recourse to any numerical valuation can be abolished within the elaboration, even if at the end we need to produce a (local) decision (that is to say a 0 or a 1).

67.6.2.2 Axiomatic Modeling of Mathematical Creativity?

With categorical tools (topoi, sketches, limits, cohomology) stressing the question of transfers, pulsation and gestures, we try to define an axiomatics for creativity in mathematics, in parallel with modeling of living systems [295] and the general method of creativity [726].

67.6.2.3 Shape Theory and Models, Cohomology, Differentials

Any mathematical problem could be presented as a question about the qualification of an object or of a relation, more precisely as the construction of a shape of an unknown object, and a seeking for a modification of any shape into a simpler one. So, very naturally the mathematical activity could be presented in terms of shape theory, rather than in terms of logic and set theory, and hence as a question of *emergence* of quality.

The analysis of shapes, and especially the research of invariants and simplifications, in fact rests on cohomological techniques. A very general presentation of cohomology is possible in terms of "Kan's extension of a Kan co-extension process". This ensures us that any cutting/gluing processes in the creative phase of mathematics could be taken into account by category theory.

We start with a *category* $\mathcal{C}$, i.e., a system of "objects" and connecting "arrows", where for consecutive arrows we suppose being given an associative and unitary composition law. Given $\mathcal{C}$, Yoneda's Lemma says that the knowledge of an object $C \in \mathcal{C}_0$ is equivalent to the knowledge of $@C$. For objects F in $\mathcal{C}^@$ and A in $\mathcal{C}$ we have $F(C) = C@F$, and we denote by $\int_\mathcal{C} F$ the *category of elements of* F, which is the category with objects the pairs (C,p) where $C \in \mathcal{C}_0$ and $p \in C@F$, a morphism from (C,p) to (C',p') being a $u : C \to C'$ in $\mathcal{C}$ such that $p'.u = p$. Then F is a gluing (inductive limit),

$$F = \varinjlim_{[C;p\in C@F]\in(\int_\mathcal{C} F)_0} @C.$$

Given a category $\mathcal{C}$, the topos of presheaves $\mathbf{Ens}^{\mathcal{C}^{\mathrm{op}}}$ on $\mathcal{C}$ is denoted by $\mathcal{C}^@$, and its objects and arrows are

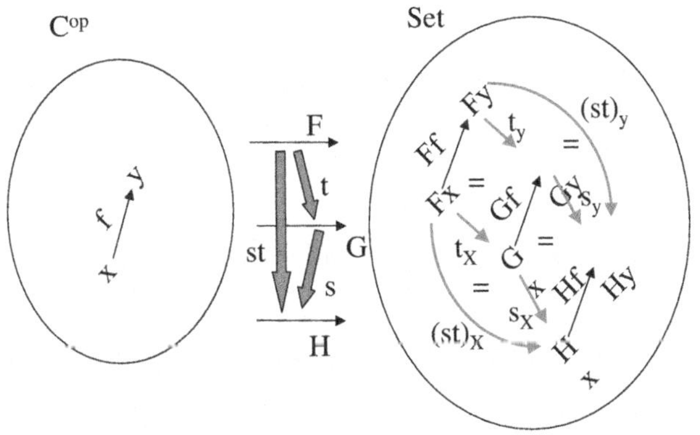

Fig. 67.14. Visualization of a topos of presheaves.

drawn as in Figure 67.14; objects are functors such as F, G, H; morphisms are natural transformations such as t, s, st.

Then we have an embedding $h_\mathcal{C} : \mathcal{C} \to \mathcal{C}^@$ given by

$$h_\mathcal{C}(C) = \hom_\mathcal{C}(-, C) = @C : D \mapsto \hom_\mathcal{C}(D, C).$$

The beginning of Yoneda's Lemma will be shown in Figure 67.15 with its set of formulas. Then we add that a natural invention will be a zig-zag as in Figure 67.16: After that we can consider the question of the shape for a given object X and a given model-functor J, as drawn in the pullback in Figure 67.17.

So the J-shape of X, that is to say J/X, can be drawn as in Figure 67.17, where X is hollowed out as a fibration q_X, where the presheaf F is re-built as a fibration s_F (Ehresmann-Grothendieck construction of the category of "elements" of F), see Figure 67.18. Then we have an isomorphism

$$F(X) = \{t : h_\mathcal{C}(X) \to F, t \text{ natural}\} = \{T : q_X \to s_F, T \text{ cartesian}\}.$$

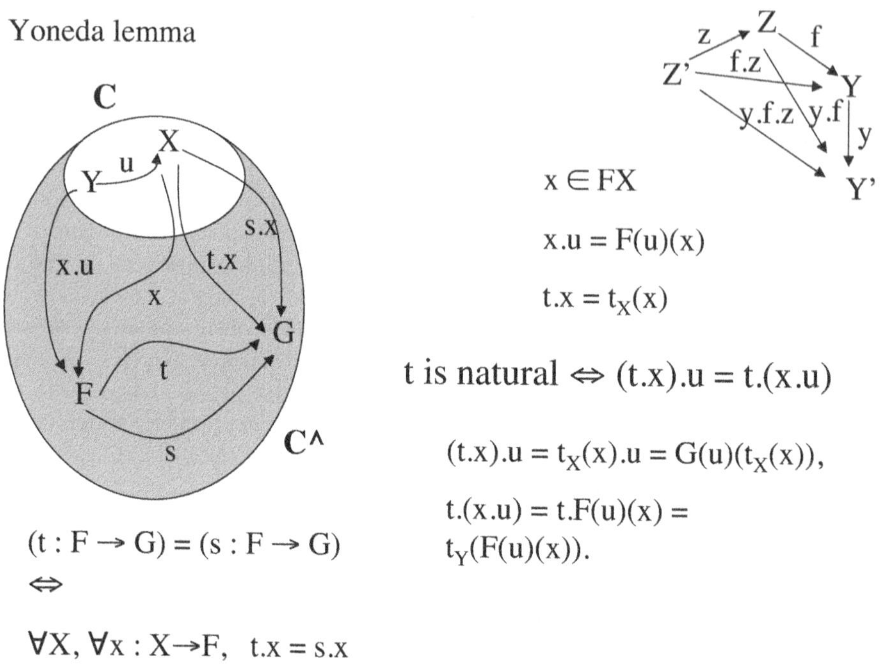

Fig. 67.15. The Yoneda Lemma and extension of a category to its category of presheaves.

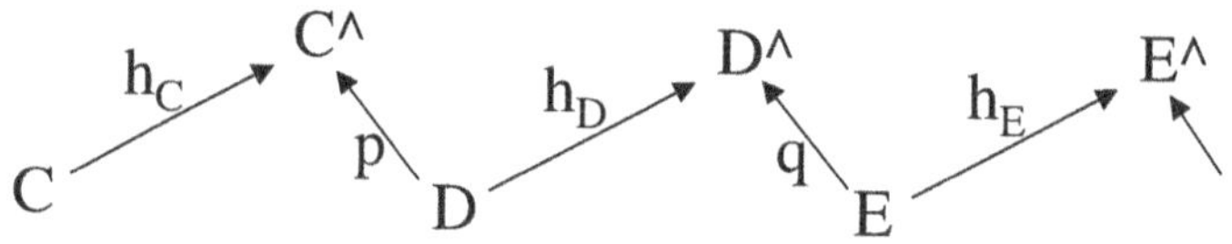

Fig. 67.16. A path of inventive modification of situations.

And finally we reach the notion of a stack (Figure 67.19), a stack on a scheme S being, as in the figure, a lax functor onto h_{Aff}/S toward the 2-category of categories (with some special conditions).

Hence if in the place of the $\mathrm{Yon}_{\mathcal{C}} = @? : \mathcal{C} \to \mathcal{C}^{@}$ we can start with a functor $J : \mathcal{M} \longrightarrow \mathcal{X}$ where $\mathcal{M}$ is thought of as the category of *known 'simple' models* M, and $\mathcal{X}$ as the category of *unknown 'complex' objects* X. Analogously to the category of elements $\int_{\mathcal{C}} F$, we consider the *J-shape of X*, which is the category $\int_{\mathcal{M}} X$—more classically denoted by J/X—with objects the pairs (M, p) where $M \in \mathcal{M}_0$ and $p : J(M) \to X$, a morphism from (M, p) to (M', p') being a morphism $u : M \to M'$ in $\mathcal{M}$ such that $p'.J(u) = p$. Let $q_X : J/X \to \mathcal{M}$ be the forgetful functor $q_X(M, p) = M$, and, if it exists, X_J the inductive limit of $J.q_X$, and $k_{X,J}$ a comparison map:

$$X_J = \varinjlim(J.q_X) = \varinjlim_{[M;p:J(M)\to X]} J(M), \qquad k_{X,J} : X_J \longrightarrow X.$$

If $k_{X,J}$ is not an isomorphism, then we consider that, with respect to J, X is an *absolute novelty*; otherwise we say that X is a *J-manifold*. Given a J-manifold X and a functor $H^* : \mathcal{X} \longrightarrow \mathcal{V}$ (e.g. cohomology), if the comparaison or differential

$$dX = d_{(H*,J)}X : \varinjlim_{[M;p:J(M)\longrightarrow X]} H^*J(M) \to H^*\Big(\varinjlim_{[M;p:J(M)\to X]} J(M)\Big)$$

is not an isomorphism, then we say that the J-manifold X has an *H^*-emergent property*.

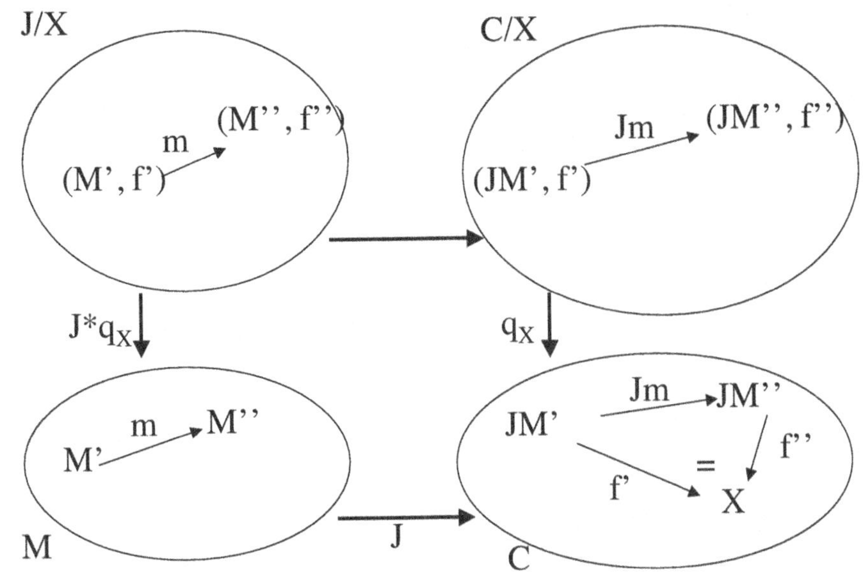

Fig. 67.17. The J-Shape of an object X.

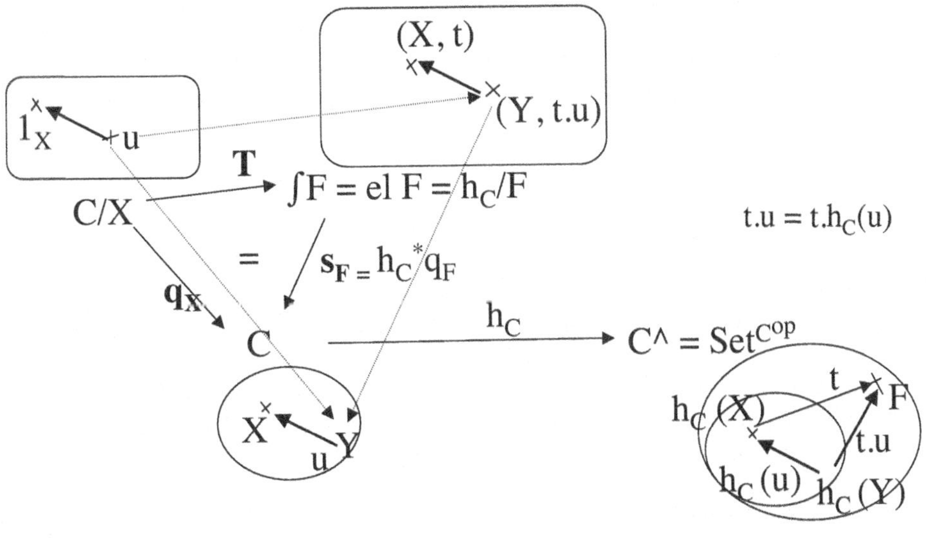

Fig. 67.18. Reformulation of Yoneda Lemma at the level of shapes and fibrations.

The expression of emergence in this way is proposed in [421] and is directly inspired by [295]. The method of inspection and extension of concept's walls, previously described in Section 67.4.3.1, could be rephrased and extended in terms of analysis and perturbations of shapes: the initial creativity moment (opening the walls) consists of choosing an inclusion functor $J_{\mathcal{A}} : \mathcal{A} \to \mathcal{C}$, and then the analysis of A in $\mathcal{C}$ with respect to $J_{\mathcal{A}}$ is—second step of creativity—the introduction of a diagram $D_{\mathcal{A}} : \Delta_{\mathcal{A}} \subset \int_{\mathcal{A}} A \overset{q_{\mathcal{A}}}{\to} \mathcal{A}$ (this introduces a perturbation of the $J_{\mathcal{A}}$- shapes towards $(J_{\mathcal{A}}.D_{\mathcal{A}})$-shapes) and the final step consists in displaying extended wall perspectives according to $D_{\mathcal{A}}$, i.e., in examining if A is a $(J_{\mathcal{A}}.D_{\mathcal{A}})$-manifold.

When the special case $\mathrm{Yon}_{\mathcal{C}} = @? : \mathcal{C} \to \mathcal{C}^{@}$ is replaced by a $J : \mathcal{M} \longrightarrow \mathcal{X}$, the functor $\mathrm{Yon}_{\mathcal{C}|\mathcal{A}} : \mathcal{C} \to \mathcal{A}^{@}$ considered in Section 67.4.3.1 has to be replaced by the functor

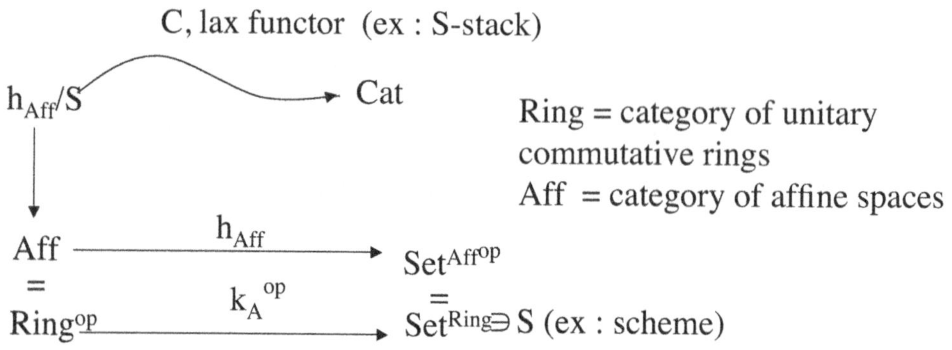

Fig. 67.19. A short presentation of the notion of stack, as a "predicate" on the shape of a scheme.

$$\mathrm{Yon}_{\mathcal{X}|_J} : \mathcal{X} \to \mathcal{M}^{@}$$

given by $X \mapsto \mathrm{Yon}_{\mathcal{X}|_J}(X) = \mathrm{Yon}_{\mathcal{X}}(X)J^{\mathrm{op}} : M \mapsto \hom_{\mathcal{X}}(J(M), X)$.

67.7 Conclusion: Categorical Presentation of Pulsations

To do mathematics is to think mathematically *and* to calculate, and these two mental activities are under the management by the mathematician of what we name "pulsation", mainly at this very moment of "creative invention". In this chapter and other papers we have explained what is this "mathematical pulsation". An old writing of Paul Valéry in 1926 about the notion of "speculation" can serve to approach the pulsation [1075, Analecta 26, p.184-185]. Valéry wrote:

> Speculation do consist to use the possible, but this possible of which I am equipped—as in prevision of variations of the environment, in order to compose them and to resist to them, in order *to wait for* them—even in order to anticipate them, in this way it can enter into the *actual*: and this is thinking!

Calculating in an open way—when we have to invent the calculus as a new system of rules for itself—is also such a speculation. To think and to compute are speculations; when in the speculation the "possible" becomes bracketed and the "unnecessary possible" we are reaching the idea of pulsation.

Clearly, creativity in mathematics needs four things: At first, 90% of eventless substitutional work in a closed well-defined known discipline. Then 9% of concentration on the impossibility of some events within this work, and 0.9% of invention derailing this ordinary game; and finally some chance. A mathematical model of creativity, and especially of creativity in mathematics, has to seek functionalities of substitutions, compositions and decompositions into a definite game with given rules and the opening of the game by modification of the rules. So a method of creation will take in charge the large thinking of analysis and synthesis, with at least these two levels (within a given closed system, and for changing systems), which allows an activation of mathematical pulsation as the heart of a living system.

The *mathematical pulsation* [412] in mathematical life arrived at a moment when an opening of mathematical multiplicities was existing, such as analogously the *multiplicity principle* is working in a memory evolutive system [295], [296]: two objects A and B in $\mathcal{C}$ are discovered to be isomorphic, whereas their constructions are done by gluings which are not comparable. Looking to A and its construction in $\mathcal{A}$, someone can forget some ingredients, reducing A to an object in $\mathcal{D}$ and adding new components, obtaining B in $\mathcal{B}$. This attitude needs to leave the environment $\mathcal{A}$ of A, to change it to $\mathcal{B}$, and so to change our frame of thinking and computing; ultimately $\mathcal{A}$ and $\mathcal{B}$ can merge into $\mathcal{C}$.

The pulsative activity is supported by a personal historical scenery or context, feeding the intuition of the mathematician with geometrical feeling, etc. The logic is only one particular technique to get proofs; the question of foundations is one special way to pretend to render absolute the game of mathematical truths.

In fact mathematical activity does not need foundation or logic; it has only to be an exact construction of paths (= proofs) in mental mathematical spaces of activity.

When we are doing mathematics on the screen of an implicit historical scenery, we are mainly taking care of exactness, i.e., of connectivity and continuity among gestures and associated transits, as it is controllable by computation of exact sequences or exact squares, and then by homotopy and cohomology. The possibility that by doing so we produce truths is in fact secondary; it is not exactly a good question for our internal active pulsation: we are looking not for truths but for paths and proofs.

This activity works by writing and reading diagrams, by transformations of diagrams, in the sense of Peirce. The effective transformations make use of 1) the incorporation of an object Z into a family or as a fiber of a map $f : Y \to X$, as $Z = f^{-1}\{x\}$, 2) the analysis of an object as a gluing of better known fragments, as a manifold, 3) the construction of spaces of figures or diagrams of a given type, and of completion of such spaces, and 4) the change of context and reduction to invariants elsewhere.

Any "mathematical thing" does not really exist; it has to be represented by an object C in a situation, i.e., in a category $\mathcal{C}$, that determines a category $\mathcal{C}/C$, such as a set E determines the powerset $\mathcal{P}(E)$ to which the study of its shape is reduced—for instance $\mathcal{C}/C$ but possibly another. (In fact to C we have to associate a structured category (a topos, a monoïdal category, etc.). That set of gestures is naturally modeled by category theory, that begins with Yoneda Lemma and the consideration of shapes.

67.8 The Hegel Group Action on a Critical Concept's Walls

Was die Wahrheit ist, ist weder das Sein noch das Nichts,
sondern dass das Sein in Nichts und das Nichts in Sein—
nicht übergeht, sondern übergegangen ist.
Ihre Wahrheit ist also
diese Bewegung des *unmittelbaren Verschwindens*
des einen in dem anderen: das Werden.
Gottfried Wilhelm F. Hegel: [453, I,Kap.1,C.a]

Summary. The present section presents a mechanism to catalyze that crucial step in a creative process where the critical concept's "walls" (referring to the common "box" metaphor) are identified and opened. We propose a concrete but generic body of six concepts and the action of a group of transformations of this body, a toolbox that should offer a set of operational perspectives on the critical concept's walls. The conceptual body and the group action are deduced from the first paragraphs of Gottfried Wilhelm Hegel's *Wissenschaft der Logik.* On this body, a group, called the *Hegel group,* acts and reflects some of the relations and operations that are hidden in Hegel's text. We then identify two technical themes from mathematical music theory, the Escher Theorem and the concept architecture of forms and denotators, as instances of this Hegel action. The Hegel action is applied to understand creative processes in two classical compositions—Beethoven's *Hammerklavier Sonata op.106*, and Liszt's *Mephisto Walzer*—but also to the creation of a small model composition.

– Σ –

67.9 Introduction

In [726], we have presented a model of musical creativity and given a number of examples, reaching from music theory to musical composition and music technology. We shall give a short description of the model's main components in Section 67.10.3 of this chapter. Although those examples confirm the validity of the model, there is one single component where the model is still abstract and far from operational in terms of concrete actions to be taken. This component can be described using the common metaphor of the "box," which has to be opened in the creative process. In our model, the box is realized by what we call a "critical concept". The decisive step is then to identify the box's "walls" and to open them. We are fully aware that there is a major debate on creativity. Our position in this context can best be traced from our book [726], in particular from Chapters 17, 18, 19, and 20. The present section however focuses on a specific methodology of acting in a creative way, not on the general debate. This section also does not claim to be a philosophical discourse, but we make use of a philosophical approach to delineate a general method to develop creativity in a very practical way.

This decisive step is what remains quite abstract in our previous work. The present section presents a mechanism that is designed to catalyze that step by offering a concrete but generic body of concepts and the action of a group of transformations of this body, a toolbox that should offer a set of operational perspectives on the critical concept's walls. As was already pointed out in [726], the comprehension of such a mechanism for creativity is not a sufficient condition for effectively producing creative results, it is only meant to be a fairly important and useful procedure for creative actions.

We are deducing the conceptual body and the group action from the first paragraphs of Gottfried Wilhelm Hegel's *Wissenschaft der Logik* [453]. This might be a logical approach since Hegel's initial dynamics in his logical architecture is in fact strongly related to the concept of the concept, i.e., to the basic structure of any concept. This is plausible since his incipit of thoughts claims to be the very beginning of the action of thinking, and in this moment the very nature of conceptualization is at stake. We shall discuss Hegel's text and deduce the conceptual body we are proposing, a body built from six concepts which we for good reasons call *Hegel's body*, it is denoted by $\mathcal{B}$. The group that acts on $\mathcal{B}$ will be called the *Hegel group*, it is denoted by $\mathcal{G}$. We give a precise definition of these objects in Sections 67.10.1 and 67.10.2.

But it is of course also highly problematic to deduce a mathematical structure such as a group action from Hegel's text since the basic situation of that incipit is far from being given a shape that could presuppose mathematical concepts for its description. We shall see that our mathematical concept framework idoes not require set theory, group theory and similar conceptual architectures. We simply use these concepts because they describe in their simplicity (which doesn't require the full mathematical formalism) what Hegel implies in his philosophical prose. It is our claim that Hegel implicitly uses some very simple operations that our small group $\mathcal{G}$ (it is a Klein four-group) comprises. We leave it to philosophers to discuss our approach in terms of what they consider being a valid argument with regard to Hegel's thought dynamics. We are also aware of the still problematic state of Hegel's text, a fact that has been discussed through the history of philosophy to the present, see for example [495]. However, we believe that our precise setup could help us avoid those well-known rhetorical deformations of dialectic argumentation which has often generated nothing more than *ex post* circumlocution of results generated by totally different methods.

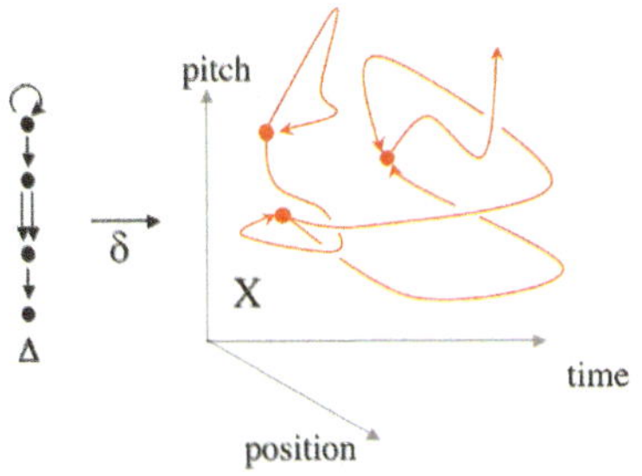

Fig. 67.20. The visual representation of a gesture whose body is a simple musical instrument space.

Shouldn't we also have a look at Hegel's writings on musical aesthetics? This section is however not focusing on Hegel's ideas about music, we only use his very primordial ideas about how we start thinking. Logically speaking, these ideas are independent of later developments in other Hegelian works. This is Hegel's own approach: The very beginnings of thought are set up in his initial sentences of [453].

Among the six basic concepts in the Hegel body, namely being, nothing, space, time, fact, and gesture, the last plays a dominant role in what follows. Although we should not presuppose higher mathematical concepts here, we believe that it is advantageous to recall the precise definition of a gesture which we have given in the mathematical theory of gestures (in music) [720, 723] (see also Chapter 61), since in the examples of this section this definition will be used.

This section is structured as follows: In Section 67.10.2 we introduce and discuss the Hegel body $\mathcal{B}$ and the Hegel group $\mathcal{G}$. In Section 67.11 we discuss the Hegel group action in the mathematical model of creativity that refers to Yoneda's lemma in category theory. Section 68 is dedicated to a Hegelian interpretation of the concept architecture of forms and denotators which has been a backbone of mathematical music theory and its computerized implementation. Section 67.13 discusses the application of the Hegel group action for the understanding of creativity in musical compositions, such as the fanfare in Ludwig van Beethoven's *"Hammerklavier" Sonata op. 106* and the incipit of Franz Liszt's *"Mephisto Walzer"*. Section 67.14 presents a small experimental composition created using the Hegel group action. Section 67.15 shortly discusses the question concerning addition symmetries of the Hegel body. We should stress for non-mathematicians that a *mathematical* category is a different concept from a philosophical category.

67.10 The Hegel Concept Group $\mathcal{G}$

In this section we first analyze Hegel's initial thought movements in his *Wissenschaft der Logik* [453, I,Kap.1,A,B,C]. This will lead to a conceptual configuration built from six components, together with a group action, introducing what will be called the Hegel group $\mathcal{G}$. Our scope here is not to interfere with philosophical debates, but to elaborate on a group structure that can be used in practical creative contexts. Nevertheless, we believe that the Hegel group structure could help understand some of the inherent dynamics in Hegel's primordial thoughts.

67.10.1 Hegel's Initial Thought Movement in *Wissenschaft der Logik*

All the following German quotations are taken from the modern German version of his writings, also referenced in [453, I,Kap.1,A,B,C]. The English quotations are taken from [454].

The Hegelian system of logic starts with these words:

A. Being (Sein)

Being, pure being, without any further determination. In its indeterminate immediacy it is equal only to itself. It is also not unequal relatively to an other; it has no diversity within itself nor any with a reference outwards. It would not be held fast in its purity if it contained any determination or content which could be distinguished in it or by which it could be distinguished from an other. It is pure indeterminateness and emptiness. There is nothing to be intuited in it, if one can speak here of intuiting; or, it is only this pure intuiting itself. Just as little is anything to be thought in it, or it is equally only this empty thinking. Being, the indeterminate immediate, is in fact nothing, and neither more nor less than nothing.

Sein, reines Sein,—ohne alle weitere Bestimmung. In seiner unbestimmten Unmittelbarkeit ist es nur sich selbst gleich und auch nicht ungleich gegen Anderes, hat keine Verschiedenheit innerhalb seiner noch nach aussen. Durch irgendeine Bestimmung oder Inhalt, der in ihm unterschieden oder wodurch es als unterschieden von einem Anderen gesetzt würde, würde es nicht in seiner Reinheit festgehalten. Es ist die reine Unbestimmtheit und Leere. Es ist nichts in ihm anzuschauen, wenn von Anschauen hier gesprochen werden kann; oder es ist nur dies reine, leere Anschauen selbst. Es ist ebensowenig etwas in ihm zu denken, oder es ist ebenso nur dies leere Denken. Das Sein, das unbestimmte Unmittelbare ist in der Tat Nichts und nicht mehr noch weniger als Nichts.

The pure being is pure indeterminacy and emptiness. Emptiness is a spatial category. This is confirmed by the statement "to be thought in it", "it", the being. The nothingness is a spatial insight: penetrating pure being results in recognizing emptiness, nothingness. The preposition "in" is opposed to "out". Both refer to a boundary of a region that we cannot understand but in a spatial way. This spatial understanding is omnipresent in conceptual architectures, such as mathematical set theory or process theory.

Therefore, to the concept of being we have to add the concept of a generic space. Such a space cannot be the concrete physical space, at this stage it is a germ of spatiality, nothing more. But it is a conceptual component of being and nothingness. And it is not only a being out there, it is the action of thinking that reifies "being", it is neither object nor subject. These are categories to be introduced later in Hegel's system.

B. Nothingness (Nichts)

Nothing, pure nothing: it is simply equality with itself, complete emptiness, absence of all determination and content—undifferentiatedness in itself. In so far as intuiting or thinking can be mentioned here, it counts as a distinction whether something or nothing is intuited or thought. To intuit or think nothing has, therefore, a meaning; both are distinguished and thus nothing is (exists) in our intuiting or thinking; or rather it is empty intuition and thought itself, and the same empty intuition or thought as pure being. Nothing is, therefore, the same determination, or rather absence of determination, and thus altogether the same as, pure being.

Nichts, das reine Nichts; es ist einfache Gleichheit mit sich selbst, vollkommene Leerheit, Bestimmungs- und Inhaltslosigkeit; Ununterschiedenheit in ihm selbst.—Insofern Anschauen oder Denken hier erwähnt werden kann, so gilt es als ein Unterschied, ob etwas oder nichts angeschaut oder gedacht wird. Nichts Anschauen oder Denken hat also eine Bedeutung; beide werden unterschieden, so ist (existiert) Nichts in unserem Anschauen oder Denken; oder vielmehr ist es das leere Anschauen und Denken selbst und dasselbe leere Anschauen oder Denken als das reine Sein.—Nichts ist somit dieselbe Bestimmung oder vielmehr Bestimmungslosigkeit und damit überhaupt dasselbe, was das reine Sein ist.

The concept of nothingness initiates being complete emptiness, again a spatial component of nothingness, shared with being. Being was emptiness when penetrated in the thinking movement, whereas nothingness is emptiness from the beginning. The conception, "Anschauen oder Denken", or "intuition of thought", determines it and therefore generates its being, the empty thought of nothingness generates its being. Nothingness is a being, and in fact, because it is emptiness, it is the pure being where Hegel started his discourse. What is important here is that both, being and nothingness, are created from each other by a movement of thoughts. Reification of each one happens through a movement of thought. This will be made explicit in the following paragraph in Hegel's text:

C. Becoming (Werden)

Pure Being and pure nothing are, therefore, the same. What is the truth is neither being nor nothing, but that being—does not pass over but has passed over—into nothing, and nothing into being. But it is equally true that they are not undistinguished from each other, that, on the contrary, they are not the same, that they are absolutely distinct, and yet that they are unseparated and inseparable and that each immediately vanishes in its opposite. Their truth is therefore, this movement of the immediate vanishing of the one into the other: becoming, a movement in which both are distinguished, but by a difference which has equally immediately resolved itself.

Das reine Sein und das reine Nichts ist also dasselbe. Was die Wahrheit ist, ist weder das Sein noch das Nichts, sondern dass das Sein in Nichts und das Nichts in Sein—nicht übergeht, sondern übergegangen ist. Aber ebensosehr ist die Wahrheit nicht ihre Ununterschiedenheit, sondern dass sie nicht dasselbe, dass sie absolut unterschieden, aber ebenso ungetrennt und untrennbar sind und unmittelbar jedes in seinem Gegenteil verschwindet. Ihre Wahrheit ist also diese Bewegung des unmittelbaren Verschwindens des einen in dem anderen: das Werden; eine Bewegung, worin beide unterschieden sind, aber durch einen Unterschied, der sich ebenso unmittelbar aufgelöst hat.

Hegel starts with a seemingly contradictory statement: Being and Nothingness are the same. And in fact, he contradicts this statement some lines later, saying that "they are not the same". This contradiction can be resolved if we view Hegel's statements as assertions of aspects of being and nothingness, not of their full "truth". We might use a geometric metaphor to illustrate this understanding. If one is positioned on a Möbius strip, it is true that it has two sides, the one where one stands, and the opposite one. But one knows that a Möbius strip has only one side. The opposite side is just a part of the front side. This apparent contradiction is resolved when one realizes that the other side is a local statement, while the sameness of the two sides is a global statement: One may walk from the first to the second local side on a global trajectory.

In this sense, sameness of being and nothingness is a global statement, while their difference is a local one. Being and nothingness are two local aspects of the same global concept. Hegel offers a clear method to perform the trajectory between being and nothingness: It is the movement that was already alluded to in the previous paragraphs when Hegel described the movement between being and nothingness in the thinking action. Now, he makes this movement explicit: "Their truth is therefore, this movement of the immediate vanishing of the one into the other: becoming." With this, to the basic concepts of being, nothingness, and space, Hegel adds the next one: becoming. It is however a delicate conceptualization since it is that movement of thought that has no subject or object yet; it is pure action. We therefore propose renaming this concept and calling it "gesture", a basic action that is not yet embedded in the dichotomy of object/subject.

In might seem that we have introduced the concept of a gesture in an arbitrary way. Let us make clear why this impression is wrong. The rationale for our conceptual choice was not to change Hegel's terminology, but first of all to solve the apparent (onto)logical contradiction between Being and Nothingness. Our discourse has in fact solved the contradiction by the introduction of a new concept (gesture) as described above. And it has done so *without* leaving classical logic in favor of some more exotic logics, such as intuitionistic, fuzzy or paraconsistent variants. Let us also remark that gestures are generically useful in artistic utterance, but see Sections 67.13.3, 67.13.4, and 67.14.

The text is moreover also specific about an aspect that every action seems to embody: time. Hegel writes "but that being does not pass over but has passed over". This reveals a time category where present and past are distinguished. Thus, we have to add the concept of time to the space concept in Hegel's setup. Finally, the statement of "being passed over" specifies a further conceptual aspect, namely that after the gestural action is established, there is a resulting fact, the transition of being into nothingness, and vice versa.

Summarizing, we have collected a sixfold conceptual anatomy, grouped into three pairs of corresponding concepts: Being/Nothingness (S/N for German Sein/Nichts), Space/Time (R/Z for German Raum/Zeit), and Gesture/Fact (G/F for German Geste/Faktum). These six conceptual "elementary particles" are shown in as vertices of an octahedron in Figure 67.21; we shall call them *fermions* in an allusion to elementary particle physics, where fermions are the particles that represent matter—as opposed to forces, which are represented by particles that are called *bosons*. We are using the famous Borromean rings in this representation to indicate that these six elementary concepts are not independent from each other. They build an irreducible body of concepts, which we call the *Hegel body*, and denote as a set (by abuse of language, since mathematical formalisms should not matter yet) by $\mathcal{B} = \{S, N, R, Z, G, F\}$.

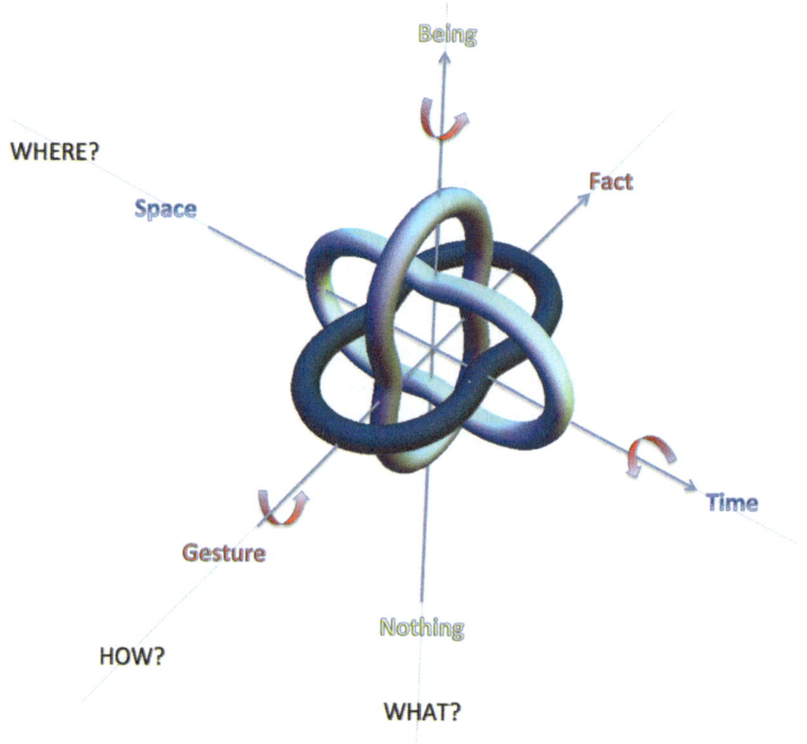

Fig. 67.21. Hegel's concept architecture, the Hegel body $\mathcal{B} = \{S, N, R, Z, G, F\}$, and the $\mathcal{G}$ group action, where the rotational axes answer the questions: how, what, where?

67.10.2 The Implicit Group Structure

The evident symmetry of the Hegel body $\mathcal{B}$ is not by case, and it is not our invention, but results from Hegel's approach as discussed above. In fact, the crucial movement between S and N is defined by the gestural becoming, G, that transforms S into N and vice versa. In this movement, there is also the result of facticity, F, that terminates the movement and puts it into its temporal past tense. This transformation may be interpreted as a symmetry of $\mathcal{B}$, namely the 180° rotation around the axis spanned by G and F, which we denote by $G@F$. In Figure 67.21, this axis corresponds to the question "HOW?"; its rotational action answers the question of how S is transformed into N and vice versa, namely by the gestural action. Using our physical metaphor, the pair $G@F$ plays the role of a boson, a force particle that moves S into N and vice versa. This interpretation is remarkable since it gives to pairs of fermions the role of bosons. In other words: The Hegel body $\mathcal{B}$ is simultaneously a body of material and of forces. The fermions are moved by bosons, and the bosons are generated by pairs of fermions. This is a philosophically essential proposition as it suspends the question of what is more elementary: movement or moved things, this is another justification of the Borromean ring visualization.

But let us first complete the transformational setup defined by the bosonic actions: We have three actions, each defined by a bosonic pair of fermions,

- the 180° rotation $G@F$ around the axis G—F, yielding the permutation $(SN)(RZ)$
- the 180° rotation $R@Z$ around the axis R—Z, yielding the permutation $(SN)(GF)$
- the 180° rotation $S@N$ around the axis S—N, yielding the permutation $(RZ)(GF)$

Together with the identity Id (all of $\mathcal{B}$ remains fixed), this defines a group

$$\mathcal{G} = \{Id, G@F, R@Z, S@N\}$$

of permutations of $\mathcal{B}$. It is evident that this commutative group verifies $x^2 = Id$ for all x, and $x \cdot y = z$ for any two different $x, y \neq Id$, where z is the third non-identity. For example, $G@F \cdot R@Z = S@N$. This group is isomorphic to the Klein four-group K_4, but we stress the equal roles of all three generators. This group of permutations is called the *Hegel group*. The orbits of the group's action are exactly the bosonic pairs.

The group structure extends the semantics of the original Hegel context, where only the action of $G@F$ on S, N are explicitly thematized. Let us therefore interpret the complete action

$$\mathcal{G} \times \mathcal{B} \rightarrow \mathcal{B}.$$

1. $G@F : S \leftrightsquigarrow N$
   This is Hegel's original movement of becoming, which he later specifies into ceasing-to-be $S \rightsquigarrow N$ and coming-to-be $N \rightsquigarrow S$.
2. $G@F : R \leftrightsquigarrow Z$
   The gestural operation maps time to space via the time-parametrization of a gesture (see also the mathematical theory of gestures [720] or Section 61.6.1). The fact as a result of a gesture recreates the time that has produced the spatial points. Let us recall here a significant statement by the great mathematician Henri Poincaré in [852]: *Localiser un objet, cela veut dire simplement se représenter les mouvements qu'il faut faire pour l'atteindre;(...) il ne s'agit pas de représenter les movements eux-mêmes dans l'espace, mais uniquement de se représenter les sensations musculaires qui accompagnent ces movements et qui ne supposent pas la préexistence de la notion d'espace.*
3. $R@Z : G \leftrightsquigarrow F$
   The time as a generator of spatial points (points as results of the pricking gesture) is embodied in the gestural movement that creates its factual results. Conversely, facts qua spatial localizations are only thought of as endpoints of a time line of a gestural movement.
4. $R@Z : S \leftrightsquigarrow N$
   Being as a thinking activity in time is annihilated to nothing when fixed in to spatial points. Conversely, points when rethought as results of the pointing action recover their temporal origin.
5. $S@N : G \leftrightsquigarrow F$
   A gesture, when taken as a being, is transformed into its resulting fact which is the nothingness that terminates the gesture. Conversely, a fact as a nothingness, when thought of as a result of an action, recovers its generating gesture.
6. $S@N : R \leftrightsquigarrow Z$
   Time, as the movement of being (recall Hegel's becoming), when frozen to nothingness, generates spatial points, endpoints. Conversely, if a point in its nothingness (it has no inner substance, so to speak) is rethought of what it brings to be, its being recreates time where the point was moved to its present location. It seems adequate to recall here Gurnemanz's lesson to Parsifal: *Du siehst, mein Sohn, zum Raum wird hier die Zeit.*

It is evident that all these operations relate to each other, and this is essential; they are not independent concepts, but define the irreducibility of the Borromean architecture and, in fact, of the Hegelian setup.

Example 86 Before we investigate more in depth the implications of this group action with regard to the creativity process, it may be useful to give a first elementary musical example of the $\mathcal{G}$ action. Let us look at the primitive action that a musician has to perform to produce a sound, hitting a key on the piano, singing a note, or blowing a tone on a trumpet. Such an action has three parts: The initial gesture moving out from the nothingness of silence to the sound production, then the sound as a product, as a being that has a factual reality, and third the termination of the sound production, the exiting gesture taking back the fact to nothingness. This is exactly what the operation $R@Z = (SN)(GF)$ does, it permutes nothingness and being as well as gesture and fact. The operation $R@Z$ first maps $N \rightsquigarrow S$ and $G \rightsquigarrow F$, creating the factual being of the sound. Then, applied in the other direction, it takes back $S \rightsquigarrow N$ and $F \rightsquigarrow G$, this corresponds to the formula $(R@Z)^2 = Id$. The orbit of $R@Z$ is what the creation of a sound realizes in terms of Hegelian action.

67.10.3 The Conceptual Box Structure

The box structure of the Hegel body $\mathcal{B}$ is shown in Figure 67.22. The box (a cube) is the dual of the octahedron defined by the six conceptual components of $\mathcal{B}$. Each component corresponds to a wall of the box. This is a perfect visualization of the principles of the creative process which we have discussed in [726]. Let us give a very short summary of such a process:

1. Exhibiting the open question
2. Identifying the semiotic context
3. Finding the question's critical sign or concept in the semiotic context
4. Identifying the concept's walls
5. Opening the walls
6. Displaying extended wall perspectives
7. Evaluating the extended walls

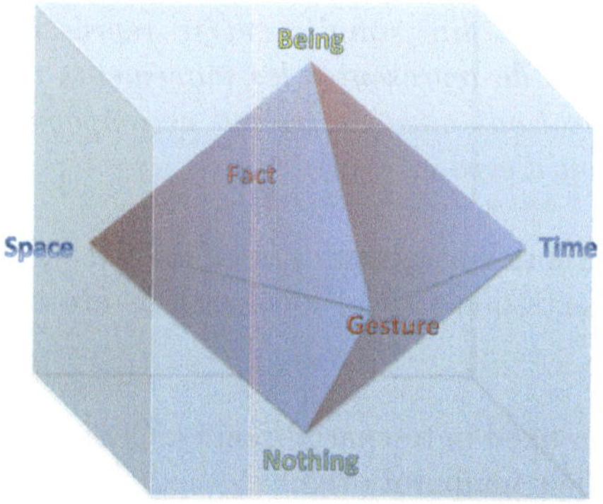

Fig. 67.22. The dual of Hegel's concept octahedron is a cube, whose six walls are associated with Hegel's basic concepts.

The metaphor of a concept's walls is now perfectly realized by the box of Hegel's body, which we may call *Hegel's box*. This coincidence is what we shall take as the starting point of our approach to creativity, namely that the critical action of identifying a concept's walls is made concrete by Hegel's box. This means that we claim that the Hegel action $\mathcal{G} \times \mathcal{B} \to \mathcal{B}$ enables a machinery that helps identify a concept's walls.

To realize this plan we however have to understand why the action $\mathcal{G} \times \mathcal{B} \to \mathcal{B}$ is a tool that could help understand and eventually soften a critical concept's walls. The first observation to this end is that in Hegel's approach, when discussing being and nothingness, he deals with a conceptual architecture that is extremely elementary. In his words, it is about emptiness, about the very beginning of conceptual thinking. This means that what he proposes, and what we have drawn from his idea, is a conceptual framework that is not yet loaded with any specific architectural details; it is *the empty canvas of conceptual construction* that we see in Hegel's text. This is the rationale that motivates us to use this conceptual canvas to investigate any critical concept in a creative process's crucial step of wall identification.

This means that using the Hegel action in the analysis of a critical concept, one might be able to identify its walls. What we called walls in the theory of creativity [726] are the concept's properties, characteristics, and specificities that circumscribe it in a more or less explicit form. In other words, what defines its inner structure, but also what delimits it from other concepts, what it is not. Therefore, using the very basics of a concept's conception (yes, this is circular, but this is essential in the beginning of conceptual organization), one should get supporting machinery dealing with the identification process of walls.

In what follows we will investigate the Hegel action from two perspectives: First, we reconsider the mathematical model of creativity as described in [726, 19.2] and [32]. Second, we analyze the architecture of forms and denotators, which have played a major role in the conceptual framework of mathematical music theory.

67.11 The $\mathcal{G}$ Action on the Yoneda Model of Creativity

Recall that in category theory, Yoneda's idea was to define a functor $Yon_{\mathcal{C}} : \mathcal{C} \to \mathcal{C}^{@}$, where $\mathcal{C}^{@}$ is the category of set-valued presheaves over the category $\mathcal{C}$, by assigning to each object A of $\mathcal{C}$ a presheaf $@A : \mathcal{C}^{opp} \to \mathbf{Ens}$ defined by $@A(X) = X@A(= \mathcal{C}(X, A))$ and for each morphism $f : A \to B$ in $\mathcal{C}$ a natural transformation $@f : @A \to @B$ given by $@f(X) : X@A \to X@B : g \mapsto f \circ g$. Yoneda's Lemma says that $Nat(@A, F) \xrightarrow{\sim} F(A) =: A@F$, for every object A of $\mathcal{C}$ and every functor F in $\mathcal{C}^{@}$. This means in particular for $F = @B$ that A and B are isomorphic if and only if their functors $@A$ and $@B$ are so. We may therefore replace the category $\mathcal{C}$ by its Yoneda-image in $\mathcal{C}^{@}$.

Although Yoneda's Lemma enables the replacement of a given category $\mathcal{C}$ by its Yoneda-image in $\mathcal{C}^{@}$, the functor $@A$ must be evaluated on the entire category $\mathcal{C}$ to yield the necessary information for its identity. The creative moment comes in here: could we not find a subcategory $\mathcal{A} \subset \mathcal{C}$ such that the functor $Yon_{|\mathcal{A}} : \mathcal{C} \to \mathcal{A}^{@} : A \mapsto @A|_{\mathcal{A}^{opp}}$ is still fully faithful? We call such a subcategory *creative*, and it is a major task in category theory to find creative categories which are as small as possible. One may even hope to find what we call an *objectively creative subcategory* for a given object A in $\mathcal{C}$, namely a creative subcategory $\mathcal{A}$ such that for this given object A in $\mathcal{C}$ there is a creative diagram D_A in $\mathcal{A}$ whose colimit C is isomorphic to A. Intuitively, a colimit of a diagram of spaces is obtained by gluing them along common subspaces; it is a generalized union operator. Taking a colimit is a natural condition since the functor $@A$ defines a big diagram whose arrows are the triples $(f : X \to Y, x \in X@A, y \in Y@A)$ with $y \circ f = x$. The colimit object C of such a diagram would ideally replace the functor $@A$ by a unique isomorphism from C to A.

In the context of the Yoneda Lemma with its creative subcategories, the generic model of creativity described in 67.10.3 looks as follows:

1. Exhibiting the open question: understand the object A
2. Identifying the semiotic context: this is the category $\mathcal{C}$ where A has been identified
3. Finding the question's critical sign or concept in the semiotic context: this is A
4. Identifying the concept's walls: this is the uncontrolled behavior of $@A$
5. Opening the walls: find an objectively creative subcategory $\mathcal{A}$
6. Displaying extended wall perspectives: calculate the colimit C of a creative diagram
7. Evaluating the extended walls: try to understand A via the isomorphism $C \xrightarrow{\sim} A$

Here is the correspondence between the Yoneda setup of creativity and the Hegel body: The pairing of G/F corresponds to the pairing object functor $@X/X$. The object is a fact, an abstract point in the category $\mathcal{C}$. The corresponding functor $@X$ enriches the factual object by the entire system of arrows that are gestural pointers to X. The bosonic action $R@Z$ maps F to G, i.e., X to $@X$. Moreover, it maps S to N in the sense that it switches from the object's identity, its simple being S, to its nothingness, its negation in the category's outside: all other objects that become the domains of the functor's arrows to X. On the other hand, the movement from $@X$ to the colimit C takes the gestural aspect back to its factual reduction, to an object C. Finally, the negation N comes back to S as a negation of the negation, the object C that re-instantiates X from its negational functor.

67.12 The Hegel Body $\mathcal{B}$ in the Concept Architecture of Forms and Denotators

Before we discuss the form and denotator concept architecture it is important to trace back this framework to the epistemological roots which define it as an application of the semiotic concept architecture set up by D'Alembert and Diderot in their Encyclopédie, see [48, 238] and Section 6.1. In their approach, an encyclopedia must comprise three characteristics: unity, completeness, and discoursivity. This means that it must realize a philosophical principle of unified presentation of knowledge. It must represent all knowledge (what we expect from a dictionary), and it must enable a discourse, a relational setup to compare its instances, the latter being given by the lexicographic ordering in a dictionary. This triple characteristic was interpreted in our form and denotator architecture by the following three characteristics: Unity was realized by the principle that concepts refer to other concepts, a recursive typology. Completeness was realized by a complete set of types of references. Discoursivity was realized by linear order and recombination of concepts.

Given these principles, a denotator is a conceptual instance in a space, called form. We refer you to Chapter 6 for details and just recall the relevant features here. A denotator has coordinates, i.e., denotators to which it refers, much as its form has coordinator forms to which it also refers. This is the recursive principle. It is the gestural aspect of this architecture. Moreover, the spatial aspect is covered by the referential typology: in topos theory, it comprises the three basic space types of limit, colimit, and powerset. Time is realized in the trajectory of gestures you have to perform to reach a denotator's recursive ingredients. Facticity is obtained when you reach the leaves of the denotator's (and the form's) recursive tree. Being is realized by the denotator's entire instance, while nothingness is realized by the linear ordering within the denotator system: the relation to what a denotator (or form) is not, what comes before and what comes after that instance.

This makes evident that the denotator and form concept architecture, which is the most general existing approach to precise conceptualization, and which has been implemented in music software with great success (see [739]), complies perfectly with the Hegel body $\mathcal{B}$. It is an open question to understand the Hegel action $\mathcal{G}$ in this situation. However, the exchange of time and space could be realized using the equivalence of breadth-first and depth-first search in forms that are built from iterated limits.

67.13 The Usage of $\mathcal{G}$ for the Dynamics of Creativity

The usage of the Hegel action for the dynamics of creativity is a multiple one. On the one hand, we can conceive it as a diagnostic tool without its necessarily acting as a generator of creative extensions. On the other hand, it can be thought of as a machine (though not a dead algorithm) that enables creative extensions. We want to discuss these two directions which, of course, are not exclusive: a good diagnosis can induce a creative extension, and the extensional spectrum can reveal quite a bit about the "patient's" health.

67.13.1 Two Preliminary Examples

Example 87 Let us give a first example of a diagnostic functionality of the Hegel action: Einstein's invention of a multiple time concept. Within our creative process scheme as displayed in Section 67.10.3, the critical concept is physical "time", time in the semiotic context of physics—not the primordial time concept which is part of the Hegel body. Let us take this concept as it was given before Einstein's in(ter)vention. When we inspect the walls of S and of N, it turns out that this concept is a singleton. It has no other copies out there, i.e., its non-being N is empty. In terms of grammar it is a singular being. Taking time and throwing it out to N by the transformation $G@F$ yields nothingness. This diagnosis generates the question "Why only one time?" Is there a physical reason for supporting this singularity of the concept? And Einstein's answer was that physical time could exist in a multiplicity of times, one for every inertial system, and that the gesture of throwing one such time instances to its nothingness in another inertial system would be realized through the famous Lorentz transformation.

Example 88 A second diagnostic example is the invention of the 3M Post-It, a creative process that has been analyzed in detail in [726, Ch.4]. Here the critical concept is "adhesive"; its inventor, 3M chemist Dr. Spencer Silver, had created a substance that did not glue as required, but only "half of it". This time the gestural wall will be inspected: What are the concept's components (recall the referential characteristic of concepts discussed in Section 68, as suggested by D'Alembert and Diderot)? One of them is that an adhesive must glue. This gluing concept's architecture in turn has a way of being that in its becoming has no further reference: it is a final fact, i.e., either gluing or not gluing, *tertium non datur*. This was exactly the point of the diagnosis that Dr. Silver learned from his friend Arthur Fry: There is no deeper reason to terminate the concept's reference tree on that final "glue" as opposed to "not glue". Gluing by 50% was introduced as a deeper conceptual reference: gluing with a percentage. The commercial success of this new adhesive proved that this creative conceptual extension was the right thing to do. In terms of denotator theory, the conceptual component of gluing that was given as a Boolean value was replaced by a real number value.

Evidently, the present state of the art is far from what one could call an expert system. But this is not surprising since the full meaning of the Hegel action must be elaborated on with respect to a variety of semiotical contexts for creative processes. However, it seems evident that a number of core questions around a concept can be built to open conceptual walls more easily.

67.13.2 The Challenge: Creating a Spectrum of Conceptual Extensions

In [726], creativity has been described as a process that takes place in a specific semiotic context. And the result of such a process is viewed as an extension of the given semiotic body. Creativity adds expression, signification and content. It is not a formal combinatorial game. Such an extension entails several critical aspects:

1. It need not be a successful extension. For example, adding a color to mathematical symbols would very probably not solve any mathematical problem. Therefore the semiotic extension might be useless for the time being, but, in the long run, it might turn out to be a good move. This means that creativity should also be judged from the global perspective of the evolution of a semiotic system. This resembles biological evolution, where a local change might show its advantage or disadvantage only after a longer period of further evolution.
2. The conceptual extension, following the Hegel action, say, need not be unique. Opening walls might create an entire "spectrum" of conceptual extensions which need not contradict each other. For example, the recent extension of counterpoint theory as described in [16] contains a variety of conceptual extensions of what are consonances and dissonances within the 12-chromatic pitch class system, but simultaneously extends to microtonal pitch systems.
3. Applying the Hegel action to a critical concept C is a manifold endeavor. For every one X of the six walls, one may create a conceptual extension $C(X)$. If one applies several extensions in a certain order, $C(X_1, X_2, \ldots X_n)$, say, it is probably not true that another extension following a permuted order, $C(X_{\pi(1)}, X_{\pi(2)}, \ldots X_{\pi(n)})$, would yield the same result.

Despite these general questions we should present a more concrete example of a conceptual spectrum created following the Hegel action. Our example relates to the operation $S@N = (RZ)(GF)$. In mathematical gesture theory, as initiated in [720, 723], one considers hypergestures, i.e., gestures $h : \Gamma \to \Delta\overrightarrow{@}X$ starting at the digraph Γ (their skeleton) and targeting the topological category $\Delta\overrightarrow{@}X$ of all gestures of skeleton Δ which target the topological category X. The Escher Theorem (see Section 62.1.2) then states that we have an isomorphism of topological categories Escher : $\Gamma\overrightarrow{@}\Delta\overrightarrow{@}X \xrightarrow{\sim} \Delta\overrightarrow{@}\Gamma\overrightarrow{@}X$. This means that we may exchange the roles of the two skeleta Γ and Δ. In other words: the gesture h which maps curve parameters to gestures *qua* points in $\Delta\overrightarrow{@}X$ can be reinterpreted as a gesture where the points now become gestures and vice versa. This is exactly what the symmetry $S@N$ does: It exchanges G and F. And it also exchanges R and Z, which in the Escher setup makes sense since the space of gestural facts in $\Delta\overrightarrow{@}X$ is transformed into the time parametrization of gestures of $\Delta\overrightarrow{@}\Gamma\overrightarrow{@}X$. The Escher procedure therefore enables us to reinterpret gestures within concepts in permuted ways, and thereby create new meanings. For example, a hypergesture defined by a line of circles can be reinterpreted as a hypergesture defined by a circle of lines. This can yield a completely new understanding of a given concept. In music theory, first species counterpoint can be viewed as a circle that connects (within the pitch class circle) the line of the cantus firmus to the line of discantus. But the Escher permutation of roles would view this counterpoint as a time line of intervals, and therefore as a completely different interpretation of what counterpoint means.

The Escher Theorem is an extremely explicit key to the Hegel action using the $S@N$ operation. It would be interesting to search for Escher-like theorems which relate to the other two Hegel actions.

67.13.3 Escher's Theorem for Beethoven's Fanfare in the *"Hammerklavier" Sonata op. 106*

We want to illustrate the creative movement as it can be interpreted using the above Escher technique as an expression of the $S@N$ operation with a concrete example: the initial fanfare of Beethoven's *Hammerklavier*

Sonata op. 106, see Figure 62.11. The fanfare's rhythmical structure and its gestural interpretation have been discussed in Section 62.4.5.

The creative action takes place when Beethoven's construction exchanges the gestural and the factual roles, following the methodology described above; but see also Figure 67.23. The factual $A \to B$ becomes the gestural part in ρ', whereas the gestural movements $A \to \rho(A)$ and $B \to \rho(B)$ become the factual parts. This is the Escher isomorphism, applied to ρ, i.e., $\rho' = \mathrm{Escher}(\rho)$. A simple topological deformation generates the second hypergesture σ. Observe that this latter hypergesture cannot be generated by direct deformation of ρ since it has a different orientation.

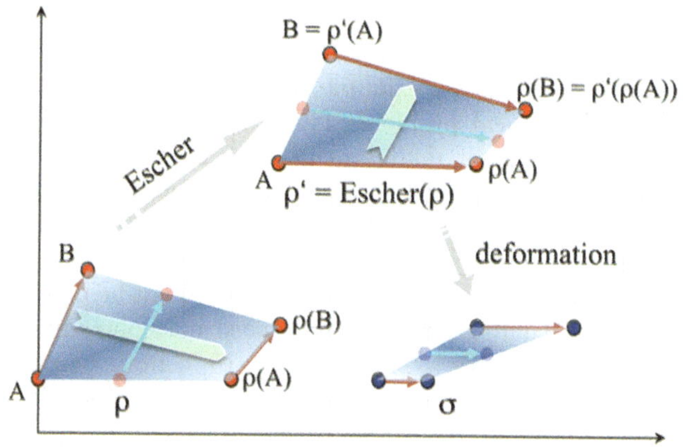

Fig. 67.23. The initial rhythmical hypergesture ρ of the fanfare is transformed into the hypergesture ρ' via the Escher isomorphism, and then deformed into the target hypergesture σ of the fanfare.

The Hegelian action underlying Beethoven's hypergesture exchanges the roles of duration and onset in the sense that a repetition of a halting gesture becomes the halting of a repetition gesture. This truly Escherian flipping movement gives the fanfare its full power. Although this example is a microscopic structure, it proves that creativity can have its germinal force in elementary compositional structures.

67.13.4 The Rotation $S@N$ as a Driving Creative Force in the Incipit of Liszt's *Mephisto Walzer* No.1

An example of the rotation $S@N = (RZ)(GF)$ can be found in the structure of the *Mephisto Waltzer* No.1 by Franz Liszt. The beginning of this composition presents a gesture of harmonic enrichment by the addition of fifths that attributes a harmonic role to the initial musical figure (a repeated tone E introduced by an *acciaccatura* $D\sharp$). The pedal note, E, is followed by a sequence of concatenated fifths in an accelerated rhythm that highlights this process. Moreover, a second compositional gesture transforms the first theme into a second one, an intensely used sequence that develops within the score.

In this composition, rhythmical and harmonic movements are linked in an inseparable way. In the first bar we have the silence, a kind of mental preparation of the initial gesture. In the Hegel group interpretation, this silence could be viewed as nothingness before being. Nothingness becomes then an integral part of the artistic work.

The initial gesture consists of an *ostinato* rhythm $\gamma = D\sharp - E - E - E$, where the last two E notes are played as an echo of the initial one, see Figure 67.24. The rhythm is presented firstly in the extended form $\gamma_0 = D\sharp - E - E - E - E - E - E$. In measures 4-5 we have the reduced sequence $\gamma = D\sharp - E - E - E$, obtained from γ_0 by an operator of horizontal reduction, $h : \gamma_0 \mapsto h(\gamma_0) = \gamma$. The following measures, 6, 10, and 12, present the ostinato γ_0, together with its second half $\gamma' = E - E - E$ (a second horizontal reduction $\gamma' = E - E - E = h'(\gamma_0)$) on a superposition of fifths: $B, F\sharp, C\sharp$. This kind of harmonic extension results from an operator of vertical completion of the fifth sequence. Measure by measure, we have a concatenation

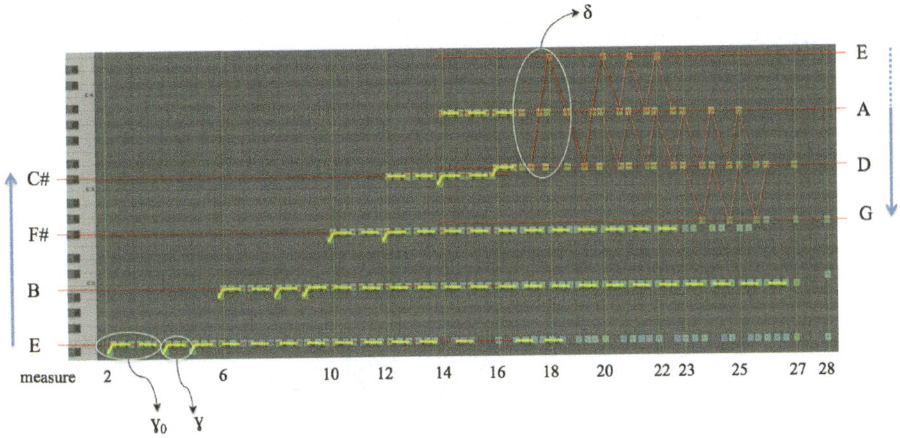

Fig. 67.24. A piano roll representation of measures 2-28 of Liszt's *Mephisto Walzer No.1*. Measures 1 and 29 are omitted since they are both *tacet* and surround the music by a nothingness of silence.

of sequences γ_0, γ, and γ'. The *factual* harmonic fifth sequence in the pitch class *space* $\mathbb{Z}_{12}$ is deployed as a *time gesture* by the ostinato rhythm. Therefore the first part of the creative process consists of an extension of the initial gesture in time, contextualizing the harmonic environment in the gesture γ. This realizes the Hegel transformation $S@N = (FG)(RZ)$.

Fig. 67.25. The ghost note C in measure 28.

Let us shortly discuss the harmonic display here. Recall that diatonic scales are defined using six consecutive fifths. This harmonic structure is a spatial one in the space $\mathbb{Z}_{12}$ of pitch classes. Although a sequence of fifths has already a temporal potentiality, the analyzed movement is still spatial (e.g. $T^7(E) = A$, where $T^7(x) = x+7$ is the fifth transposition). In our case, the measures 2-14 show a first ascending fifth sequence $E-B-F\sharp-C\sharp$, followed by a second descending sequence $A-D-G$, yielding a complete total of $G-D-A-E-B-F\sharp-C\sharp$ that defines D-major. Liszt has succeeded in defining the harmonic basis of D-major in a rhythmically triggered gesture. In measure 18, the chord with dynamics "***f** marcato*" is composed as a symmetrical part around E, i.e., $D-A-E-B-F\sharp$. The couple $E-E$ is then the inferior and superior limit of this chord. This again confirms the inversion symmetry of D-major around E.

Following the key signature (three $\sharp$), the pedal note E could appear as the dominant of A-major. However, this is not the case, since A-major does not appear until measure 111. In measures 2-27, E plays the role of the second degree of D-major. Then, in turn, in measure 28 D appears as the second degree of C-major; in fact, almost unexpectedly appears a C in measure 28 that is confirmed in the following measures 30-34. The B of the left hand (measure 27) moves towards C (measure 28) as a leading tone. This movement suggests the effect of a "ghost note" on the superior melodic line: $D \to C$, see Figure 67.25. Summarizing, after the long preparation of $E \to D$ as $II \to I$ in D-major, we have another interpretation (this time virtual) of $II \to I$ as $D \to C$. Tone E has still a relevant role for C-major, being the symmetric pitch of the tonic with respect to the C-major inversion symmetry around D, and it continues to act as a pedal in the reprise of the initial gesture (measure 35).

In measures 17-18, the rhythmical and harmonic developments are completed by a *melodic creation*. It is the birth of a melodic motive $\delta = D-A-E-A$ (with an initial rest rhythm, and a sequence of quaver-quaver-crochet-quaver). In Figure 67.24 the notes of δ and their rarefaction are highlighted by red lines. Measures 20-26 present a progressive rarefaction of this fragment. Gesture δ will be then modified by transposition and variation of intervals in measures 93-94, generating δ_1, see Figure 67.26. In measure 97 we will have a fusion of the head of γ ($D\sharp - E$) and the tail of δ_1. This hybrid is followed by a new sequence of fifths, where E is effectively the dominant of A. A new version of δ, named δ_2, affirms clearly the tonality of A-major.

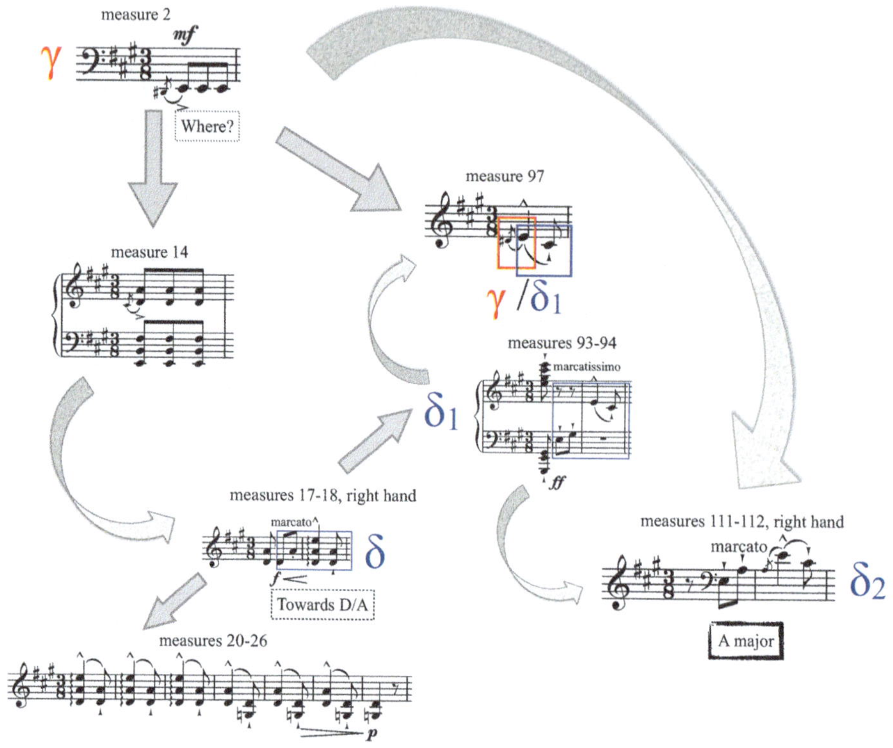

Fig. 67.26. In the *Mephisto Waltzer* by Franz Liszt, the first gesture γ, is progressively repeated and translated in superposed fifths. The main theme, δ, comes from these fifths played this time melodically. A transition gesture with the head of γ and the tail of δ leads to δ_1 and then to δ_2, the varied theme in A-major.

Finally, in measure 29 we have the silence, symmetrically with the silence in measure 1. It is a gesture from the nothing to the being, and back to the nothing of silence, full of potentiality and tension. Rarefaction of tones corresponds to a dilation of rhythmical figures.

Summarizing, in the incipit of *Mephisto Walzer*, two creative gestures are evident:

1. rhythmic-harmonic deployment of γ_0 and its two parts γ, γ' in the circle of fifths of D-major,
2. birth of δ through a deforming gesture of γ within its repositioning on the circle of fifths (rhythmic/harmonic variation).

Thus, the rhythmic structure unfolds through time the spatial relations of harmony. The factuality of abstract points is unfolded in a gestural transformation in time, realizing the Hegelian creative action $S@N = (RZ)(GF)$. The rhythm is the generative force of the transformations $R \to Z$, $F \to G$. The action of deforming γ into δ is a movement generated by the rhythm and by the mapping of the circle of fifths into time.

67.14 An Experimental Composition

Let us try to compose a little musical piece using the Hegel model. It will be obtained, via $S@N$ transformations, as a little melody from an elementary gesture, see Figure 67.27.

The creativity model that we have defined can potentially describe all mechanisms of artwork production. In the previous examples, the interest of analyzing pages of Beethoven and Liszt was given by the evidence of the construction principle. In these examples is not only evident the artistic "fact" but also the path of its creation. *Mephisto Waltzer* does not start with the fifths already superposed and with the

Fig. 67.27. The little composition *The Hegel Song* has been composed starting form an elementary gesture and its variations constructed via the rotation $S@N = (RZ)(GF)$. This composition shows that fundamental compositional processes, variation and thematic development, can be easily obtained using the Hegel action.

δ theme that dominates, but Liszt prepared the entry of the main theme, revealing the very mechanism of construction.

One could observe that, sometimes, the artist can produce an artwork without making all steps explicit: It is the case of "sudden inspiration." This case also complies with our creativity model since inspiration could be correlated to a time compression of steps within the Hegel scheme. These steps, even reduced to instants, must be defined.

A score contains a musical fact. Musicians' educated hands can realize such a fact via appropriate gestures. Playing fingers are like a dancer, who moves from one point to another: Dance is not characterized by the targeted points, but by the trajectory to reach them. Recall Poincaré's citation in Section 67.10.2. Gestures define the dynamics to reach these points. The choice of time and space (meter/*tempo* and keys), explicit in the musical score, contains all the required information to allow the hands' movements. Then, if we want to create a different musical fact, we may deform the used gesture, i.e., connecting it through a hypergestural line to a new gesture. A way to deform a gesture is by changing meter/tempo (Z) and notes (R). Time is the driving force, while notes are the 'target' and thus the factual endpoints of any gesture. If a gesture terminates in an endpoint, the following gesture—that germinates from the preceding one at this endpoint—is connected to it via a hypergesture relation.

In our example of a little composition, realized using the Hegel group scheme, we will use meter/*tempo* and pitch indications to allow the pianist to create and annihilate the small beings commonly called "musical phrases." It is true that a lot of music is written with meter/*tempo* and pitch; however, in this case we intentionally start from a primitive gesture, named γ, without any precise meter/*tempo* indication, to represent the zero level of the pianist's action. A sequence of rotations around the S—N axis will generate the permutations (RZ) and (GF), and we will obtain a little melody.

In the case of a pianist—as already mentioned in the discussion of Liszt's *Mephisto Walzer*, where two notes are hit in the beginning—the primitive gesture here, γ, is a simple hitting of a key with one finger. This maps the spatial fact of a note symbol to a gesture in time. Repeating the same note means doubling this elementary gesture: 2γ. This repetition expresses time in a simple movement with a new characteristic of the note gesture as opposed to the timeless note symbol: γ can be positioned at different times. Some repetitions of γ and 2γ yield $n\gamma$. Until now, there is no quantification of time, and the gesture is localized on the same key. This unarticulated sequence of gestures must be connected and modified in order to generate a meaningful musical composition. Connected structures in music are realized via connected gestures. The same gesture, if repeated at regular intervals (time Z, in a 2/4 meter in our example) and shifted at particular points (space R), leads us to some musical facts, different but related by a common origin. This yields the step $n\gamma \mapsto n\gamma(t)$; pitch space is deployed along time.

Until now, we have only couple of quavers; by doubling gestures we obtain a group of four notes, γ^2. In order to enrich the musical discourse, we can deform this simple scheme of four repeated notes. A different choice of spatial endpoints, an R transformation, deforming the hand gesture form, modifies the *ribattuto* notes into a more articulated sequence; thus $R : \gamma^2 \mapsto \tilde{\gamma}^2$. For example, the sequence G, G, G, G becomes $C\sharp, B, D, B$. Such sequences are typical for Bach's keyboard compositions; perhaps its simplicity and universality derives from these simple deformations of primitive gestures.

In Figure 67.27, γ^3 is obtained from $\tilde{\gamma}^2$ by time modification $4/4 \mapsto 3/4$, and $\tilde{\gamma}^{3-}$ from $\tilde{\gamma}^3$ by the suppression of the second note of each couple, realized by a 'jumping' hand gesture. Transition t from $\tilde{\gamma}^{3-}$ to γ^4 is again a change of meter, $3/4 \mapsto 3/8$. Transition from γ^4 to γ^{4+} is a transformation of time (rests) to space, filling up empty time. Transition from γ^{4+} to $\tilde{\gamma}^{4+}$ is a spatial contour preserving deformation. Tansition $\tilde{\gamma}^{4+}$ to γ^5 is a third change of meter, $3/8 \mapsto 6/8$. Finally, we will then use a Z transformation of $\tilde{\gamma}^5$ to modify some group of three quavers into a pointed crochet (γ'^5). In the new sequence we have a little *cantabile* melody. The transformations applied are a Hegelian-gestural equivalent of variational and developing strategies used by composers. In this way we have just completed a little musical clockwork, *The Hegel Song* as displayed in Figure 67.28.

67.15 Still More Symmetries? Future Developments

Reviewing the Hegel group $\mathcal{G}$, one might be tempted to extend it to the full symmetry group of the Hegel body $\mathcal{B}$, i.e., the automorphism group of this tetrahedron. There are two types of such automorphisms:

(1) Movements, elements of the special orthogonal group $SO(3, \mathbb{R})$ of 3D space, such as the 120° rotation around the axis through the centers of the triangles R, G, S and Z, F, N, see Figure 67.29.

(2) Automorphisms in the orthogonal group $O(3, \mathbb{R})$ with determinant -1, such as the inversion

$$-Id = (GF)(RZ)(SN)$$

which exchanges all our conceptual pairs. Why should one reject the first-case automorphisms that are not in $\mathcal{G}$? One reason could be that they have no fixpoints, or, in other words: They are not bosonic actions generated by a selection of fermions, such as $R@Z$. A reason for avoiding automorphisms with determinant -1 (the case (2) above) could be that a change of orientation of the Hegel body $\mathcal{B}$ could be forbidden because human conceptualization is fundamentally using orientation, whatever that could mean in this embryonal state of mind.

We are not aware of any philosophical interpretation of additional symmetries, but the intrinsic geometry of $\mathcal{B}$ might create new aspects that could not have been conceived without this geometric rendering. Together with the problem of proving Escher-type theorems for (SN) and (RZ), this is a subject of future investigations.

The Hegel Song

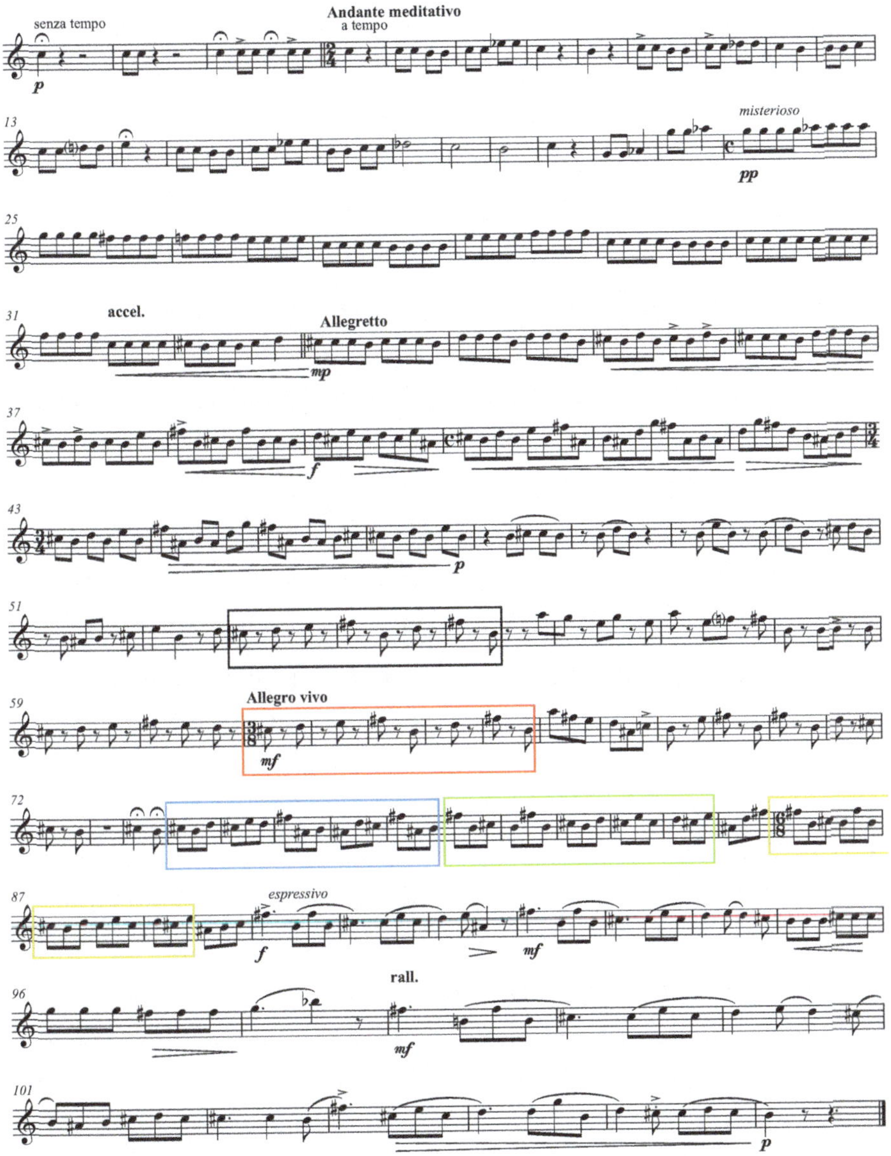

Fig. 67.28. The experimental composition *The Hegel Song*, created by Maria Mannone. Black: γ^{3-}, red: γ^4, blue: γ^{4+}, green: contour preserving deformation, yellow: metrical change $3/8 \mapsto 6/8$.

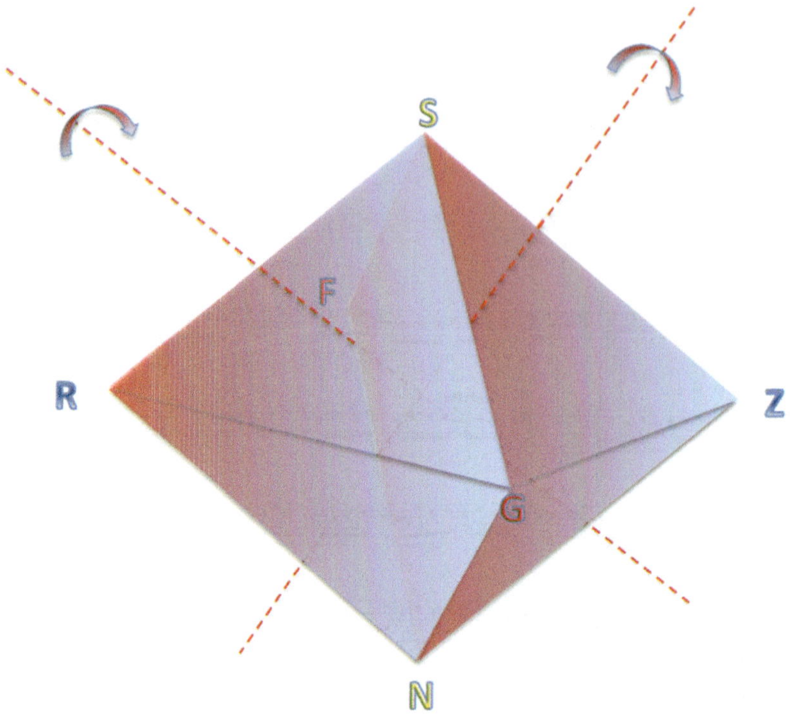

Fig. 67.29. Additional symmetries of the Hegel body might be considered, but at present, no philosophical interpretation is given.

Part XVII

Concept Architectures and Software for Gesture Theory

68

Forms and Denotators over Topological Categories

Summary. This chapter introduces the concept architecture of forms and denotators for gesture theory. It also discusses a Galois theory of concepts in the case of denotators over the category $\mathbf{Mod}^@$.

– Σ –

68.1 The General Topos—Theoretical Framework

Summary. We discuss some basic properties of the category **TopCat** of small topological categories.

– Σ –

68.1.1 The category TopCat of Small Topological Categories

The category of (small) topological categories **TopCat** is the category $Int(\mathbf{Top})$ of categories internal to **Top** (See Section J.4.2). The category **Top** of topological spaces is a full subcategory of **TopCat** if we take the morphism set of a topological space X to be X^2 with the product topology.

Lemma 54 *If $C = (C_0, C_1)$ is a topological category with object set C_0 and morphism set C_1, then C_0 can be identified with its image $e(C_0) \subset C_1$ with its induced topology.*

Clearly, open sets in C_1 induce open sets in the image $e(C_0)$. Conversely, if $U \subset e(C_0)$ is open for the topology on C_0, then its inverse image $d^{-1}(U)$ (d is the domain map) is open in C_1 and we have $d^{-1}(U) \cap e(C_0) = U$. Therefore every open in $e(C_0)$ stems from an open set in C_1.

Proposition 67 *If $C = (C_0, C_1)$ is a topological category and X is a topological space, then there is a bijection*

$$\mathbf{TopCat}(C, X) \xrightarrow{\sim} \mathbf{Top}(C_0, X) : f \mapsto f|C_0,$$

where to the left we identify X with its topological category.

Clearly, this map is an injection since the knowledge of the object map determines the morphism map as the morphism sets on X are singletons. Let us show that it is also surjective. Take a continuous map $g : C_0 \to X$ and an open set $U \subset X^2$. Then its inverse image $V = (g^2)^{-1}(U) \subset C_0^2$ is open. But its inverse image in C_1 is the open set $(d, c)^{-1}(V)$, where $(d, c) : C_1 \to C_0^2$ is the continuous map defined by the domain and codomain maps d, c.

Proposition 68 *The category* **TopCat** *is finitely complete and cocomplete.*

G. Mazzola et al., *The Topos of Music III: Gestures*, Computational Music Science,
https://doi.org/10.1007/978-3-319-64481-3_13

We restrict our proof to completeness since the dual statement follows the same lines of proof. We use the criterion from Appendix Section G.2.2, Proposition 111 and show that **TopCat** has products and equalizers. If $C = (C_0, C_1), D = (D_0, D_1)$ are topological categories, their product category, together with the product topologies, defines a product in **TopCat**. Let $a, b : C \rightrightarrows D$ be two morphisms, i.e., continuous functors. This means that we have the following commutative diagram (with the left and right vertical maps, respectively) for both, the morphism and object maps $a = (a_0, a_1), b = (b_0, b_1)$:

$$\begin{array}{ccc} C_0 & \xrightarrow{e_C} & C_1 \\ a_0 \downarrow\downarrow b_0 & & a_1 \downarrow\downarrow b_1 \\ D_0 & \xrightarrow[e_D]{} & D_1 \end{array}$$

But the two equalizers $\Delta(a_0, b_0), \Delta(a_1, b_1)$ qua topological spaces also define a topological category, as one easily verifies.

This implies that the category $\mathcal{C}^{©}$ of continuous presheaves over a category $\mathcal{C}$, i.e., $F : \mathcal{C}^{opp} \to$ **TopCat** (not only **Ens**-valued, but **TopCat**-valued) is also finitely complete and cocomplete, the limits and colimits being taken pointwise, as known from set-valued presheaves.

68.2 Forms and Denotators

Summary. Forms and denotators for topological categories, to be used for gesture theory, are introduced.

– Σ –

Forms and denotators for topological categories and gestural spaces can be defined much as they have been defined for module categories. However, there is an important difference, namely the fact that the category of gestural presheaves $\overrightarrow{\mathbf{TopCat}}^{©}$ is not a topos, it is finitely complete and cocomplete, but there is no subobject classifier.[1]

Instead of Ω, the power type needs to take the set-theoretical powerset functor 2^F, which evaluates to a topological space with the indiscrete topology. According to Proposition 67, any natural transformation $G \to 2^F$ is possible if the object map is defined set-theoretically since every such map is automatically continuous with the indiscrete topology on 2^F.

Also, the "gesturally representable" functors $@X$ are continuous presheaves that evaluate at objects (Σ, A) in **Digraph** × **TopCat** to $\Sigma \overrightarrow{@}_A X$, see Section 62.7.

Definition 119 *A gestural form or G-form*

F is a quatruple $F = (NF, TF, CF, IF)$ where

(i) NF is a string of ASCII characters; it is called the name *of F and denoted by $N(F)$.*
(ii) TF is one of the symbols
   1. *G-***Simple***,*
   2. *G-***Power***,*
   3. *G-***Limit***,*
   4. *G-***Colimit***;*

   it is called the type *of F and denoted by $T(F)$.*
(iii) CF is one of the following objects according to the previous symbols:
   A. *For G-***Simple***, CF is a topological category X,*
   B. *for G-***Power***, CF is a gestural form,*
   C. *for G-***Limit** *and G-***Colimit***, CF is a diagram* **D** *of gestural forms;*

[1] The usual construction of Ω in topos theory only identifies full topological subcategories if the sieve functor is given the indiscrete topology.

it is called the coordinator *of F and denoted by* $C(F)$. *The diagram* $\mathbf{D}$ *is a diagram of continuous functors* $Fun(F_i)$, *as defined in (iv), for a family* $(F_i)_i$ *of gestural forms.*

(iv) IF is a monomorphism of functors $IF : Fu \rightarrowtail X$ *in* $\overrightarrow{\mathbf{TopCat}}^{©}$, *with this data:*

1. *For G-***Simple**, $X = @X$,
2. *for G-***Power**, $X = 2^{Fun(CF)}$,
3. *for G-***Limit**, $X = lim(\mathbf{D})$,
4. *for G-***Colimit**, $X = colim(\mathbf{D})$;

it is called the identifier *of F and denoted by* $I(F)$, *whereas its domain Fu is called the* space (functor) *of F and denoted by* $Fun(F)$. *The codomain of the identifier is called the* frame space *of the form.*

To denote a gestural form F, we inherit the notation of the module-theoretic setup and add the identifier below the arrow:

$$Name \underset{Identifier}{\longrightarrow} Type(Coordinator). \tag{68.1}$$

The definition of a gestural denotator is completely analogous to the definition of a classical denotator, see Section 6.3.1.

68.3 Mathematics of Objects, Structures, and Concepts

Summary. This section deals with the large scale change of mathematical objectives that shape the development of theoretical frameworks.

– Σ –

We start with the classical focus on selected mathematical objects, such as integers in number theory, special symmetry groups in Galois theory, complex numbers in analysis and algebra, or elliptic functions in complex analysis. This focus was dominant until—roughly—the second decade of the 20th century, when the interest in types of structures became the focus of mathematicians such as Emmy Noether or Emil Artin.

One may see the invention of category theory by Eilenberg and Mac Lane and the Bourbaki enterprise as the climax of this structural style of mathematical research. Category theory was created following the insight that many concrete objects share structural principles that would yield theorems on structures qua meta-objects of the mathematical reality.

Grothendieck had completely incorporated this style in his advances in functorial algebraic geometry. However, with his approach to the Weil conjectures, and then more radically in his proposal of a motivic unification of cohomology theories, the topic had changed from structural catechism to the question concerning the right concepts to solve a type of problem, or, in the case of motives, even the question of conceptual unification of structurally related theories. This was no longer a structural problematic but a conceptual one: The search for good structures, not the application of structurally determined entities.

68.4 Galois Theory of Concepts

Summary. This section deals with those background strategies which have profiled this book's overall concern. It relates to what [519] qualifies as the mathematical knowledge by reason from concept construction. Our approach to this constructivist perspective shares the nature of Galois theory: to understand new concepts as extensions of given concept frameworks via specific 'equations' and their Galois groups. The "formal ontology" of our approach is described in the language of form semiotics over given topoi, see Section G.5.3.

– Σ –

68.4.1 Introduction

The switch from the original title "Geometrie der Töne" of [682] to the title "The Topos of Music" of [714] is not only due to the topos-theoretic generalization of the structural setup of mathematical music theory; it more radically testifies to a change of methodological principles. The former book was meant as a mathematical description of spaces, models, and theorems about musical objects and facts.

Here, modules, categories and geometric spaces were just a language to restate musical subjects in precise terms of prefigured mathematics. In this framework, the construction of conceptual entities was not a theme per se, but a prescientific activity as is standard in other applications of mathematics to determined fields of knowledge, such as physics, economics, or psychology.

The progressive infiltration of what they now call mathematical music theory (à défaut de mieux, but what in fact is much more than just a special type of music theory, namely a geometric logic of concepts, theory, and performance) by conceptual construction issues was induced by the programming work in the context of the RUBATO® project from 1992 to 2002 as conducted by the author in collaboration with Oliver Zahorka, Thomas Noll, Jörg Garbers, Stefan Göller, Stefan Müller, and Gérard Milmeister. This research taught us that music needs universal concept architectures. We learned this on the level of object-oriented programming, but such engineering procedure is everything less than a proof, and this is why the theoretical counterpart in [714] (see Part II) had to be developed to build the mathematical formalism backing those programming activities.

In this chapter, we describe the presently most complete version of this mathematical formalism. It has two main components: the first of them is a basis of mathematical objects which we take for granted from classical mathematical theory. To guarantee all necessary structural and logical constructions required in musical conceptualization, this basis must be a topos $\mathcal{E}$. Typically, this topos is the topos **Ens** of sets, or the topos $\mathbf{Mod}^{@}$ of presheaves over the category **Mod** of modules and diaffine maps. The latter category is sufficiently powerful to englobe all usually envisaged structures in mathematical music theory and is therefore central to the exposition in this book. However, performance theory in its generalization to gestural dynamics needs topoi that are related to differentiable structures. Therefore the limitation to a single topos is neither preconized from practice nor mandatory in theory.

The second component provides us with the mechanism for concept construction. This mechanism is a recursive one in the sense that already construed concepts are used to build new ones by virtue of universal tools, such as limits, colimits, and power objects. It is not clear whether other than topos-theoretic "universal" constructions are required for future developments. We admit however a fundamental extension of classical recursion: Our recursive construction process includes circular concepts, i.e., the objects X being under construction are defined by use of universal tools when applied to—among others—these same objects X.

Although in such a context existence theorems are crucial, we do not stress the mathematical aspect. Rather it is our concern to communicate the fundamentally philosophical relevance of this methodology. Kant characterizes the mathematical method by its constructivist nature: A concept must be understood from its construction and not by pure philosophical meditation [519, A713-B741]. This construction is of a significant nature: A new concept is defined as a solution of a defining 'equation'. Here, equation means that we have to solve functional correspondences between determined domains of concepts, coupled with equations of topos-theoretic nature: limits, colimits, and power objects.

The point is that such solutions do not exist within the given concept domains; they generate proper conceptual extensions, such as the extension of the real numbers to the complex numbers, stemming from a solution of the equation $X^2 + 1 = 0$, which over the reals is impossible. We have learned to handle such extensions as plain solution spaces, i.e., algebraic field extensions. But they are in fact conceptual extensions which were brought to life under hard existential struggles. We claim that any fundamental mathematical progress is due to conceptual extensions which enable solutions of hitherto unsolvable 'equations'. Recall that one of the more recent dramatic events in this development was Deligne's solution of the Weil conjectures in view of the generalized concept of a topological space as proposed and developed by Grothendieck.

We contend that conceptual extensions are precisely what Grothendieck in his autobiographic *Récoltes et semailles* [398] alludes to when explaining his method of smoothening and eventually dissolving the hard surface of a coconut in tepid water under the sun's patient warmth. What happens is that a manifest identity—the coconut's firm shape—is being dissolved under the osmose of its negation, not under the brute force destruction of its identity. Let us look at what is outside the coconut, let us confront its epidermis with what is beyond the object's boundary, and we shall understand what is within the coconut. In his Logic [453, 2. Chapter, A.b)], Hegel, citing Spinoza, states that all determination is by an affirmative negation, and that something is a negation of negation. It is obvious that the solution of Grothendieck's coconut problem is precisely this insight: Something is the dissolution of its conceptual epidermis, the solution of its conceptual negation (what is impossible within a given concept?) in the negation of what it excludes. To put it in French, it is all about "dévissage de l'identité".

It may seem that such philosophical far-out mysteries are not what formal and effective science is about, but this is erroneous: Once we have understood the conceptual epidermis, the boundary of a concept's power, we can transcend it and offer solutions to the present conceptual limitations, solutions which help overcome the inherent limitations. Recall that Galois' solution of old questions—such as the trisection of an angle by use of ruler and compass—is in fact the result of a thorough analysis of the conditions for such a solution, showing that any solution space has properties not shared with a specific ruler- and compass-aided construction method.

In other words, Galois shows that any conceptual extension by ruler and compass must contradict the targeted extension by trisection. To our mind it is not by chance that Grothendieck's more recent research (his unpublished manuscript *La Longue Marche à travers la Théorie de Galois*, written in 1981) is about 'great Galois unification theories' (unifying Galois and fundamental groups, but see [947] for more details).

The following technical sections should be viewed in this light in order to understand why we so strongly insist on conceptual extensions and on the related Galois theory.

68.4.2 Form Semiotics

We refer to Appendix Section G.5.3 regarding the development and musicological motivation for the structure of a form semiotic. Here, we want to give a slightly more elegant definition of a form semiotic. To this end, we suppose being given a topos $\mathcal{E}$ with subobject classifier Ω, together with a subcategory $\mathcal{R}$ such that the Yoneda map $@? : \mathcal{E} \to \mathcal{E}^{@}$ into the topos of presheaves over $\mathcal{E}$ yields a fully faithful functor $@? : \mathcal{E} \to \mathcal{R}^{@}$ if the presheaves are restricted to $\mathcal{R}$. Recall that this is the case, for example, for any presheaf topos $\mathcal{E} = \mathcal{C}^{@}$ over a category $\mathcal{C}$, together with its full subcategory $@\mathcal{C}$ of represented $\mathcal{C}$-objects $@X = Hom(?, X)$ (this *is* Yoneda's Lemma), or for the topos $\mathcal{E} = \mathbf{Ens}$, together with any singleton subcategory $\mathcal{R} = Sing(S)$ defined by any singleton $S = \{s\}$. For reasons stemming from the context of mathematical music theory introduced in [714], we call such a subcategory an *address subcategory* of $\mathcal{E}$, and its objects are called addresses. Denote by $Mono(\mathcal{E})$ the subcategory of monomorphisms of $\mathcal{E}$, where it is understood that objects in a category are identified with their identity morphisms. We further need a set $\mathcal{T} = \{\mathbf{Simple}, \mathbf{Limit}, \mathbf{Colimit}, \mathbf{Power}\}$ of four type symbols; the meaning of this set will become clear in the following definition. Finally, given a set Y, we need the set $Dia(Y/\mathcal{E})$ of finite diagrams : $D \to \mathcal{E}$, where D is a diagram scheme (a quiver, see Appendix Section C.2.2), the vertexes $d, e, f, \ldots$ being elements of Y, where for each pair e, f of vertexes, the arrows $i : e \to f$ are identified by positive natural numbers $i = 1, 2, 3, \ldots$, i.e., an arrow is a triple (e, f, i). Finally, we denote by $Dia^*(Y/\mathcal{E})$ the disjoint union $\mathcal{E} \sqcup Dia(Y/\mathcal{E})$. By abuse of language, we call the elements of $\mathcal{E}$, when embedded in $Dia^*(Y/\mathcal{E})$, trivial diagrams (in fact, in previous definitions of form semiotics, we used a special denotator to graph trivial diagrams, and the present formalism is just a more direct restatement of that artifical setup).

Definition 120 *Given the type set $\mathcal{T}$, a topos $\mathcal{E}$, an address subcategory $\mathcal{R}$, a set $\mathcal{D}$, the elements of which are called* denotators, *and a set $\mathcal{F}$, the elements of which are called* forms, *a* form semiotic *is the data of a* type *map $T : \mathcal{F} \to \mathcal{T}$, a* diagram *map : $\mathcal{F} \to Dia^*(\mathcal{F}/\mathcal{E})$, an* identifier *map $Id : \mathcal{F} \to \mathcal{E}$, a* coordinate *map $C : \mathcal{D} \to \mathcal{E}$, a* denotator name *map $DN : \mathcal{D} \to \mathcal{D}$, a* form name *map $FN : \mathcal{F} \to \mathcal{D}$, and a* denotator form *map $DF : \mathcal{D} \to \mathcal{F}$. These data are required to have the following properties:*

(*i*) *A form* F *is uniquely determined by its name* $fn = FN(F)$, *its identifier* $id = Id(F)$, *its type* $t = T(F)$, *and its diagram* $d = (F)$. *We therefore also denote a form* F *by the* DenoteX[2] *symbol* $fn : id.t(d)$ *and also write* $F \sim fn : id.t(d)$ *to indicate that* F *is determined by its four images.*

(*ii*) *A form's identifier* $Id(F)$ *is supposed to be in* $Mono(\mathcal{E})$; *we call the domain* $dom(Id(F))$ *the form's* space *and denote it by* $space(F)$, *whereas the codomain* $codom(Id(F))$ *is called the form's* frame (space) *and is denoted by* $frame(F)$.

(*iii*) *The domain* $dom(C(D))$ *of the coordinate* $C(D)$ *of a denotator* D *is supposed to be an address* $a = A(D)$ *within* $\mathcal{R}$; a *is called the* denotator's address.

(*iv*) *The coordinate codomain* $codom(C(D))$ *of a denotator* D *is the space of the denotator's form* $F(D)$. *The image* $Id(F(D)) \circ C(D)$ *is also called the* frame coordinate of D.

(*v*) *A denotator* D *is uniquely determined by its name* $dn = DN(D)$, *its form* $f = DF(D)$, *and its coordinate* $c = C(D)$, *with address* a. *We therefore also write down a denotator by its DenoteX symbol* $dn : a@f(c)$ *and also write* $D \sim dn : a@f(c)$ *to indicate that* D *is determined by these data. Logically, the address is superfluous since it is already contained in the coordinate; the stress of this important information is however advantageous for immediate recognition.*

(*vi*) *If a form's type is simple, i.e.,* $T(F) =$ **Simple**, *then its diagram* (F) *is an address* A. *This address equals the frame space* $frame(F)$ *of* F.

(*vii*) *If* $T(F) =$ **Power**, *then* (F) *is a diagram with one vertex form* G, (F) *has no arrows, we have* $(F)(G) = space(G)$, *and* $frame(F) = \Omega^{space(G)}$.

(*viii*) *If* $T(F) =$ **Limit**, *then for all vertexes* G_ι *of* (F), *we have* $(F)(G_\iota) = space(G_\iota)$, *and* $frame(F) = \lim(F)$.

(*ix*) *If* $T(F) =$ **Colimit**, *then for all vertexes* G_ι *of* (F), *we have* $(F)(G_\iota) = space(G_\iota)$, *and* $frame(F) = colim(F)$.

Let us give some comments on this definition with respect to the previous approaches as documented in [714] and in the present book, of course.

Remark 24 In reasonable programming contexts, it is assumed that the names of forms are keys, i.e., that the application FN is injective. In this case, a diagram can also be considered with vertexes being form names instead of forms. The denotator names are also denotators now, instead of being elementary character strings. This approach is a substantial enrichment compared to the usually string-oriented naming technique, as described, for example, in [942]. This generalization enables us to work with global and more complex name spaces. In particular, this enables us to assume that like form names, denotator names are also keys, i.e., that DN is injective. And it opens the path to more structured name space concepts by use of forms which are specially designed for name management.

For example, if $\mathcal{E} = \mathbf{Mod}^{@}$ and $\mathcal{R} = @Mod$, we may consider the monoid $\mathbb{Z}$-algebra $\mathbb{Z}\langle UNICODE\rangle$ over the set $UNICODE$ of Unicode symbols and its represented presheaf $@\mathbb{Z}\langle UNICODE\rangle$ address object. Then we have the form $NF \sim fn : Id.\mathbf{Simple}(Dg)$, where $Dg = @\mathbb{Z}\langle UNICODE\rangle$, Id is the identity on Dg, and $fn \sim fn : 0@FN(C)$, the coordinate $C : \mathbb{Z} \to \mathbb{Z}\langle UNICODE\rangle : 0 \mapsto$ "$NameForm$" representing the corresponding natural transformation $@\mathbb{Z} \to @\mathbb{Z}\langle UNICODE\rangle$. Observe that the denotator fn is its proper name denotator. To this initial naming tool, we may add any name denotator $n \sim n : 0@NF(C : 0 \mapsto$ Anyname) with $n \neq fn$, and Anyname $\in \mathbb{Z}\langle UNICODE\rangle$, for example Anyname $=$ "$3.Violin + 4.Piano$". On this construction level, we have a single form, and a number of denotators for this form, each of them being zero-addressed, and its own name denotator, i.e., it essentially identifies to its coordinate value Anyname.

Remark 25 *In former setups for form semiotics, we had included the type "Synonymy" in order to allow plain changes of form names. This feature is easily realized by use of type* **Limit**, *and a one-point diagram, similarly to the diagram used for* **Power** *type. Therefore one is dispensed from this type.*

[2] DenoteX is an ASCII-based denotation language for denotators. An EBNF specification of DenoteX is available from `http://www.ifi.unizh.ch/mml/musicmedia/downloads.php4`.

Remark 26 There is a subtlety in the diagram definition which one should observe. In a sloppy language it is possible to have diagrams of objects in a category such that one and the same object appears in more than a single vertex. This is correct since diagram schemes are charged with the "indexing" job. Here, we do not have arbitrary diagram schemes, their vertexes must stem from the form set $\mathcal{F}$. This is not an impoverishment of mathematical expressivity, it is just the strict duty to use nothing except what is given in the form semiotic in order to comprise all features without being forced to invent new names and symbols on the flight.

Once such a system is implemented in a programming environment, such discipline pays. Therefore, in order to place the same form on different vertexes, one has to produce a number of isomorphic copies, i.e., synonymous forms (in the above sense) with the same spaces. Or else, we may enrich the form names in order to be able to define as many forms as needed via rich name spaces. In practice, we do however often refer to "copies of a given form" without specifying these naming accents.

Remark 27 In practice, lists are very useful. There is no list type in our setup, but one may easily mimic it as follows. We are working in the presheaf topos $\mathbf{Mod}^{@}$ and its address category $@\mathbf{Mod}$, where we often identify modules with their representable functors, if no confusion is likely. Suppose that we want to have a list length $n = 0, 1, 2, 3, \ldots \in \mathbb{N}$, the denotators of which have the form F. This is achieved by a form $L_n(F)$ of limit type and a discrete diagram without arrows, consisting of n copies of F. This enables us to introduce a list form for all list lengths up to N, say. We take the colimit of all the forms $L_n(F), n = 0, 1, 2, 3, \ldots N$.

A more elegant way without using an indetermined number of cofactors for our list form works as follows: Consider the form

$$List(F) : Id.\mathbf{Colimit}(Item(F), Terminal),$$

where $Item(F) : Id.\mathbf{Limit}(F, List(F))$ and where $Terminal : Id.\mathbf{Simple}(\mathbb{Z})$. Here we denote the vertex forms for limits and colimits if we deal with discrete diagram schemes. The form $Terminal$ is made for terminating the list entries and writing down the list's length. This however is not a complete definition since we do not know whether such a form exists! This definition is circular, and the existence of the corresponding presheaves must be proven. This is in fact true (see Section G.2.2), but the proof uses more than finite completeness.

Despite the missing list type, the given types can be used to define more general list forms in the sense that general index sets can be used. To this end, let I be an index set, together with a linear ordering relation $<$. Call a subset $J \subset I$ an initial interval iff there is either $x \in I$ such that $J = \{\iota | \iota < x\}$ or $J = I$. Suppose further that F is a given form over $\mathbf{Mod}^{@}$, $@\mathbf{Mod}$, and that we want to define a form $List(I, F)$ such that its elements at address module A are precisely the "lists" $(l_\iota)_J$, i.e., the sequences of $A@space(F)$-elements for initial intervals J of I. To this end, consider the family $[\iota] : Id.\mathbf{Simple}(@0)$ of simple forms with the zero module 0 over the zero ring as a trivial diagram, $Id = Id_{@0}$, and name keys $[\iota]$ over $NameForm$. Let $X(I, 0) : [\iota] \mapsto @0, \iota \in I$, be the (generally infinite, but our topos is complete, not only finitely complete) diagram of these simple forms. Consider the 'pure list' form $[I] : Id.\mathbf{Colimit}(X(I, 0))$, Id being the identity on the frame space colim $X(I, 0)$. Observe that for any address module A, we have $A@\text{colim}\, X(I, 0) \xrightarrow{\sim} I$ since 0 is final in $\mathbf{Mod}$. We then define the auxiliary form $G : Id.\mathbf{Limit}([I], F)$ and finally set

$$List(I, F) : Id.\mathbf{Power}(G),$$

with the identifier $Id : space(List(I, F)) \to 2^{space(G)} \subset \Omega^{space(G)}$ the subpresheaf whose value at address A is the set of subsets $L \subset A@space(G) \xrightarrow{\sim} I \times A@space(F)$ such that the first projection $pr_1 : A@space(G) \to I$ is injective on L, and $pr_1(L)$ is an initial interval of I. Moreover, this construction yields the lexicographic linear ordering relation among lists if F bears a linear ordering, see Section 6.8 for orderings on denotators.

The subject of a circular form definition is a first example of the Galois problem in defining concepts: We are given a determined stage of a form semiotic and would like to add new forms by specific properties, as for example the above list property. Besides this existence problem, we also would like to know how many solutions we may expect, and whether they have an influence on the denotators which will live in these new forms.

68.4.3 The Category of Form Semiotics

Evidently, the problem of successive extensions of form semiotics must be accessed by use of a tool for comparing form semiotics: We need to know when a semiotic is an extension of a given form semiotic, what an isomorphism between form semiotics is, and so forth. In other words, we have to introduce morphisms between form semiotics. We shall therefore elaborate on the categorical aspect which has been sketched in Section G.5.3.1. The type set $\mathcal{T}$ being fixed once for all (in our context), a form semiotic Sem is given by four objects, $\mathcal{E}, \mathcal{R}, \mathcal{F}, \mathcal{D}$, and seven maps, $T, , Id, DN, FN, DF$. We write

$$Sem = (\mathcal{E}, \mathcal{R}, \mathcal{F}, \mathcal{D}; T, , Id, DN, FN, DF)$$

for these data. For the following definition we consider the subset $Dia^s(\mathcal{F}/\mathcal{E})$ of $Dia^*(\mathcal{F}/\mathcal{E})$ consisting of $\mathcal{E}$ and of those diagrams $: D \to \mathcal{E}$ such that for every vertex form $F \in D$, we have $(F) = space(F)$. The maps and Id may then be given the codomain $Dia^s(\mathcal{F}/\mathcal{E})$ instead of $Dia^*(\mathcal{F}/\mathcal{E})$.

Definition 121 *Given two form semiotics*

$$\begin{aligned} Sem_1 &= (\mathcal{E}_1, \mathcal{R}_1, \mathcal{F}_1, \mathcal{D}_1; T_1, {}_1, Id_1, DN_1, FN_1, DF_1) \\ Sem_2 &= (\mathcal{E}_2, \mathcal{R}_2, \mathcal{F}_2, \mathcal{D}_2; T_2, {}_2, Id_2, DN_2, FN_2, DF_2), \end{aligned}$$

a morphism $f : Sem_1 \to Sem_2$ is a triple of maps $f = (f_\mathcal{F}, f_\mathcal{D}, f_\mathcal{E})$ with the following properties:

(*i*) *We have two set maps $f_\mathcal{F} : \mathcal{F}_1 \to \mathcal{F}_2$, $f_\mathcal{D} : \mathcal{D}_1 \to \mathcal{D}_2$ and a logical*[3] *morphism of topoi $f_\mathcal{E} : \mathcal{E}_1 \to \mathcal{E}_2$ which preserves addresses.*

(*ii*) *They commute with all seven maps of the respective forms, more precisely: $t_1 = t_2 \circ f_\mathcal{F}$, $DN_2 \circ f_\mathcal{D} = f_\mathcal{D} \circ DN_1$, $DF_2 \circ f_\mathcal{D} = f_\mathcal{F} \circ DF_1$, $FN_2 \circ f_\mathcal{F} = f_\mathcal{D} \circ FN_1$. The morphism f induces a map $f^s : Dia^s(\mathcal{F}_1/\mathcal{E}_1) \to Dia^s(\mathcal{F}_2/\mathcal{E}_2)$, to be defined below, such that $Id_2 \circ f_\mathcal{F} = f^s \circ Id_1$ and ${}_2 \circ f_\mathcal{F} = f^s \circ {}_1$.*

Here is the definition of the critical map f^s. Whenever no confusion is likely, we omit the indexes $\mathcal{F}, \mathcal{D}, \mathcal{E}$ of f. On the topoi $\mathcal{E}_1, \mathcal{E}_2$, it is the given logical morphism. On diagrams, we have this construction: Let $X : S \to \mathcal{E}_1$ be diagram in $Dia^s(\mathcal{F}_1/\mathcal{E}_1)$. We define a diagram $f^s(X) = X' : S' \to \mathcal{E}_2$. Its quiver S' has these data: The vertex set is the image $S' = f(S)$. For a given vertex couple $A, B \in S'$ the arrow are all arrows (U, V, i) in S such that $f(U) = A, f(V) = B$. We enumerate these arrows by natural indexes $j = 1, 2, 3, \ldots$ and denote these indexes[4] by $j = j(U, V, i)$. With this, the new diagram X' sends an arrow $j = j(U, V, i)$ to the morphism $f(X(i)) : space(A) \to space(B)$. This follows from the axiom $Id_2 \circ f_\mathcal{F} = f^s \circ Id_1$, and therefore, form spaces and frames commute with f, i.e., for any form F, we have $f^s(Id(F) : space(F) \to frame(F)) = Id(f(F)) : space(f(F)) \to frame(f(F))$.

Now, if F is simple, the diagram and frame maps coincide, and diagram commutation with f^s means commutation of frames. If F is of limit type, the conservation of the limit $frame(F) = \lim(F)$ under the logical f means precisely $frame(f(F)) = \lim f^s((F))$; analogously for colimit and powerset type. Of course, a number of conditions upon a morphism f are intertwined, but this is not the place to discuss a minimal set of conditions. More important is the following fact:

Sorite 12 *Given three form semiotics, Sem_1, Sem_2, Sem_3, and two morphisms, $f = (f_\mathcal{F}, f_\mathcal{D}, f_\mathcal{E}) : Sem_1 \to Sem_2$, $g = (g_\mathcal{F}, g_\mathcal{D}, g_\mathcal{E}) : Sem_2 \to Sem_3$, the factorwise composition $g \circ f : Sem_1 \to Sem_3$ is a morphism of form semiotics. This composition is associative, and the identity is such a morphism. Call $ForSem$ the category of form semiotics with these morphism data.*

In Section G.5.3.2, the category $ForSem$ is used to propose global form semiotics by the usual gluing procedure. Here, we are interested rather in the local extension problem.

[3] It preserves finite limits, exponentials, and subobject classifiers, see [639, p.170]; to be clear, we also require here that it preserve colimits.

[4] This indexing function is however only defined up to permutations, but in this theory, quivers are only considered modulo permutations of the arrow numbers for given vertex couples, as limits and colimits are invariant under these permutations.

68.4.4 Galois Correspondence of Form Semiotics

Given a topos $\mathcal{E}$, together with an address category $\mathcal{R}$, we denote by $\emptyset(\mathcal{E}, \mathcal{R})$ the *empty form semiotic* with $\mathcal{F} = \mathcal{D} = \emptyset$. An automorphism of $\emptyset(\mathcal{E}, \mathcal{R})$ is just an (automatically logical) automorphism of $\mathcal{E}$ which preserves addresses, i.e, induces an automorphism on the subcategory of addresses, denote by $Aut(\mathcal{E}, \mathcal{R})$ the group of these automorphisms. Let Φ be a subgroup of $Aut(\mathcal{E}, \mathcal{R})$. Given a form semiotic S over $\mathcal{E}, \mathcal{R}$, call $Aut_\Phi(S)$ the *group of automorphisms of a semiotic S over Φ*, i.e., the automorphisms which have elements of Φ as underlying topos morphisms. In particular, for the trivial group $\Phi = Id_{\mathcal{E}}$, we write $Aut_{\mathcal{E}}(S)$ and call this the *group of automorphisms of S over $\mathcal{E}$*. For any subsemiotic R of S over $\mathcal{E}$, we consider the subgroup $Gal_{\mathcal{E}}(S/R) \subset Aut_{\mathcal{E}}(S)$ consisting of those automorphisms which leave R pointwise fixed. Conversely, given a subgroup $H \subset Aut_{\mathcal{E}}(S)$, there is a unique maximal subsemiotic $Sem(H)$ of S over $\mathcal{E}$, which is left pointwise fixed under H, much in the same way as there is a maximal subfield of a field which is left pointwise fixed under a given group of automorphisms of the given field. This defines a *Galois correspondence*

$$Sub_{\mathcal{E}}(S) \underset{Sem}{\overset{Gal}{\rightleftarrows}} Sub(Aut_{\mathcal{E}}(S))$$

between the set $Sub_{\mathcal{E}}(S)$ of subsemiotics of S over $\mathcal{E}$ and the set $Sub(Aut_{\mathcal{E}}(S))$ of subgroups of $Aut_{\mathcal{E}}(S)$.

Example 89 Suppose that we are given a form semiotic S over $\mathbf{Mod}^@$, $@\mathbf{Mod}$ with form set $\mathcal{F}_S$ denotator set $\mathcal{D}_S$ with the above defined name form NF with name value "$NameForm$". Suppose further that we are given a fixed diagram $X \in Dia^s(\mathcal{F}_S/\mathbf{Mod}^@)$. We may define two new forms, $U \sim PN : Id.\mathbf{Limit}(X), U' \sim PN' : Id.\mathbf{Limit}(X)$, as follows: The identifiers are the identities on the frames. The names are *by definition* two different zero-addressed denotators, $PN \sim PN' : 0_\mathbb{Z}@NameForm(c), PN' \sim PN : 0_\mathbb{Z}@NameForm(c)$, of the given name form NF, represented by its name value. These denotators are their mutual name denotators. As to the coordinates c, c', we consider two situations:

In the first, we set $c = c' : 0_\mathbb{Z} \to \mathbb{Z}\langle UNICODE\rangle$, which means that the only difference of names resides in the declaration that $PN \neq PN'$. Therefore, the form semiotic $S(U, U'; PN, PN')$ defined as the extension of S by the two new forms and the new names evidently has the automorphism f over S which exchanges forms and names, i.e., $f(U) = U', f(U') = U, f(PN) = PN', f(PN') = PN$. This situation means that we have a purely conceptual automorphism in $Gal_{\mathcal{E}}(S(U, U'; PN, PN')/S)$ without touching any topos-theoretic 'basic' data.

The second situation is the same for everything except that $c \neq c' : 0_\mathbb{Z} \to \mathbb{Z}\langle UNICODE\rangle$. In this case, suppose that the values $c(0), c'(0)$ differ from the given name values in S. Then we consider the following construction of an automorphism in any category $\mathcal{C}$ as follows: Suppose that we are given an automorphism ϕ of an object A of $\mathcal{C}$. By definition, the automorphism $[\phi] : \mathcal{C} \xrightarrow{\sim} \mathcal{C}$ leaves all objects fixed. On the morphisms, we have four cases:

- $Hom(A, A)$ is left pointwise fixed.
- For $X \neq A$, we have $Hom(X, A) \xrightarrow{\sim} Hom(X, A) : g \mapsto \phi \circ g$, whereas
- $Hom(A, X) \xrightarrow{\sim} Hom(A, X) : g \mapsto g \circ \phi^{-1}$.
- For $X, Y \neq A$, $Hom(X, Y)$ is left pointwise fixed.

We now apply this construction to $\mathcal{C} = \mathbf{Mod}^@, A = @\mathbb{Z}\langle UNICODE\rangle$ and $\phi = @\mathbb{Z}\langle u\rangle$, $u \in \mathfrak{S}(\langle UNICODE\rangle)$, the symmetric group of words over $UNICODE$. Using the transposition f, this defines an automorphism in $Aut_{Aut(\mathbf{Mod}^@, @\mathbf{Mod})}(S(U, U'; PN, PN'))$.

Example 90 Suppose again that we are given a form semiotic S with form set $\mathcal{F}_S$ denotator set $\mathcal{D}_S$, over $\mathbf{Mod}^@$, $@\mathbf{Mod}$, and with the above defined name form NF with name value "$NameForm$". Given an identity morphism $Id : S \xrightarrow{\sim} S$ in $\mathbf{Mod}^@$, we introduce two forms, $NF_i : Id.\mathbf{Limit}(X_i), i = 1, 2$, with diagrams $X_1 : NF_2 \mapsto S, X_2 : NF_1 \mapsto S$ and *different* name denotators $NF_1 \sim NF_2 : 0_\mathbb{Z}@NameForm(c : 0 \mapsto w), NF_2 \sim NF_1 : 0_\mathbb{Z}@NameForm(c : 0 \mapsto w)$ having the same name coordinate c. Then the exchange of forms and corresponding names NF_i defines an element of $Gal_{\mathbf{Mod}^@}(S(NF_1, NF_2)/S)$. Here, we have used the circular definition of forms by mutual reference.

Proposition 69 *Every finite group is isomorphic to an automorphism group of type* $Aut_G(S)$ *for a form semiotic* S *over* $\mathbf{Mod}^@$, $@\mathbf{Mod}$, *and* $G \subset Aut(\mathbf{Mod}^@, @\mathbf{Mod})$.

It suffices in fact to consider any group $G \subset \mathfrak{S}(\langle UNICODE \rangle)$ of word permutations and to construct name denotators with name values that are permuted under G over the form *"NameForm"*, which are their proper names.

Problem 2 *The main problem of this Galois theory is to investigate the relation between conceptual constructions of forms and denotators and their associated automorphism groups and the existence problem of such concepts. This problem is evidently related to the underlying topoi and address categories. In view of this latter constraint, the question is also about how far concept constructions are a function of these topoi or else of a generic character, i.e., independent of the given topoi.*

The problem is likewise manageable for regular extensions, i.e., for a set of forms and denotators, the names and diagrams of which depend on an already existing form semiotic, whereas for circular, i.e., non-regular, extensions, very little is known. For example, as already mentioned above, one has some results concerning the existence of specific presheaves, see Proposition 115 in Section G.2.2.1. Concerning existence problems involving circularity, we suggest you review what Paul Finsler introduced in [321] and what Peter Aczel calls Hyperset Theory [5, 87, 813], together with associated techniques for the solution of set equations with circularity conditions as presented in [813].

69

The Rubato Composer Architecture

Summary. In this chapter, we give a short presentation of the RUBATO® Composer software environment. It is the basis for the subsequent chapters about the BigBang rubette.

– Σ –

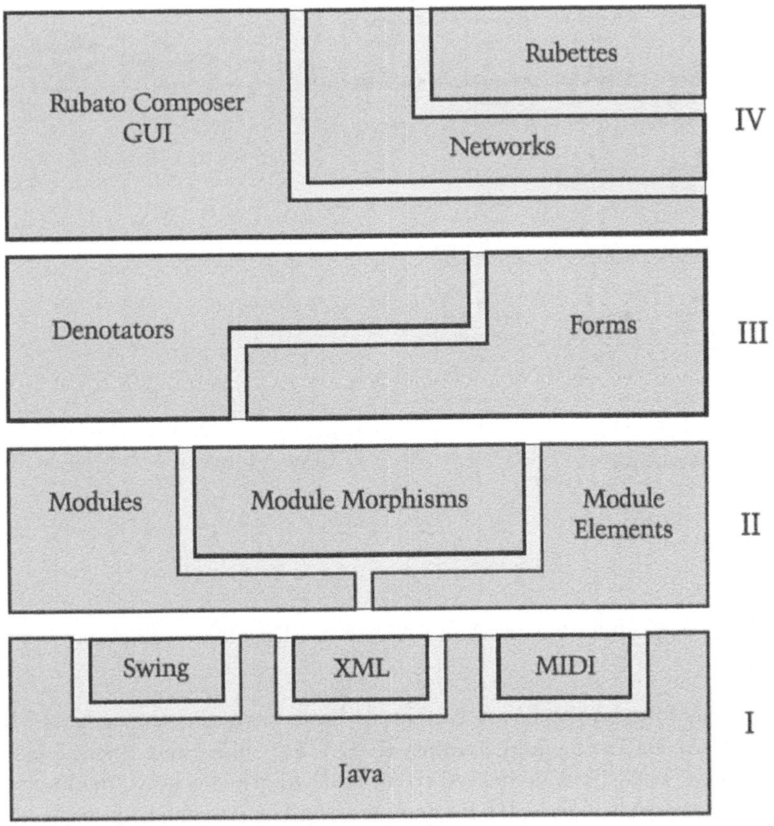

Fig. 69.1. The software architecture of RUBATO® Composer.

As opposed to initial rubette concepts for RUBATO®, rubettes can now be managed in a totally flexible way, the classification into composition, performance, analysis, and math rubettes is obsolete. The only persistent

G. Mazzola et al., *The Topos of Music III: Gestures*, Computational Music Science,
https://doi.org/10.1007/978-3-319-64481-3_14

essential architectural element in the collaboration of rubettes is that they communicate by an exchange of denotators. A detailed discussion of this new environment is documented in Gérard Milmeister's book [739].

69.1 The Software Architecture

The software architecture of RUBATO® Composer is shown in Figure 69.1.

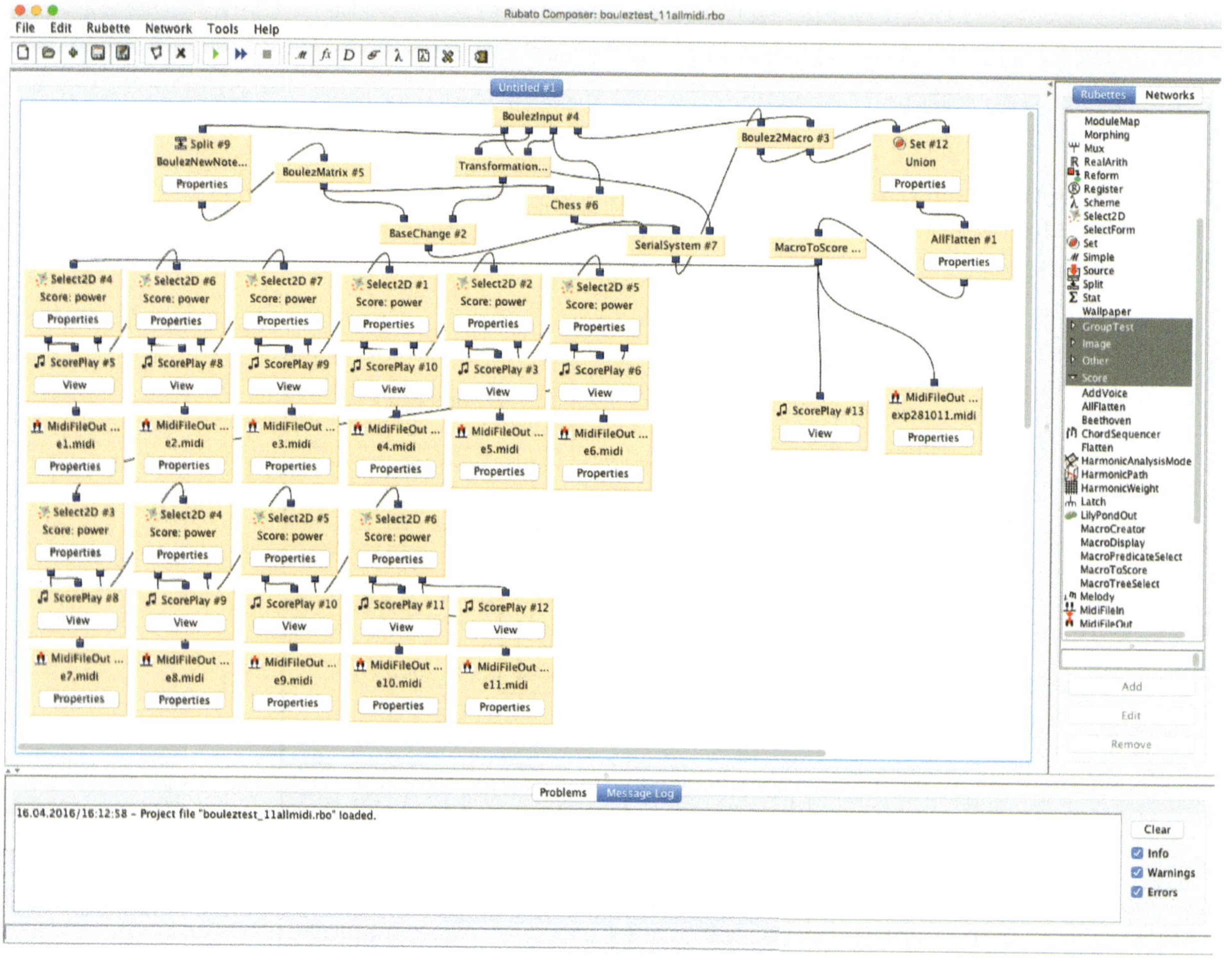

Fig. 69.2. A network of rubettes in RUBATO® Composer.

This architecture has four layers: Layer I includes basic Java classes, such as MIDI classes for MIDI management, XML classes for saving and loading RUBATO® files, and Swing classes for the graphical user interface management. Layer II is dedicated to all mathematical classes, dealing with modules, module elements, and module morphisms. Layer III implements the classes which describe denotators and forms, and layer IV implements rubettes as well as the GUI for their interaction, which is called the network GUI. The network GUI is shown in Figure 69.2.

While earlier RUBATO® versions had a limited display and interaction of rubettes, RUBATO® Composer displays a GUI where any existing rubette, shown to the right of the network surface in Figure 69.2, can be activated by dragging it into the network surface. Rubettes can be added to RUBATO® Composer at any time in a rubette plugin directory.

Once a number of rubettes are instantiated in the network surface, they can be connected to build a functional network of such components. Every rubette has a number of input and output ports. The user may connect any output port to a number of input ports of other rubettes. To guarantee a reasonable functionality, only one connection can terminate at an input port. Special rubettes are available which load or save MIDI files.

69.2 The Rubette World

At present, dozens of rubettes are available. They are mainly designed for compositional purposes, including generalized counterpoint tools according to our counterpoint theory from Part VII, a software which is discussed in detail in [16]. A new version of the HarmoRUBETTE is also avaliable (see Section 69.2.2), and in particular Florian Thalmann's BigBang rubette that is discussed in the following chapters. Unfortunately, a MeloRUBETTE is not available yet, and the powerful PerformanceRUBETTE is also a *desideratum* at this time. See `www.rubato.org` to see all presently available implementations and additional documentation.

A rubette has, besides its input and output ports, the option to display properties, which access the processual parameters, as well as a view dialog for visual display and interaction, see Figure 69.3. Some rubettes are described in Chapter 71.

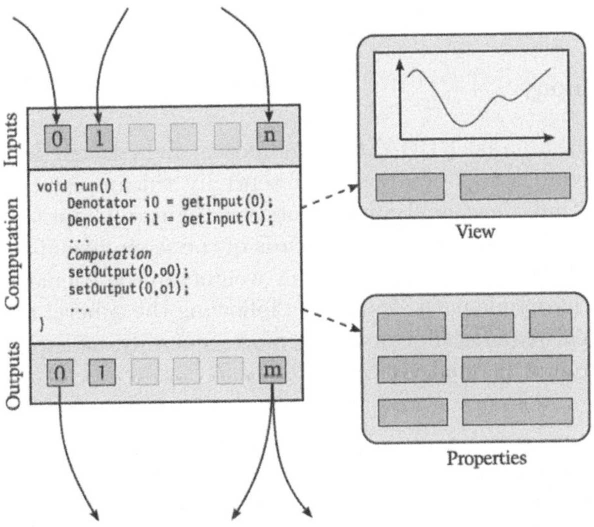

Fig. 69.3. The functionality of a rubette.

69.2.1 Rubettes for Counterpoint

The contrapuntal composition method for all of the six strong dichotomy classes has been implemented by Julien Junod in the RUBATO® Composer software; see Figure 69.4 It shows a simple network for random counterpoint generation. The details are described in [16]. Junod's seven rubettes allow for the generation of counterpoint compositions in any one of the six strong dichotomy classes. In the network shown in Figure 69.4, the BoolyWorldRUBETTE selects the strong dichotomy, the MidiFileInRUBETTE (#1) provides the cantus firmus, the BollyCarloRUBETTE creates a random counterpoint, the DeCounterpointiserRUBETTE creates the discantus and cantus firmus components of the internal calculation forms, and the VoiceMergerRUBETTE puts the two voices together. The two MidiFileInRUBETTEs (#7, #8) input additional note parameters for the cantus firmus and discantus, respectively.

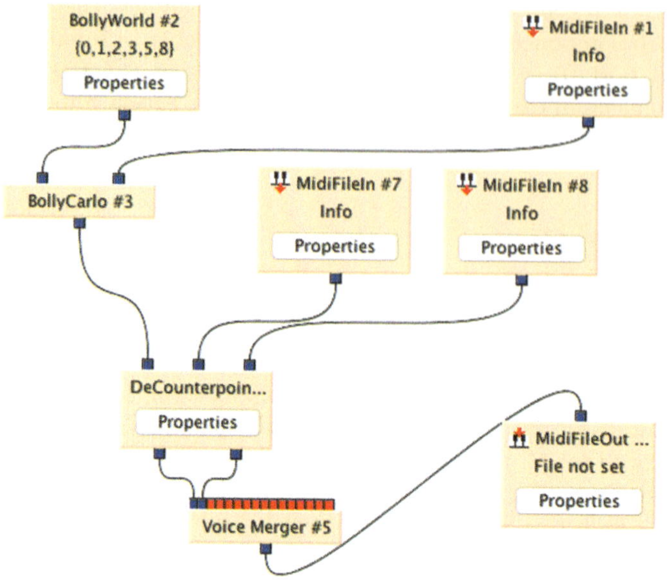

Fig. 69.4. A simple network for random counterpoint generation.

69.2.2 Rubettes for Harmony

A series of rubettes on the Java-based RUBATO® Composer environment for harmonic analysis has been implemented by Ruhan Alpaydin, see [23] for details. Essentially, this implementation splits the original HarmoRUBETTE's functionality into functional components; see Figure 69.5 for the corresponding network. The ChordSequencerRUBETTE creates the sequence of chords of the score given from the MidiFileInRUBETTE, the HarmonicWeightRUBETTE calculates the harmonic weights in the Riemann matrix, and the HarmonicpathRUBETTE calculates a harmonic path, everything following the general method as described in Section 41.3. However, some analytical parameters can be set more generally, as explained in [23]. The calculation of note weights has not been implemented yet.

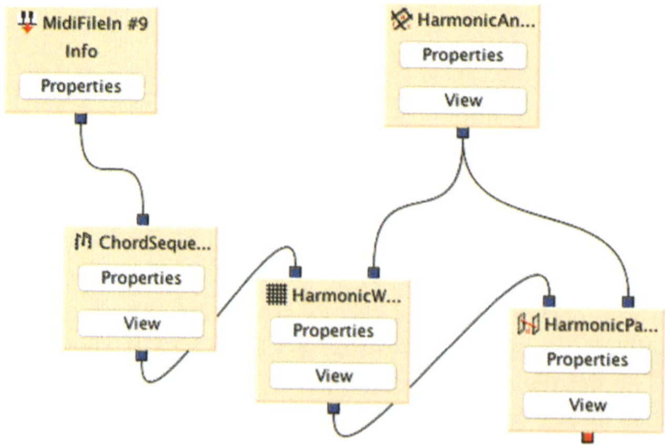

Fig. 69.5. Alpaydin's network for harmonic analysis.

69.2.3 MetroRubettes

Ruhan Alpaydin has also implemented an attractive MetroRubette on the Java-based RUBATO® Composer environment. Besides the metrical weight graph, one may select any set of maximal local meters and view their positions in the given score; see Figure 69.6.

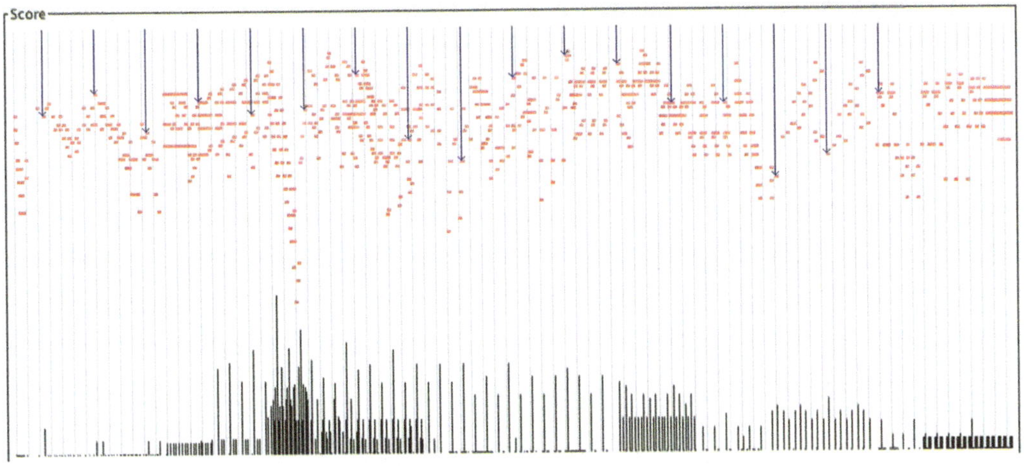

Fig. 69.6. Alpaydin's interface for metrical analysis, showing a maximal local meter together with the score and the metrical weight graph.

70

The BigBang Rubette and the Ontological Dimension of Embodiment

The *BigBang* rubette is a gestural music visualization and composition tool that was developed with the goal to reduce the distances between the user, the mathematical framework, and the musical result. In its early stages, described for instance in [1043, 1045], it enabled defining, manipulating, and transforming *Score* denotators using an intuitive visual and gestural interface. Later on, it was generalized for transformation-theoretical paradigms based on the ontological dimension of embodiment, consisting of facts, processes, and gestures, and the communication between these levels [1042].

BigBang is a regular Rubato Composer rubette, as introduced in the previous chapter, with a variable number of view windows using which users can easily create denotators by drawing on the screen, visualize them from arbitrary perspectives, and transform them in a gestural way. On a higher level, they can interact with a visualization of their compositional or improvisational process, and even gesturalize the entire process in various ways. *BigBang* simply has one input and one output and now accepts almost any type of denotator as an input, to be visualized and interacted with. Figure 70.1 shows the rubette, incorporated in a *Rubato Composer* network, and one of the rubette's view windows including the so-called facts and graph views.

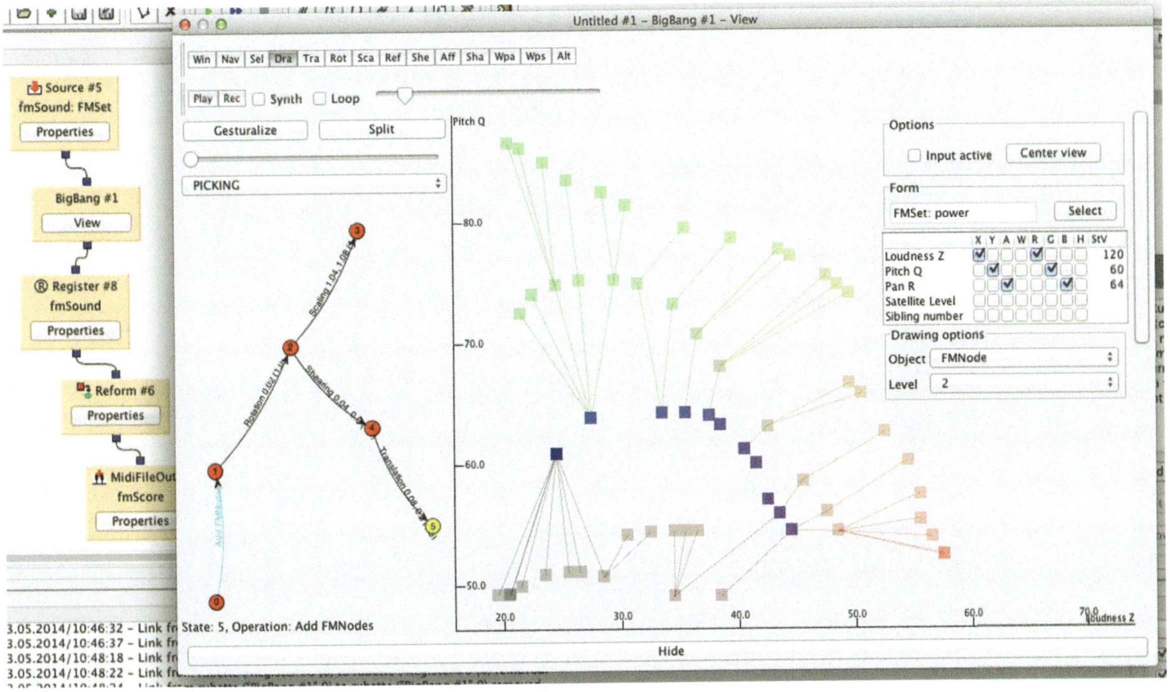

Fig. 70.1. A network including the *BigBang* rubette and its view next to it.

G. Mazzola et al., *The Topos of Music III: Gestures*, Computational Music Science,
https://doi.org/10.1007/978-3-319-64481-3_15

All this is made possible through an architecture based on the three levels of embodiment. The *facts view*, the large area on the right, visualizes the musical objects or facts, which are simply denotators in their coordinate space. It allows for different views of these objects, configurable using the grid of checkboxes on the right. The smaller area on the left, the *process view*, visualizes the graph of the creation process of the music, in a similar way to graphs in transformational theory. Each arrow corresponds to an operation or transformation, while each node corresponds to a state of the composition. The graph can also be interacted with and used as a compositional or improvisational tool itself. Finally, the *gestures*, topic of this volume, are visualized in the facts view, whenever the represented musical objects are interacted with. Any operation or transformation performed in *BigBang* is immediately and continuously sonified and visualized and can be reconstructed at any later stage.

Apart from this functionality, *BigBang* behaves just like any other rubette: whenever the user presses on *Rubato Composer*'s run button, *BigBang* accepts a denotator, either adding it to the one already present or replacing it (depending on the user's settings), and sends its previous denotator to the next rubettes in the network. The rubette can be duplicated, which copies the graph in the process view along with any denotators created as part of the process. This way, users can include a *BigBang* with a defined process in other parts of the network, or other networks, and feed them with different inputs, while the process remains the same. Finally, as any rubette, *BigBang* can be saved along with the network, which again saves processes and corresponding facts.

Before we get to the most relevant aspect of *BigBang* in the context of this book, the gestures, we will introduce the fundamental concepts underlying the software, without which its high-level gestural capabilities would not be possible. The following chapters are structured according to the three levels of embodiment. Chapter 71 will deal with the facts view, explain how arbitrary musical objects can be visualized, how they can be sonified, and how they are represented within *BigBang*. Chapter 72 is devoted to processes and explains what operations and transformations are available, how they can be applied, and how they are represented in *BigBang*'s process view. Finally, in Chapter 73, we will explain how the user's gestures are formalized, i.e. mapped onto transformations and operations, and how the resulting processes can be gesturalized again, so that users can see and hear their composition's evolution in a continuously animated way. These chapters deal mainly with conceptual matters. A more detailed description of the architecture and implementation can be found in [1042].

Check for updates

71

Facts: Denotators and Their Visualization and Sonification

The facts, or objects, that the rubette *BigBang* in RUBATO® Composer deals with are denotators, which can be considered points in the spaces defined by their forms, as introduced in Chapter 6. So far, we have only seen a small portion of the variety of forms that can be defined in RUBATO® *Composer*. However, any conceivable musical or non-musical object can be expressed with forms and denotators, many of them just with the category of modules $\mathbf{Mod}^{@}$. The most recent version of *BigBang* was made compatible with as many forms as possible, even ones that the users may spontaneously choose to define at runtime. In order to handle this as smoothly as possible, we had to find a suitable way of representing denotators within the rubette, which we call `BigBangObjects`.[1] In this chapter, we describe how this works.

71.1 Some Earlier Visualizations of Denotators

In order to understand the evolution of *BigBang*'s visualization system it will be helpful to look at some earlier attempts at visualizing denotators. Several dissertations were based on an implementation of denotators and forms. Stefan Göller's had visualization as its main focus and Gérard Milmeister's included a number of smaller visualization tools.

71.1.1 Göller's PrimaVista Browser

The goal of Göller's dissertation was to visualize denotators "in an active manner: visualization as navigation" [372, p.55]. The result was the sophisticated *PrimaVista Browser*, implemented in Java3D, that featured a three-dimensional visualization in which users could browse denotators in first-person perspective. *PrimaVista* could be customized in many ways using a virtual device, the *Di*, shown in Figure 71.1 [372, p.107].

PrimaVista was capable of representing any type of zero-addressed $Mod^{@}$ denotator as a point or a set of points in $\mathbb{R}^3$ while preserving both order and distance of the original data structure as well as possible. **Limit** and **Colimit** denotators of any dimensionality and their nested subdenotators were folded in a two-step process, first into $\mathbb{R}^n$ then into $\mathbb{R}^3$. Thereby, for any denotator d the mapping $Fold : F(d) \to \mathbb{R}^3$ had to be injective. The first step of this process mapped the values of the **Simple** denotators found in the given denotator hierarchy, regardless of their domain, into $\mathbb{R}^n$ by injecting or projecting each of the individual values into $\mathbb{R}$. A matrix defined which denotator dimensions were mapped into which of the n dimensions of the real codomain space, allowing for both multiple mappings and merging mappings. A so-called *greeking* procedure made sure that only denotator values up to a certain level of hierarchical depth were taken into account, which enabled dealing with circular structures. The second step of the process consisted in folding

[1] Every object that literally exists as a Java object in *BigBang*'s code will be written in verbatim font here, even if we just define them conceptually here.

G. Mazzola et al., *The Topos of Music III: Gestures*, Computational Music Science,
https://doi.org/10.1007/978-3-319-64481-3_16

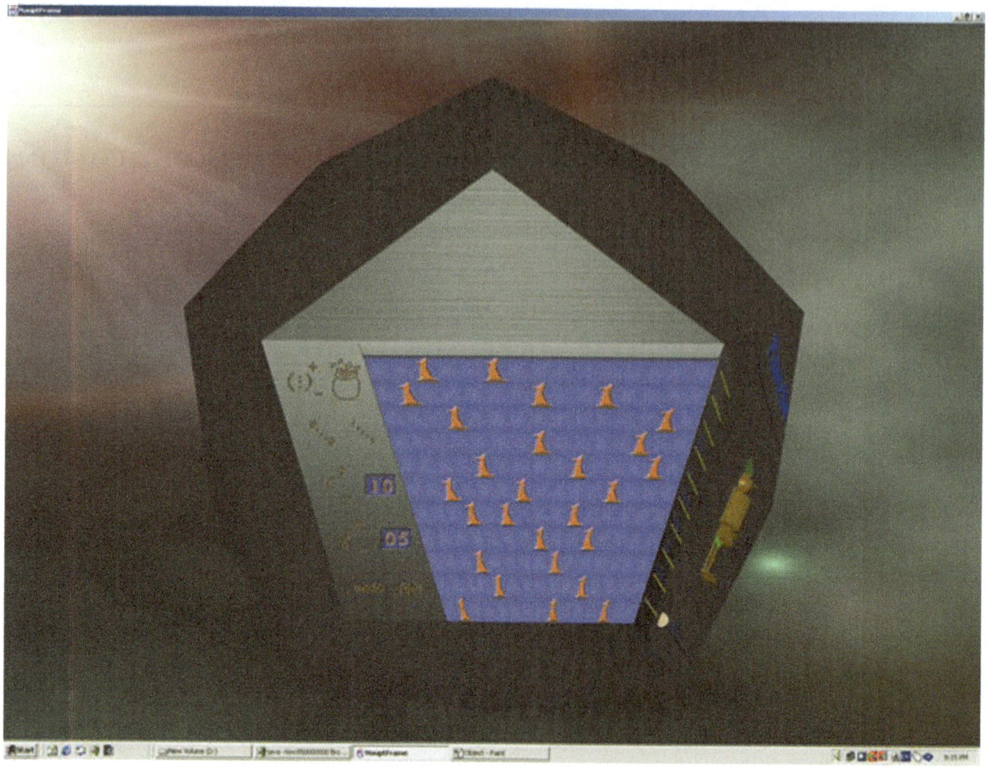

Fig. 71.1. The *Di* of Göller's *PrimaVista* browser.

the obtained $\mathbb{R}^n$ vectors into $\mathbb{R}^3$ by privileging specified dimensions and folding the remaining ones to the mantissa, the decimal digits after the comma.

Göller discussed adventurous ways of visualization, replacing the points in $\mathbb{R}^3$ with complex three-dimensional objects the parts of which he called *satellites*, not to be confused with satellites as they are defined in this context,[2] each of them representing additional characteristics of the represented denotators. Each of Göller's satellites is characterized by the following variable visual parameters: position (x, y, z), rotation vector (rx, ry, rz, α), scale (sx, sy, sz), color $(red, green, blue)$, texture, sound $(pitch, loudness, instrument, sysex)$ [372, p.77]. The most complex object finally implemented is the Pinocchio satellite shown in Figure 71.2. Göller even suggests some satellites may be moving in time to represent parameters such as frequency. This feature was, however, finally not implemented. Another feature not implemented was a generalization of the musical score, where each satellite is associated with sounds that would be played when intersected with a plane, or more generally an algebraic variety, moving in time [372, p.84-5]. Finally, Göller discusses the concept of so-called *cockpits*, where an object's subsatellites become actuators in the form of levers, buttons, or knobs, through which users can change the underlying denotator [372, p.95]. Again, this was not implemented within the scope of his thesis. In addition to this, Göller envisioned ways of transforming and manipulating objects that are similar to the ones of the *BigBang* rubette [372, p.123f].

There are several issues with Göller's approach, some of which explain the difficulties that arose when trying to implement the ideas. First, the folded spaces pose problems of ambiguity in visualization and especially transformation. If one dimension of $\mathbb{R}^3$ represents several denotator dimensions at the same time and the user starts transforming the denotator, it is not intuitively deducible from the visible movement how the denotator values are affected. Representation is often ambiguous, where differences in dimensions folded to the mantissa become only subtly visible and often visually indistinguishable from a simple projection.

[2] See Sections 71.2 and 71.3, where satellites are defined as elements of sub-powersets of a denotator. Also, in Göller's work, there are only two levels: the main satellite and its subsatellites.

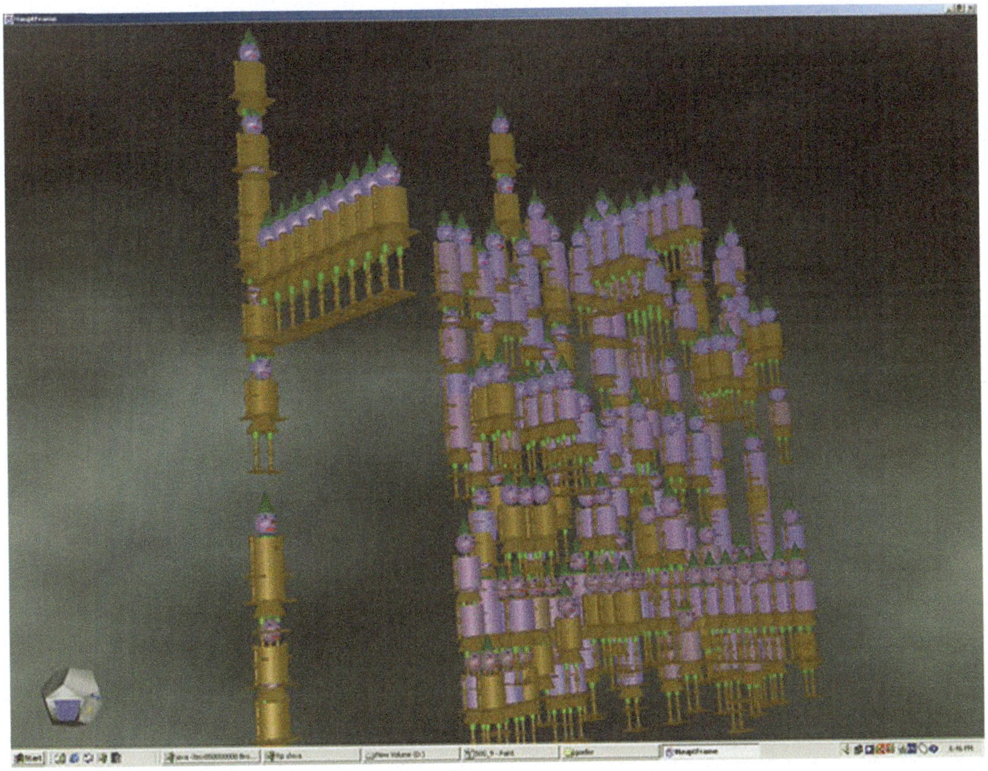

Fig. 71.2. A denotator visualized in *PrimaVista* using Pinocchios (satellites) of varying size and differently positioned extremities (subsatellites).

Second, several simplifications of the denotator concept were made to enable representation within this model. Göller does not, for instance, consider higher-dimensional **Simple** forms, such as ones using modules based on $\mathbb{R}^2$ or $\mathbb{C}$. Third, he mainly visualizes denotators on the topmost level, thereby assuming that it consists of a **Power** [372, p.63]. The BigBang rubette offers solutions to several of these problems, as discussed later.

71.1.2 Milmeister's ScorePlay and Select2D Rubettes

Even though the focus of Milmeister's work lay in building the basic mathematical framework as well as the interface of *RUBATO® Composer*, some of his rubettes offer visualizations of denotators of both general and specific nature. The *ScorePlay* rubette limits itself to *Score* denotators and represents them in piano roll notation. It simply visualizes a *Score* and enables users to play it back at a variable tempo and using different built-in MIDI instruments. It does not allow for any interaction with the represented notes.

The *Select2D* rubette represents any incoming **Power** or **List** denotator as points projected to a customizable two-dimensional coordinate system, the axes of which can be freely associated with any **Simple** denotator somewhere in the denotator hierarchy. Users can then select any number of these points by defining polygons around them (Figure 71.3). The rubette then outputs the subdenotators associated with these points as one runs the network.

Milmeister's rubettes provide several improvements over Göller's software while being more limited in other ways. *ScorePlay* only accepts denotators of one form and visualizes them rigidly. However, its visualization is minimal and based on a standard immediately understandable by the user, which Göller's might not always be. *Select2D*, in addition to **Power** denotators, also accepts **List** denotators, which were only introduced in Milmeister's work [739, p.105]. Furthermore, it is able to represent more types of **Simple** denotators than Göller's, more precisely ones containing free modules over any number ring except for $\mathbb{C}$. Nevertheless, higher-dimensional **Simple** coordinates and product rings can again not be represented.

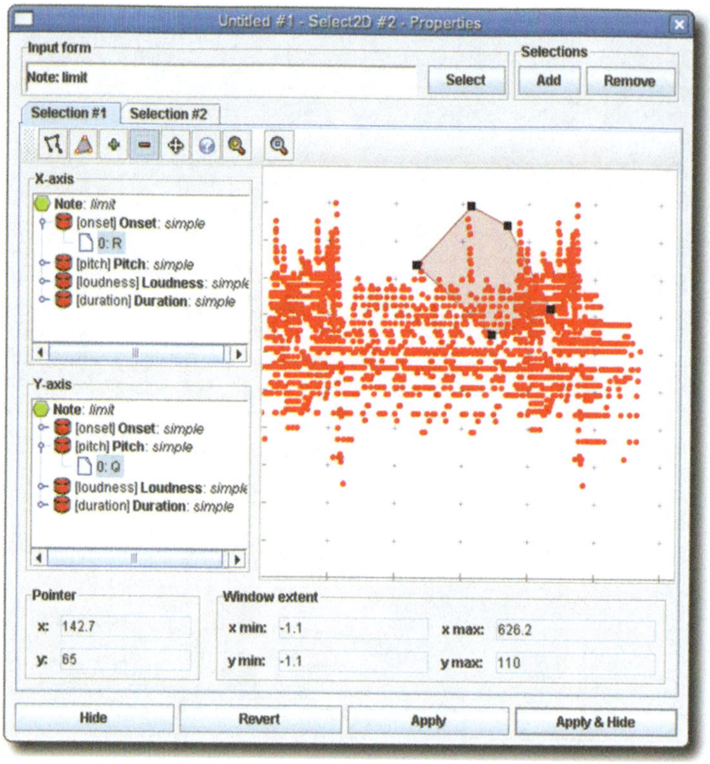

Fig. 71.3. The *Select2D* rubette showing a *Score* denotator on the *Onset* × *Pitch* plane.

Furthermore, the rubette's visualization capabilities do not exceed the representation of points projected to a two-dimensional coordinate system.

71.2 An Early Score-Based Version of BigBang

Initially, the *BigBang* rubette was designed for a small set of score-related denotators. The first version allowed users to handle *Scores* and *MacroScores* and was developed before in the context of an independent research project at the University of Minnesota [1043]. *MacroScore* is a conceptual extension of the form *Score* which we casually defined earlier. It brings hierarchical relationships to *Notes* by imitating the set-theoretical concept of subsets.[3] The form is defined in a circular way, as follows:

$$\begin{aligned}&MacroScore : .\mathbf{Power}(Node),\\&Node : .\mathbf{Limit}(Note, MacroScore),\\&Note : .\mathbf{Limit}(Onset, Pitch, Loudness, Duration, Voice)\end{aligned}$$

Each *Node* associates thus a *Note* with a set of again *Nodes*, each of which again contains a *Note* and a set, and so on. In short, with this construction, each *Note* of a *MacroScore* has a set of so-called *satellites* on a lower hierarchical level. We could go on infinitely, but in order to stop at some point, we give some of the *Nodes* empty sets, thus no satellites. The idea behind this form is that in music, we not only often group objects together and wish to treat them as a unity, but also establish hierarchies between them. A trill, for

[3] This complies with Graeser's notion of counterpoint as "a set of sets of sets of notes", cited in Section 13.1.

instance, consists of a main note, enhanced by some ornamental subnotes.[4] A simplified trill denotator could be defined as follows:

$$\begin{aligned}
&shakeWithTurn : @MacroScore(mainNode),\\
&mainNode : @Node(mainNote, ornamentalNotes),\\
&mainNote : @Note(\ldots),\\
&ornamentalNotes : @MacroScore(\\
&\quad upNode, midNode, upNode, midNode, lowNode, midNode),\\
&upNode : @Node(upNote, emptySet),\\
&upNote : @Note(\ldots),\\
&emptySet : @MacroScore(),\\
&\ldots
\end{aligned}$$

What is crucial to the notion of satellites is that their values are defined relatively to the ones of their anchor. So if for instance the *mainNote* defined above has *Pitch* 60 and its satellite *upNote Pitch* 61, the latter in fact obtains a *Pitch* of 1. If another had *Pitch* 58 it would be defined as −2. This way, if we transform the anchor, all its satellites keep their relative positions to it.

Later on, another form was added to *BigBang*'s vocabulary, *SoundScore*, which combines frequency modulation synthesis with the *MacroScore* concept. Each note, in addition to having satellites, can have modulators which modulate its frequency and change its timbre [1045]. Again, modulators have a relative position to their carrier and would be transformed with it. The form is defined as follows:

$$\begin{aligned}
&SoundScore : .\mathbf{Power}(SoundNode),\\
&SoundNode : .\mathbf{Limit}(SoundNote, SoundScore),\\
&SoundNote : .\mathbf{Limit}(Onset, Pitch, Loudness, Duration, Voice, Modulators),\\
&Modulators : .\mathbf{Power}(SoundNote)
\end{aligned}$$

Denotators of these forms are all based on the same five-dimensional space spanned by the **Simple** forms *Onset*, *Pitch*, *Loudness*, *Duration*, and *Voice* and can thus be visualized the same way. The early *BigBang* rubette did this using a generalized piano roll representation, as we will explain later on.[5] In sum, all of the objects the early *BigBang* rubette dealt with were essentially notes.

71.2.1 The Early BigBang Rubette's View Configurations

The visualization principle of the BigBang rubette [1043, p.4-5] combines elements of both Göller's and Milmeister's models, but focuses on a minimalist appearance aiming towards simplicity and clarity. It generalizes the piano roll notation also used in the ScorePlay rubette (see Section 71.1.2). Notes are represented by rectangles on a two-dimensional plane, just as in a piano roll. However, already in early versions of BigBang, the visual elements of the piano roll were separated from their original function so that they could be arbitrarily assigned to the symbolic dimensions of the represented score denotator. This is reminiscent of the ways Göller's subsatellites could be assigned to any folded denotator dimensions (Section 71.1.1) or of the spacial representation of Milmeister's *Select2D* rubette (Section 71.1.2). A similar method of visualizing was also available in *presto*®'s local views (see Section 50).

In order to do this we defined a set of six visual parameters

$$N = \{X\text{-}Position, Y\text{-}Position, Width, Height, Opacity, Color\}$$

[4] In a similar way, Schenkerian analysis describes background harmonic progressions enhanced by ornamental foreground progressions, which could be represented with *MacroScores* as well. However, we may find forms that are better suited, as will be discussed below.

[5] Piano roll is a standard in music software.

corresponding to the visual properties of piano roll rectangles, along with a set of six note parameters

$$M' = \{Onset, Pitch, Loudness, Duration, Voice, SatelliteLevel\},$$

which corresponds to the **Simple** denotator in *Scores* with the exception of *SatelliteLevel*, which was used to capture the hierarchical level of satellite notes in *MacroScores* and *SoundScores*. We then defined a *view configuration* to be a functional graph $V \subset N \times M'$. This ensures that each screen parameter $n \in N$ is associated with at most one musical parameter $V(n)$ that defines its value, as well as that V does not need to include all $n \in N$. View parameters not covered by V obtain a default value that can be defined by the user. The traditional piano roll notation could be produced by selecting the following pairing (shown in Figure 71.4):

$$V_1 = \{(X\text{-}Position, Onset), (Y\text{-}Position, Pitch), (Width, Duration)\}.$$

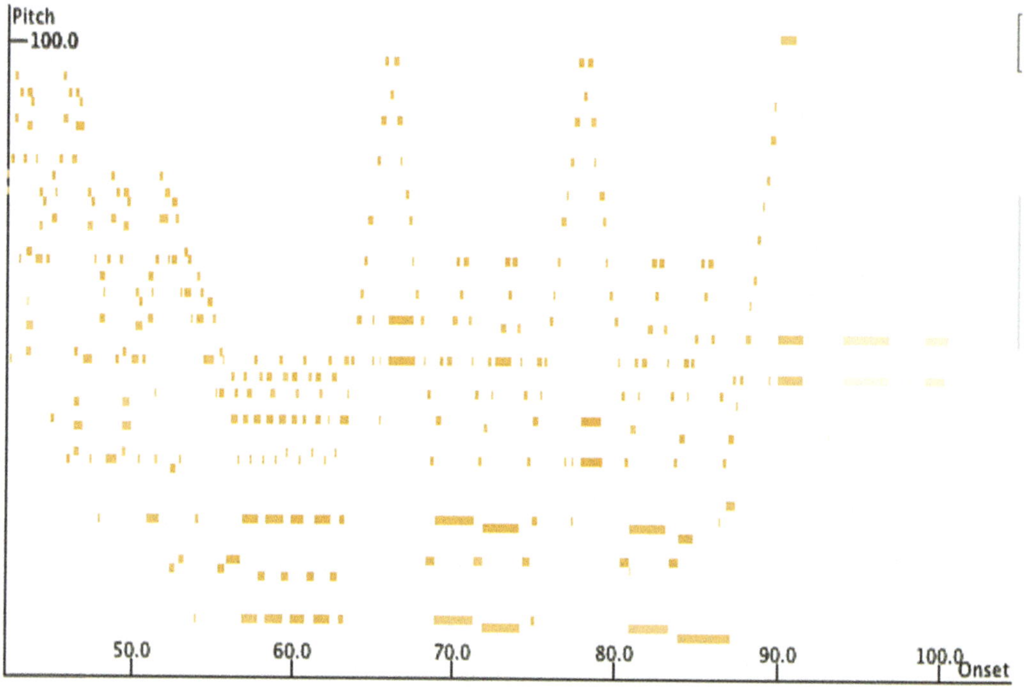

Fig. 71.4. The early *BigBang* rubette showing a *Score* in piano roll notation.

An enhanced version of the piano roll that often appears in software products also uses opacity and color:

$$V_2 = \{(X\text{-}Position, Onset), (Y\text{-}Position, Pitch), (Opacity, Loudness),$$
$$(Width, Duration), (Color, Voice)\}.$$

The possibility of arbitrary pairings, however, also enables more adventurous but possibly also interesting view configurations, such as the following (Figure 71.5):

$$V_3 = \{(X\text{-}Position, Onset), (Y\text{-}Position, Loudness), (Width, Pitch),$$
$$(Color, Onset), (Height, Loudness)\}.$$

Experimenting with such view configurations may be especially valuable for analysis and may lead to a different understanding of given musical data sets.

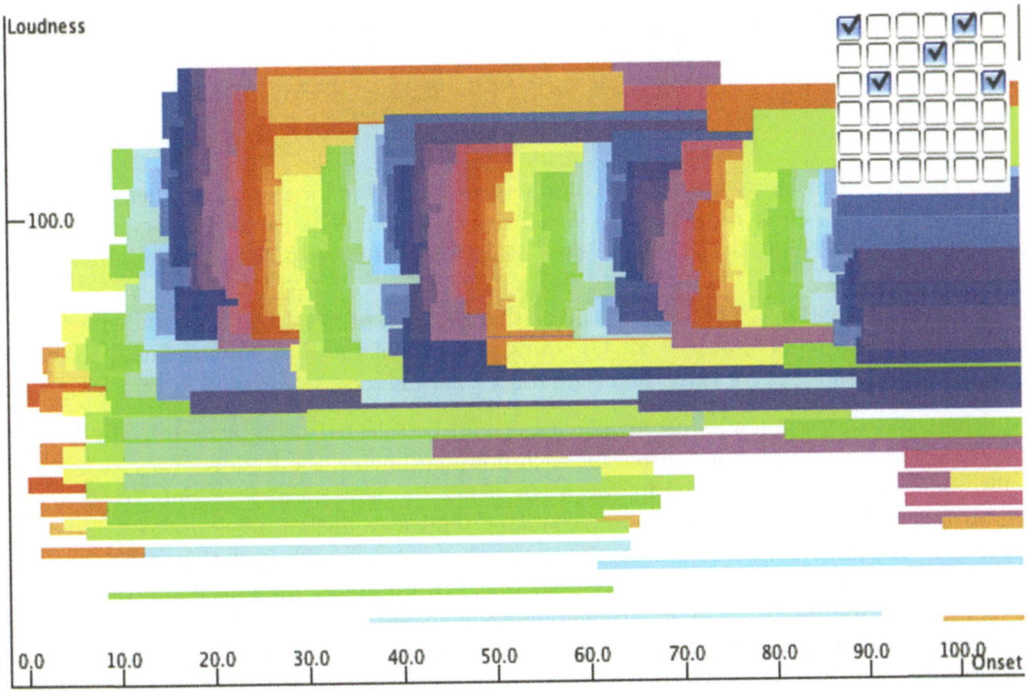

Fig. 71.5. The early *BigBang* rubette visualizing a *Score* in a more experimental way.

Every view parameter can be customized at runtime. Depending on the represented note parameter, it can be useful to ensure that a screen parameter's value does not exceed a specific value range. For example it may look more clear when the rectangle's heights are limited in a way that their areas do not intersect, just as with piano roll notation. Thus, for each $n \in N$, we optionally define min_n and max_n, the minimal and maximal screen values. We then have two options to define the way note parameters are mapped to the screen parameters.

1. If we choose the conversion to be *relative*, the minimal and maximal values of the given note parameters $min_m, max_m, m \in M'$, are determined for the actual score, and then mapped proportionally so that the note with min_m is represented by min_n and the note with max_m by max_n. For this, we use the formula

$$v_n = \frac{v_m - min_m}{max_m - min_m}(max_n - min_n) + min_n,$$

where v_n is the screen value for the note value v_m.

2. On the other hand, *absolute* mapping means that every value with $v_m < min_n$ or $v_m > max_n$ is mapped to a new value, while all other values stay the same, i.e., $v_n = v_m$. For absolute mapping, we have two choices. In *limited* mapping, the values that surpass the limits are given the min_n and max_n values, respectively. The following formula is used:

$$v_n = \begin{cases} min_n, \text{ if } v_m < min_n \\ max_n, \text{ if } v_m > max_n \\ v_m \text{ otherwise.} \end{cases}$$

For *cyclical* mapping, we use the formula

$$v_n = \begin{cases} (v_m \mod (max_n - min_n)) + min_n, \\ \qquad \text{if } v_m < min_n \text{ or } v_m > max_n \\ v_m \text{ otherwise.} \end{cases}$$

This mapping type can be useful for the color screen parameter for example, where it is reasonable to cycle through the color circle repeatedly to visualize a specific note parameter interval, such as an octave in pitch, or a temporal unit, as shown in Figure 71.5, where color visualizes a time interval of length 24, i.e., six 4/4 measures.
With absolute mapping it is possible to leave either or both of the **Limits** as undefined. Accordingly, we assume $min_n = -\infty$ or $max_n = \infty$. Of course, if none of the limits are defined, the visible screen parameters correspond exactly with the original note parameters.

At runtime, the view window's current pairings could be selected using a *matrix of checkboxes* with a column for each screen parameter and a row for each note parameter, see Figure 71.5.

Satellite relations can be displayed in two ways. First, the note parameter *SatelliteLevel*, mentioned above, can be assigned to any arbitrary visual parameter. This way, anchor notes are associated with integer value 0, first-level satellites with 1, and so on. On the other hand, satellite relations may also be displayed as lines between the centers of two note objects so that every note has lines leading to each of its direct satellites, as shown in Figure 71.6.[6] As mentioned above, since all anchors and satellites in *MacroScore* and *SoundScore* denotators are notes, they can be represented in the same space.

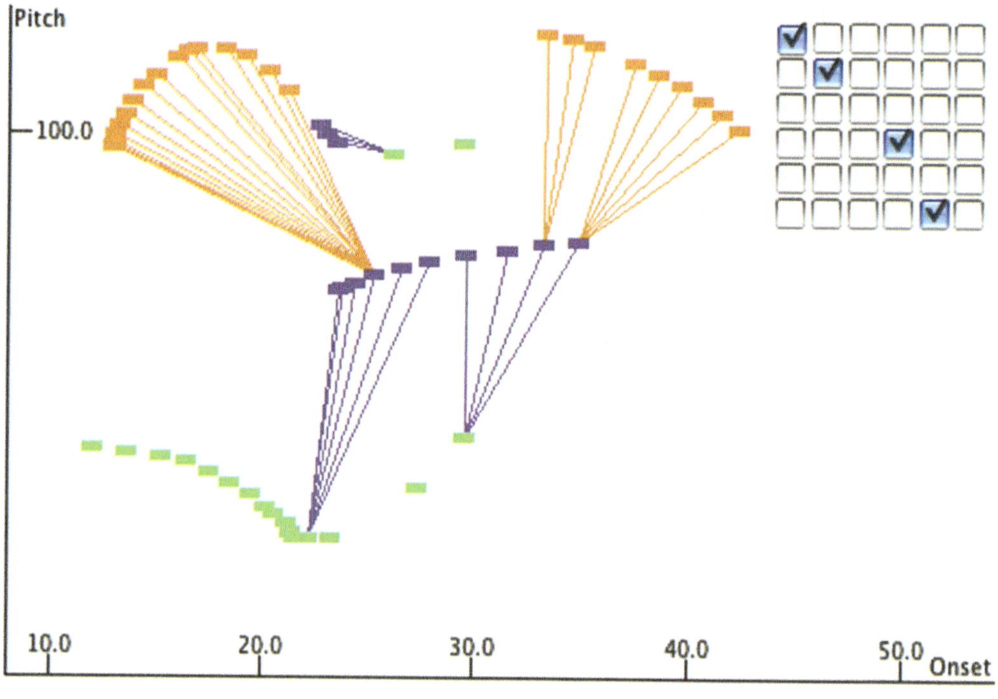

Fig. 71.6. The early *BigBang* rubette showing a *MacroScore* with two levels of satellites.

[6] This is a notion of satellites significantly different from Göller's (Section 71.1.1). While Goeller uses the term to denote movable parts of objects and represents them as denotators, but here we use it to speak of circular denotator structures (also see Note 2).

71.2.2 Navigating Denotators

Users can navigate this two-dimensional space not only by changing their view of the space by choosing different note parameters for the x and y view parameters, but also by changing their viewpoint by scrolling the surface and zooming in and out without limitations. This is similar to Göller's *PrimaVista*, but using two instead of three dimensions. However, users can also open several of these views simultaneously and choose different perspectives on the composition. This is especially valuable when performing transformations in one view while observing how the composition is affected from the other perspective.

71.2.3 Sonifying Score-Based Denotators

In early *BigBang*, denotators could not only be visualized but also sonified. Even though this may be done using another, specialized rubette such as *ScorePlay*, we decided to include this functionality within *BigBang*. The main reason for this was the gestural interaction concept, where immediate auditive feedback is key, in order to evoke a sense of continuity of motion. Users have to be able to judge musical structures by ear while they are creating them, and the use of an external rubette would have slowed down the process. A second reason was that many of the possible musical structures in early *BigBang* were micro-tonal, for which MIDI feedback, as implemented in *ScorePlay*, is unsuited since it is strictly chromatic.[7] The extension of *BigBang* for *SoundScores* was another reason, for now timbre was part of the musical objects and had to be judged while it was defined.

Since all the structures dealt with in early *BigBang* were *Score*-based denotators, sonification was rather straightforward. All the objects that had to be played were *Notes* that existed in the same space. They were simply played back in time, giving the user control over tempo. The microtonal and frequency modulation structures of *SoundScores* made it necessary for a synthesizer to be used. For each note, a synthesizer object, a so-called `JSynNote` footnoteJSyn is the name of the synthesizer framework we decided to use, as will be explained later on. was created by converting symbolic time, pitch, and loudness into the physical parameters time, frequency, and amplitude.

Outside *BigBang*, *MacroScores* usually have to be converted into *Scores* in order to be played back, a process called *flatten* (see next chapter). In early *BigBang*, this happened in the background, since it would have significantly slowed down the composition process. Satellites were simply converted into additional `JSynNotes` accordingly. Modulators in *SoundScores*, however, became modulators of `JSynNotes`. There were two options for playing back modulators: either their temporal parameters were ignored and they simply played whenever their carrier was playing, or they only modulated their carrier according to their own onset and duration. In the latter case, users had to make sure the anchor notes were playing at the same time as their modulators, but they also had the chance to create temporally varying configurations of modulators for a single note.

71.3 BigBangObjects and Visualization of Arbitrary $Mod^{@}$ Denotators

Despite its customizability, the view concept of the early *BigBang* rubette was first designed to represent *Score*, *MacroScore*, and *SoundScore* denotators, which are all based on the same musical space: (**Power** of...) **Power** of **Limit**. There, the view concept has proven its viability, compared to other concepts such as the ones discussed above. Now how can this concept be generalized so that *BigBang* can accept any denotator?

In this section we describe how we can do this for a major part of the spaces available in Milmeister's version of RUBATO® Composer, which are all based on elements of the topos $Mod^{@}$ over the category of modules, as described earlier.[8] The number of denotator types capable of being represented by the new

[7] The use of pitch bend is an option for monophonic material, but limited as soon as several notes have to be bent in different ways.

[8] This procedure is also described in [1048, p.3], as well as briefly in [1047].

BigBang rubette is significantly higher than the two comparable modules *PrimaVista* and *Select2D*. Nevertheless, for the time being we restrict ourselves to 0-addressed denotators and focus on number-based modules. We exclude both modules based on polynomial rings and ones based on string rings, since their visualization may differ markedly and will be left to future projects.

71.3.1 A Look at Potential Visual Characteristics of Form Types

As a starting point we need to reflect on the role of the five form types **Simple**, **Limit**, **Colimit**, **Power**, and **List** and the way they can best be visualized. Each of these types implies another visual quality that may be combined with the others. These qualities in early *BigBang Scores* were shown as clusters of rectangles (**Power**) within a coordinate system (**Limit**) of five axes (**Simple**), which could in turn be variably shown as any of six visual dimensions (x-position, y-position, width, height, color, opacity). Three of the five form types are involved here. The **Simple**s in a *Note* are based on free modules on a number ring and can thus easily be represented by one number axis or one of the other visual properties. However, *Rubato Composer* allows for many more types of **Simple**s, each of which must be considered here.

71.3.1.1 Simple Denotators

Simple denotators are crucial to a system of visualization, since they are the only denotators that stand for a specific numerical value in a space. Basically, every form that will be used in a practical way should contain **Simple**s. This despite the fact that it is possible to conceive more pathological forms, such as the circular form that describes sets as sets of sets:

$$Set : .\mathbf{Power}(Set).$$

Such forms will be of little use in our context, since anything to be represented and especially transformed needs to contain specific numerical values. We can thus declare a first rule here:

Rule 1 *In our system denotators will only be represented if they contain at least one* **Simple** *denotator somewhere in their structure.*

With the system, **Simple** denotators over the following modules can be represented:

Free Modules over Number Rings

The most straightforward type of modules are the free modules based on number rings such as $\mathbb{Z}, \mathbb{Q}, \mathbb{R}$, or $\mathbb{C}$. Elements of the first three are typically represented along an axis, whereas ones of the last on a two-dimensional Cartesian system. For modules $\mathbb{Z}^n, \mathbb{Q}^n, \mathbb{R}^n$, and $\mathbb{C}^n$ an n-dimensional or $2n$-dimensional system of real axes will be appropriate.

Furthermore, as shown in Section 71.2.1, as long as all values of a specific denotator are known and finite, dimensions of free modules over number rings can equally be represented by other visual parameters, such as an object's width, height, color, etc. Elements of the free module over $\mathbb{C}$, for instance, could convincingly be represented as width and height of objects.

Quotient Modules

For free modules over quotient modules such as $\mathbb{Z}_m = \mathbb{Z}/m\mathbb{Z}$, $\mathbb{Q}_m = \mathbb{Q}/m\mathbb{Z}$, $\mathbb{R}_m = \mathbb{R}/m\mathbb{Z}$, and $\mathbb{C}_m = \mathbb{C}/m\mathbb{Z}$ we choose a manner of representation that corresponds to the one introduced in the previous section, where values are simply projected on a one- or two-dimensional coordinate system. However, instead of being potentially infinite, the axes maximally show the numbers of the interval $[0, m[$, which makes zooming out beyond this point impossible. This works in an analogous way for other view parameters that do not allow cyclical representation, such as width, height, and opacity. Of the defined visual parameters, only color allows for cyclical representation, in analogy to the color circle. Again, for $\mathbb{Z}^n_m, \mathbb{Q}^n_m, \mathbb{R}^n_m$, and $\mathbb{C}^n_m$, the system can be extended to be n-dimensional or $2n$-dimensional.

Modules over Product Rings and Direct Sums of Modules over Quotient Modules

Representation is straightforward for direct sums of any of the quotient modules discussed so far. Each of the factors is independently associated with one of the view parameters. For example, for $\mathbb{Z} \times \mathbb{R}_7$ we might choose to represent the $\mathbb{Z}$ part with the x-axis and the $\mathbb{R}_7$ part with color.

Remark: More general modules which are not derived from direct sums of such quotient modules are not yet dealt with.

71.3.1.2 Limit Denotators

The fact that **Limit**s are products or conjunctions makes them always representable in the conjoined space, i.e., the cartesian product of the spaces of their factors. The most simple case is a **Limit** of **Simple** denotators, just as with our common *Score* denotators. *Notes* can be represented in $Onset \times Pitch \times Loudness \times Duration \times Voice$. This is even possible if the same subspaces appear in several times. For instance, if we define a form

$$Dyad : .Limit(Pitch, Pitch),$$

its denotators are representable in the space $Pitch \times Pitch$. This also works in cases where the factors of a **Limit** are not directly **Simple**.

71.3.1.3 Colimit Denotators

Colimits, disjunctions or coproducts, are again representable in the product space of their cofactors, even if they then typically do not have defined values in all of the product's dimensions. For "missing" dimensions, we set standard values, so that the denotators are represented on a hyperplane in the entire product space. The case where cofactors share common subspaces is especially interesting, since these subspaces will always be populated.

An example will clarify this: we assume a form *EulerScore* which consists of *EulerNotes* and *Rests*, which share the **Simple** forms *Onset* and *Duration*. The product space of all cofactor spaces is $Onset \times EulerPitch \times Loudness \times Duration$. While *EulerNotes* fill the entire space, *Rests* are simply represented on the $Onset \times Duration$ plane. Thus, even though *EulerNotes* and *Rests* are actually separated by a coproduct, both can be shown in the same space.

71.3.1.4 Power and List Denotators

Power and **List** forms define sets and ordered sets of distinct objects on any hierarchical level. In practice we typically encounter them on the topmost level as for instance with all the forms supported by early *BigBang*, *Score*, *MacroScore*, and *SoundScore*. However, it is also conceivable that they occur only at lower levels, as in Mariana Montiel's more detailed score form, which is defined as [760]

$$\begin{aligned} Score' : .\mathbf{Limit}(&BibInfo, Signatures, Tempi, Lines, GeneralNotes, \\ &GroupArticulations, Dynamics). \end{aligned}$$

There, **Power**s appear in almost all the coordinator forms, but not at the top level. In this case we can for instance see all *BarLine* or *Slur* denotators that appear on lower levels as indirect satellites of our main $Score'$.

Power denotators can always be represented as a set of points in the space of their coordinate. An *EulerScore*, for instance, can be shown as a cloud of objects in the *EulerNoteOrRest* space described above. **List** denotators can be shown the same way, however, at the expense of the order of their elements, for it may contradict the spatial organization. In any case, **Power** and **List** forms are in fact the main constructs that define the discrimination of distinct visual objects. Wherever they occur, we have the opportunity to define as many elements as we would like.

71.3.2 From a General View Concept to BigBang Objects

From these characteristics we can imply that all we need to have for a representation of any denotators is a conjunction of the **Simple** spaces and a visualization of clouds of objects within them. These objects can be represented in just the same way as the ones in the generalized piano roll described above, as multidimensional rectangles. Whenever a denotator enters *BigBang*, the visualization space is reset based on its form, and users have the possibility to select any form space and start drawing objects, as will be described below.

We arrive at the core part of our generalization. In short, the representation of any arbitrary denotator relies on the fundamental difference between the various types of compound forms, **Limit**, **Colimit**, and **Power**. We propose a novel system of classification that generalizes the previous notion of anchors and satellites, based on occurrences of **Power** denotators. For this, we maintain the following rules:

Rule 2 The general visualization space consists of the cartesian product of all **Simple** form spaces appearing anywhere in the anatomy of the given form. For instance, if we obtain a *MacroScore* denotator of any hierarchical depth, this is *Onset* × *Pitch* × *Loudness* × *Duration* × *Voice*.

Rule 3 Any **Simple** form X the module of which has dimension $n > 1$ is broken up into its one-dimensional factors $X_1, \ldots, X_n$. The visual axes are named after the dimension they represent, i.e. X_n, or X if $n = 1$.

Rule 4 If the same **Simple** form occurs several times in a **Limit**, it is taken to occur several times in the product as well. For instance, the product space of *Dyad* is *Pitch* × *Pitch*. However, if the same **Simple** form occurs at different positions in a **Colimit**, this is not the case. For instance, **Colimit** of *Pitch* and *Pitch* results in the space *Pitch*. This renders the space more simple, but we also lose some information. This loss can be regained thanks to an additional spatial dimension, *cofactor index*, as described under Rule 7.

Rule 5 **Power** or **List** denotators anywhere in the anatomy define an instantiation of distinct visual objects represented in the conjoined space. Objects at a deeper level, i.e., contained in a subordinate **Power** or **List**, are considered *satellites* of the higher-level object and their relationship is visually represented by a connecting line. For example, *SoundScore* objects formerly considered modulators are now visually no different from regular satellites.

Rule 6 Given a view configuration, the only displayed objects are denotators that contain *at least one* **Simple** form currently associated with one of the visual axes. However, if an object is a satellite and one of the **Simple** forms associated with the axes occurs anywhere in its parental hierarchy, it is represented at exactly that value.

Rule 7 If there is an occurrence of either **Colimits** or *satellites*, additional dimensions are added to the ones defined in Rules 1-3. For **Colimit** we add *cofactor index*, and for satellites *sibling index* and *satellite level*.[9] These dimensions are calculated for each object and can be visualized in the same manner as the other ones. For instance, associating satellite level with y-position facilitates the selection of all denotators at distinct positions of the satellite hierarchy.

We call the objects defined by these rules `BigBangObjects`. They are not only visual entities, but they are the entities that the *BigBang* rubette deals with in every respect. All operations and transformations available in *BigBang* are applied to sets of `BigBangObjects`, as we will see in the next chapter. The consequence is that we simplify the structure of forms and denotators significantly, so that if we, for instance, are handling denotators of a form defined as **Limit** of **Limit** of **Limit** and so on, we can treat it as a single object. New objects are broken up only if there are **Powers** or **Lists** in the hierarchy. We can thus for instance claim that in *BigBang* we assume that

$$\mathbf{Limit}(A, \mathbf{Limit}(B, \mathbf{Limit}(C, D))) = \mathbf{Limit}(A, B, C, D).$$

71.3.2.1 Implications for Satellites

One of the main innovations of these definitions is a new notion of the concept of satellites. Previously, the term was uniquely used to describe *Notes* in a *MacroScore* that are hierarchically dependent on other

[9] Already present in the early *BigBang*, as seen in Section 71.2.1.

Notes. For instance, the analogous construction of *Modulators* in *SoundScores* was not referred to in this way; neither was the relationship represented the same way as satellites are [1045]. Following Rule 5 above, *Modulators* are now equally considered satellites and represented in precisely the same way. Another new aspect of this is that now satellites do not technically have to have a shared space with their anchor. For instance, if we define

$$\begin{aligned} &HarmonicSpectrum : .\mathbf{Limit}(Pitch, Overtones),\\ &Overtones : .\mathbf{Power}(Overtone),\\ &Overtone : .\mathbf{Limit}(OvertoneIndex, Loudness),\\ &OvertoneIndex : .\mathbf{Simple}(\mathbb{Z}), \end{aligned}$$

Overtones do not have a *Pitch* themselves, but merely a *Loudness*. Because of Rule 6, however, if we choose to see *Loudness* × *Pitch* as the axes of a view configuration, the *Overtones* are represented above their anchor. An example visualization of this form will be shown below.

Above, we discussed how satellites and modulators were defined relatively to their anchors (Section 71.2). This can also be generalized for the new notion of satellites. We add another rule:

Rule 8 Given a **Simple** form F, every denotator $d_i : @F$ in a satellite `BigBangObject`, i being its index in case the satellite contains several denotators of form F, is defined in a relative way to $d_i@F$ in its anchor, if there is such a denotator.

For instance, if we define

$$\begin{aligned} &MacroDyad : .\mathbf{Power}(DyadNode),\\ &DyadNode : .\mathbf{Limit}(Dyad, MacroDyad), \end{aligned}$$

the first *Pitch* in each satellite *Dyad* is defined relatively to the first *Pitch* in its anchor, and the second *Pitch* in each satellite relatively to the second *Pitch* in the anchor. On the other hand, in a *HarmonicSpectrum* none of the satellites share **Simple** denotators with their anchor and are thus defined absolutely.

71.3.3 New Visual Dimensions

The facts view of the new *BigBang* maintains all the features of the early *BigBang* and can still be navigated the same way as described in Section 71.2.2. However, the newest version allows independent zooming in and out horizontally and vertically, when the shift or alt keys are pressed. It also features some additional view parameters. There is now an option to use, instead of hue (*Color*) values, RGB values for color, in a similar way as that in *PrimaVista*. This adds more visual variety at the expense of the cyclical nature of hue, and is especially beneficial when working with data types other than musical ones, such as images. The new view parameters vector thus looks as follows:

$$N' = \{X\text{-}Position, Y\text{-}Position, Width, Height, Alpha, Red, Green, Blue\}.$$

In the future, more visual characteristics can easily be added, such as varying shapes, texture, or a third dimension.

The former note parameters, in turn, now called *denotator parameters*, include *SiblingNumber* and *ColimitIndex* where appropriate and vary according to the input or chosen form. The former identifies the index of a denotator in its **Power** or **List**, whereas the latter refers to an index based on all possible combinations of **Colimit** coordinates. For instance, for an object form

$$\mathbf{Colimit}(\mathbf{Colimit}(X_0, X_1), \mathbf{Colimit}(X_2, X_3, X_4), X_5),$$

where $X_0, ..., X_5$ are any other forms not containing **Colimits**, we get six possible configurations: an object containing X_0 gets index 0, one containing X_1 gets 1, and so on.

For *EulerScore* denotators, for instance, the entire denotator parameters look as follows:

$$M_{EulerScore} = \{Onset, EulerPitch1, EulerPitch2, EulerPitch3, \\ Loudness, Duration, ColimitIndex\}.$$

The three-dimensional space of *EulerPitch* is broken up into its three constituent dimensions, and *ColimitIndex* is added, with two potential values: 0 for *EulerNoteOrRests* containing an *EulerNote*, 1 for *EulerNoteOrRests* with a *Rest*.

71.4 The Sonification of BigBangObjects

As seen above in Section 71.2.3, in the early *BigBang* rubette, sonification was relatively straightforward, since all objects that had to be dealt with existed in the same five-dimensional space. For the new *BigBang*, this concept had to be generalized as well. For users to be able to sonify a multitude of denotators, even ones they define themselves, the sonification system had to become more modular.

Our solution generalizes the `JSynNotes` described above into `JSynObjects`, which can contain any number of a set of standard musical parameters. Each `BigBangObject` is converted into a `JSynObject`, by searching for occurrences of these musical parameters anywhere in their anatomy. Any parameters necessary for a sounding result subsequently obtain a standard value. For instance, if we play back a **Simple** denotator *Pitch*, a `JSynObject` is created with a standard *Loudness*, *Onset*, and *Duration*, so that it is audible. Especially *Onset* and *Duration* are relevant in this case. The standard values assigned for temporal parameters are chosen such that the object plays continuously for as long as the denotator is being played. This is particularly interesting when the denotator is transformed, which results in continuously sounding microtonal sweeping.

`JSynObjects` can also have multiple pitches, in order to work with denotators such as *Dyad*, as defined in Section 71.3.1.2, or other user-defined types that might describe chords, and so on. Some of the recognized simple forms so far are all note parameters (*Onset*, *Pitch*, *Loudness*, *Duration*, *Voice*), as well as *BeatClass*, *ChromaticPitch*, *PitchClass*, *TriadQuality*, *OvertoneIndex*, *Rate*, *Pan*, and *OperationName*. *Rate* replaces *Onset* by defining the rate at which a `JSynObject` is repeatedly played, *OperationName* distinguishes between frequency modulation, ring modulation, and additive synthesis, and *TriadQuality* adds an appropriate triad above each *Pitch* in the object, assuming that they are root notes. Some of the other forms are discussed below, along with examples of the visualization of their denotators.

Another recent addition is the option of having everything played back through MIDI, either with Java's internal MIDI player, or by sending live MIDI data to any other application or device, via IAC bus or MIDI interface. MIDI is event-based and thus problematic for playing continuous objects without temporal parameters. There are two solutions to this problem implemented so far. Either, objects are repeated continuously at a specified rate, or a note-off event is only sent when an object is replaced by another. In the latter case, note ons are only sent again once a denotator is transformed.

In order to play back the composition in *BigBang*, users can press the play button in the lower toolbar. If the denotator has a temporal existence, i.e., it contains *Onsets*, it can be looped, where the player automatically determines the loop size to be the entire composition. In addition to this, any musical denotator in *BigBang* can be played back by using an external MIDI controller such as a keyboard controller. Each MIDI key of such a controller triggers one performance of the denotator, i.e., a one-shot temporal playback, a loop, or a continuous playback, depending on the denotator. Middle C (60) corresponds to the visible denotator, while all other keys trigger transposed versions, e.g. a half step up for 61, etc. This is especially practical when designing sounds, i.e., denotators without *Onset* or *Duration*, such as the *HarmonicSpectrum* form defined above. This way, users can design sounds and immediately play the keyboard with them, just as with a regular synthesizer.

For the future, this system of sonification could be extended in order to work in a similar way to view configurations. For now, whenever a new **Simple** form is introduced that should be sonified in a novel way, the system has to be adjusted accordingly. With a free association of any **Simple** form with a sonic parameter, just as is done for the visual system, users can experiment with spontaneously performing parameter exchanges, or with sonifying non-musical forms.

71.5 Examples of Forms and the Visualization of Their Denotators

In this chapter, we have discussed what the objects on *BigBang*'s factual level are and how they are visualized and sonified. It is now time to give some specific examples of forms that can potentially be defined and show how their denotators are visualized. Sonification will have to be left to the readers to try themselves. Anything we feed the new BigBang rubette will be analyzed and visualized as described above. Users may also select a form within *BigBang* upon which the facts view is cleared and they may simply start drawing denotators, as will be described in the next chapter.

We will start with some simple constructs from set theory, move to tonal constructs, and finally give some examples from computer music and sound design.

71.5.1 Some Set-Theoretical Structures

The most basic construct to be represented is necessarily a single **Simple** denotator. For instance, if we input a *Pitch*, the space is merely one-dimensional, but it can be represented in various visual dimensions simultaneously. Figure 71.7 shows the pitch $middleC : @Pitch(60)$ – C4 is MIDI pitch 60 – being represented in every possible visual dimension in RBG mode,[10] however reasonable this may be. X, y, width, height, alpha, red, green, and blue, all represent the value 60, depending on the min_m, max_m defined (see Section 71.2.1).

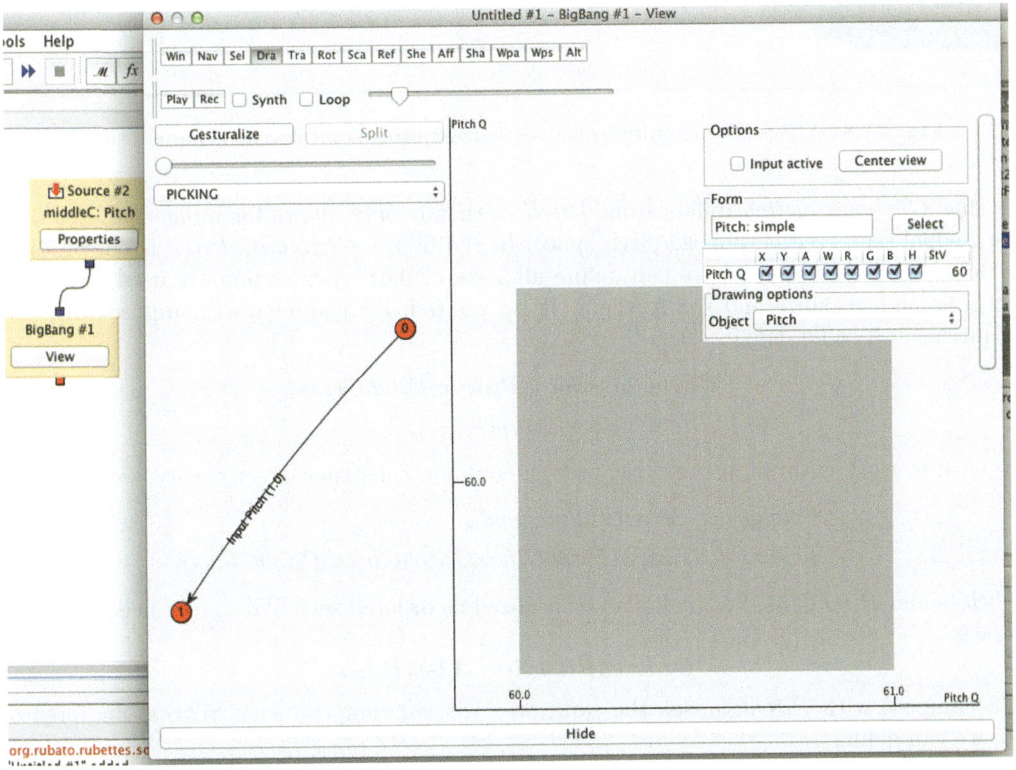

Fig. 71.7. The new *BigBang* rubette visualizing a *Pitch* denotator in every visual dimension.

[10] Explained in Section 71.3.3.

For a **Power** of a **Simple**, we get a cloud of values. Figure 71.8 shows an example of

$$PitchSet : .\mathbf{Power}(ChromaticPitch),$$
$$ChromaticPitch : .\mathbf{Simple}(\mathbb{Z}).$$

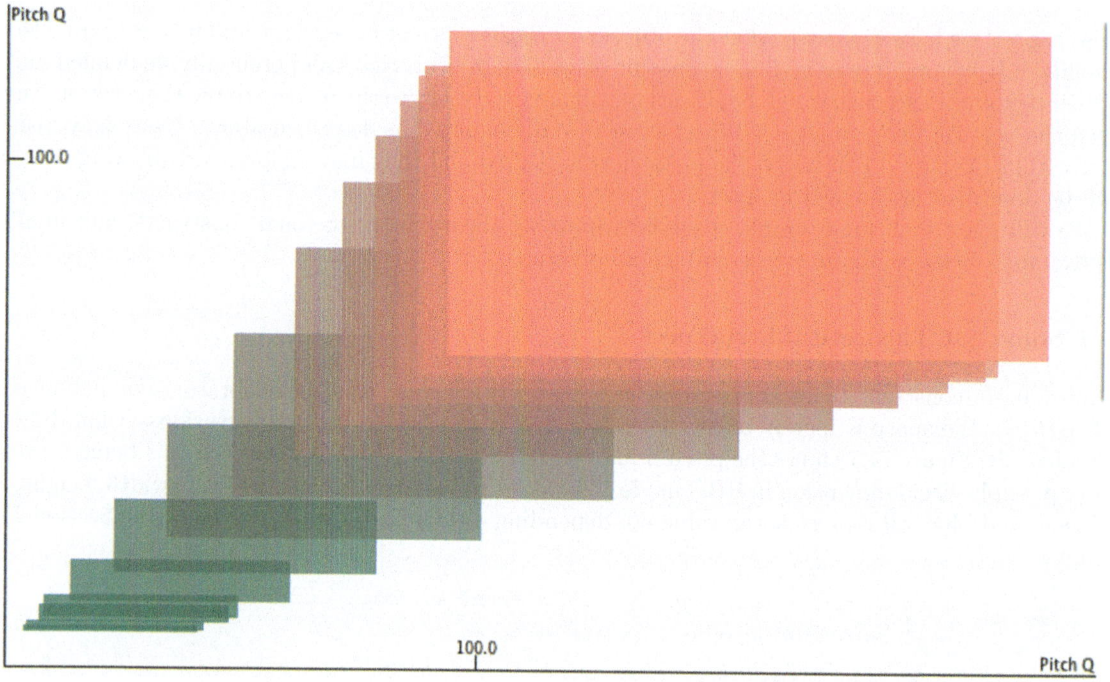

Fig. 71.8. A *PitchSet* simultaneously visualized using several visual characteristics.

Note that *ChromaticPitch* differs from *Pitch* in that it only allows for integer values, which models the Western equal-tempered chromatic pitch space. In the figure, *ChromaticPitch* is shown on both axes, color, width and height. This way, we can define all sorts of datatypes commonly used in music theory or sound synthesis and visualize and sonify them. If we wanted, for instance, to compose with pitch classes instead of pitches, we could define

$$PitchClassSet : .\mathbf{Power}(PitchClass),$$
$$PitchClass : .Simple(\mathbb{Z}_{12}).$$

If we wish to work with pitch-class trichords, a common construct in set theory, we can define

$$Trichords : .\mathbf{Power}(Trichord),$$
$$Trichord : .\mathbf{Limit}(PitchClass, PitchClass, PitchClass).$$

PitchSets and *PitchClassSets* can also be realized as ordered sets. We simply need to replace **Power** with **List**, e.g.

$$OrderedPitchSet : .\mathbf{List}(Pitch).$$

In order to compose with *PitchClasses* the same way we can compose with *Scores*, i.e., create temporal structures, we can define

$$PitchClassScore : .\mathbf{Power}(PitchClassNote),$$
$$PitchClassNote : .\mathbf{Limit}(Onset, PitchClass, Loudness, Duration, Voice),$$

which is then visualized as shown in Figure 71.9.

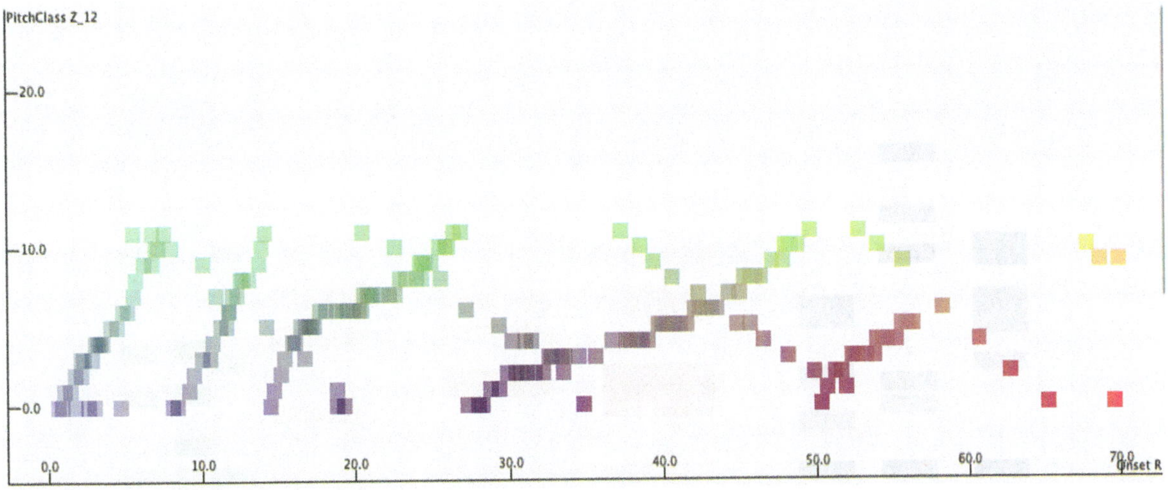

Fig. 71.9. A *PitchClassScore* drawn with ascending and descending lines to show the cyclicality of the space.

71.5.2 Tonal and Transformational Theory

The next few examples imitate spaces and constructs from transformational theory and traditional music theory.[11] For a model of triads, as they are often used in transformational theory, we define

$$Triad : .\mathbf{Power}(Pitch, TriadQuality),$$
$$TriadQuality : .\mathbf{Simple}(\mathbb{Z}_4),$$

where Quality stands for one of the four standard qualities in tonal music: diminished, minor, major, and augmented.

More generally, a simplified notion of chord progressions can be implemented as follows:

$$Progression : .\mathbf{List}(Chord),$$
$$Chord : .\mathbf{Limit}(Onset, PitchSet, Loudness, Duration),$$

assuming that all members of a chord have the same temporal and dynamic qualities. In so doing, the pitches of a chord are actually satellites and thus also visualized this way, as can be seen in Figure 71.10. From there, we can also define hierarchical chord progressions the same way as we did above for *Score* or *Dyad*. For instance, we can define

$$MacroProgression : .\mathbf{List}(ChordNode),$$
$$ChordNode : .\mathbf{Limit}(Chord, MacroProgression).$$

This way, each chord can have ornamental progressions, just as we know it from Schenkerian theory. If a main progression is transposed, its ornamental progressions, defined in a relative way to them, are transposed with it. The next chapter will clarify what this means.

Figure 71.11 shows an example of the depiction of **Colimits**. It shows a denotator of a form similar to *EulerScore*, but with regular *Pitch* and an additional *Voice* parameter, thus simply using regular *Notes* and *Rests*. In the image we see that all the rests are depicted at *Pitch* 0, since they do not contain a *Pitch*. If we chose to depict the denotator on the *Onset* × *Duration* plane, the rests would also be shown in two dimensions.

A final example illustrates a way we can introduce rhythmical relationships other than using *Onset*. If we write

[11] Some of them were described in [1048].

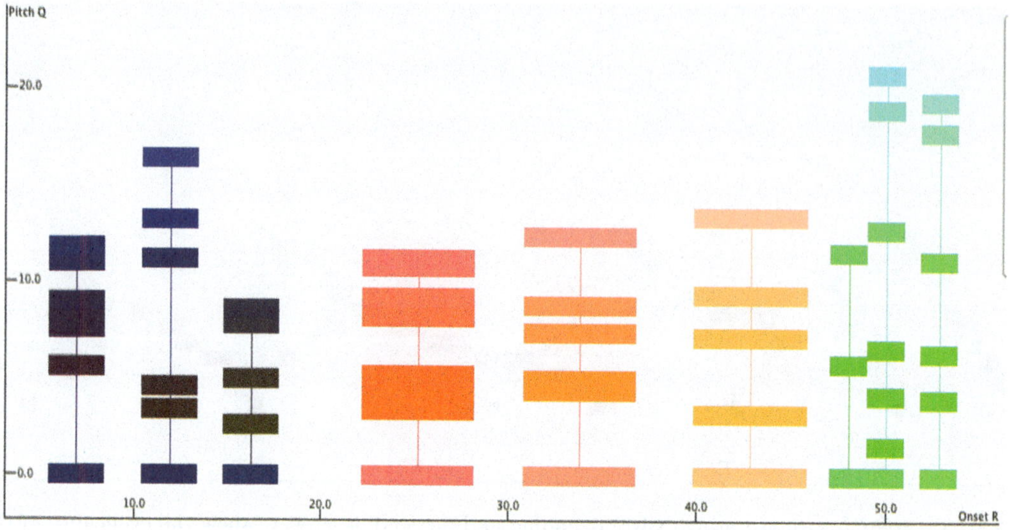

Fig. 71.10. A *Progression* where pitches adopt the visual characteristics of their anchor chord.

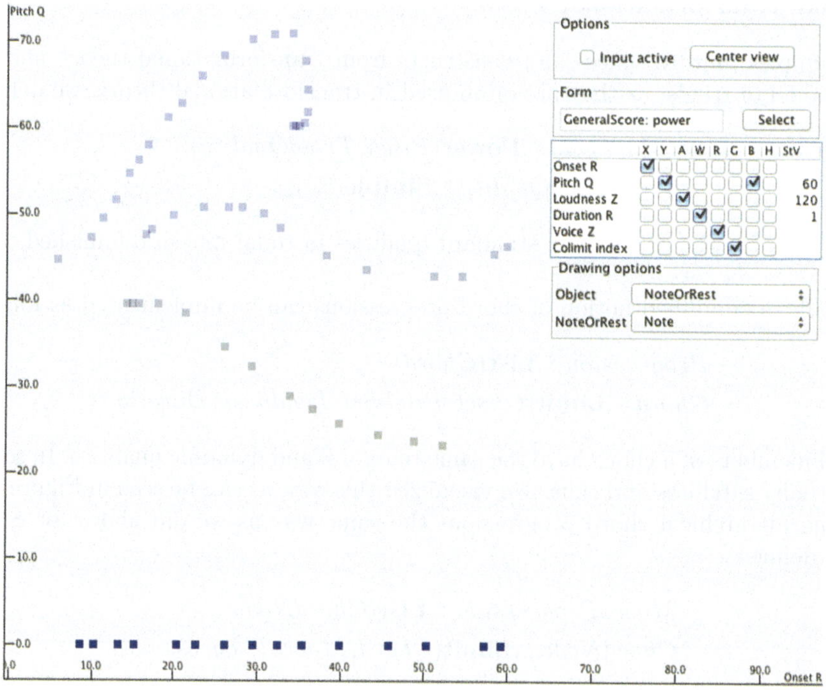

Fig. 71.11. A *GeneralScore* with some *Notes* and *Rests* shown on the *Onset* × *Pitch* plane.

$$\begin{aligned}&Texture : .\mathbf{Power}(RepeatedNote),\\&RepeatedNote : .\mathbf{Limit}(Pitch, Loudness, Rate, Duration),\\&Rate : .\mathbf{Simple}(\mathbb{R})\end{aligned}$$

we obtain a set of notes that are repeated at a certain rate, altogether forming a characteristic *Texture*.

71.5.3 Synthesizers and Sound Design

Finally, here are some examples of forms that allow for more sound- and timbre-oriented structures. Some of the forms shown in Section 71.5.1 could be considered to be sound-based forms as they may be seen as somewhat related to additive synthesis, but we can go much farther than that.[12]

For instance, we can define

$$Spectrum : .\mathbf{Power}(Partial),$$
$$Partial : .\mathbf{Limit}(Loudness, Pitch).$$

This models a constantly sounding cluster based on only two dimensions. Since it is not using *ChromaticPitch* but *Pitch*, the cluster can include any microtonal pitches. Figure 71.12 shows an example of a *Spectrum*. If we, however, wanted to define a spectrum that only allows for harmonic overtones, this form would not be well suited, as we would have to meticulously arrange each individual pitch so that it sits at a multiple of a base frequency. Instead, we could simply use the form already introduced above, *HarmonicSpectrum* (Section 71.3.2.1). Figure 71.13 shows an example denotator of a set of harmonic spectra, defined as

$$HarmonicSpectra : .\mathbf{Power}(HarmonicSpectrum).$$

Since satellites (*Overtones*) and anchors (*HarmonicSpectrum*) do not share **Simple** dimensions, they can only be visualized if one **Simple** of each is selected as an axis parameter, here *Pitch* × *OvertoneIndex*. However, as we will see in the next chapter, they can both be transformed in arbitrary ways on such a plane. These are examples of the simplest way of working with additive synthesis in *BigBang*. All oscillators are expected to be based on the same wave form and a phase parameter is left out for simplicity. This is also the case for the following examples.

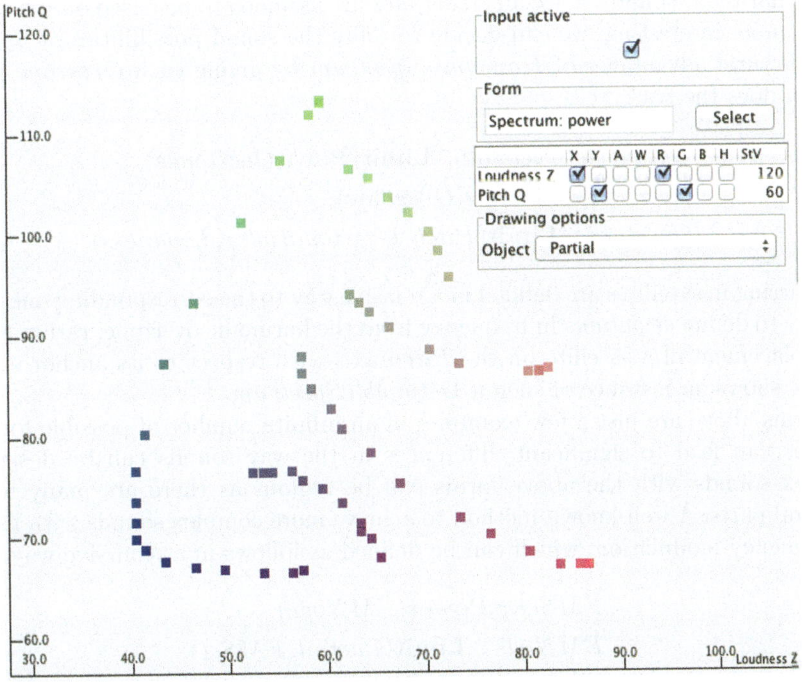

Fig. 71.12. A *Spectrum* shown on *Loudness* × *Pitch*.

[12] Some of these constructions were described in [1047].

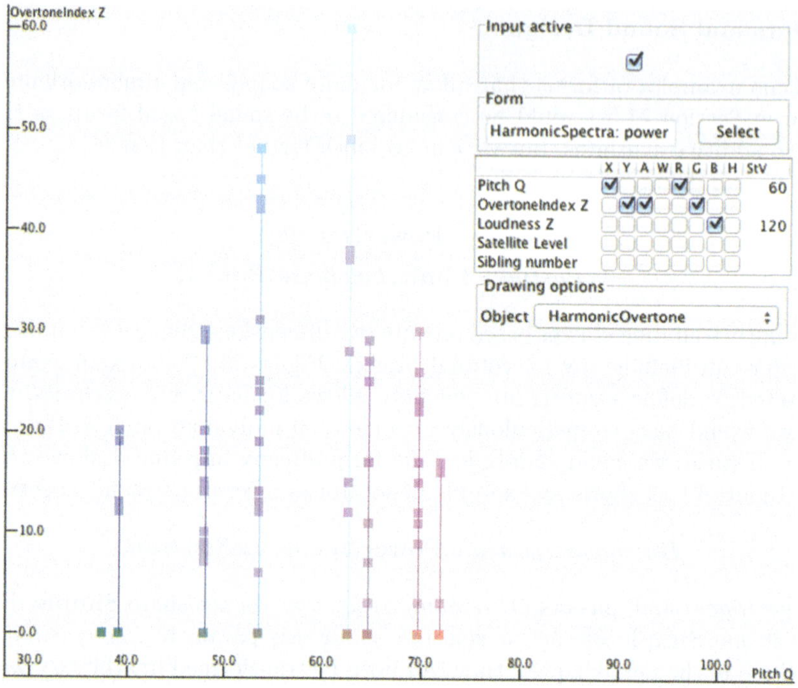

Fig. 71.13. A constellation of eight *HarmonicSpectra* with different fundamental *Pitches* and *Overtones*.

Even though the previous form leads to more structured and visually appealing results, we limited ourselves to purely harmonic sounds, since all *Overtones* are assumed to be based on the same base frequency *Pitch*. To make it more interesting, we can decide to unite the sound possibilities of *SoundSpectrum* with the visual and structural advantages of *HarmonicSpectrum* by giving each *Overtone* its own *Pitch*. The following definition does the trick:

$$\begin{aligned}
&DetunableSpectrum : .\mathbf{Limit}(Pitch, Overtones),\\
&Overtones : .\mathbf{Power}(Overtone),\\
&Overtone : .\mathbf{Limit}(Pitch, OvertoneIndex, Loudness).
\end{aligned}$$

Since values reoccurring in satellites are defined in a relative way to the corresponding ones of their anchor, we get the opportunity to define *deviations* in frequency from the harmonic overtone, rather than the frequencies themselves. A displacement of a satellite on the *Pitch* axis with respect to its anchor enables us to detune them. Figure 71.14 shows an instance of such a *DetunableSpectrum*.

The three forms above are just a few examples of an infinite number of possible forms. Slight variants of the above forms can lead to significant differences in the way sounds can be designed. For instance, generating complex sounds with the above forms can be tedious as there are many ways to control the individual structural parts. A well-known method to achieve more complex sounds with much fewer elements (oscillators) is frequency modulation, which can be defined as follows in a recursive way:

$$\begin{aligned}
&FMSet : .\mathbf{Power}(FMNode),\\
&FMNode : .\mathbf{Limit}(Partial, FMSet),
\end{aligned}$$

with *Partial* as defined above. Examples as complex as the one shown in Figure 71.15 can be created this way. Frequency modulation, typically considered highly unintuitive in terms of the relationship of structure and sound [199], can be better understood with a visual representation such as this one. All carriers and modulators are shown respective to their frequency and amplitude and can be transformed simultaneously and

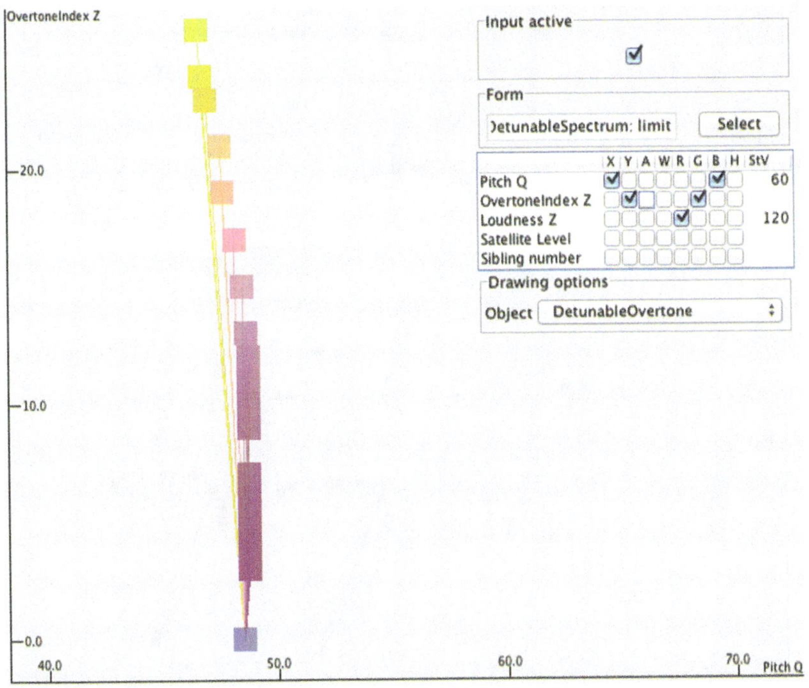

Fig. 71.14. An instance of a *DetunableSpectrum*, where the fundamentals of the *Overtones* are slightly detuned.

parallelly, which has great advantages for sound design compared to old-fashioned skeuomorphic synthesizers and applications.

In order to include other synthesis models, we can define

$$GenericSound : .\mathbf{Limit}(Oscillator, Satellites, Operation),$$
$$Oscillator : .\mathbf{Limit}(Loudness, Pitch, Waveform),$$
$$Satellites : .\mathbf{List}(GenericSound),$$
$$Operation : .\mathbf{Simple}(\mathbb{Z}_3),$$
$$Waveform : .\mathbf{Simple}(\mathbb{Z}_4),$$

where *Operation* represents the three synthesis operations for additive synthesis, ring modulation, and frequency modulation. For each anchor/satellite relationship, we can choose a different operation. Each *Oscillator* also has its own *Waveform*, here a selection of four varying ones, for instance sine, triangle, square, and sawtooth.[13] Sounds designed this way can immediately be played with by using a keyboard controller, as seen in Section 71.4.

Finally, we can also combine multiple forms into higher-level forms that contain several objects. For instance, a **Limit** of *SoundSpectrum* and *Score* allows us to create compositions containing both constantly sounding pitches and notes with *Onsets* and *Durations*. We simply need to define

$$SpectrumAndScore : .\mathbf{Limit}(Spectrum, Score).$$

Figure 71.16 shows an example of such a composition. This way, any number of synthesis methods and musical formats can be combined to higher-level forms and can be used simultaneously in *BigBang*.

These examples show how much structural variety we can create by just using a small given set of **Simple** forms, and how their visualization can help us understand the structures. All of them can directly

[13] A slightly different *GenericSound* form is described in [1048].

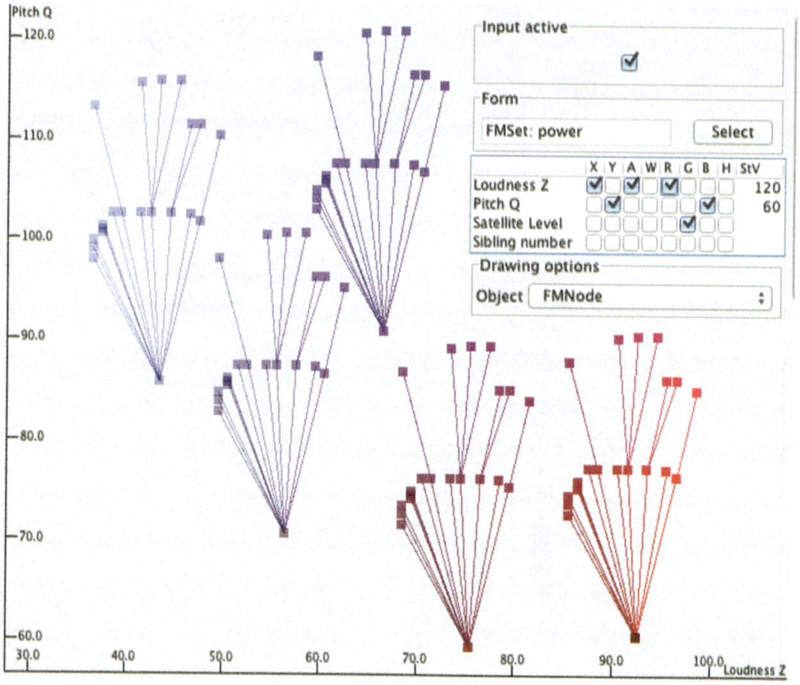

Fig. 71.15. An *FMSet* containing five carriers all having the same modulator arrangement, but transposed in *Pitch* and *Loudness*.

be sonified, even while we are building the denotators. Most importantly, such forms can be defined at runtime in *Rubato Composer* and they can immediately be used in *BigBang*. In addition to musical data types, as in the examples here, one can define forms describing any kind of fact. For example, we programmed rubettes that read image files (*ImageFileIn*), translate them into forms, and make them available to transformation in *BigBang*, before being exported again or converted into musical objects by other rubettes.

In the next chapter we will discuss how such objects, once their form is defined, can be created, manipulated, and transformed in *BigBang*. For this, we need to examine how the *BigBang* rubette implements the level of processes.

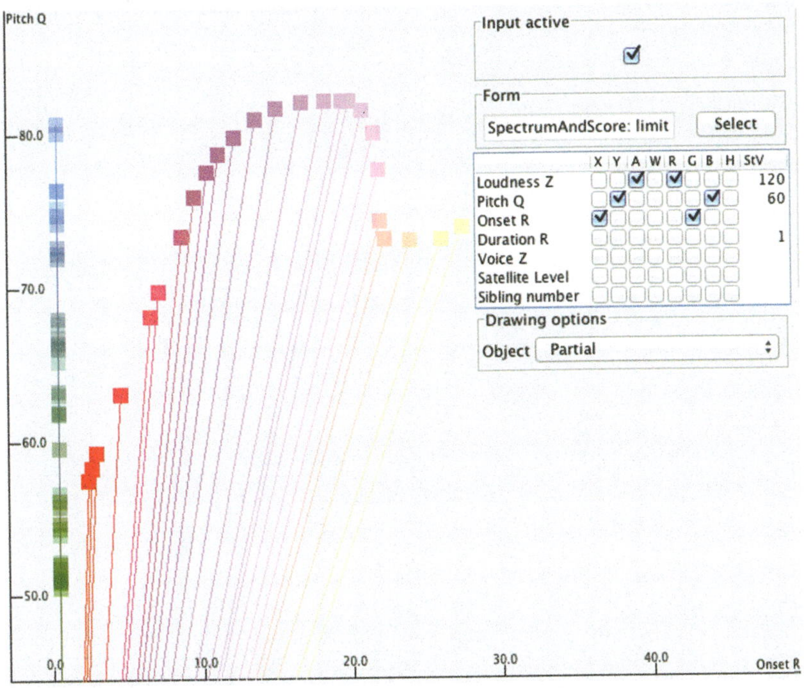

Fig. 71.16. A composition based on a **Limit** of a *SoundSpectrum* (*Pitches* at *Onset* 0) and a *Score* (*Pitches* with *Onsets*).

72

Processes: BigBang's Operation Graph

The main idea behind the *BigBang* rubette is to give composers and improvisers a way to use the software RUBATO® *Composer* in a way that is more intuitively understandable, more spontaneous, and more focused on audible results than on the mathematical underpinnings. After discussing the types of facts available in *BigBang* we need to examine how we can create them and what we can do with them. From earlier in this book we now know that both of these activities, making and manipulating, are instances of processes. *BigBang* keeps track of these processes in a more sophisticated way than other musical software, especially ones dedicated to symbolic structures.

RUBATO® *Composer* itself already allows users to create processes by building networks of rubettes. Why does *BigBang* need its own processes? There are several significant differences between the processes of RUBATO® *Composer* and other systems such as *Max/MSP* and the ones of *BigBang*.

1. *BigBang*'s visualization of processes is the dual graph of the rubette networks in RUBATO® *Composer*. Its focus on transformations as arrows is closer to the way we imagine processes, as seen in the first part of this book. RUBATO® *Composer*'s representation of denotators traveling through connecting lines imitates the physical reality of electric signals traveling through cables and has hardly anything to do with our imagination and representation of the mathematical constructs, e.g. diagrams of morphisms in category theory, or transformational graphs.
2. *BigBang* emphasizes spontaneity and a quick work process. Rather than representing a definite composition process, its processes represent experimental stages, are easily mutable, and allow for the creation of alternatives.
3. Its processes focus on a simple vocabulary of operations and transformations that can be combined to create larger structures in a transparent way.
4. The processes in *BigBang* are always directly connected to facts, and the user has the chance to observe and interact with both, facts and processes, simultaneously while composing or improvising. In RUBATO® *Composer*, facts remain hidden for large parts of the process.
5. *BigBang* focuses on processes that can be directly connected to gestures. For many features of RUBATO® *Composer* this would be difficult to do.
6. Most importantly, its processes represent the workings of the rubette itself. Users can use it just the way they use Macro Rubettes,[1] which encapsulate entire RUBATO® *Composer* networks within them. For instance, they can define a compositional process in *BigBang*, duplicate the rubette, and use it in multiple contexts by adjusting the process as needed. Thereby they can even decide to send denotators of any other form and see what the rubette yields.

Despite these differences, there is a chance that someday RUBATO® *Composer* may be extended to adopt some of *BigBang*'s principles, so that they can be used on a higher level and in a greater variety of ways.

Underlying the representation of processes in *BigBang* is a notion of a composition or improvisation as a dynamic rather than static entity. Rather than seeing the composition as a definite fact, we see it as a

[1] See [739, p.237].

G. Mazzola et al., *The Topos of Music III: Gestures*, Computational Music Science,
https://doi.org/10.1007/978-3-319-64481-3_17

conjunction of a set of musical input objects along with a processual graph into which these objects are fed, similar in this respect to transformational theory. Figure 72.1 shows a diagram of this scheme.[2] This way, the only existing facts are the initial facts, such as input denotators, and everything else can be produced by the transformations in the graph. Any stage of the composition process, i.e., any node of the process graph, can be dynamically generated using the factualizing procedure first introduced in Chapter 71. Internally, denotators are never saved at every step of the composition, but always generated dynamically.

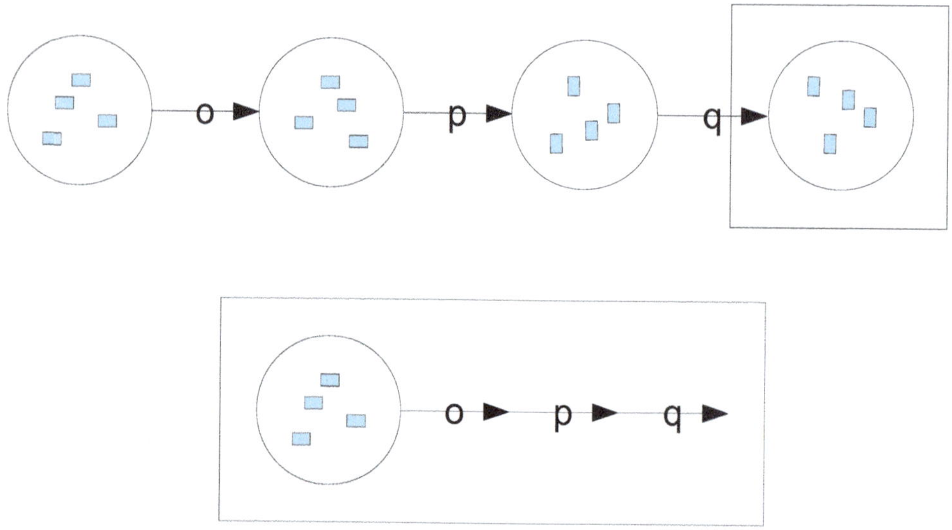

Fig. 72.1. A factual notion of a composition, above, versus a dynamic notion, below.

In this chapter, we will first introduce the available operations and transformations and describe what they do. Then we will describe how factualizing works in *BigBang*. Finally, we will discuss the way processes are visualized and how users can interact with the visualizations.

72.1 Temporal BigBangObjects, Object Selection, and Layers

Before performing an action, we need to be able to decide which objects are going to be affected by the action we wish to perform. In *BigBang*, users can make selections in the facts view by drawing rectangles around objects. During this process, they can arbitrarily change their perspective. For instance, it may be tedious to select all visible second-level satellites in a complex *MacroScore* composition. By selecting *SatelliteLevel* as one of the visible axes, the composition will be shown as a number of levels (Figure 72.2). The user can then simply draw a rectangle around the objects shown on level two to select all satellites there.

Whenever the user performs an operation, it will be applied to all selected objects. When selecting all objects, the operation or transformation will be applied to as many objects as were selected, even if the input of the rubette changes. For instance, suppose we input a piece with twelve notes into *BigBang*, select all of them, and then transform them. If we then input another composition with many more notes, the transformation will be applied only to the twelve first notes of the new composition.

72.1.1 Selecting None and Lewin's Transformation Graphs

If *none* of the objects are selected, an operation or transformation will always be applied to *all* objects present at the respective state, however many there may be. This has a major advantage, besides speeding

[2] These concepts were first discussed in [1046, 730].

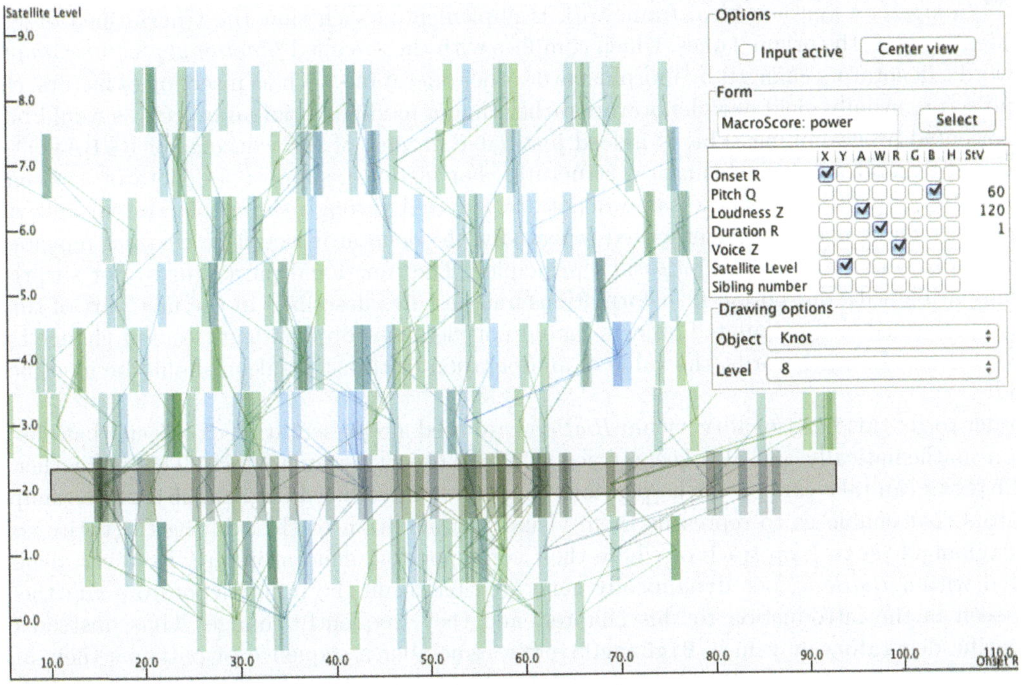

Fig. 72.2. A *MacroScore* with its second level of satellites being selected. Note that the y-axis is *SatelliteLevel* to facilitate the selection. The x-position and colors of the objects hint at their chaotic arrangement on the *Onset* × *Pitch* plane.

up the compositional process of users who like to work with the full set of musical objects. Users may often be in the situation where they know the actions they would like to perform without being sure what the objects are that they will be working with, especially if they plan on using *BigBang* as a rubette in different contexts, as described under Point 6 above. In this case, they may decide to not select any objects, which means that whatever they send the *BigBang* rubette, however many notes or sounds, the operations they defined will be applied to them. This way, they can experiment by applying the same process to as many different inputs as they like.

This establishes an interesting connection to Lewin's theories. In his distinction between networks and graphs, the nodes of the former are associated with specific objects, while the nodes of the latter are not. *BigBang* precisely models the latter. We can design transformation graphs without having a specific application in mind, and then subsequently make them into networks when we send *BigBang* specific denotators.

72.1.2 The Temporal Existence of BigBangObjects

One of the major problems faced when implementing *BigBang* emerged from the problem of object selection. The idea of selecting something with a specific identity, and possibly even assigning it some characteristics such as visibility or audibility as described below, and moreover expecting that it remains selected even when transformed, is hardly compatible with mathematical language. We have already discussed some aspects of this problem in the beginning of this discussion of *BigBang* with respect to functions and the anti-cartesian notion of transformation. What we saw there is that a function, mathematically speaking, does not really "move" its argument into the value, but rather associates a value, a new, different object, to the argument.

When dealing with computer software such as the *BigBang* rubette, we expect an entirely different behavior. When we select something, transform it, and continuously observe it during the transformation, we assume it to still be selected at any stage of the transformation, and thus the selected objects to maintain the same identity. However, since denotators are mathematical objects, this is not evident.

RUBATO® *Composer*'s mathematical framework is implemented such that the transformed objects have a different identity from the original ones, which complies with the so-called *functional programming* paradigm. In other words, denotators changed by morphisms or other operations such as insertion of factors, changing of values, and so on, usually yield new denotators rather than a modified original object, as would be expected in object-oriented programming. This is indeed justifiable in view of the workings of RUBATO® *Composer* and in view of denotators as mathematical structures. For instance, suppose a denotator was sent through several rubettes sequentially and the same denotator traveled through the whole system and was altered by each rubette. If the network was executed repeatedly, the denotator would be changed repeatedly, which does not comply with RUBATO® *Composer*'s principles. The functional paradigm is, not surprisingly, entirely incompatible with our notion transformations and gestures described in the first part of this book. In *BigBang*, transformed or manipulated objects should not yield new object identities, but change the ones we chose to transform. In other words, the value of an operation or transformation should be identified with its argument.

In order to get this functionality within *BigBang*, we need a representation of objects that observes what is going on mathematically and also keeps track of which object became which. This is another task that `BigBangObjects` can take care of. In the previous chapter we introduced `BigBangObjects` as simplifications of denotators that enable us to represent them visually. In addition to this, during the entire composition process, `BigBangObjects` keep track of where their corresponding denotators are. However, denotators are never saved within *BigBang*, but dynamically generated based on the input denotators and the operation graph as seen in the introduction to this chapter, and they frequently change. Thus, instead of remembering specific denotators or values, `BigBangObjects` remember a sequence of paths as their history – in the RUBATO® *Composer* framework, both forms and denotators are referred to by paths, for the topmost denotator identifying the object, at any state of the composition. Since every `BigBangObject` is either the top-level object or an element of a **Power** or **List**, this path usually ends with an index in a **Power** or **List**.[3] Figure 72.3 shows how this works.

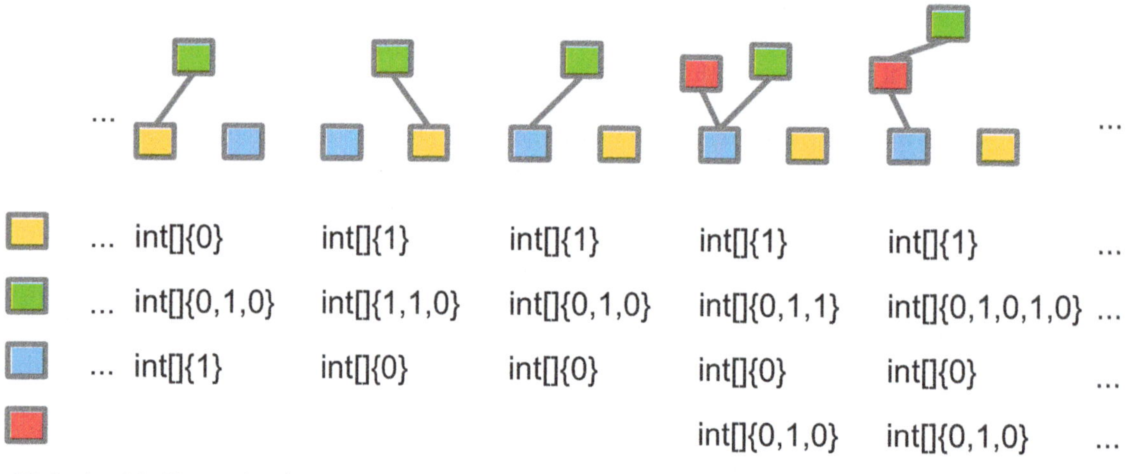

Fig. 72.3. A table illustrating how BigBangObjects keep track of their location. Each column is a state of a simple composition process with an *FMSet*. The rows are what each of the objects save: a path for each of the states the object exists at, pointing to the denotators corresponding to the objects (*FMNodes*) are at, at the respective state. Note that all paths are assigned according to the x-axis here (*Loudness* in *FMSet*). A *MacroScore* with its second level of satellites being selected. Note that the y-axis is *SatelliteLevel* to facilitate the selection. The x-position and colors of the objects hint at their chaotic arrangement on the *Onset* × *Pitch* plane.

[3] In RUBATO® *Composer*, **Powers** are sorted automatically and every one of their elements can thus be referred to by a definite index.

72.1.3 BigBang Layers

But `BigBangObjects` can do more. Composers often think in groups of musical objects, representing different voices, logical parts, or recording tracks. One of the characteristics of *BigBang*'s facts view is that it does not distinguish between such tracks in a spatial way. This leads to a significant gain in space. However, because of this, working in different parts can become tedious. However, with just the facts view, if users wanted to process several parts individually, they would have to repeatedly select each individual group of objects. In order to simplify this, we introduced so-called layers, which correspond to tracks in sequencers, except for that they are not strictly tied to specific voices or instruments.

In early *BigBang*, layers were realized on the level of forms. All *SoundNotes* had an additional **Simple** denotator *Layer*, which represented the index of the layer on which the note was present. *SoundNotes* could be moved from layer to layer either by being transformed or with a specific menu function, and layers could be made invisible, and inactive, which rendered the notes unselectable [1045]. This way of implementation had two disadvantages. First, the layers would only work for forms that contained the *Layer* **Simple** form. *MacroScores* and *Scores* could thus not be represented on layers. Second, the *Layer* form was of no use outside of *BigBang*, which hardly justified its appearance on the level of forms.

For the new generalized *BigBang* we had to find another solution. Layers were one of the main reasons for the introduction of `BigBangObjects` as temporal structures that were described in the previous section. We decided that each `BigBangObject` can be part of as many so-called `BigBangLayers` as needed. Users can add new layers, move objects to specific layers, or add them to additional ones. Each `BigBangLayer` can be made inactive (unselectable), invisible, and/or inaudible, which affects all of its objects. If an object is audible on at least one layer, it is audible. The same is true for visibility, and activeness. Figure 72.4 shows an example of an *FMSet* with active, inactive, inaudible, and selected layers.

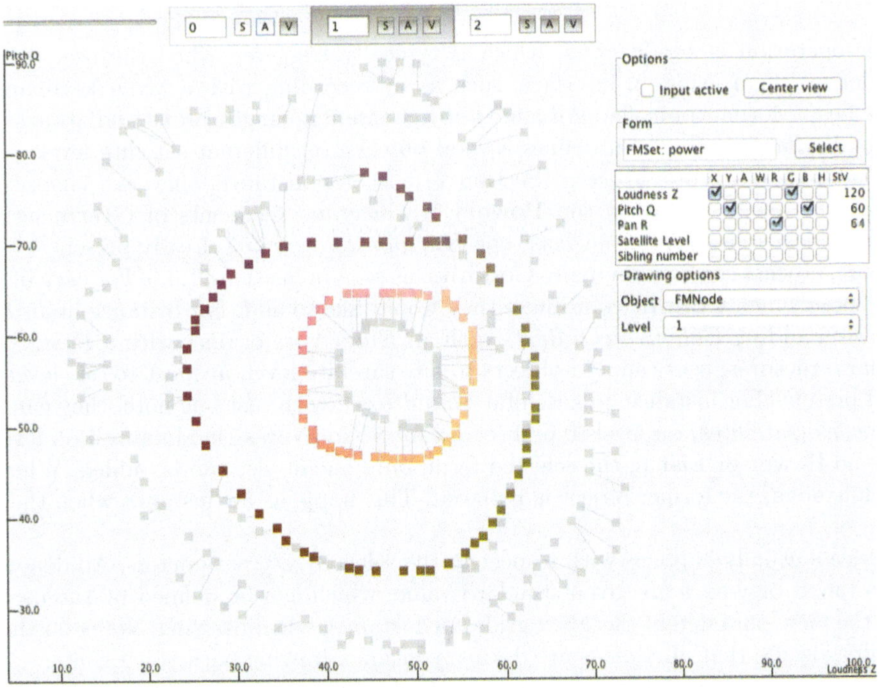

Fig. 72.4. An *FMSet* distributed on three layers, represented by the rectangluar areas at the top. Layer 0 is inactive and inaudible (its *Partials* in the facts view are greyed out), layer 1 is active and selected (its *Partials* are darkened), and layer 2 is active, but not selected (normal bright color).

72.2 Operations and Transformations in BigBang

This section introduces all activities that are part of the graph represented in *BigBang*'s process view. In *BigBang*, we distinguish between two types of processes, operations and transformations.[4] *Operations* are all activities that affect *BigBang*'s denotators, such as creating denotators, adding satellites, deleting denotators, etc. *Transformations* are special operations that include all activities that can be formulated as morphisms in the category of modules $\mathbf{Mod}^{@}$. Almost all operations and transformations are defined relatively to the x/y coordinates currently selected. This way, every dimension of the visible denotators can be manipulated. Since the way of interaction with *BigBang* is gestural for most of the operations, the next chapter, which deals with gestures, will explain how this works.

Almost all operations, except for the ones that add objects to the composition, are applied to a selection of `BigBangObjects` and keep references to these objects, rather than to denotator paths, as they did in earlier *BigBang*. This has major advantages in case operations are modified, removed, or inserted, as will be discussed in Section 72.3.

72.2.1 Non-transformational Operations

We do not consider all activities available in *BigBang* as part of the compositional or improvisation process. Only activities that change the denotator and the musical structure represented by the rubette are included in *BigBang*'s operation graph. For example, when the musician decides to select notes, and move them from layer to layer, or pushes the play button, denotators are left unchanged. This section considers all operations that are not transformations.

72.2.1.1 AddObjects and DeleteObjects

The most basic operation is `AddObjects`, which is typically triggered when the user draws objects onto the screen or defines them using an interface, such as by recording with a MIDI keyboard or by defining points with the Leap Motion controller. All added objects are `BigBangObjects` and always comply with the form selected in *BigBang*. If the form defines several objects, on different satellite levels, users can choose which ones to add. In *HarmonicSpectra* (Section 71.5.3), for instance, they can choose between adding *HarmonicSpectrum* (elements of the top **Power**) or *Overtone* (elements of *Overtones*). Note that the latter cannot be added unless there is at least one *HarmonicSpectrum* already present.

Furthermore, objects can contain many **Colimit**s, as seen in Section 71.3.3. For each of those **Colimits**, users need to choose which **Colimit** coordinate they would like to add. For instance in an *EulerScore* they can choose between adding *EulerNoteOrRests* with an *EulerNote* or ones with a *Rest*.

For circular structures, uses can add objects to any satellite level, at most to one level higher than the maximum level present. For instance, in case of a *SoundScore* with just one note, they can choose to either create *SoundNodes* (satellites) on levels 0 or 1, or create *SoundNotes* (modulators) on level 1.

If there is no **Power** or **List** in the selected form, only one object can be added. Whenever users keep performing `AddObjects`, the former object is replaced. This happens for instance when the selected form is *Pitch* or *Note*.

Adding objects usually happens with respect to the selected x/y parameters. All denotator parameters not assigned to the x- or y-axes are given standard value, which can be defined by the user, in the column to the right of the view parameters checkbox grid. For instance, when drawing *Notes* on the *Onset* × *Pitch* plane, we can first decide that all *Voices* are 0, then continue drawing with voice 1, etc.

The `DeleteObjects` operation simply removes all objects currently selected.

[4] Even though Lewin made the same distinction, our notion of operations greatly differs from what Lewin defined them to be. Rather than being more specific—bijective transformations—they are more general and denote the entire set of activities available in *BigBang* that change the contained denotators.

72.2.1.2 InputComposition

Instead of adding denotators in any of the above ways, users can also input denotators by means of the rubette's input, which creates an `InputComposition` operation. For this, they need to connect the rubette to another rubette that outputs a denotator, for instance *MidiFileIn*, and run the RUBATO® *Composer* network. If there is already a denotator in *BigBang*, the incoming one is of the same form, and the top-level denotator is a **Power** or a **List**, the elements of the incoming denotator are added to the denotator already present. Otherwise, the entire denotator is replaced. For instance, users can add as many other *Scores* to a *Score* as they wish, which leads to a large union, in exactly the same way as can be done with the *Set* rubette. If they do not wish to do that, but replace the *Score* already present, they have to select and modify the `InputComposition` operation, as will be described in Section 72.3.2.

72.2.1.3 BuildSatellites and Flatten

When users want to build hierarchical structures, they can simply add objects by drawing on a higher satellite level, as seen above. However, they can also use the `BuildSatellites` operation, usually triggered through a pop-up menu, with which they can add objects that are already present in the composition as satellites to other objects, if the form contains **Powers** with the same coordinate form, at several places. This way, for example, with an *FMSet* denotator with several top-level *Partials*, users can add some of these *Partials* as modulators to another one of the *Partials*.

The opposite operation is called `Flatten` [715]. It adds all selected satellites to the **Power** or **List** that contain their respective anchors, e.g. it changes first-level *FMSet* modulators into simple additive oscillators.

72.2.1.4 Shaping

`Shaping` allows users to change the values of the visible objects. It is based on two given denotator values u, v, for instance *Onset*, *Pitch*, and a number of real number pairs $(u_1, v_1), \ldots, (u_n, v_n) \in \mathbb{R}^2$. For each `BigBangObject` in the composition, if its u value is close to any u_n, its v value is assigned v_n. In practice, this is used in *BigBang*'s shaping mode, where users can click-and-drag as in drawing mode on the x/y plane, and every object that is close to the x value is assigned the y value of the current drawing location.[5] This can be helpful especially with mouse drawing, where users are limited to drawing in two dimensions. After drawing on one plane (`AddObjects`), they can switch into shaping mode and define more dimensions the same way. Figure 72.5 shows a *Score* composition that was drawn on the *Onset* × *Pitch* plane, before being shaped in the *Onset* × *Loudness* plane, where for each *Onset*, a new *Loudness* was assigned. These varying *Loudnesses* are now represented with hue color values, red being both the loudest and quietest, green being *mp* and blue *mf*.

72.2.1.5 Wallpaper Operations

Wallpapers were introduced as generalizations of *Presto Ornaments* [1041]. In simplified terms, for a wallpaper we need a *motif* m of coordinates of **Powers** or **Lists**, a *grid* of morphisms $f_1, \ldots, f_n$ and corresponding *ranges* $r_1, \ldots, r_n$ with $r_k = (r_k^{min}, r_k^{max}) \in \mathbb{Z}^2$ and $r_k^{min} \leqslant r_k^{max}$. The first wallpaper dimension then results from the repeated application of $f_1(\ldots f_1(m))$ and creating the union of all copies. $1 + r_1^{max} - r_1^{min}$ determines the number of copies of m we get. If $r_1^{min} \leqslant 0 \leqslant r_1^{max}$, m itself is included. The next dimension, if there is one, is then produced by f_2, r_2, applying f_2 to all copies of m resulting from the first dimension, then f_3, r_3 to all results of f_2, r_2 and so on.[6] Figure 72.6 shows an instance of a two-dimensional wallpaper in early *BigBang*.

In *BigBang*, the morphisms of a wallpaper are limited to its transformations, i.e., affine morphisms, to be defined in Section 72.2.2. However, *BigBang* adds a functionality that the *Wallpaper* rubette was not

[5] An early version of this mode was introduced in [1045].

[6] For more details, see [1041, p.33f].

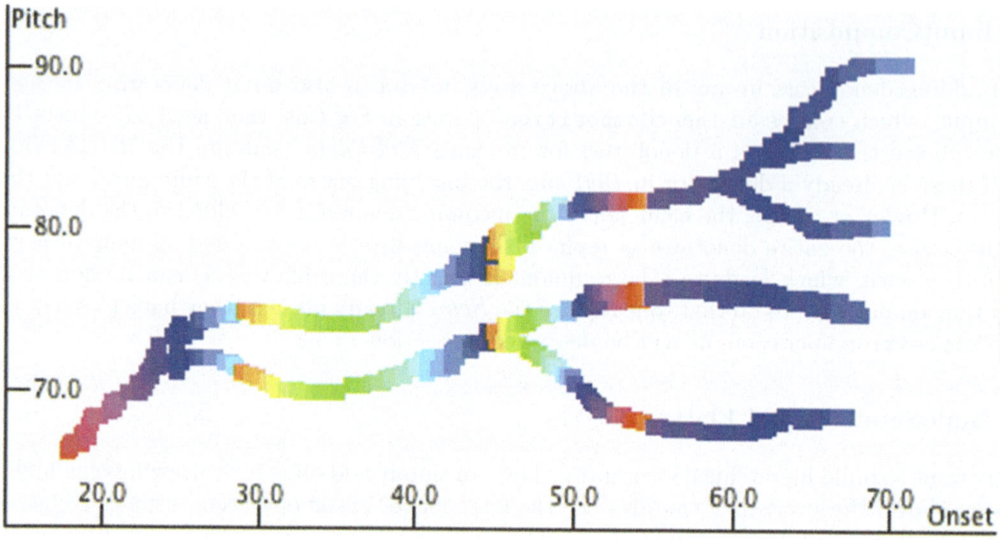

Fig. 72.5. A composition drawn in *Onset* × *Pitch* with a shaped third dimension represented by color.

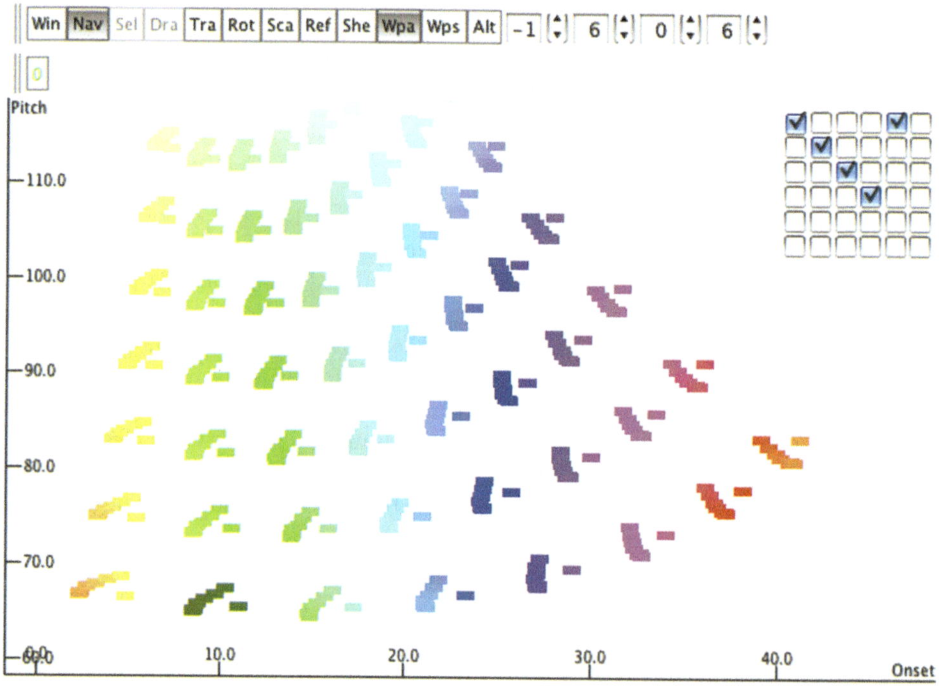

Fig. 72.6. A two-dimensional wallpaper in early *BigBang*.

capable of. The motif is no longer the entire composition, as for the *Wallpaper* rubette, but the selection the user has made. This way, elements on any level of the denotator anatomy can simultaneously participate in a wallpaper. For instance, we can generate a regular structure of *FMSet* denotators including both, carriers and modulators. Each mapped modulator is added to its original carrier.

Also, each dimension of the wallpaper can be composed of several two-dimensional *BigBang* transformations, each of them performed on arbitrary x/y planes. This way, when working with a *Score*, a single dimension can for instance consist of a translation on the *Onset* × *Pitch* plane, followed by a shearing on the *Onset* × *Loudness* plane, which results in each copy of the motif being transposed in pitch, time, *and* loudness.

Two operations regulate the creation of wallpapers in the new *BigBang*, `AddWallpaperDimension` and `EndWallpaper`. Whenever `AddWallpaperDimension` is performed for the first time, all objects that are selected at the time are taken to be the motif, and all following transformations constitute the first dimension of a wallpaper. Each additional performance of `AddWallpaperDimension` adds another dimension, again constituted by all following transformations, while being based on the same motif. Finally, `EndWallpaper` ends the wallpaper and goes back to normal transformation mode.

72.2.1.6 Alteration

The last operation available in the current version of *BigBang* is `Alteration` and corresponds to the functionality of the *Alteration* rubette, a generalization of part of *Presto OrnaMagic* [1041, p.36f]. In *BigBang*, alteration consists in deforming a set of objects O_1 gradually towards another set of objects O_2. In *BigBang*, as with wallpaper motifs, both of these compositions can be selected using the selection tool and do not have to include the entire denotator. Again, compared to the *Alteration* rubette, where both inputs have to be **Power** denotators the direct elements of which are altered, in *BigBang* the sets of objects can include objects on different anatomical levels of the composition. For every object in $o_i \in O_1$, alteration finds the nearest object in $o_j \in O_2$, based on spatial distance, and moves the values of o_i towards the ones of o_j by a given degree. These degrees work as follows: for a degree of 0% the object o_i stays unchanged whereas for 100%, o_i becomes o_j.

Alteration can be performed simultaneously for as many of the denotator parameters as the user would like. For instance, if we merely alter *Pitch* in a *Score*, we get the tonal alteration familiar from music theory. If we alter just *Onset*, we get a generalized version of quantizing, familiar from sequencer systems. However, if we alter every **Simple** denotator in *Score*, we get intermediary compositions between O_1 and O_2. In the new *BigBang*, users can define two alteration degrees dg_1, dg_2 that act according to the denotator parameter currently associated with the x-axis. For instance, if we alter a *Score* while looking at the *Onset* × *Pitch* plane, dg_1 defines the degree by which the object with the earliest *Onset* is altered, while dg_2 designates the degree for the object with the latest *Onset*. If we switch to *Pitch* × *Onset*, dg_1 concerns the object with the lowest *Pitch* and dg_2 the one with the highest. All degrees for the intermediary objects are interpolated linearly. Figure 72.7 shows an example from early *BigBang* where a *Score* is altered with $dg_1 = 0\%, dg_2 = 100\%$.

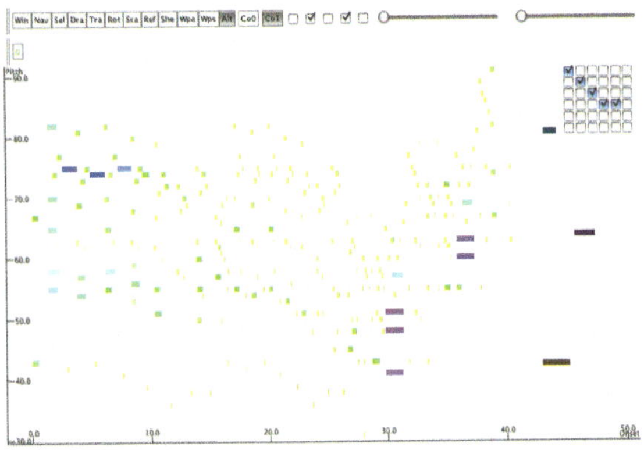

Fig. 72.7. A *Score* alteration in early *BigBang*. (a) shows the unaltered *Score*

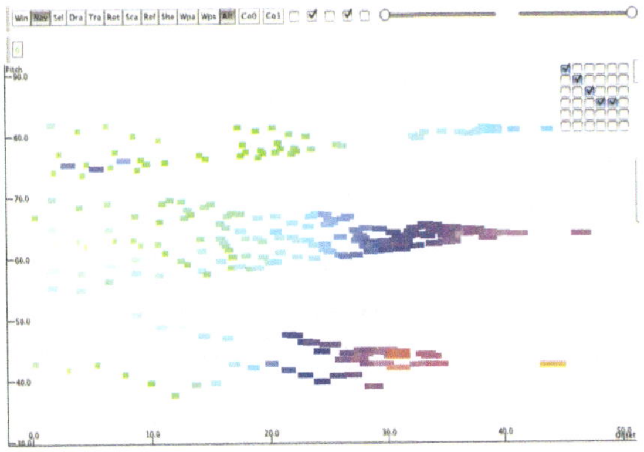

Fig. 72.8. A *Score* alteration in early *BigBang*. (b) *Pitch* and *Duration* are altered with $dg_1 = 0\%$ and $dg_2 = 100\%$

72.2.2 Transformations

In the context of RUBATO® *Composer*, transformations are directly based on morphisms between denotators, as we defined them above. *BigBang* allows for five different kinds of geometric transformations on the visible x/y plane, `Translation`, `Rotation`, `Scaling` (= dilation), `Shearing`, and `Reflection`, which take advantage of a lemma that says that any multi-dimensional affine transformation can be described as a concatenation of such two-dimensional geometrical transformations. These transformations are typically applied with a gestural interface, as will be described in detail in the next chapter. For example, when using a multi-touch interface, users can directly define an `AffineTransformation` based on combined dilation, rotation, and translation with two fingers, or all five transformations with three fingers [1044].

Regularly, transformations replace the selected objects with transformed versions. However, users also have a choice to perform so-called *copy-and-transform*, a generalized version of copy-and-paste, which adds the transformed objects to the given **Power** denotator, while keeping the originals. `Translation` with copy-and-transform yields classical copy-and-paste. Figure 72.9 shows a composition made with several copy-and-transforms.

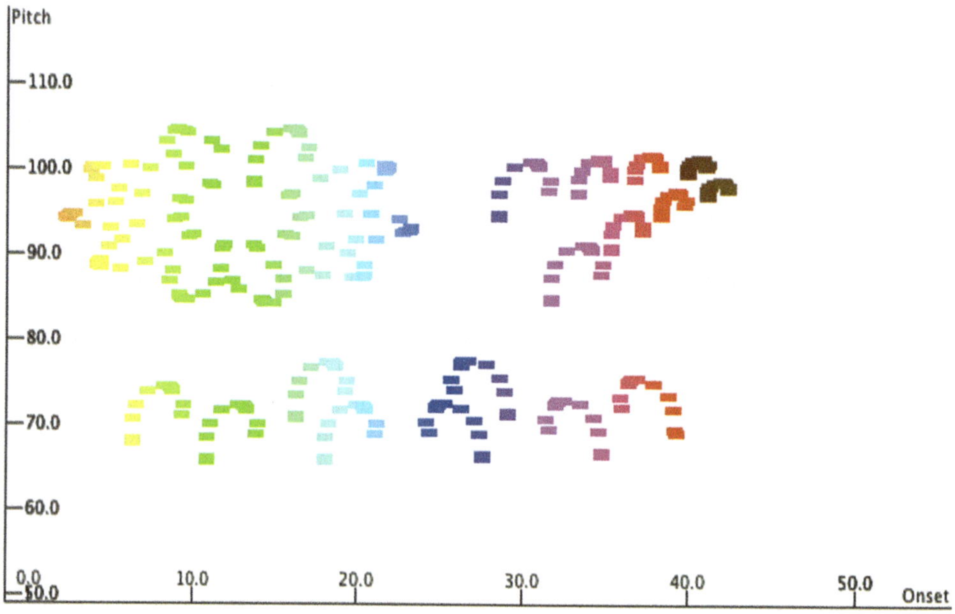

Fig. 72.9. A small composition made with copy-and-translate (bottom left), copy-and-rotate (top left), copy-and-scale (top right), and copy-and-reflect (bottom right).

In contrast to earlier versions, in the new *BigBang* rubette all transformations can be applied to any selection of `BigBangObjects`, on whatever anatomical level of the denotator they are, as seen above with wallpapers and alterations. This makes it possible for anchors and satellites to be transformed simultaneously, which leads to interesting results. Since satellites are designed to keep their relative position to their anchor when the anchor is transformed, a simultaneous transformation of anchors and satellites leads to satellites transforming doubly, once along with their anchor and once themselves.

Futhermore, even different objects in a **Colimit** can be transformed together in shared dimensions. For instance, in an *EulerScore* composition, we can transform *Notes* and *Rests* simultaneously if either *Onset* or *Duration* is (or both are) associated with the x- and y-axes.

Since the view parameters can be freely assigned in *BigBang*, objects also need to be able to be transformed when only one of their parameters is associated with one of the visual axes. If this is the case, objects are represented on the respective axis and the transformation, defined in two dimensions, acts on the objects as though they were located in two-dimensional space. However, the results remain projections on the axis at any time.

72.2.2.1 Transformation in Arbitrary Spaces

Even though what the new *BigBang* rubette does in terms of operations and transformations may appear straightforward, from a theoretical point of view it is trickier than expected. In this section, we will briefly illuminate one of the solutions we found in order to deal with the potentially infinite number of object types that *BigBang* can handle.

Most importantly, the transformative system needed to be adjusted in order to transform more general types of **Simple** denotators, and not just *Note* parameters. For this, we could build on a procedure that allows for mapping denotators by arbitrary morphisms that Florian Thalmann defined in his master's thesis and used in the context of the *Wallpaper*, *Alteration*, and *Morphing* rubettes [1041, p.32f]. Here we describe the necessary extensions and generalizations.

In the original procedure, the goal was to map a **Power** denotator d by a morphism f, even if the modules of its **Simple** denotators do not match the domain of f. This was done by inserting auxiliary *injection*, *projection*, and *casting* morphisms on both sides of f in order to adapt it to the chosen **Simple** morphisms.

We assume $f : V \to W$ to be any kind of affine or non-affine morphism where V and W are products of arbitrary modules $V = V_1 \times \ldots V_s$ and $W = W_1 \times \ldots W_t$. In the original procedure we tacitly assumed these modules to be one-dimensional free modules over the number rings $\mathbb{Z}, \mathbb{Q}, \mathbb{R}$ or $\mathbb{C}$. With the new extended repertoire of denotators, including **Limit**, **Colimit**, and **Power** denotators based on **Simple** denotators on more-dimensional free modules as well as modules based on product rings, we needed to make some adjustments.

Assuming that we would like to map values of a given denotator $d : A@F$, where F is any form containing **Simple** forms, we again define two sequences, $G. = (G_1, \ldots G_s)$ and $H. = (H_1, \ldots H_t)$, their cardinality corresponding to domain dimension s and codomain dimension t of f. However, as opposed to the earlier procedure, their elements G_j and H_k are not **Simple** forms but either component modules of one- or more-dimensional free modules over a certain ring, or factors of direct sum modules or modules over a product ring. More formally, $G_j, H_k \in R_F$, where R_F is the set of all module components or factors throughout the denotator tree. There are significant differences between the set S_F introduced earlier [1041, p.31] and R_F. Not only does R_F contain modules and not simple forms, but it may also contain several instances of the same type of component module or factor module, unless it is contained at the same position in a different instance of the same **Simple** form. There is thus no function analogue to $SA(S, d)$ involved.

Due to the fact that we now allow more-dimensional **Simple** denotators, we also need more auxiliary morphisms. In addition to i_j, p_k, g_j and h_k in the earlier procedure [1041, p.33], we define two additional sequences of projection and injection morphisms, $\mathfrak{p}_m$ and $\mathfrak{i}_n$, which leaves us with the following collection of morphisms:

- the initial *projection* morphisms $\mathfrak{p}_1, \ldots \mathfrak{p}_s$ with $p_j : M_{G_j} \to G_j$,
- the initial *casting* morphisms $g_1, \ldots g_s$ with $g_j : G_j \to V_j$,
- the initial *injection* morphisms $i_1, \ldots i_s$ with $i_j : V_j \to V$ with $i_j(v) = v' = (0, \ldots v, \ldots 0)$, where v is at the jth position of v',
- the final *projection* morphisms $p_1, \ldots p_t$ with $p_k : W \to W_k$ with $p_k(w) = w_k$ for $w = (w_1, \ldots w_t)$,
- the final *casting* morphisms $h_1, \ldots h_t$ with $h_k : W_k \to H_k$, and
- the final *injection* morphisms $\mathfrak{i}_1, \ldots \mathfrak{i}_t$ with $p_k : H_k \to M_{H_k}$.

In these definitions, M_{G_j} and M_{H_k} stand for the modules of which the G_j and H_k are components or factors. They of course do not have to be pairwise different, since several of the elements of $G.$ and $H.$ might be different components or factors of the same modules.

We then define a ϕ'_f analogously to ϕ_f [1041, p.33]:

$$\phi'_f(d, (G_j)) = f(i_1 \circ g_1 \circ \mathfrak{p}_1 \circ A@M_{G_j}(d) + \ldots i_s \circ g_s \circ \mathfrak{p}_s \circ A@M_{G_j}(d)).$$

Finally, we define

$map_f(d)$ as a copy of d,
where every module M_{H_k} is replaced by the sum of $\mathrm{i}_k \circ h_k \circ p_k \circ \phi'_f(d,(G_j))$,
and the injected projection of every component or factor M_i of M_{H_k}
with $M_i \neq H_k$.

72.3 BigBang's Process View

When they are performed, all of the operations and transformations described above are added to *BigBang*'s process view. In this section, we discuss how processes are visualized and how users can interact with them.

72.3.1 Visualization of Processes

As seen above, the process view shows a directed graph, which we call this graph *operation graph*, since it contains all operations performed, including transformations, and since its node values are not defined in an absolute way and thus resemble Lewinian transformation graphs rather than networks (see Section 72.1.1).

Whenever a new *BigBang* rubette is created or the user decides to start over by selecting a new form to work with, the operation graph is reset, which means that it merely consists of one node, labelled 0. For every operation performed, the graph obtains a new arrow, labelled with the operation, and a new node, representing the so-called `CompositionState` after the execution of the operation. Composition states are identified with unique increasing numbers, the highest of them representing the state last added. As long as the user merely interacts with the facts view, the graph grows as a linear sequence of arrows and nodes. Figure 72.10 shows such a simple linear graph including an `AddObjects` operation followed by all five geometric transformations.

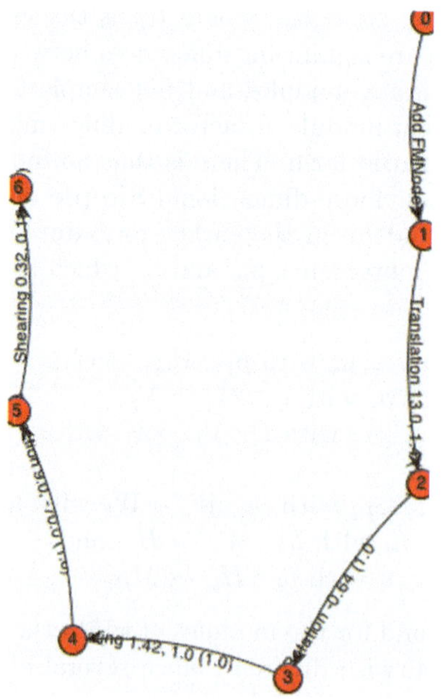

Fig. 72.10. A *BigBang* operation graph showing a linear composition process.

72.3.2 Selecting States and Modifying Operations

Users can interact with the graph by selecting its nodes, which immediately updates *BigBang's* composition to the one at the corresponding state, in both the facts view and the sonification. This way, the users can compare and contrast different states of their composition process, and evaluate them. Every time a state is selected, the shortest path between state 0 and the selected state is calculated and the corresponding facts are dynamically generated, which corresponds to the factualizing procedure described in Chapter 71.

When an operation is selected, the corresponding screen tool—a schematic representation of the operation as described in the next chapter—is shown, and users have the opportunity to modify the operation. Any state can be selected during this procedure and the consequences of the modification are shown for that state. This enables composers to change past decisions in their composition process, while observing their present composition, much in the fashion of Boulezian *analyse créatrice*, where composers use an analytical process to find other compositions in the neighborhood of theirs.[7]

Transformations are modified by dynamically changing the transformational parameters, e.g. the rotation angle or center for a `Rotation`, or the scale factors of a `Scaling`, which will be described in Section 73.1.4. Operations can have more distinct consequences. For instance, with modifying `AddObjects`, users can entirely replace the objects they were working with. The same composition process following the selected operation will then be applied to the new objects. The same applies to `InputComposition`. If the user selects such an operation before running the RUBATO® *Composer* network, the operation's composition is replaced. If no such operation is selected, a new `InputComposition` operation is created at the end of the graph or the selected composition state.

This is where the definition of operations with none of the objects selected becomes interesting, as described in Section 72.1.1. If a user replaces the entire input of *BigBang*, the entire composition process will be applied to all objects, no matter how many of them there are.

Now what happens when operations are modified that later operations depend on? If, for instance, we modify a `Rotation` by a 180 degrees, all concerned objects' denotator paths may change, since they are based on lexicographical sorting, especially in **Power**. In early *BigBang*, this would have led to major problems, since all operations were directly based on denotator path references. In the current version, as mentioned in Section 72.2, operations keep references to `BigBangObjects` instead, which dramatically simplifies the case. In the case of the modified rotation, all paths the `BigBangObjects` refer to are changed. All operations following the rotation can then dynamically obtain the actual paths from the objects when updating the denotator, i.e., during factualization.

72.3.3 Alternative and Parallel Processes

In addition to the linear processes described in Section 72.3.1, there are currently two more process types, alternative and parallel processes.

If users select a state other than the latest composition state and perform an operation, the operation is added to the graph by building a fork at the selected state, building an *alternative process*. This way, users can experiment by building processes that share an initial part, but then continue individually. Such alternative composition states can again be selected and are immediately visualized and sonified accordingly. Figure 72.11 shows such a graph generating two alternative wallpapers starting from the same input material.

Parallel processes are created when an operation is selected in the graph. Then, any new operation performed is added as a parallel arrow to the selected operation, starting and ending at the same states. This is the only way operations are added to the graph without adding a new composition state. Logically, parallel operations are no different from sequential operations at the current time. They are executed in their order of addition, since conflicting situations might arise with a parallel execution, especially with non-commutative transformations. However, as we will see in the next chapter, they differ from sequential operations in the way they are gesturalized (Section 73.2). Furthermore, they can be a good choice for composers to group their operations in order to get fewer composition states, if they are composing on a

[7] The notion of neighborhoods was introduced in [718], based on *analyse créatrice* in [140].

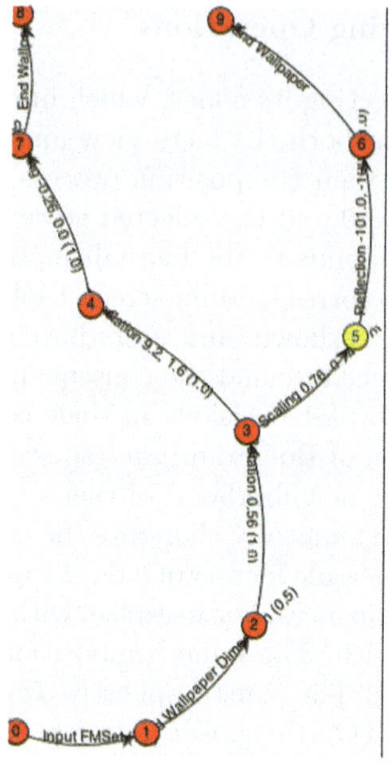

Fig. 72.11. An operation graph with two alternative processes.

meta-level (see Section 74.2). Figure 72.12 shows a graph including three parallel geometric transformations, followed by a parallel reflection and alteration.

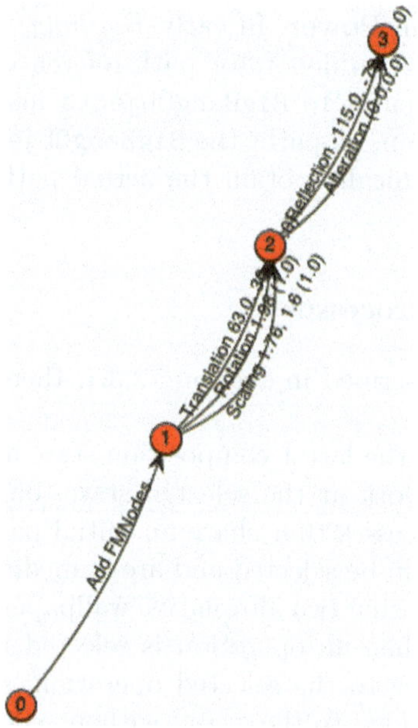

Fig. 72.12. An operation graph with parallel processes.

Currently, parallel operations can only be added as directly parallel to one operation. In the future, however, there will be the possibility to create higher-level operations that are executed simultaneously with several lower-level operations. This is especially attractive for gesturalization, where higher-level operations would be animated much slower than lower-level operations. Internally, *BigBang* already supports the definition of such processes.

72.3.4 Structurally Modifying the Graph

So far we described ways in which operations can be added to the graph simply by performing them. There are additional ways in which users can interact with *BigBang's* operation graph.

72.3.4.1 Removing Operations

At any stage of the composition process, users can decide to remove an operation in the graph using a popup menu. When an operation is removed, all other operations are still executed and applied to the same selection of `BigBangObjects`. However, there is a chance that the concerned `BigBangObjects` are not there anymore, if the removed operation is for instance an `AddObjects` operation. This is why all operations are always applied to all of their objects there, while all others are ignored.

Removing operations is an especially attractive solution to a problem that early *BigBang* had with its undo/redo system. Since its architecture was facts-based, as shown in the top half of Figure 72.1, non-invertible transformations such as projections were impossible to be undone. With the new, process-based *BigBang*, any type of operation can be undone without problems.

72.3.4.2 Inserting Operations

Users can also insert an operation at any state, by simply deciding where to insert and by selecting objects and executing an operation as usual. This replaces the selected state node by two nodes, and connects them with an arrow representing the new operation.

72.3.4.3 Splitting Operations

Any operation can be split into two operations by indicating a ratio between 0 and 1 at which the operation should be split. This can be done using a slider, as described in the next chapter. Thereby, the operation arrow is replaced by two arrows and an intermediary node. For instance, if `Rotation` is split with angle α at ratio 0.4, this results in two subsequent `Rotations` with the same center, the first with 0.4α and the second with 0.6α.

72.3.5 Undo/Redo

BigBang's operation graph represents the composition or improvisation process. The possibility of interacting with it in the above ways may be seen as a replacement of traditional undo/redo functionality in software. However, on top of this, *BigBang* has a regular undo/redo system that works on the level of graph interaction. It allows users to undo and redo any activity of adding operations to and removing them from the graph. This is important for an even faster and more flexible way of interaction. For instance, if users decide to remove an operation early in the process and dislike the effect, they can bring it back using a standard key combination.

73

Gestures: Gestural Interaction and Gesturalization

We have so far seen that the *BigBang* rubette allows users to visualize and sonify facts, and create and manipulate them using processes. In the previous chapter, we also discussed that the only structures that *BigBang* represents internally are processes, only one of which refers to facts in the form of denotators (`InputComposition`). All other facts are generated dynamically, whenever an operation is added or modified. In order to offer an intuitive way of interacting with the software, we need yet another level: gestures.

BigBang builds on the gestural principles described in Section 61. There are two ways in which gestures come into play with *BigBang*, the ones that are performed by the user when applying operations, and the ones that are recreated from processes. With the former, anything composers and improvisers do within *BigBang* is immediately audible and most operations can be performed in continuous ways, using continuous physical controllers such as a mouse, a multi-touch surface, or a Leap Motion controller. All operations are accessible through a minimal amount of actions or gestures, designed to be understandable to any user, even ones without a mathematical background.

Nevertheless, what *BigBang* saves are not the gestures as such, but their processual abstractions. From the point of view of computer science, this is an infinitely more economical solution than saving every temporal state of a gesture. The latter would be possible to implement with current computers, but it is not yet conceivable in terms of denotators and will thus be left to further projects.[1] Thus, the second way gestures are available in *BigBang* is by turning processes back into gestures, in the form of an animated composition history that can again be used for compositional purposes.

All this corresponds to the communication scheme between the levels of embodiment introduced in Chapter 71. The two types of gestures correspond to the inputs of the formalizing procedure and the outputs of the gesturalizing procedure. In this chapter, we explain how both formalizing and gesturalizing is implemented in *BigBang*.

First, an overview of gestural possibilities will be helpful. Table 73.1 lists all operations currently available in *BigBang* and shows whether or not their definition occurs in a gestural way (first type of gesture) and whether or not they are gesturalizable (second type of gesture). While several of the operations are not defined in a gestural way, most of them are gesturalizable. All transformations are both defined gesturally and gesturalizable. The two last columns will be discussed in Section 73.1.4.

73.1 Formalizing: From Gestures to Operations

In this section, we discuss the ways gestures are used to define operations, more precisely how controller gestures are mapped into appropriate operations and transformations. We thereby move from the most simple supported gestural interface to more complex ones. The standard gestural controller is the computer mouse. It was a design principle that almost everything in *BigBang* can be done in a satisfying way using a mouse.

[1] In order to do this properly within RUBATO® *Composer*, we have to extend its vocabulary to include constructs in the category of topological spaces, as used in the definition of gestures (see Section 61).

G. Mazzola et al., *The Topos of Music III: Gestures*, Computational Music Science,
https://doi.org/10.1007/978-3-319-64481-3_18

operation	defined gesturally	gesturalizable	modifiable	range (in $\mathbb{R}$)
`AddObjects`	yes	yes	yes	[0,1]
`DeleteObjects`	no	yes	yes	[0,1]
`InputComposition`	no	yes	yes	[0,1]
`BuildSatellites`	no	yes	yes	[0,1]
`Flatten`	no	yes	yes	[0,1]
`Shaping`	yes	yes	yes	[0,1]
`AddWallpaperDimension`	no	no	yes	[0,2]
`EndWallpaper`	no	no	no	no
`Alteration`	yes	yes	yes	[0,1]
`Translation`	yes	yes	yes	[0,2]
`Rotation`	yes	yes	yes	[0,2]
`Scaling`	yes	yes	yes	[0,2]
`Shearing`	yes	yes	yes	[0,2]
`Reflection`	yes	yes	yes	[0,2]
`AffineTransformation`	yes	yes	yes	[0,2]

Table 73.1. *BigBang*'s operations and their gestural capabilities.

Other currently supported interfaces include multi-touch surfaces, the Leap Motion controller, and various MIDI controllers. Gestural devices vary significantly in the dimensionality of their topological space, the number of recognized parameters in this space, as well as the potential interdependency of the parameters based on physical limitations.

In our case, the mouse recognizes one point in $\mathbb{R}^2$, multi-touch a number of points in $\mathbb{R}^2$ (maximally 10 per user), and the Leap Motion twelve points and twelve vectors in $\mathbb{R}^3$. What do gestures look like in these spaces? For the mouse, for instance, we can define a simple click-and-drag gesture $g : \uparrow \rightsquigarrow f$ where $\uparrow$ is the arrow digraph with $\uparrow = \bullet \to \bullet$ and $f : I \to \mathbb{R}^2$. Since we will always deal with single click-and-drag gestures below, we will simply identify the gestures by defining f.

73.1.1 Modes, Gestural Operations, and the Mouse

Complying with our principles for operation-based gestural systems, we decided the mouse operations in *BigBang* would be atomic gestures with as few clicks and movements as possible, so that they can be quickly applied, in an improvisational and potentially virtuosic way [1043, p.3]. Most gestural operations can be defined with a click-and-drag gesture. In order to distinguish the different operations from each other we did not implement a recognition system, but defined a number of `Modes` in which the program can be, one for all gestural operations (see Table 73.1), plus one each for `AddWallpaperDimension` and `EndWallpaper`.[2] These modes are accessible through buttons in the top toolbar, but will soon be made accessible through keyboard shortcuts when using a mouse, or even a MIDI foot controller when working with two-handed gestural interfaces,[3] in order to keep the hands focused on gestures. Most modes for gestural operations have a corresponding `DisplayTool`, which represents the ongoing operation in a schematic way as a reference for the user. Figure 73.1 shows the tool displayed in `Shearing` mode, which consists of a grey square representing the original state and a clear parallelogram representing the sheared version of the square.

[2] There are also modes for non-operational activity, e.g. *Navigation* mode and *Selection* mode.
[3] As suggested in [1049].

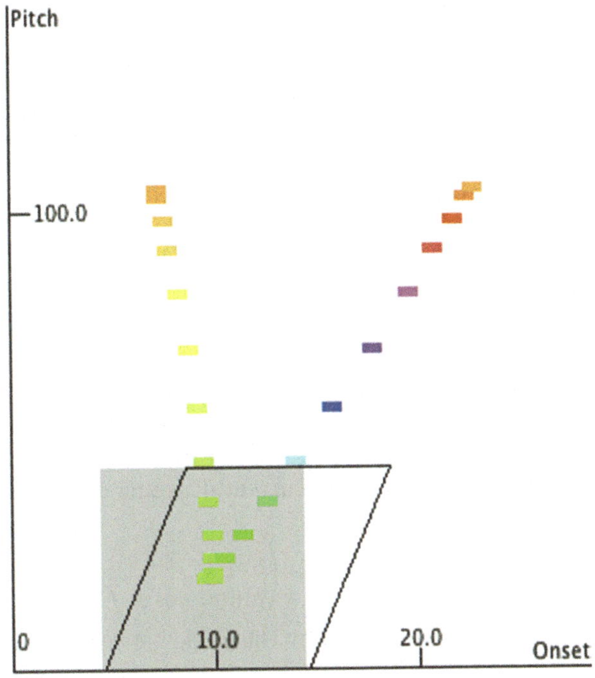

Fig. 73.1. The shearing *DisplayTool* shown while a copy-and-shear is performed.

73.1.1.1 Gestural Transformations

The most interesting case of gestural control is transformations. In this section, we describe in a mathematical way in which the user gestures are transformed into gestures on the canonical topological space of affine[4] morphisms in $\mathit{Aff}_2(\mathbb{R})$, and finally how we obtain the transformation morphism m that will be applied to the denotators represented by the selected objects.

All transformations in *BigBang* are currently two-dimensional affine transformations, which can be expressed as

$$y = Ax + b, \text{ with } A = \begin{pmatrix} a_{11} & a_{12} \\ a_{21} & a_{22} \end{pmatrix} \text{ and } b = \begin{pmatrix} b_1 \\ b_2 \end{pmatrix}.$$

For every mouse gesture, we find a gesture in $\mathit{Aff}_2(\mathbb{R})$ by a gesture morphism or throw morphism, which we defined earlier as a pair (u, v) with $u : \Gamma \to \Delta$ and $v : X \to Y$ such that $h \circ u = \vec{v} \circ g$ (Sections 61.5 and 74.2). Since on both sides we deal with the digraph with two edges and an arrow—this is what a simple click-and-drag gesture corresponds to—and thus $\Gamma = \Delta$, we can assume that $u(\gamma) = \gamma$ is the identity morphism on digraphs. All we thus need to do is define a $v : \mathbb{R}^2 \to \mathit{Aff}_2(\mathbb{R})$ for each transformation type.

In *BigBang*, each point of a click-and-drag mouse gesture is simply describable by two coefficients, λ_x, λ_y, which represent motion along the x- and y-axes of the currently selected view configuration (see Section 71.2.1). Thus, λ_x, λ_y can stand for any of the denotator parameters. For simplicity, we assume here that the scales of the two axes directly correspond to the scales of the denotator parameters. In practice, however, depending on the currently selected zoom level, we need an additional conversion algorithm.

All transformations also depend on location, except for translation in our case. For this, users define an additional center point $c = (c_x, c_y)$. Unless indicated otherwise, the center is automatically defined by the initial click of the click-and-drag gesture. In order to execute an affine transformation relative to a center, we first need to translate by $-c$, then apply $Ax + b$ and finally translate back by c. This can be packed into a simple constant. If we assume $y = Ax + b$, then

[4] $\mathit{Aff}_2(\mathbb{R})$ denotes the affine endomorphism set $\mathbb{R}^2@\mathbb{R}^2$.

$$y_c = y + (1 - A)c.$$

In the following discussion, we will omit this from the formulas for simplicity and only define $Ax + b$, even though c is always considered in *BigBang*.[5]

Translation

The most simple case is translation, where we can simply map the mouse space to the linear coefficient b. For this we define

$$v_T(\lambda_x, \lambda_y) = \begin{pmatrix} 1 & 0 \\ 0 & 1 \end{pmatrix} x + \begin{pmatrix} \lambda_x \\ \lambda_y \end{pmatrix}.$$

Rotation

Rotation is the only transformation that needs more than a click-and-drag gesture. First, users need to select a center around which to rotate, by simply clicking anywhere in the facts view. The center affects the rotation as seen above. Then a click-and-drag gesture decides the rotation angle. Here, we need more than λ_x, λ_y. Two points (x_1, x_2) and (y_1, y_2) are the starting and current dragging or ending points of the click-and-drag gesture. We map as follows:

$$v_{Ro}(x_1, x_2, y_1, y_2) = \begin{pmatrix} \cos\phi & -\sin\phi \\ \sin\phi & \cos\phi \end{pmatrix} x,$$

where ϕ is the angle around the center c determined by the angle between the straight line from c to (x_1, x_2) and the one from c to (y_1, y_2).

Scaling

For scaling, the initial click of the click-and-drag motion defines the center. λ_x, λ_y, defined by the dragging distance, determine the so-called *scale factors*:

$$v_{Sc}(\lambda_x, \lambda_y) = \begin{pmatrix} \lambda_x & 0 \\ 0 & \lambda_y \end{pmatrix} x.$$

If the shift key is pressed, we set $\lambda_y = \lambda_x$ to allow for equal scaling in both dimensions.

Shearing

Shearing works as the scaling does, where λ_x, λ_y define the *shearing factors*, where λ_x shears horizontally and λ_y vertically. We get the following formula:

$$v_{Sh}(\lambda_x, \lambda_y) = \begin{pmatrix} 1 & \lambda_x \\ \lambda_y & 1 \end{pmatrix} x.$$

Purely horizontal or vertical shearing can be performed by pressing the shift key during the click-and-drag gesture. If $\lambda_x \geqslant \lambda_y$, we set $\lambda_y = 0$, else $\lambda_x = 0$.

[5] For instance, for a scaling by λ_x, λ_y around c (see below) with

$$A = \begin{pmatrix} \lambda_x & 0 \\ 0 & \lambda_y \end{pmatrix} \text{ and } b = 0$$

we obtain

$$(1 - A)c = \begin{pmatrix} 1 - \lambda_x & 0 \\ 0 & 1 - \lambda_y \end{pmatrix} c = \begin{pmatrix} (1 - \lambda_x)c_x \\ (1 - \lambda_y)c_y \end{pmatrix}$$

and thus

$$v_{Sc} = \begin{pmatrix} \lambda_x & 0 \\ 0 & \lambda_y \end{pmatrix} + \begin{pmatrix} (1 - \lambda_x)c_x \\ (1 - \lambda_y)c_y \end{pmatrix}.$$

Reflection

Reflection is slightly more complex. The click-and-drag gesture determines the reflection axis rather than the positions of objects and reflected images, which could be done as well. We thus obtain

$$v_{Re}(\lambda_x, \lambda_y) = \begin{pmatrix} \frac{\lambda_x^2 - \lambda_y^2}{\lambda_x^2 + \lambda_y^2} & \frac{2\lambda_x \lambda_y}{\lambda_x^2 + \lambda_y^2} \\ \frac{2\lambda_x \lambda_y}{\lambda_x^2 + \lambda_y^2} & \frac{\lambda_y^2 - \lambda_x^2}{\lambda_x^2 + \lambda_y^2} \end{pmatrix} x.$$

Reflection is the only transformation that cannot be performed in a purely gestural way. We will see below that it can be easily gesturalized by interpolating through a projection on the axis. Here, however, we had to find a different solution. As soon as the initial click and a slight dragging motion is performed, the objects are abruptly reflected. However, as the user continues dragging, the axis is adjusted in a gestural way until a satisfying result is found.

Transformations in Wallpapers

The initial/final operations that frame the execution of a wallpaper are not gestural. `AddWallpaperDimension` simply decides that all following transformations, until `EndWallpaper` occurs, will be part of the wallpaper, and the two operations are executed by a simple click on the corresponding mode buttons. However, the way a wallpaper grows is always gestural, since the user applies regular transformations, executed as just described. For instance, if after an `AddWallpaperDimension` operation we start translating, we gesturally perform as many subsequent translations as determined by the range of the wallpaper dimension. Every transformation we perform afterwards has a similarly gestural effect.

73.1.1.2 Other Gestural Operations

In addition to the transformations just described, there are also other operations that can be considered gestural.

Drawing

For drawing with the mouse, which happens with the same click-and-drag gesture as above and triggers `AddObjects`, we can define a gesture morphism that, instead of going into the topology of affine transformations, directly reaches a topological space defined by the two denotator parameters associated with the x/y view parameters. For instance, if we draw *EulerNotes* on the *Onset* × *EulerPitch*1 plane (see Section 71.3.3), we can create a gesture morphism with u as above and $v : \mathbb{R}^2 \to \mathbb{Q} \times \mathbb{Z}$, if *Onset* is defined over $\mathbb{Q}$ and *EulerPitch* over $\mathbb{Z}$.

For each λ_x, λ_y, if we assume again a correspondence of view and denotator parameters as in Section 73.1.1.1, drawing defines an object with x/y parameters λ_x, λ_y. In reality, even though such a gesture morphism defines an infinite number of objects, only a finite number are created due the discrete nature of mouse movements (pixel by pixel) in combination with a purposeful time constraint that limits the number of objects drawn each second. However, by zooming in the facts view, objects can be created as closely together as necessary.

In sum, drawing could be considered the most gestural of all operations, since, for objects in **Powers** or **Lists**, *BigBang* does not only remember the last state, as it is true for transformations, but creates and remembers all objects in order reached along the path. We will see later on that this has implications for gesturalizing, since we do not have to reconstruct a gesture but can in fact use this trace for gesturalizing.

Shaping

Shaping works in a similar way to drawing, in terms of how it can be defined gesturally. The image space of the topological part v of the gesture morphism is also two-dimensional. However, while the second dimension is also the denotator space associated with the y-axis parameter, the first dimension is a discrete space defined by the set of all present values of the denotator parameter associated with the x-axis space.

Nevertheless, shaping remembers all shaping locations as elements of $\mathbb{R}^2$, which makes one shaping gesture applicable to any denotator, if for instance the input composition changes, or more objects are inserted at an earlier stage of the process.

Alteration

When performing an alteration, users have gestural control over the alteration degrees dg_1, dg_2 (see Section 72.2.1.6) over two sliders in the top toolbar, which can be considered one-dimensional gestural controllers. Initially, both degrees are 0, which means that we see and hear the unchanged composition. Then, as soon as the sliders are moved, the composition O_1 moves continuously towards O_2.

The configuration with two sliders makes it impossible with the mouse to control both degrees at the same time. However, this could be solved in the future using other controllers or a two-dimensional "slide field" instead of sliders, as in many sequencing softwares.

73.1.1.3 Non-gestural Operations

Several operations are not defined in a gestural way. Deleting, building satellites, and flattening are all based on a selection of objects and happen at once, as described earlier on, upon a menu or keyboard command. For all of these, gestural versions are conceivable, but only partially implemented. For instance, deleting is possible in a gestural way when selecting an `AddObjects` operation and holding the shift key while clicking-and-dragging. This way, users can *undraw* notes previously drawn. In a similar way, instead of adding satellites, users can draw satellites simply by entering drawing mode and selecting the satellite level on which they would like to draw (described in Section 72.2.1.1). Despite their limited gesturality, these operations are all gesturalizable, as we will explain below.

The two framing wallpaper operations, as seen above, are the only operations that are neither gestural nor gesturalizable. They are simply discrete events with structural consequences for denotators and thus also need to be part of the process graph.

73.1.2 Affine Transformations and Multi-touch

When using controllers other than the mouse, users also have the chance to directly define more general affine transformations. Such transformations combine all geometrical ones defined above. Before discussing how this works, we will briefly summarize how commonly used transformational multi-touch gestures work.

Multi-touch devices typically support the three gestural types *drag*, *pinch*, and *twist*, shown in Figure 73.4, which are all executed using two fingers and which correspond to the geometrical transformations translation, scaling, and rotation. Drag works the same way as the mouse gesture defined above, with the difference that λ_x, λ_y are determined by the average position of the two fingers. In contrast, the gestural space of the other two gestures is not directly determined by finger position, but by a certain relationship between the two fingers used. For pinch, the distance between the two fingers determines a gesture on a one-dimensional topological space, and for twist it is the angle at which the fingers are placed that defines the topological space. Both gestures could thus be independently expressed as $g : I \to \mathbb{R}$.

Since these three parameters are all defined independently they can be used simultaneously. We can thus define a four-dimensional gesture $g' : I \to \mathbb{R}^4$ the components of which are $\lambda_x, \lambda_y, \lambda_p$ and λ_t for x- and y-position, pinch, and twist, with which we can simultaneously translate, scale, and rotate.

In an earlier paper, we generalized these gestures for two-dimensional affine transformations by adding a third finger [1044]. We defined three fingers $f_i = (p_i^s, p_i^e)$ with finger indices $i = 1, 2, 3$ with starting point

Fig. 73.2. The three most common two-dimensional multi-touch gestures: (a) drag.

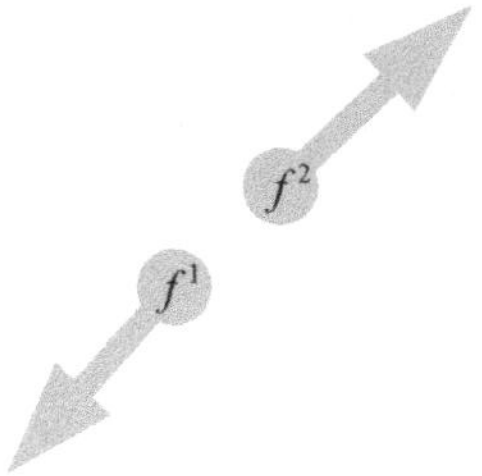

Fig. 73.3. The three most common two-dimensional multi-touch gestures: (b) pinch.

Fig. 73.4. The three most common two-dimensional multi-touch gestures: (c) twist.

p_i^s and intermediary or ending point p_i^e, along with four vectors $v^j = p_2^j - p_1^j$, $w^j = p_3^j - p_1^j$ with $j = s, e$. Figure 73.5 visualizes these components. We also define the $d_i = p_i^e - p_i^s$ and $\hat{v}_j = \frac{v_j}{|v_j|}$. We then obtain the following gestural transformation parameters:

- For two fingers, the *translation* component is defined by $\frac{d_1+d_2}{2}$,
- the *scaling* component is $\frac{v^e}{v^s}$, and
- the *rotation* component is $\arccos\left(\frac{v^e}{|v^e|} \cdot \frac{v^s}{|v^s|}\right)$.
- In addition to the above parameters, we obtain the *shearing* parameter, which is the projection length $|(d_3 \cdot \hat{v}^e)\hat{v}^e|$, and
- the *reflection* component defined by the projection length of $|(d_3 \cdot \hat{u}^e)\hat{u}^e|$, where $\hat{u}^e$ is a vector perpendicular to v^e.

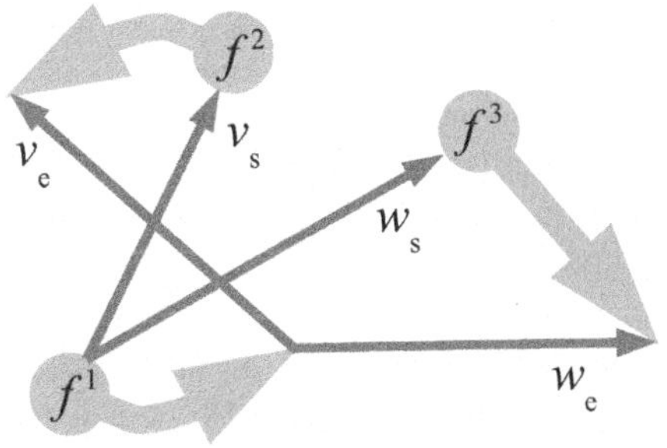

Fig. 73.5. The components resulting from a three-finger gesture.

With this, all geometrical transformations available in *BigBang* can be performed simultaneously. However, if we just wish to perform a reflection, we can hold fingers f_1 and f_2 steady at a distance, which can be seen as the reflection axis, and then move the third finger in a motion perpendicular to this axis (Figure 73.7 (a)). For a shearing, f_3 should move in parallel to the f_1-f_2-axis (Figure 73.7 (b)).

If we forget about the components just defined, we can move the three fingers around freely and perform any conceivable two-dimensional affine transformation, of course limited by physical constraints.

73.1.3 Dynamic Motives, Sound Synthesis, and Leap Motion

The most complex controller currently supported by *BigBang* is the Leap Motion controller. It can be used for precisely the same things as multi-touch, including two-dimensional affine transformations [1049, p.4]. *BigBang* thereby recognizes up to three fingers, projects them onto the plane perpendicular to the user's

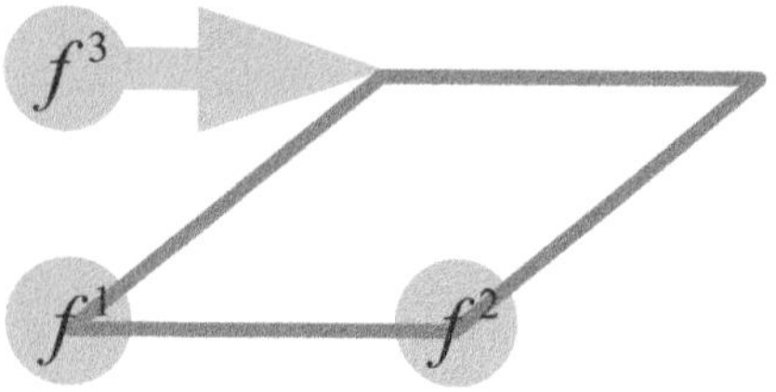

Fig. 73.6. The two three-finger gestures for (a) shearing.

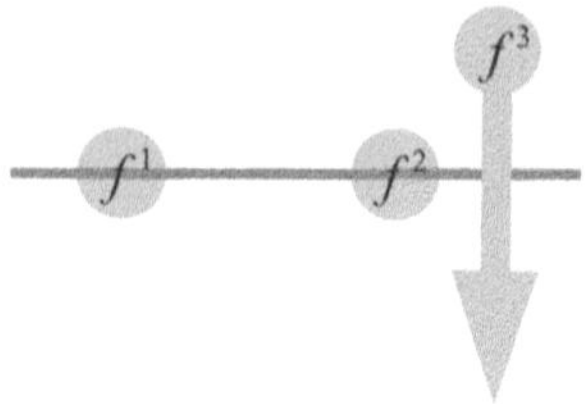

Fig. 73.7. The two three-finger gestures for (b) reflection.

viewing direction (perpendicular to the z-axis at $z = 0$), and uses a procedure similar to the ones described in the previous section to perform two-dimensional affine transformations.

However, here we will be concerned with another functionality, in order to show the gestural possibilities of *BigBang*. Leap Motion can also be used to draw objects. In contrast to the procedure described above, where each gestural position generates an object, there is also the possibility to create and replace objects. When using the Leap Motion we treat each fingertip as a denotator and map the (x,y,z) location of each finger using a linear scaling into the coordinate system represented as currently displayed by the *BigBang* rubette. Whenever the fingers move around the corresponding denotators are adjusted, which provides an immediate visual and auditive feedback. From there, we have the option to capture the currently defined denotators and keep adding new ones using the same method. If we use all three dimensions of the Leap Motion space, capturing is only possible with an external trigger (such as a MIDI trigger). To avoid the use of an external trigger the user can decide to use only two dimensions for drawing (preferably $x \times y$) and the third dimension for capturing, whenever a certain threshold, e.g. the plane perpendicular to the z-axis at $z = 0$, is crossed.

Figure 73.8 shows a situation where the modulators of a carrier in an *FMSet* are defined using Leap Motion. Their arrangement directly corresponds to the fingertips in space, as can be verified visually. Compared to drawing with a mouse or another device, this method has significant advantages. The user can quickly compose complex musical structures while being able to smoothly preview each step until satisfied. Furthermore, the user can also easily edit musical objects added earlier in the process in the same continuous way, which has many musical applications. The high precision of the Leap Motion makes this method just as accurate as using a mouse or trackpad.

One of the most useful musical applications of this way of generating objects is to go back to editing the `AddObjects` operation, in the manner described in Section 72.3.2, after several transformations have been performed. This way, users can gesturally redefine the motif that was transformed, and the entire following composition process is immediately applied to every gesturally changed state of the motif. This is especially interesting when the transformations consist in copying-and-transforming, which can yield an entire composition created from the same motif. Even more powerful is the use of the wallpapers to transform a motif, where the motif can virtually be grabbed by the user and moved around, upon which the entire wallpaper moves accordingly. Figure 73.9 shows an example of such a wallpaper, where the motif has a recognizable hand shape defined by the user.

Instead of defining motifs in a composition or improvisation, users can also design sounds when choosing appropriate forms. For instance, while playing the keyboard, the positions of the fingers over the Leap Motion controller can be directly mapped to carrier oscillators or frequency modulators, as shown in Figure 73.8, and each hand movement changes their parameters. Furthermore, in a similar way, the user can create sounds and transform them gesturally in any of the geometrical transformation modes. This way, instead of changing simple parameters in a linear way as with commonly available synthesizer interfaces, multiple parameters can be changed in a complex way, for instance by manipulating both frequency and amplitude of hundreds of oscillators around a defined sound center.

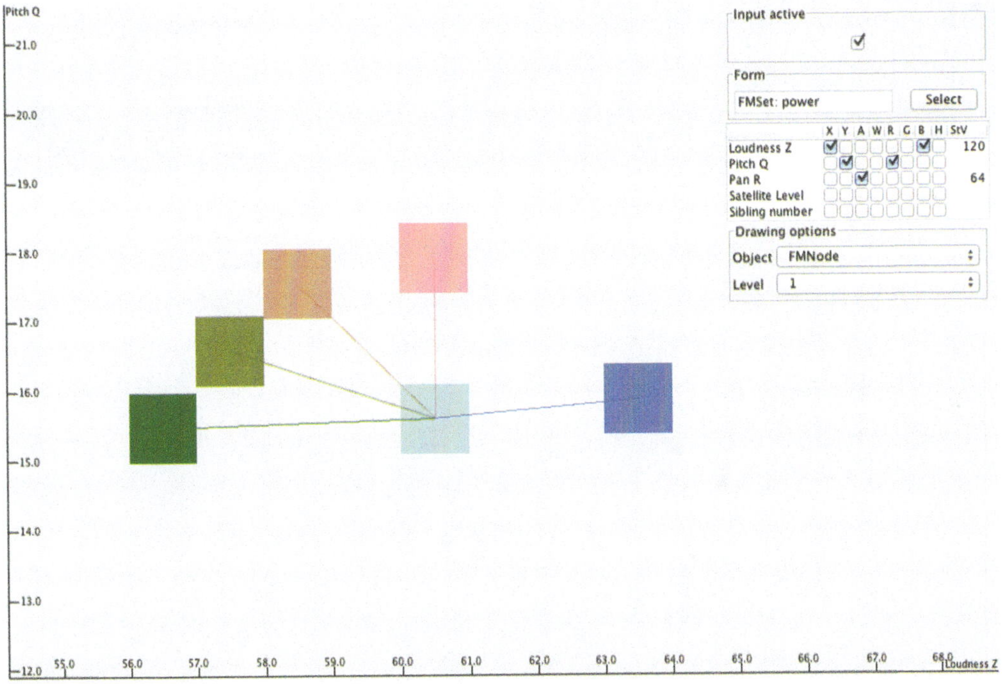

Fig. 73.8. An *FMSet* denotator consisting of a carrier and five modulators defined by the fingertips of the user.

Depending on form and transformation choice, users have almost infinite possibilities of dynamically mapping their gestural parameters to musical parameters. While translations map one-to-one, the other transformations have the potential to map a simple gesture to several parameters.

73.1.4 Recording, Modifying Operations and MIDI Controllers

Finally, here is a fourth way of controlling *BigBang* in a gestural way. Several types of MIDI controllers were made available, including keyboard controllers, mixing controllers, and combined ones. Keyboard controllers can be used to record MIDI notes into *BigBang*, by converting them not only into *Score* denotators but into any denotators containing *Loudness* (from velocity), *Pitch*, or temporal **Simple** forms, in a similarly versatile way to that of the playback function discussed in Section 71.5.3. For instance, when working with a *Spectrum*, the temporal parameters of the MIDI input are ignored, while *Onset*/*Pitch* objects are added in a way similar to that of drawing mode.

While the above does not conform with the conditions for gestural control (note on/off events cannot be considered continuous), there are other uses of MIDI that are more gestural. The knobs and sliders on many devices send control changes that are gestural, even in a discrete space ($g : I \to \mathbb{Z}$). Currently, such control changes are mapped to the modification of operations and transformations. All knobs and sliders are assigned to the operations in the order in which they were added to the graph. For instance, the 16 knobs of the *E-MU Xboard* are assigned to the 16 first operations, regardless of whether they occur in linear, alternative, or parallel processes. Since for each controller the control change assignments may vary, they all have to be configured individually.

For each control change, the sent values, integers in $[0, 127]$, are mapped to real numbers within $[0, 2]$, where 0 corresponds to the identity, 1 to the original operation, and 2 to double the operation. The latter value is then used to replace the operation's values or morphism by a new one found at the corresponding point on the gesture. How this is done will be discussed in detail in the next section. For now, an example will suffice: if the modified operation is a rotation by 45°, MIDI value 31 will be mapped to 0.5, and will thus modify the angle to 22.5°, whereas $127 \to 2$ will lead to 90°. Almost all operations can be modified this

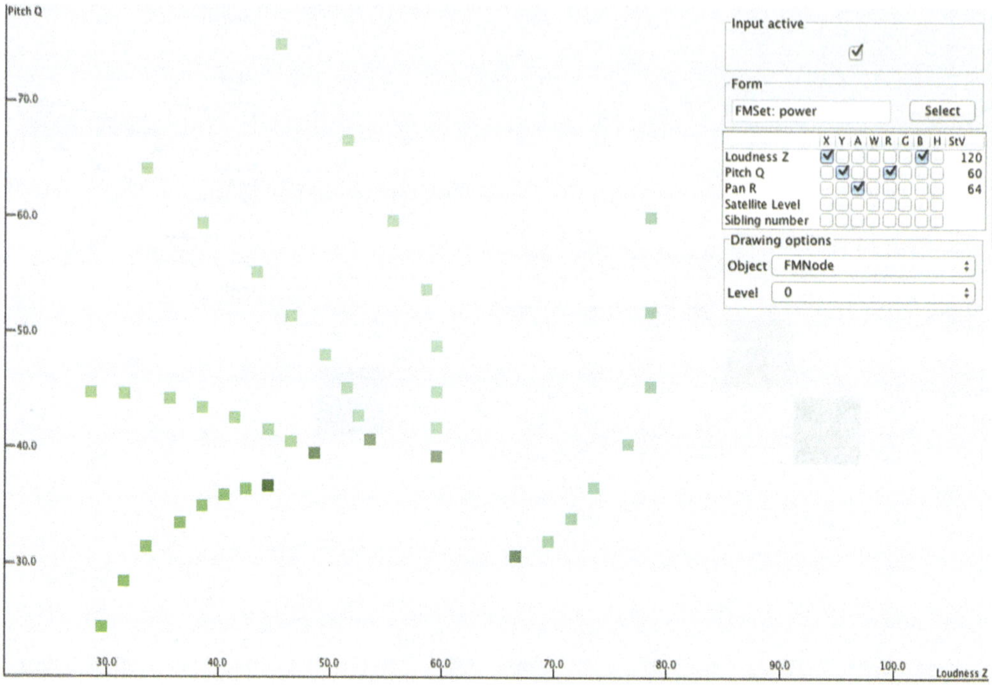

Fig. 73.9. A wallpaper with a motif defined by the fingers of a hand.

way. However, only some, including all transformations, can be extended to double their amount. Table 73.1 shows which operations can be modified and lists all the ranges.

73.2 Gesturalizing and the Real BigBang: Animated Composition History

When we started designing the *BigBang* rubette, we chose *BigBang* as the working name for the prototype, because of the innumerable possibilities it brings to RUBATO® *Composer*. Meanwhile, this name has gained an initially unexpected and highly appropriate new meaning. The operation graph recounts the evolution of a sounding universe, which shows many parallels to the evolution of our physical universe. An initial group of musical objects expands and multiplies by being copied and transformed into a highly complex musical structure based on rules of symmetries. The ultimate functionality in *BigBang* is a gestural animation of this evolution, from the initial compositional actions to the actual state. In this section, we describe how this second type of gesture can be created.

As seen above, *BigBang* saves processes rather than gestures. From these processes, we can not only generate facts, but also turn the processes back into gestures. The construct of a gesture ensures continuous and unidirectional motion in its topological space, by anchoring it in the interval I. Animating a gesture is thus straightforward: we just need to gradually interpolate on I, which gives us a sequence of points in the topological space, be it affine transformations ($A\!f\!f_2(\mathbb{R})$) or any other structure. However, how do we get gestures from processes, which consist of merely a point in the corresponding topological space? We will start by answering this for transformations, and then move on to operations.

73.2.1 Gesturalizing Transformations

We saw that what *BigBang* keeps from the gestures performed by the user when transforming are merely the ending points in the topological space, the final morphisms. However, it also remembers which type of transformation the user was executing, which is helpful for reconstructing a standard gesture. As above, in the following definitions we ignore center c. In reality, we keep c constant during the entire gesture.

73.2.1.1 Translation

In Section 73.1.1.1, we saw that translations merely consist in the b-part of $Ax + b$. Thus, for a given translation

$$x + \begin{pmatrix} b_1 \\ b_2 \end{pmatrix},$$

all we need to do to create a gesture is define $g : I \to \mathit{Aff}_2(\mathbb{R})$ as follows:

$$g(i) = x + \begin{pmatrix} ib_1 \\ ib_2 \end{pmatrix},$$

for $i \in I$.

73.2.1.2 Rotation

For rotations, we interpolate on the angle ϕ and calculate the appropriate element of $\mathit{Aff}_2(\mathbb{R})$ as above. We thus define

$$g(i) = \begin{pmatrix} \cos i\phi & -\sin i\phi \\ \sin i\phi & \cos i\phi \end{pmatrix} x.$$

73.2.1.3 Scaling

For a scaling

$$\begin{pmatrix} a_{11} & 0 \\ 0 & a_{22} \end{pmatrix} x$$

we define

$$g(i) = \begin{pmatrix} 1 + i(a_{11} - 1) & 0 \\ 0 & 1 + i(a_{22} - 1) \end{pmatrix} x.$$

73.2.1.4 Shearing

For a shearing

$$\begin{pmatrix} 1 & a_{12} \\ a_{21} & 1 \end{pmatrix} x$$

accordingly

$$g(i) = \begin{pmatrix} 1 & ia_{12} \\ ia_{21} & 1 \end{pmatrix} x.$$

73.2.1.5 Reflection

Finally, we interpolate a reflection

$$\begin{pmatrix} a_{11} & a_{12} \\ a_{21} & a_{22} \end{pmatrix} x$$

by traveling through a projection on the reflection axis by doing the following:

$$g(i) = \begin{pmatrix} ia_{11} + (1 - i) & ia_{12} \\ ia_{21} & ia_{22} + (1 - i) \end{pmatrix} x.$$

73.2.1.6 Affine Transformations

The procedure for reflections also works for any affine transformation, since we interpolate between the identity matrix and any arbitrary matrix. What is missing in this formula is the b part, which we can deal with as discussed in the translation section. Thus:

$$g(i) = \begin{pmatrix} ia_{11} + (1-i) & ia_{12} \\ ia_{21} & ia_{22} + (1-i) \end{pmatrix} x + \begin{pmatrix} ib_1 \\ ib_2 \end{pmatrix}.$$

However, the problem with this is that somewhere on the way, we might encounter singular projections that may not be musically optimal. In a former paper, we suggested the use of Bruhat standardized transformations in order to decompose affine transformations into their geometrical parts, and finally gesturalize on these obtained parts, which leads to more satisfying musical results [730].

73.2.1.7 Gesturalizing Beyond the Transformation

As mentioned in Section 73.1.4, transformations cannot only be gesturalized in the interval $[0, 1]$ but even beyond it. The $i \in I$ in the formulas in this section can simply be replaced by an $r \in \mathbb{R}$, for which we get an extended gesture of infinite length. If we keep $r \in [0, 2]$, we get what we described above, and we can obtain exaggerated versions of the transformations, of up to double the amount.

73.2.2 Gesturalizing Other Operations

Almost all other operations can be gesturalized as well (see Table 73.1). We thereby distinguish between so-called `ObjectBasedOperations` that operate on a single set of objects. They include `AddObjects`, `DeleteObjects`, `InputComposition`, `BuildSatellites`, `Flatten`, and `Shaping`. For these operations, we interpolate on the number of objects the operation manipulates. We define a function $o : I \to [0, n]$, n being the number of objects, and $o(i) = in$. For instance, if a `BuildSatellites` operation adds 30 objects as satellites of any anchors, for $i = 0.2$ it only adds the first six objects. For operations that remember the order of their objects, such as `AddObjects`, the objects are manipulated in order. This leads to an accurate reconstruction of a drawing gesture, as described above.

Two other operations can be gesturalized in a different way, even though during gesturalization `AddWallpaperDimension` is ignored, since a wallpaper only evolves through its transformations. However, `AddWallpaperDimension` can be modified in the way described in Section 73.1.4. Then, $[0, 2]$ is simply used to adjust the upper range r^{max} of the wallpaper dimension, i.e., $o'(i) = i * r^{max}$ becomes the modified upper range. Finally, for `Alteration`, gesturalization affects the two alteration degrees dg_1 and dg_2. Thereby, $o''(i) = (idg_1, idg_2)$ is the modified pair of alteration degrees.

73.2.3 Using Gesturalization as a Compositional Tool

In *BigBang*, pressing on the *Gesturalize* button in the upper part of the process view initiates a gesturalization of the shortest path that connects composition state 0 and the currently selected state in the operation graph, or the last added state if no state is selected. Each gesturalizable operation along the way is gesturalized until the current state is reached. Users can specify an arbitrary duration for each operation. At each point in time, the current state is visualized and sonified as described above. Parallel operations are gesturalized simultaneously, despite their logical succession.

Even though gesturalizing can be used to reconstruct the composition process, it can become part of the composition itself. For instance, composers can design continuously evolving textures, by defining continuously sounding objects such as *FMSets*, transforming them in various ways, and finally creating a temporal structure by selecting various durations for the transformations. This way, the gesturalized structure becomes the actual composition.

This can also be done in a more improvisational way, by using a slider at the top of the process view. The space of the slider represents the entire gesturalization and by moving the slider back and forth, users can continuously travel back and forth in the compositional evolution, while hearing the respective temporal states.

74

Musical Examples

The new *BigBang* rubette offers many possibilities of creating music, due to the great variety of forms that can be defined. We already presented some simple ideas of forms in Section 71.5. In this section, we introduce some of innumerable slightly larger musical examples created in the course of writing the code of *BigBang* and the corresponding thesis. These examples illustrate a variety of composition techniques and types of musical results possible with *BigBang*. All examples are available for listening on SoundCloud, and some of the more performative ones can be found on YouTube, at the addresses indicated below.

74.1 Some Example Compositions

This section explores some of the compositional possibilities of *BigBang*, i.e., preparing music outside of musical time that can later be played back, recorded, or performed. More spontaneous and real-time methods of creating music with *BigBang* will be discussed in the next section.

74.1.1 Transforming an Existing Composition

Form *Score*
Graph four states, three sequential operations
Techniques inputing a composition, transforming, modifying
Output *BigBang* synth with sine wave oscillators, slightly post-processed
Link `http://www.soundcloud.com/bigbangrubette/k003`

Instead of creating denotators from scratch, there are many ways in which existing compositions can be used to create strikingly different musical results. This example is part of a series of variations based on Sonatas by Domenico Scarlatti, all of them using composition procedures that are as simple as possible. Here, we input Scarlatti's *K003* into *BigBang* via a *MidiFileIn* rubette, thus in *Score* form; then we stretched and compressed it in time and pitch, respectively (`ScalingTransformation`), and finally transposed it down several octaves (`TranslationTransformation`). The resulting graph therefore consists of three sequential operations (see Figure 74.1). Using the option of modifying transformations, we found the range we envisioned, resulting in a pulsating bass sound emerging from the beating based on the close frequencies after the pitch compression. The clicking noise, resulting from a chosen short attack time of the *BigBang* synthesizer, preserves the rhythmical qualities of the Scarlatti. The visualization of the final result (Figure 74.1) was partially created due to aesthetic decisions. However, it shows the composition on the $Onset \times Pitch$ plane, where the close *Pitch* range is visible (the vertical middle of the blocks), around MIDI pitch 24, which corresponds to $C1$ or approximately 32 Hz.

G. Mazzola et al., *The Topos of Music III: Gestures*, Computational Music Science,
https://doi.org/10.1007/978-3-319-64481-3_19

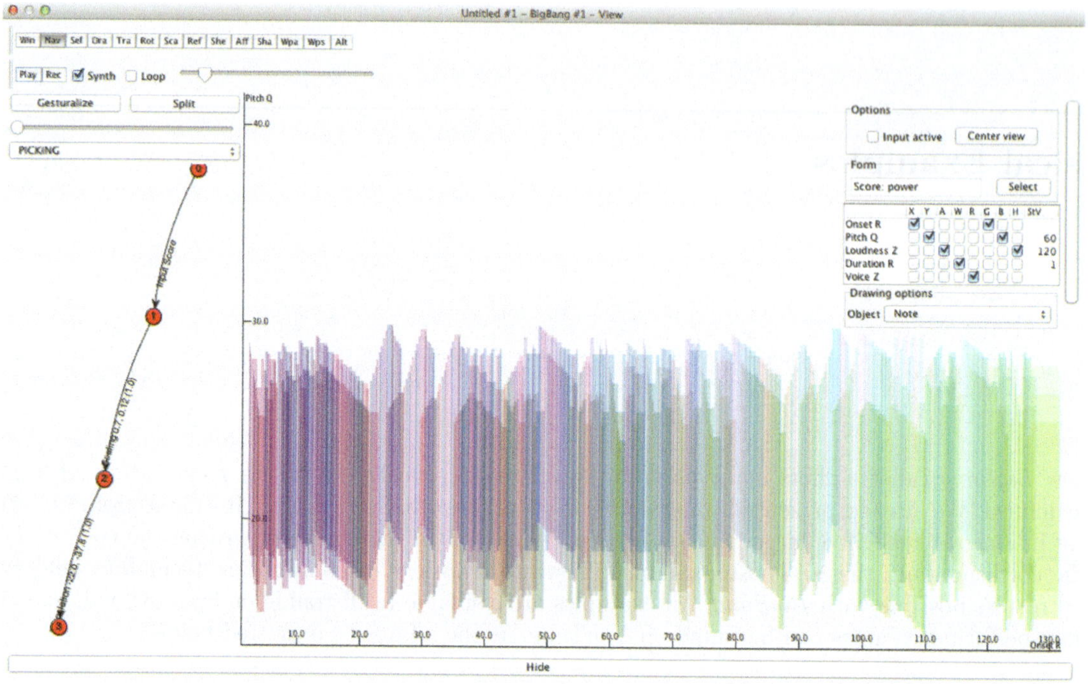

Fig. 74.1. A transformation of Scarlatti's Sonata *K003* resulting in a pulsating bass sound.

74.1.2 Gesturalizating and Looping with a Simple Graph

Form *Score*
Graph two states, two parallel operations
Techniques looping, gesturalizing
Output MIDI to Ableton Live, Guitar-Jazz preset
Link `http://www.soundcloud.com/bigbangrubette/k002`

Another piece part of the Scarlatti series, this example uses *BigBang*'s gesturalizing function. Its graph consists of merely two states, between which we find two parallel operations. Again, the original (*K002*) enters through a *MidiFileIn* rubette, resulting in an `InputCompositionOperation`. Then, by selecting the operation and performing a counterclockwise rotation by 180 degrees, we add a parallel `RotationTransformation`, which results in a minimal graph with two states and two operations (Figure 74.2). When gesturalized, these two operations occur simultaneously (see Section 72.3.3), so that the composition simultaneously grows note by note, and gradually rotates. During gesturalization, the composition is played back in loop mode, where in this case, the loop grows longer and longer, and outputs through MIDI directly to Ableton Live, where it is played back by a guitar sound. In order to find a musical result, we experimented with operation duration and tempo, settling on a gesturalization time of 200 seconds at a pace around two to three times as fast as the tempo the sonata is often played at. The resulting piece has a strong improvisational and gestural quality, where the motivic content is gradually developed and grows larger and larger. A contrapuntal effect reminiscent of group improvisation emerges due to the variety of pitch ranges produced by the counterclockwise rotation, which are well captured by the lower end of the guitar sound. In the end, the musical material converges towards the retrograde inversion of the Scarlatti, modulating increasingly slowly, and culminating in a congenial closure. Figure 74.2 shows the piece shortly after midway through the gesturalization.

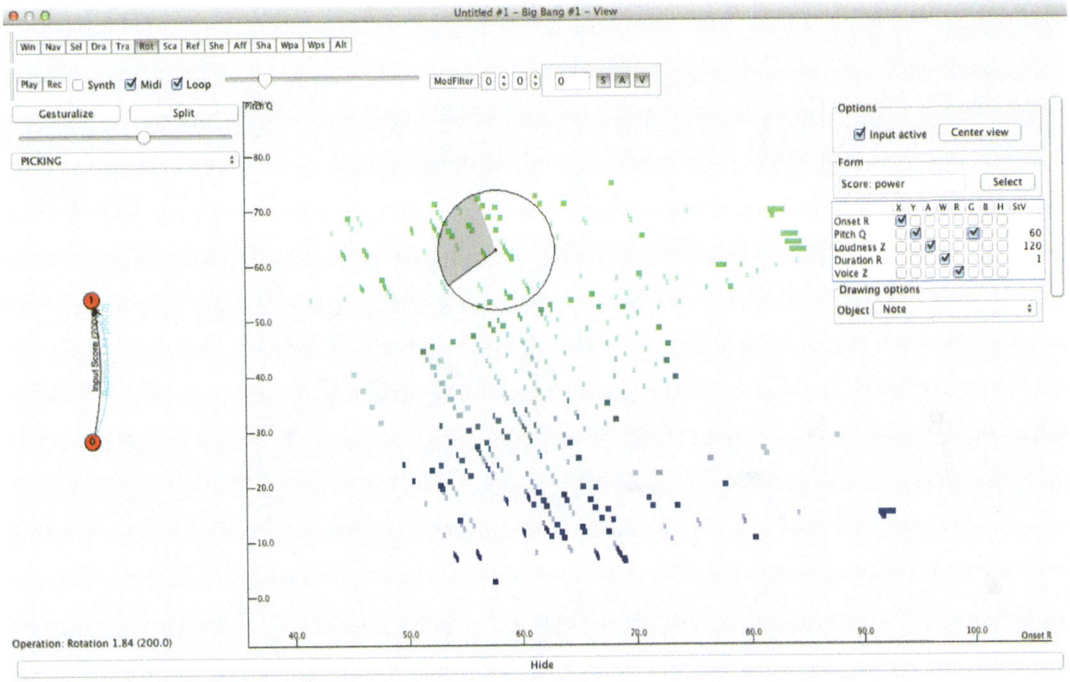

Fig. 74.2. A growing and rotating Scarlatti *K002* during gesturalization.

74.1.3 Drawing UPIC-like Motives and Transforming

Form *PanScore*
Graph long sequential graph
Techniques drawing, shaping, copy-and-transforming
Output *BigBang* synth with sine wave oscillators, post-processed
Link `http://www.soundcloud.com/bigbangrubette/upic`

In comparison with Xenakis's *UPIC* system [652], *BigBang* has several advantages, the two most important of which are that composers can work with arbitrary musical object types, and that they can transform these objects. This example makes use of the latter, while keeping a similar data type as was used with Xenakis's system, *Score*, however, an enhanced version that allows for stereo panning, which we called *PanScore*. The example is based on a drawn structure with ramifications similar to, for instance, parts of Xenakis's *Mycenae Alpha*. Since drawing can only occur in two dimensions at a time, we used shearing transformations as well as the shaping operation to process the drawn structure in dimensions other than *Onset* and *Pitch*, here mainly *Pan*. Then, we multiplied the initial motive and partial motives by using various copy-and-transform operations (Section 72.2.2), a simple and intuitive way of ensuring motivic unity in a piece. Figure 74.3 shows the score on the UPIC-typical *Onset* × *Pitch* plane. Figure 74.4 shows the results of the shaping and shearing on the *Onset* × *Pan* plane, with the same color distribution as in Figure 74.3. Note that the images do not contain the original graph, which was of linear nature and almost confusingly long, for the first result was saved (using the *Register* and *Source* rubettes) and worked on in several sessions. However, the original graph was of a linear nature. The result is a microtonal spectral composition that reiterates the initial motive in increasingly contracted and cut out versions, evolving from a single voice to about 45. The result was post-processed in Ableton Live, with some reverb, equalizing, and compression.

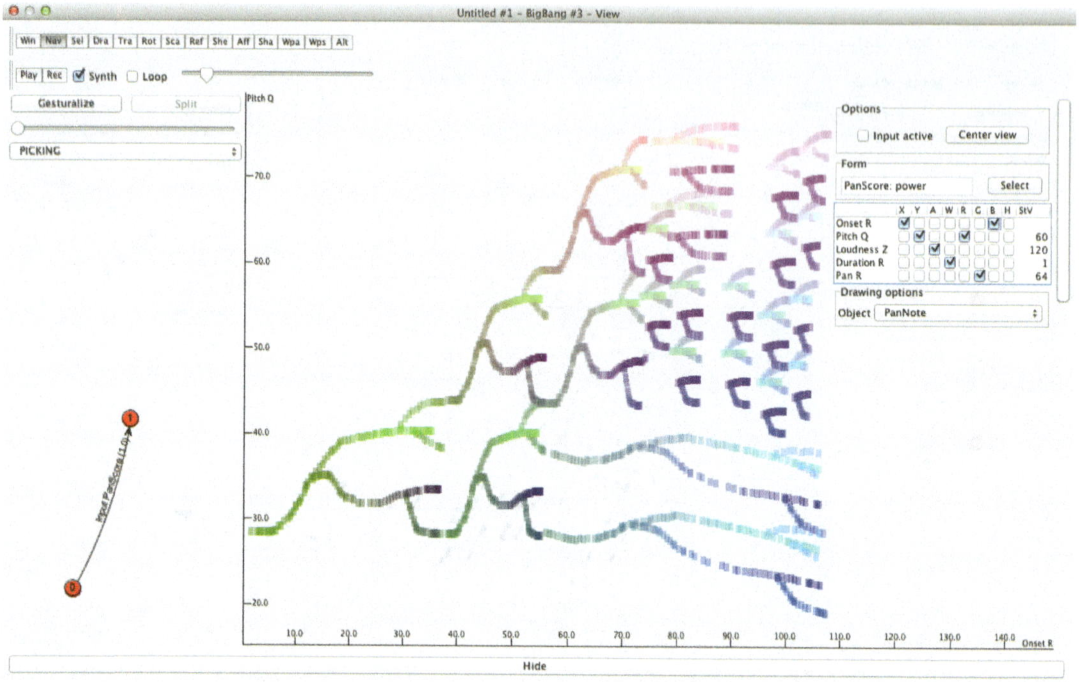

Fig. 74.3. The *Onset* × *Pitch* plane of the *UPIC*-like composition.

74.1.4 Drawing Time-Slices

Form *PanScore*
Graph just a drawing operation
Technique drawing, changing preset values
Output *BigBang* synthesizer with triangle waves, post-processed
Link `http://www.soundcloud.com/bigbangrubette/slices`

This example illustrates another technique of drawing in several dimensions. Instead of switching to other planes and shaping and transforming drawn motives, as described in the previous example, it is also possible to determine the values in the dimensions absent from the x/y plane by entering standard values into the boxes to the right of each denotator value row in the view parameters table (right-hand side of *BigBang* interface). In this case, we drew a *PanScore* on the *Pan* × *Pitch* plane, while manually entering *Onsets* and *Durations*. This way, starting with $Onset = 0$ and $Duration = 2$, we were able to draw overlapping slices of the same duration, by increasing *Onset* step by step, and by drawing increasingly many *PanNotes* in approximate concentric circles. For the second part of the piece, we drew single longer pitches, all of them at $Onset = 12$, first with $Duration = 12$, then decreasing the duration step by step, which resulted in a gradual disappearing of the pitches of the final chord. The piece is played back using triangle wave oscillators and mastered in Live. Figure 74.5 shows the drawing plane, whereas Figure 74.6 shows the resulting slices in a temporal representation, the colors being kept the same for both representations, in order to show the respective missing spatial dimension.

74.1.5 Converting Forms, Tricks for Gesturalizing

Form *Texture*
Graph several sequential and parallel operations
Technique reforming, identities, immediate operations, gesturalizing, selecting states
Output MIDI to Ableton Live, string ensemble

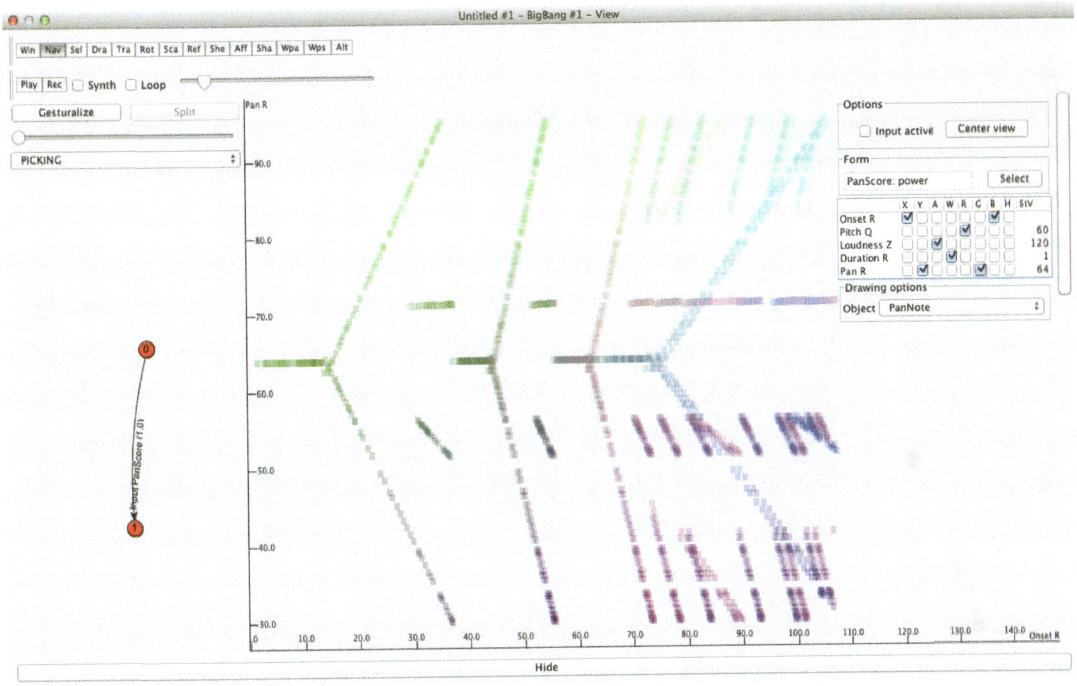

Fig. 74.4. The *Onset* × *Pan* plane of the *UPIC*-like composition.

Link `http://www.soundcloud.com/bigbangrubette/textures`

This example is slightly more complex. It uses a network with three rubettes. The *Texturalize* rubette converts a *Score* into a *Texture* (introduced in Section 71.5.2) based on analytical values. It gathers all pitches present in the *Score* and for each *Pitch* p, remembers the number of occurrences o_p, the average duration d_p, as well as the average loudness l_p. The output of the *Texturalize* rubette is then a *Texture* with a *RepeatedNote* for each p, with a *Rate* based on o_p, a *Duration* based on d_p. and a *Loudness* based on l_p. The *Texture* can thus be seen as a scrambled but regularized version of the input piece, with the same average tone material, resembling the textures of the American Minimalists.

The current example makes use of a *Texture* based on a part of a live performance of Frédéric Chopin's *Ballade Op. 23* in *G* minor, at the indication *agitato* and then *sempre più mosso*. The fact that it is a live performance leads to a great variety of durations and dynamic values, as opposed to the notated score, which is particularly interesting when converted into a *Texture*. Figure 74.7 shows the facts at the initial stage (composition state 1).

The first part of the example, played by a string ensemble, is based on a gesturalization of various transformations of the original texture, which results in slowly changing rates, durations, loudnesses, and pitches of the texture's notes. In addition to the evolving parts, we also wanted to include parts where the current texture is resting. Currently, the trick to do this is to insert an identity transformation, for instance, a translation by 0, as in the example, and assign the transformation a gesturalization duration. Another trick is used in the beginning of the piece, where we did not want the texture to gradually build up when gesturalized. For this, we assigned the `InputCompositionOperation` a duration of 0.

The entire trajectory of the first part reaches three stable states, one at the original texture, one at a lower, quieter, and more legato retrograde inversion of the original, and one at a variation that is louder, faster, staccato, and that consists of an extended pitch space. These variations of the original texture come about using parallel transformations, which are equally gesturalized, with the effect of two gradual textural changes between the three static parts. The first transition consists of a translation down in loudness, a rotation by 180 degrees on the loudness/pitch plane, and a scaling in duration in order to make the notes

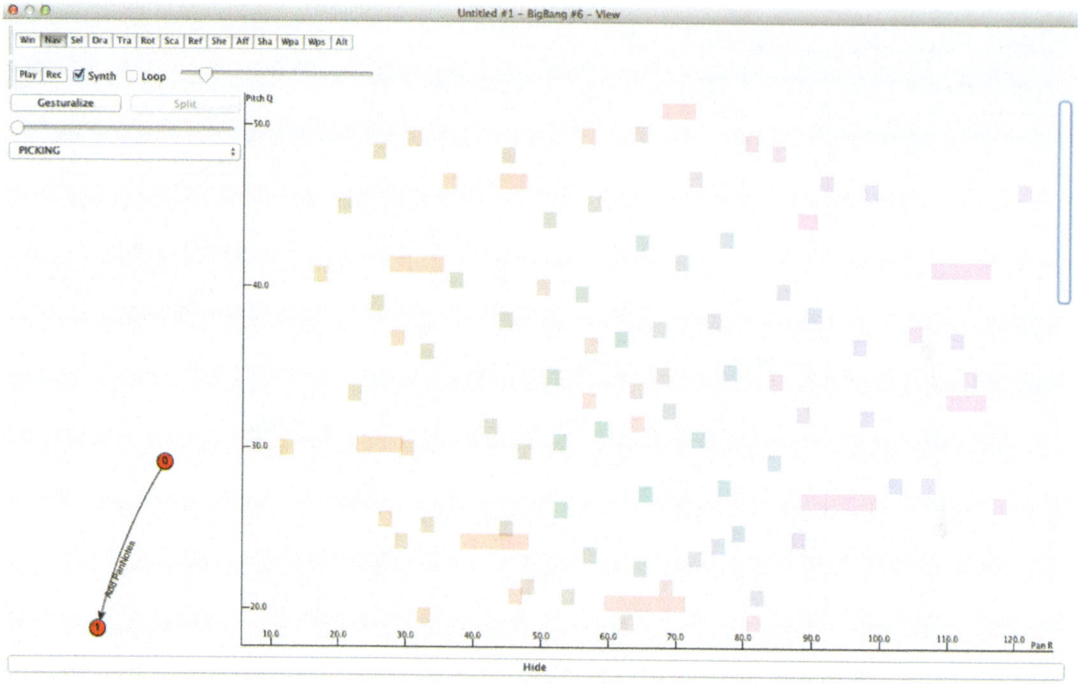

Fig. 74.5. The *Pan* × *Pitch* plane on which drawing took place.

longer. The second transition consists of a translation up in loudness, a scaling and a translation on the rate/duration plane, and a scaling on the loudness/pitch plane in order to expand both dynamic range and pitch range. At the end of the first part, the piece jumps back to the original texture, which was done manually by selecting composition state 1.

The second part of the piece, realized by pizzicato strings, makes use of a technique that rather fits into the part on improvisation and performance with *BigBang*. Using the same graph, we recorded the strings playing while we jumped from composition state to composition state, using the number keys of the computer keyboards. This way, the strings freely jump back and forth between the three textural states and remain static for various amounts of time.

74.1.6 Gesturalizing a Spectrum

Form *Spectrum*
Graph many sequential and parallel transformations
Technique drawing, selecting and transforming groups of objects
Output *BigBang* synthesizer with sine waves, post-processed with ring modulation
Links `http://www.soundcloud.com/bigbangrubette/spectrum4`
`http://youtu.be/JlIpj0lKYUc`

Gesturalization was illustrated in the examples in Sections 74.1.2 and 74.1.5. However, the two forms used there, *Score* and *Texture*, define denotators with a distinct temporal existence, through the **Simple** forms *Onset*, *Duration*, and *Rate*. As described in Section 71.4, if none of these forms are present, the musical objects sound constantly when played back. This is especially interesting when they are gesturalized, which results in microtonal glissandi. In this example, we used the *Spectrum* form to create a slowly evolving spectral texture. The gestural result starts out by gradually adding *Partials*, in the order we defined them. After that, the *Partials* are transformed either sequentially or in parallel, in pairs or as a whole, resulting from differently timed transformations based on various selections of *Partials*. Figure 74.8 shows a moment during one of the sequential scalings of a pair.

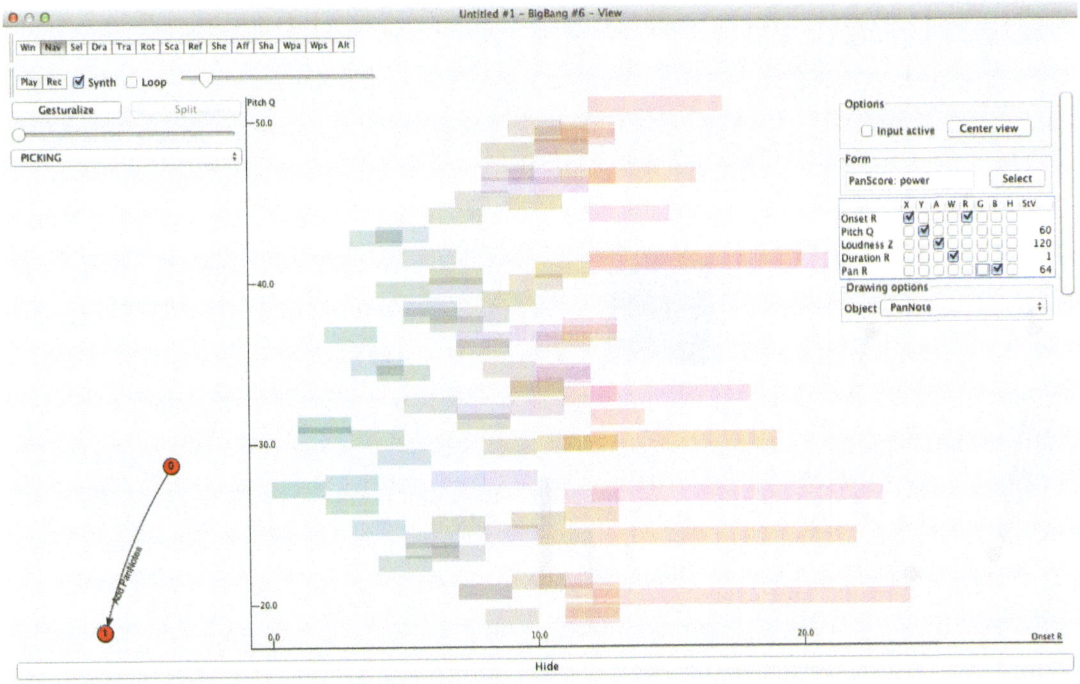

Fig. 74.6. The resulting slices seen on the *Onset* × *Pitch* plane.

74.1.7 Using Wallpapers to Create Rhythmical Structures

Form *Score*
Graph a wallpaper with a few sequential transformations
Technique creating regular structures using wallpapers
Output MIDI to GarageBand
Links `http://www.soundcloud.com/bigbangrubette/wallpapers`

One of the main uses of ornamental structures made with the *OrnaMagic* module in the *presto*® software, following Mazzola, was the creation of regular drum patterns. Mazzola's *Synthesis* composition [681] makes wide uses of transformationally reiterating drum patterns created this way. With *BigBang*, such structures can be created in a much less tedious way, as this small example shows. Two short random drum motives are created using the *Melody* rubette, input into *BigBang*, and iteratively transformed with two different wallpapers, as shown in Figure 74.9. This way, we obtain slowly altering drum patterns, the first one translated, sheared, and scaled, and the second one translated and scaled.

74.2 Improvisation and Performance with BigBang

Even though many of the aspects of the compositional examples seen so far were created in spontaneous ways, we now turn to a discussion of more momentary ways of creating music with *BigBang*. Several aspects of the requirements for gestural control are ideal for more performative and improvisatory ways of making music. For instance, the fact that at all times, when working with *BigBang*, there is immediate auditory and visual feedback inspires musicians to experiment and spontaneously react to the outcome of their actions. Also, the support of various physical interfaces enables users to handle *BigBang* like an instrument.

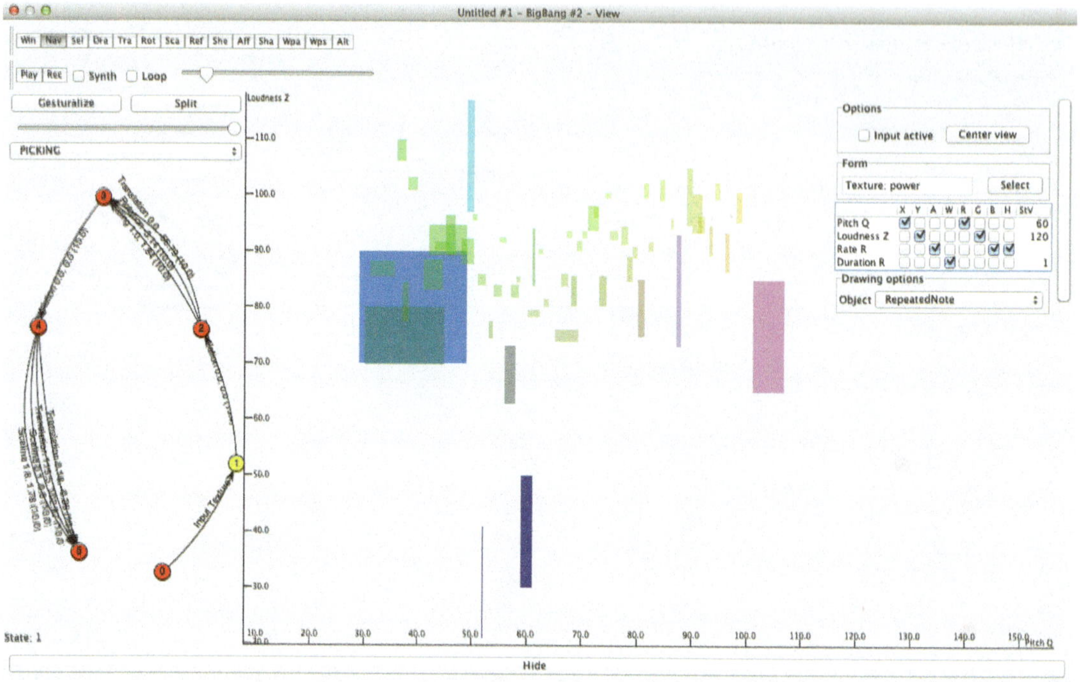

Fig. 74.7. The facts view shows the *Texture* at state 1, with rate and duration represented by height and width, respectively. The process view shows the graph generating the entire composition.

74.2.1 Improvising by Selecting States and Modifying Transformations

Form *Spectrum*
Graph sequential and alternative transformations
Technique selecting states with number keys, modifying transformations with MIDI controller knobs
Output *BigBang* synth with sine waves, postprocessed
Link `http://www.soundcloud.com/bigbangrubette/selections`

The first example simply consists of a drawn widely panned *Spectrum* with a predefined transformation graph containing alternative paths. As seen at the end of the previous section, the number keys on the computer keyboard can be used to directly select states in rapid succession. In this example we did this to add a rhythmic quality, and we also used the knobs on a MIDI keyboard to spontaneously modify the transformations. The result is a changing spectral texture, each moment of which is based on a similar sound structure. Figure 74.11 shows state 5 of the process, from where the two alternative transformations fork off.

74.2.2 Playing Sounds with a MIDI Keyboard and Modifying Them

Form *FMSet*
Graph a simple succession of transformations
Technique MIDI keyboard triggering and modifying
Output *BigBang* synth with different wave forms into Ableton Live
Link `http://www.soundcloud.com/bigbangrubette/designs`

The harmonic material in the previous example is inherent in the constellation of the *Spectrum* and its transformations. However, with *BigBang* it is also possible to regard its current contents as sonic material with which harmonic structures can be built by interaction with a MIDI keyboard, as described in Section 73.1.3. This example illustrates how *BigBang* can be used for sound design, by defining a few *FMSet* structures

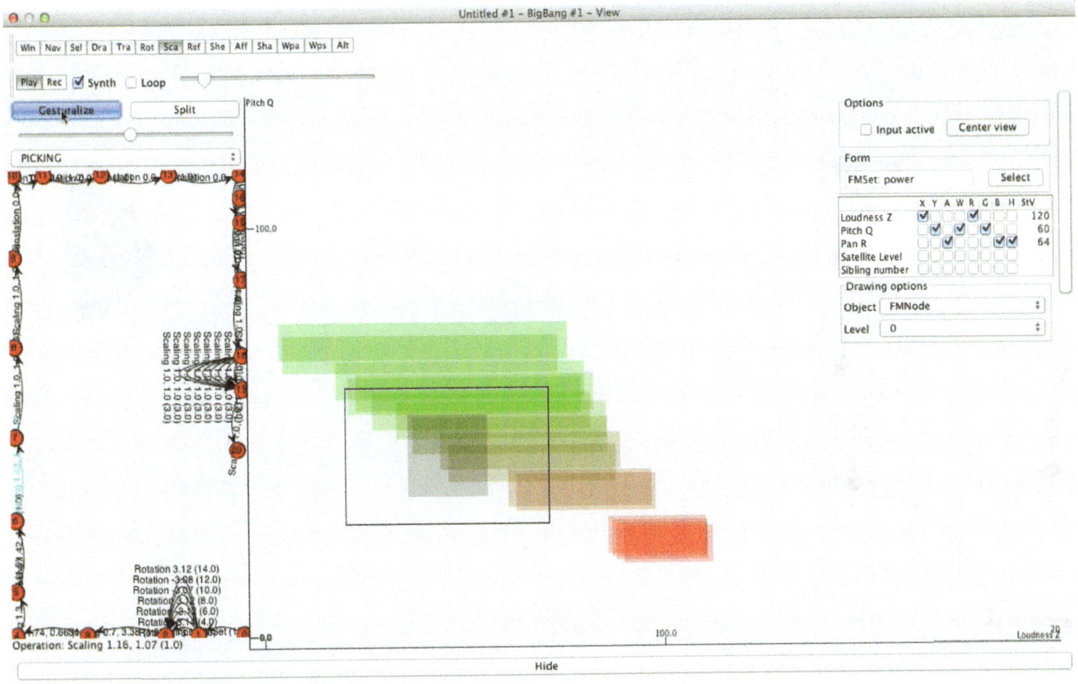

Fig. 74.8. The *Spectrum* during gesturalization. For a video, see the link above.

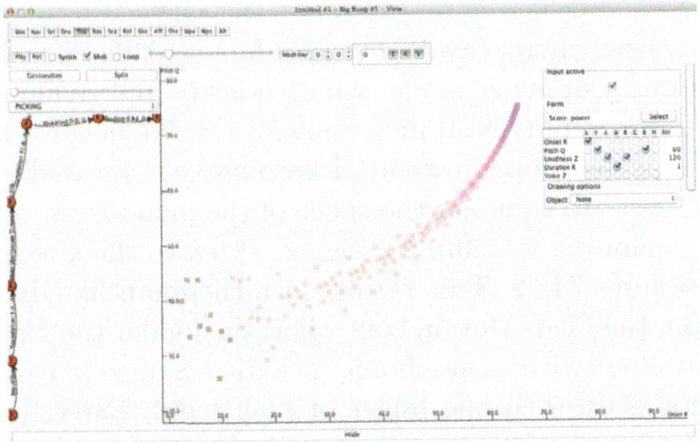

Fig. 74.9. The two wallpaper patterns in this example.

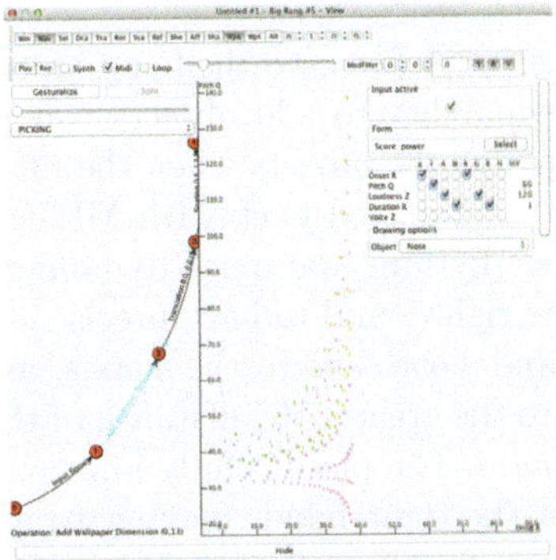

Fig. 74.10. The two wallpaper patterns in this example.

and modifying them. All sounds except the drum sounds were created in *BigBang* and played back using a MIDI keyboard, where each key plays a chromatic transposition of the current sound. Using the control change knobs of the keyboard, we modified the sounds while playing, which resulted in the various sweeping sounds in this example. An example of such an FM sound structure was shown in Figure 71.15.

74.2.3 Playing a MIDI Grand Piano with Leap Motion

Form *Spectrum*
Graph just an `AddObjectsOperation`
Technique drawing with Leap Motion
Output MIDI to a Steinway Grand Piano with the PianoDisc system
Link `http://youtu.be/ytGcKfhzF2Q`

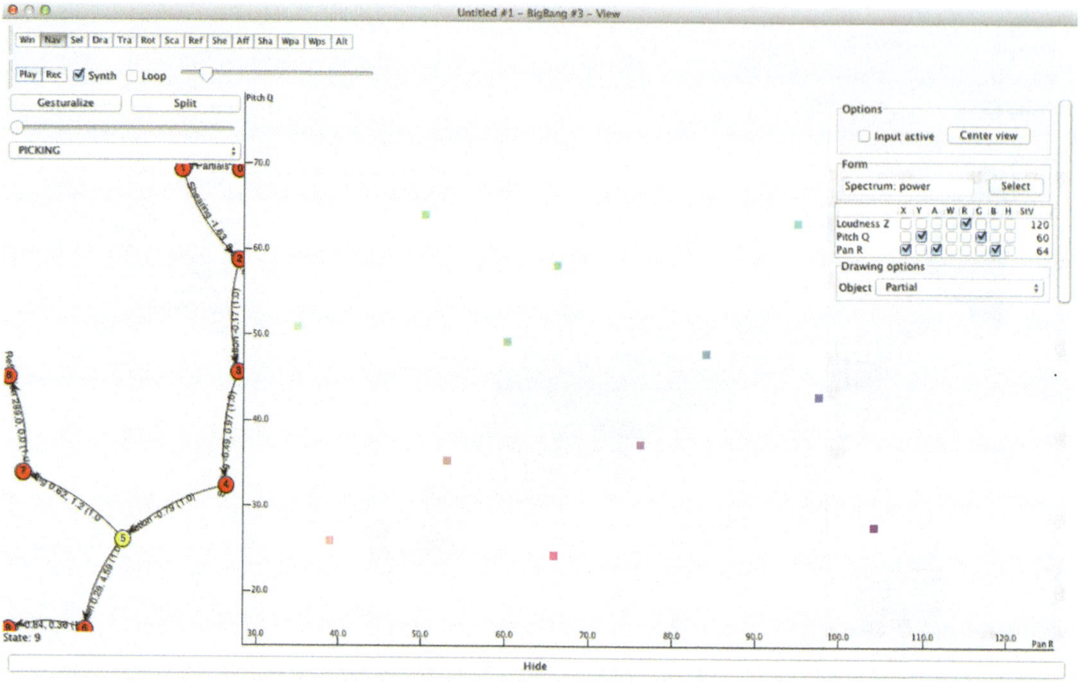

Fig. 74.11. The *Spectrum* and process that form the basis of this brief improvisation.

In this example we used the Leap Motion controller newly made available to *BigBang* [1049]. Specifically, we use the capability to add objects using Leap Motion in drawing mode, which quickly and dynamically adds and replaces the objects when the fingers move, described in detail in Section 73.1.3. We decided to use a *Spectrum* form played back with MIDI as quickly repeated notes instead of keeping keys pressed, which allows for fast rhythms and quick dynamic changes. In order to simulate the space of the piano keys – higher pitches to the right – and to have precise control over dynamics, we simply assigned *Pitch* to the x-axis view parameter, and *Loudness* to the y-axis, as shown in Figure 74.12. This results in a theremin-like dynamic setting, where the greater the distance of the hands from the Leap Motion controller, the louder the *Partials* of the *Spectrum*. The piece is fully improvised and starts out with monophonic melodic gestures played with one finger of the right hand, moving to a contrapuntal part with one finger of each hand. Later, we add more and more fingers to each hand, sometimes playing in parallel, sometimes independently, culminating in a part with increasingly energetic and fast gestures, and at an increasing distance from the Leap Motion, culminating in a loud and choppy part of thrown gestures.

74.2.4 Playing a MIDI Grand Piano with the Ableton Push

Form *Spectrum*
Graph four sequential transformations
Technique aftertouch dynamics, manual gesturalizing using the Ableton Push
Output MIDI to a Steinway Grand Piano with the PianoDisc system
Link `http://youtu.be/n2Pi281XZP4`

As opposed to the previous example, the simple setup of which enabled maximal freedom of determining the structure of the piece and its tonal material spontaneously, this example shows how it is also possible to improvise with some deliberately prepared material. The piece uses the same form, *Spectrum*, with the same rate of MIDI note repetition, but it also includes a simple graph that predefines intervallic and transformative gestural material. The transformed entity is a four-note motive of purely harmonic nature,

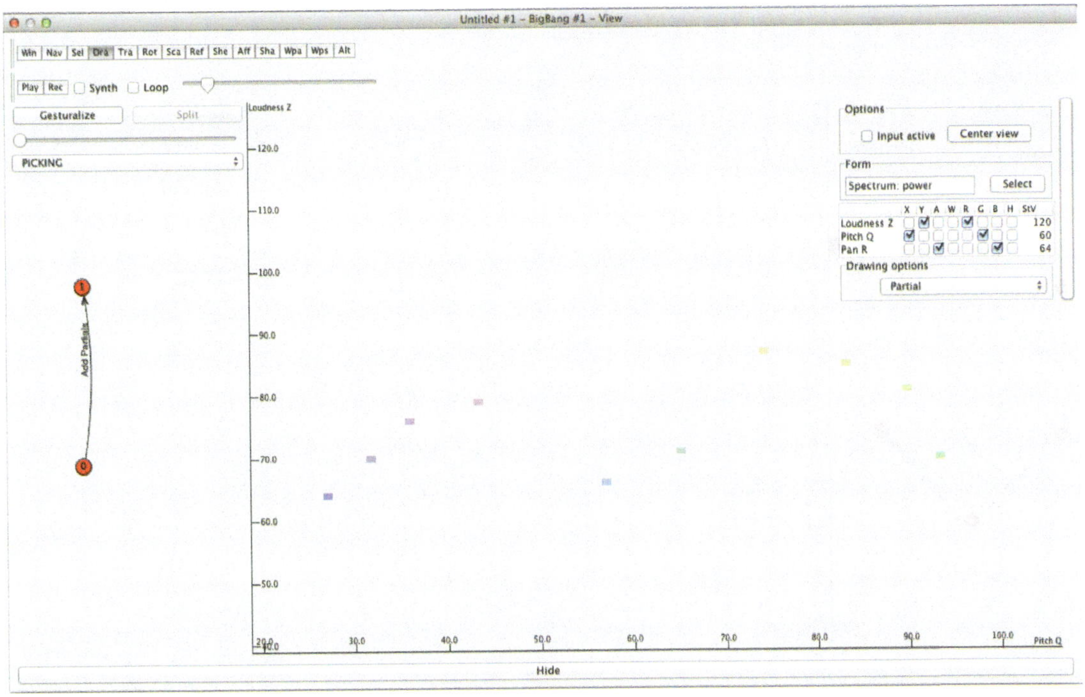

Fig. 74.12. Two piano hands drawn with Leap Motion on the *Pitch* × *Loudness* plane.

due to the *Spectrum*'s atemporal quality. Figure 74.13 shows the initial motive and the graph. First, the motive is rotated by about 180 degrees, then scaled, mainly in pitch (about 4.5 times larger), then rotated counterclockwise by slightly less than 90 degrees, and finally scaled to an intervallic and dynamic unison.

After defining the simple graph, we decided to perform the piece using the dynamic capabilities of the Ableton Push controller. The Push offers pads sending note on/off messages, which we mapped to played back versions of the four-note motive, in a similar way to that in which the sounds were triggered by the MIDI keyboard in the example in Section 74.2.2. More importantly, the Push's pads send out highly precise monophonic aftertouch control changes, which we used to control the dynamics of the played back motives. Furthermore, we mapped the Push's large touch strip to the *BigBang*'s manual gesturalizing slider, which allowed me to freely move back and forth through the various transformed versions of the motive, both continuously by sliding and discretely by tapping the strip, which jumps to the corresponding position of the gesturalization (this way users can not only select the composition states represented by the nodes of the graph, but any other intermediary state!). In the beginning of the improvisation, we gradually add more pitches by gesturalizing the `AddObjectsOperation`; then we spontaneously move through different states of the motive's transformational path, reacting to the temporary constellations by playing them in different ways on the pads. In the end, the motive disappears in a unison produced by the ultimate scaling.

74.2.5 Improvising with 12-Tone Rows

Form *Score*
Graph a sequential transformation graph with a few identities
Technique gesturalizing and looping, aftertouch dynamics with the Ableton Push
Output MIDI to a Steinway with PianoDisc
Link `http://youtu.be/n1RQimytD2A`

This last example uses a similar setup as the previous one, playing the MIDI Grand with the Push. However, it is even more deliberately prepared, using automatic gesturalization, and links to the second

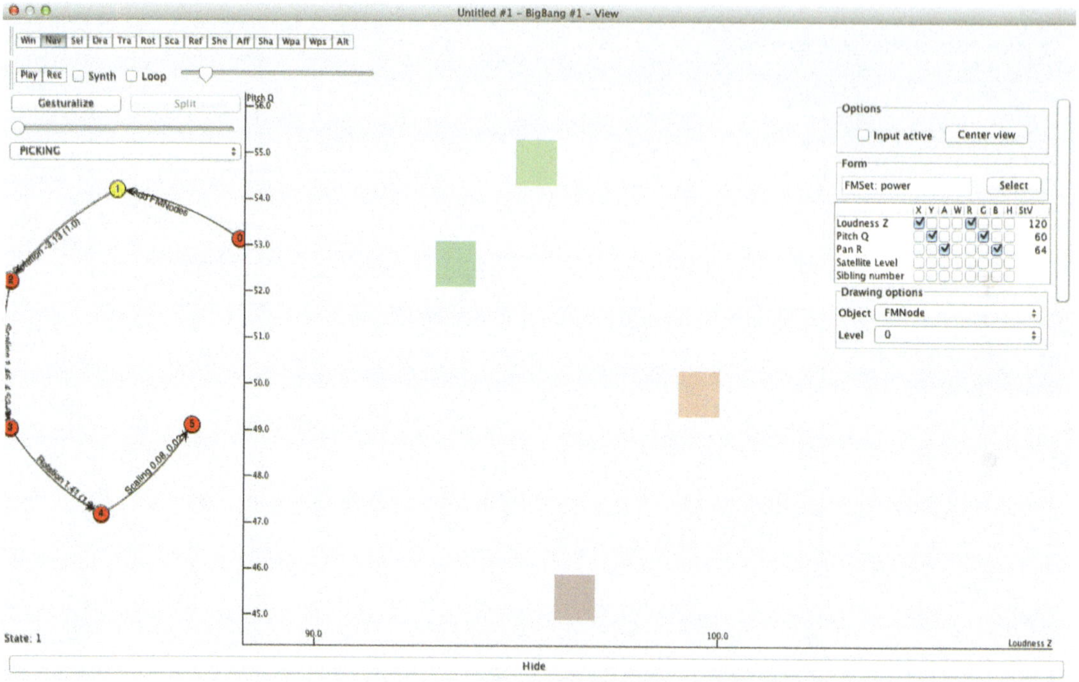

Fig. 74.13. The initial motive and the simple sequential graph.

musical example presented here (Section 74.1.2) in its use of the *Score* form in combination with looping while the graph is being gesturalized. Here, the base material is a simple twelve-tone row, generated by another rubette we recently created, the *NTone* rubette, which generates rows of N equidistant tones, microtonal if necessary, within a specific interval of I semitones (here, $N = 12$ and $I = 12$). The twelve-tone row is then transformed in simple ways (Figure 74.14 shows the original row and the graph). It is first compressed (scaled) in *Onset*, which results in a faster version, then expanded in *Onset*, leading to a staccato version. At this point, we inserted an identity scaling in order to achieve a static moment in gesturalization, as suggested in Section 74.1.6. After that, the row gradually assumes its retrograde inversion form by being rotated by 180 degrees, again stays static, is inverted by a reflection, again stays static, and finally disappears in a single pitch (scaling to 0 in both onset and pitch). In the performed version, again each pad of the Push triggers a transposed version of the twelve-tone row, again dynamically modulated by aftertouch. In this case, we could fully focus on playing the Push, since gesturalization was predetermined and happened automatically.

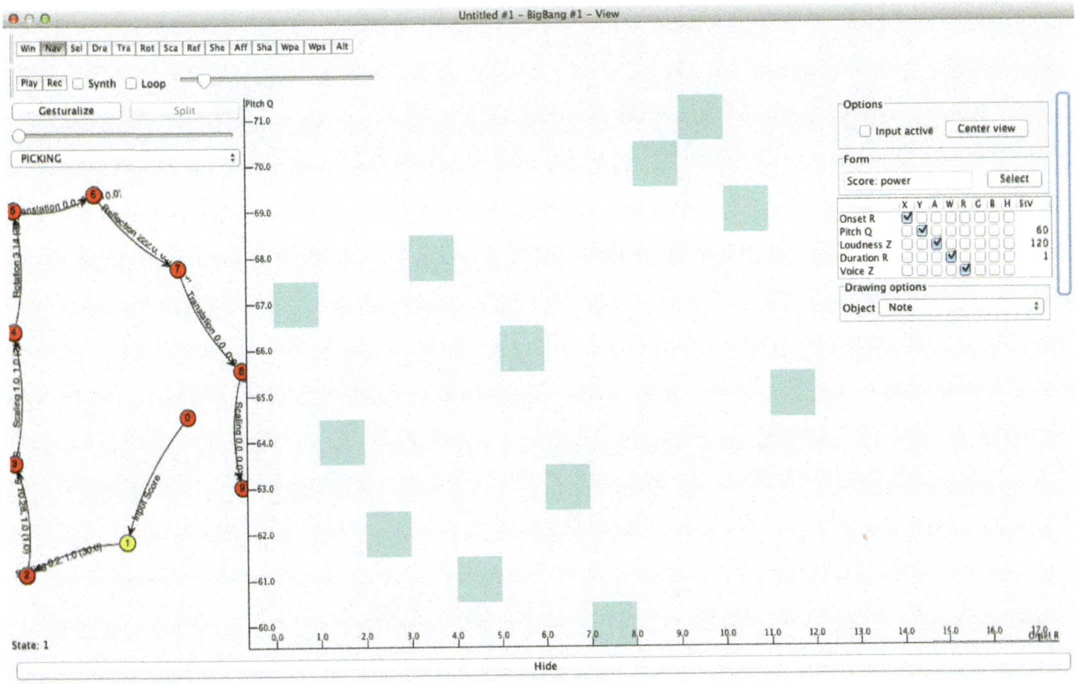

Fig. 74.14. The sequential graph with the initial state selected, showing the original twelve-tone row.

Part XVIII

The Multiverse Perspective

75

Gesture Theory and String Theory

Summary. This chapter introduces the creative relationship between mathematical gesture theory and physical string theory.

– Σ –

So far, we have discussed the mathematical theory of gestures, a formalized approach to the well-known phenomenon of gestural dynamics in music (and other arts). This theory was developed starting in 2002 by one of the authors (Mazzola), and a first version of the new formalism was published in a joint paper with Moreno Andreatta [720] in 2007. The philosophy of that approach is that instead of considering point-like notes (points in mathematical modules of musical parameters), one would envisage systems of curves as elementary objects in music. These objects could be a description of physical movements of human limbs (fingers, hands, arms, the tongue, etc.) or more abstract movements such as a melodic line in a musical parameter space, or a curve within a space of spectral sound parameters.

This shift of musical conceptualization also meant in Grothendieck's spirit a shift from categorical points, i.e., elements of $A@F$ for a module A and a functor F, to a different type of points, namely elements of $\Gamma@\overrightarrow{X}$, i.e., Grothendieck points in the category of digraphs with values in spatial digraphs $\overrightarrow{X}$. These points are special Grothendieck points, but at the same time generate a different optic: the spaces of curves are a priori of a continuous nature, and gestures are therefore a priori focusing on systems of continuous curves. In musical gesture theory, we could therefore state that the most elementary objects, the "atomic" points, are continuous curves. Grothendieck's shift from ordinary points to points with general addresses now is turned around to a situation where the relativity of such addresses is replaced by a thoroughly continuous perspective. There is nothing more elementary than continuous curves. Recall that since Euclid, a point is "what has no parts". In musical gesture theory, curves and their gestural combinations are exactly those indecomposable conceptual atoms.

We stress this philosophical fact because such a shift of the concept of an atomic entity also happened with the physical string theory that is being developed since 1960 (after a complex prehistory going back the first two decades of the 20th century). Here, point-like elementary particles are replaced by strings, i.e., curves that move in space-time. Similarly to musical gestures, strings are atomic concepts, they have no proper parts *qua* physical objects. This is also the reason why this evidently parallel setup for music was recognized from the beginning and mentioned in [720]: musical gesture theory was conceived as a *musical string theory*, see also Chapter 61.2.

Physical string theory has been (dis)qualified as being ugly, and there were complaints that not enough effort had been made to develop other theories. In the musical realm, there is some parallelism to that, less in the sense that musical gesture theory is considered as being ugly, but in that it is without doubt far too complex when compared to traditional mathematics in music theory. One now has to shift away from elementary algebra and combinatorics to topology, topos theory, algebraic topology, homology theory and other far out mysteries. Yes, musical gesture theory is ugly in the sense of overwhelming complexity. It is clear that gestures are important in musical practice, above all to performers, but it is not less clear that a

G. Mazzola et al., *The Topos of Music III: Gestures*, Computational Music Science,
https://doi.org/10.1007/978-3-319-64481-3_20

mathematical theory of such dynamic configurations is beyond the given competence of music theorists. If musical set theory or its evolution to transformational theory is taken as a reference theory, gesture theory appears as a veritable "diabolus in musica".

It might be by case that music and physics are sharing this new paradigm of "string atoms", but it is without doubt that in both contexts this approach yields a number of fundamental insights that no other existent theory can provide. This parallelism is very probably also a manifestation of Hegel's "Weltgeist", the simultaneous awaking of a consciousness in different localities of *fundamental* human knowledge.

76

Physical and Musical Multiverses

Summary. We shortly discuss the question of unicity in music and physics, a question that in physics has been virulent since the advent of string theory, but which in music has been relevant since the approach to music via individual compositions at the end of the Middle Ages.

– Σ –

The present state of theoretical physics is dominated by string theory (see also our discussion of this theory in Chapter 75 and Appendix K.2). It is however not a valid theory, it only provides us with presently the best ideas of how gravitation and quantum physics could be reconciled.

One of the irritating consequences of string theory is that it provides us not with one string "landscape" of this universe (i.e. solution of its equations), but with a quantity of the range of 10^{450}. This is quite difficult for the theory as such, but it also opens up questions that are deeper: Are we living on a single universe? If not, in what sense can a variety of universes exist? What is the overall framework of such a variety? What would it mean that they exist "simultaneously" since the would have their own space-time each? Can a multiplicity of solutions of string theory define a higher universal reality? And for whom? Although physicists speculate on such varieties, it is in any case a big question how human mathematical constructions of string theory might describe multiple universes. The question is about the 'transuniversal' validity of mathematics, about a transcendental reality that might cover the reality of 10^{450} Universes.

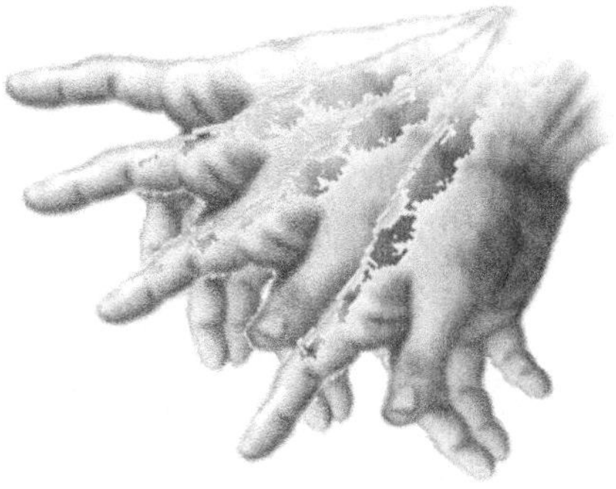

Fig. 76.1. The creation of a multiverse.

Such a multiplicity of realities however questions any naive unicity claims after the abolishment of geocentricity, chronocentricity, and ratiocentricity. Can deocentricity still be defended? It can, but the price is high. How could we humans remain so exquisitely important as inhabitants of one out of 10^{450} cosmic landscapes? The SETI (Search for ExtraTerrestrial Intelligence) enterprise would turn out to be a very local affair.

The confrontation of music and physics that is mediated via gestural paradigms does not thematize only multiverses in physics, but also in music. When introducing gestures to music, every concept of the traditional theory melts into a manifold of gestural vibrations. To begin with, classical music theory is softened to enable many transformations that were unthinkable without gestures. Tonal modulation theory now can be developed between tonalities of completely different isomorphism classes, see Chapter 80. Counterpoint can now be thought with an infinity of consonant and dissonant intervals, see Chapter VII. And gestural performance theory (finally!) offers a melting pot of operators of an infinity of different performance strategies that emerge from the world-sheet paradigm, the hypergestural connectivity of symbolic and physical performance gestures, see Chapter 78.

In music, gestures seem to produce and connect many worlds of music, they seem to offer a communicative basis to a genuine multiplicity of thinking and making music. For example, in the documentary

G. Mazzola et al., *The Topos of Music III: Gestures*, Computational Music Science,
https://doi.org/10.1007/978-3-319-64481-3_21

movie *Teak Leaves at the Temples* that was produced in 2006 in Indonesia, featuring Mazzola's free jazz trio (Mazzola on piano, Heinz Geisser on percussion, Sirone Norris Jones on bass) and many Indonesian ensembles, gestures for the musical language were essential and helpful in the intercultural communication with local Indonesian musicians.

But there is more to say about multiverses in music. Music in its practice and theory seems to be genuinely driven by multiplicities, ambiguities, and manifolds of thoughts and feelings, sometimes as a sign of ill-defined conceptualization ("inversion", "cadence", and "tonality" do have multiple meanings). But there is a substantial aspect that transcends plain deficiency, it is a basic existential style. *Music is not the simply indeterminate, but the multiply determined.*[1] This style cannot be described and understood in the language of discrete thought units such as is represented by abstract algebra and category theory *alone.* It is the elastic conceptual tools of algebraic topology, topological categories, differential and algebraic geometry, and corresponding homology theories that can cover these exigencies of fundamental continuity. It is important here to cite Fernando Zalamea's statement about contemporary mathematics (between 1950 and 2000) in [1149, p. 271-272]: "*Multiplicity* everywhere underlies contemporary transit, and the objects of mathematics basically become *webs and processes.*" Of course, we would add: "*and gestures,*" the innermost position in the dimension of embodiment. And it is equally important to stress that multiplicity with an inner logical coherence need not result in a juxtaposition of independent and isolated items, as is beautifully illustrated by the Möbius band with its intriguing but consistent multiplicity of local and only local orientations.

[1] In the Preface of Mazzola's book *Geometrie der Töne* [682], he wrote: "Musik erweist sich in diesem Unternehmen als das im Vieldeutigen Bestimmte und darin als Gegenstand, welcher den Paradigman heutiger Mathematik in natürlicher Weise entspricht."

77

Hesse's Melting Beads: A Multiverse Game with Strings and Gestures

Summary. A critical review of Hermann Hesse's idea of a *Glass Bead Game* in the light of recent developments in mathematics, music theory, and theoretical physics is presented. The common denominator of these new dynamics is the shift from Wittgenstein's world of rigid facts to an ocean of elastic gestures. In such a soft architecture of knowledge production, the ultimate principle of uniqueness as conceived in the idea of a singular universe breaks down to a multiverse, a multiplicity of worlds that terminates the historical breakdowns of uniqueness principles from geocentricity (Copernicus) to anthropocentricity (Darwin), chronocentricity (Einstein), and ratiocentricity (computers). We discuss contributions from eminent mathematicians Alexander Grothendieck and Yuri Manin, theoretical physicist Edward Witten, music theorist David Lewin, and philosophers Tommaso Campanella, Paul Valéry, Gilles Châtelet, Jean Cavaillès, and Charles Alunni. We complement their positions with our own contributions to topos-theoretical concept architectures and theories in gestural music theory, and offer realizations, both by means of gestural composition software and with examples from contemporary free jazz. The chapter concludes with a reconsideration of the game concept as a synthesis of artistic and scientific activity in the light of gestural fluidity.

– Σ –

77.1 Review of Hesse's Glass Bead Game

Hermann Hesse published his novel *Das Glasperlenspiel* in 1943 in Zurich [463] and was accordingly awarded the Nobel Prize in Literature in 1946. Hesse comments on the novel's substance [463]: "It (the glass bead game) represented a symbolic form of seeking for perfection, an approach to that Mind which beyond all images and multiplicities is one within itself—in other words, to God."

A precise description of this glass bead game does not exist. But a passage [464] describes to some degree what the game is about: "Under the shifting hegemony of now this, now that science or art, the game of games had developed into a kind of universal language through which the players could express values and set these in relation to one another. Throughout its history the game was closely allied with music, and usually proceeded according to musical or mathematical rules. One theme, two themes, or three themes were stated, elaborated, varied, and underwent a development quite similar to that of the theme in a Bach fugue or a concerto movement. A game, for example, might start from a given astronomical configuration, or from the actual theme of a Bach fugue, or from a sentence out of Leibnitz or the Upanishads, and from this theme, depending on the intentions and talents of the player, it could either further explore and elaborate the initial motif or else enrich its expressiveness by allusions to kindred concepts. Beginners learned how to establish parallels, by means of the game's symbols, between a piece of classical music and the formula for some law of nature. Experts and masters of the game freely move the initial theme into unlimited combinations."

The combination of mathematics and music was inspired by the work of music theorist Hans Kayser (1891-1964), who also lived in Switzerland when Hesse was designing his novel, and whose work is a Neo-Pythagorean mathematical theory of music [526].

G. Mazzola et al., *The Topos of Music III: Gestures*, Computational Music Science,
https://doi.org/10.1007/978-3-319-64481-3_22

77.2 Frozen Glass Beads of Facticity

Hesse's approach is clearly interdisciplinary. The technical character of the game, although not explicated in detail, is a combinatorial one. This follows from Hesse's enthusiasm for the famous Chinese *I Ching*, that yarrow stalk oracle and cosmology which results from the $64 = 2^6$ combinations of six broken or unbroken lines (represented by yarrow stalks). This combinatorial cosmology gives the user a corresponding number of cosmological treads. It is not creative, but strictly selective. Hesse's combinatorial perspective is also evident from his enthusiasm for Gottfried Wilhelm Leibnitz's calculus of logic, that idea of a perfect language where its conversations would be reduced to an exchange of formal logical formulas [287].

This approach to reality as a combinatorial setup that can be controlled by formal logical calculus is the ontology we know from Ludwig Wittgenstein's *Tractatus Logico-Philosophicus* [1138], whose first sentence reads: "Die Welt ist alles, was der Fall ist." ("The world is everything that is the case.") This point of view is utterly reductionist. What is the case means: what has been made and is there now. Once for ever? And who made it, and how, this does not matter. It is a perspective on existence as a collection of frozen objects. Take it or leave it, but you are only the user, much as in the Medieval cosmology where humans could just observe God's world without any option of active intervention. A recent book [1120] confirms this view, describing a computer program which implements those combinatorial options in the spirit of *I Ching* and Leibniz. Hesse in fact also stresses that his glass bead game is a search for God, that hidden universal spirit which connects all those diversities at the phenomenal surface.

In this style, Hesse's *Glass Bead Game* idea has been realized through my own work in mathematical music theory in 1985 [670]. It is however not a Neo-Pythagorean theory, but relies on modern mathematics (such as module theory, category theory, and algebraic topology). We want to give a short overview of such work to make evident its power, but also the fundamental limitation of the combinatorial method.

In our approach, we started from Beethoven's famous *Hammerklavier* Sonata Op. 106. We applied mathematical models of tonal modulation and paradigmatic classification of melodies (the latter in the spirit of Jean-Jacques Nattiez [782]) to a thorough analysis of Beethoven's harmonic and motivic construction. In the vein of Hesse's sketch of the game, we derived a mathematical formula that would describe his musical operations. This formula is the symmetry group $Sym(C\sharp^{-7})$ of the diminished seventh chord $C\sharp^{-7} = \{c\sharp, e, g, a\sharp\}$. For the composition of a new sonata, we then exchanged this chord for an equally famous combination of pitches, the augmented triad $C\sharp^{+} = \{c\sharp, f, a\}$, and then used its symmetry group $Sym(C\sharp^{+})$ to construe all modulations and the motivic germs of the new sonata, which means that we threw back the mathematical formula to the musical realm. The resulting sonata *L'essence du bleu* has been published, including a CD recording [708]. The result could look like a real product of creativity. But it is simply a restatement of Beethoven's thoughts (viewed through the microscope of our mathematical analysis) with changes of corresponding structures, mutatis mutandis. We don't qualify this type of game as a truly creative one. There is no box, whose walls are being opened to an unknown space outside, rather are we opening a door of one box (music) to the neighboring box (mathematics).

In what follows we shall explain how and in which domains of creative knowledge production Hesse's glass bead game can be morphed into a less passive, factual, and rigid activity.

77.3 The Revolution of Functors

A first conceptual revolution occurred in the new mathematics that was introduced by Samuel Eilenberg and Saunders Mac Lane with their theory of categories in 1945 [637]. They analyzed the basic objects of mathematics, namely points, those objects that have no parts according to the classical approach by Euclid: *punctus est cuius pars nulla est.* For categories, points were replaced by new elementary objects: arrows. Category theorists considered points as a result of a pointer action, *punctus* is the result or head of an arrow that pricks its target. The new concept of an arrow has three parts: its head and its tail, which are connected by a shaft that symbolizes a pricking movement from tail to head. This seemingly harmless change introduced the point of departure of a conceptual movement, not only its pricked result.

With this dramatic change, the subject, the point of departure of an arrow, was introduced and thereby the inclusion of the point of view or "address" from which the target point was addressed. In Alexander Grothendieck's (1928-2014) algebraic geometry [395], this paradigmatic revolution introduced a completely new conceptualization, a kind of relativity theory in geometry. This theory eventually led to the solution of hard problems, such as the Weil and Fermat conjectures, by Pierre Deligne (1974) and Andrew Wiles (1995), respectively.

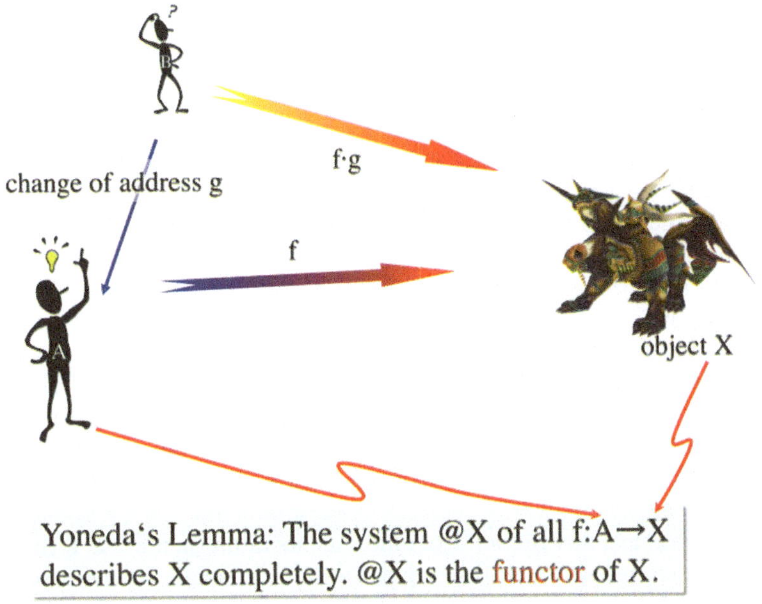

Fig. 77.1. Yoneda's Lemma replaces Whatness by Suchness.

The big program was therefore to understand algebraic geometry as a science of relative points. The strongest general argument for this revolution was a simple, but powerful lemma, a result introduced by the Japanese computer scientist Nobuo Yoneda in 1954. He could prove that a space X can be completely described by its functor, i.e., by the system of all arrows $f: A \to X$ where the address A runs over all possible choices. This functor, which we denote by $@X$, replaces the space X as such by its system of variable arrows. More precisely, the lemma states that two spaces X and Y are isomorphic if and only if their functors $@X$ and $@Y$ are isomorphic as functors, see Section G.2.

This result allows mathematicians to replace spaces (and in fact any objects of general categories) by their functors. In other words, the spaces *qua* objects with their intrinsic identity are replaced by their functors which are systems of relative behavioral perspectives from variable addresses. In philosophical terms: the *whatness* of X is replaced by the *suchness* of the functorial entities $@X$. Mathematics is no longer interested in the abstract identities X but only in their behavioral systems, the functors $@X$, see Figure 77.1.

This revolution generated a new paradigm of conceptual manipulation: Abstract spaces became only relevant via their behavior under specific, but arbitrarily variable addresses. This implies that now, a mathematical object was conceived as a distributed behavior, its identity was now reified as a collaborative system of addressed perspectives. With respect to Hesse's original God-oriented approach, we now see that divine whatness being replaced by a "God functor" which is realized in its not necessarily representable howness that our collaboration experiences in a distributed identity.

In mathematical music theory, we have composed a piece, which is in fact a recomposition of Pierre Boulez's *structures pour deux pianos I*, following the functorial methodology [726] and applying Boulez's idea of a creative analysis. Our analysis of Boulez's original composition followed György Ligeti's analysis [611] which we reinterpreted as being a functorial procedure. Of course, Boulez did not know category theory when he composed that piece, but in the light of Yoneda's Lemma, his compositional strategy was in perfect congruence with modern functorial methods. We then implemented our functorial analysis in the rubato composer software and were able to recompose new variants of that piece by canonical address changes [726]. It is not an exaggeration to understand Boulez's structures as *the historically first genuinely functorial composition.*

77.4 Gestures in Philosophy and Science

Despite the distributed identity of functorial mathematics, its collaborative style was still unsatisfactory and mechanical. Let us explain why. In the functorial system $@X$ of a space X, we are given the set $A@X$ of arrows $f : A \to X$ for every address A. The communication from $A@X$ to $B@X$ for two addresses A, B is defined from a morphism $m : B \to A$ by composition of arrows, $m@X : A@X \to B@X : f \to f \circ m$. The set map $m@X$ is however an abstract function in the sense of logician Gottlob Frege, who introduced the modern function concept of mathematics. The arrow between sets $m@X : A@X \to B@X$ is only intuitive, there is no movement from elements $f \in A@X$ to elements $f \circ m \in B@X$. Fregean functions are "teleportations", the arguments f "disappear" and "reappear" as values $f \circ m$. The French mathematician and philosopher Gilles Châtelet has criticized this Fregean tromperie [189, 190]: "The function gives only the form of the transit from one external term to another external term, whereas the act exhausts itself in its result." A function is similar to an industrial plant: Input — Black Box — Output. The great mathematician Henri Poincaré accordingly stated [852]: "Localiser un objet en un point quelconque signifie se représenter le mouvement (c'est-à-dire les sensations musculaires qui les accompagnent et qui n'ont aucun caractère géométrique) qu'il faut faire pour l'atteindre." He was aware that the gestural origin of geometry is beyond formally mathematical mechanisms known at his time.

Let us make clear the radical abstraction from movements in mathematical formulas. Take a rotation in the real three-space. This movement of the space's point around an axis is represented by a 3×3 matrix M in linear algebra. But the matrix M does not show any information about the axis or the rotation's angle. You have to work quite hard to exhibit such an axis from M. The catchword is eigenvalues and corresponding eigenvectors (that could play the role of rotational axes), and the rotation's angle—if an eigenvector of a rotation can be found at all—is still another problem to be solved by linear algebra. In other words: The mathematical representation of movements by matrices is a radical abstraction, a compactification of a movement in a formula, a kind of Fregean prison, where the original movement has been encapsulated. If we were to relate music to mathematics we could state here that they relate formulas to gestures, but in opposite directions: While mathematics compactifies gestures to formulas, music unfolds formulas to gestures.

Let us now focus on the gestural ontology which has been hidden to this date even in contemporary functorial mathematics. For an excellent treatise on the history of gestures in European philosophy, we refer to Jean-Claude Schmitt's *La raison des gestes dans l'Occident médiéval* [946], see also Section 57.2. He has given the most complete and important contribution to a history of the concept, philosophy, social and religious roles of gestures during the early centuries of our modern Western culture. Recall that he exhibits the first (and still one of the best) definition of a gesture, given by Paris-based theologist Hugues de Saint-Victor (1096-1141) [946, p. 177]: "Gestus est motus et figuratio membrorum corporis, ad omnem agendi et habendi modum."[1] Observe that in Saint-Victor's definition, the specification *ad omnem agendi et habendi modum* is not semiotic, but merely describes the generic modality of action and being.

A gesture is a presemiotic concept, it does not automatically mean a thing. The pointer gesture is presemiotic also since Saussurean structuralist semiotics is built upon the pointer from signifiant to signifié. A pointer is not a sign, but a *prequisite* to any semiotic concept architecture.

In the 20th century, gesture philosophy was above all developed by French philosophers, linguists, and mathematicians. Their works also differ from the Anglo-Saxon linguistic philosophy of gestures that has been developed by Adam Kendon and David McNeill [530, 741], see also Section 57.7. They focus on gestures

[1] "Gesture is the movement and figuration of the body's limbs with an aim, but also according to the measure and modality proper to the achievement of all action and attitude."

that are co-present in linguistic utterances, and from this perspective, their concept of a gesture is strictly semiotic: gestures are special signs that support the building of linguistic syntagms and contents. And they are always related to the body's actions; no more abstract concept of a gesture, such as a gesture in a musical melody or a thought gesture, is addressed. We don't follow this rather restrictive conceptual line in more detail and refer you to Section 57.7.

The French tradition of gesture philosophy is characterized by the thesis that gestures constitute a proper ontology that is independent of, and typically precedes, semiotic systems, it is presemiotic.

Ahead of his time, French mathematician and philosopher Jean Cavaillès in 1938 stated a core property of gestures that bypasses any semiotic basis [181, p. 178]: "Understanding is catching the gesture and being able to continue."

Cavaillès' dancing thought (also shaped in Pierre Boulez's reflection on gesture in music [141]) was in fact stated with respect to mathematical theories, and as such it was one of the very first principles of gestural embodiment in mathematics, an idea now quite fashionable through the work of George Lakoff and Rafael Núñez [570] but also anticipated in Châtelet's observation [189] that the Fregean concept of a function in mathematics is a dramatic (and questionable) abstraction.

Gestures—except when "tamed" by social codes—are not signs in a semiotic environment. They are not a realization of Ferdinand de Saussure's classical signification process.

Summarizing, we learn that gestures are in general understood as pertaining to a proper ontology that is not subordinate to semiotic lines of thought. In particular, the dominant French diagrammatic philosophy exhibits a sharp dichotomy between "wild" and "tamed" gestures, the former being independent or antecedent of semiotic realms, while the latter serving semiotic purposes as special types of signs. Conceptual creativity is exhibited in the layer of wild gestures. The communicative characteristic of (wild) gestures stresses their "howness" as opposed to their substantial "whatness". Gestures are understood in their behavior, not in their absolute being (such as Kant's Ding an sich).

It is astonishing that despite the sensational success of Grothendieck's "mathematical relativity theory", there has been some work in the direction of replacing abstract Fregean functions and their formalism in category theory by gestural concepts. In *Categorical Gestures* [723], Mazzola started an investigation of the possibility of enriching Yoneda's Lemma by gestures, and the ultimate goal would be to replace the present abstract foundational entities of mathematics, such as sets or arrows, by gestures.

In a extraordinary interview with the *Notices of the American Mathematical Society* [642], the prominent mathematician Yuri Manin states his vision of future foundations of mathematics: "Instead of sets, clouds or discrete elements, we envisage some sort of vague spaces, which can be very severely deformed, mapped one to another, and all that while the specific space is not important, but only the space up to deformation. If we really want to return to discrete objects, we see continuous components, the pieces whose form or even dimension does not matter. (...) I am pretty strongly convinced that there is an ongoing reversal in the collective consciousness of mathematicians: the right hemispherical and homotopical picture of the world becomes the basic intuition, and if you want to get the discrete set, then you pass to the set of connected components of a space defined only up to homotopy. (...) That is, the Cantor points become continuous components, or attractors, and so on almost from the start. Cantor's problems of the infinite recede to the background: from the very start, our images are so infinite that if you want to make something finite out of them, you must divide them by another infinity. (...) I see in this an analogy with the rebuilding of pragmatic foundations in terms of a category theory and homotopic topology."

77.5 Gesture Theory in Music

In music philosophy, music theory, and performance research, gestures have been playing a role of conceptual enrichment for a long time, but a full-fledged theory of musical gestures has still been delayed, probably also because of the difficulty of an epistemologically valid conceptualization of gestures. Let us give a summary of some important gestural perspectives in the science and art of music. See also Chapter 60.

Already in Eduard Hanslick's determination of musical content as "tönend bewegte Formen" [438], not just forms, but forms that are moved in a sounding manner, or moved in sound for short, the formal aspect, the formula, of a cadence, for example, is not sufficient to generate content. The form(ula) needs to be moved, it is deployed in a gestural dynamics. And Hanslick illustrates his idea with the kaleidoscope, a dynamical arrangement of forms that receive their aesthetical value in a self-referential internal relationship.

In the 20th century, German music theorist Wolfgang Graeser applied the mathematical theory of symmetry groups to restructure Bach's *Art of Fugue*, but while observing dancers who interpreted Bach's *Goldberg Variations*, switched from abstract symmetries to what he called "Körpersinn", the sense of embodiment in music [388]. His book project on this topic could not be accomplished as Graeser felt totally lonesome with such advanced ideas and committed suicide in 1928 at the age of 22.

In the theory of musical performance, Theodor Wiesengrund Adorno in 1946 wrote an essay about performance which gave strong arguments for the gestural essence in performance [8]. He followed Paul Valéry's famous dictum: "C'est l'exécution du poème qui est le poème." In this essay Adorno argues that "the idea of performance pertains to music as such and isn't an accidental attribute", and here is his analysis of the gestural basis of performance: "Correspondingly the task of the interpreter would be to consider the notes until they are transformed into original manuscripts under the insistent eye of the observer; however not as images of the author's emotion—they are also such, but only accidentally—but as the seismographic curves, which the body has left to the music in its gestural vibrations." Adorno argues for what I had called "the score as a repertory of frozen gestures." He does not argue for the emotional message of gestures, rather, he argues for their nature as "vibrating" bodily utterances. At first sight, this may look overly materialistic and far from the symbolic meaning of musical creation, but Adorno insinuates a spiritual component in the gestural dynamics. This perspective is in fact supported by the very history of score notation. Originally, scores encoded the gestural hints in the graphemes of Medieval neumes. These graphemes then successively morphed to the present notation, which has abstracted neumatic threads to discrete point symbols.

Adorno's student, Renate Wieland, and her fellow scholar Jürgen Uhde make the teacher's approach more explicit and apply it to the their system of piano performance [1067]. She makes clear that gestures are abstractions from concrete actions, however they remain geometric entities in some more generic space. Wieland also argues that the emotional connotation in music originally is e-motion, out-movement, and so the gestural transmutation is not an artificial construct, but the restatement of the original phenomenon.

The crucial but still underestimated role of gestures in performance has been described in a beautifully clear way by American composer and music critic Roger Sessions [972].

On the level of music performance in technology, Manfred Clynes with his sentograph and Johan Sundberg and McAngus Todd with their performance software and gesture-driven concepts of the final retard in music state in particular that "the performance and perception of tempo/musical dynamics is based on an internal sense of motion." Similar approaches to cognitive models of gestures in music are shared by Marcelo Wanderley, Claude Cadoz, and Marc Leman [168, 371], see also Section 60.5.

Coming from a different position, namely music theory, the great American music theorist David Lewin introduced in 1987 the gestural perspective in his seminal book [605]. Well, nearly, since the theory and the textual representation are more complex. Lewin's book describes what is now called "transformational theory", later adapted by his student Henry Klumpenhouwer to become K-nets. Such networks replace an 'amorphous' set of tone objects by a diagram, where the tone objects are placed at the diagram's vertices, while the diagram's arrows designate (affine) transformations mapping tone objects into each other.

Lewin argues against what he calls the "cartesian thinking", which observes musical objects as res extensae. Opposed to this passive attitude, Lewin suggests that transformations between musical points (such as pitch classes, for example) are the new path to pursue. Lewin's formalism and his wording are different; however, they show what Lewin is targeting: at a gestural theory of music. It would be very interesting to investigate Lewin's text with that subtext of gestural thinking in mind, since he repeatedly uses this metaphor in a speaking way.

We should also recall that Cecil Taylor, the monstre sacré of free jazz piano, describes his approach to creative improvisation with these words: "I try to imitate on the piano the leaps in space a dancer makes." This is a completely gestural concept, and Taylor comments (in a documentary DVD "All the Notes" by Christopher Felver) on score reading that deciphering those symbols takes away most of the energy that is needed for creativity.

Summarizing, we recognize that there are numerous approaches to music and performance theory that stress the primordial role of gestures in their conception as presemiotic components of a deeper understanding of music.

Let us now give a short overview of the mathematical theory of gestures which we have developed since 2002. It is a theory of expressions that point in a complex way to gestures, those exotic "animals of human communication" that we try to understand in their dynamical behavior. It is in no way an attempt to turn gestures as such into signs. In this section, mathematical prerequisites are required, and a reference to introductory texts [720] is recommended. This approach is inspired by Saint Victor's definition, but it gives it a formal shape that enables the development of an interesting concept architecture and the proof of powerful theorems. This theory was developed as a theory of musical gestures to capture many of the problems that have been mentioned above. However, this mathematical formalism is also applicable to more general contexts of gestural utterance, such as dance and painting.

To begin with, we propose a semiotic setup that attempts to generate a precise formalism of gestural structures. Observe that we do not attempt to recast gestures as signs, but as entities that may be expressed as mathematical structures. In terms of Hjelmslev semiotics, this amounts to establishing a gesture semiotic $GestSem$ that has an expressive level $Ex(GestSem)$, which realizes a mathematical theory of gestures, i.e., a classical system $Ex(GestSem) = MathGest$ within the semiotic of mathematics that offers a set of "expressions" or "forms" that point to gestures, the content level $Ct(GestSem) = Gestures$ of gestures proper. The signification level $Sg(GestSem)$ is not restricted to any particular gesture, meaning that we may set $Sg(GestSem) = GestSem$. This means that the semiotics of gestures is its own signification level, that its expressive level is a mathematical theory of gestural forms, and that gestures remain disclosed from metatheoretical specifications, gestures are not signs. We now give a definition of the pointer gesture.

A simple sign in $GestSem$ would be the pointer P. We could set $Ex(P) : \uparrow \to C(I)$, the map which sends the digraph $\uparrow$ (with one arrow that connects two different points) to the identity curve $Id : I \to I$; we would define $Ct(P) =$ the gesture of pointing, and finally let $Sg(P) = P$, the autoreference. One could then also declare the middle signification level in Hjelmslev's model to be identified with $GestSem$, thereby replacing the simple pointer by any gestural structure.

77.6 A Remark on Gestural Creativity

Musical creativity has always been a mysterious business since it is not only creativity as such, but moreover remains largely disclosed from verbalization. Music is created in a non-verbal process. But if we look at creativity in general, it is largely presemiotic since it deals with opening boxes, transcending given languages and formal environments. Creativity is a semiotic generator [726, Chapter 22.3.2]. Therefore, the gestural layer is quite canonical as a candidate for understanding and performing musical creativity.

In *Flow, Gesture, and Spaces in Free Jazz* [721], we have described musical creativity in jazz, especially in free jazz, using these ideas. It is about making the rules (in the sense of Alunni), negotiating them in a gestural interplay, and establishing a distributed identity as a rotational movement around the axis of a distributed identity. This is what Ornette Coleman meant when asking for making the music, and not the background, in the liner notes of his famous LP *Free Jazz.* He wanted to step away from the reproduction of others' ideas, and to make music without copying an already given template. That's why Cecil Taylor imitates a dancer's leaps in his creative piano universe, he would not want to imitate given forms. It's his dancing thoughts that he aims at.

77.7 Gestures and Strings

See also Chapter 75. The softening of creative knowledge production in a gestural approach softens not only the conceptual framework, it also softens the ddisciplinary barriers. Soft knowledge cannot be limited by traditional disciplinary walls. Such a transgression of disciplinary limits is in fact observed in our theory of hypergestures, especially for the example of counterpoint, see Section 61.14.1. The Escher Theorem 38 enables an exchange of roles in hypergestural combinations. External gestural skeleta can be exchanged with internal skeleta. A loop of lines can be transformed into a line of loops. This gestural insight is in fact also observed in physical string theory, where in S-duality, strings can be reinterpreted as branes, and gauge particles as bosons, see Appendix Section K.2 and [440].

This string-theoretic connection has been realized in a third application of mathematical gesture theory in music that relates to the investigation of the complex transitional process between reading a score in its symbolic realm and performing it in physical reality [648]. The model here represents the score data as symbolic gestures and the performed events as physical gestures, and then connects these two components by a hypergesture that is constructed following the Lagrangian formalism from physical string theory. This model enables a detailed analysis of the artistic potential that shapes the connection between symbolic and physical gestural utterance.

77.8 Playing the Multiversed Game in a Pre-semiotic Ontology

If we collect and summarize the dramatic changes in the basic conditions of Hesse's Glass Bead Game, we can state that it has these new features:

- It is played by a distributed identity of collaborators.
- It is not following given rules but creates them on the spot and according to an interplay of equivalent partners, in a presemiotic layer.
- The collaborative interplay is made by gestural exchanges, by a comprehension that does not follow templates, but is built upon the repercussion of gestural dynamics.

The conceptual framework is successful in its creation of a rotational energy around the axis of a distributed identity. Success is possible and addressed, but it does not result from given criteria; it is established in a distributed harmony without the time-space invariance of traditional laws. These characteristics redefine the glass bead game of "melting glass beads" with the following consequences:

- The gestural ontology is not auxiliary or preliminary to a factual layer of reality. It has its own persistent reality that does not serve what might become the case later on.
- The incessant gestural remaking of rules and concepts eliminates the world's unicity, completing the historical suspensions of geocentricity (Copernicus), anthropocentricity (Darwin), chronocentricity (Einstein), and ratiocentricity (Turing), and adding the *suspension of factocentricity.*
- Creativity is no longer delegated to arcane divinities. It has also become a radically human endeavor. Creation is no longer limited to God's initial Big Bang.
- The transdisciplinary parallel between music and physics (see also Chapter 76) shows these parallelisms:
  - Hypergestures — Strings,
  - Escher Theorem — Duality,
  - Works (typically 10^{37} 72-element motives) — Universes (10^{450} string theory landscapes),
  - Communication via gesture interaction — Interaction via exchange of bosonic strings.

78

Euler-Lagrange Equations for Hypergestures

Summary. This chapter deals with a model from mathematical physics of string theory that describes the transition from symbolic reality to physical reality of musical gestures. We demonstrate, using multidimensional Fourier theory and Green functions, that the physical gesture can be viewed as a function of a potential and the symbolic gesture. The role of this potential is however not fully understood to date, but the idea is that it should encompass artistic rationales, together with physical components.

– Σ –

78.1 The Problem in Performance Theory with the Physical Nambu-Goto Lagrangian

Refer to Appendix K.2 for the Nambu-Goto action for strings in physics. In the mathematical theory of gestures, string world-sheets correspond to hypergestures h in $\uparrow \overrightarrow{@} \uparrow \overrightarrow{@} X$, where X is the four-dimensional space-time with its Minkowski metric. The Nambu-Goto action is based on the surface $A(h)$ formula of the world-sheet of h, together with the factor $-\frac{T_0}{c}$ that guarantees the energy$\times$time dimension of the action.

In the musical situation, we have a number of characteristic differences from physics. To begin with, the reparametrization invariance that is given by the surface formula is not required in music. A gesture is not defined modulo reparametrization, but essentially relies on the chosen parametrization. Reparametrization is a topic dealt with in hypergesture theory, where (generalized) homotopy covers this transformational relation. Second, the musical world-sheet has to connect the symbolic gestures that are defined by the score's symbols to the physical gestures in performance. This creates a major problem for the naive physical approach since the symbolic gesture would allow for infinite velocity. This obstruction requires a different approach to velocity, more precisely, to time. The third problem is that the Lagrangian density is a function of a scalar T_0 which represents the mass density in physics. But it is not evident that this quantity should also be a scalar in the musical situation. It could be that the inertia in the musical world-sheet is a function of the direction, which would enforce a vectorial T_0. Let us now discuss an approach to a musical Lagrangian action.

To begin with, we are given a musical world-sheet $M : I^2 \to \mathbb{R}^{p.} : (x, y) \mapsto M(x, y) = (s(x, y), t(x, y))$ that maps the two sheet parameters x, y to a point in a parameter space $\mathbb{R}^{p.}$ defined by a sequence $p. = p_1, p_2, \ldots p_k$ of physical and cognitive/symbolic time parameters. We may think of a very elementary situation of a single mass point (one finger movement). The first $k-1$ parameters designate space coordinates of performance, such as the key position on a piano, level above the instrumental interface, etc., whereas the last parameter is the physical time of the performer's movement. We have denoted the space values by $s(x, y)$, whereas the time value is denoted by $t(x, y)$. For the space needed for a hand's gestures, refer to Section 78.2.3.

To simulate Lagrangian density $\mathcal{L}(x, y)$, we need to simulate kinetic energy first. The naive approach to velocity is $\frac{ds}{dt} = \partial_x s \frac{dx}{dt} + \partial_y s \frac{dy}{dt}$, where $\frac{dx}{dt}, \frac{dy}{dt}$ don't exist and must be replaced by $\frac{1}{\partial_x t}, \frac{1}{\partial_y t}$, respectively, i.e.,

G. Mazzola et al., *The Topos of Music III: Gestures*, Computational Music Science,
https://doi.org/10.1007/978-3-319-64481-3_23

$\frac{ds}{dt} = \frac{\partial_x s}{\partial_x t} + \frac{\partial_y s}{\partial_y t}$. To simplify the discussion, let us suppose for a moment that the two summands are orthogonal in the given metric of the spatial component. Then the energy density would be $\mathcal{E}(x, y) = \frac{\mu}{2}(|\frac{\partial_x s}{\partial_x t}|^2 + |\frac{\partial_y s}{\partial_y t}|^2) = \frac{\mu}{2}(\frac{|\partial_x s|^2}{\partial_x t^2} + \frac{|\partial_y s|^2}{\partial_y t^2})$. The big problem here is that the denominators $\partial_x t^2, \partial_y t^2$ could vanish, and in fact $\partial_y t$ will vanish for $x = 0$, the initial symbolic gesture, if infinite velocity is required by symbolic data. See also Figure 78.1.

Fig. 78.1. The musical world-sheet hypergesture from a symbolic (back) to a physical (front) gesture.

A first solution to this problem could be to require not the vanishing of denominators $\partial_x t^2, \partial_y t^2$ only, but the vanishing of inertia μ, too, and this in a vectorial sense, i.e., introducing a vector $\mu = (\mu_x, \mu_y)$ such that we have $\mathcal{E}(x, y) = \frac{1}{2}(\mu_x \frac{|\partial_x s|^2}{\partial_x t^2} + \mu_y \frac{|\partial_y s|^2}{\partial_y t^2})$, but with finite limit values $\frac{\mu_x}{\partial_x t^2}, \frac{\mu_y}{\partial_y t^2}$. Although this solution yields a reasonable $\mathcal{E}(x, y)$, the way back to reasonable values of $t(x, y)$ is problematic as it requires a delicate analysis of the vectorial inertia $\mu = (\mu_x, \mu_y)$.

A more elegant, and philosophically and physically more reasonable, solution would be to question the concept of time. In physics, it is known that complex instead of real time can help solve singularity problems of the Big Bang model. Such an approach would mean defining a time function $t(x, y)$ with complex values: $t(x, y) = t_r(x, y) + it_i(x, y)$, $t_r(x, y)$ a real value that the physical performance needs, and $t_i(x, y)$ an imaginary value. This means that we now have a world-sheet $M : I^2 \to \mathbb{C}^{p.}$, where in particular the last time coordinate is complex. We have discussed imaginary time in several previous publications (e.g. [725, 4.12] and [726, 22.3]) in the context of the problem of artistic presence, where physical duration of the "now" vanishes, whereas the cognitive time expands to comprise all the artistic dynamics. It is therefore natural to think of imaginary time in performance, and to reify this parameter in a mathematically concrete way in the world-sheet model of gestural performance.

78.1.1 Complex Time and Descartes' Dualistic Ontology

This is just a small addendum to a possible ontological enrichment which complex time in its cognitive presence could entail. Viewing the classical space-time with complex time, it is represented as a five-dimensional real vector space $ST = \mathbb{C} \oplus \mathbb{R}^3$ that is a sum of the four-dimensional "real" space-time $RST = \mathbb{R} \oplus \mathbb{R}^3$ and the "imaginary" space-time $IST = i\mathbb{R} \oplus \mathbb{R}^3$ that intersect in the space component $\mathbb{R}^3$. Descartes, in his *Principia philosophiae* [263] describes the three substances of being: res extensa, res cogitans, and God. The human existence is comprised of res extensa and res cogitans, the latter being strongly associated with consciousness, and their mysterious interaction. It is not clear where this interaction should happen (Descartes' idea of the pineal gland being the crossing locus is too naive). And this dualism also poses the problem of the causality of such interaction. In view of our complex time approach, we would at least have the following elements of a potential solution: The total space-time ST is a sum $ST = RST + IST$ of two 4-dimensional subspaces that intersect in the "spatial" space: $RST \cap IST = \mathbb{R}^3$, see also Figure 78.2.

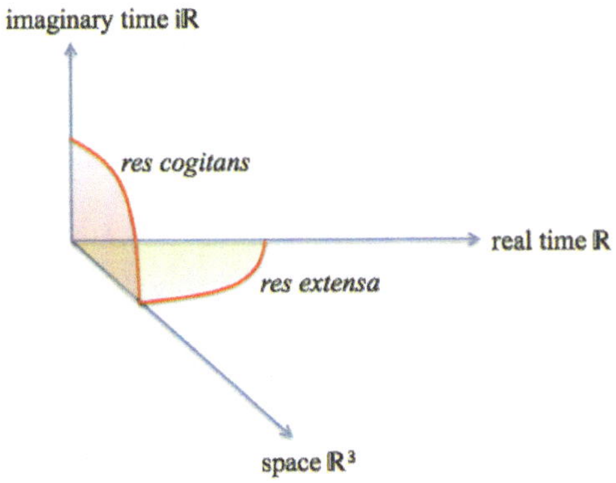

Fig. 78.2. Descartes' duality in a space-time with complex time.

The total dynamics is an affair of ST, but it has two 4-dimensional components, the classical "real" physical RST space-time, and the new "imaginary" IST space-time. If we identify the dynamics of res extensae with RST and the dynamics of res cogitantes with IST, we envisage two ontologies with different and independent time dimensions, but which are connected spatially on $RST \cap IST$, which is a spatial ontology that could be metaphorically (and nothing more!) associated with the space of the pineal glans.

This approach in principle opens the possibility of an extended physical dynamics that could include causal or other interactions between RST and IST. These different spaces would no longer be disjoint geometrically and causally. We are however not aware of any physical theory that would describe the interactions in ST that exceed those in RST—except the cosmological approach to the Big Bang singularity by Stephen Hawking, where no hypothesis is made that would include human cognitive dynamics.[1] In the following discourse, we shall describe an explicit transitional space from imaginary to real time, and thereby from IST to RST with intermediate stages in ST that are both, real and imaginary, much as general complex numbers have non-vanishing real and imaginary coordinates. At this point, we do not have a thorough understanding of these intermediate stages of musical reality.

78.2 Lagrangian Density for Complex Time

For the above reasons, we would have to restate the formula with a "complex" velocity, i.e., $\mathcal{E}(x,y) = \frac{1}{2}(\mu_x \frac{|\partial_x s|^2}{|\partial_x t|^2} + \mu_y \frac{|\partial_y s|^2}{|\partial_y t|^2})$. The advantage of such an approach is that now both $\partial_x t$ and $\partial_y t$ needn't vanish although the real part of these complex numbers might vanish. Imposing the existence of a lower bound $q > 0$ such that $q \leqslant |\partial_x t|$ and $q \leqslant |\partial_y t|$, we get rid of the singularity problem of infinite speed, in fact, this condition means that both velocities, $\frac{1}{|\partial_x t|}$ and $\frac{1}{|\partial_y t|}$, are limited from above by $\frac{1}{q}$. Such a condition is well known in physics, namely as the limit by the speed of light, only that here we work in a time reality that comprises the speed of thoughts, too, not only the speed of moving fingers. We now propose a Lagrangian density defined by

[1] Hawking didn't answer repeated emails by one of the authors (Mazzola) concerning a possible cognitive reality of imaginary time.

$$\mathcal{L}(x,y) = \frac{\mu}{2}\left(\frac{|\partial_x s|^2}{|\partial_x t|^2} + \frac{|\partial_y s|^2}{|\partial_y t|^2}\right) - U(s(x,y)),$$

where $U(s(x,y))$ is a potential, but not only the usual physical potential (the gravitational field). It will also comprise still unknown artistically defined potential contributions.

We shall see below that the above limitation of $|\partial_x t|, |\partial_y t|$ to a stronger condition, namely that map $t : (x,y) = x + iy \mapsto t_r(x,y) + it_i(x,y)$, is complex analytic (see Appendix J.5), hence *conformal*, also a standard situation in string theory [1159, 22.4]. Under this condition, using the canonical sesquilinear form $\langle , \rangle$ on $\mathbb{C}^n$, we first have $\frac{ds}{dt} = \frac{\partial_x s}{\partial_x t} + \frac{\partial_y s}{\partial_y t} = \frac{1}{|t'|^2}(\partial_x s \overline{\partial_x t} + \partial_y s \overline{\partial_y t})$. Then, taking into account that the space coordinates are supposed to be real numbers,

$$\begin{aligned}\langle \frac{ds}{dt}, \frac{ds}{dt}\rangle &= \frac{1}{|t'|^4}(\langle \partial_x s\overline{\partial_x t} + \partial_y s\overline{\partial_y t}, \partial_x s\overline{\partial_x t} + \partial_y s\overline{\partial_y t}\rangle) = \\ &\frac{1}{|t'|^2}(|\partial_x s|^2 + |\partial_y s|^2) + \frac{1}{|t'|^4}(\langle \partial_x s, \partial_y s\rangle(\overline{\partial_x t}\partial_y t + \overline{\partial_y t}\partial_x t)).\end{aligned}$$

But since $\partial_x t = -i\partial_y t$ because t is analytic, $\overline{\partial_x t}\partial_y t + \overline{\partial_y t}\partial_x t = 0$, and the above hypothesis of orthogonality of $\partial_x s$ and $\partial_y s$ is absorbed by the analyticity hypothesis.

78.2.1 The Lagrangian Action for Performance

The Lagrangian action is $S = \int_y \int_x \mathcal{L}(x,y)$, and we have to calculate the spatial[2] variation δS, defined from a variation $M_\delta(x,y) = M(x,y) + \delta M = (s(x,y), t(x,y)) + (\delta s, 0)$, to obtain the Euler-Lagrange equations. We focus on the kinetic energy part, the potential variation is as usual. With the scalar product $\langle , \rangle$ we have

$$\begin{aligned}&\frac{\mu}{2}\frac{|\partial_x(s+\delta s)|^2}{|\partial_x(t+\delta t)|^2} = \\ &\frac{\mu}{2}\frac{|\partial_x s|^2 + |\partial_x \delta s|^2 + 2\langle \partial_x s, \partial_x \delta s\rangle}{|\partial_x t|^2} = \text{(mod quadratic order in } \delta) \\ &\frac{\mu}{2}\frac{|\partial_x s|^2}{|\partial_x t|^2} + \mu\frac{\langle \partial_x s, \partial_x \delta s\rangle}{|\partial_x t|^2}.\end{aligned}$$

Using the derivative $\partial_x(\langle \frac{\partial_x s}{|\partial_x t|^2}, \delta s\rangle) = \langle \partial_x(\frac{\partial_x s}{|\partial_x t|^2}), \delta s\rangle + \langle \frac{\partial_x s}{|\partial_x t|^2}, \partial_x \delta s\rangle$, the variation becomes

$$\mu(\partial_x(\langle \frac{\partial_x s}{|\partial_x t|^2}, \delta s\rangle) + \partial_y(\langle \frac{\partial_y s}{|\partial_y t|^2}, \delta s\rangle) - \langle \partial_x(\frac{\partial_x s}{|\partial_x t|^2}), \delta s\rangle - \langle \partial_y(\frac{\partial_y s}{|\partial_y t|^2}), \delta s\rangle),$$

and since variation of s vanishes at the endpoints, and using the fundamental lemma of the calculus of variations (see Appendix J.9), we get the Euler-Lagrange equation

$$-\nabla U = \mu(\partial_x(\frac{\partial_x s}{|\partial_x t|^2}) + \partial_y(\frac{\partial_y s}{|\partial_y t|^2})),$$

i.e.,

$$-\nabla U = \mu(\frac{\partial_x^2 s}{|\partial_x t|^2} - 2\frac{\langle \partial_x t, \partial_x^2 t\rangle_{\mathbb{R}}}{|\partial_x t|^4}\partial_x s + \frac{\partial_y^2 s}{|\partial_y t|^2} - 2\frac{\langle \partial_y t, \partial_y^2 t\rangle_{\mathbb{R}}}{|\partial_y t|^4}\partial_y s),$$

with $\langle , \rangle_{\mathbb{R}}$ being the real part of $\langle , \rangle$, which is the usual real scalar product in $\mathbb{R}^2$. Using the usual derivative symbols in physics, $\dot{s} = \partial_y s$, $\ddot{s} = \partial_y^2 s$, $s' = \partial_x s$, $s'' = \partial_x^2 s$, etc., we get

[2] This choice is made in view of the hypothesis that $t(x,y)$ is complex analytic, a condition that would be violated by general variations.

$$-\nabla U = \mu(\frac{s''}{|t'|^2} - 2\frac{\langle t', t''\rangle_{\mathbb{R}}}{|t'|^4}s' + \frac{\ddot{s}}{|\dot{t}|^2} - 2\frac{\langle \dot{t}, \ddot{t}\rangle_{\mathbb{R}}}{|\dot{t}|^4}\dot{s}).$$

This complex formula can be simplified with the following assumption, namely that $t' \perp_{\mathbb{R}} t''$. This is equivalent to the condition that $|t'|^2$ is only a function of y. The orthogonality $t' \perp_{\mathbb{R}} t''$ is a special case of the condition that for an analytical function g, we have $g \perp_{\mathbb{R}} g'$, and this means that g'/g is imaginary. But this is evidently equivalent to the differential equation $0 = \partial_x(|g|^2)$. And this means that $|g|^2$ is only a function of y. Applying this result to $g = t'$, our claim follows.

In a first approximation, we neglect the term $\dot{s}$ and discuss the simplified Euler-Lagrange equation $-\nabla U|t'|^2 = \mu(s'' + \ddot{s})$. Now, given a differentiable map $f : \mathbb{R}^n \to \mathbb{R}^n$, we have the kth Jacobian $J^k(f)$, which is the $n \times n$-matrix whose entries are the partial derivatives $\partial^k_{x_j} f_i$ on row i and column j. Let $\Delta = (1, 1)$ be the diagonal vector in $\mathbb{R}^2$. Then the simplified Euler-Lagrange equation is

$$-\nabla U|t'|^2 = \mu J^2 s\Delta,$$

a remarkable formula since the Jacobian and the diagonal vector also appear in the definition of the performance vector fields in performance theory for symbolic and physical events, see Chapter 33. Evidently, we could reintroduce the vectorial $\Delta_\mu = (\mu_x, \mu_y)$ and write the Euler-Lagrange formula as

$$-\nabla U|t'|^2 = J^2 s\Delta_\mu.$$

Let us focus on the Euler-Lagrange equation with a scalar density μ. It is in fact a Poisson equation (see Appendix J.10) in each spatial factor: $\frac{-1}{\mu}\partial_{s_i} U|t'|^2 = \Delta s_i$ for spatial coordinate s_i, Δ now being the Laplace operator $\partial_x^2 + \partial_y^2$, or, in vectorial writing,

$$\frac{-|t'|^2}{\mu}\nabla U = \Delta s.$$

Supposing that ∇U is given as a function of x, y, the solution $s_i(x, y)$ for each spatial coordinate s_i of this Poisson equation uses Green's function G (see Appendix J.10) and is an integral of the form (everything in space vectors)

$$s = P(\frac{|t'|^2}{\mu}\nabla U) + S(\partial s)$$

where $\partial s = \sum(-1)^i s_i$ is the homological boundary of the singular cube s, and P, S are linear functions. The quotient $\frac{|t'|^2}{\mu}$ looks like a factor for temporal versus material density. In this formalism, the component $s_0(y) = s(0, y)$ is fixed: it is the given symbolic gesture defined by the score. The component $s_1(y) = s(1, y)$ is the physical gesture of the performer. It must comply with physical laws and can be calculated separately. Components $s_2(x) = s(x, 0), s_3(x) = s(x, 1)$ are not given a priori, except for their start and end points. The potential U is not given either a priori. *The artistic program is therefore to investigate the variety of curves* s_2, s_3 *and of the potential* U *in terms of artistic determinants.*

It is worthwhile to add a short remark on the possible causal relations that occur during the transition from imaginary time (the vertical line state of $x = 0$ in Figure 78.3) to real time (the horizontal line state of $x = 1$ in Figure 78.3). A specific moment in the evolution of the pianist's gesture is given by the value of y. The initial moment is for $y = 0$, the final one is for $y = 1$. The evolution at time $y = s$ is traced along the quarter circle $a(s)$ connecting the imaginary vertical time line and the real horizontal time line, see Figure 78.4. We recognize that a gestural moment $a(s)$ cannot be changed in a real time $\tau > t$ after $a(s)$'s intersection with the real time axis at t. In particular, the initial moment cannot be changed after the initial real time. The only moments that admit a change at time t are those $a(\sigma)$ which intersect the real time line after t, they are shown as open intermediate states at time t in Figure 78.4. However, these open states have been shaped quite a lot until real time t (the part of $a(\sigma)$ to the left of the vertical line at real time t). Therefore only the part of $a(\sigma)$ to the right of t can be used to terminate the shaping of $a(\sigma)$. This fact generates a kind of causality in the shaping process of the pianist's hypergesture.

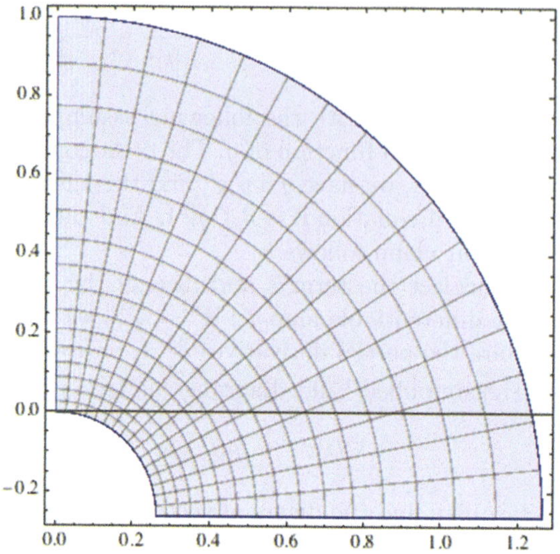

Fig. 78.3. A time map for the function $t(x, y)$ described in the text.

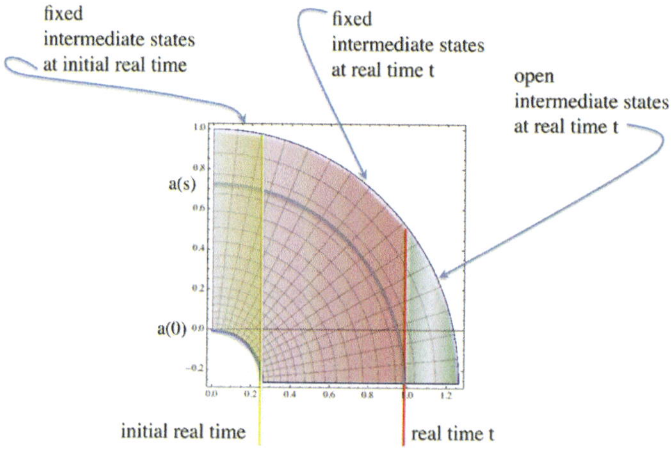

Fig. 78.4. The causal relations during the hypergestural evolution in complex time.

78.2.2 The World-Sheet of Complex Time

The previous discussion is built on the hypothesis that $t' \perp \dot{t}$ is the derivative of analytic $t(x, y)$. This means that the Jacobian

$$Jt(x,y) = \begin{pmatrix} \partial_x t_1 & \partial_y t_1 \\ \partial_x t_2 & \partial_y t_2 \end{pmatrix} = (t', \dot{t})$$

has orthogonal columns. See Figure 78.3 for a time function

$$t(x,y) = -0.2624i + \frac{1}{3.8104} e^{\frac{1}{2}\pi i(1-x) + \frac{\pi y}{2}},$$
$$3.8104 \approx e^{\pi/2} - 1, \quad 0.2624i \approx e^{\frac{1}{2}\pi i}/(e^{\pi/2} - 1),$$

mapping the unit imaginary time interval at $x = 0$ to the unit real time interval at $x = 1$. The vector field defined by the y-derivative $\dot{t}$ is shown in Figure 78.5.

This situation is completely analogous to the one encountered for the calculation of performance from performance vector fields, see Section 33.2.2. The time function $t(x, y)$ can described by the function $Jt(x, y)$

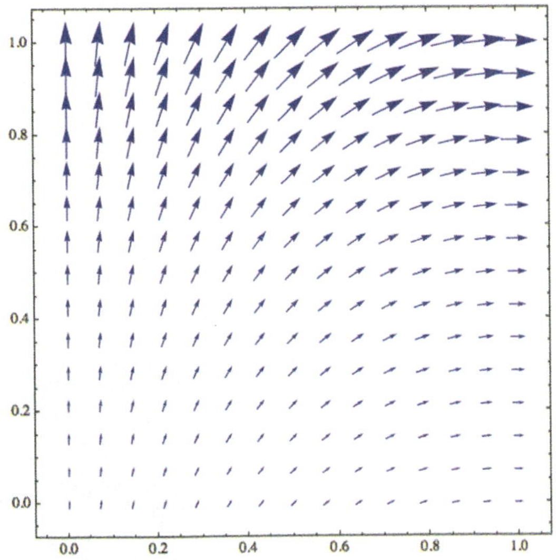

Fig. 78.5. The y-derivative $\dot{t}$.

and adequate initial conditions. The *(complex) time performance field* here is generated by the inverse Jacobian, acting on the diagonal vector Δ (known from performance theory),

$$\mathfrak{Z}^t_{(x,y)} = Jt(x,y)^{-1}\Delta,$$

while the initial time performance defines the values of t on selected initial arguments; see Figure 78.6 for an example of such a time performance vector field for the above time function $t(x,y)$.

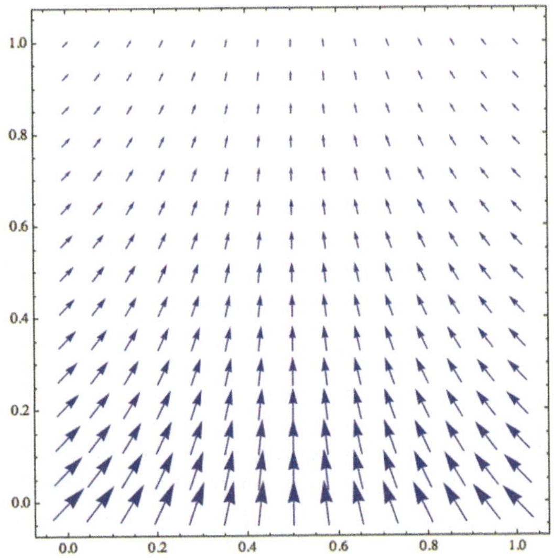

Fig. 78.6. A time performance vector field.

This figure illustrates a typical situation for the initial symbolic gesture, i.e., $x = 0$. Here, the field is diagonal, meaning that the vertical partial derivative $\partial_y t_1$ for real time t_1 vanishes, in other words, the physical velocity would be infinity. But the vertical partial derivative $\partial_y t_2$ of imaginary time has positive length, time only elapses in the mental direction, which is a natural requirement in the initial symbolic position of our world-sheet. At the terminal position, for $x = 1$, the mental time stands still while the physical time takes over the control of the movement, the performance field is rotated by 90°.

The terminology of "initial" vs. "terminal" for values $x = 0$ vs. $x = 1$ is slightly misleading since there is no causal background for these terms. It is not clear whether the symbolic reality is pushing into a physical one, or whether the movement is a finality towards physics. Both could be happening. This becomes even more evident if we view the passage in the opposite direction, $x : 1 \to 0$. This could happen if an improviser is creating a score from played music, in fact a common approach.

In view of the complexity of this time performance field, it is an open question whether the world-sheet should/could be defined by a Lagrangian action in the sense of an economy of thoughts in creativity.

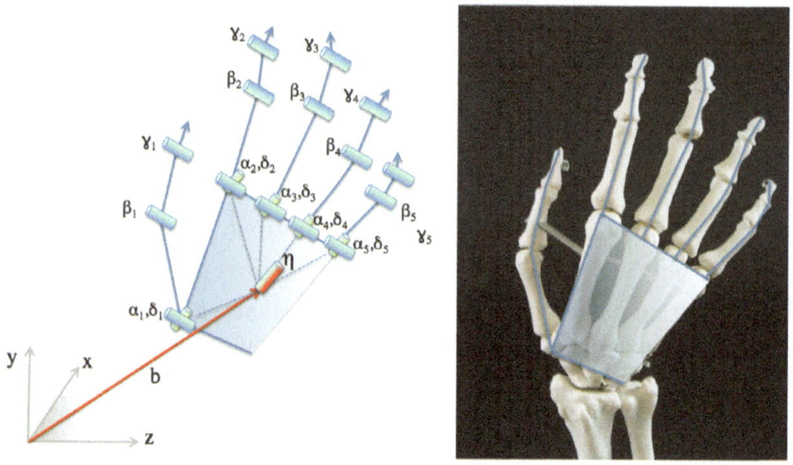

Fig. 78.7. The spatial coordinates of a hand.

78.2.3 The Space for a Hand's Gestures

The hand's dynamic has two aspects, a spatial and a temporal one. Refer to Figure 61.2 for the spatial aspect. As shown in Figure 78.7, a hand's position is defined by four angles per finger, a vector $b \in \mathbb{R}^3$ for the position of the carpus center, two real numbers for the orientation of the carpus plane, and one real number for the rotational position of the carpus plane around the normal vector through the center of the carpus plane, a total of 26 real numbers. Denote this position by $s \in \mathbb{R}^{26}$. We may also assume that for every position, a small open neighborhood is also a set of possible (however small) variations, which means that the spatial information of the hand's dynamic is a connected open set $W \subset \mathbb{R}^{26}$. Is W simply connected?

78.2.4 The World-Sheet for a Simple Case

We have calculated the solution of the world-sheet for a simple movement of one finger: push the key, lift the finger, keep it up, go down again on that key, keep it on the key. The symbolic gesture is the linear edgy curve to the left in Figure 78.15, the physical gesture is the smooth curve to the right. The potential is taken as a test case function $\nabla U(s(x, y)) = (x + 4y - 12)e^{-\pi y/6}$, and $\mu = 1$. In this case we have considered equal time intervals for each part of the splines, as we will describe in Section 78.2.5. We have also calculated the world-sheet for more complex potentials, as shown in Figure 78.9.

78.2.5 The Elementary Gesture of a Pianist

The elementary gesture of the pianist is the movement of a finger that presses a key. In general, elementary gestures in music are conceptually similar, following the general idea of *arsis—thesis*, passing through the

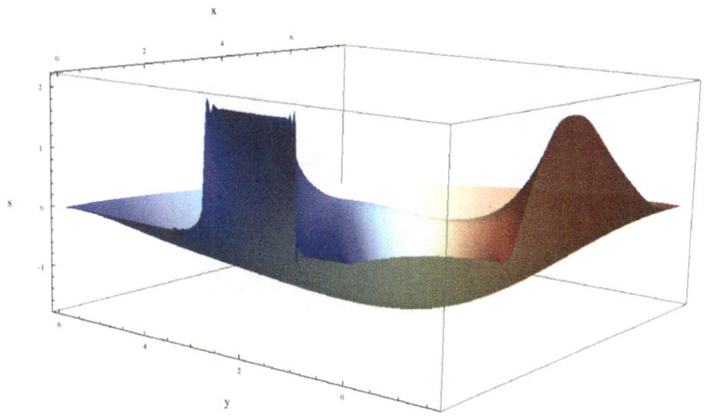

Fig. 78.8. The world-sheet for a simple up-down movement of one finger.

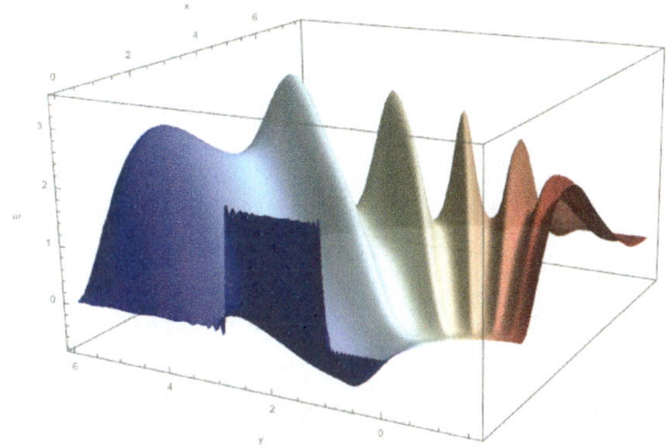

Fig. 78.9. The world-sheet for a simple up-down movement of one finger with a more complex potential.

attack gesture of a conductor [791], or the breath with inspiration-exhalation of singers. It is the general idea of a preparation by an ascending movement, and the accomplishment of a gesture by a descending movement. The completion of gesture is signaled by a third part, again an ascending movement, that prepares for a new gesture, in a chain of gestures—the simplest hypergesture.

Here we focus on the formal description of the elementary piano gesture, representing velocity by spline interpolation to describe variable velocity. Initially the finger—which we schematize as a massive point—is at rest on a pressed key. It then accelerates moving up, moves at constant speed for a time interval, decelerates, then reaches a distance from the keyboard (*arsis* movement) where it is at rest. It then starts to move again, with identical motion, but in opposite direction, until the key is completely pressed.

We choose cubic polynomial splines. For each part of the movement we use three different interpolations: cubic, constant velocity, and cubic again. If we consider the cubic polynomial $v(t) = at^3 + bt^2 + ct + d$, to find the first spline, representing the increasing velocity from zero to the maximal value, we have to solve the following system of equations:

- velocity zero at time $t_0 \Rightarrow at_0^3 + bt_0^2 + ct_0 + d = 0$
- acceleration zero at time $t_0 \Rightarrow 3at_0^2 + 2bt_0 + c = 0$
- maximal velocity at $t_1 \Rightarrow W = at_1^3 + bt_1^2 + ct_1 + d$
- zero acceleration at $t_1 \Rightarrow 0 = 3at_1^2 + 2bt_1 + c$.

The absolute value of acceleration is maximal at the center of the first and the third spline. Solving the system we obtain $v_y^1(t_0, t_1, t, W) = \frac{(t-t_0)^2(2t+t_0-3t_1)W}{(t_0-t_1)^3}$. Proceeding in an analogous way, we obtain the velocity for the first part of the motion, a massive point that starts from the key and reaches height H:

$$v_y^1(t_0, t_1, t, W) = \frac{(t - t_0)^2(2t + t_0 - 3t_1)W}{(t_0 - t_1)^3} \text{ if } t_0 < t < t_1,$$
$$v_y^2(t, W) = W \text{ if } t_1 < t < t_2,$$
$$v_y^3(t_2, t_3, t, W) = -\frac{(t - t_3)^2(2t - 3t_2 + t_3)W}{(t_2 - t_3)^3} \text{ if } t_2 < t < t_3.$$

To complete the vertical gesture with the *thesis* part, we also consider the descending motion:

$$v_y^4(t_3, t_4, t, W) = -\frac{(t - t_3)^2(2t + t_3 - 3t_4)W}{(t_3 - t_4)^3} \text{ if } t_3 < t < t_4,$$
$$v_y^5(t_4, t_5, t, W) = -W \text{ if } t_4 < t < t_5,$$
$$v_y^6(t_5, t_6, t, W) = \frac{(t - t_6)^2(2t + t_6 - 3t_5)W}{(t_5 - t_6)^3} \text{ if } t_5 < t < t_6,$$

where W is the maximal vertical velocity. The complete graph of velocity is shown in Figure 78.10.

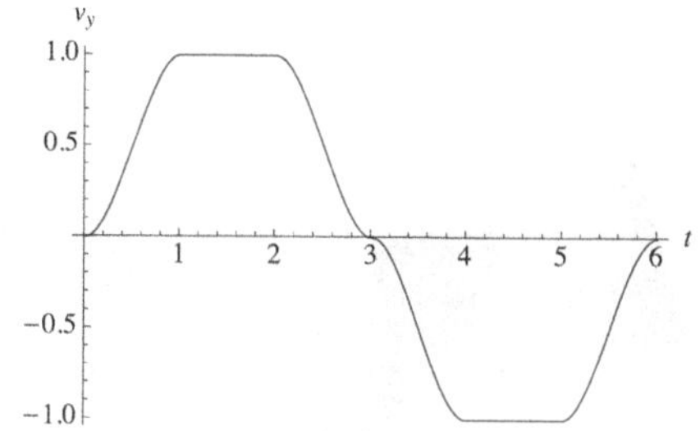

Fig. 78.10. Vertical velocity of piano primitive gesture, in the raising and lowering part.

Artistic parameters are related to these results. For example, acceleration is related to force via the well-known $F = ma$, and force determines the loudness of the sound. The maximum values of acceleration have a superior physical limit given by the following equation (with $t' = \frac{t_0+t_1}{2}$):

$$\frac{d}{dt}\left[v_y^1(t_0, t_1, t, W)\right]_{t'} = \frac{d}{dt}\left[\frac{(t - t_0)^2(2t + t_0 - 3t_1)W}{(t_0 - t_1)^3}\right]_{t'} = \frac{3}{2}\frac{W}{(t_0 - t_1)} \leqslant \frac{F}{m}.$$

Summarizing, even in this simple case, we have the variety of parameters:

$$t_0, t_1, \ldots t_6, W, H, m.$$

The vertical displacement is obtained by integration of the previous speed functions,

$$\begin{aligned}
&AscendingPosition[t_0, t_1, t_2, t_3, t, W, H_0] = H_0 + \\
&\text{If } t < t_1, ((t - t_0)^3 \frac{(t + t_0 - 2t1)W}{2(t_0 - t_1)^3}, \\
&\text{If } t < t_2, -(t_0 W)/2 + (t_1 W)/2 + (t - t_1)W, \\
&\text{If } t < t_3, -(t_0 W)/2 + (t_1 W)/2 + (-t_1 + t_2)W - \\
&\frac{1}{2(t_2 - t_3)^3}(t - t_2)(t^3 - t_2^3 - tt_2(t_2 - 4t_3) + \\
&4t_2^2 t_3 - 6t_2 t_3^2 + 2t_3^3 - t^2(t_2 + 2t_3))W,
\end{aligned}$$

and by the following ones for the descending movement (H_f is the highest vertical position):

$$DescendingPosition[t_0, t_1, t_2, t_3, t_4, t_5, t_6, t, W, H_0, H_f] = H_f +$$
$$\text{If } t < t_4, -(t-t_3)^3 \frac{(t+t_3-2t_4)W}{2(t_3-t_4)^3},$$
$$\text{If } t < t_5, (t_3W)/2 - (t_4W)/2 - (t-t_4)W,$$
$$\text{If } t < t_6, (t_3W)/2 - (t_4W)/2 - (-t_4+t_5)W+$$
$$\frac{1}{2(t_5-t_6)^3}(t-t_5)(t^3 - t_5^3 - tt_5(t_5-4t_6)+$$
$$4t_5^2t_6 - 6t_5t_6^2 + 2t_6^3 - t^2(t_5+2t_6))W.$$

We then define the complete function Moto (in Mathematica) for the vertical movement depending on time (Figure 78.11).

```
Moto[t0_,t1_,t2_,t3_,t4_,t5_,t5_,W_,H0_,Hf_] :=
 If[t < t3, AscendingPosition[t0,t1,t2,t3,t,W,H0],
  If[t < t6,
   DescendingPosition[t0,t1,t2,t3,t4,t5,t6,t,W,H0,Hf]]].
```

We make the following choice of parameters: $t_0 = 0, t_1 = 1, t_2 = 2, t_3 = 3, t_4 = 4, t_5 = 5, t_6 = 6, W = 1, H_0 = 0, H_f = 2$, and therefore $a = 6$, $b = 6$ for the rectangular domain of Poisson equation. With this parameter choice we obtain the right curve on Figure 78.15.

The complete graph is obtained, as said above, solving the Poisson equation with a test potential. By variation of these parameters, we can describe a huge amount of piano touches. These concepts are useful for understanding not only general ideas about music performance, but in particular can be used to avoid vagaries in performance didactics.

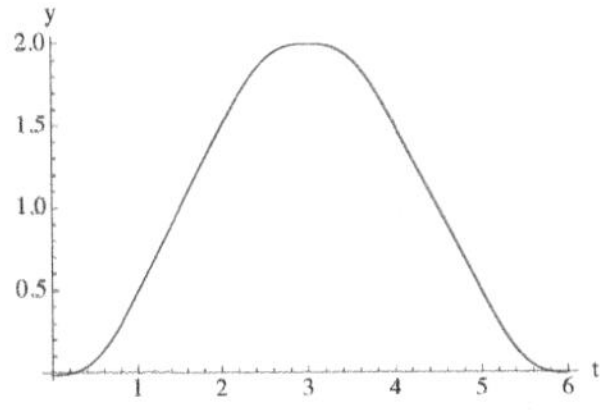

Fig. 78.11. Vertical movement of piano elementary gesture, in the raising and lowering part.

78.2.6 The Overarching Framework Between Note Performance and Gesture Performance

In this section, we want to recapitulate the overarching performance framework where gestural world-sheets are embedded. To begin with, the symbolic gesture s_0 in the homological boundary ∂s is supposed to be fixed, also projecting to the symbolic note score $p_{symbol.}(s_0) = \sigma_0$, a local composition in the traditional space $\mathbb{R}^{EHLD...}$ of symbolic parameters. The note performance map $\wp : \mathbb{R}^{EHLD...} \to \mathbb{R}^{ehld...}$ into the space of physical note parameters is also supposed to be given according to the rationales that control this construction, see also Chapter 44.7. This determines the note-level performance $\sigma_1 = \wp(\sigma_0)$, and we suppose that a physical gestural performance s_1 in ∂s yields a note-level performance $p_{phys.}(s_1) = \sigma$ that lies in a small open neighborhood $O(\sigma_1)$ of σ_1. In other words, $s_1 \in OW := p_{phys.}^{-1}(O(\sigma_1)) \cap W$, where $W \subset \mathbb{R}^p$ is the open subset of physical gesture parameters defined by the geometrically possible positions, as presented for a pianist's hand in Section 78.2.3.

We suppose for the moment that for each selection $s_1 \in OW$, the two boundary gestures s_2, s_3 are calculated by a given formula. This means that, using the solution s of the Poisson equation $\frac{-1}{\mu}\nabla U|t'|^2 = \Delta s$, the action $S = \int_y \int_x \mathcal{L}(x,y)$ is determined by the choice of potential U and of a physical gesture s_1, see Figure 78.12.

The action surface $S(U, s_1)$ now opens the discussion framework for the effective selection of U and s_1. It seems reasonable to look for couples (U, s_1) with locally minimal $S(U, s_1)$, as shown in Figure 78.12, since we want the total energy to perform a physical gesture to be (locally) minimal. On the other hand, we also want the potential U to be not only physically, but also aesthetically meaningful. *This condition is the main open question for the time being.* Musically speaking, it means that we have an entire manifold of physical hand gestures (the points in OW) that may realize given physical note scores (the points in $O(\sigma_1)$), but we also want the energy strategy to shape the world-sheet hypergesture from symbolic s_0 to physical s_1 to be taken into account, i.e., the kind of transformational effort from thinking to making this piece of music.

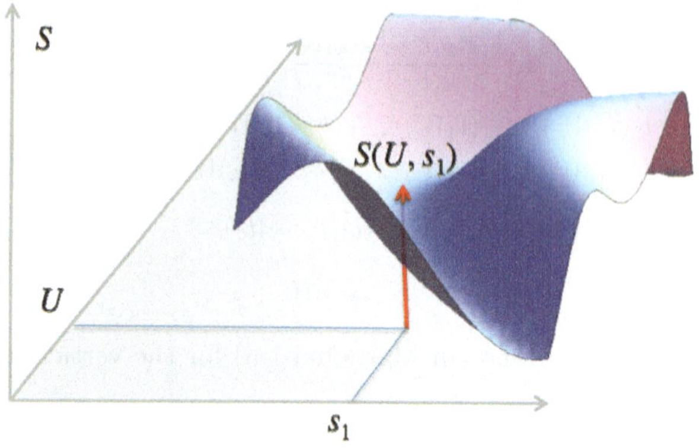

Fig. 78.12. Local action minima.

Here is what we are hoping for: The surface $S(U, s_1)$ is a candidate for exhibiting more precise relations between potentials U and physical gestures s_1. Ideally, one should be able to find a functional subset F of that surface, i.e., a subset whose projection to the $U - s_1$ plane defines a function $U \mapsto s_1(U)$, and thereby generates a causally interpretable relation of a cause U to its effect s_1, see Figure 78.13. But even a non-functional graph as a projection would be advantageous for a better understanding of how potentials can "cause" physical gestures.

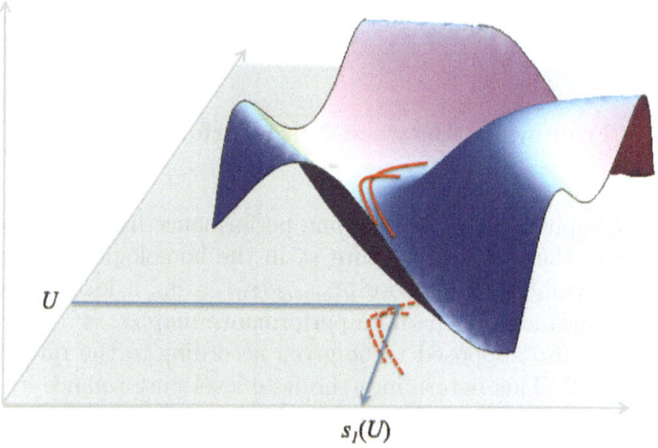

Fig. 78.13. Functional graphs of local action minima.

78.2.7 Examples of Functional Relations Between Potential and Physical Gesture

For the technical reason of explicit calculations, we shall use for the solutions of Poisson equations that describe the Euler-Lagrange equations in Section 78.2.1 the more general rectangular boundaries a, b instead of the $1, 1$ that were used in Appendix J.10. We shall use here the time function $t(x, y)$ defined in Section 78.2.2, and its derivative module

$$|t'|^2 = 0.17e^{\frac{\pi y}{b}}. \tag{78.1}$$

To simplify calculations, we shall also set the density $\mu = 1$. In our context, the potential Φ in the Poisson equation $\Delta s + \Phi = 0$ of Appendix J.10 is realized by the gradient expression $\frac{|t'|^2}{\mu}\nabla U = |t'|^2\nabla U$. We further suppose that the gradient of U is given as a direct function of x, y, which is a special requirement since a priori U is only given as a function of space $s(x, y)$.

78.2.7.1 Solving the Poisson Equation for Rectangular Boundary conditions

First, suppose the potential is zero. The solution is the sum of four integrals of type shown in Equation J.2, each corresponding to the values of w on one of the four sides of the domain rectangle. We suppose here that w vanishes on the two horizontal sides $y = 0, y = b$. Denote $s_0(y) = s(0, y), s_1(y) = s(1, y)$.

Then we have to calculate these two integrals, I_1 for the side $x = 0$, and I_2 for the opposite side $x = a$. Setting $q_m = \frac{m\pi}{b}$, and we have

$$I_1 = \frac{2}{b} \sum_{m=0}^{\infty} \frac{\sin(q_m y)\sinh[q_m(a-x)]}{\sinh(q_m a)} \int_0^b s_0(\eta)\sin(q_m\eta)d\eta \tag{78.2}$$

$$I_2 = \frac{2}{b} \sum_{m=0}^{\infty} \frac{\sin(q_m y)\sinh[q_m x]}{\sinh(q_m a)} \int_0^b s_1(\eta)\sin(q_m\eta)d\eta. \tag{78.3}$$

The integral here, for s_0, evaluates to

$$\int_0^b s_0(\eta)\sin(q_m\eta)d\eta = \int_0^b \mathit{Symbolic}[t, 0, 1, 2, 3, 3, 3, 4, 5, 5, 6, 2]\sin(q_m\eta)d\eta$$

where η stands for symbolic time, and the function *Symbolic* (used for calculations with Mathematica) describes the symbolic gesture, a hat function shown in Figure 78.19. The second integral, for s_1, calculates the physical gesture, where η now is the real time component,

$$\int_0^b s_1(\eta)\sin(q_m\eta)d\eta = \int_0^b \mathit{Moto}[0, 1, 2, 3, 3, 4, 5, 6, t, 1, 0, 2]\sin(q_m\eta)d\eta.$$

The function *Moto* (used for calculations with Mathematica) was defined in Section 78.2.5 and is shown in Figure 78.11.

Let's now introduce the potential. The general world-sheet solution with potentials, according to Appendix J.10, is the sum of the solution with zero potential and specified solutions on the boundary and the solution with zero boundary values and general potential,

$$s(x, y) = (I_1 + I_2 + \mathit{Part}_1 + \mathit{Part}_2), \tag{78.4}$$

where Part_1 is the integral for $y \leqslant \eta \leqslant b$, and Part_2 is the integral for $0 \leqslant \eta \leqslant y$:

$$\mathit{Part}_1 = 8 \sum_{m=0}^{\infty} Q(m, x, y) \int_0^a d\xi \int_y^b d\eta \sin\left(\frac{m\pi\xi}{a}\right) \sinh\left[\frac{m\pi}{a}(b-\eta)\right] \nabla U(\xi,\eta)|t'|^2$$

$$\mathit{Part}_2 = 8 \sum_{m=0}^{\infty} Q(m, x, b-y) \int_0^a d\xi \int_0^y d\eta \sin\left(\frac{m\pi\xi}{a}\right) \sinh\left[\frac{m\pi}{a}\eta\right] \nabla U(\xi,\eta)|t'|^2,$$

setting $Q(m, x, y) = \frac{\sin(p_m x)}{m\sinh(p_m b)}\sinh(p_m y)$.

As discussed previously, we will not consider the zero term in Fourier series because the denominator of the Green function vanishes.

78.2.7.2 Three Examples of Potentials

The simplest choice for the gradient is inspired by the inverse of $|t'|^2$, i.e., $\nabla U = e^{-\pi y/b}$. In this trivial case, the effect of the gradient is just a little deformation due to the constant factor 0.17 from formula 78.1, see Figure 78.14. Next, we take a simple potential, whose gradient is given by

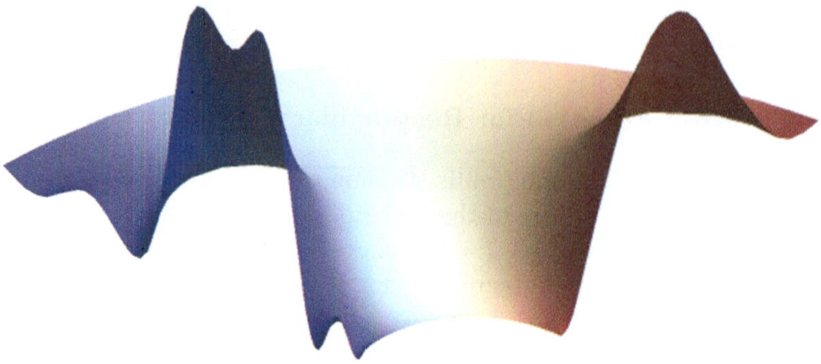

Fig. 78.14. World-sheet (10 terms in the Fourier series) for the very simple potential, inverse of $|t'|^2$.

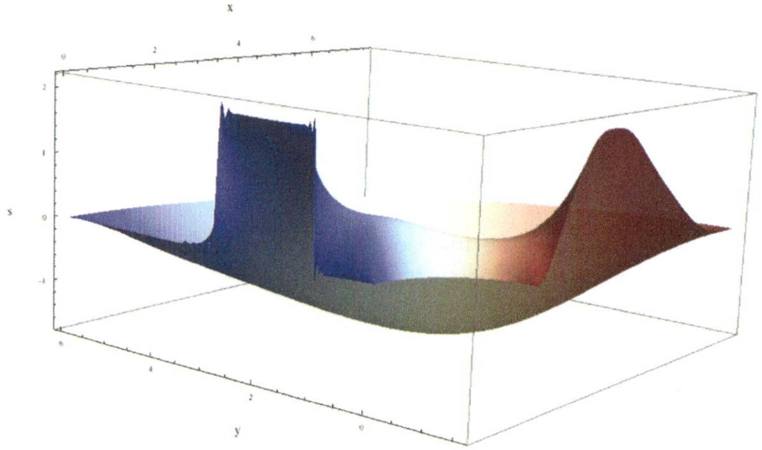

Fig. 78.15. World-sheet for a simple potential as defined in Equation 78.5.

$$\nabla U(s(x,y)) = (x + (y-3)4)e^{\frac{-\pi y}{6}}. \tag{78.5}$$

In this case we have a visible deformation of the shape of the world-sheet, see Figure 78.15.

The third example is a less simple gradient formula:

$$\nabla U(s(x,y)) = 0.6\left[(-y-1)^3 + (y+5)^2\right]\sin(x^6)e^{\frac{-\pi y}{b}}e^{\frac{-x}{6}}. \tag{78.6}$$

The shape of the world-sheet is shown in Figure 78.16. The conceptual interpretation of this case and of the previous one will be given in Section 78.2.14. However, we can immediately think of the different shape of the graph as a measure of the 'artistic effort' to reach a physical realization of the symbolic gesture.

78.2.7.3 Examples of Lagrangian Action

Given the solution of the Euler-Lagrange equation, we are able to find the effective expression of the potential. In fact, we have

$$U(s(x,y)) = \int_0^x d\xi\, \nabla U(s(\xi,y))\,\partial_\xi s(\xi,y) + \int_0^y d\eta\, \nabla U(s(x,\eta))\,\partial_\eta s(x,\eta). \tag{78.7}$$

For concrete calculations, we use the gradient of the potential $e^{-\pi y/b}$ given in Section 78.2.7.2, for technical reasons due to the size of calculation. The result is shown in Figure 78.17.

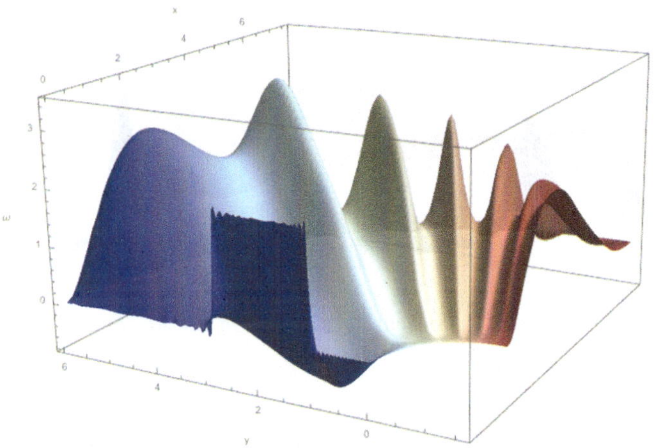

Fig. 78.16. World-sheet (100 points Fourier series) for the gradient $\nabla U(s(x,y))$ given in Equation 78.6.

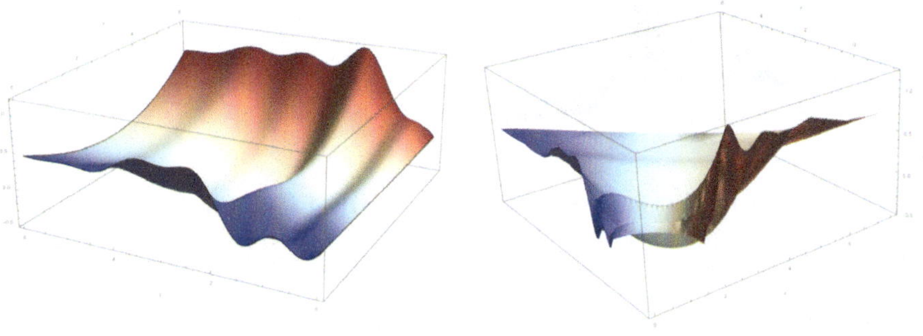

Fig. 78.17. Visualization of a potential, shown both in rectangular domain (left) and in deformed rectangular domain (right). The potential is obtained from the gradient discussed in Section 78.2.7.2 and from the world-sheet with 10-point Fourier series.

Following our discussion in Section 78.2, the Lagrangian is defined as

$$\mathcal{L}(x,y) = \frac{\mu}{2}\left(\frac{|\partial_x s|^2}{|\partial_x t|^2} + \frac{|\partial_y s|^2}{|\partial_y t|^2}\right) - U(s(x,y)), \tag{78.8}$$

where $s = s(x,y)$ is the world-sheet space defined in Equation 78.4, and $U(s(x,y))$ is the potential defined in Equation 78.7. The Lagrangian density is visualized in Figure 78.18.

Here, $|\partial_x t(x,y)|^2 = 0.17e^{\pi y/b}$ and $|\partial_y t(x,y)|^2 = 0.17e^{\pi y/b}$. With the same choice of a simple gradient as explained in Section 78.2.7.2, we obtain the graph of Figure 78.19. Observe that in the left part of Figure 78.19, two high points in the Lagrangian correspond to the points with "infinite" real velocity for the symbolic gesture, as shown in the right part of Figure 78.19.

The action is defined as the integral of the Lagrangian with respect to the world-sheet's parameters x, y, where y parametrizes complex time.

$$S = \int_x \int_y \mathcal{L}(x,y). \tag{78.9}$$

As we are fixing the initial score data, i.e., the symbolic gesture s_0, as well as the zero gestures for $y = 0$ and $y = b$, the action is a function of the physical gesture s_1 and the potential U. We want to investigate whether these two variables, s_1, U, are related in a functional way. To this end, we have restricted our domain of analysis to some specific gestures and potentials. In particular, we have analyzed physical gestures

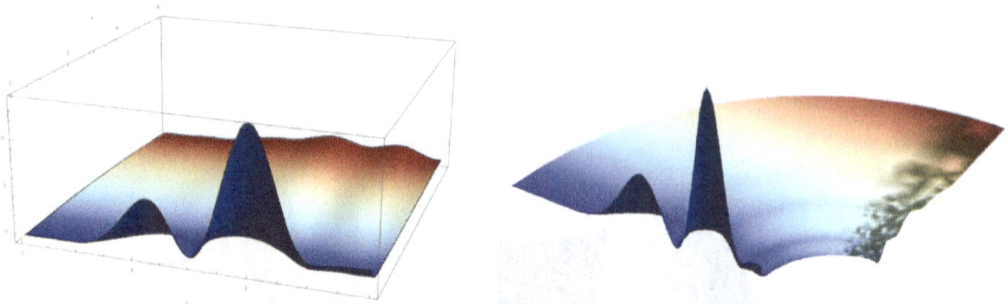

Fig. 78.18. The Lagrangian density from the potential discussed in Section 78.2.7.2, shown in both rectangular (left) and deformed rectangular (right) domains.

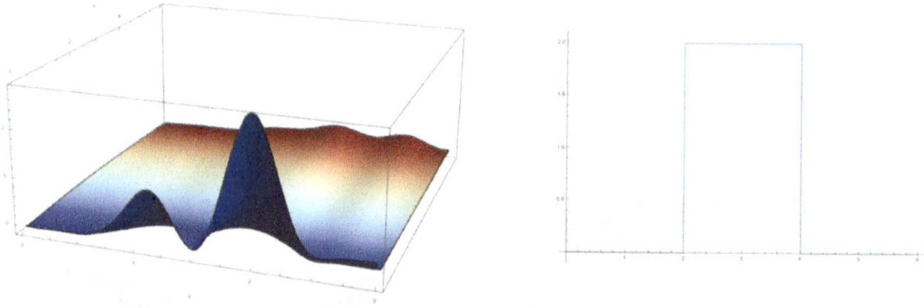

Fig. 78.19. Left: detail of higher points in the symbolic gesture of the Lagrangian density. Right: the correspondence with time points 2 and 4 with "infinite real velocity" in symbolic gesture time.

corresponding to matrix-like strategies of "diagonal" and "co-diagonal" vertical velocities, as shown in Figure 78.20. The chosen labels are 00, 11, 22 for diagonal and 20, 11, 02 for co-diagonal elements.

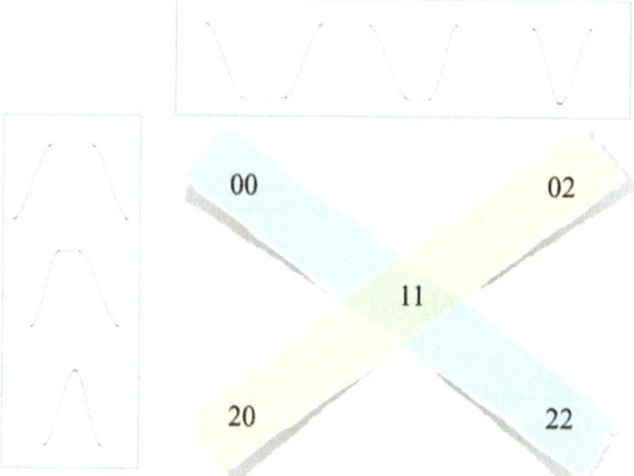

Fig. 78.20. A choice of diagonal and co-diagonal vertical velocities for physical gestures, and the labels of the corresponding gestures. From $0 \to 2$, there are also increasing intervals of zero velocity between positive and negative parts. In particular, in the calculation of gestures of the co-diagonal, the maximal velocity has been modified, in order to have the same integral for each part, and keep the starting and ending points of gestures at the position of a pressed key.

While considering points in the co-diagonal, we took care of changing the maximal velocity to have the same integral for the positive and negative parts. In fact, we require that the motion of the pianist's finger start from 0, the level of pressed key, and return to zero, again the pressed key. Regarding the potentials,

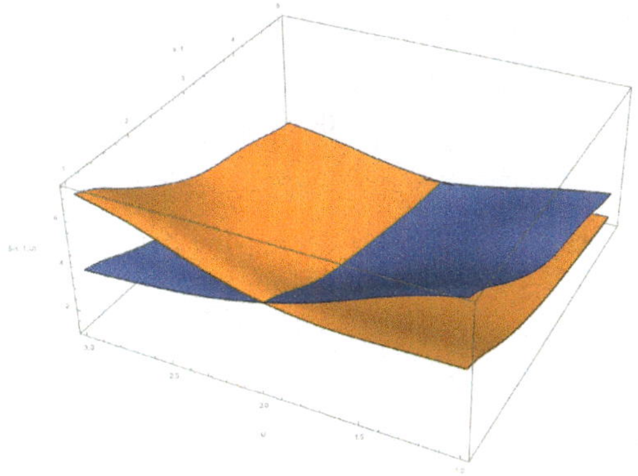

Fig. 78.21. Action for diagonal and co-diagonal physical gestures s_1, with close choices of the (gradient of potential) U.

we started from the simplest case described in Section 78.2.7.2, i.e., the gradient of the potential equals to $e^{-\frac{\pi y}{b}}$, and we moved to close functions, more precisely: $e^{-\frac{3.57y}{b}}$, $e^{-\frac{4y}{b}}$, $e^{-\frac{4.5y}{b}}$, $e^{-\frac{5y}{b}}$, indicated using the labels 0, ... 4, respectively. Of course, to evaluate the Lagrangian and the action it was necessary to calculate the explicit expression of potentials as described in formula 78.7. With a little abuse of notation, we used the labels $0, \ldots 4$ to directly indicate the potentials, and not their gradients. The elements 00, 11, 22, 20, 02 have been evaluated for each expression of the potential.

The results are shown in Figure 78.21, where the surface has been obtained using a polynomial interpolation of third order.

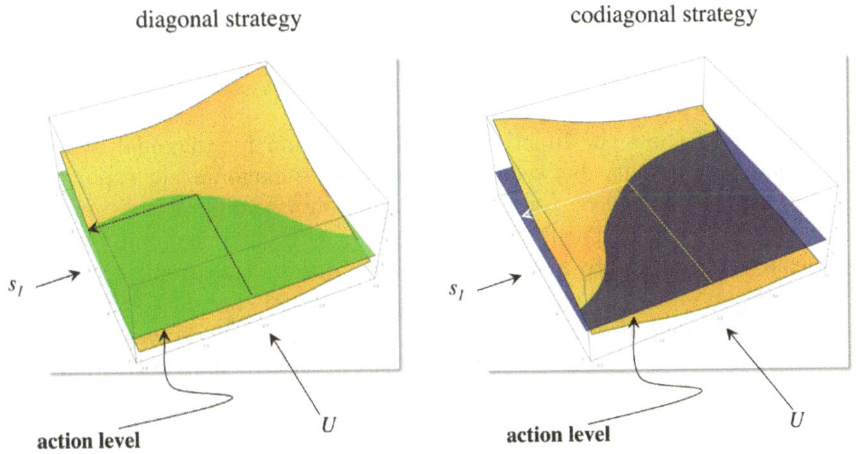

Fig. 78.22. Functionality of diagonal (left) and co-diagonal (right) elements of the action.

From the chosen parameters, it appears that, for diagonal elements:

- fixed U, for increasing indices of s_1 the action is raising,
- fixed s_1, for increasing indices of U the action is decreasing;

and for co-diagonal elements:

- fixed U, for increasing indices of s_1 the action is decreasing,
- fixed s_1, for increasing indices of U the action is decreasing.

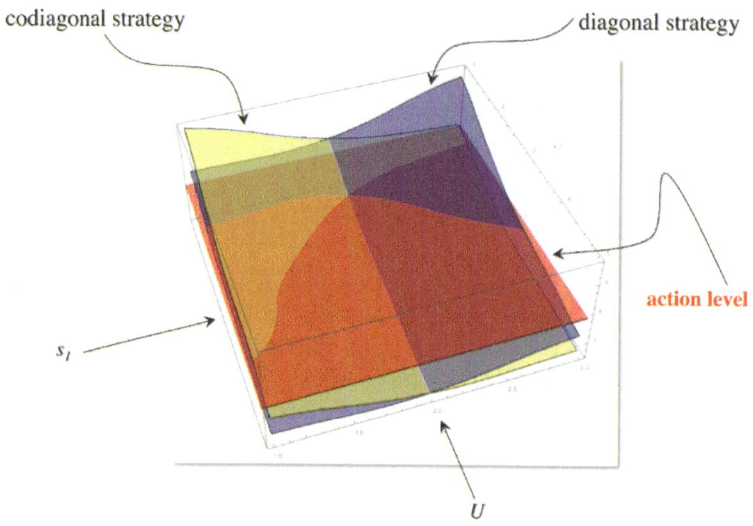

Fig. 78.23. Comparison between the functionality of diagonal and co-diagonal elements of the action.

Figures 78.22 and 78.23 show the functionality of the action for diagonal and co-diagonal elements separately and together. The duality between diagonal and co-diagonal corresponds to the situation of symmetry versus non-symmetry between the complementary movements of *arsis* and *thesis*. It is a crucial problem for all musical performances. The first strategy allows an immediate communication, from conductor to performer, and from a performer to another performer, due to the predictability of the gesture. The second strategy generates an effect of surprise, supplying additional content to the artistic communication. A situation of non-symmetry can also be found in the connection point between two completely different gestures. Now the challenge is to describe artistic varieties by the characterization of functional action strategies.

78.2.8 Calculus of Variations for the Physical Gesture

The question of a functional relation between potential U and physical spatial gesture $s_1(y) = s(1, y)$ can also be approached using a double variational argument. We have used the variational principle to calculate a (local) minimum of the action $S = \int_y \int_x \mathcal{L}(x, y)$. But this was performed with fixed boundary curves: the symbolic gesture s_0, the physical gesture s_1, and the two x-parametrized curves at $y = 0, 1$. Fixing these curves as well as the symbolic gesture, we have an action that is minimized for every given physical curve s_1. Let us write it as $S(s_1)$. Now, this function is minimal for given s_1, but its value is not minimal if we are allowed to vary s_1. We could therefore calculate the vanishing variation $\delta S(s_1) = 0$ when varying s_1. This gives us a physical gesture s_1 where $S(s_1)$ is (locally) minimal. This generates a functional relation $U \mapsto s_1$. As we have the explicit formulae for the solution $s(x, y)$ for every fixed boundary condition from Appendix J.10, we can effectively calculate the variation $\delta S(s_1) = 0$.

Remark 28 At present, we don't know whether this double variational calculus yields the same result as a variational calculus where we don't fix the physical gesture, but only three other boundary curves, and the look for vanishing variation of variable physical gesture. Clearly, the double variational method yields a solution that is weaker than the direct method. It could be equal if we can decompose every variation (with non-fixed physical gesture) into a double variation that first fixes the physical gesture and then adds its variation.

78.2.9 A First Solution. World-Sheet Potentials Determine a Pianist's Gesture: Calculus of Variations and Fourier Analysis

Let us first recall the context of the calculations we want to perform to prove a functional relation between potential U and symbolic and physical gesture. The Lagrangian density is defined by

$$\mathcal{L}(x,y) = \frac{\mu}{2|t'|^2}(|\partial_x s|^2 + |\partial_y s|^2) - U(x,y).$$

with a "material" density μ and a potential U. Here the time function t is supposed to be complex analytic: The argument (x, y) is viewed as a complex number $x+iy$, and $t'(x,y) = \partial_x t(x,y)$. In the following discussion we shall simplify the formalism setting $\mu = 1$ and fixing the time derivative to $|t'|^2 = 0.17e^{\pi y}$, a formula that is derived from the complex analytic function

$$t(x,y) = -0.2624i + \frac{1}{3.8104}e^{\frac{1}{2}\pi i(1-x)+\frac{\pi y}{2}},$$
$$3.8104 \approx e^{\pi/2} - 1, \quad 0.2624i \approx e^{\frac{1}{2}\pi i}/(e^{\pi/2} - 1),$$

mapping the unit imaginary time interval at $x = 0$ to the unit real time interval at $x = 1$.

The variational calculus, when applied to the Lagrangian action

$$S = \int_y \int_x \mathcal{L}(x,y),$$

yields the Euler-Lagrange equation, which we simplify in a first approximation as described in Section 78.2.1 to a Poisson equation:

$$\frac{-|t'|^2}{\mu}\nabla U = \Delta s.$$

Using the calculus of Green functions [768], it was possible to provide solutions of the space function s. This worked by fixing the values of s on the world-sheet's boundary that has four parts[3]: the symbolic gesture[4] $s_0 \in \uparrow \overrightarrow{@} M_{\mathbb{C}}$ for $x = 0$, the physical gesture $s_1 \in \uparrow \overrightarrow{@} M_{\mathbb{C}}$ for $x = 1$, the initial transition line for $y = 0$, and the final transition line for $y = 1$. These last two straight lines s_2, s_4 are completely determined by the initial and final values of s_0, s_1. Therefore the space function s is determined by the potential U, and the time function t, the symbolic and physical gestures.

The space function s has the following form:

$$s(x,y) = I_1(x,y) + I_2(x,y) + P_1(x,y) + P_2(x,y).$$

Here, setting $q_m := m\pi$ and, for $m = 0$, $\frac{\sin(q_0 y)}{\sinh(q_0)} := 1$,

$$I_1 = 2\sum_{m=0}^{\infty} \frac{\sin(q_m y)\sinh[q_m(1-x)]}{\sinh(q_m)} \int_0^1 s_0(q_m \eta)d\eta$$

[3] Recall that $M_{\mathbb{C}} = W \times \mathbb{C}$, where W is the spatial domain of the hand described in Section 78.2.3.

[4] We write $[n]$ for the line digraph with $n+1$ vertices $0, 1, \ldots n$ and one arrow from vertex i to vertex $i+1$ for $i = 0, 1, \ldots n-1$. The arrow digraph a special case: $\uparrow = [1]$.

is a function of the symbolic gesture s_0 only (neither of s_1, nor of U). Further,

$$I_2 = 2\sum_{m=0}^{\infty} \frac{\sin(q_m y)\sinh(q_m x)}{\sinh(q_m)} \int_0^1 s_1(q_m \eta) d\eta$$

is a function of the physical gesture s_1 only (neither of s_0, nor of U). The contributions P_1, P_2 relate to the potential (but not to s_0, s_1):

$$P_1 = 8\sum_{m=0}^{\infty} \frac{\sin(q_m x)}{m\sinh(q_m)} \sinh(q_m y) \int_0^1 d\xi \int_y^1 d\eta \sin(m\pi\xi)\sinh\left[m\pi(1-\eta)\right] \nabla U(\xi,\eta) 0.17 e^{\pi\eta}$$

$$P_2 = 8\sum_{m=0}^{\infty} \frac{\sin(q_m x)}{m\sinh(q_m)} \sinh[q_m(1-y)] \int_0^1 d\xi \int_0^y d\eta \sin(m\pi\xi)\sinh(m\pi\eta) \nabla U(\xi,\eta) 0.17 e^{\pi\eta}$$

In this approach there is no information about how the three variables s_0, s_1, U relate to each other; s_0 and s_1 are two gestures that until now could be exchanged in the given formalism. In the present model of creative performance they are, however, characterized, depending on the degree of smoothness, as less—at the limit of performative impossibility—for the symbolic gesture and more for the physical gesture. It would be desirable to understand how—if at all—the physical gesture could be in a functional relation to the symbolic gesture and the potential.

Here, the potential is thought to be an artistic operator, similar to the performance operators in classical performance theory 44.7. Artistically, the situation is quite substantial. Composers, while writing a score, that is s_0, think of future performance and the characteristics of the performers' gestures s_1. They take into account what can be called an artistic potential, both of performers and of composers. In common language, what is called the *quid* of artistic creative performance, that blend of expressivity, precision, character, passion, and originality, could be identified with a hidden and "ineffable" potential U. It might be ineffable, but it is not un-calculable.

We start from the very general Poisson equation, without imposing any constraint on smoothness of gestures. Our approach is structured as follows.

- We first use the solution $s(x, y)$ of the above Euler-Lagrange equation, but now we calculate the action S with this function and view it as a function of s_1 with fixed parameters s_0, U. As such, we calculate its variation $\delta_{s_1} S = 0$ when varying s_1. This means that after having calculated s using the Euler-Lagrange equation—that is a consequence of the variational calculus for s—we now vary the physical gesture and get a second type of Euler-Lagrange equation.
- Next, we analyze this second Euler-Lagrange equation, which in fact, using a Fourier coefficient argument, turns out to be an infinite sequence $C^*(m) = 0$ of equations for $m = 0, 1, 2, \ldots$
- Thirdly, the equations $C^*(m) = 0$ are interpreted as an infinite system of linear equations with a non-singular transformation matrix (we work with Schauder bases in functional analysis [921]). The solution of this system turns out to be the system of Fourier coefficients of s_1 with respect to a non-standard basis, in other words, the function s_1 can be calculated from s_0 and U.

The complexity of this solution is however not an obstruction for the explicit calculus of s_1, it is completely explicit (no existential obscurities using the axiom of choice arise). As such, it opens up a new field of operator research for gestural performance.

The following sections are organized as follows: We first calculate everything with vanishing potential, $U = 0$. Using this calculus, we then redo everything with a general potential. This is advantageous since many formulae are more transparent when split into zero potential and non-zero potential expressions.

78.2.10 The Calculus with Vanishing Potential

In this section, we set $U = 0$, and therefore $P_1 = P_2 = 0$. We let $\alpha = \frac{1}{2 \cdot 0.17}$, $\alpha' = 0.17 \cdot 8\pi$ and abbreviate $Q_p = \frac{q_p}{\sinh(q_p)}$, with $Q_0 = 1$. Observe that $Q_p = Q_{-p}$.

78.2.10.1 The Variational Calculus on s_1

With the above formula for the Lagrangian, we have

$$\begin{aligned}
\mathcal{L} &= \alpha\frac{A^2+B^2}{e^{\pi y}} \text{ with}\\
A &= \partial_x I_1 + \partial_x I_2,\\
B &= \partial_y I_1 + \partial_y I_2,\\
\partial_x I_1 &= -2\sum_{m=0}^{\infty} \sin(q_m y)\cosh(q_m(1-x))Q_m \int_0^1 s_0 \sin(q_m\eta)d\eta,\\
\partial_x I_2 &= 2\sum_{m=0}^{\infty} \sin(q_m y)\cosh(q_m x)Q_m \int_0^1 s_1 \sin(q_m\eta)d\eta.
\end{aligned}$$

Therefore

$$\begin{aligned}
A &= 2\sum_{m=0}^{\infty} \sin(q_m y)Q_m\left(\cosh(q_m x)\int_0^1 s_1\sin(q_m\eta)d\eta - \cosh[q_m(1-x)]\int_0^1 s_0\sin(q_m\eta)d\eta\right),\\
B &= 2\sum_{m=0}^{\infty} \cos(q_m y)Q_m\left(\sinh(q_m x)\int_0^1 s_1\sin(q_m\eta)d\eta + \sinh[q_m(1-x)]\int_0^1 s_0\sin(q_m\eta)d\eta\right).
\end{aligned}$$

We now vary the physical gesture, replacing s_1 by $s_1+\varepsilon W$, and deriving $S(\varepsilon) = S(s_0, s_1+\varepsilon W)$ with respect to ε at $\varepsilon = 0$:

$$\delta S = \frac{d}{d\varepsilon}S(\varepsilon)|_{\varepsilon=0} = \frac{d}{d\epsilon}\int_{x,y}\mathcal{L}(\epsilon)|_{\epsilon=0} = \int_{x,y}\frac{d}{d\epsilon}\mathcal{L}(\epsilon)|_{\epsilon=0}.$$

The derivative of the Lagrangian is

$$\frac{d}{d\varepsilon}\mathcal{L}(\varepsilon) = \alpha\frac{2A\frac{dA}{d\epsilon}+2B\frac{dB}{d\varepsilon}}{e^{\pi y}} = \frac{2\alpha}{e^{\pi y}}\left(A\frac{dA}{d\varepsilon}+B\frac{dB}{d\varepsilon}\right)$$

with

$$\begin{aligned}
\frac{d}{d\varepsilon}A &= 2\sum_{m=0}^{\infty}\sin(q_m y)Q_m\cosh(q_m x)\int_0^1 W\sin(q_m\eta)d\eta,\\
\frac{d}{d\varepsilon}B &= 2\sum_{m=0}^{\infty}\cos(q_m y)Q_m\sinh(q_m x)\int_0^1 W\sin(q_m\eta)d\eta.
\end{aligned}$$

Then

$$\frac{d}{d\epsilon}\mathcal{L}(\epsilon)|_{\epsilon=0} = \frac{2\alpha}{e^{\pi y}}2\sum_{m=0}^{\infty}\left[A\sin(q_m y)\cosh(q_m x) + B\cos(q_m y)\sinh(q_m x)\right]Q_m\int_0^1 W\sin(q_m\eta)d\eta.$$

This entails

$$\begin{aligned}
0 = \delta S &= \int_x\int_y d_\epsilon\mathcal{L}(\epsilon)|_{\epsilon=0}\\
&= 4\alpha\sum_{m=0}^{\infty}\int_x\int_y\frac{Q_m}{e^{\pi y}}\left[A\sin(q_m y)\cosh(q_m x)+B\cos(q_m y)\sinh(q_m x)\right]\int_0^1 W\sin(q_m\eta)d\eta\\
&= 4\alpha\sum_{m=0}^{\infty}\int_0^1 W\sin(q_m\eta)d\eta\int_x\int_y\frac{Q_m}{e^{\pi y}}\left[A\sin(q_m y)\cosh(q_m x)+B\cos(q_m y)\sinh(q_m x)\right]\\
&= 4\alpha\sum_{m=0}^{\infty}C(m)\int_0^1 W\sin(q_m\eta)d\eta\\
&= 4\alpha\int_0^1 W\left(\sum_{m=0}^{\infty}C(m)\sin(q_m\eta)\right)d\eta,
\end{aligned}$$

where we set

$$C(m) = Q_m \int_x \int_y \frac{1}{e^{\pi y}} \left[A \sin(q_m y)\cosh(q_m x) + B\cos(q_m y)\sinh(q_m x)\right].$$

But by the fundamental lemma of variational calculus, this entails

$$\sum_{m=0}^{\infty} C(m)\sin(q_m \eta) = 0$$

for all $\eta \in [0,1]$. This terminates the first step of our calculations.

78.2.10.2 The Fourier Calculus

This last infinite sum is the Fourier representation of a function of frequency $\frac{1}{2}$. But we only know that the function vanishes for $[0,1]$, half the period. This is half the information to conclude that all $C(m) = 0$. If we now require that $C(2n+1) = 0$ for all n, the equation is the Fourier representation of a function with frequency 1. Therefore, by Fourier's theorem, we have $C(2n) = 0$ for all n. This means that we have

$$0 = \int_x \int_y \frac{1}{e^{\pi y}} \left[A \sin(q_m y)\cosh(q_m x) + B\cos(q_m y)\sinh(q_m x)\right] \quad \text{for all } m,$$

which terminates the second step in our analysis.

Let us now analyze the vanishing condition for the $C(m)$. Writing them explicitly yields these equations for $m = 0, 1, 2, \ldots$:

$$0 = \sum_{p=0}^{\infty} Q_p \int_x \int_y \frac{1}{e^{\pi y}} \left\{\sin(q_p y)\sin(q_m y)\cosh(q_m x)[1]_{x,p} + \cos(q_p y)\cos(q_m y)\sinh(q_m x)[2]_{x,p}\right\},$$

where

$$[1]_{x,p} = \cosh(q_p x)\int_0^1 s_1 \sin(q_p \eta)d\eta - \cosh[q_p(1-x)]\int_0^1 s_0 \sin(q_p \eta)d\eta$$

and

$$[2]_{x,p} = \sinh(q_p x)\int_0^1 s_1 \sin(q_p \eta)d\eta + \sinh[q_p(1-x)]\int_0^1 s_0 \sin(q_p \eta)d\eta.$$

Next, we use the goniometric formula

$$\sin(a)\sin(b) = 1/2[\sin(a + \pi/2 - b) + \sin(a - \pi/2 + b)]$$

to derive

$$\begin{aligned}
\sin(q_p y)\sin(q_m y) &= \frac{1}{2}(\cos(q_{p-m} y) - \cos(q_{p+m} y)),\\
\cos(q_p y)\cos(q_m y) &= \frac{1}{2}(\cos(q_{p-m} y) + \cos(q_{p+m} y)).
\end{aligned}$$

Inserting this in the above formula we get

$$\begin{aligned}
0 = \sum_{p=0}^{\infty} \frac{Q_p}{2} \int_x \int_y \frac{1}{e^{\pi y}} \{&(\cos(q_{p-m} y)\cosh(q_m x) - \cos(q_{p+m} y)\cosh(q_m x))[1]_{x,p}\\
&+(\cos(q_{p-m} y)\sinh(q_m x) + \cos(q_{p+m} y)\sinh(q_m x))[2]_{x,y}\}.
\end{aligned}$$

This is rewritten as

$$0=\sum_{p=0}^{\infty}\frac{Q_p}{2}\int_x\int_y\frac{1}{e^{\pi y}}\left[\cos(q_{p-m}y)R_{x,m,p}+\cos(q_{p+m}y)S_{x,m,p}\right],$$

with

$$\begin{aligned}R_{x,m,p}&=\cosh(q_m x)[1]_{x,p}+\sinh(q_m x)[2]_{x,p},\\ S_{m,p,x}&=-\cosh(q_m x)[1]_{x,p}+\sinh(q_m x)[2]_{x,p},\end{aligned}$$

which yields

$$0=\sum_{p=0}^{\infty}\frac{Q_p}{2}\left[\int_y\frac{\cos(q_{p-m}y)}{e^{\pi y}}\int_x R_{x,m,p}+\int_y\frac{\cos(q_{p+m}y)}{e^{\pi y}}\int_x S_{x,m,p}\right].$$

We have

$$\int_0^1 dy\,\frac{\cos(q_{p\pm m}y)}{e^{\pi y}}=\frac{e^{-\pi}(e^{\pi}-1)}{\pi(1+(m\mp p)^2)},$$

and setting $L=\frac{e^{-\pi}(e^{\pi}-1)}{\pi}$, we get

$$0=\sum_{p=0}^{\infty}\frac{Q_p}{2}L\left(\frac{1}{1+(m-p)^2}\int_x R_{x,m,p}+\frac{1}{1+(m+p)^2}\int_x S_{x,m,p}\right).$$

To calculate the integral of $R_{x,m,p}$, we have

$$\begin{aligned}R_{x,m,p}&=\cosh(q_m x)[1]_{x,p}+\sinh(q_m x)[2]_{x,p}\\ &=\cosh(q_m x)\left(\cosh(q_p x)\int_0^1 s_1\sin(q_p\eta)d\eta-\cosh[q_p(1-x)]\int_0^1 s_0\sin(q_p\eta)d\eta\right)+\\ &\quad\sinh(q_m x)\left(\sinh(q_p x)\int_0^1 s_1\sin(q_p\eta)d\eta+\sinh[q_p(1-x)]\int_0^1 s_0\sin(q_p\eta)d\eta\right)\\ &=\cosh(q_m x)\cosh(q_p x)\int_0^1 s_1\sin(q_p\eta)d\eta-\cosh[q_p(1-x)]\cosh(q_m x)\int_0^1 s_0\sin(q_p\eta)d\eta+\\ &\quad\sinh(q_m x)\sinh(q_p x)\int_0^1 s_1\sin(q_p\eta)d\eta+\sinh[q_p(1-x)]\sinh(q_m x)\int_0^1 s_0\sin(q_p\eta)d\eta\\ &=(\cosh(q_m x)\cosh(q_p x)+\sinh(q_m x)\sinh(q_p x))\int_0^1 s_1\sin(q_p\eta)d\eta+\\ &\quad(-\cosh[q_p(1-x)]\cosh(q_m x)+\sinh[q_p(1-x)]\sinh(q_m x))\int_0^1 s_0\sin(q_p\eta)d\eta.\end{aligned}$$

Therefore

$$m=p=0\text{ implies }\int_x R_{x,m,p}=0,$$

$$m=p>0\text{ implies }\int_x R_{x,m,p}=\frac{1}{Q_{2m}}\int_0^1 s_1\sin(q_p\eta)-\frac{1}{Q_m}\int_0^1 s_0\sin(q_p\eta),$$

$$m\neq p,\ m+p>0\text{ implies }\int_x R_{x,m,p}=\frac{1}{Q_{p+m}}\int_0^1 s_1\sin(q_p\eta)-\left(\frac{1}{q_p\cdot Q_m}+\frac{1}{q_m\cdot Q_p}\right)\int_0^1 s_0\sin(q_p\eta).$$

To calculate the integral of $S_{x,m,p}$, we have

$$\begin{aligned}
S_{x,m,p} &= -\cosh(q_m x)[1]_{x,p} + \sinh(q_m x)[2]_{x,p} \\
&= -\cosh(q_m x)\left(\cosh(q_p x)\int_0^1 s_1 \sin(q_p\eta)d\eta - \cosh[q_p(1-x)]\int_0^1 s_0 \sin(q_p\eta)d\eta\right) + \\
&\sinh(q_m x)\left(\sinh(q_p x)\int_0^1 s_1 \sin(q_p\eta)d\eta + \sinh[q_p(1-x)]\int_0^1 s_0 \sin(q_p\eta)d\eta\right) \\
&= \cosh(q_m x)\cosh(q_p x)\int_0^1 s_1 \sin(q_p\eta)d\eta - \cosh[q_p(1-x)]\cosh(q_m x)\int_0^1 s_0 \sin(q_p\eta)d\eta + \\
&\sinh(q_m x)\sinh(q_p x)\int_0^1 s_1 \sin(q_p\eta)d\eta + \sinh[q_p(1-x)]\sinh(q_m x)\int_0^1 s_0 \sin(q_p\eta)d\eta \\
&= (-\cosh(q_m x)\cosh(q_p x) + \sinh(q_m x)\sinh(q_p x))\int_0^1 s_1 \sin(q_p\eta)d\eta + \\
&(\cosh[q_p(1-x)]\cosh(q_m x) + \sinh(q_m x)\sinh[q_p(1-x)])\int_0^1 s_0 \sin(q_p\eta)d\eta.
\end{aligned}$$

Therefore

$$m = p \text{ implies } \int_x S_{x,m,p} = -\int_0^1 s_1 \sin(q_p\eta) + \cosh(q_m)\int_0^1 s_0 \sin(q_p\eta),$$

and

$$m \neq p \text{ implies } \int_x S_{x,m,p} = -\frac{1}{Q_{m-p}}\int_0^1 s_1 \sin(q_p\eta) + \left(\frac{1}{q_{-p}\cdot Q_m} + \frac{1}{q_m \cdot Q_{-p}}\right)\int_0^1 s_0 \sin(q_p\eta).$$

78.2.10.3 The Non-singular Matrix

This calculation yields the equation

$$0 = \sum_p Q_p Z_{m,p}$$

with

$$Z_{m,p} = \frac{1}{1+(m-p)^2}\int_x R_{x,m,p} + \frac{1}{1+(m+p)^2}\int_x S_{x,m,p}.$$

For $m = p = 0$ we have $Z_{0,0} = 0$. For $m = p > 0$, we have

$$\begin{aligned}
Z_{m,m} &= \left(\frac{1}{Q_{2m}} - \frac{1}{1+4m^2}\right)\int_0^1 s_1 \sin(q_m\eta) + \left(\frac{-1}{Q_m} + \frac{\cosh(q_m)}{1+4m^2}\right)\int_0^1 s_0 \sin(q_m\eta) \\
&= E_{m,m}\int_0^1 s_1 \sin(q_m\eta) + F_{m,m}\int_0^1 s_0 \sin(q_m\eta),
\end{aligned}$$

where $E_{m,m} = \frac{1}{Q_{2m}} - \frac{1}{1+4m^2}$ and $F_{m,m} = \frac{\cosh(q_m)}{1+4m^2} - \frac{1}{Q_m}$. For $m \neq p$, we have

$$Z_{m,p} = E_{m,p}\int_0^1 s_1 \sin(q_p\eta) + F_{m,p}\int_0^1 s_0 \sin(q_p\eta),$$

with

$$E_{m,p} = \frac{1}{1+(m-p)^2}\frac{1}{Q_{m+p}} - \frac{1}{1+(m+p)^2}\frac{1}{Q_{m-p}}$$

and

$$F_{m,p} = \frac{1}{1+(m+p)^2}\left(\frac{1}{q_{-p}\cdot Q_m} + \frac{1}{q_m \cdot Q_{-p}}\right) - \frac{1}{1+(m-p)^2}\left(\frac{1}{q_p \cdot Q_m} + \frac{1}{q_m \cdot Q_p}\right).$$

We therefore have for every m this linear equation in the variables $X_p = Q_p \int_0^1 s_1 \sin(q_p\eta)$:

$$0 = \sum_{p=1}^{\infty} Q_p \left(E_{m,p} \int_0^1 s_1 \sin(q_p \eta) + F_{m,p} \int_0^1 s_0 \sin(q_p \eta) \right). \tag{78.10}$$

In order to solve this equation, one needs to verify that the matrix $E_N = (E_{m,p})_{1 \leqslant m,p \leqslant N}$ is non-singular as $N \to \infty$. We have not verified this as a mathematical theorem, but the evaluation of $Det(E_N)$ yields a curve that does not vanish for $1 \leqslant N \leqslant 100$, as shown in Figure 78.24.

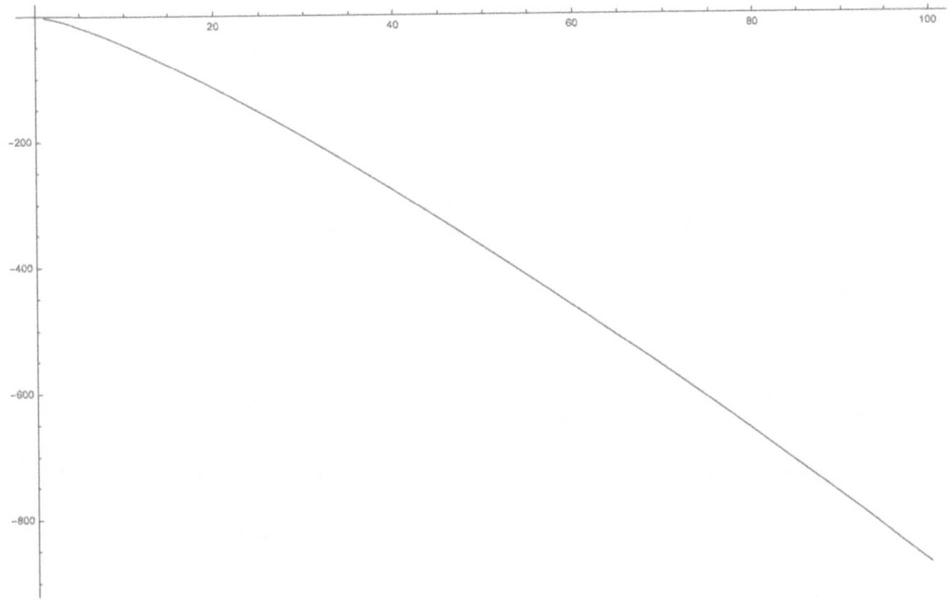

Fig. 78.24. The logarithm of the determinant $Det(E_N)$ for the values $1 \leqslant N \leqslant 100$.

78.2.10.4 A Second Fourier Calculus

The non-singularity of the matrix E_N implies that we can calculate the unknown values $X_p = Q_p \int s_1 \sin(q_p \eta)$ and therefore the integrals $\int_0^1 s_1 \sin(q_p \eta) = \int_0^1 s_1 \sin(\pi p \eta)$ for all $p = 1, 2, \ldots$ These integrals are akin to the function's s_1 Fourier coefficients, i.e., we have $\int_0^1 s_1 \sin(\pi p \eta) = \int_0^1 s_1 \sin(2\pi r \eta)$ for even $p = 2r$ and frequency $\nu = 1$. These numbers are half of what we need, the other half would be the integrals $\int_0^1 s_1 \cos(2\pi r \eta)$. This information is what must be contributed by the odd integrals $\int_0^1 s_1 \sin(\pi(2r+1)\eta)$. We have not succeeded in proving that the odd multiples $\sin(\pi(2r+1)\eta)$ together with the even ones $\sin(2\pi r \eta)$ form a complete Schauder base. We would have to prove that replacing the cosine functions $\cos(2\pi r \eta)$ by the odd multiple functions $\sin(\pi(2r+1)\eta)$ defines a base together with the even multiples $\sin(2\pi r \eta)$. However, we have checked a number of numerical data on the determinants of cosine representations of the odd multiples, and found that these determinants don't vanish. We conjecture that we indeed have a Schauder basis.

Proposition 4. *Assuming that our Schauder basis conjecture holds, the physical gesture s_1 is determined by the argument(s) s_0 (and U) in the case of a vanishing potential $U = 0$.*

78.2.11 The Calculus with General Potential

We now stop over to the situation of a general potential. Recall from the introduction that the space function is $s(x, y) = I_1(x, y) + I_2(x, y) + P_1(x, y) + P_2(x, y)$, the potential being an argument only in the last two formulae P_1, P_2 that are also made explicit in the introduction.

78.2.11.1 The Variational Calculus of s_1 with Potential

To calculate the Lagrangian

$$\mathcal{L}(x,y) = \frac{\mu}{2|t'|^2}(|\partial_x s|^2 + |\partial_y s|^2) - U(x,y)$$

we need the partial derivatives

$$A^* = A + A',\ A' = \partial_x P_1 + \partial_x P_2,$$
$$B^* = B + B',\ B' = \partial_y P_1 + \partial_y P_2$$

with the quantities A, B being defined in Section 78.2.10.1. We have

$$\partial_x P_1 = 8\sum_m \frac{\cos(q_m x)}{\sinh(q_m)}\pi \sinh(q_m y)\int_0^1 d\xi \int_y^1 d\eta \sin(q_m\xi)\sinh[q_m(1-\eta)]\,\nabla U(\xi,\eta)0.17e^{\pi\eta}$$

$$\partial_x P_2 = 8\sum_m \frac{\cos(q_m x)}{\sinh(q_m)}\pi \sinh[q_m(1-y)]\int_0^1 d\xi \int_0^y d\eta \sin(q_m\xi)\sinh(q_m\eta)\nabla U(\xi,\eta)0.17e^{\pi\eta}$$

$$\partial_y P_1 = 8\sum_m \frac{\sin(q_m x)}{\sinh(q_m)}\pi \cosh(q_m y)\int_0^1 d\xi \int_y^1 d\eta \sin(q_m\xi)\sinh[q_m(1-\eta)]\,\nabla U(\xi,\eta)0.17e^{\pi\eta} -$$
$$8\sum_m \frac{\sin(q_m x)}{m\sinh(q_m)}\sinh(q_m y)\int_0^1 d\xi \sin(q_m\xi)\sinh[q_m(1-y)]\nabla U(\xi,y)0.17e^{\pi y}$$

$$\partial_y P_2 = 8\sum_m \frac{\sin(q_m x)}{m\sinh(q_m)}\sinh[q_m(1-y)]\int_0^1 d\xi \sin(q_m\xi)\sinh(q_m y)\nabla U(\xi,y)0.17e^{\pi y} -$$
$$8\sum_m \frac{\sin(q_m x)}{\sinh(q_m)}\pi \cos[q_m(1-y)]\int_0^1 d\xi \int_0^y d\eta \sin(q_m\xi)\sinh(q_m\eta)\,\nabla U(\xi,\eta)0.17e^{\pi\eta}.$$

Let us calculate the term $A^* = A + A'$, where $A = A(s_0, s_1)$, $A' = A'(U)$, and $B^* = B + B'$, where $B = B(s_0, s_1)$, $B' = B'(U)$. Setting $\alpha' = 0.17\cdot 8\pi$, we have

$$\begin{aligned}
A^* &= A + \\
&\alpha' \sum_{m=0}^{\infty}\left[\frac{\cos(q_m x)}{\sinh(q_m)}\sinh(q_m y)\int_0^1 d\xi \int_y^1 d\eta \sin(q_m\xi)\sinh[q_m(1-\eta)]\,\nabla U(\xi,\eta)e^{\pi\eta} + \right.\\
&\left.\frac{\cos(q_m x)}{\sinh(q_m)}\sinh[q_m(1-y)]\int_0^1 d\xi \int_0^y d\eta \sin(q_m\xi)\sinh(q_m\eta)\nabla U(\xi,\eta)e^{\pi\eta}\right]\\
B^* &= B + \\
&\alpha' \sum_{m=0}^{\infty}\left[\frac{\sin(q_m x)}{\sinh(q_m)}\cosh(q_m y)\int_0^1 d\xi \int_y^1 d\eta \sin(q_m\xi)\sinh[q_m(1-\eta)]\,\nabla U(\xi,\eta)e^{\pi\eta} - \right.\\
&\frac{1}{\pi}\left(\frac{\sin(q_m x)}{m\sinh(q_m)}\sinh(q_m y)\int_0^1 d\xi \sin(q_m\xi)\sinh[q_m(1-y)]\nabla U(\xi,y)e^{\pi y}\right) -\\
&\frac{\sin(q_m x)}{\sinh(q_m)}\cos[q_m(1-y)]\int_0^1 d\xi \int_0^y d\eta \sin(q_m\xi)\sinh(q_m\eta)\nabla U(\xi,\eta)e^{\pi\eta} +\\
&\left.\frac{1}{\pi}\left(\frac{\sin(q_m x)}{m\sinh(q_m)}\sinh[q_m(1-y)]\int_0^1 d\xi \sin(q_m\xi)\sinh(q_m y)\nabla U(\xi,y)e^{\pi y}\right)\right].
\end{aligned}$$

We then consider the derivative

$$\frac{d\mathcal{L}(\varepsilon)}{d\epsilon} = \alpha\frac{2A^*\frac{dA^*}{d\epsilon} + 2B^*\frac{dB^*}{d\epsilon}}{e^{\pi y}},$$

where obviously $\frac{dA^*}{d\epsilon} = \frac{dA}{d\epsilon}$ and $\frac{dB^*}{d\epsilon} = \frac{dB}{d\epsilon}$. Therefore we have

$$\frac{d\mathcal{L}(s_0, s_1 + \varepsilon W, U)}{d\epsilon} =$$
$$\frac{\alpha}{e^{\pi y}}\left[2(A(s_0, s_1 + \varepsilon W) + A'(U))\frac{dA(s_0, s_1 + \varepsilon W)}{d\epsilon} + 2(B(s_0, s_1 + \varepsilon W) + B'(U))\frac{dB(s_0, s_1 + \varepsilon W)}{d\epsilon}\right].$$

This yields the derivative at $\varepsilon = 0$,

$$\frac{d\mathcal{L}(\varepsilon)}{d\epsilon}|_{\varepsilon=0} =$$
$$\frac{4\alpha}{e^{\pi y}}\left[(A + A')\sum_{m=0}^{\infty}\sin(q_m y)Q_m\cosh(q_m x)\int_0^1 W\sin(q_m\eta)d\eta +\right.$$
$$\left.(B + B')\sum_{m=0}^{\infty}\cos(q_m y)Q_m\sinh(q_m x)\int_0^1 W\sin(q_m\eta)d\eta\right].$$

Then the variation of the action is $0 = \delta S = \int_x\int_y \frac{d\mathcal{L}(\varepsilon)}{d\epsilon}|_{\varepsilon=0}$, whence by the fundamental lemma of variational calculus, we have

$$0 = \sum_{m=0}^{\infty}\sin(q_m\eta)\int_x\int_y\frac{Q_m}{e^{\pi y}}((A + A')(\sin(q_m y)\cosh(q_m x)) + (B + B')(\cos(q_m y)\sinh(q_m x)))$$

for all η. By the argument that views the factors of $\sin(q_m\eta)$ as Fourier coefficients, we again have a system of equations for all m:

$$0 = C^*(m) = C(m) + \int_x\int_y\frac{Q_m}{e^{\pi y}}(A'\sin(q_m y)\cosh(q_m x) + B'\cos(q_m y)\sinh(q_m x))$$

with the $C(m)$ as defined earlier in Section 78.2.10.1. The second summand

$$C'(m) = \int_x\int_y\frac{Q_m}{e^{\pi y}}(A'\sin(q_m y)\cosh(q_m x) + B'\cos(q_m y)\sinh(q_m x))$$

is only a function of the potential U and of m. The symbolic and physical gestures s_0, s_1 don't appear here. This implies that all the calculations that were made for the zero potential to separate s_1 as a function of s_0 are still valid, however now with a functionality that includes U, too. This is easily checked when reviewing the calculations in Section 78.2.10.2. Therefore we have—modulo the above conjecture—this result:

Theorem 7. *Assuming that our Schauder base conjecture holds, the physical gesture s_1 is determined by the arguments s_0 and U for any potential U.*

In view of the above explicit calculations, this functional relation can be made more precise. If we represent the gestures s_0, s_1 as functions on $[0, 1]$ by their Fourier coefficients

$$a_r(s_1) = \int_0^1 s_1\sin(2\pi r\eta),\ b_r(s_1)\int_0^1 s_1\cos(2\pi r\eta)$$

and

$$u_r(s_0) = \int_0^1 s_0\sin(2\pi r\eta),\ v_r(s_1)\int_0^1 s_0\cos(2\pi r\eta)$$

as vectors $F_1 = F(s_1) = (a_r, b_r)_r, F_0 = F(s_0) = (u_r, v_r)_r$, then, using Equation 78.10, we have an equation

$$E(F_1) = K(F_0) + C(\nabla U)$$

with non-singular linear operators E, K on the subspace $T \subset \mathbb{R}^{\mathbb{N}}$ of convergent series, and a linear operator C on vector fields with values in T. Writing it as

$$F_1 = E^{-1}K(F_0) + E^{-1}C(\nabla U) = M(F_0) + D(\nabla U)$$

for invertible operator M and linear operator D, we may think of the physical gesture F_1 as being generated from an already given intermediate physical gesture F_1', where

$$F_1' = M(F_0) + D(\nabla U'),$$

and then replace F_0 by

$$M^{-1}(F_1') - M^{-1}D(\nabla U'),$$

whence

$$F_1 = M(F_0) + D(\nabla U) = MM^{-1}(F_1') - MM^{-1}D(\nabla U') + D(\nabla U) = F_1' - D(\nabla(U - U')) = F_1' + D(\nabla f),$$

where f is a potential difference function representing the passage from F_1' to F_1. This setup enables us to view the hypergestural world-sheet process as being driven by a performance operator on gestures, and specifically driven by gradients of potential functions.

78.2.12 Solution of the Differential Equation Using 2D Fourier Series

The above method used the Poisson differential equation as a simplified approach, however with a general potential. In this section, we want to discuss solutions of the general equation

$$-\nabla U = \mu\left(\frac{\Delta s}{|t'|^2} - 2\frac{\langle \dot{t}, \ddot{t}\rangle_{\mathbb{R}}}{|\dot{t}|^4}\dot{s}\right)$$

with a determined time function

$$t(x, y) = -0.2624i + \frac{1}{3.8104}e^{\frac{1}{2}\pi i(1-x)+\frac{\pi y}{2}},$$
$$3.8104 \approx e^{\pi/2} - 1, \quad 0.2624i \approx e^{\frac{1}{2}\pi i}/(e^{\pi/2} - 1).$$

With this function, the above differential equation becomes

$$-\frac{\nabla U|t'|^2}{\mu} = \Delta s - \pi\partial_y s.$$

The temporal derivative in our example is $|t'|^2 = \frac{\pi^2}{4\times 3.8104^2}e^{\pi y} \approx 0.1699e^{\pi y}$. We now proceed to a solution of this equation using 2D Fourier series. To ease notation, we set $|t'|^2 = \frac{1}{r}e^{-by}$, i.e., $r = 1/0.1699, b = -\pi$, $-\frac{\nabla U(x,y)}{\mu} = V(x, y)$ and $V^*(x, y) = V(x, y)\frac{1}{r}e^{-by}$, whence $V(x, y) = V^*(x, y)re^{by}$. We also need a better symbol for the world-sheet function for the following formulae. We set $s(x, y) = sheet(x, y)$. With these notations, we get the differential equation

$$V^* = \Delta sheet - \pi\partial_y sheet.$$

We now represent functions on the unit square I^2 by functions that are 2-periodic in both coordinates x, y, i.e., frequency is 1/2 in both directions. This is necessary in order to get any functions on I^2. The Fourier representation of such a 2-periodic function $f(x, y)$ is

$$f(x, y) = \sum_{n,m\geqslant 0}^{\infty} (f^{cc}_{n,m}\cos(\pi nx)\cos(\pi my) + f^{cs}_{n,m}\cos(\pi nx)\sin(\pi my) +$$
$$f^{sc}_{n,m}\sin(\pi nx)\cos(\pi my) + f^{ss}_{n,m}\sin(\pi nx)\sin(\pi my)).$$

This means that we represent f by the system $f^{\cdot\cdot}_{\cdot\cdot} = (f^{cc}_{n,m}, f^{cs}_{n,m}, f^{sc}_{n,m}, f^{ss}_{n,m})_{n,m}$ of its coefficients.

We have two systems of coefficients: the system $V^{\cdot\cdot}_{\cdot\cdot}$ for the potential V, and the system $V*^{\cdot\cdot}_{\cdot\cdot}$ for the product $V*$ of V with the time function derivative $|t'|^2 = \frac{1}{r}e^{-by}$. All our calculations will use $V*^{\cdot\cdot}_{\cdot\cdot}$, following the above differential equation. This is very useful because we may then have constant coefficients for $\Delta sheet$ and $\partial_y sheet$. This simplifies calculations considerably. We however need to control the relation between $V*^{\cdot\cdot}_{\cdot\cdot}$ and the original potential system $V^{\cdot\cdot}_{\cdot\cdot}$. It turns out that this relation is a linear isomorphism. More precisely, suppose we are given the system $V*^{\cdot\cdot}_{\cdot\cdot}$, then, we can take the Fourier representation of $V*$:

$$V*(x,y) = \sum_{n,m\geqslant 0}^{\infty} (V*^{cc}_{n,m}\cos(\pi nx)\cos(\pi my) + V*^{cs}_{n,m}\cos(\pi nx)\sin(\pi my) + V*^{sc}_{n,m}\sin(\pi nx)\cos(\pi my) + V*^{ss}_{n,m}\sin(\pi nx)\sin(\pi my)),$$

and get V by

$$V(x,y) = \sum_{n,m\geqslant 0}^{\infty} (re^{by}V*^{cc}_{n,m}\cos(\pi nx)\cos(\pi my) + re^{by}V*^{cs}_{n,m}\cos(\pi nx)\sin(\pi my) + re^{by}V*^{sc}_{n,m}\sin(\pi nx)\cos(\pi my) + re^{by}V*^{ss}_{n,m}\sin(\pi nx)\sin(\pi my)).$$

The Fourier coefficients of V then are sums of integrals of type

$$\int_{x=0}^{2}\int_{y=0}^{2} re^{by}\left[V*^{cc}_{n,m}\cos(\pi nx)\cos(\pi my)\cos(\pi rx)\cos(\pi sy) + V*^{cs}_{n,m}\cos(\pi nx)\sin(\pi my)\cos(\pi rx)\cos(\pi sy) + V*^{sc}_{n,m}\sin(\pi nx)\cos(\pi my)\cos(\pi rx)\cos(\pi sy) + V*^{ss}_{n,m}\sin(\pi nx)\sin(\pi my)\cos(\pi rx)\cos(\pi sy)\right]$$

or similar formulae replacing cos by sin, etc. This evidently implies that we have a linear isomorphism $L: V*^{\cdot\cdot}_{\cdot\cdot} \xrightarrow{\sim} V^{\cdot\cdot}_{\cdot\cdot}$.

Let us now investigate the solutions—in terms of Fourier coefficients—of our differential equation $V* = \Delta sheet - \pi\partial_y sheet$. Comparison of Fourier coefficients yields the following.

For a pair n,m, we consider the four-dimensional spaces $V*^{\cdot\cdot}_{n,m}$ and $sheet^{\cdot\cdot}_{n,m}$ of coefficients of $V*$ and $sheet$ of index n,m. Then we have a linear map $D_{n,m}: sheet^{\cdot\cdot}_{n,m} \to V*^{\cdot\cdot}_{n,m}$, where

$$D_{n,m} = \begin{pmatrix} E_{n,m} & 0 \\ 0 & E_{n,m} \end{pmatrix}$$

is a matrix with 2×2 blocks

$$E_{n,m} = \pi^2\begin{pmatrix} -(n^2+m^2) & -m \\ m & -(n^2+m^2) \end{pmatrix}$$

of determinant $d(n,m) = \pi^4((n^2+m^2)^2 + m^2)$, that vanishes only for $n=m=0$. The map $D_{n,m}$ determines the relations of coefficients, i.e., for any n,m we must have $v_{n,m} = D_{n,m}(sheet_{n,m})$ for coeffient quadruples $v_{n,m} \in V*^{\cdot\cdot}_{n,m}$ and $sheet_{n,m} \in sheet^{\cdot\cdot}_{n,m}$.

For $n=m=0$ this means that coefficients $v_{0,0} = 0$ for the potential $V*$, and that $sheet_{0,0}$ are arbitrary for the world-sheet $sheet$. For any other indices n,m, the four-dimensional coefficient spaces correspond under the isomorphism $D_{n,m}$. Also observe that the two-dimensional summands $V*^{c\cdot}_{n,m}$ and $sheet^{c\cdot}_{n,m}$ correspond under the isomorphism $E_{n,m}$; same for $V*^{s\cdot}_{n,m}$ and $sheet^{s\cdot}_{n,m}$.

This result is however not what we want, because we also want the symbolic and physical gestures $Symbolic(y), Physical(y)$ to be the boundary values of $sheet$ for $x=0, x=1$, respectively. This means that we require

$$Symbolic(y) = sheet(0,y) = \sum_m\left[\left(\sum_n sheet^{cs}_{n,m}\right)\sin(\pi my) + \left(\sum_n sheet^{cc}_{n,m}\right)\cos(\pi my)\right]$$

and

$$Physical(y) = sheet(1,y) = \sum_m \left[\left(\sum_n (-1)^n sheet^{cs}_{n,m} \right) \sin(\pi m y) + \left(\sum_n (-1)^n sheet^{cc}_{n,m} \right) \cos(\pi m y) \right].$$

This means in particular, that these conditions are only imposed upon the two-dimensional spaces $V{*}^{c.}_{n,m}$ and $sheet^{c.}_{n,m}$, the other summands are free.

Let us give the 2-periodic Fourier representations of symbolic and physical gestures:

$$Symbolic(y) = \sum_m \sigma_m \sin(\pi m y) + \tau_m \cos(\pi m y),$$

$$Physical(y) = \sum_m \sigma^*_m \sin(\pi m y) + \tau^*_m \cos(\pi m y).$$

This entails the equations

$$\sigma_m = \sum_n sheet^{cs}_{n,m},\ \sigma^*_m = \sum_n (-1)^n sheet^{cs}_{n,m},\ \tau_m = \sum_n sheet^{cc}_{n,m},\ \tau^*_m = \sum_n (-1)^n sheet^{cc}_{n,m}.$$

Let us see what these equations imply for possible V^* coefficients. Let us start with $m = 0$. Then we have

$$\sigma_0 = \sum_n sheet^{cs}_{n,0},\ \sigma^*_0 = \sum_n (-1)^n sheet^{cs}_{n,0},\ \tau_0 = \sum_n sheet^{cc}_{n,0},\ \tau^*_0 = \sum_n (-1)^n sheet^{cc}_{n,0}.$$

We know that $sheet^{cs}_{0,0}, sheet^{cc}_{0,0}$ are arbitrary. Moreover, we have $sheet^{cs}_{n,0} = \frac{-1}{\pi^2 n^2} V{*}^{cs}_{n,0}$ and $sheet^{cc}_{n,0} = \frac{-1}{\pi^2 n^2} V{*}^{cc}_{n,0}$ for $n \neq 0$.

Let us now calculate the coefficient for a Fourier approximation for $m = 0$ and until $n = N$. The above formulae entail the matrix equations

$$\begin{pmatrix} \sigma_0 \\ \sigma^*_0 \end{pmatrix} = sheet^{cs}_{0,0} \begin{pmatrix} 1 \\ 1 \end{pmatrix} + M_0 \begin{pmatrix} V{*}^{cs}_{1,0} \\ V{*}^{cs}_{2,0} \\ V{*}^{cs}_{3,0} \\ \ldots \\ V{*}^{cs}_{N,0} \end{pmatrix}$$

with the matrix

$$M_0 = \frac{-1}{\pi^2} \begin{pmatrix} \frac{1}{1^2} & \frac{1}{2^2} & \frac{1}{3^2} & \ldots & \frac{1}{N^2} \\ \frac{-1}{1^2} & \frac{(-1)^2}{2^2} & \frac{(-1)^3}{3^2} & \ldots & \frac{(-1)^{N+1}}{N^2} \end{pmatrix}$$

that has rank 2. This means that the subspace of vectors $\begin{pmatrix} V{*}^{cs}_{1,0} \\ V{*}^{cs}_{2,0} \\ V{*}^{cs}_{3,0} \\ \ldots \\ V{*}^{cs}_{N,0} \end{pmatrix}$ that yield $\begin{pmatrix} \sigma_0 \\ \sigma^*_0 \end{pmatrix}$ is the coset $Sing + Ker(M_0)$ of a single solution $Sing$ of the above equation and the $(N-2)$-dimensional kernel of M_0. Together with the condition $V{*}^{cs}_{0,0} = 0$, we have an $(N-3)$-dimensional coset $V(\sigma_0, \sigma^*_0, s^{cs}_{0,0}) \subset (V{*}^{cs}_{n,0})_{n \leqslant N}$ that describes all potential V^* solutions for Fourier coefficients σ_0, σ^*_0 and world-sheet value $sheet^{cs}_{0,0}$. The same result—mutatis mutandis—holds for the coefficients of the cosine part of the symbolic and physical gestures: $V(\tau_0, \tau^*_0, sheet^{cc}_{0,0}) \subset (V{*}^{cc}_{n,0})_{n \leqslant N}$.

For $m > 0$ and $n \leqslant N$, we have the equation

$$\begin{pmatrix} \sigma_m \\ \sigma_m^* \end{pmatrix} = M_m \begin{pmatrix} V{*}^{cs}_{0,m} \\ V{*}^{cs}_{1,m} \\ V{*}^{cs}_{2,m} \\ \dots \\ V{*}^{cs}_{N,m} \\ V{*}^{cc}_{0,m} \\ V{*}^{cc}_{1,m} \\ V{*}^{cc}_{2,m} \\ \dots \\ V{*}^{cc}_{N,m} \end{pmatrix}$$

with $u(n, m) = -(n^2 + m^2)$ and the rank 2 matrix

$$M_m = \frac{1}{\pi^2} \begin{pmatrix} \frac{u(0,m)}{d(0,m)} & \frac{u(1,m)}{d(1,m)} & \frac{u(2,m)}{d(2,m)} & \cdots \frac{u(N,m)}{d(N,m)} & \frac{0}{d(0,m)} & \frac{-1}{d(1,m)} & \frac{-2}{d(2,m)} & \cdots \frac{-N}{d(N,m)} \\ \frac{u(0,m)}{d(0,m)} & \frac{(-1)^1 u(1,m)}{d(1,m)} & \frac{(-1)^2 u(2,m)}{d(2,m)} & \cdots \frac{(-1)^N u(N,m)}{d(N,m)} & \frac{0}{d(0,m)} & \frac{(-1)^2 1}{d(1,m)} & \frac{(-1)^3 2}{d(2,m)} & \cdots \frac{(-1)^{N+1} N}{d(N,m)} \end{pmatrix}$$

deduced by inversion of the matrix $E_{n,m}$. This data defines a 2-codimensional coset

$$V(\sigma_m, \sigma_m^*) \subset (V{*}^{cs}_{n,m})_{n \leqslant N} \oplus (V{*}^{cc}_{n,m})_{n \leqslant N}.$$

The same holds—mutatis mutandis—for the cosine coefficients:

$$V(\tau_m, \tau_m^*) \subset (V{*}^{cs}_{n,m})_{n \leqslant N} \oplus (V{*}^{cc}_{n,m})_{n \leqslant N}.$$

We therefore have a complete set of conditions for the potential V^* to verify the symbolic and physical boundary conditions.

78.2.12.1 Functional Dependence of the Physical Gesture

It is now easy to describe the functional dependence of the physical gesture for given potential V^* and symbolic gesture $Symbolic(y)$.

Let us discuss this topic for the generic case $m > 0$ and the sinusoidal components σ_m, σ_m^*. If the physical coefficients σ_m are given; this means that we have a linear form equation

$$\sigma_m = \frac{1}{\pi^2} \begin{pmatrix} \frac{u(0,m)}{d(0,m)} & \frac{u(1,m)}{d(1,m)} & \frac{u(2,m)}{d(2,m)} & \cdots \frac{u(N,m)}{d(N,m)} & \frac{0}{d(0,m)} & \frac{1}{d(1,m)} & \frac{2}{d(2,m)} & \cdots \frac{N}{d(N,m)} \end{pmatrix} \begin{pmatrix} V{*}^{cs}_{0,m} \\ V{*}^{cs}_{1,m} \\ V{*}^{cs}_{2,m} \\ \dots \\ V{*}^{cs}_{N,m} \\ V{*}^{cc}_{0,m} \\ V{*}^{cc}_{1,m} \\ V{*}^{cc}_{2,m} \\ \dots \\ V{*}^{cc}_{N,m} \end{pmatrix}$$

that defines a one-codimensional coset $C(\sigma_m) \subset (V{*}^{cs}_{n,m})_{n \leqslant N} \oplus (V{*}^{cc}_{n,m})_{n \leqslant N}$. Within this coset of the potential's parameters, the value of σ_m^* is given by the linear form

$$\sigma^*_m = \frac{1}{\pi^2}\begin{pmatrix} \frac{u(0,m)}{d(0,m)} & \frac{(-1)^1 u(1,m)}{d(1,m)} & \frac{(-1)^2 u(2,m)}{d(2,m)} & \ldots & \frac{(-1)^N u(N,m)}{d(N,m)} & \frac{0}{d(0,m)} & \frac{(-1)^1 1}{d(1,m)} & \frac{(-1)^2 2}{d(2,m)} & \ldots & \frac{(-1)^N N}{d(N,m)} \end{pmatrix} \begin{pmatrix} V^{*cs}_{0,m} \\ V^{*cs}_{1,m} \\ V^{*cs}_{2,m} \\ \ldots \\ V^{*cs}_{N,m} \\ V^{*cc}_{0,m} \\ V^{*cc}_{1,m} \\ V^{*cc}_{2,m} \\ \ldots \\ V^{*cc}_{N,m} \end{pmatrix}.$$

This functional relation has as arguments all vectors in the one-codimensional coset

$$C(\sigma_m) \subset (V^{*cs}_{n,m})_{n\leqslant N} \oplus (V^{*cc}_{n,m})_{n\leqslant N}$$

defined above by σ_m.

78.2.13 Parallels Between Performance Operators for Scores and for Gestures

It is important now to investigate the possible relationship between performance operators for scores and for gestures. Recall from Figure 61.4 on page 913 that we relate the score performance to gestural performance by a commutative diagram that involves the projections of symbolic and physical gestures to score symbols and sound events, respectively. In this setup, we recall an important type of performance operator, described under the title of "basis operators" in [714, Section 39.7.1].

Basis operators describe the transition in the stemmatic unfolding of score performance from a given "mother" performance vector field $\mathfrak{Z}_0$ to a "daughter" field $\mathfrak{Z}_1$ using a weight function Λ (typically derived from harmonic, rhythmical, or melodic analysis). It has the formal shape

$$\mathfrak{Z}_1 = \mathfrak{Z}_0 - L_{\mathfrak{Z}_0}\Lambda . i_V Dir,$$

where Dir is a linear endomorphism of the underlying parameter space and i_V is the embedding of V in the total space of $\mathfrak{Z}_0$. In view of the fact that the Lie derivative of weight Λ is in fact a linear operator on the gradient $\nabla\Lambda$, we recognize that the gestural (relative) operator $F_1 = F_1' + D(\nabla f)$ is formally similar to the basis operator. Both involve the previous performance (field $\mathfrak{Z}_0$ or gesture F_1', respectively) and an additive correction by a linear operator on the weight function's gradient:

$$\begin{aligned} \mathfrak{Z}_1 &= \mathfrak{Z}_0 - L_{\mathfrak{Z}_0}\Lambda . i_V Dir, \\ F_1 &= F_1' + D(\nabla f). \end{aligned}$$

We should add that evidently the 2D Fourier analysis of the solution of the general elliptic differential equation also yields the same type of functional dependence: The physical gesture is a linear function of the gradient of the potential.

It remains to be shown that the two weight or potential functions can be related to each other to yield corresponding transition modes on both levels of the commutative diagram of Figure 61.4.

78.2.13.1 Some Detailed Calculations Regarding the Variational Calculus of s_1 with Potential

Describe in detail the contribution $C(\nabla U)$, and then also the formulae for a more general time function $|t'|^2$. The contribution $C(\nabla U)$ is determined by the quantity $C'(m)$ that depends only on the potential's gradient, not on the symbolic and physical gestures. Here is what it looks like, but recall that this is the formula with our special time function $|t'|^2 = 0.17e^{\pi y}$:

$$C'(m) = \frac{\alpha'}{2}\sum_p \frac{Q_p}{q_p}\left(\mathcal{H}^+_{m,p}\mathcal{I}_1 + \mathcal{H}^-_{m,p}\left(\mathcal{I}_2 + \mathcal{I}_3\right)\right),$$

where

$$\mathcal{H}^+_{m,p} = \int_0^1 dx \cos(p\pi x)\cosh(m\pi x) = \frac{p\cosh(m\pi)\sin(p\pi) + m\cos(p\pi)\sinh(m\pi)}{\pi(m^2+p^2)}$$

$$\mathcal{H}^-_{m,p} = \int_0^1 dx \sin(p\pi x)\sinh(m\pi x) = \frac{m\cosh(m\pi)\sin(p\pi) - p\cos(p\pi)\sinh(m\pi)}{\pi(m^2+p^2)}$$

and

$$\begin{aligned}
\mathcal{I}_1 &= \int_0^1 dy \frac{\sin(m\pi y)}{e^{\pi y}}\Big(\sinh(\pi p y)\int_0^1 d\xi \int_y^1 d\eta \sin(p\pi\xi)\sinh[\pi p(1-\eta)]\nabla U(\xi,\eta)e^{\pi\eta} + \\
&\qquad \sinh[\pi p(1-y)]\int_0^1 d\xi\int_0^y d\eta \sin(p\pi\xi)\sinh(\pi p\eta)\nabla U(\xi,\eta)e^{\pi\eta}\Big)\\
\mathcal{I}_2 &= \int_0^1 dy \frac{\cos(m\pi y)}{e^{\pi y}}\Big(\cosh(\pi p y)\int_0^1 d\xi \int_y^1 d\eta \sin(p\pi\xi)\sinh[\pi p(1-\eta)]\nabla U(\xi,\eta)e^{\pi\eta} - \\
&\qquad \cosh[\pi p(1-y)]\int_0^1 d\xi\int_0^y d\eta \sin(p\pi\xi)\sinh(\pi p\eta)\nabla U(\xi,\eta)e^{\pi\eta}\Big)\\
\mathcal{I}_3 &= \frac{1}{\pi}\int_0^1 dy \frac{\cos(m\pi y)}{e^{\pi y}}\Big(\sinh(\pi p y)\int_0^1 d\xi \sin(p\pi\xi)(-\sinh[\pi p(1-y)])\nabla U(\xi,y)e^{\pi y} + \\
&\qquad \sinh[\pi p(1-y)]\int_0^1 d\xi \sin(p\pi\xi)\sinh(\pi p y)\nabla U(\xi,y)e^{\pi y}\Big).
\end{aligned}$$

When we use a general time function $|t'(x,y)|^2$, the coefficients of Equation 78.10 and its generalization for non-vanishing potential have the following shape. The coefficients $Q_p \cdot E_{m,p}$ become

$$\mathcal{I}^1_{m,p} = \int_x\int_y \frac{\cosh(q_p x)\cosh(q_m x)}{|t'(x,y)|^2}\sin(q_p y)\sin(q_m y) + \int_x\int_y \frac{\sinh(q_p x)\sinh(q_m x)}{|t'(x,y)|^2}\cos(q_p y)\cos(q_m y),$$

while the coefficients $Q_p \cdot F_{m,p}$ become

$$\mathcal{I}^2_{m,p} = \int_x\int_y \frac{\cosh[q_p(1-x)]\cosh(q_m x)}{|t'(x,y)|^2}\sin(q_p y)\sin(q_m y) + \int_x\int_y \frac{\sinh[q_p(1-x)]\sinh(q_m x)}{|t'(x,y)|^2}\cos(q_p y)\cos(q_m y),$$

and the coefficients $C'(m)$ become

$$\begin{aligned}
C'(m) = \sum_{p=0}^{\infty}\Bigg(&\int_x\int_y\left(\mathcal{G}^{1\,(x,y)}_{p,m} + \mathcal{G}^{5\,(x,y)}_{p,m}\right)\int_0^1 d\xi \sin(q_p\xi)\int_y^1 d\eta \sinh[q_p(1-\eta)]\nabla U(\xi,\eta)|t'(\xi,\eta)|^2 + \\
&\int_x\int_y\left(\mathcal{G}^{3\,(x,y)}_{p,m} + \mathcal{G}^{6\,(x,y)}_{p,m}\right)\int_0^1 d\xi \sin(q_p\xi)\int_0^y d\eta \sinh(q_p\eta)\nabla U(\xi,\eta)|t'(\xi,\eta)|^2 + \\
&\int_x\int_y \mathcal{G}^{2\,(x,y)}_{p,m}\int_0^1 d\xi \sin(q_p\xi)\sinh[q_p(1-y)]\nabla U(\xi,y)|t'(\xi,y)|^2 + \\
&\int_x\int_y \mathcal{G}^{4\,(x,y)}_{p,m}\int_0^1 d\xi \sin(q_p\xi)\sinh(q_p y)\nabla U(\xi,y)|t'(\xi,y)|^2\Bigg),
\end{aligned}$$

where the integrands $\mathcal{G}^i = \mathcal{G}^{i\,(x,y)}_{p,m}$, for $i = 1, ...6$, are, respectively:

$$\mathcal{G}^{1\,(x,y)}_{p,m} = \alpha^* \cos(q_p x)\cosh(q_m x)\sinh(q_p y)\sin(q_m y)|t'(x,y)|^{-2},$$
$$\mathcal{G}^{2\,(x,y)}_{p,m} = -\frac{1}{q_p}\alpha^* \sin(q_p x)\sinh(q_m x)\sinh(q_p y)\sin(q_m y)|t'(x,y)|^{-2},$$
$$\mathcal{G}^{3\,(x,y)}_{p,m} = \alpha^* \cos(q_p x)\cosh(q_m x)\sinh[q_p(1-y)]\sin(q_m y)|t'(x,y)|^{-2},$$
$$\mathcal{G}^{4\,(x,y)}_{p,m} = \frac{1}{q_p}\alpha^* \sin(q_p x)\sinh(q_m x)\sinh[q_p(1-y)]\cos(q_m y)|t'(x,y)|^{-2},$$
$$\mathcal{G}^{5\,(x,y)}_{p,m} = \alpha^* \sin(q_p x)\sinh(q_m x)\cosh(q_p y)\cos(q_m y)|t'(x,y)|^{-2},$$
$$\mathcal{G}^{6\,(x,y)}_{p,m} = -\alpha^* \sin(q_p x)\sinh(q_m x)\cosh[q_p(1-y)]\cos(q_m y)|t'(x,y)|^{-2}.$$

78.2.14 Complex Time and the Artistic Effort

Intuitively, more complicated choices for the gradient in Equations 78.5 and 78.6 suggest to us the idea of different levels of artistic difficulty. What can be described as *difficult* in art? It's the possibility of converting thought into physical realization. The movements of the hand of a pianist or of the paint-brush of a painter or the choice of the words for a writer are some examples.

Fig. 78.25. Dante Alighieri.

Composers follow an inverse mechanism. If s/he first improvises, s/he starts from his or her thought (imaginary time) to realize a performance (physical time) and later to write a score (again imaginary time). The score will be used by other musicians to play the music that the composer has written (imaginary → physical time). If the composer writes music without playing it, s/he makes an internal transition inside the imaginary component. The writing of a score is an intermediate step before the physical realization. Otherwise, the thought of a composer can already be seen as a collection of physical gestures, that are frozen in a score (imaginary time). Writing a score actually means writing a collection of signs that will be used by the performer as an indication of what variety of gestures move in. The score is a way of fixing on a sheet of paper an indication of movement. The goal to reach, by the composer, is in fact not the realization of a score, but the final performance of his or her music. That is the main difference between music composition and drawing or sculpture: the score is not the final step, but an intermediate one.

When there is a musician who plays his or her music and another who writes his or her notes under dictation, we return to symbolic reality, and thus to the imaginary component, and the circle is completed.

Sometimes the transition imaginary → real time is difficult or even impossible. It is the case of a musician who cannot play the piano because s/he does not know the *technique*, or because a particular composition is too difficult to play, and so on.

78.2.15 Opening the Aesthetic Question that Is Quantified in Lagrange Potentials

Regarding the artistic fight with a deaf material, let us recall Dante's saying in his *Divine Comedy* [242, Paradiso, I Canto, vv. 127 - 132] that

Vero è che, come forma non s'accorda / molte fiate a l'intenzion de l'arte,
perch'a risponder la materia è sorda, / così da questo corso si diparte
talor la creatura, c'ha podere / di piegar, così pinta, in altra parte;

(It is true that, as form resists / many times the intention of the artist,
because its matter in response is deaf, / so it moves away from this path
sometimes the creature with bending power / from innate attitude, in another way;)

To approach this struggle in a more quantitative and precise way, we have demonstrated that the complex time approach yields a dynamical framework where the least Lagrange action can be solved by Poisson equation methods using Green functions for the corresponding Euler-Lagrange equations.

This is a satisfactory result. However, it is nothing but a necessary initial study to approach the following types of musical problems:

- How can potentials be defined from the aesthetic point of view?
- What is the variety of physical gestures for given symbolic gestures and potentials?
- How can the minimal action be understood as an effort of thinking in the making of musical creativity?
- What are the neurophysiological correlates of our model, e.g. with regard to a possible role of mirror neurons for gestural processing?

78.2.16 A Musical Composition by Maria Mannone Realized Using These Ideas

The analysis of piano gestures can be used also to compose new musical compositions. It is possible to start from improvisation and later to freeze the gestures into symbolic indications. It is also possible to start from a simple symbolic gesture, such as the up-down one already considered, choosing some cases from the physical-related gestural variety, and using them as *themes* for gestural variations. This idea has been used to compose the original piano piece *Three Musical Gestures*, built upon a progressive deformation and extension of primitive gestures with different values of initial parameters.

The piece is divided into three movements, with a total of twelve variations. The three parts are titled *Staccato leggero*, *Legato* and *Staccato violento*, respectively.

78.2.16.1 First Movement

The first movement, *Staccato leggero*, is composed by six variations. It starts with only gestures without any movement, and then (variation 1, m. 1-15) has isolated staccato single notes or chords, see Figure 78.26. The idea is taken from Asian music, where the single-gesture movement is clearly defined. However, the repetition of single-gesture movements constitutes a sequence, and a sequence is the basic idea for Western music. If single-movement can be visualized as a circle, and a sequence as an arrow, the composition arrow-circle gives us the structure of a complete musical thought.

In the second variation (m. 17-29) the repetition of a single note is followed by the repetition of a pair of notes, the kernel of any melodic sequence. Upon this basis is built a melody, played by the right hand, that ends with a suspension (see Figure 78.26, m. 28-29).

Variation 3 (m. 31-43) starts with a more classic pattern, with slurs between two notes, where the second one is staccato, see Figure 78.26.

More rhythmical variety is introduced in variation 4 (m. 44-64), see Figure 78.27.

The hands' movement is becoming wider and faster, see Figure 78.28. In this variation, the rhythmical and melodic variety is introduced as a consequence of progressive gestures deformations.

The sequence staccato-staccato-legato-staccato is the starting point of variation 5 (m. 66-79), more tonal in the beginning, see Figure 78.29.

A cluster-like sequence in measure 80 introduces variation 6 (m. 80-91), where a pentatonic scale on the black keys of the piano is used. The last fragment is isolated and repeated when other clusters (obtained as enrichment of the notes of the last fragments) abruptly interrupt the sequence, ending the first movement, see Figure 78.31.

78.2.16.2 Second Movement

The second movement starts again with isolated gestures in variation 7 (m. 92-110): complex chords cluster where the notes are progressively released, leaving only a fifth, see Figure 78.32.

The interval of fifth is used as theme also in variation 8 (m. 112-123), a *legato* sequence of fifths, see Figure 78.33.

Fig. 78.26. Score of *Three Musical Gestures*, including variations 1 (m. 1-15), 2 (m. 17-29), and the beginning of 3 (m. 31-34, out of m.31-43).

Variation 9 (m. 124-129), more dissonant, presents a group of isolated chords, always played with a very legato technique, see Figure 78.34.

Variation 10 (m. 131-148) presents a simple *arpeggiato* in 6/8 tempo, see Figure 78.35. The sequence ends with a progressive rarefaction of the arpeggiation, and the repetition of more and more short fragments thereof. However, the unit is given not only by the pitches but also by the required gesture, always strongly

Fig. 78.27. Beginning of variation 4 (m. 44-53).

Fig. 78.28. Fragment of variation 4 (m. 56-61).

legato. The performer can repeat these soft movements—as though caressing the piano—also at the end of the movement.

78.2.16.3 Third Movement

A completely different piano gesture is needed at the beginning of the third movement, *Staccato violento*, with variation 11 (m. 150-160), see Figure 78.36. There is also here a single note as starting point, but the indication of staccato, jointly with *fortissimo* and *Agitato* tempo, requires a completely different approach to the keyboard. Also the choice of the register (mostly lower in the keyboard, contra-posed with the higher notes of *Legato* movement), jointly with a faster and heavier gesture, contributes to creating stronger resonances

Fig. 78.29. Beginning of variation 5 (m. 66-67, out of m. 66-79).

Fig. 78.30. Beginning of variation 6 (m. 80-85, out of 80-91).

Fig. 78.31. End of variation 6 (m. 86-91).

in the piano. Now the movement is faster and the touch stronger and even metallic at some points. Musical phrases are short and always abruptly interrupted. The sound will be loud and powerful.

Variation 12 (m. 162-180) (see Figure 78.36) has the character of an improvisation, with an alternating of fast and improvised *fortissimo* movements and dramatical variation of intensity in tremolos. The final gesture (see Figure 78.37) must be the largest and most evident of the entire piece.

Fig. 78.32. Beginning of variation 7 (m. 92-99, out of 92-110).

Fig. 78.33. Beginning of variation 8 (m. 112-115, out of 112-123).

Fig. 78.34. Variation 9 (m. 124-129).

78.3 Global Performance Hypergestures

The following section deals with global hypergestures in performance. The general setup of global gestures is described in Section 66.5. Refer to that text before delving into the mathematical discussion of our topic in Section 78.4.2.

78.3.1 The Musical Situation: An Intuitive Introduction

To get off the ground, we first consider some elementary movements of the pianist's hand. However, this is an excessive simplification of physical reality of musical performance for two main reasons: First, because we will impose restricted conditions on gestures; second, because in general even simple gestures are not isolated but musically (and therefore gesturally!) connected to other gestures within a wider context.

Isolated "micro-gestures" are actually the first object of study and training of musicians. When musicians are practicing a piece of music, they can start by isolating a difficult gesture, repeating and correcting it until its quality is satisfactory for performance. Successively, they do the same exercise but now connecting well the micro-gestures among each other. Looking at a musical score, it is usually straightforward to exhibit the splice points between any two adjacent simple micro-gestures.

A complete musical performance is thus realized as a superposition of a number of "local" micro-gestures to create "global" gestures. In our theory of world-sheets of performance hypergestures, gluing together two

Fig. 78.35. Ending of variation 10 (m. 140-148, out of 131-148).

different (hyper)gestures means gluing together their corresponding local world-sheets to obtain a global world-sheet. The conditions that allow this union will be explained later.

To understand why the superposition of gestures is a topic of crucial importance for musical performance, we want to consider some simple examples of musical patterns for keyboard, see Figure 78.38.

In the examples from (a) to (e) of Figure 78.38, the blue square highlights the first gesture, and the yellow square the second one. The common part is indicated by their intersection (light green).

In example (a) the first gesture is given by the repetition of the vertical movement of a finger on the A key, alternated with the rests. The second gesture is a variation of the first one, where the pattern of rest-note has been substituted by rest-note-note-note, with an articulation in the group of three notes. The first note of the group is played with a vertical movement on the keyboard following the rest; thus the connection with the previous and simplest movement is evident.

Example (b) shows a very common combination of a repeated gesture consisting in the alternation of fingers on two notes, in a musical *trillo*, and its resolution, with the same rhythm, but with other notes (and fingers) involved. When pianists practice trills, they have to use the same speed for the embellishment and for its resolution, in a continuous deformation of the gesture.

Example (c) contains the superposition of a melodic pattern and a fragment of descending scale. A melodic pattern is also present in example (d), where the movement for playing the last two notes is rhythmically deformed from a dotted quaver and a sixteenth to two quavers, and later to a triplet with the repetition of the first note.

The last example (e) starts with a fragment of descending scale with the rhythm of a triplet, and continues with the repetition of the last triplet alternated with its melodic variation.

The *global character* of these examples means that we have to play two micro-gestures, but not independently; they are interlocked by a common sub-gesture, which implies that they have to be played under the condition that yields the same gesture on their common restriction. The performance of the two micro-gestures would be different if played independently.

78.4 Categorical Gestures and Global Performance Hypergestures

78.4.1 Categorical Gestures: The Case of Potentials

We used a potential function $U : M \to \mathbb{R}$ which, solving the Euler-Lagrange equation, yielded the hypergesture in its world-sheet function $st : I^2 \to M_{\mathbb{C}}$. The arguments for st are, as recalled in the introduction, the potential U together with the symbolic and physical gestures. We keep the time function aside for this

Fig. 78.36. Variation 11 (m. 150-160) and beginning of variation 12 (m. 162-163, out of 162-180).

Fig. 78.37. Ending of variation 12 (m. 176-180, out of 162-180).

example. If we denote by $A = \mathcal{C}^\infty(M, \mathbb{R})$ and $B = (\uparrow \overrightarrow{@}_A M_\mathbb{C})^2$ the topological (space) categories whose points are for A the functions $U : M \to \mathbb{R}$ and for B the pairs $(w_s : \uparrow \to M_\mathbb{C}, w_p : \uparrow \to M_\mathbb{C})$, we can view this information as a specification of the potential U, the symbolic gesture w_s, and the physical gesture w_p needed to specify the world-sheet hypergesture st. This means that we may view a hypergesture $h \in \uparrow \overrightarrow{@}_A \overrightarrow{@}_B M_\mathbb{C}$ as follows: for every argument $(x, U) \in I \times A$, we have a gesture $g_U(x) \in \uparrow \overrightarrow{@}_B M_\mathbb{C}$, which in turn means that we have a point $g_\alpha(x)_\beta(y) \in M_\mathbb{C}$ for every $y \in I$, $\alpha \in A$, and $\beta \in B$. We therefore have a function value $g_{U,w_s,w_p}(x, y) \in M_\mathbb{C}$, which is exactly what we need to have a world-sheet as a function of the three parameters U, w_s, w_p.

Fig. 78.38. Five examples of simple global phenomena occurring while practicing elementary piano gestures. Each example shows two parts (shown in a blue and a yellow rectangle) of the total gesture, intersecting in a common subgesture (shown as a green rectangle).

78.4.2 The Mathematics of Global Performance Hypergestures

In the previous discussion of performance world-sheets, we have always been focusing on hypergestures $h \in \uparrow \overrightarrow{@} \uparrow \overrightarrow{@} X$, where X is the four-dimensional space-time. This is the elementary case of most simple skeleta (twice $\uparrow$), one (with parameter y) for the space-time gesture, the other (with parameter x) for the "logical" unfolding of a gesture. But the most frequent situation will not be as simple as this, and the most evident generalization is to consider a more general space-time-related skeleton Σ (*Sigma* for space-time), and then hypergestures with the logical unfolding $\uparrow$ digraph, i.e., $h \in \uparrow \overrightarrow{@} \Sigma \overrightarrow{@} X$. The simplest generalization is obtained by dividing the original $\uparrow$ into a linear sequence $\uparrow^n$ of concatenated arrows as, for example, in the one-finger movement down on the key, remaining there for the note's duration, and then moving up again. But ramifications may also occur for more complex hand gestures.

A second generalization regards the logical unfolding $\uparrow$ digraph. Why should this one become more complex? This question regards the nature of the logical unfolding. Is it possible to have a more complex unfolding? After all, the initial symbolic gesture has to end up in a single physical one. Yes, but it is not mandatory to realize this transition in a single-arrow gesture. The logical unfolding might very well be a concatenation of a number of partial unfoldings, referring to the linear sequence skeleton $\uparrow^n$ in the sense that only the initial and final gestures are symbolic or physical, but a number of intermediate stations are conceived, where the musician has not fully thought about the performative transformation. This might then also take place in purely imaginary time and only switch to physical time in the last arrow of $\uparrow^n$. But it might also happen that the final physical gesture is the convergent result of a number of simultaneous initial symbolic gestures, or even a number of intermediate ramified gestures that have been created starting from a single symbolic initial gesture. We therefore have to envisage world-sheets defined by hypergestures $h \in \Lambda \overrightarrow{@} \Sigma \overrightarrow{@} X$ (Λ for "Logical").

The global hypergesture topic refers to the space-time skeleton Σ. It means that we are given a covering Σ^I of Σ by a set $I = \{\Sigma_\iota \subset \Sigma\}$ of non-empty sub-digraphs of Σ. We work in the Escher-Theorem-transformed hypergesture space $\Sigma \overrightarrow{@} \Lambda \overrightarrow{@} X$, and correspondingly in the hypergesture spaces $\Sigma_\iota \overrightarrow{@} \Lambda \overrightarrow{@} X$. The global setup is defined by a family $h_\iota \in \Sigma_\iota \overrightarrow{@} \Lambda \overrightarrow{@} X$ and isomorphisms $\phi_{\iota,\kappa} : h_\iota | \Sigma_\iota \cap \Sigma_\kappa \xrightarrow{\sim} h_\kappa | \Sigma_\iota \cap \Sigma_\kappa$ for every pair $\iota \neq \kappa$ of indices with $\Sigma_\iota \cap \Sigma_\kappa \neq \emptyset$.

This general configuration initiates a number of critical questions.

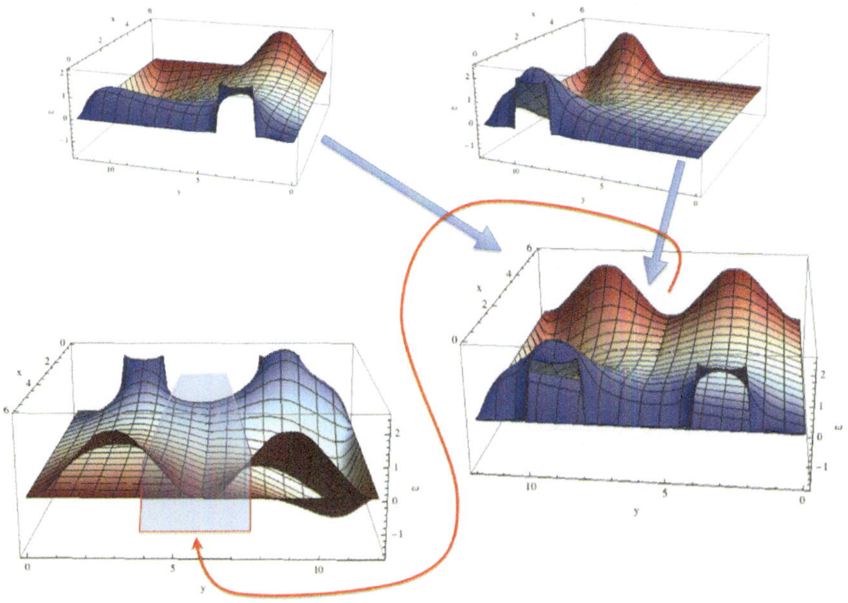

Fig. 78.39. Two partial hypergestures differ in their intersection.

Question 1. To begin with, we now have to face the construction of world-sheets for charts h_ι, defined on general skeleta Σ_ι, and not just simple arrows $\uparrow$. This means that the Lagrange formalism will no longer work for the simple square domain I^2 as before. We are however in the happy situation of having defined integrals for more general hypergestures than simple products of $\uparrow$'s (cubic chains), see Chapter 64. The action integral calculation for this situation is however not trivial anymore.

Question 2. Suppose that one succeeds in calculating an Euler-Lagrange equation for each chart h_ι; this is not what we really need since the variational calculus cannot be executed for each chart independently of the others. We have to perform a global variational calculus in the sense that the restrictions of variations to intersections $h_\iota|\Sigma_\iota \cap \Sigma_\kappa$, $h_\kappa|\Sigma_\iota \cap \Sigma_\kappa$ must correspond under the given transfer isomorphisms $\phi_{\iota,\kappa}$. Figure 78.39 shows that the world-sheets for two partial hypergestures differ in their intersection and also from the restriction of the total hypergesture to these parts.

Question 3. If we have to perform local variational calculi on charts, the given potentials, densities, and time world-sheets must also be global data, i.e., global potentials, densities, and time world-sheets.

Question 4. The definition of global potentials and densities is clear, but the concept of a global time-sheet is less trivial. It is not even clear what a local time world-sheet should be for a non-trivial skeleton Σ_ι. It could be a gluing of conformal mappings on each arrow along their common lines (in the simple case of a $\uparrow$-type logical unfolding), as shown in Figure 78.40 for the gluing of three local conformal time sheets for a "Y"-shaped skeleton. The passage from local to global time-sheets is not problematic here since no Lagrange action calculations are necessary for time. But, of course, the Poisson equations in the global case refer to norms of $|t'|^2$ of time derivatives.

78.5 World-Sheet Hypergestures for General Skeleta

In this section, we want to present a simple but characteristic musical example that involves more general skeleta for hypergestures to describe the transition from symbolic to physical reality in piano performance. We start with a simple original composition by one of the authors (Mannone) as shown in Figure 78.41. It is remarkable that this compositional fragment was created for the sake of our mathematical discourse, but this discourse turned out to be also a useful tool for musical creativity.

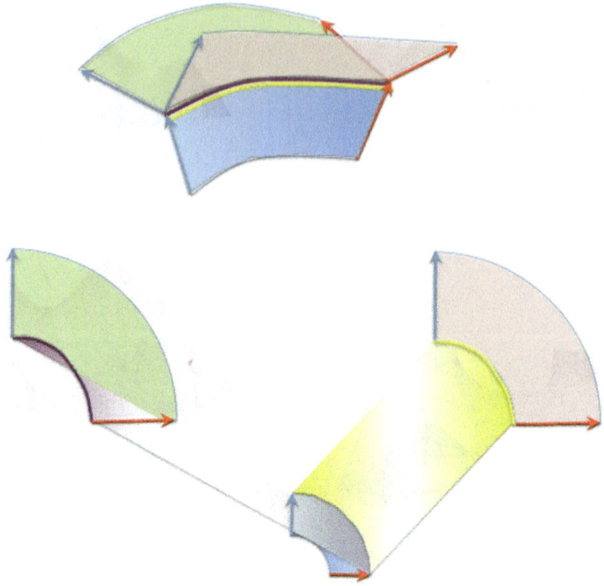

Fig. 78.40. Gluing three conformal time world-sheets for a "Y"-shaped space-time skeleton.

Fig. 78.41. An original score showing the ramification of a symbolic gesture for right and left hands.

The score shows a single voice in measures 1 and 2, and then splits into a right and left hand branch. This can be formalized using the disjoint union $\Gamma_{L,R} = [2]_L \sqcup [2]_R$ of two copies of the digraph[5] $[2] = \bullet_0 \to \bullet_1 \to \bullet_2$, one for the right, one for the left hand. The middle vertex 1 is the end of the second note and the beginning of the third. This gesture $g_{L,R} : \Gamma_{L,R} \to \overrightarrow{M_{\mathbb{C}}}$ has its values in the space-time $M_{\mathbb{C}}$ of full hand parameters in time. So its curves are two groups of left and right hand curves, each one being articulated in the middle by vertex 1 values. To simplify the discussion, we want to consider the coordinates of the left and right indexes. This yields a projection $p_{Index} : M_{\mathbb{C}} \to M_{Index,\mathbb{C}}$. And the index coordinates are now identical until the second note of the third measure. This is formally described by the following commutative diagram, where $\Gamma_{L+R} = [2]_L \sqcup_{[1]} [2]_R$ is the "Y"-shaped digraph obtained by gluing $[2]_L$ with $[2]_R$ along the initial segment $[1] = \bullet_0 \to \bullet_1$; we write $t : \Gamma_{L,R} \to \Gamma_{L+R}$ for the associated projection. The gesture g_{L+R} of left and right index movements is obtained through this projection.

$$\begin{array}{ccc} \Gamma_{L,R} & \xrightarrow{g_{L,R}} & \overrightarrow{M_{\mathbb{C}}} \\ {\scriptstyle t}\downarrow & & \downarrow{\scriptstyle p_{Index}} \\ \Gamma_{L+R} & \xrightarrow[g_{L+R}]{} & \overrightarrow{M_{Index,\mathbb{C}}} \end{array}$$

[5] We write $[n]$ for the line digraph with $n+1$ vertices $0, 1, \ldots n$ and arrows from vertex i to vertex $i+1$ for $i = 0, 1, \ldots n-1$. The arrow digraph a special case: $\uparrow = [1]$.

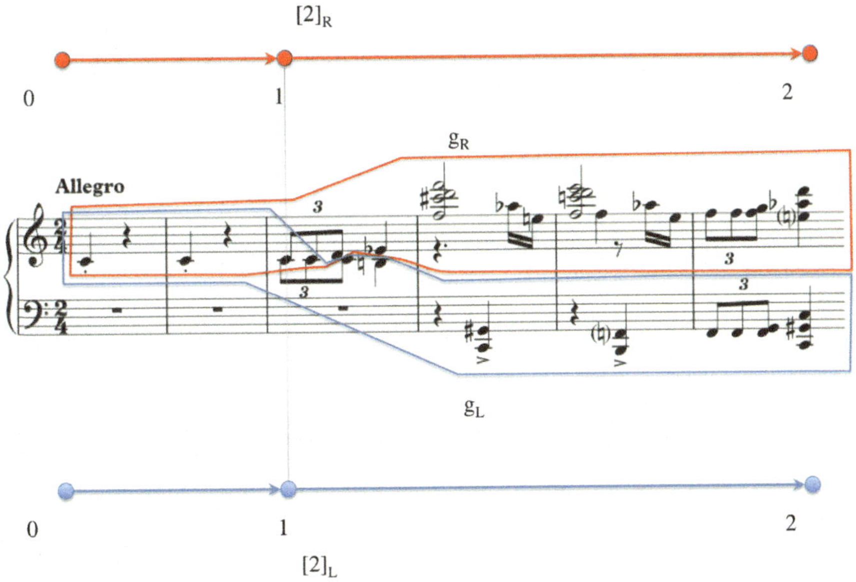

Fig. 78.42. The score's ramification using a gesture with skeleton $\Gamma_{L,R}$.

The symbolic gesture g_{L+R} may now be deformed into a physical gesture by a hypergesture $g_{L+R}(?) : \uparrow \overrightarrow{@}_A \Gamma_{L+R} \overrightarrow{@}_B M_{Index,\mathbb{C}}$, where $g_{L+R}(0)$ is the initial symbolic and $g_{L+R}(1)$ is the final physical gesture, and where the parameter sets A and B are the potential and the symbolic and physical gestures, as in our previous discussion. Figure 78.43 shows this situation.

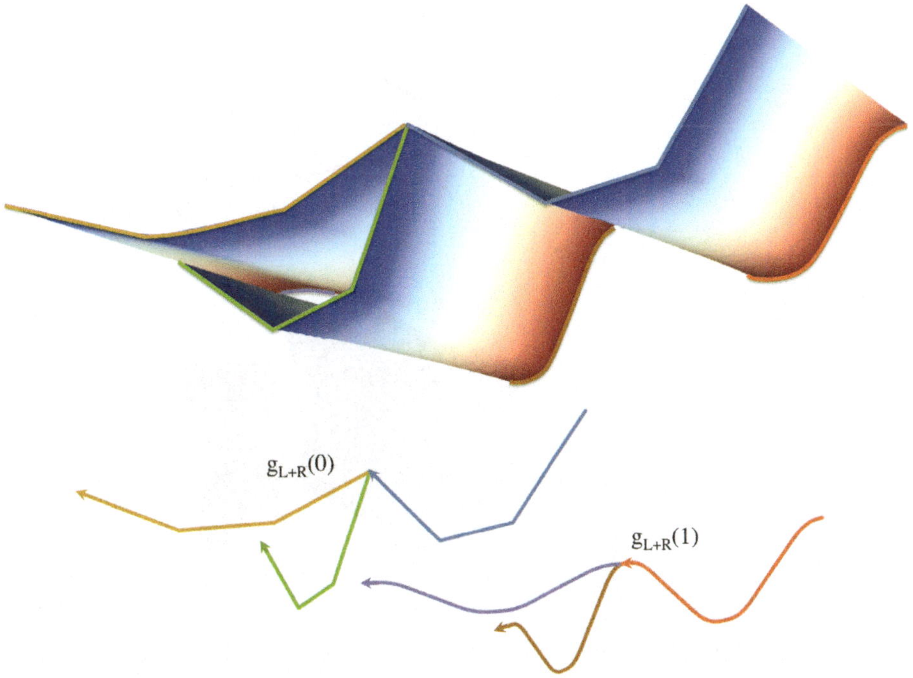

Fig. 78.43. The score's ramification using a gesture with skeleton Γ_{L+R}.

The generalization of skeleta can also occur in the second digraph, which was only a simple line graph $\uparrow$ to this point. Musically speaking, it may happen that the development of a physical gesture in performance takes different directions after an initial unique unfolding, until system time $x = 1/2$, say. Figure 78.44 shows such a splitting in performance. This means that we now have the world-sheet of "Y" shape as shown in

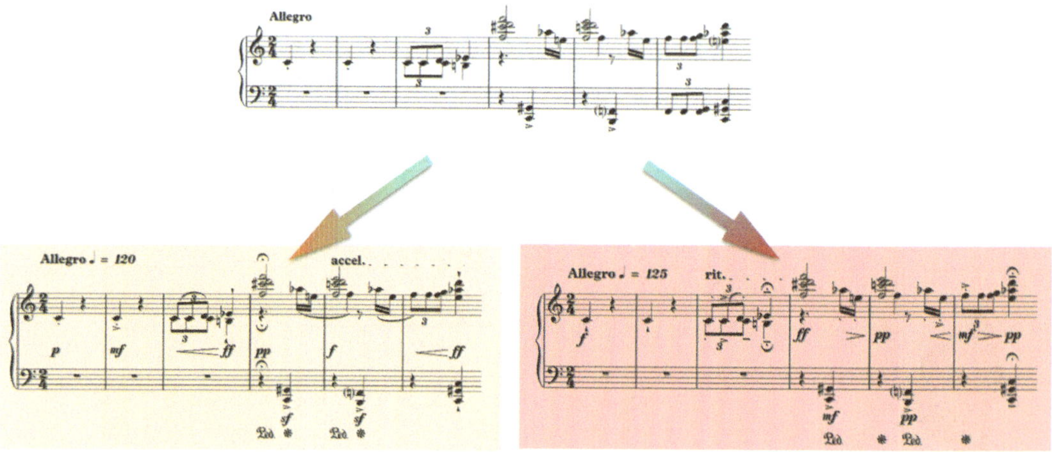

Fig. 78.44. The score's performance ramification using a hypergesture with skeleton Γ_{L+R}.

Figure 78.43 being split into a two-branch shape. This situation is shown in Figure 78.45.

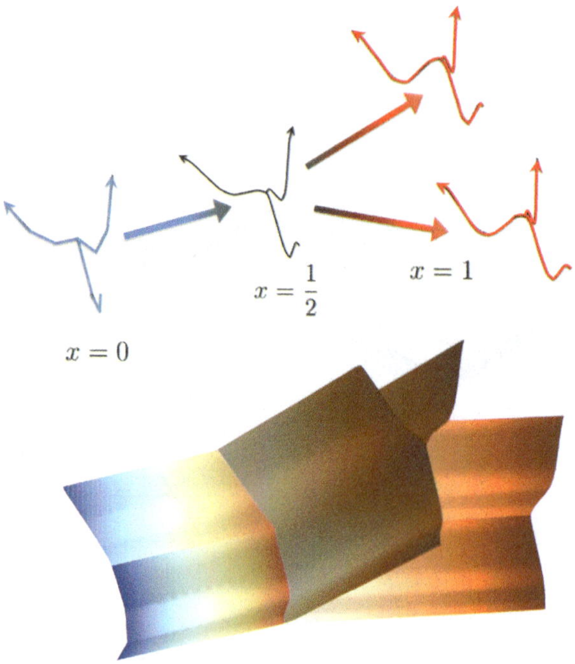

Fig. 78.45. The score's world-sheet ramification with skeleton Γ_{L+R}.

78.6 A Global Variational Principle for the Lagrange Formalism

The globalization of gestures is not only a formalism for the kinetic aspect of hypergestures, it also deals with the Euler-Lagrange equations that describe corresponding world-sheets. And these equations involve the "system parameters" of potential, initial symbolic, and final physical gestures, as described above. When globalizing gestures, this also means globalizing these system parameters. Let us have a look at the potentials for global gestures.

In the previous notations, if Γ^G is a global gesture and G_ι, G_κ are two charts, the hypergesture $g_\iota : G_\iota \to \overrightarrow{\Sigma^I}$ must coincide with the hypergesture $g_\kappa : G_\kappa \to \overrightarrow{\Sigma^I}$ on the intersection $G_\iota \cap G_\kappa$ (modulo automorphisms of Σ^I). This imposes a condition upon the family $(U_\iota)_\iota$ of local potentials, since they cannot be independent from each other. They must induce the same solution of the Euler-Lagrange equation $\frac{-|t'|^2}{\mu}\nabla U = \Delta s$ on the intersections, i.e., $\nabla U_\iota = \nabla U_\kappa$ for intersecting charts. We may interpret this condition in terms of Čech cohomology. To this end, consider the space $A = \mathcal{C}^\infty(M, \mathbb{R})$ as above in Section 78.4.1. Define the following real vector spaces as cochain spaces: $C^0 = A$, $C^1 = \bigoplus_\iota A$, and $C^2 = \bigoplus_{\iota,\kappa \text{ non-empty intersections}} \mathscr{X}(M)$, where $\mathscr{X}(M)$ is the space of $\mathcal{C}^\infty$ vector fields over M, and $C^i = 0$ in all other cases. Define

$$\begin{aligned}
\partial_0 &: C^0 \to C^1 : U \mapsto (U_\iota = U)_\iota,\\
\partial_1 &: C^1 \to C^2 : (U_\iota)_\iota \mapsto (\nabla(U_\iota - U_\kappa))_{\iota,\kappa},\\
\partial_i &= 0 \text{ in all other cases.}
\end{aligned}$$

With this setup, the compatibility condition for a family (U_ι) of local potentials means that $(U_\iota) \in Z^1(C^*)$, i.e., (U_ι) *must be a cocycle of this Čech cohomology.* If we have such a cocycle of local potentials, we may solve the Euler-Lagrange equations on each chart following the variational calculus for local Lagrangian densities.

Part XIX

Gestures in Music and Performance Theory, and in Ethnomusicology

Check for updates

79

Gesture Homology for Counterpoint

Summary. The purpose of this chapter is to review our contrapuntal model such that the group-theoretical contrapuntal symmetries are reinterpreted in the framework of topology, where continuity can be addressed. In particular, we shall interpret the set of intervals as being a topological category. We shall then develop a theory of hypergestures in such a category and investigate the first singular homology group associated with hypergestures. It will turn out that the above conditions defining contrapuntal symmetries can be restated in terms of topology and its associated homology of hypergestures.

– Σ –

79.1 Summary of Mathematical Theory of Counterpoint: What It Is About and What Is Missing

Let us first summarize what our mathematical model of counterpoint is about and why we believe it needs to be reinterpreted in terms of algebraic topology. A detailed description of the model, including proofs, can be found in Part VII. The mathematical model of counterpoint relates to the classical Fuxian canon with its rules, and it deals with the first species[1]: describing which consonant intervals are allowed to succeed which consonant intervals. Our model does not refer to psychological rationales and also not to acoustical arguments; neither would explain those rules. This is well known, since, for example, the fourth is dissonant in the first species, a fact which contradicts the acoustical theories of consonances. Our model yields a set of interval successor rules that is very similar to the Fux rules. In particular, the forbidden parallels of fifths result from this model and are not a consequence of psychological arguments.

The rules of admitted successors of a given consonant interval are deduced by the following setup. We first model intervals as dual numbers $\xi = a+\epsilon k \in \mathbb{Z}_{12}[\epsilon]$, where $k \in K = \{0,3,4,7,8,9\}$ is a consonant interval quantity in the ring $\mathbb{Z}_{12}$ of pitch classes. We denote by $K[\epsilon]$ the set of consonant intervals (consonances), while $D[\epsilon]$ is the set of dissonant intervals (dissonances), where $D = \mathbb{Z}_{12} \backslash K$ is the set of dissonant interval quantities. Given a consonance ξ, we select *contrapuntal symmetries* g in $G = \overrightarrow{GL}(\mathbb{Z}_{12}[\epsilon])$, the group of affine automorphisms of $\mathbb{Z}_{12}[\epsilon]$. These are by definition those automorphisms such that

1. $\xi \in g(D[\epsilon])$,
2. $g \circ AK_\xi = AK_\xi \circ g$,
3. $g(K[\epsilon]) \cap K[\epsilon]$ is a maximal set with the first two properties.

In this definition, AK_ξ is the unique affine automorphism AK_ξ which leaves the intervals at cantus firmus pitch class a of ξ invariant (see Section 30.2.2), and recall that $AK_\xi^2 = Id$. The idea is that following classical ideas about contrapuntal motion (see [924]), the tension between successive intervals is to move between consonances and dissonances. Since this is impossible, we simulate this desired motion as a transition

[1] In his doctoral thesis [15], Octavio A. Agustín-Aquino extended this model to second species counterpoint.

G. Mazzola et al., *The Topos of Music III: Gestures*, Computational Music Science,
https://doi.org/10.1007/978-3-319-64481-3_24

from *deformed* dissonances to *deformed* consonances, where the deformation is given by the dichotomy $g(K[\epsilon])/g(D[\epsilon])$, i.e., the interval ξ is a deformed dissonance that is succeeded by a deformed consonance. The second requirement in the above rules means that the autocomplementarity AK_ξ is also one of the deformed dichotomy. The maximality requirement means that we want to reach every possible deformed consonance with a maximal set of choices.

This model however suffers from the same conceptual deficiency as David Lewin's famous statement in [605] about the "characteristic gesture" when "getting from s to t". There is no gesture whatsoever in his theory, the continuous gestural movement from s to t is a fiction. Lewin's transformations are strictly algebraic, no continuous curves connecting s to t are defined. In our model we recognize a similar deceptive intuition, namely when we speak about those deformed interval sets. The concept of a deformation suggests a continuous movement that successively morphs an interval set into a deformed result. In our setup, similar to Lewin's, we only have symmetries, transformations which don't share any continuous character.

The scope of this chapter is to review our contrapuntal model such that the group-theoretical contrapuntal symmetries are reinterpreted in the framework of topology, where continuity can be addressed. In particular, we shall interpret the set of intervals as being a topological category. We shall then develop a theory of hypergestures in such a category and investigate the first singular homology group associated with hypergestures as developed in [635]. It will turn out that the above conditions defining contrapuntal symmetries can be restated in terms of topology and its associated homology of hypergestures.

What is the meaning of this result for music theory? It is comparable to the meaning of a result that would enable us to consider Lewin's transformations as gestures, not only metaphorically speaking, but as real gestures in a topological space. Such a result—and ours is of this type—opens up a topological music theory, a theory that understands music-theoretical rules as consequences of a topological concept framework that invokes homology of hypergestures. We believe that this initializes a considerable paradigmatic shift in music-theoretical thinking. But it also initializes thoughts about a more intimate relationship between thinking and making music, the latter being strongly connected to gestural embodiment.

79.2 Hypergestural Singular Homology

This chapter does not deal with general hypergestural singular homology, but we focus on the classical context of cubic singular homology [635], meaning that the gestures are all curves $f : \nabla \to K$, i.e., continuous functors on the simplex category ∇, with values in a topological category K. The simplex category ∇ is essentially the unit line, enriched with the morphism pairs $(x, y), x \leqslant y$, see also Section 62.1.1. Viewed as a categorical gesture [723], such a curve is the representation of the gestural *skeleton* $\uparrow$ as a curve f in the *body* of the category K.

Cubic homology is based upon categories $\uparrow @ \uparrow @ \ldots \uparrow \overrightarrow{@} K$ of n-fold hypergestures in K for the sequence $\uparrow, \ldots \uparrow$ of n copies of the one-arrow digraph $\uparrow$ (two vertices, t (tail) and h (head), and one connecting arrow a from t to h). The difference from classical cubic homology is that here we look at curves which are also functors, not only continuous. The basic tool of hypergesture homology is Escher's Theorem 2, Section 62.1.2, which states that we have an isomorphism of topological categories $\Gamma_1\Gamma_2 \ldots \Gamma_k \overrightarrow{@} K \xrightarrow{\sim} \Gamma_{\pi(1)}\Gamma_{\pi(2)} \ldots \Gamma_{\pi(k)} \overrightarrow{@} K$ for any sequence of digraphs $\Gamma_1, \Gamma_2, \ldots \Gamma_k$ and any permutation $\pi \in S_k$ in the symmetric group S_k. The Escher Theorem is used to define the homological boundary operator $\partial_n z$ of an n-chain z, which in the case of cubic homology by definition is an R-linear combination of hypergestures in the n-fold hypergesture category $\uparrow^n \overrightarrow{@} K = \uparrow @ \uparrow @ \ldots \uparrow \overrightarrow{@} K$, R being a commutative unitary ring. This operator goes from the free R-module $C_n(R, K)$ of n-chains to $C_{n-1}(R, K)$, with the starting space $C_0(R, K)$ being the free R-module over the objects of K, and $C_n(R, K) = 0$ for negative n. Since one proves that $\partial^2 = 0$, one has $Im(\partial_{n+1}) = B_n \subset Ker(\partial_n) = Z_n$, and the (cubic) homology module $H_n(R, K) = Z_n/B_n$ is defined. Elements of Z_n are called *cycles*, elements of B_n are called *boundaries*. Refer to Chapter 63 for technical details of the general hypergestural setup.

For the general theory of hypergestures, the yoga of homology is that it formally represents a relationship between neighboring layers of hypergestures over a given topological category. In this chapter we shall study

$H_1(R, K)$ for any ring R and for topological categories K related to continuous actions of topological groups on topological spaces.

79.3 A Classical Example of a Topological Category from Counterpoint

We consider a special topological groupoid ${}^G X$ for a continuous left action $G \times X \to X$ of a topological group G on a topological space X. The groupoid ${}^G X$ is defined as follows. It has as objects elements of X, and as morphisms the triples $(f, x, y) \in G \times X \times X$ such that $f(x) = y$. The topology on ${}^G X$ is induced from the product topology on $G \times X \times X$. A first immediate example from musical set theory is the indiscrete topological spaces $X = \mathbb{Z}_{12}$, $G = T^{\mathbb{Z}_{12}} \rtimes \mathbb{Z}^\times$, the pitch classes on which the full affine group acts.

The following example is our test category since it relates intimately to counterpoint theory. The topological space is the space $X = \mathbb{Z}_{12}[\epsilon]$ of contrapuntal intervals of pitch classes. The topology is defined as follows. We select a consonant interval $\xi = a + \epsilon k$ and consider the unique affine automorphism AK_ξ of X which leaves the intervals at cantus firmus pitch class a of ξ invariant (Section 30.2.2), and recall that $AK_\xi^2 = Id$. The Kuratowski closure operator of our topology (see also Section H.1) is defined by $\overline{Y} = Y \cup AK_\xi(Y)$. This means that closed subsets $Y \subset X$ are the AK_ξ-invariant subsets. This also implies that $Y \subset X$ is open iff it is closed. In particular, if $\eta \in X$ is an interval, its closure is the subset $\overline{\eta} = \{\eta, AK_\xi(\eta)\}$. We call this topology the *AK_ξ-topology*. It has no closed points, and the set $K[\epsilon] = \{a + \epsilon b | a \in \mathbb{Z}_{12}, b \in K\}$ ($K = \{0, 3, 4, 7, 8, 9\}$ is the set of consonant interval quantities) of all consonant intervals is dense.

The topological group in this example is $G = \overrightarrow{GL}(\mathbb{Z}_{12}[\epsilon], AK_\xi)$, the group of AK_ξ-topology continuous affine automorphisms of $\mathbb{Z}_{12}[\epsilon]$. The following lemma shows what it means for an affine automorphism $g : \mathbb{Z}_{12}[\epsilon] \xrightarrow{\sim} \mathbb{Z}_{12}[\epsilon]$ to be AK_ξ-continuous.

Lemma 55 *An affine automorphism $g : \mathbb{Z}_{12}[\epsilon] \xrightarrow{\sim} \mathbb{Z}_{12}[\epsilon]$ is AK_ξ-continuous iff it commutes with AK_ξ.*

Proof. If g commutes with AK_ξ, then, if $Y \subset X$ is AK_ξ-invariant, we have $g^{-1}(Y) = g^{-1}(AK_\xi(Y)) = AK_\xi(g^{-1}(Y))$ since g commutes with AK_ξ iff g^{-1} does so. Therefore g is continuous. Conversely, if g is continuous, taking the open set $\overline{\eta}$, we get $g^{-1}(\overline{\eta}) = \{g^{-1}(\eta), g^{-1}(AK_\xi(\eta))\}$, which has two elements and is open, meaning that it has the shape $\overline{\tau}$. Now, if $g^{-1}(\eta) = \tau$, then $g^{-1}(AK_\xi(\eta)) = AK_\xi(\tau) = AK_\xi(g^{-1}(\eta))$ and g^{-1} commutes with AK_ξ, so also g commutes with AK_ξ. Else, if $g^{-1}(\eta) = AK_\xi(\tau)$ and $g^{-1}(AK_\xi(\eta)) = \tau$, then $AK_\xi(g^{-1}(\eta)) = \tau = g^{-1}(AK_\xi(\eta))$, and commutation is again true, QED.

This elementary topological fact is interesting since the commutation condition is exactly what is required for a so-called contrapuntal symmetry, see Definition 95 in Section 31.1. This supports the hope that counterpoint theory could be restated in terms of topology or even in terms of algebraic topology and singular homological algebra.

We now take the group $G = \overrightarrow{GL}(\mathbb{Z}_{12}[\epsilon], AK_\xi)$ with the compact-open topology. Since $\mathbb{Z}_{12}[\epsilon]$ is locally compact (it is even finite), the composition of continuous automorphisms is continuous, and we get a topological group. Moreover, we have the following.

Lemma 56 *With the above notations, the canonical group action $m : G \times X \to X$ is continuous.*

Proof. It suffices to show that $m^{-1}(\overline{\eta})$ is open for the smallest open sets $\overline{\eta}$. But $m^{-1}(\overline{\eta}) = \{(g, \zeta) | g(\zeta) \in \overline{\eta}\}$. In the compact-open topology of G, we have the open sets $[\zeta, \eta] = \{g | g(\zeta) \in \overline{\eta}\}$ since ζ is the unique element of the compact singleton $\{\zeta\}$. This means that $m^{-1}(\overline{\eta}) = \{(g, \zeta) | g \in [\zeta, \eta]\}$. But we evidently also have $[\zeta, \eta] = [AK_\xi(\zeta), \eta]$; therefore, $m^{-1}(\overline{\eta}) = \bigcup_{\zeta \in X} [\zeta, \eta] \times \overline{\zeta}$, an open set in the product topology, QED.

This terminates the construction of the topological category

$${}^G X = {}^{\overrightarrow{GL}(\mathbb{Z}_{12}[\epsilon], AK_\xi)} \mathbb{Z}_{12}[\epsilon]$$

of contrapuntal intervals. In what follows, we shall also consider full subcategories Z of this ${}^G X$. The essential difference is that in those cases it will not be so that any group action $g.\eta$ on an object η of Z will automatically yield another object in Z. For some such subcategories we can nonetheless prove that they are in fact also groupoids ${}^G X$. Here is the lemma that enables this special situation:

Lemma 57 *With the above notations, if $\eta \in \mathbb{Z}_{12}[\epsilon]$ is an interval, then the full subcategory on the open set $\overline{\eta}$ is a groupoid defined by the group $G(\eta)$ generated by the automorphisms that define morphisms of $\overline{\eta}$.*

Proof. The automorphisms of $\overline{\eta}$ are generated by 1) AK_ξ 2) $Aut(\eta)$. But since AK_ξ and morphisms on ${}^{\overrightarrow{GL}(\mathbb{Z}_{12}[\epsilon],AK_\xi)}\mathbb{Z}_{12}[\epsilon]$ commute by Lemma 56, $G(\eta)$ consists of $Aut(\eta)$ and of the products $Aut(\eta) \circ AK_\xi$, QED.

In the context of continuous groupoids, the following lemma is useful and often used without special mention:

Lemma 58 *If $f : \nabla \to C$ is a curve, then for $0 < \lambda \leqslant 1$ the concatenation curve $f_\lambda = Id_{f(1)} \cdot_\lambda f$, defined by $f_\lambda(\mu) = f(\mu/\lambda)$ for $\mu \leqslant \lambda$ and $f_\lambda(\mu) = f(1)$ for $\mu > \lambda$ with the evident transition morphisms, admits a morphism of curves $f \to f_\lambda$. We also have a morphism ${}_\lambda f \to f$ for the opposite concatenation ${}_\lambda f = f \cdot_\lambda Id_{f(0)}$, defined by the analogous construction. These morphisms between curves are isomorphisms if f is a curve of isomorphisms (all morphisms of the functor are isomorphisms).*

Proof. We define a morphism of curves $q : f \xrightarrow{\sim} g_\lambda$ as follows. For $\mu \leqslant \lambda$, we map $f(\mu) \to g_\lambda(\mu) = f(\mu/\lambda)$ by the morphism given from f. For $\mu > \lambda$, we map $f(\mu) \to g_\lambda(\mu) = f(1)$ by the given morphism from f. The proof of the second statement works in complete analogy, and if f is a curve of isomorphisms, evidently, the curve morphisms are both isomorphisms, QED.

A general remark on continuous curves $f : \nabla \to {}^G X$ is important for the above topological category of contrapuntal intervals. If $f(0) = \eta$, then the whole image $f(\nabla)$ must be in the full subcategory induced on $\overline{\eta}$ since the inverse image $f^{-1}(\overline{\eta})$ is a non-empty closed and open subset of ∇, which can only be all of ∇.

79.3.1 Generators of $H_1({}^G X)$ for a Groupoid ${}^G X$ Defined by a Group Action

We want to describe a set of generators (over any ring R) of the first homology module $H_1({}^G X)$ deduced from the standard (i.e., cubic) hypergesture configuration $\uparrow\uparrow \overrightarrow{@}^G X$. Here are two standard cycle types in $Z_1({}^G X)$:

Definition 122 • *Loop curves $Loop(x, f) = x \circlearrowleft f$ are defined by loops f starting and ending at objects x.*
• *Pairs of different curves $f \neq g$ from x to $y \neq x$ with opposite signs define these parallel cycles:*

$$Para(x, y, f, g) = x \underset{-g}{\overset{f}{\rightrightarrows}} y\,.$$

We shall prove the following theorem from a sequence of lemmata and corollaries:

Theorem 43 *For the standard hypergesture configuration $\uparrow\uparrow \overrightarrow{@}^G X$ of a continuous groupoid ${}^G X$, the homology module $H_1({}^G X)$ is generated by loop curve cycles in $Z_1({}^G X)$.*

Lemma 59 *For any three curves $g, h, k : \nabla \to {}^G X$ in this configuration*

$$\begin{array}{ccc} z & & w \\ g \uparrow & & \uparrow k \\ x & \xrightarrow[h]{} & y \end{array}$$

there exists a hypergesture $t \in \uparrow\uparrow \overrightarrow{@}^G X$ such that $\partial t = g - h - k + l$, with $z \xrightarrow{l} w$ as a fourth curve.

Proof. The construction of t goes in two steps. First, we construct a hypergesture $t_1 \in \uparrow\uparrow \overrightarrow{@}^G X$ with $t_1(0) = g$ and $t_1(1) = Id_y$, such that its Escher-corresponding hypergesture starts at h and ends in a gesture m from z to y. Then we define a second hypergesture $t_2 \in \uparrow\uparrow \overrightarrow{@}^G X$ with $t_2(0) = Id_z$ and $t_2(1) = k$, whose Escher-corresponding hypergesture starts at m and ends at l, a gesture from z to w as required by our theorem. Concatenating t_1 with t_2 yields a hypergesture that starts at g and ends at k, and whose Escher-corresponding hypergesture starts at h and ends at l. This defines the desired hypergesture t. Hypergesture t_1 has the gesture $t_1(\lambda) = {}^{\lambda}g_{h(\lambda)}$ at curve parameter $\lambda \in I$, where ${}^{\tau}g(\mu) = g(\tau + (1-\tau)\mu)$ is the partial gesture of g from τ to 1. And t_2 has gesture $k^{\lambda}_{m(\lambda)}$ at curve parameter $\lambda \in I$, where k^{τ} is the partial gesture of k starting at 0 and ending at τ. The gesture morphisms $t_i(\lambda, \mu) : t_i(\lambda) \to t_i(\mu), i = 1, 2$, for $\lambda \leqslant \mu$ are evident.

Corollary 28 *Every curve* $x \xrightarrow{g} y$ *in* ${}^{G}X$ *can be reversed modulo loops, i.e., there is a curve* $y \xrightarrow{h} x$ *such that* $g = -h + \partial t + 2Loop(x, Id_x)$.

Proof. Attaching two copies of $Loop(x, Id_x)$ to g at x generates the configuration

$$\begin{array}{ccc} y & & x \\ {\scriptstyle g}\uparrow & & \uparrow{\scriptstyle Id_x} \\ x & \xrightarrow[Id_x]{} & x \end{array}$$

which solves our problem in view of Lemma 59.

This corollary enables us to reverse any curve direction and to focus on cycles with any curve directions we like.

Corollary 29 *In the category* ${}^{G}X$*, for any two cycles* $Loop(x, f), Para(x, y, g, k)$*, there is a loop* $Loop(y, l)$ *such that*

$$Loop(x, f) - Loop(y, l) \equiv Para(x, y, g, k) \ (mod\ B_1)$$

In particular, any parallel cycle is equivalent to a difference of two loops modulo B_1.

Lemma 60 *In* ${}^{G}X$*, modulo parallel arrow cycles* $Para(x, y, f, g)$*, every cycle in* Z_1 *is equivalent to a cycle without multiple arrows.*

Proof. If we have two curves c, d between points x, y, Corollary 28 enables us to suppose the curves have same direction. Then, if these curves come up with multiplicities γ, δ, we can subtract the γ-fold of the parallel cycle $Para(x, y, c, d)$ from the given cycle and only curve d remains with scalar $\delta + \gamma$, QED.

And here is the proof of Theorem 43. Modulo loops, we may suppose that a given cycle $z \in Z_1$ has no multiple arrows or loops. Take a longest closed path p within z. Taking a sequence of three curves in p, we may suppose that they have the directions described in Lemma 59. If the factor of the middle curve is λ, we may subtract from p the boundary $\lambda\partial t$ for the t generated in Lemma 59. This yields a new closed path p' which is shorter than p. It is evident that the curve introduced by ∂t does not yield new paths having the length of p. This means that now p' has fewer longest closed paths. We may continue this procedure until in the resulting cycle z' no closed paths of length more than 2 are left, applying Lemma 60 if necessary. Finally, following Corollary 29, the parallel cycles can be replaced by differences of loops, QED.

If we take the set of all loop cycles $Loop(x, f)$, they are evidently linearly independent. But are there relations among such loops modulo B_1? The next proposition describes relations among loop cycles modulo B_1.

Lemma 61 *Let $C(K)$ be the set of curve-connected components*[2] *of a topological category K, and let $(x_c)_{c \in C(K)}$ be a family of objects, one from each component $c \in C(K)$. Then for $K = {}^{G}X$, every cycle $z \in Z_1$ can be represented modulo B_1 by a linear combination of loop cycles $Loop(x_c, f)$.*

Proof. We know from Lemma 60 that z is represented by a linear combination of loop cycles $Loop(x_z, f_z)$. Suppose that x_z is in the connected component c containing x_c. Now, choose a curve $x_z \xrightarrow{l} x_c$. Then there is a hypergesture t defined by the configuration

$$\begin{array}{ccc} x_c & \xrightarrow{Loop(x_c,f)} & x_c \\ {\scriptstyle l}\uparrow & & \uparrow{\scriptstyle l} \\ x_z & \xrightarrow[Loop(x_z,f_z)]{} & x_z \end{array}$$

according to Lemma 59 such that we have $\partial t = Loop(x_c, f) - Loop(x_z, f_z)$, and therefore $Loop(x_c, f)$ represents $Loop(x_z, f_z)$ on x_c, QED.

Lemma 62 *With the above notations, every loop cycle $Loop(x_c, f)$ at x_c is equivalent to the constant identity cycle $Id_c = Loop(x_c, Id_{x_c})$, which implies that*

$$H_1({}^{G}X) = \bigoplus_{c \in C({}^{G}X)} R.Id_c^*,$$

Id_c^* being the class of Id_c.

Proof. If we repeat the construction of a hypergesture $t \in \uparrow\uparrow \overrightarrow{@}^{G} X$ from Lemma 43 with any loop cycle h at x and the identity loop cycle, we get this diagram

$$\begin{array}{ccc} x & & x \\ {\scriptstyle h}\uparrow & & \uparrow{\scriptstyle Id_c} \\ x & \xrightarrow[h]{} & x \end{array}$$

and the constructed top horizontal gesture is also h, i.e., $\partial t = h - Id_c$. Since there are no morphisms between different components, the sum is direct, QED.

79.4 The Meaning of H_1 for Counterpoint

The next step concerns the calculation of homology according to curve connected components $C(K)$ of topological categories. Evidently, since there are no curves connecting different components, the homology module $H_1(K)$ is the direct sum of the homology modules $H_1(c)$ of components $c \in C(K)$. In our contrapuntal example of Section 79.3 above, the components are those minimal open two-element sets $\overline{\eta} = \{\eta, AK_\xi(\eta)\}$. Each curve must live in one of these minimal open sets. But these sets bear the indiscrete topology. So the full subcategories $Z = \overline{\eta}$ they induce are exactly of the type we assumed in the preceding results. Calculations of homology can therefore be performed on each of these small subcategories. We may take the representatives $\eta \in K[\epsilon]$, the set of consonant intervals, and we then get the homology

$$H_1({}^{\overrightarrow{GL}(\mathbb{Z}_{12}[\epsilon], AK_\xi)}\mathbb{Z}_{12}[\epsilon]) = \bigoplus_{\eta \in K[\epsilon]} H_1(\overline{\eta}).$$

The last step consists in showing that the classes Id_c^* don't vanish. And we may also suppose that we are working in a fixed curve component, call it W_c.

[2] Observe that curves are special topological paths in that they must be functors, but not all paths need to be functorial.

Lemma 63 *With the notation of Lemma 62, every class Id_c^* is non-vanishing modulo B_1.*

Proof. We have to show that an equation $Id_c = \sum_{i=1}^{k} \lambda_i \partial t_i$ is impossible. We may wlog suppose that all hypergestures t_i are defined at object c; otherwise define a continuous functor $e : W_c \to End(c)$ by conjugation. Take a morphism $f_z : z \to c$ for each object z of W_c and map a morphism $t : z \to z'$ to $e(t) = f_w \circ t \circ f_z^{-1}$. As homology commutes with continuous functors, the original equation is conserved, replacing the hypergestures t_i by hypergestures in $End(c)$. Every boundary ∂t_i is of shape $\partial t_i = a_i + b_i - c_i - d_i$ for curves in C_1. So we have the equation $Id_c = \sum_{i=1}^{k} \lambda_i(a_i + b_i - c_i - d_i)$. But the elements a_i, b_i, c_i, d_i are all in the canonical basis of C_1. Therefore we may define a linear map f sending all a_i, b_i, c_i, d_i different from Id_c to a curve $y \neq Id_c$ and leaving Id_c fixed. If no $y \neq Id_c$ exists, then the boundaries vanish, which contradicts our equation. It is immediate that the f-image of such a boundary is either $\pm(Id_c - y)$ or $\pm 2(Id_c - y)$. But no linear combination of such elements can be Id_c. Therefore Id_c does not vanish modulo B_1, QED.

We finally get this description of the first homology module:

Theorem 44 *For a topological category ${}^{G}X$ that is defined as the groupoid of a group action, the first homology module of is the direct sum*

$$H_1({}^{G}X) = \bigoplus_{c \in C(Z)} R.Id_c^*$$

of one-dimensional summands $R.Id_c*$. We therefore have $rk(H_1({}^{G}X)) = card(C({}^{G}X))$.

Corollary 30 *We have*

$$H_1({}^{\overrightarrow{GL}(\mathbb{Z}_{12}[\epsilon], AK_\xi)}\mathbb{Z}_{12}[\epsilon]) \xrightarrow{\sim} R^{K[\epsilon]},$$

and for a contrapuntal symmetry g,

$$dim_R(H_1(\overline{g(K[\epsilon]) \cap K[\epsilon]})) = card(g(K[\epsilon]) \cap K[\epsilon]).$$

In mathematical counterpoint theory, Part VII, the intersection $g(K[\epsilon]) \cap K[\epsilon]$ is the set of consonances that must be maximized as a candidate of target intervals starting from the consonance $\xi \notin g(K[\epsilon])$. According to Corollary 30, this maximality is also the maximality of the first homology module of that intersection's closure $\overline{g(K[\epsilon]) \cap K[\epsilon]}$. Therefore the *set-theoretical* maximality in fact carries over to a maximality of a *linear dimension of a homology module*. This is what we have been looking at.

79.5 Concluding Comments

This result is the consequence of a tricky puzzle between the translation of the commutativity condition for contrapuntal symmetries into a topological statement (they are continuous) and the possibility of reducing homology cycles modulo boundaries such that dimensions are boiled down to a single one per curve component. Observe that despite the small number, two, of elements in such a component $\overline{\eta}$, it contains infinitely many curves. It is therefore not trivial to learn that all of these curves contribute to one and only one homology module dimension.

The crucial point in replacing the maximal set cardinality condition of the previous model by a maximal homology dimension condition lies in the fact that such a dimension is defined even if the underlying set is infinite. This means that this topological approach enables such counterpoint models even if the sets of consonances are infinite. The maximality only relates to the dimension of a module, not its set-theoretical cardinality. This situation may occur when stepping to the colimit of infinite microtonal towers [15], where we are dealing with infinite interval sets.

80

Modulation Theory and Lie Brackets of Vector Fields

Summary. In a recent book [16], we have opened the discussion of a hypergestural restatement of mathematical counterpoint theory. The present chapter aims at a discussion in the same vein of the classical mathematical modulation theory [682, 670]. The present approach to modulation theory is based on the idea that degrees in the start tonality are interpreted as gestures that move to degrees (qua gestures) of the target tonality by means of hypergestures. This means that the symmetries relating tonalities in the classical setup are replaced by hypergestures that connect gesturally interpreted degrees. The present hypergestural model solves the problem, but it opens more questions than it answers in the sense that the construction of hypergestures that replace the classical inversion symmetries is by no means unique.

– Σ –

80.1 Introduction

The present chapter aims at a gesturally driven discussion of the classical mathematical modulation theory [682, 670]. Following that approach, it can be proved that tonal modulation as described by Arnold Schönberg[1] [948] can be modeled using symmetries S between scales underlying the involved tonalities. For example, to modulate from C-major to F-major, Schönberg proposes the three modulation degrees II_F, IV_F, VII_F. These degrees also come out from the mathematical model, where the C scale is mapped to the F scale using the inversion symmetry $S = T^9.-1 = U_{e/f}$ between e and f. The mathematical model yields exactly Schönberg's modulation degrees in all cases where he describes a direct modulation, namely for fourth and fifth circle distances 1,2,3,4.

The present approach is based on the idea that degrees in the start tonality are interpreted as gestures that move to degrees (qua gestures) of the target tonality by means of hypergestures. This means that the symmetries relating tonalities in the classical setup are replaced by hypergestures that connect gesturally interpreted degrees.

The present hypergestural model solves the problem, but it opens more questions than it answers in the sense that the construction of hypergestures that replace the classical inversion symmetries is by no means unique. We are still in search of a theory that might generate natural "minimal action" hypergestures in the sense of Hamilton's variational principle in mechanics. In fact, the classical modulation model was driven by the idea of elementary fermion particles in physics, interacting via bosons that materialize interaction forces. The hypergestural restatement would view symmetry-corresponding degrees $X, S(X)$ as musical fermions connected via a boson hypergesture $h : X \to S(X)$. More precisely, the homological boundary $\partial h = (S(X) - X, -h_1^\square)$ has the first component $S(X) - X$ as the difference of the involved fermions, whereas the second component $-h_1^\square$ is the boson deduced from the face operator $?^\square$ acting on the Escher-inverted h_1 of h, but

[1] There are two Arnolds here: the original Austrian Schönberg, and the Americanized Schoenberg. Harmony was written by Schönberg.

G. Mazzola et al., *The Topos of Music III: Gestures*, Computational Music Science,
https://doi.org/10.1007/978-3-319-64481-3_25

see [727] for details. An intuitive illustration in Figure 80.1 shows this situation, where X is given as a pitch class gesture $C \to B, B \to A, C \to A$; $S(X)$ is given by the gesture $C^* \to B^*, B^* \to A^*, C^* \to A^*$,; and the hypergesture h deforms X to $S(X)$ along the lines from A to A^* etc., whereas the Escher-inverted perspective h_1 consists of the hypergesture deforming the line $C \to C^*$ to $A \to A^*$, the line $C \to C^*$ to $B \to B^*$, and the line $B \to B^*$ to $A \to A^*$.

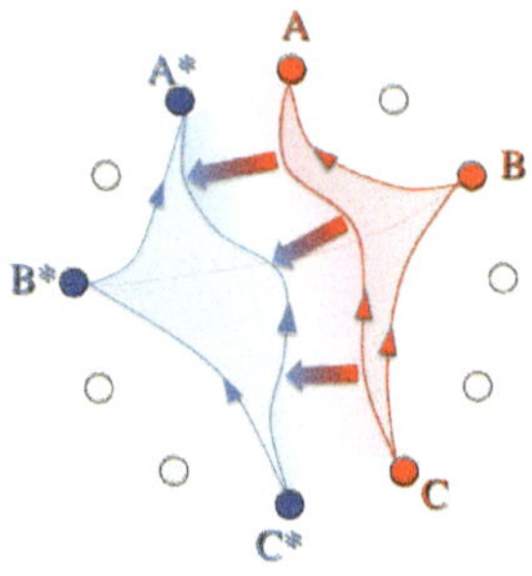

Fig. 80.1. For a pitch class gesture X, with curves $C \to B, B \to A, C \to A$, the target gesture $S(X)$ is given by the gesture $C^* \to B^*, B^* \to A^*, C^* \to A^*$, and the hypergesture h deforms X to $S(X)$ along the lines from A to A^*, B to B^*, C to C^*.

The general procedure will be as follows: We first model gestures and hypergestures in the topological space $\mathbb{R}^2$, where the usual pitch class set $\mathbb{Z}_{12}$ is embedded on a circle. We then look at triadic degrees X of pitch class points, which are represented as gestures of lines connecting these points. Next, we construct vector fields on $\mathbb{R}^2$ whose integral curves give rise to hypergesture curves that deform the gestural degrees. Then we discuss cadences of such triadic degrees and their behavior under hypergestural deformation. We shall prove that for a specific choice of such vector fields, the inversion symmetries used in the classical model map pitch classes x into pitch classes living in the same integral curve as x. Next we consider the trajectories of the curves of the Escher-inverted perspective and calculate energy integrals of such curves. Under the condition of non-vanishing energy, we can then exhibit the admitted degrees. These integrals refer to Stokes' theorem, and we therefore need to think about Stokes' theorem for hypergestures. Its statement and proof are found in the concluding sections of this paper and might be of more general interest.

80.1.1 Short Recapitulation of the Classical Model's Structure

The classical model is described in Section 27.1, we only give a short and not exhaustive recapitulation thereof here. For a modulation from major tonality X to major tonality Y, the triadic modulation degrees (in the sense of Schönberg) in Y are calculated by means of a modulation quantum Q, which is a set of pitch classes. Its intersection $Y \cap Q$ is, by construction, the union of the modulation degrees. This modulation quantum is defined by a number of properties:

1. Q has an inner symmetry that transforms X to Y.
2. For a given cadential set J of Y, all degrees of J are subsets of Q.
3. The intersection $Q \cap Y$ is rigid, i.e., it has no nontrivial inner symmetries (in the group of transpositions and inversions).
4. The quantum Q is minimal with the first two properties.

The motivation of such a quantum is that, by (i), it "materializes" a symmetry qua "force" that transforms X to Y, that, by (ii), it is rich enough to determine Y by a cadence, and that, by (iii), the symmetry of Q that transforms X to Y is uniquely determined by Q; and (iv) is a Hamilton principle: we want Q to be minimal with the first two properties. Observe that this setup does not guarantee the existence of modulation quanta. The modulation Theorem 30, in Section 27.1.4, for 12-tempered tuning guarantees the existence of such quanta. This theorem is also valid for just tuning, see Section 27.1.7, but in the present chapter we focus on 12-tempered tonalities.

80.2 Hypergestures Between Triadic Degrees That Are Parallel to Vector Fields

As said above, we embed the set $\mathbb{Z}_{12}$ of pitch classes in a circle as a part of the real plane $\mathbb{R}^2$. A triadic degree, and more generally any pitch class set X, is enriched by a system of differentiable curves $l(x, y)$ from x to y (or vice versa), one for every *unordered* pair x, y of different points in X; the selected direction is irrelevant (why will be seen later); the corresponding gesture is denoted by $\overrightarrow{X}$. An example is shown in Figure 80.1 for the set $X = \{A, B, C\}$. If such a gesture has skeleton Σ, it is an element of $\Sigma\overrightarrow{@}\mathbb{R}^2$. We shall consider hypergestures $h \in\uparrow \overrightarrow{@}\Sigma\overrightarrow{@}\mathbb{R}^2$ that connect two pitch class set gestures (of same skeleton) $\overrightarrow{X}, \overrightarrow{Y}$, i.e., $\overrightarrow{X} = h(0), \overrightarrow{Y} = h(1)$. We shall now look at vector fields $\mathcal{X}$ on $\mathbb{R}^2$ that are smooth enough to have integral curves, fields that are locally Lipschitz, to be precise. For every point $x \in \mathbb{R}^2$, there is a uniquely defined maximal integral curve $\int_x \mathcal{X} : D(x) \to \mathbb{R}^2$, defined on the open domain $D(x) \subset \mathbb{R}$, starting at x, i.e., $\int_x \mathcal{X}(0) = x$, and $T(\int_x \mathcal{X})(\lambda) = \mathcal{X}(\int_x \mathcal{X}(\lambda))$ for all parameters $\lambda \in D(x)$ of $\int_x \mathcal{X}$.

Definition 123 *Given a hypergesture $h \in \uparrow \overrightarrow{@}\Sigma\overrightarrow{@}\mathbb{R}^2$ connecting $\overrightarrow{X} = h(0)$ to $\overrightarrow{Y} = h(1)$, we say that it is* parallel to a vector field $\mathcal{X}$ *if for every point x in X, there is a function $f : I \to D(x)$ of the unit interval $I = [0, 1]$ into the domain $D(x)$ such that the $\uparrow$-gesture $h_1(x)$ of h_1 starting at x has values $h_1(x)(\lambda) = \int_x \mathcal{X}(f(\lambda))$ for all $\lambda \in I$.*

The terminology is justified for such a differentiable function f since its tangent function Tf evaluates to vectors parallel to the vector field's vectors. The following lemma enables us to construct parallel hypergestures from curves on vertices of a pitch class set X.

Lemma 64 *Given a vector field $\mathcal{X}$, a pitch class set X with a gesture $\overrightarrow{X}$ in $\Sigma\overrightarrow{@}\mathbb{R}^2$, and a pitch class set Y such that for every point $x \in X$, there is a curve $f_x : I \to D(x)$ such that $\int_x \mathcal{X}(f(1)) =: y(x)$ defines a bijection $X \xrightarrow{\sim} Y$, then there is a hypergesture $h \in \uparrow \overrightarrow{@}\Sigma\overrightarrow{@}\mathbb{R}^2$ connecting $\overrightarrow{X}$ with a gesture $\overrightarrow{Y}$ that is parallel to $\mathcal{X}$.*

The critical point here is the question whether we can find vector fields that connect degrees X, Y that are symmetric images of each other, i.e., $Y = S(X)$ for a symmetry S connecting two tonalities, by parallel hypergestures.

80.3 Lie Brackets Generate Vector Fields That Connect Symmetry-Related Degrees

In this section we define vector fields associated with pairs of tonalities and which fulfill the conditions explained above. Although such vector fields can be defined for quite general situations of tonality pairings, we want to restrict our attention to the pairing of two tonalities that are one fourth apart from each other, and we may choose the concrete situation of C-major and F-major. For each such tonality T, which we identify with its scale for this special discussion, we define a vector field $\mathcal{X}_T$ that is motivated by the unique inner symmetry S_T of T. For $T = C$ this is the inversion $S_C = U_d$; for $T = F$, it is $S_F = U_g$. To have a simple representation of symmetries and fields, we choose a labeling of the pitch classes in $\mathbb{Z}_{12}$ such that $0 = d, 1 = d\#, 2 = e, 3 = f, 4 = f\#, 5 = g, 6 = g\#, 7 = a, 8 = a\#, 9 = b, 10 = c, 11 = c\#$. With this notation, and 0 being on top, and 3 to the right of the circular representation (as with normal time visualisation), the symmetry S_C is the reflection at the vertical diameter through the pitch class circle. We now represent this reflection as a movement in horizontal direction from left to right, thinking of a 180°-rotation in $\mathbb{R}^3$. This can be represented by a vector field $\mathcal{X}_C(x, y) = (\cos(y)\cos(x), 0)$. Similarly, for tonality F, we define its vector field $\mathcal{X}_F$ as being the clockwise rotation of $\mathcal{X}_C$ by $5\pi/6$. More generally, if R is a nonsingular linear transformation of $\mathbb{R}^2$, we construct a vector field X^R from X by $X^R(x) := R(X(R^{-1}(x))$. Then we have $Y = X^R$ if R is the clockwise rotation by $5\pi/6$. Figure 80.2 shows these fields in a graphic generated by Mathematica software.

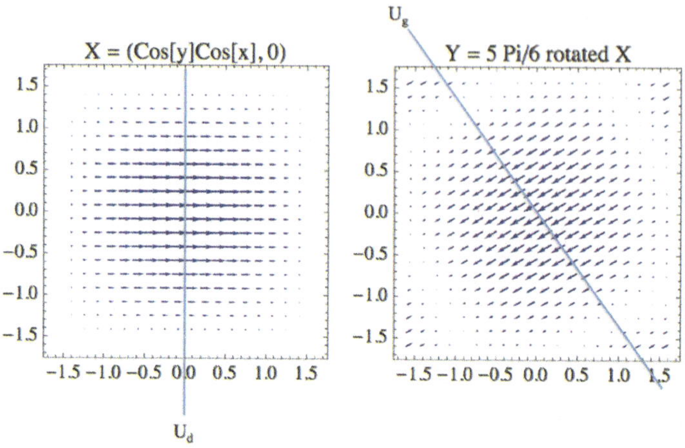

Fig. 80.2. Field X is $\mathcal{X}_C$ (left), field Y is $\mathcal{X}_F$ (right).

The next step is mathematically well defined, but we actually don't know why it works. To generate the field which will eventually guide the hypergestural lines, we consider the Lie bracket $[\mathcal{X}_C, \mathcal{X}_F]$ of the fields associated with the inner symmetries S_C, S_F. They are defined using the fact that vector fields are in one-to-one correspondence with derivations on functions, and then using the fact that the commutator of such derivations is again a derivation. Lie brackets are very important in mathematical physics, and in particular in Lagrangian and Hamiltonian mechanics. See [2] for the calculus of Lie brackets and its application to mechanics. Here is the explicit formula for this Lie bracket:

$$
\begin{aligned}
&[\mathcal{X}_C, \mathcal{X}_F](x, y) = \\
&\qquad (-(1/2)\sqrt{3}\cos((\sqrt{3}x)/2 + y/2)\cos(y)\cos(x/2 - (\sqrt{3}y)/2)\sin(x) - \\
&\qquad 1/2\cos(x)\cos((\sqrt{3}x)/2 + y/2)\cos(x/2 - (\sqrt{3}y)/2)\sin(y) + \\
&\qquad \cos(x)\cos(y)(3/4\cos(x/2 - (\sqrt{3}y)/2)\sin((\sqrt{3}x)/2 + y/2) + \\
&\qquad 1/4\sqrt{3}\cos((\sqrt{3}x)/2 + y/2)\sin(x/2 - (\sqrt{3}y)/2)), \\
&\qquad \cos(x)\cos(y)(1/4\sqrt{3}\cos(x/2 - (\sqrt{3}y)/2)\sin((\sqrt{3}x)/2 + y/2) + \\
&\qquad 1/4\cos((\sqrt{3}x)/2 + y/2)\sin(x/2 - (\sqrt{3}y)/2)))
\end{aligned}
$$

The integral curve display of this field is shown in Figure 80.3. This field has two remarkable properties which we want to list as a proposition:

Proposition 70 *With the above notations, the Lie bracket field $[\mathcal{X}_C, \mathcal{X}_F]$ has the following properties:*

1. *The twelve pitch class points are contained in three closed integral curves: C_b through $\{b, a\#\}$, C_c through $\{c, c\#, d, d\#, f\#, g, g\#, a\}$, and C_e through $\{e, f\}$.*
2. *The curves C_b, C_c and C_e are symmetrical with respect to the modulator symmetry $U_{e/f}$ that maps C to F in the sense that every pitch class p in its integral curve C_b, C_c or C_e is mapped to the $U_{e/f}(p)$ that is contained in the* same *integral curve.*
3. *If R is the 180°-rotation in $\mathbb{R}^2$, we have $[\mathcal{X}_C^R, \mathcal{X}_F^R] = [\mathcal{X}_C, \mathcal{X}_F]^R = [\mathcal{X}_C, \mathcal{X}_F]$.*
4. *If $R = U_{e/f}$ then $\mathcal{X}_F = -\mathcal{X}_C^R$, and we have $-[\mathcal{X}_C, \mathcal{X}_F]^R = [\mathcal{X}_C, \mathcal{X}_F]$. The latter is also true if R is the reflection orthogonal to $U_{e/f}$. These formulas mean intuitively that the two reflection axes that are visible in the left part of Figure 80.3 transform the Lie bracket field into its negative.*

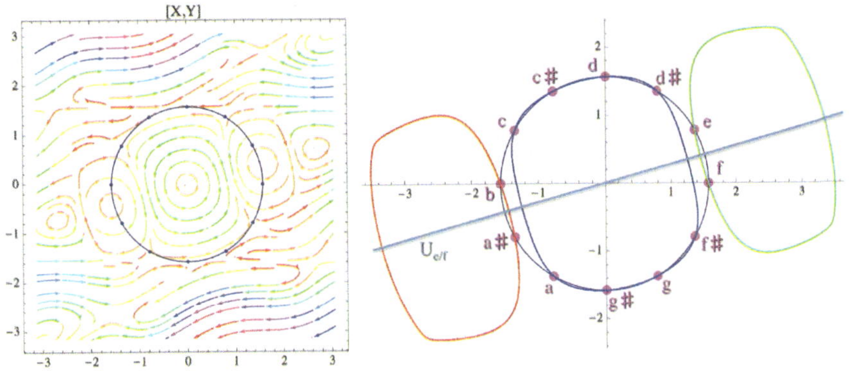

Fig. 80.3. The global display of the integral curves of the Lie bracket field $[\mathcal{X}_C, \mathcal{X}_F]$ (left), and (right) the three closed integral curves comprising all twelve pitch classes, and this in such a way that the modulator transformation $U_{e/f}$ from C to F maps pitch classes into pitch classes within the same integral curves.

Recall that if $J(X)$ denotes the Jacobian of a vector field X on $\mathbb{R}^2$, then $[X,Y] = J(Y)X - J(X)Y$. Property (iii) is evident since for the 180°-rotation R, $X^R = -X$ and $Y^R = -Y$, whence $[X^R, Y^R] = [-X, -Y] = [X, Y]$. The equation $[\mathcal{X}_C, \mathcal{X}_F]^R = [\mathcal{X}_C, \mathcal{X}_F]$ follows immediately from the Jacobian formula. The last property in (iv) follows from (iii) and the first part of (iv). To prove it, we need two easy auxiliary results about Lie brackets. The first result relates to the Jacobian of a vector field $XT(x) := X(T(x))$ deduced from a non-singular linear transformation T on $\mathbb{R}^2$. We have $J(XT)(x) = J(X)(T(x))T$. Using this result, if R is a linear involution ($R^2 = Id$), then we have $[X, -X^R] = -[X, -X^R]^R$. Property (iv) now follows from this last result since in our case, $\mathcal{X}_F = -\mathcal{X}_C^R$ for $R = U_{e/f}$. Property (ii) follows from property (i) and property (iv). Property (i) of the Lie bracket field is not evident. We don't know why the twelve pitch classes are grouped in just three integral curves that are invariant under $U_{e/f}$. We have no mathematical proof of this proposition in the sense that we were not able to calculate symbolically (with explicit formulas) those three symbolic integral curves C_b, C_c, C_e and to prove that the subsets of pitch classes are precisely contained in those curves. Also, Mathematica did not yield a solution using its `DSolve` function; our result is obtained using the numerical integration function `NDSolve`, QED.

Using this proposition, we can now find hypergestures h, parallel to $[\mathcal{X}_C, \mathcal{X}_F]$ that map degrees of C-major or more general pitch class sets to symmetry-connected degrees or pitch class sets, respectively, in F-major. In fact, referring to the notations of Lemma 64, given a pitch class set X in C, we can find by Proposition 70 a curve $f_x : I \to D(x)$ for every $x \in X$, such that $\int_x [\mathcal{X}_C, \mathcal{X}_F](f(1)) = S(x)$ defines a bijection with the symmetric pitch class set $Y = S(X)$. Therefore, by Lemma 64, there is a hypergesture h, parallel to $[\mathcal{X}_C, \mathcal{X}_F]$, that maps X to $S(X)$.

80.4 Selecting Parallel Hypergestures That Are Admissible for Modulation

The next step consists of the selection of "good" hypergestures for the intended modulation. To this end, we look at the hypergestures $h_{x,y}$ obtained from the above parallel hypergestures h when restricting them to the single curves $l(x,y)$ in $\overrightarrow{X}$, deformed under $h_{x,y}$ to curves $l(S(x), S(y))$ that define $\overrightarrow{S(X)}$. Such a deformation hypergesture consists of a (smooth) curve of curves $h_{x,y}(\lambda), \lambda \in I$, whose endpoints x_λ, y_λ are all moving within one of the three integral curves C_b, C_c and C_e, and $h_{x,y}(0) = l(x,y), h_{x,y}(1) = l(S(x), S(y))$; see Figure 80.4 for an example; starting at a curve from $l(c,e)$ and ending at curve $l(S(c) = a, S(e) = f)$, the intermediate curves $h_{x,y}(\lambda)$ all move along the integral curves C_c, C_e with their endpoints.

Definition 124 *With the above notation, such a hypergesture $h_{x,y}$ from curve $l(x,y)$ to curve $l(S(x), S(y))$ is called non-singular if for every parameter $\lambda \in I$, the gesture $h_{x,y}(\lambda)$ is not a loop.*

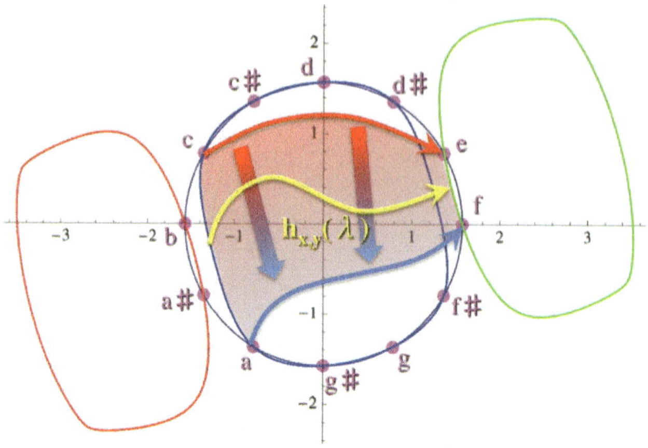

Fig. 80.4. Starting at a curve from $l(c,e)$ and ending at curve $l(S(c)=a, S(e)=f)$, the intermediate curves $h_{x,y}(\lambda)$ move along the integral curves C_c, C_e with their endpoints.

Although this definition looks only geometric, it has an interpretation in terms of the energy function. Suppose that $E(x,y)$ is a differentiable potential function on $\mathbb{R}^2$. Then we may consider the usual line integral $\int_{h_{x,y}(\lambda)} dE$, expressing the work to move from $h_{x,y}(\lambda)(0)$ to $h_{x,y}(\lambda)(1)$ under the given potential E. If we suppose that Stokes theorem can be proved for hypergestures, we have $\int_{h_{x,y}(\lambda)} dE = \int_{\partial h_{x,y}(\lambda)} E = E(h_{x,y}(\lambda)(1)) - E(h_{x,y}(\lambda)(0))$. This latter vanishes if the curve $h_{x,y}(\lambda)$ is a loop. The converse is not true, but we can enforce the converse if we find enough potentials such that the vanishing of the integral for all these potentials implies that the curve is a loop. In fact, in our situation it is possible to find two simple potentials, $E_C(x,y) = x$ and its clock-wise rotation by $5\pi/6$, $E_F(x,y)$ (similarly to the vector field construction). Evidently, $h_{x,y}(\lambda)$ is a loop if and only if $d(\lambda) := (\int_{h_{x,y}(\lambda)} dE_C)^2 + (\int_{h_{x,y}(\lambda)} dE_F)^2 = 0$. This will be our condition for an *admissible (parallel) hypergesture h from pitch class set X to $S(X)$*, namely that *all of its curve sub-hypergestures $h_{x,y}, x \neq y$, are non-singular.* The Stokes theorem can in fact be proved for hypergestures; we refer you to Chapter 64 for a thorough discussion of a hypergestural Stokes theorem.

In the classical modulation model, one looks at all minimal cadential sets of triadic degrees, see Section 26.2.1. Here they are:

$$J_1 = \{II, III\}, J_2 = \{II, V\}, J_3 = \{III, IV\}, J_4 = \{IV, V\}, J_5 = \{VII\}.$$

One then considers the S-transformed cadential sets. These involve all degrees, II, III, IV, V and VII (in both scales, C and F, since S switches II_C to V_F, III_C to IV_F, IV_C to III_F, V_C to II_F, and VII_C to VII_F). We have this proposition:

Proposition 71 *For every triadic degree*

$$X_C = I_C, II_C, III_C, IV_C, V_C, VI_C, VII_C,$$

there is a non-singular parallel hypergesture h_{X_C} from $\overrightarrow{X_C}$ to $\overrightarrow{S(X_C)}$ for the Lie bracket field $[\mathcal{X}_C, \mathcal{X}_F]$.

The proof of this proposition is an easy verification. Therefore each triadic degree can be connected hypergesturally to its symmetric counterpart. However, if we look at the cadential sets and the pitch class sets they define by union of their degrees, such as $\cup(J_1) := II_C \cup III_C$, such a connection is no more possible in general for corresponding gestures. Here are the obstructions, and Figure 80.5 visualizes the singular situation for the hypergestural movement:

- For $J_1 = \{II_C, III_C\}$, the hypergesture h has a singular part for the curve $l(a,g)$ ($a \in II_C, g \in III_C$) that maps to $l(c,d)$.

- For $J_2 = \{II_C, V_C\}$, the hypergesture h has a singular part for the curve $l(a, g)$ ($a \in II_C, g \in V_C$) that maps to $l(c, d)$.
- For $J_3 = \{III_C, IV_C\}$, the hypergesture h has a singular part for the curve $l(a, g)$ ($a \in IV_C, g \in III_C$) that maps to $l(c, d)$.
- For $J_4 = \{IV_C, V_C\}$, the hypergesture h has a singular part for the curve $l(a, g)$ ($a \in IV_C, g \in V_C$) that maps to $l(c, d)$.

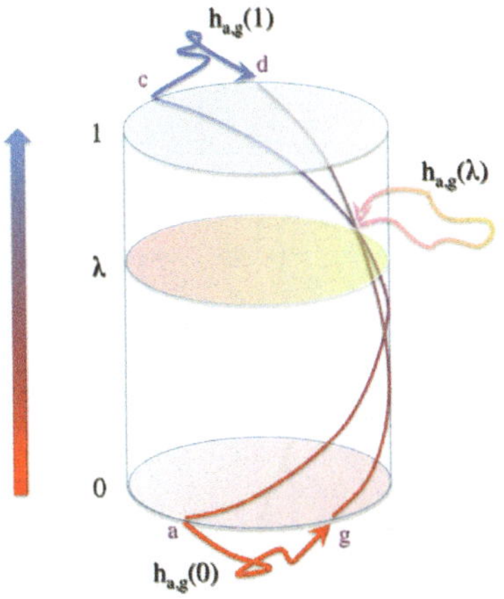

Fig. 80.5. The hypergesture from curve $h_{a,g}(0) = l(a, g)$ to curve $h_{a,g}(1) = l(d, c)$ enforces an intermediate singular loop position.

Therefore the only admissible hypergestural transformation is that from $\overrightarrow{VII_C}$ to $\overrightarrow{VII_F}$. This is the selection we find using the present hypergestural arguments. Then, going back to the construction of the modulation quantum in the classical model, we have to look at the intersection $F \cap (VII_C \cup VII_F) = F \cap \{b, d, f, e, g, a\#\} = \{d, f, e, g, a\#\} = II_F \cup IV_F \cup VII_F$, and the latter is exactly the set of modulation degrees described in the classical model and by Schönberg.

This model also works for the fifth circle modulation from C to G. It is an easy exercise to go through all steps for this movement, and we get the classical modulation degrees III_G, V_G and VII_G as in the classical case.

80.5 The Other Direct Modulations

For other modulation types to more distant fourth circle tonalities, from C to $E_\flat$, say, we propose the following hypergestural construction. We factor the movement into fourth circle steps, e.g. C to F, then F to $B_\flat$, then $B_\flat$ to $E_\flat$, then $E_\flat$ to $A_\flat$. The corresponding integral curves through the twelve pitch classes are shown in Figure 80.6.

But this is not factorizing the modulation steps, i.e., we only factor the hypergestural curves and then exhibit those hypergestures which have non-singular steps. Figure 80.7 shows such a factorization for the hypergesture moving e to d in a modulation $C \to B\flat$. The first part of the curve moves e to f on the closed integral curve C_e, the second part of the curve moves f to d on F_c. We shall realize this model for all fourth circle relations $C \to B_\flat, C \to E_\flat, C \to A_\flat$ (and of course for the corresponding fifth circle relations $C \to D, C \to A, C \to E$). The result will again yield the same modulation degrees as with the classical model.

The precise setup for modulations $C \to B_\flat, C \to E_\flat, C \to A_\flat$ is that we look for sequences of admissible parallel hypergestures. Denote by $S_C, S_F, S_{B_\flat}, S_{E_\flat}$ the four inversions mapping $C \to F, F \to B_\flat, B_\flat \to$

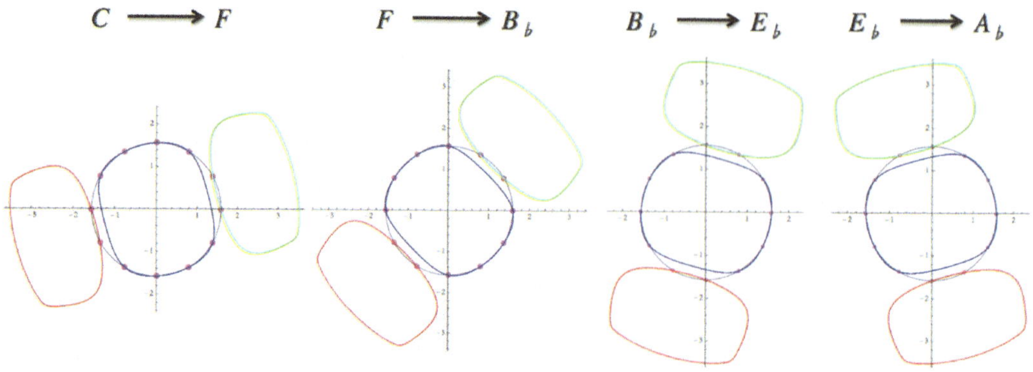

Fig. 80.6. The four closed integral curves for fourth circle modulations starting from $C, F, B_\flat, E_\flat$.

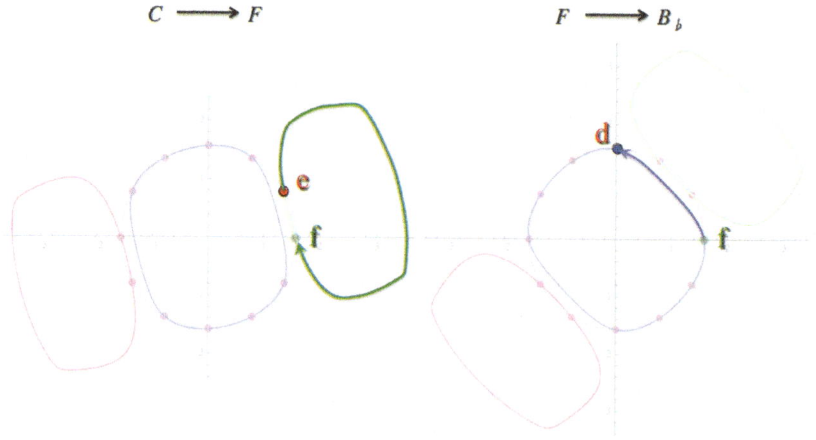

Fig. 80.7. The hypergestural curve from e to d factors through f on two integral curves, C_e and F_c, while a direct movement is not possible for any of these closed integral curves.

$E_\flat, E_\flat \to A_\flat$. For example, for $C \to E_\flat$, and for a set X of pitch classes in C, we look for a sequence of admissible parallel hypergestures h_C, h_F and $h_{B_\flat}$ where h_C connects X to $S_C(X)$, h_F connects $S_C(X)$ to $S_F(S_C(X))$, and $h_{B_\flat}$ connects $S_F(S_C(X))$ to $S_{B_\flat}(S_F(S_C(X)))$, the last being the target set in $E_\flat$. The concatenation $h = h_{B_\flat} \circ h_F \circ h_C$ of these three hypergestures is what we call an *admissible parallel hypergesture connecting a gesture* $\overrightarrow{X}$ *to* $\overrightarrow{S_{B_\flat}(S_F(S_C(X)))}$.

Let us make an example to understand the special character of such concatenations. We again look at the above concatenation $h = h_{B_\flat} \circ h_F \circ h_C$, and we start with a pitch class set $X = V_C$. We are looking for three admissible parallel hypergestures $h_{B_\flat}, h_F$ and h_C that connect V_C to $II_{E_\flat} = S_{B_\flat}(S_F(S_C(V_C)))$. Figure 80.8 shows that this is possible. The only non-trivial step is the first hypergesture; we have shown to the left the non-singularity of this hypergesture.

With this approach we now look at cadence sets $J_1, \ldots J_5$ in C which (more precisely, as above: the unions of their members, e.g. $\cup(J_1) = II_C \cup III_C$, etc.) can be connected by admissible parallel hypergestures to corresponding cadence sets in the target tonality. If such hypergestures between cadence set J_k in C and cadence set J'_l in the target tonality exist, we proceed as before: We take the union $(\cup(J_k)) \cup (\cup(J'_l))$ and check whether their intersection $T \cap (\cup(J_k)) \cup (\cup(J'_l))$ with the target tonality T is rigid. The difference with the classical algorithm is that we don't check for minimality anymore. This condition has been taken care of by the distinguished hypergestural connection described by the integral curves of the Lie bracket vector fields. Minimality seems to be taken care of by the hypergestural transformation. The result is this:

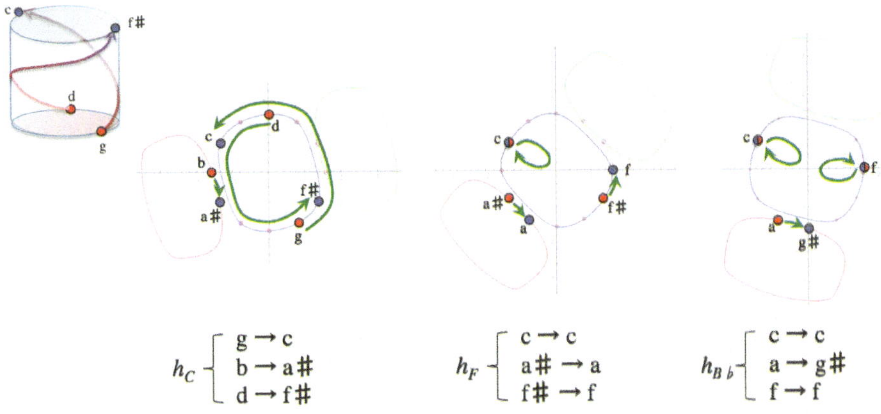

Fig. 80.8. The concatenation of three admissible parallel hypergestures, connecting $V_C = \{g, b, d\}$ to $II_{E_\flat} = \{f, g\#, c\}$.

Proposition 72 *With the above procedure, when applied to all fourth and fifth circle modulations for up to four circles, the resulting modulation steps coincide with the steps calculated in the classical model (coinciding with Schönberg's steps).*

The proof (which we omit here) is lengthy, but easy; one has to go through all possible admissible parallel hypergestures and then calculate the modulation steps as described above.

81

Hypergestures for Performance Stemmata

Summary. In the context of performance stemmata, different hypergestures correspond to different strategies of deformation from mother to daughter performances. We discuss and classify types of such strategies using topological obstructions in terms of singular hypergesture homology. This gives hypergesture homology a nice interpretation in terms of human performance practice.

– Σ –

81.1 Motivation, Terminology, and Previous Results

In Part VIII, a general theory of musical performances was developed as a background for the performance rubette of the software RUBATO® [689], also described in Part X. According to this approach, musical performance can be understood as a process of unfolding a primary "mother" performance into a tree of "daughter" performances according to a successively refined shaping of intermediate performances and applying *performance operators* that are typically specified according to given *weight functions* derived from rhythmical, motivic or harmonic analyses (other analyses not being excluded, these three types merely reflect the status quo of software implementation). It is straightforward that this evolutionary performance tree, termed *stemma*, is in fact a *local network* in the sense of [716] and [728], the transition processes playing the role of morphisms in local networks.

Although performance theory models the processual level of performative genealogy, it is not sufficiently explicit for the unfolding of performance when rehearsed by a musician. The transition from one level of sophistication to the next daughter level is a discrete "beaming" action without intermediate stages. Unfortunately, all research dealing with performance models is, to our knowledge, limited to this processual methodology. This is also the reason why our references are quite limited: there is nothing that we would be able to refer to when leaving this processual approach. But refer to Gerhard Widmer's overview of some important approaches to performance theory [1119]. Humans do not rehearse in this way, they tend to approximate a refined performance by successive, continuous deformation of a mother performance to a new daughter performance. In [728, 2.3.1] a topological approach to such a continuous deformation has been set up in terms of categorical gesture theory. In this paper that approach is made more concrete with regard to the specific performance operators involved in stemmatic deployment. It has been shown in Section 39.7 that an important class of performance operators involves the Lie derivative $L_{צ}w$ of *weight function* w with respect to a *performance vector field*[1] צ. We shall describe the performance hypergestures related to such operators.

[1] The beautiful Hebrew letter צ, "tsadeh", was introduced by Mazzola in 1992 for performance vector fields, motivated by the German "Tempo-Stimmungs-Feld", performance vector fields for tempo and tuning. The abbreviation "TS" for such fields later (in [714]) muted to the somewhat artificial notation $\mathfrak{Ts}$ because Hebrew letters were not available to the author. As this problem is solved now, we are able to use it and to put the old symbol to retirement.

G. Mazzola et al., *The Topos of Music III: Gestures*, Computational Music Science,
https://doi.org/10.1007/978-3-319-64481-3_26

In this context, different hypergestures correspond to different strategies of deformation from mother to daughter performances. We discuss and classify types of such strategies using topological obstructions in terms of singular hypergesture homology. This gives hypergesture homology a nice interpretation in terms of human performance practice.

To provide the reader with a more down-to-earth access to the mathematically and conceptually complex approach exposed in this paper, we shall present and discuss a concrete example in Section 81.5 after having worked through the theoretical setup.

81.1.1 Performance Stemmata and Performance Gestures of Locally Compact Points

Before we investigate the gestural aspects of stemmatic deployment, we want to recall the setup of stemma theory of performance as exposed in detail in Chapter 38. A performance stemma (LPS, also called *local performance score* in Section 35.4) is the formal description of a diagram $LPS : S \to PerCell$ over a tree digraph S (the *stemma tree*) whose values at vertices s in S are *cellular hierarchies* $\mathfrak{h}_s$, i.e., functors $\mathfrak{h}_s : H_s \to PerCell$ on a space category H_s with values in the category $PerCell$ of *performance cells*. The arrows $o : t \to s$ of S are then mapped to functions $LPS(o) : \mathfrak{h}_t \xrightarrow{\mathfrak{h}_o} \mathfrak{h}_s$ that map mother hierarchies $\mathfrak{h}_t$ to daughter hierarhices $\mathfrak{h}_s$ applying performance operators $\mathfrak{h}_o$, i.e., $\mathfrak{h}_o(\mathfrak{h}_t) = \mathfrak{h}_s$.

Let us recall these terms. The category $PerCell$ has performance cells as objects. A performance cell is a 5-tuple $C = (K, R, \mathfrak{Z}, I, \wp_I)$ consisting of a local composition $K \subset \mathbb{R}^{X.} \approx \mathbb{R}^n$, the cell's *kernel*, where $X. = X_1 X_2 \dots X_n$ is a sequence of n real-valued musical parameters; a *frame* R, i.e., a cube $R = [a_1, b_1] \times \dots [a_n, b_n] \subset \mathbb{R}^{X.}$, containing K; a locally Lipschitz-continuous *performance vector field* $\mathfrak{Z} : R \to TR$, where TR denotes the tangent bundle of R; an *initial set* $I \subset R$ such that for every element $k \in K$, the maximal integral curve $i(k)$ of $\mathfrak{Z}$ through k cuts I, and an initial performance $\wp_I : I \to PS$, with codomain a corresponding physical parameter space PS over $\mathbb{R}^n$, such that for any point $k \in K$ and any two points $a = i(k)(\alpha), b = i(k)(\beta)$ in I, we have $\wp_I(b) - \wp_I(a) = (\alpha - \beta).\Delta$, where $\Delta = (1, \dots 1)$ is the diagonal vector in the $\mathbb{R}^n$ underlying PS. Such a cell defines a well-defined performance map $\wp : K \to PS$ by $\wp(k) = \wp_I(l) - \alpha.\Delta$, if $i(k)$ hits $l \in I$ at curve parameter α.

A morphism $p : C_1 \to C_2$ in $PerCell$ is a standard coordinate projection of underlying real spaces $p : \mathbb{R}^{n_1} \to \mathbb{R}^{n_2}$ such that $p(K_1) \subset K_2$ and $p(R_1) \subset R_2$, p induces a morphism of vector fields $\mathfrak{Z}_1 \to \mathfrak{Z}_2$, and every point of $p(I_1)$ can be reached via integral curves of $\mathfrak{Z}_2$ from the initial set I_2. This concept guarantees that the performance maps commute, i.e., $p \circ \wp_1 = \wp_2 \circ p$. A hierarchy is a morphism $\mathfrak{h} : H \to PerCell$ starting from a space category H, whose objects are a set of symbolic musical parameter spaces $U = \mathbb{R}^{X.}$, such as, for example, $U = \mathbb{R}^{EHLD}$, where E is onset, H pitch, L loudness, and D duration. The category must be a lattice, i.e., it is closed under finite unions and intersections of the sets defining the parameter sequences, and have a maximal element $Top(H)$ (the *top space* of H). For $U = \mathbb{R}^{X.}, V = \mathbb{R}^{Y.}$ the morphisms $U \to V$ are those space couples with partial sequences $Y. \subset X.$.

For example, the default piano hierarchy has spaces

$$\mathbb{R}^{EHLD}, \mathbb{R}^{EHL}, \mathbb{R}^{EHD}, \mathbb{R}^{ELD}, \mathbb{R}^{EH}, \mathbb{R}^{EL}, \mathbb{R}^{HL}, \mathbb{R}^{ED}, \mathbb{R}^{E}, \mathbb{R}^{H}, \mathbb{R}^{L}.$$

Typically, D is not an independent parameter since duration is a function of onset. The *fundamental*, i.e., minimal, spaces $\mathbb{R}^E, \mathbb{R}^H$ and $\mathbb{R}^L$ of this hierarchy are performed via one-dimensional tempo, intonation, and intensity vector fields, respectively. The hierarchy functor $\mathfrak{h}$ must map a space $U = \mathbb{R}^{X.}$ from H to a performance cell $\mathfrak{h}(U)$ which has its kernel in U, and the projections defined by the morphisms $U \to V$ must define performance cell morphisms.

In [728] we have given the example of performance stemmata as an illustration of gestures of locally compact points, see also Section 65.3.1. Our locally compact topological categories were $\mathcal{C} = \mathbf{Frame}_{X.}$ of n-dimensional frames defined over the parameter sequences $X.$ of length n. These were interpreted in the sense that all dimensions of this space refer to a musical parameter given by a fixed choice $X.$ of n parameters. The category $\mathbf{Frame}_{X.}$ has the inclusions of frames as morphisms, and its topology is defined by the Euclidean metric of the representation of frames as points in $\mathbb{R}^{2n}$. We look at a particular presheaf $F_{X.} : \mathbf{Frame}_{X.}^{opp} \to \mathbf{Top}$ of $\mathcal{C}^1$ vector fields, i.e., if c is a n-dimensional frame for $X.$, then $F_{X.}(c) = \{v : c \to$

$Tc|v$ is a $\mathcal{C}^1$ vector field$\}$, morphisms being transformed to restrictions of vector fields. Since c is compact, the metric $(v,w) = max_c(d(v(x),w(x))$ for the Euclidean distance d on vectors in $\mathbb{R}^n$ defines a locally compact topological space, i.e.,

$$F_{X.} : \mathbf{Frame}_{X.}^{opp} \to \mathbf{LCTop},$$

where **LCTop** denotes the topological category of locally compact Hausdorff topological spaces. This presheaf $F_{X.}$ will play a crucial role in the construction of hypergestures of locally compact points over $\mathbf{Frame}_{X.}$ as described in [728]. The role of $F_{X.}$ is to control the stemmatic unfolding of performance vector fields $\mathfrak{S}$, appearing in performance cells of hierarchies of performance stemmata LPS, under the action of specific performance operators $LPS(o) : \mathfrak{h}_s \xrightarrow{\mathfrak{h}_o} \mathfrak{h}_t$. The example in [728] focused on frames and performance fields, although we have to consider entire performance *cells* in stemma theory. We want however to focus on the two components frame and performance field, and suppose that the other components are automatically defined according to a standard algorithm. We now focus on the spatial hierarchy $\mathfrak{h}_s : H_s \to \int_{End(F)}^{lc}$ associating with selected spaces U over parameter sequence $X.$ of H_s performance vector fields $\mathfrak{S}_s(U) : c_U \to Tc_U$ in $c_U@F_{X.}$.

The target category of the spatial hierarchy $\mathfrak{h}_s$ was defined in [728], but for the sake of not torturing the reader with those generalities about topological categories, we want to give a short-hand definition of $\int_{End(F)}^{lc}$ here. The integral sign is used in category theory to designate categories of points. In our case, we are looking at the topological category $\mathbf{Frame}_{X.}$ of frames in a space specified by the sequence $X.$ of parameter names, as introduced above. When talking about points in $\mathbf{Frame}_{X.}$, one addresses contravariant functors (presheaves) $F : \mathbf{Frame}_{X.}^{op} \to \mathbf{LCTop}$ which are continuous for the underlying topologies of these topological categories. Since our functor F evaluates to sets of $\mathcal{C}^1$ vector fields on frames, it is a candidate for such a presheaf.

In this context, a point is an element $x \in F(c)$ for a frame object c. Using Yoneda's Lemma, stating that $F(c) \xrightarrow{\sim} Nat(@c, F)$, we may identify such a point by the associated natural transformation $x : @c \to F$. The category of such points (all for one and the same F, all for the frames in $\mathbf{Frame}_{X.}$) has these morphisms from point $v_1 : @c_1 \to F$ to point $v_2 : @c_2 \to F$. They are pairs $(\alpha : c_1 \to c_2, \phi : F \to F)$ such that the diagram

$$\begin{array}{ccc} @c_1 & \xrightarrow{v_1} & F \\ {\scriptstyle\alpha}\downarrow & & \downarrow{\scriptstyle\phi} \\ @c_2 & \xrightarrow{v_2} & F \end{array}$$

commutes. Here, ϕ is an endomorphism of F on tangent bundles. This data defines the category $\int_{End(F)}^{lc}$. The index $End(F)$ refers to the endomorphisms ϕ of F, while the exponent lc refers to the locally compact target spaces, a condition for obtaining a topological category, as mentioned above. It is shown in [728] that such a category $\int_{End(F)}^{lc}$ is in fact a topological category.

81.2 Gestures with Lie Operators in Stemma Theory

In [728] performance operators were described that relate to the situation described in the above commutative diagram; see also Section 65.3.1. This diagram is precisely what we have explained at the end of the preceding section as being a morphism in the category $\int_{End(F)}^{lc}$. This means that the α-restricted field v_1 over c_1 is given by $\phi^{-1} \circ v_2 \circ \alpha = v_1$.

Here we want to make this concrete and look for gestures and hypergestures related to such operators. We start on a cellular hierarchy $\mathfrak{h} : H \to PerCell$ which is supposed to be constructed in a previous performance process. Our operator will be of Lie type, a so-called *basis specialization* operator (see Section 39.7.1). Lie type operators seem to play a crucial role in performance theory. This operator acts on a performance field $\mathfrak{S}_W$ defined on a space W of H. We suppose that $W = U \oplus V$, where U is in H (but V not necessarily in H). For example, in the default piano hierarchy, we could take $U = \mathbb{R}^E, V = \mathbb{R}^D, W = \mathbb{R}^{ED}$, where $\mathbb{R}^D$ is not in the piano hierarchy.

This operator uses an analytical weight, i.e., a $\mathcal{C}^2$ function $\Lambda : c_U \to \mathbb{R}$. Such functions are available from musical analysis; in the RUBATO® software they are provided by analytical rubettes, see Chapter 41. Denote by $\mathcal{F}_n(c_U)$ the real vector space of $\mathcal{C}^n$-functions on c_U. Performance theory is not limited to given weights, but also uses linear combinations to shape performance fields. For example, the one-dimensional tempo field may be shaped using a weighted sum $0.3\Lambda_1 + 0.7\Lambda_2$ of a melodic weight Λ_1 and a harmonic weight Λ_2, meaning that we are working in the vector space $\mathbb{R}\Lambda_. \subset \mathcal{F}_n(c_U)$ generated by a family $\Lambda_.$ of analytical weights.

The Lie-type basis specialization operator we are using here also uses an affine "directional" endomorphism $Dir : V \to V$ and is defined by the formula

$$\mathfrak{Z}_{W,\Lambda,Dir} = \mathfrak{Z}_W - L_{\mathfrak{Z}_U}\Lambda.i_V Dir$$

where $\mathfrak{Z}_U$ is the performance field on U, acting on Λ as a derivation via its Lie representation, and $i_V : V \to W$ is the embedding map. Since U and V share no common parameters, $(\mathfrak{Z}_{W,\Lambda,Dir})_U = \mathfrak{Z}_U$. The operator only changes vector field components in V. It is immediate that this operator is continuous and linear in $\mathfrak{Z}_W$ and additive in Λ in the sense that

$$\mathfrak{Z}_{W,\Lambda_1+\Lambda_2,Dir} = (\mathfrak{Z}_{W,\Lambda_1,Dir})_{\Lambda_2,Dir}.$$

Since $\mathfrak{Z}_{W,0,Dir} = \mathfrak{Z}_W$, this means that $\mathfrak{Z}_{W,-\Lambda,Dir}$ results from the inverse operator to $\mathfrak{Z}_{W,\Lambda,Dir}$; the operator is an automorphism of the tangent bundle as required in our general setup. This means that $\mathfrak{Z}_{W,\Lambda,Dir}$ corresponds to the inverse $\phi^{-1}_{\Lambda,Dir}$ of the operator

$$\phi_{\Lambda,Dir}(\mathfrak{Z}_W) = \mathfrak{Z}_{W,-\Lambda,Dir} = \mathfrak{Z}_W + L_{\mathfrak{Z}_U}\Lambda.i_V Dir.$$

We therefore have a special case of the general setup: The daughter field is generated by the frame restriction $c_1 \subset c_2$ and the weight Λ together with the directional endomorphism via the basis specialization operator. These facts allow for gestural constructions as follows: If $\gamma : I \to \mathcal{F}_2(c_U)$ is a continuous curve of $\mathcal{C}^2$-weights, starting at $\gamma(0) = 0$ and ending at $\gamma(1) = \Lambda$, then $\gamma^*(t) = \mathfrak{Z}_{W,\gamma(t),Dir}$ deforms the mother field $\mathfrak{Z}_W$ to the daughter field $\mathfrak{Z}_{W,\Lambda,Dir}$.

81.3 Connecting Stemmatic Gestures for Weights and Performance Fields

We now want to construct gestures of stemmata defined by basis specialization operators. The stemma starts at a given hierarchy $\mathfrak{h} : H \to PerCell$ as above, using the space configuration $W = U \oplus V$ for hierarchy spaces U, W. The stemma starts on the frame c_W and is defined on a digraph of subframes of c_W as follows:

Calling $c_W = c_0$, we start with a sequence $c_{00}, \ldots c_{0k_0} \subset c_0$ of mutually disjoint subframes. In performance theory this procedure defines a split of the composition into relevant, mutually disjoint, subcompositions, for example into left hand and right hand parts, and then into four periods of left or right hand in an $AABA$ form, and then for each period into one subframe for each measure, etc. For each such subframe c_{0i} we repeat the construction by a sequence $c_{0i0}, \ldots c_{0ik_{0i}} \subset c_{0i}$ of mutually disjoint subframes, etc. This defines a stemmatic tree digraph c_S of frames and subframes, where we write $c_t \to c_s$ if c_s is a direct subframe of c_t in our construction (s, t denote sequences of indices used to define the tree).

Given c_S, we take a finite sequence $\Lambda^s_. = \Lambda^s_1, \Lambda^s_2 \ldots \Lambda^s_{l_s}$ of analytical weights for each vertex frame c_s, except the top frame c_0. The role of such a sequence is to create a daughter performance on c_s from the performance on the stemmatic predecessor c_t of c_s. This setup is classical in the sense that the initial weight is 0 while the final one is $\sum_i \Lambda^s_i$. Accordingly the initial (mother) performance field (over W) is $\mathfrak{Z}_t$, the field inherited from c_t and restricted to the subframe c_t, and the final (daughter) field is $\mathfrak{Z}_{t,\sum_i \Lambda^t_i,Dir}$, for a directional endomorphism Dir, which we suppose is chosen once, forever to ease the discourse here.

This setup is however not a gestural one since no continuous curves from mother to daughter performance are defined. To this end, we define a domain of parameters that are available for a continuous deformation in gestural curves. For each pair $c_t \to c_s$, we take the cube I^{l_s} and allow for weights of form $\Lambda^s(x) = \sum_i x_i \Lambda^s_i$ for $x \in I^{l_s}$. We further allow for continuous deformation of frames for $\xi \in I$, defining $c_t(\xi)$ to be the frame of parameter ξ on the straight line from c_t to c_s in their representation in $\mathbb{R}^{2n}$, $n = dim(W)$. This defines an $l_s + 1$-dimensional parameter space $p(s) = I^{l_s+1}$, the first l_s dimensions being assigned to parameter sequences x for weights and one parameter ξ for inter-frame positions. We always start at $(x, \xi) = (o, o) \in p(s)$ and terminate at $(x, \xi) = (\Delta, 1) \in p(s)$, $\Delta = (1, 1, \ldots 1)$. We do however always, for all intermediate performances between c_t and c_s, restrict the kernels K_t to the kernel K_s defined by the final restriction $c_s \subset c_t$.

To generate gestures, we first introduce a topological space of parameters, the *parameter stemma*. It is the colimit $P_S = \text{colim}\ p(s)$ of the topological spaces $p(s)$ which are glued together as follows: If $c_t \to c_s$ is an arrow of c_S, then $(\Delta, 1) \in p(t)$ is identified with $(0, 0) \in p(s)$. We intuitively replace each arrow $c_t \to c_s$ in c_S by the cube $p(s)$, see Figure 81.1.

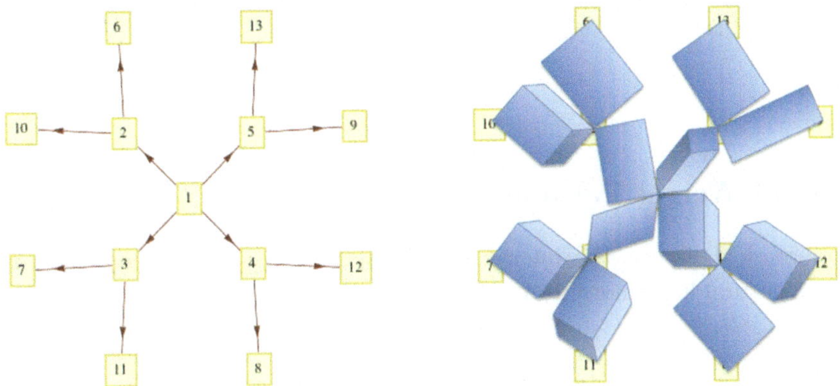

Fig. 81.1. Left: a stemmatic digraph c_S; right: a parameter stemma space P_S for c_S. Observe that it is not necessary to have growing dimensions of the parameter cubes when stepping down from the mother performance. In fact, it could happen that an early stage needs more weights than a later one. For example, the final performance operator might only work very delicately on some local shaping of loudness as a function of a single rhythmical weight, whereas an initial shaping might use melodic, harmonic and rhythmical weights at once.

Next we turn P_S into a topological category. Its objects are the points of P_S, and the morphisms $p \to q$ are those pairs (p, q) such that if p is in cube $p(t)$ then q is in a cube $p(s)$ of a vertex s equal to or following t in the stemmatic tree. Suppose we have chosen a family of weights $\Lambda^s_\cdot$ for each vertex $s \neq 0$ of the stemma, call this choice Λ^S. We have a topological functor

$$\int \Lambda^S : P_S \to \int^{lc}_{End(F)}$$

defined as follows. Suppose that a point $p = (x, \xi) \in P_S$ is in the cube $p(s)$. It is then connected to the tree's source c_0 by a unique path $c_0 \to c_1 \to \ldots c_t \to c_s$, where c_t is the predecessor of c_s. Then we set $\Lambda^S(p) = x.\Lambda^s_\cdot + \sum_{i=1}^t \Delta \Lambda^i_\cdot$ with the notation $x.\Lambda^s_\cdot = \sum_i x_i \Lambda^s_i$. We then set

$$\int \Lambda^S(p) = \mathfrak{Z}_{W,\Lambda^S(p),Dir}$$

together with its intermediate frame $c_t(\xi)$ between c_t and c_s defined by ξ. Formally speaking, this is the point $\int \Lambda^S(p) : @c_t(\xi) \to F$. If $p \to q$ is a morphism in P_S, the associated morphism

$$\int \Lambda^S(p) \to \int \Lambda^S(q)$$

is the basis specialization operator associated with the difference of weights $\Lambda^S(q) - \Lambda^S(p)$.

We are in principle interested in gestures in $\int^{lc}_{End(F)}$, but there are several reasons for defining such gestures via the functor $\int \Lambda^S$. To begin with, our performance operators are defined using linear combinations of previously calculated analytical weights, therefore the coefficients of such linear combinations are a natural data set. Second, calculating homology of hypergestures in $\int^{lc}_{End(F)}$ is difficult since the Lie derivative can produce uncontrollable functions due to the local gradients of weights with respect to the local vectors of $\mathfrak{Z}_U$. Even though the map $\int \Lambda^S$ can be generically injective, it will not be open.

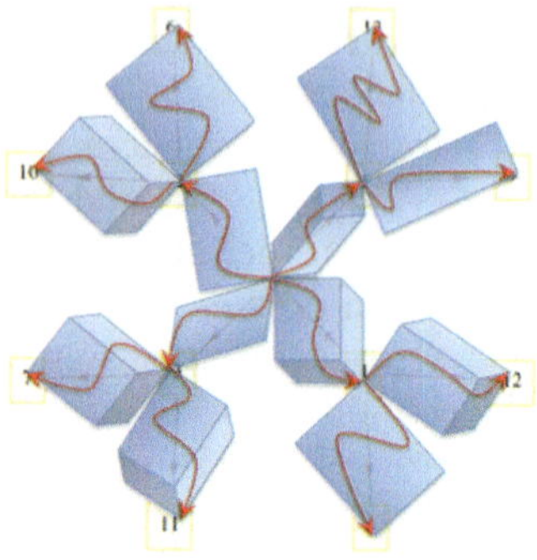

Fig. 81.2. A parameter stemma gesture composed of curves connecting initial and final points for weight operators.

81.4 Homology of Weight Parameter Stemmata

For these reasons we shall discuss the integer homology $H_*(P_S, \mathbb{Z})$ of hypergestures in P_S rather than $H_*(\int^{lc}_{End(F)}, \mathbb{Z})$. The natural homomorphism

$$H_*(\int \Lambda^S) : H_*(P_S, \mathbb{Z}) \to H_*(\int^{lc}_{End(F)}, \mathbb{Z})$$

associated with $\int \Lambda^S$ however connects simpler homology over the parameter stemma with more difficult homology of locally compact points.

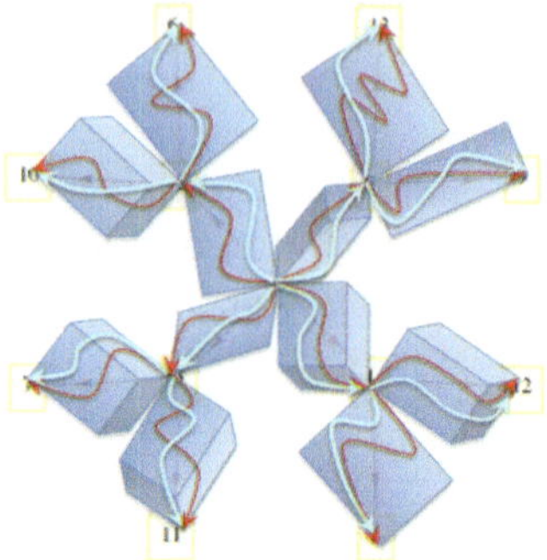

Fig. 81.3. Two gestures in the same parameter stemma.

We investigate gestures that represent a stemmatic unfolding from the primary mother performance to the ramifications of the tree's leaf daughters. Such a gesture has the shape of a digraph morphism $g : c_S \to \overrightarrow{P_S}$, where $\overrightarrow{P_S}$ denotes the digraph of curves in the topological category P_S [723], but with the boundary condition that the vertices of c_S map to the gluing points in P_S; more precisely, if $f : c_t \to c_s$ is an arrow in c_S, then $g(f) : \nabla \to P_S$ is a curve in $p(s)$ starting at 0 and terminating at Δ, see Figure 81.2. Each such curve represents a deformation path from a mother to a daughter performance, and this is—musically speaking—a continuous trajectory of the local rehearsal process. If we consider two such gestures $g_1, g_2 : c_S \to \overrightarrow{P_S}$, as shown in Figure 81.3, the question arises whether they are essentially the same procedure or not. A natural condition for such an equivalence would be that they are homotopic, i.e., initial and final values of a hypergesture $h \in \uparrow \overrightarrow{@} c_S \overrightarrow{@} P_S$.

To understand this notation, recall that the $\Delta \overrightarrow{@} K$ denotes the topological category of gestures on a digraph Δ with values in a topological category K. Therefore, $\uparrow \overrightarrow{@} c_S \overrightarrow{@} P_S$ denotes the topological category of gestures on digraph $\uparrow$ with values in the category $c_S \overrightarrow{@} P_S$ of gestures on the digraph c_S with values in P_S. In other words, we are looking for gestures that have their vertex values in common and are equivalent modulo boundaries of hypergestures. The former condition of course implies that their difference is a 1-cycle of singular homology.

This means that we are not interested in all hypergestures in this singular homology, but only in the subspaces generated by the above ones having their vertex values in common. The formal setup is therefore as follows. Let us look at the singular chain spaces involved in this homology. In the generalized homology theory for hypergestures (Chapter 63) we have the following boundary homomorphism diagram:

$$\mathbb{Z} \uparrow \overrightarrow{@} c_S \overrightarrow{@} P_S \xrightarrow{\partial_2} \mathbb{Z} c_S \overrightarrow{@} P_S \oplus \mathbb{Z} \uparrow \overrightarrow{@} P_S \xrightarrow{\partial_1} \mathbb{Z} P_S$$

with

$$\partial_2(g) = g_0^\square - g_1^\square = (g(1) - g(0), -g_1^\square)$$

on a hypergesture $g \in \uparrow \overrightarrow{@} c_S \overrightarrow{@} P_S$ where $g_0^\square \in \mathbb{Z} c_S \overrightarrow{@} P_S, -g_1^\square \in \mathbb{Z} \uparrow \overrightarrow{@} P_S$, and with

$$\partial_1(h, k) = h^\square + k^\square = h^\square + k(1) - k(0).$$

on a pair of gestures $h \in c_S \overrightarrow{@} P_S, k \in \uparrow \overrightarrow{@} P_S$. Here $?^\square$ is the face operator. It generalizes by a recursive formula the classical face operator to arbitrary digraphs. For example, if c_S is the digraph shown to the left Figure 81.1, then the face operator yields

$$\begin{aligned} g^\square = {} & 5401944 g[1] + 55044(g[2] + g[3] + g[4] + g[5]) - \\ & 80142(g[6] + g[7] + g[8] + g[9] + g[10] + g[11] + g[12] + g[13]). \end{aligned}$$

In the following discussion, we want to specialize the homological setup to gestures and hypergestures that are of interest in the performance shaping. We are looking at gestures $h \in c_S \overrightarrow{@} P_S$ with fixed values on the vertices of c_S, i.e., $h|V_{c_S} = \xi$, meaning that for all vertices v of c_S, $h(v)$ is the gluing point in P_S corresponding to v. Denote by $\mathbb{Z} c_S \overrightarrow{@}_\xi P_S$ the subgroup of chains generated by such gestures in $\mathbb{Z} c_S \overrightarrow{@} P_S$. Similarly, we restrict the second chain space $\mathbb{Z} \uparrow \overrightarrow{@} c_S \overrightarrow{@} P_S$ to the subspace $\mathbb{Z} \uparrow \overrightarrow{@} c_S \overrightarrow{@}_\xi P_S$, and we have the restricted boundary map

$$\mathbb{Z} \uparrow \overrightarrow{@} c_S \overrightarrow{@}_\xi P_S \xrightarrow{\partial_2} \mathbb{Z} c_S \overrightarrow{@}_\xi P_S \oplus \mathbb{Z} \uparrow \overrightarrow{@} P_S \xrightarrow{\partial_1} \mathbb{Z} P_S.$$

The Escher Theorem that is needed in the definition of boundary maps here establishes an isomorphism of topological categories,

$$\uparrow \overrightarrow{@} c_S \overrightarrow{@}_\xi P_S \xrightarrow{\sim} c_S \overrightarrow{@}_{\xi *} \uparrow \overrightarrow{@} P_S,$$

where ξ_* is the gesture derived from ξ by replacing its values by constant loops.

Let us now calculate the homology in this context, more precisely the homology group $H_1^*(c_S)$ generated by the cycles of differences $h_1 - h_2$ of gestures $h_1, h_2 \in c_S \overrightarrow{@}_\xi P_S$. Call Z_1^* the group of cycles (subgroup of Z_1) generated by these differences. We have to calculate the boundary image group B_1^* stemming from the second chain group $\mathbb{Z} \uparrow \overrightarrow{@} c_S \overrightarrow{@}_\xi P_S$, and then its intersection with Z_1^* to derive $H_1^*(c_S)$.

If $g \in \uparrow \overrightarrow{@} c_S \overrightarrow{@}_\xi P_S$, then we have $\partial_2(g) = (g(1) - g(0), -g_1^\square) = (g(1) - g(0), \eta)$, where η is constant since it only depends on ξ. Moreover, one easily sees that $(0, \eta) = \partial_2(\xi_*)$. Therefore the ∂_2-boundaries of gestures are linear combinations of $(0, \eta)$, the boundary of ξ_*, and the differences $(h_1 - h_2, 0)$ of gestures that are related to each other by a homotopy given by a hypergesture. Let us now define a basis of the space Z_1^* which takes care of such homotopy: Partition the set of all $h \in c_S \overrightarrow{@}_\xi P_S$ into homotopy classes $C_i, i \in J + 1$. Choose one representative h_i for every homotopy class i. Take this family of generators: for every class i take all $h \in i, h \neq h_i$, and then the family of differences $(h - h_i, 0)_h$. Then choose one representative h_0 of any class 0 and also take the family of differences $(h_i - h_0, 0)_i, i \neq 0$. Finally, take the vector $(0, \eta)$. Then Z_1^* is generated by these families. The reason for this choice is that the two families are linearly independent ∂_1 cycles, and any difference $(h - h', 0)$ is contained in the free group they generate. In fact, if h is in homotopy class i, h' in class j, then $h - h' = (h - h_i) + (h_i - h_0) - (h_j - h_0) - (h' - h_j)$. This implies that all boundaries are in the subgroup generated by the families $(h - h_i, 0)_h$ and by $(0, \eta)$. Therefore the homology group $H_1^*(c_S) = Z_1^*/(B_1^* \cap Z_1^*)$ is free of rank J, proving the formula:

$$H_1^*(c_S) \xrightarrow{\sim} \mathbb{Z}^{\oplus J}$$

Let us terminate this homological calculation by exhibiting the number J of homotopy classes in the stemmatic tree. Recall that we musically wanted to identify equivalent strategies of rehearsals related to the given stemma. For every arrow $f : c_t \to c_s$ in c_S, we are given a cube $p(s)$ of weight parameters. We are selecting curves in these cubes, reaching from 0 to the diagonal vector $\Delta = (1, \ldots 1)$ introduced above. The musical obstructions in such a cube could come from certain forbidden weight parameter combinations. For example, it could be impossible or not desired to play an intermediate weight combination of a melodic weight Λ_1 and a rhythmical weight Λ_2 with weight parameters 0.3 and 0.4 for these weights, respectively. Then we would have a hole in the plane I^2 defined by these two weights, and therefore not all curves would be homotopic. Suppose that for each $p(s)$, some obstructions are defined and generate a subspace $p^*(s)$ with $\pi(s)$ homotopy classes of curves from 0 to Δ. Then the total number of homotopy classes of the parameter stemma space is $\pi_S = \prod_s \pi(s)$ and $rk(H_1^*(c_S)) = \pi_S - 1$. In other words:

$$H_1^*(c_S) \xrightarrow{\sim} \mathbb{Z}^{\oplus \pi_S}.$$

81.5 A Concrete Example

The mathematical and conceptual complexity of the previous calculations and reflexions should be illustrated by a concrete example in order to enable the reader to realize the connections of this theory to practical aspects of performance.

We choose the composition *Träumerei* T, the seventh piece in Robert Schumann's op. 15, *Kinderszenen.* It consists of four eight-measure periods $T = A, A', B, A''$ (including the repetition A' of the first period A). Each period is split into two four-measure phrases: $A = A_1, A_2, A' = A'_1, A'_2, B = B_1, B_2, A'' = A''_1, A''_2$. We shall use a stemmatic digraph c_S that corresponds to this ramification. It is the same digraph we used to exemplify the general theory in Figure 81.1, with this correspondence of nodes: $T = 1, A = 2, A' = 3, B = 4, A'' = 5, A_1 = 6, A_2 = 10, A'_1 = 7, A'_2 = 11, B_1 = 8, B_2 = 12, A''_1 = 9, A''_2 = 13$. We shall focus on the parameter sequence $X. = E, D$, with parameter spaces $W = \mathbb{R}^{ED}, U = \mathbb{R}^E, V = \mathbb{R}^D$, where W and U are in the default piano hierarchy, whereas V is not. This means that our performance shaping in this example deals with tempo (relating to $Tempo(E) = \mathfrak{T}_E$ at onset E) and articulation (relating to the D-component of the tempo-articulation field $\mathfrak{T}_{E,D}$; pay attention, there is no independent articulation performance field $\mathfrak{T}_D$ since articulation is a function of onset in our setup).

The frames on our parameter space are the objects of the frame category $\mathbf{Frame}_{E,D}$. We have thirteen frames $c_1, c_2, \ldots c_{13}$, corresponding to the parameter rectangles in $\mathbb{R}^{ED}$ defined by the parts $T, A_1, \ldots, A''_2$, with their inclusion morphism, respectively, as follows: $c_2, c_3, c_4, c_5 \to c_1$, $c_6, c_{10} \to c_2$, $c_7, c_{11} \to c_3$, $c_8, c_{12} \to c_4$, $c_9, c_{13} \to c_5$.

Next we suppose that the preliminary analytical work has provided us with three weights: a rhythmical weight $\Lambda_r(E)$, a melodic weight $\Lambda_m(E)$, and a harmonic weight $\Lambda_h(E)$, each weight being a C^1 function of onset E only. This is what the RUBATO® software effectively calculates, see Chapter 41. As is standard in performance theory, it may happen that we don't apply the given weights, but derived ones to be able to express what the performative shaping addresses. Let us suppose for example that the four periods c_2, c_3, c_4, c_5 are shaped in their tempi using weights derived from rhythmical and melodic weights. Suppose we want to shape the mother tempo $\mathfrak{T}_E$ on the four periods $c_i, i = 2, 3, 4, 5$, by a factor $\gamma(E) = \Lambda_r(E) + \Lambda_m(E)$, yielding four daughter tempi $\gamma(E)\mathfrak{T}_E|_{c_i}, i = 2, 3, 4, 5$, on c_2, c_3, c_4, c_5. In order to achieve this deformation of tempo by use of the Lie operator, we take the new weight function $\Lambda(E) = \int \frac{1-\gamma}{E}$, and we take the directional morphism $Dir = Id_E$. Then we get the desired formula $\gamma(E)\mathfrak{T}_E = \mathfrak{T}_E - L_{\mathfrak{T}_E} \Lambda.Id_E$.

Suppose now that we have shaped the performance cells for the four periods according to the above (or some other) Lie operator approach. Let us then look at the shaping procedure of one of the leaves of the stemma digraph c_S to see the homological situation more concretely. We know from harmonic analysis (see for example Alban Berg's famous analysis [112]) that the first phrase B_1 of the B period involves modulatory movements from F major to G minor, $B_\flat$ major, and D minor, back to F major. We therefore want to shape articulation (as said above: the duration component of the field $\mathfrak{T}_{ED}$) using information pertaining to the melody weight Λ_m as well to the harmonic weight Λ_h. We start at the performance field $\mathfrak{T}^{c_4}_{ED}$ given on the B period. At the end of a gesture g leading from the B period (on c_4) to the B_1 phrase (on c_8), we have the B_1 performance field $\mathfrak{T}^{c_8}_{ED} = \mathfrak{T}^{c_4}_{ED} - L_{\mathfrak{T}^{c_4}_{ED}}(\Lambda_m + \Lambda_h)i_D Id_D$ (the directional morphism being taken as identity to make things simpler). According to our general setup, this gesture is parametrized by a gesture in the cube I^3 for three parameters: ξ for the shrinking of c_4 to c_8, and λ, μ for the mixed combination $\lambda \Lambda_m + \mu \Lambda_h$.

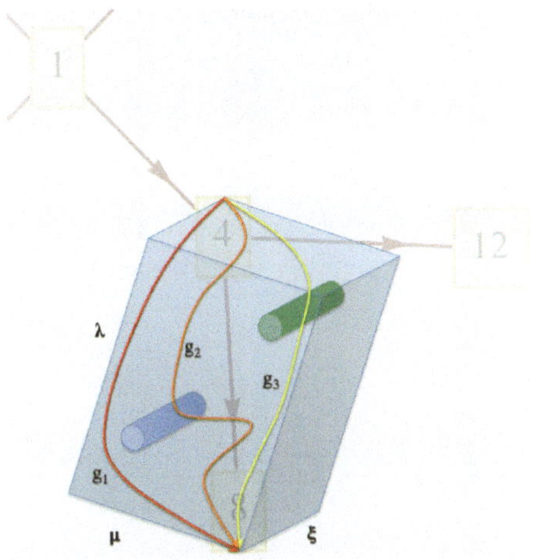

Fig. 81.4. Three gestures in homotopically inequivalent positions with respect to two parametric holes.

The critical object is the gesture g that moves from $(0,0,0)$ in the cube to the $\Delta = (1,1,1)$ value of our three parameters ξ, λ, μ, i.e., $g(0) = (0,0,0), g(1) = \Delta$. Refer to Figure 81.4 for the following discussion. The movement of a gesture g has a meaning for the performer, which is that he/she starts from the performance that is defined at value $(0,0,0)$, i.e., the period's previous performance, and now steps to the refined performance of B_1, using the two weights, melodic and harmonic, to reshape articulation. In a naive approach, one would guess that whatever the pianist does in this creation is essentially the same; the process is to just introduce successively articulations that are shaped by mixed melodic and harmonic weights. But when the pianist tries to play according to those weights, it may happen that he/she cannot play any linear combination $\lambda \Lambda_m + \mu \Lambda_h$. In Figure 81.4, we have inserted two holes in the cube, where there are regions of λ, μ parameter combinations, which cannot be played. Why not? Because it may be impossible for the pianist to articulate according to a strong melodic contribution versus a weak harmonic one, or vice versa, a strong harmonic combined with a weak melodic contribution. In other words, the pianist may only be capable of moving from $(0,0,0)$ to Δ on a curve that does not include such unbalanced contributions of the two weights. In Figure 81.4, we have drawn three gestures, g_1, g_2, g_3, where the first and third are impossible since they move through regions of unbalanced contributions, g_1 below the left hole, or g_3 above the right hole. Only the second curve g_2 is sufficiently balanced to be playable by the pianist. But this is a situation of homotopy: The two holes define three homotopy classes, each being represented by one of the three gestures g_1, g_2, g_3. The contribution to the homology group rank of this cube is dual: There are three classes of continuous transgressions targeting at a defined performance of phrase B_1, but we have only one good gesture homotopy class, the two others are bad ones.

81.6 A Final Comment

Singular homology of hypergestures offers a first precise classification of types of rehearsal strategies in musical performances if they are built upon stemmatic deployment using Lie-type performance operators. Although this looks quite abstract, the gestural expressivity could be discussed in a rigorous manner without being detached from practical consequences. The homotopy concept within spaces of weight parameters is a rather intuitive account of rehearsal strategies.

Check for updates

82

Composing and Analyzing with the Performing Body

Summary. In this chapter, we tackle both analysis and composition as reciprocal processes to investigate the performer's body. We argue that performers, more specifically, the performers' bodily gestures, are key to the critical understanding and the creation of music. This chapter contains three parts: first, we will investigate the concept of embodied musical gestures through a range of inter-disciplinary scholars, ultimately defining a concept that is useful and fruitful in discussing performance. In the second part, we will use Toru Takemitsu's *Rain Tree Sketch II for Piano* (1994) as a testing ground for analyzing with the performative body as the starting point. And lastly, we will discuss how composing with performative gestures in my composition *Sheng* (2016) for piano, audience's smartphones, and fixed audio playback elicits the cross-modal, inter-sensory nature of embodied musical gestures.

Indeed, the concept of embodied musical gesture has the potential to dissolve the artificial fractures between the activities of thinking, creating, and doing. Analyzing and composing with the performing body do away with this mind-body split, offering refreshing and generative insights that do justice to the physical nature of music making.

– *Σ* –

Theorists are thinkers, composers are creators, performers are doers. Over-simplified as they are, these designations nevertheless sum up our perception of each of these musical sub-disciplines in singular descriptors. Each of these three distinctive roles in the practice and study of classical music has its own exclusive preoccupations: theorists write about music, composers create it, and performers concretize it in real time. However, a silent power differential lurks in the background that renders the task of performers—the doers—as less serious, even less intelligent. Composers generate original works as scores that are played by performers and analyzed by theorists; theorists produce writings that provide insights that inspire new compositional aesthetics and performative interpretations. Meanwhile, the majority of the output of performers—performances—is a lot more ephemeral, vanishing only hours into thin air after it begins. Partially as a consequence of these differences in output formats, composers and theorists are regarded to be at the forefront of creative and critical enquiry respectively; in contrast, the contribution of performers in pushing the envelope of musical innovation does not seem to be quite on par.

Traditionally, performers in this tripartite relationship are seen to be purely at the receiving end: they simply play the composers' works, assisted by theorists who hold up illuminating torches that shine forth meaning. While an analyst may rely on recordings of performances to get an impression of her work at hand and a composer may ask for a performer's feedback on the playability of her piece, performers themselves play a secondary role in conceiving of music. Indeed, performers are merely doers; they are laborers whose work is primarily done through their practical, physical skills.

Yet, to think that performers cannot contribute to analytical and creative enquiry is to entirely miss the point of music making. Without performers, a score may as well be an abstract piece of art. Without enlivening music through the arms, hands, fingers, and mouths of performers, theorists can only hope to conjure up sounds in their heads. Performers, with their skilled bodies, should be at the front and center

G. Mazzola et al., *The Topos of Music III: Gestures*, Computational Music Science,
https://doi.org/10.1007/978-3-319-64481-3_27

of analysis and creation. Theorists and composers must move away from conceiving of their tasks as mind-focused in order to address the performative, bodily connection that has traditionally been missing.

Surely, since Susan Cusick famously decried the "mind/body problem" in music theory and musicology in the 1990s, the field of performance-related musical analysis has slowly gained momentum [227]. Take, for instance, Judith Lochhead and George Fisher's article "Analyzing from the Body," which inspired the title of this chapter, where they approach two contrasting works for clarinet and piano by Joan Tower and Johannes Brahms from a performative perspective [323]. Or Elizabeth Le Guin's analysis of Baroque composer Luigi Boccherini's cello sonatas from the point of view of herself as a performer, which offers fresh, groundbreaking insights into the understanding of this body of work, under-appreciated precisely because of the inadequacies of traditional analyses [592]. Conversely, a movement of contemporary composers is focusing on the compositional potentiality of the performer's bodily gestures by giving explicit, often unusual bodily instructions. Celeste Oram's *Toccata/Bruise* (2016) for piano and Mark Applebaum's *Aphasia* (2010) for an actor performing in sign language are two such examples where deliberate, instructive performative gestures foreground the corporeality of the performer on stage.

In this chapter, we tackle both analysis and composition as reciprocal processes to investigate the performer's body. We argue that performers, more specifically, the performers' bodily gestures, are key to the critical understanding and the creation of music. This chapter contains three parts: first, we will investigate the concept of embodied musical gestures through a range of inter-disciplinary scholars, ultimately defining a concept that is useful and fruitful in discussing performance. In the second part, we will use Toru Takemitsu's *Rain Tree Sketch II for Piano* (1994) as a testing ground for analyzing with the performative body as the starting point. And lastly, we will discuss how composing with performative gestures in my composition *Sheng* (2016) for piano, audience's smartphones, and fixed audio playback elicits the cross-modal, inter-sensory nature of embodied musical gestures.

Indeed, the concept of embodied musical gesture has the potential to dissolve the artificial fractures between the activities of thinking, creating, and doing. Analyzing and composing with the performing body do away with this mind-body split, offering refreshing and generative insights that do justice to the physical nature of music making.

82.1 Gesture: A Sign or a Totality?

To begin with, the embodied musical gesture is a musical event (a figure, a phrase, or a contiguous group of sounds) that necessarily has a bodily correlate. As a concept, the embodied musical gesture has recently drawn a body of inter-disciplinary scholars together in addressing music in two fundamentally novel ways: as an experience that is not just sonic but full-bodied, and as a discipline that cannot be adequately understood through a purely mental-cognitive perspective.[1] In other words, the musical experience is both inter-sensory and corporeal. This pique in interest in how the whole body is implicated in musical meaning coincides with the ever-growing field of neuro-physiological research in how humans perceive the world. One such example of the intersection between scientific and music studies is the use of mirror neuron research by musicologist Arnie Cox in [218, pp.45-59] to defend the "mimetic hypothesis." Cox theorizes that because of mirror neuron action, audiences understand music not only in terms of thoughts and emotions, but also through various levels of sympathetic resonances in the body while witnessing (or imagining) the corporeal performer.

This idea of the embodied gesture is not new; in fact, it resounds with the foundational thoughts of Maurice Merleau-Ponty seventy odd years ago, in his seminal *Phenomenology of Perception* [749] in 1945. Merleau-Ponty's belief that "motility [is] basic intentionality"—that intentions as heard originate from intentions as movement—was ground-breaking for its time [749, p.154]. Shunning the well-established cartesian split between the body and mind that had dominated philosophical thinking for centuries, Merleau-Ponty picked up these two segregated halves to complete the human person again: he regards the body, rather

[1] I distinguish the "mental-cognitive" approach from that of just a cognitive one, since the latter would imply the involvement of the body, given recent research in the field of embodied cognition.

than merely the disembodied mind, as central to how humans understand the world.[2] Through explicit or implicit bodily actions, not mere mental activity, meaning is created. In addition to rejecting the cartesian mind-body split, Merleau-Ponty also argues that bodily experiences of the world are not subdivided into sensory components such as sight, touch, hearing, and so forth. Rather, a "thing" is experienced as an "inter-sensory entity," where all our senses communicate with one another. In [749, p.166], Merleau-Ponty writes (emphasis mine):

> And in so far as my hand knows hardness and softness, and my gaze knows the moon's light, it is as a certain way of linking up with the phenomenon and communicating with it. Hardness and softness, roughness and smoothness, moonlight and sunlight, present themselves in our recollection, not preeminently as sensory contents but *as a certain kinds of symbiosis*, certain ways the outside has of invading us and certain ways we have of meeting this invasion, and memory here merely frees the framework of the perception from the place where it originates.

Merleau-Ponty conceives of perception as a rich, inter-connected sensorium of the body. In light of this, the experience of music does not merely involve hearing but also, for instance, feeling the thick *texture* of a fully orchestrated chord, sensing the *highs and lows* of range, seeing the *bright* tone color of the flute, or soaking up the *warmth* of a clarinet's chalumeau register. The language with which we speak of music already points towards this inter-sensory experience. Such an experience is not merely metaphorical but real, just as our memories of height, temperature, sight, and touch become enmeshed in the "symbiosis" of a fully corporeal experience.

To follow this line of thinking, in recent decades, related notions of embodied musical gestures have been explored in different contexts, from David Lidov's semiotics of gestures to Robert S. Hatten's work on tropes in Beethoven's music [446, 610]. In their examination of musical gesture, musicologists and theorists have traditionally embraced a semiotic approach. While David Lidov's system of gestures is based on Peircian semiotics such as sinsigns (singular gestures) and legisigns (repeated, recognizable gestures), Hatten uses topics that signify extra-musical meaning to create emergent tropes with new meanings. Useful as these are, a purely semiotic approach to gestures faces some limitations.

By its very nature, this "signification model" splits the musical experience into the body and its affects—resurrecting Cusick's mind-body dilemma. The body performs a gesture, the mind then interprets this as a sign invested with meaning. However, as Cox points out in his mimetic hypothesis, mirror neuron research suggests that we understand music at an even more elementary, neuro-muscular level [218]. It decentralizes the experience of music away from the central mind and semiotic interpretations, and moves it to more peripheral, sympathetic resonances with the performer's hands, arms, vocal apparatus, and torso.

Recent scholarship on music and emotion raises an even more fundamental problem with the signification model. Lawrence Zbikowski's "Music, Emotion and Analysis" [1153] suggests that music and musical gestures do not consist of the specific, categorized emotions on which much of current gesture analysis is based. Rather, gestures consist of generic feelings that are emotively uncategorized. This implies that, under the signification model, gestures signify emotions to listeners only upon explicit suggestions and contextualization (for example, in the title of the work, concert program notes, assumed knowledge about the composer and style, etc.). While the semiotic approach yields fruitful insights into contextualized meanings, in studying these gesture-emotive signifiers within explicitly embedded contexts, are we ignoring these more elementary, generic, and uncategorized feelings?

In an effort to address this conundrum I have turned to the more recent domain of performance studies. In *Agency and Embodiment*, Carrie Noland [795, pp.66-77] provides valuable insights into these uncategorized feelings through a close study of Bill Viola's memorable video installation, *The Quintet of the Astonished.* In the making of this video, Viola directed five actors to perform a sequence of emotions in direct succession. He then slowed the video to the point where the micro-facial expressions, twitches, and contortions of the faces between stereotypical emotive expressions were foregrounded.

[2] Maurice Merleau-Ponty, regarded as the founder of embodiment philosophy, sees the bodily experience as the starting point of human knowledge and understanding, see [750].

Noland characterizes these uncategorized facial expressions and gesticulations that are in between emotive states as gesture "deviances." Not categorically emotive, these gestural deviances go between emotive states; that is, they still contain in them some affective information that cannot be easily codified. By drawing on the agency of embodied beings within culture through a Merleau-Pontyian point of view, she concludes that cultural change stems precisely from our kinesthetic awareness of these gestural deviances. Instead of a model of humans as repetitious robots who perform perpetually the same stereotypical gestures, she favors the kinesthetic awareness of these interstitial, uncodified gestural deviances that enables the forging of new meanings.

Noland [795, p.75] refers to these uncategorized feelings or gesticulations as "vitality affects," a term borrowed from child psychologist Daniel Stern in direct contrast to codified, emotive "categorical affects." Vitality affects are primal ("vital" to life) and pertain to the pre-emotive kinetic and energetic qualities of touch, movement, and sound. In child development theory, as a child grows in a social environment, some of these vitality affects are instilled with meaning and thus become categorical affects. But for adults, Noland argues that by being kinesthetically aware of vitality affects, they can push their way into the culturally inscribed sphere.

Noland's vitality affects have groundbreaking implications on the field of music analysis. Her insistence that new meanings are created by their *kinesthetic awareness* of vitality effects implies that the responsibility of analysis now lies with those who are able to reflect upon and become aware of these vitality effects in the body. In other words, Noland's argument can be reinterpreted in a musical context as a call for a bodily-based approach to analysis. Those who are most qualified to undertake this task, who have the most intimate bodily knowledge of a piece of music, are exactly those who are traditionally left out of the field of analytical enquiry: the performers. In the analysis below, I will thus take this viewpoint of the performer with privileged information: analyzing from my self-awareness of performative gestures the feelings of tension, relaxation, direction, and the quality of movements, as a starting point to investigate Takemitsu's *Rain Tree Sketch II.*

The study of the vitality effects through a performer's kinesthetic awareness is even more crucial in understanding contemporary music. While the signification model could be effective in studying old works such as those of Beethoven and Brahms, that cannot be said of new music. Contemporary composers are at the forefront of forging new meanings that are uncodified, and are often difficult to situate within a clear context from which a signification model needs to be based. A performance-based analysis of *Rain Tree Sketch II* below reveals structural information that can only be manifested through such an intimate bodily knowledge of the work.

82.2 A Gesture-Based Structural Reading in *Rain Tree Sketch II* by Toru Takemitsu

Written in 1992, Toru Takemitsu's *Rain Tree Sketch II*—dedicated to his beloved teacher Olivier Messiaen—counts as his final piano composition. The title alludes to the short stories of Takemitsu's friend and poet Kenzaburo Oe about the rain tree, a miraculous ancient tree that stores rainwater in its many leaves and releases droplets well after a storm has passed. Takemitsu draws his musical and aesthetic inspiration from both his Western music training and his Japanese heritage. This is most apparent in his collection of writings about the relationship of his compositional aesthetic to Japanese philosophy and his works that combine Western and traditional Japanese instruments [1032].

Like much of Takemitsu's piano music, *Rain Tree Sketch II* contains many stops and starts; three or four measures of sound are often followed by a measure of rests throughout the piece. As such, the sound and the in-between stillness are embodied by the performer with a predictable alternating bodily sequence: after a configuration of movement finishes, the body is held in poise, after which another configuration of movements begins again.

As a frequent performer of his music, I would describe Takemitsu's music, in particular *Rain Tree Sketch II*, as physical, more so than it is "intellectual" (like Babbitt) or "abstract" (like Webern). That is, the success of a piece of music hinges most crucially on how well I execute choreographed movements. For instance, I argue that finishing a gesture by holding my body absolutely still in the rests before commencing the next gesture is just as important (or even more so) as playing all the right notes. While other composers' music might involve plenty of intellectual reflection, interpretation, and physical trial and error in order to make the piece "work" as a whole, *Rain Tree Sketch II* seems to lend itself to a "structural playing." The structure and the progression towards the piece's climax seem to simply play themselves out during a performance.

Consequently, I will take the performative body as a starting point to investigate two structurally important moments in the piece, the "peak" and the "climax" (explained below). Specifically, I will consider these two moments from the point of view of performative gestures. By a performative gesture, I mean a choreography of movements that has an intentional start and end. I will consider how these performative gestures interact with certain recurring motifs in the piece to sculpt the peak and the climax.

Bars	1	23	25	49	59	92
Sections	A		B		A′	coda
Processes	I	(peak)	II	(climax)		

Fig. 82.1. Formal sections and structural processes in *Rain Tree Sketch II.*

In an ABA-Coda form that could be described as a free-ternary form, the piece features two structural processes (Figure 82.1). Figure 1 shows these two structural processes in relation to the formal ternary sections of the piece. As Figure 82.1 shows, Process I spans the "A" section of the ternary form, while Process II starts at the "B" section and lasts until the end of the piece.

These two processes each have a high point: a local "peak" (measure 23) in Process I and a global "climax" (measure 49) of the piece in Process II. In the performance of the piece, I feel a distinction between these two high points: whereas the peak is a tensing up of the whole body that feels unresolved and fleeting, the climax is accompanied by a deeply peaceful feeling with a sense of visceral unwinding. The embodiment of these structural points can be further explored by taking a closer look at two particular motifs in the piece and how they are gesturally manifested.

82.2.1 Process I: Synergy of Mirroring and Parallel Gestures

A note of clarification before we proceed: a motif is defined by two or three pitches that recur recognizably throughout the piece. While a motif is defined by pitch and rhythm, it goes through different transformations including changes in texture, tempo, dynamics, etc. Consequently, motifs are *played* differently, that is, they elicit different gestures.

One such motif that recurs saliently throughout the piece is motif i, defined by the $A5-D6-C\#6$ figure that first appears (measure 1), which supplies the principal material for Process I (Figure 82.2, circled). In its first appearances (measures 1 and 2 at pitch; transposed in measures 4-5), motif i, with its upward then downward pitch pattern, affords a subtle circular gesture in the right hand.

However, the different periodicities across the four vertical layers restrict the hands to relatively bound positions; there is a certain sense of physical conflict between the layers. To play the top layer, my right hand is naturally inclined to rest and recoil on the keybed at the end of motif i. However, the slurred pairs below it do not permit me to do this; my right hand must keep undulating up and down for three times per measure. In conflict with the right hand, which is already on a delicate fulcrum trying to compromise two independent layers, the left hand is performing acrobatics of its own. The repeating $C\#5-B\flat 4-F\#4$ figure calls for a natural circling-in wrist motion that repeats two times per measure; however, this circling motion is executed not without effort. Tucked into this circular motion, I must fit the irregular, longer notes of $D4$

and $E4$ below them. All to be executed in a soft dynamic with a "celestially light" touch, the beginning's embodiment is one of fastidious hand balance, full of compromises and tension within its limited, carefully negotiated circular motion.

Fig. 82.2. *Rain Tree Sketch II*, measures 1-7, motif *i* circled. Copyright 1992 by Schott Music Co. Ltd., Tokyo, All Rights Reserved, Used by permission of European American Music Distributors Company, sole U.S. and Canadian agent for Schott Music Co. Ltd., Tokyo.

This balancing act thankfully does not last for long: after a bar of stillness in measure 8, motif *i* changes registration and is treated chordally in two hands (Figure 82.3). Motif *i* appears in this guise for the rest of Process I. While the two hands were working against each other in different periodicities in the beginning, here, the hands work together with the same articulation, where a slurred pair is followed by a tenuto-accent in both hands. Moreover, while the right hand moves *up* from $A4$ to $D5$ and then *down* to the $C\#5$ to play motif *i*, the left hand does the exact mirror image, moving *down* from $B\flat3$ to accommodate the fifth finger on the $D3$, then finally *up* to the $F\#3$ on the third finger. Thus, the two hands make a mirroring, circling-in gesture. This is a highly relaxing and satisfying motion which exploits the symmetry of the hands. In contrast to the bounded and careful hand undulations of the first seven measures, in measures 9 and 12, the hands are liberated to perform a mirroring circular gesture that releases the tension from the beginning. The *tenuto* accent on the third chord allows me to relax and lean immediately into the keybed, transferring my arm weight into the keys. In addition, while the first seven measures are in the upper part of the keyboard that destabilizes the torso, here I can relax not only my arms but also my whole trunk, as I play in the part of the keyboard that is straight in front of me. Motif *i* thus becomes a circling-in, mirroring, weightful, and relaxed gesture of the two hands and whole upper body.

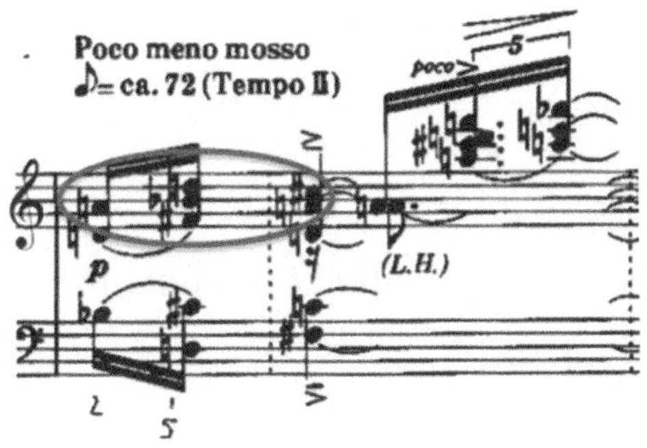

Fig. 82.3. Measure 9, motif *i* circled. Copyright as in Figure 82.2.

In contrast to motif *i*'s transformation into a mirroring, circular gesture, motif *ii* emerges soon after: motif *ii* first appears in measure 15 (Figure 82.4). This motif, the figure comprising of *C*#5 – *A*5 – *G*#6 sixteenth notes in the right hand and *A*3 – *D*4 – *F*4 in the left, recurs time and time again as a kind of afterthought, almost always appearing after a longer gesture and leading to a measure of rests (other appearances include measures 21, 24, 28, and 33 in Process I).

Fig. 82.4. Measure 15, motif *ii*. Copyright as in Figure 82.2.

Motif *ii* balances out the mirroring circular gesture of motif *i* of measure 9: both hands move in a similar parallel motion upwards. Instead of motif *i*'s end that is a weightful relaxation into the keybed, motif *ii* drifts upwards weightlessly with a *diminuendo molto*, ending up with arms in the air above the keyboard. However, the weightlessness betrays the effort in the body: while motif *i* relaxes, the off-centered end position of the arms to the far right of the keyboard in motif *ii* requires much effort in the shoulder and upper arm region to maintain its poise in the following measure of stillness (measure 16). Figure 82.5 summarizes the shapes and qualities of movement in the gestural renditions of motifs *i* and *ii*, in measures 9 and 15 respectively.

Given these two gestures with contrasting bodily tensions and relaxations, shapes, and relations to gravity, we now approach our first moment of enquiry: the peak in measure 23. While the two gestures are played out in succession up until this point, immediately before the peak in measure 22, these two archetypal gestures are synthesized (Figure 82.6). In the two slurred pairs of chords in measure 22, the hands play in contrary motion outwards in a *mirror* gesture, a trait of motif *i*. This mirror motion is emphasized by crescendi, as if lunging outwards towards the distal part of the hands. However, on a macroscopic level, the arms are moving *upwards* in *parallel* motion, characteristically belonging to motif *ii*. The combination of the mirror and parallel upward gestures lacks certain traits of the original motifs though; while the mirror gesture was also circular in motif *i*, this is not the case here, as we only hear the first two notes of motif *i* ($A5 - D5$ in the top voice in the first beat of measure 22 instead of $A4 - D5 - C\#5$). The torso tension is also more akin to that of motif *ii*; the body moves upwards towards the far right of the keyboard and stays off-center for motif *ii*'s afterthought gesture in measure 24. Thus, in Process I, we experience a move towards some kind of synergy between contrasting gestures. However, this attempt to combine the two gestures feels incomplete; the body is unresolved by holding onto motif *ii*'s prior tension, yearning for a counter-balance to the preponderance of right-sided tension.

	Motif *i* (measure 9)	**Motif *ii* (measure 15)**
Hand shape	mirroring, circular	parallel, upwards
Feeling of weight	heavy	light
Torso tension	relaxed	high degree of tension towards the right

Fig. 82.5. Bodily attributes of motif *i*'s and motif *ii*'s gestures.

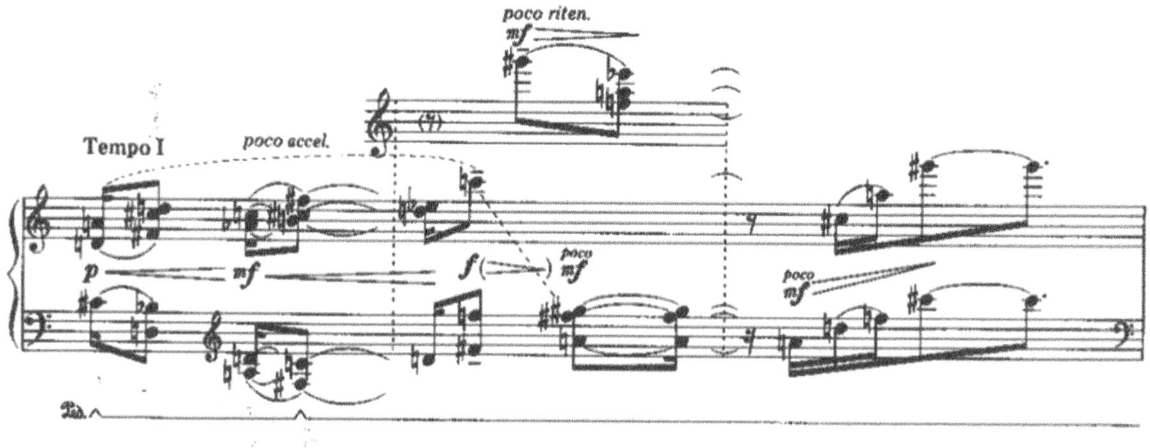

Fig. 82.6. Measures 22-24, peak. Copyright as in Figure 82.2.

82.2.2 Process II: Towards Relaxation, Balance, and Weightfulness

Process II starts the journey towards its own high point, the piece's climax, with a renewed bodily comportment, beginning in measure 35 (Figure 82.7). From the outset, the performer faces a different gestural strategy. In Process I, the hands work in tandem in both the mirroring, circular gesture of motif *i* and the parallel upward gesture of motif *ii*. That is, the hands simultaneously scoop outwards in a contrary motion, or travel up the keyboard in similar motion at the same time. At the start of Process II, the hands do cooperate, but are also independent of each other: they play in canon, moving in a parallel, but staggered motion (Figure 82.7). This interdependence of the hands gives the body a fresh approach, as if calling for a cooperation that also respects the individuality of the two hands.

Fig. 82.7. Measures 35-37. Copyright as in Figure 82.2.

However, this fresh, gestural innovation does not come without a price; throughout Process II, a higher demand of energy is placed on the performer's body. While in Process I, gestures are mostly flanked by pedaled, still resonances indicated as full measures of rests, in Process II, there is no such respite, with one gesture moving straight to another. Here, the performer is allowed only a meagre few seconds of recovery in the long held notes at the end of each gesture (for instance, in measure 36 of Figure 82.7), before moving onto the next gesture. The effortful gestures and transitional still recoveries create a macrocosmic rhythm undergirding the piece that is akin to throbbing. This throbbing rhythm between movement and stillness is slow in Process I, with a long recovery time after almost every gesture. In contrast, Process II witnesses a quickening of this rhythm, with less and less recovery time until the gestures are juxtaposed one after the other as they head toward the climax in measure 48. The effortful gestures and recovering stillnesses thus supply within the piece a general buildup of an embodied rhythmic energy towards the climax.

This arduousness in quickly shifting from one gesture to the next in Process II is coupled with yet another bodily demand: an involvement of the body that is not just hands and arms, but the full body. The lead-up to the climax entails a rocking gesture that moves the bodily engagement from the distal regions—the fingers—toward the central torso (Figure 82.8). In measure 39 the $F4 - D4 - F4$ figure moves in a quick to-and-fro rocking gesture. The fast sixteenth-note movements within a small space of whispery dynamics ensure that the articulations are limited largely to the fingers. However, the answering phrase in measures 45-46 slows down to eighths and dotted eighths, allowing the arm to initiate its rocking motion. Finally, the next measure engages the whole torso: the catapulting from the accented chord up ($G\#4$ in the top voice of the right hand) to a high register launches the entire torso toward the right end of the keyboard, lifting up the chest. Following quickly, the rapid movement from the high register back down to the weighty middle-register chords materializes in a sudden, quick jerk back to the left, including downward abdominal force. In quick, successive movements the body rocks to the right and up and then to the left and down. The spatiality of the music (the registers of the piano) and the weight of the gestures produce two distinct dimensions (right/left and up/down).

The lead-up to the climax engages progressively more and more of the body: from the distal, to the intermediate, to the central. This increase in physical energy and expansion of bodily involvement prepares the performer for an outward spatial expansion in the climax in measures 48-49 (Figure 82.9). Here, a fragment of motif *ii*—a $D4 - F4$ melodic fragment from the left hand—is treated chordally, in the middle of the keyboard. Instead of being parallel, upward moving, and light (see Figure 82.5), it is now treated with motif *i*'s quality: contrary (mirroring) in motion as the two hands move to emphasize both of the fifth fingers, heavy with an accent. Pitch-wise, we hear motif *ii*, but in physicality, it adopts all of motif *i*'s qualities. The bodily tension also assumes that of motif *i*; I can both relax into the keys with my arms as well as sit with a stable trunk, playing at the center of the keyboard. This gestural adoption of motif *i*'s qualities by motif *ii*'s

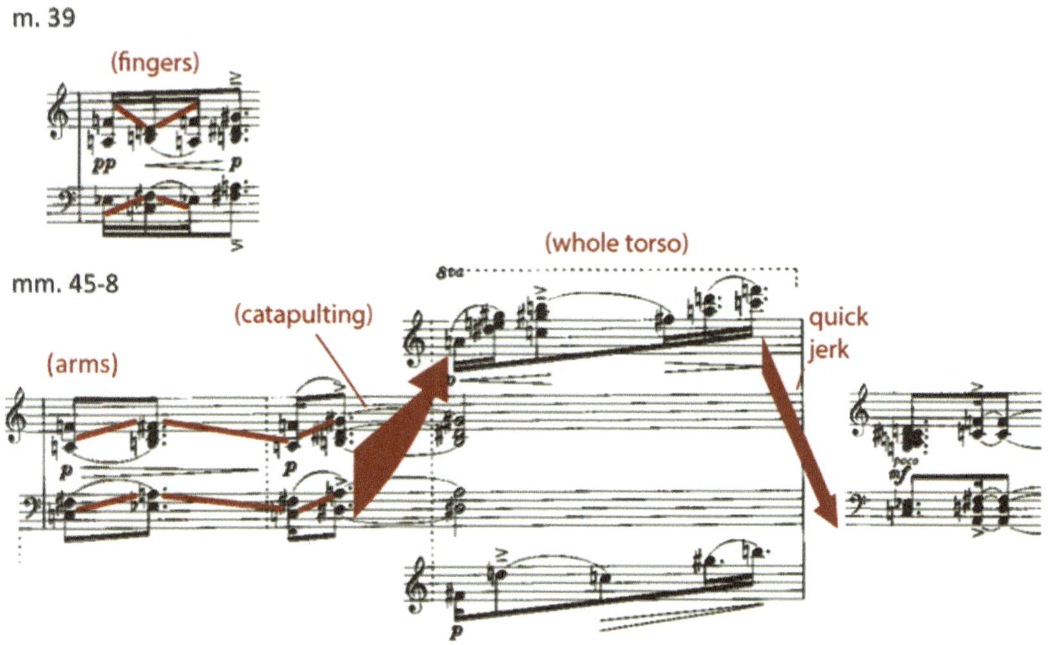

Fig. 82.8. Rocking motion, from distal to central. Copyright as in Figure 82.2.

pitches is tweaked with several improvements. Firstly, while motif *i* is played with a circling *inwards* motion, here, right after playing motif *ii*, the arms shoot *out* symmetrically towards the extremes of the keyboard in measure 49. Thus, instead of contracting inwards as in motif *i*, the arms expand, and expand generously. Secondly, in contrast to the off-centeredness of the torso at the peak in Process I, the symmetry with which the arms move out at the climax assumes a centering balance. Bodily expansion and gestural synergy thus need not be compromised by bodily tension; instead they can be accompanied by relaxation, balance, and weight.

The gestural transformations of motifs *i* and *ii* thus drive the two processes towards their respective high points. While both of these moments share a similar attempt to merge two contrasting motifs, the kinesthetic feeling or physical experience cannot be more contrasting. A performative, gesture-based analysis reveals the bringing together of differences in two different ways: through bodily tension and upward striving in Process I, and through balance and bodily expansion in Process II, the latter being by far the more satisfying of the two. What can be said about this structural experience of *Rain Tree Sketch II* in terms of Takemitsu's compositional aesthetic?

Indeed, the idea of merging opposing qualities resonates with Takemitsu's own philosophy of reconciliation, as expressed in his collection of writings, *Confronting Silence* [1032, p.81]:

> That rich world of sound around me... those are the sounds that I should have the courage to let live within my music. To reconcile those diverse, sometimes contradictory, sounds around us, that is the exercise we need in order to walk that magical and miraculous road we call life.

The climactic reconciliation of opposing gestural qualities in *Rain Tree Sketch II* emerges not through negation or domination but rather through unification within an expanded body. This kind of harmonious reconciliation resonates with Takemitsu's peaceful attitude toward the contradictions inherent in human beings and nature, in music from East and West—where one element does not dominate another but both live together in a state of "naturalness." [1032, pp.59-67] The interdependence of opposites—light and heavy, circular and parallel, balanced and off-centered—is played out also in terms of the fundamental bodily opposition between left and right in the parallel, staggered motion at the very start of process II. This

Fig. 82.9. Measures 48-49. Copyright as in Figure 82.2.

parallel, staggered motion is heard again immediately after the climax in measures 51 and 54, reminding us of the opposite but complementary nature of the hands (Figure 82.10).

Fig. 82.10. Measure 51. Copyright as in Figure 82.2.

At the very last two measures, this parallel, staggered motion is stretched out even further, such that the left hand's energy seems to extend to the right hand, in one continuous flow (Figure 82.11). The completion of the piece with this unified gesture certainly gives testament to these philosophical concerns that advocate for unity achieved by bringing together complementary opposites.

Fig. 82.11. Measures 75-6. Copyright as in Figure 82.2.

In this brief analysis of how a performer's body is engaged both as a whole and as arm-based gestures, the structure of the piece—the crucial moments and their kinesthetic experiences—is unveiled in ways that traditional, even semiotic-based analyses could never achieve.

These kinesthetic feelings can be viewed in light of Takemitsu's compositional aesthetic, giving way to a nuanced and culturally sensitive interpretation. From a reverse perspective, embodied performative gestures can also be the focus of compositional procedures. While composers traditionally pay painstaking attention to the translation of sounds onto scores, a gesture-oriented approach refocuses musical creation to attend directly to the act of performance

82.3 The Last Leg of a Bodily Journey

As a composer, I have explored embodied gestures through another medium, one that is perhaps not as rigorous as analysis but is nonetheless illuminating and generative. *Sheng* is my composition involving piano, audience's smartphones, and fixed audio playback that grew out of this very research into performance-based gestural analysis. It only counts as one piece of a larger, more ambitious multi-work project: a collaboration of music, art, and technology called "Synaesthesia Playground" with fourteen other creators. In the rest of this chapter, I will be discussing how the compositional process of *Sheng* shines light on the physical-affective-sonic makeup of an embodied musical gesture, as well as its cross-modal and shared, inter-corporeal nature.

Before we launch into a discussion of *Sheng*, some background about "Synaesthesia Playground" is in order. In this project, I led an interdisciplinary team of six composers, two visual artists, five software developers, and two fashion designers to create six multimedia works for a piano recital (performed by me). As suggested by the project's title, the theme of inter-sensoriality is explored. The idea of inter-sensoriality stems from the Merleau-Pontyian concept of perception as a "symbiosis" of the senses as discussed in Section 82.1, and more specifically, the notion that the musical experience involves the full body and an intermingling of the senses. The multimedia component in "Synaesthesia Playground" is thus not a superfluous addition to the sound, but a crucial and integrative aspect that highlights the synergistic nature of musical perception. More specifically, there are two major visual installations in the project. These two installations examine the concept of the body and the dichotomy between its interiority and exteriority. The first installation is a video projection by Celeste Oram and Takefumi Ide that is cast onto the body of the piano, reimagining the piano as a living organism with a "skin" (Figure 82.12). Figure 82.12 shows a still of the video projection for the very first piece of the recital, *Toccata and Bruise* by Celeste Oram. This video projection is featured throughout the entire first half of the recital, unifying the works with a connected visual element. In contrast,

Fig. 82.12. "Epidermis of the piano" in Oram's Toccata and Bruise.

the installation in the second half of the recital is not on the body of the piano, but on that of the pianist (Figure 82.13).

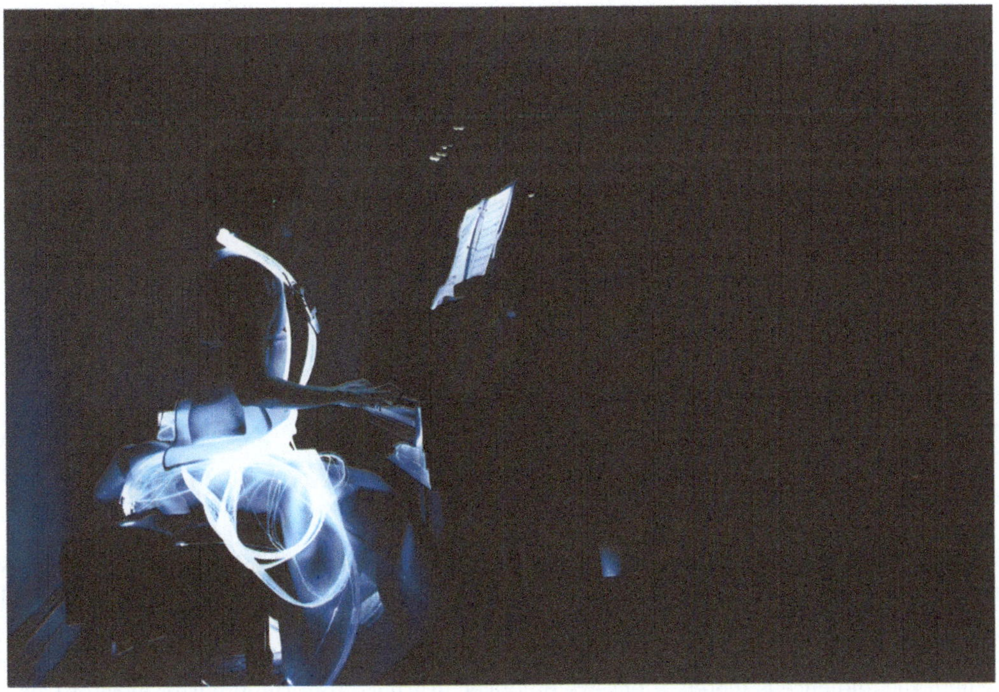

Fig. 82.13. *Bio Lux* by Nagasawa, Hui, in Arnold Batt-Rawden's *Love Spiral.*

As Figure 82.13 shows, this is a luminescent performance attire made of optical fibers, called *Bio Lux* by Nobuho Nagasawa, Bopha Hul and Troy Arnold. Changing colors and pulsating to my heartbeat, breathing, and movements, the attire renders the inner workings of my body visible, virtually flipping my viscera inside-out. Thus, the skin or "epidermis of the piano" is featured in the first half of the recital while the "viscera of the pianist" is made visible in the second half. These two visual installations that explore the dichotomy between the exteriority of the piano and the interiority of the pianist define the general thematic movement of the two halves, as seen in the recital program (Figures 82.14 and 82.15). As Figures 82.14

Program

First half: A journey into the visceral

Toccata and Bruise — Celeste Oram
[skin:touch]

Iyalāmai — Anne-Sophie Andersen
[voice of feminist struggle]

Missa di Glossa — Sidney Boquiren
[spirituality:core]

Visuals: **Piano epidermis**
by Takafumi Ide and Celeste Oram

Intermission

Fig. 82.14. "Synaesthesia Playground" recital program (first half).

and 82.15 reveal, each half of the recital comprises of three musical works. Altogether, the six works form an overarching recital narrative: that of a journey from the exterior of the body to the interior. Each work deals with a particular bodily experience and leads to the next; together, they create a gradual movement inwards until reaching the very interior viscera of the body. However, this bodily journey is not solely one of my own, where the audience is merely a passive onlooker. Rather, from the point of view of Cox's "mimetic hypothesis," as I perform, the audience resonates with me such that this bodily journey is experienced first hand by all who witness the recital. Seen from this perspective, I literally take the audience on a journey inside their own bodies.

Second half: From internal to collective bodies

Love Spiral — Andrew Batt-Rawden
[heart] — with Ben Hinchley and Pierre Depaz
**Fiber optic attire responds and pianist plays to her live heart-beat*

Lux Venit — Daniel Weymouth
[breath] — with Al Petersen and Pierre Depaz
**Fiber optic attire responds to the energy and breath of the pianist*

Sheng — Jocelyn Ho
[inter-bodies] — with Drew Petersen, Al Petersen and Pierre Depaz
**Collective improvisation with audience's mobile phones*

Visuals: **Luminescent fiber optic attire "Bio Lux"**
by Nobuho Nagasawa with Bopha Hul and Troy Arnold,
assisted by Roy Shilkrot
Concert attire: 'Musculi' by HUL ARNOLD

All works are world premieres

Fig. 82.15. "Synaesthesia Playground" recital program (second half).

A closer look at the program in Figures 82.14 and 82.15 shows that the first half of the recital, "epidermis of the piano," is a movement from the very superficialities of the body—the sense of touch—through the vocal apparatus, to the inner spiritual core. Skin, voice, and core: these bodily experiences are encapsulated respectively in the three works: *Toccata and Bruise* by Celeste Oram, *Iyalāmai* by Anne Sophie Andersen, and *Missa de Glossa* by Sidney Boquiren. After this active, inward-moving journey, the second half of the recital, "viscera of the pianist," dwells in the interior viscera for a while. The first two pieces feature the intimate physiology of the performer's body: in *Love Spiral* by Andrew Batt-Rawden, I play to the continuously changing tempo set by my own heartbeat (detected by an Arduino pulse sensor), while in *Lux Venit*, the music and live sound playback responds to my breathing (detected by a microphone). These two pieces thus explore the organic inner workings of the body. However, after this inward journey, the recital narrative finally shoots outwards towards an interactive, collective experience. The last piece, *Sheng*, ties all the elements of the journey together, exploring interiority and exteriority simultaneously with the concept of inter-corporeality through performer-audience interactions.

82.3.1 *Sheng* for Piano, Smartphones, and Fixed Playback

Sheng is a collective structured improvisation involving the audience's mobile phones as makeshift instruments, with one guiding improviser (me) at the front. At its heart is a vision to extend performative gestures to the audience not only as bodily resonance with the performer but as active, shared participation. *Sheng* is the Chinese word for "life"; specifically in this composition, it refers to three life-giving elemental aspects—metal, water, and air—that are universally shared, according to traditional Chinese philosophy. The piece is divided into three sections of almost identical names—Metal, Underwater, and Air—that contain vastly different soundscapes on the fixed playback. Throughout the three sections, the audience (whom I will refer to as participants) and the pianist play sounds that pertain to the elemental aspect of each section. While the piece is mostly improvisational, both the pianist and the participants do have some rules to follow. The pianist improvises to broad guidelines: Metal concentrates on extended techniques on the metallic strings of the piano, Underwater foregrounds pulsating pitches ("wobbles" that come and go), and Air sees ascending pitch structures that reach up to the top four prepared pitches on the piano. Throughout the performance, the pianist is asked to respond to the sounds of the participants, and the colors and pulsations of *Bio Lux* (preprogrammed to the fixed audio playback). She also has control over when a section changes to the next via an interface panel. The participants, in contrast, receive instructions on their smartphones in real-time, which ask them to play various gestures in relations to other participants, the pianist, and their own bodily comportments.[3]

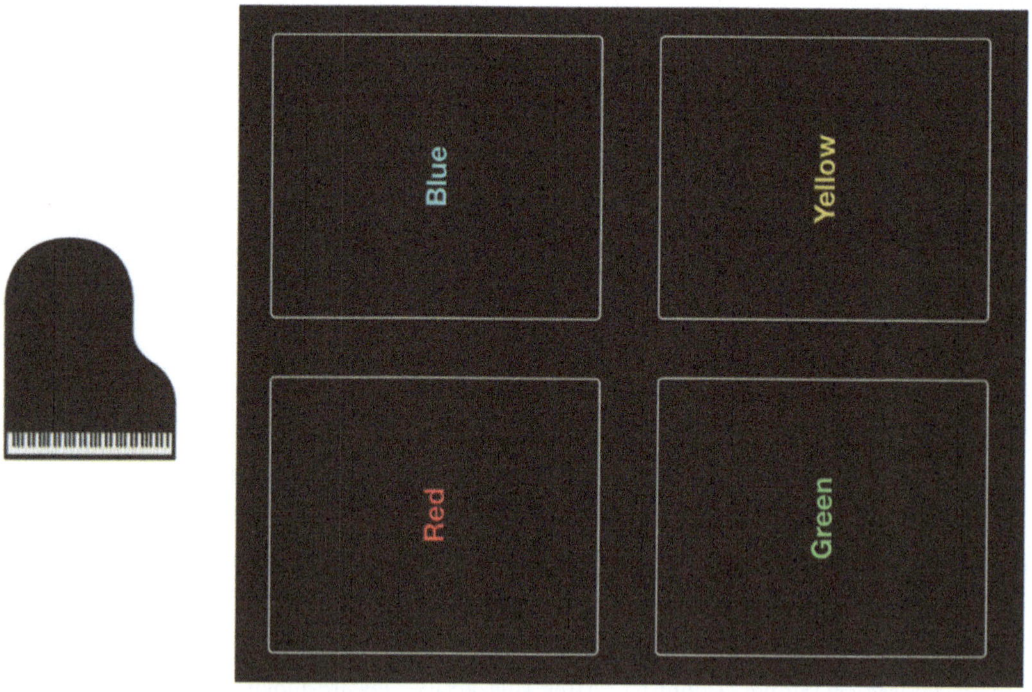

Fig. 82.16. Layout of audience quadrant divisions in *Sheng*.

Among the inter-connected elements of the piece's set up—the pianist wearing *Bio Lux*, the participants with their smartphones, and the audio playback—I will focus the discussion here on the participants, since

[3] A video of a performance of *Sheng* in the premiere of Synaesthesia Playground at Stony Brook University can be found here: `https://www.youtube.com/watch?v=tL6i9NRFBHM`

embodied gestures are most creatively explored through composing and performing with smartphone gestures. Before launching into the discussion, however, the spatial logistics and technical infrastructure of the piece need to be addressed. Spatially, the audience is divided into four quadrants in relation to the piano—left upper, right upper, left lower, and right lower—as shown in Figure 82.16. These four quadrants are given color labels (blue, red, yellow, and green) for convenience. Before the performance begins, participants are asked to join the local area network with their smartphones and identify the quadrant in which they are located on their devices. This division allows the participants in the four quadrants to receive different instructions on their phones, which specify when and how to perform various physical gestures with their smartphones (shake them, slash them, tickle the screen, etc.) in order to play sounds. An example of this is in Figure 82.17, which shows a screen capture of a particular instruction. Here, the participant is asked to be aware of his internal body (swallowing saliva) in order to perform a gesture, upon which a "popping" sound will be played. These on-screen instructions change every couple of minutes or so, thus functioning as a kind of "score" for the audience.

82.3.2 Cross-modality of Gestures

One might ask, why use gestures when one can simply play a sound via a button on the smartphone? Composing with smartphone gestures in *Sheng* is a way to explore the connection between sound and movement. In programming gestures to map to certain sounds on the smartphone, I have converted a task-oriented, everyday tool to become an interim, makeshift musical instrument. The design of this makeshift instrument is in the mapping between a gesture with a particular sound. While on the piano (or any standard instrument), gesture-sound mappings are mostly fixed (for instance, one can only play a middle C in so many ways), as a composer/instrument designer for the smartphone, I can create countless and arbitrary mappings. However, not all mappings are equal; the design task of mapping a gesture to a sound involves careful consideration of cross-modal bodily experiences to construct an intuitive match between the physical, affective, and the sonic. In other words, the intuitive complementation of certain gesture-sound mappings depends on how a particular movement is usually felt, executed, seen, heard, or experienced in everyday life by the target audience/user. The question of complementarity of a gesture-sound mapping, in fact, already has an informal designation in computer game-speak. A game has "juice" if it has inter-sensory (audio, haptic, etc.) elements that make on-screen events realistic, exciting or interactive; for instance, an explosion may correspond with a shake in the controller, or a "woop" sound with a jump by an on-screen character. In *Sheng*, the compositional process consequently involves constantly drawing on my own bodily experiences of how a physical motion might feel and sound. For instance, a metallic, slashing sound is programmed to a diagonal, downward fast motion, while a shivering, fluttering sound is programmed to a shaking motion in the first section Metal. In contrast, the minimal tapping motion in Figure 82.17 is connected to an internal swallowing motion. This instruction appears in the second section, Underwater, in which the fixed audio playback creates a watery, submerged impression.[4] The participant's swallow is connected to the "popping" sound that suggests an underwater air bubble or an ear-pop, immersing him in the submerged experience through his own body. The process of creating effective gesture-sound mappings thus draws on the bodily history of an audience through cross-modal references.

82.3.3 Learning the Smartphone Instrument

With newly programmed sound-gesture mappings on their phones, the audience is in a sense learning a new instrument as the piece progresses. To investigate the implications of learning the smartphone-as-musical-instrument, one could benefit from a closer look at the process of instrumental skill acquisition. Acquiring instrumental skills involves training a specific set of physical movements that successfully produce sounds for which one is intently listening. These sounds are linked to certain experiences such as affects, emotions, and imaginative cross-modal analogies (one only needs to listen to any work by Impressionist composer Claude Debussy to sense the connection between his pieces' evocative titles and sound-worlds). In other words, to

[4] This section could be found $2'30''$ into the video.

Fig. 82.17. Smartphone screen capture for a particular instruction in *Sheng*.

learn an instrument is to acquire a set of physical-affective-sonic connections that pertain to the peculiarities of the instrument. Obviously, these connections may vary between individuals because of differences in body types; nevertheless, learned pedagogical traditions ensure some commonalities among those who play the same instrument. For instance, for pianists, legato and warm tones are associated with flexible wrists and supple fingers, while loud and violent passages are associated with a steadying of the whole torso, which forms the solid basis for forceful and quick changes in the arms. Furthermore, style plays a big factor in determining this multi-facted connection: the light legato touch demanded in an early classical work such as a Mozart slow movement tends to be different from a richer legato touch in, say, a Brahms Intermezzo. Thus, to learn an instrument is to acquire a habitus, a specific set of bodily techniques in which are ingrained physical-affective-sonic connections influenced by stylistic assumptions and tendencies.

In *Sheng*, the learning of the smartphone-as-musical-instrument does not start from scratch as it would with a standard instrument; rather it is disruptive of its already prescribed usage. This prescribed usage is task-driven and heavily dependent on the visual sense: just think of the hand-eye—or finger-eye—association needed while one scrolls through her Twitter feed with a flick of her finger, eyes darting up and down to catch all the nuggets of information. Information retrieval through the coordination between finger-based gestures and the visual sense, whether the information is communicative through chat apps or knowledge-based through web browsers, is at the heart of mobile device usage. Even when the information to be retrieved is sonic (for example, songs on a streaming audio app such as Spotify or Pandora), the finger-based gestures are connected to the visual information displayed on the screen, and are performed before the sonic information is given; once the song starts to play, one can easily close the lid of the device, sit back or do unrelated tasks.

In *Sheng*, I am disrupting the participants' learned finger-eye associations and information-driven stylistic use of the smartphone. Instead, one could say that a new "style," or a new attitude, for the smartphone's usage is created, one that is sonically focused and creative. While another stylistic use might be that of video games (disruptive of information-oriented usage but nevertheless mainly visual), here in *Sheng*, the smartphone is used as a musical instrument in an immersive, interactive concert setting. The gestures in *Sheng* are not only finger-based but also involve arm and upper body motions. During the performance, physical-affective-sonic connections are thus inscribed onto the participant's body such that her smartphone becomes full, intuitive extension of her bodily being. In other words, learning the smartphone-as-musical-instrument involves suspending its mundane, everyday associations to use it in a creative, full-bodied, affective, sonically oriented, and interactive way.

82.3.4 Kinesthetic Awareness and Modes of Listening

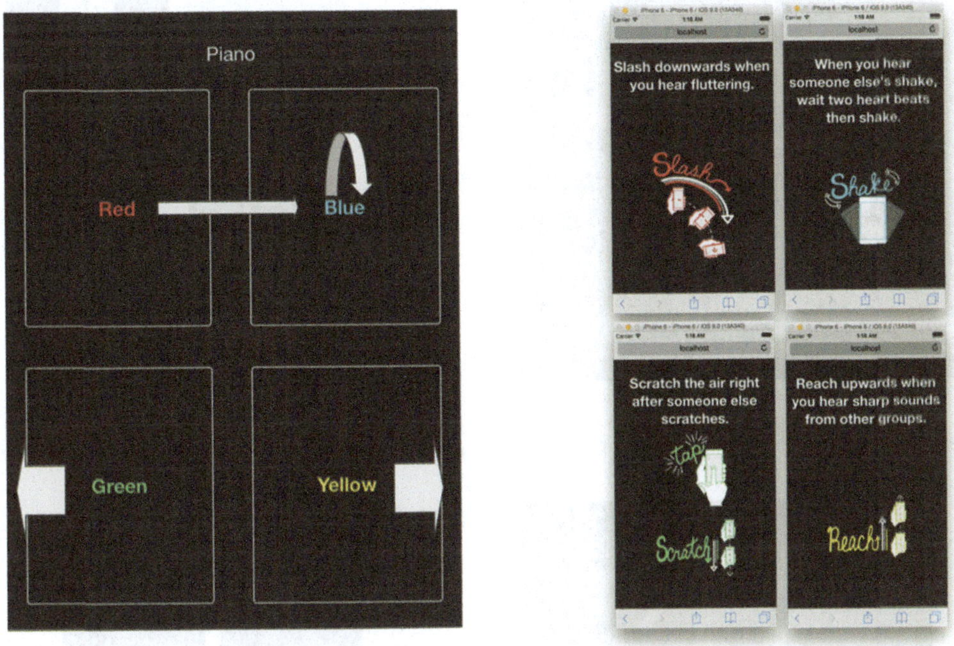

Fig. 82.18. Example of instructions (right) and audience dynamics (left) in first section, Metal.

To exploit this concept of the smartphone as instrument, the instructions I have devised exploit two crucial abilities required to learn any instrument: kinesthetic awareness and flexible listening. When one learns an instrument such as the piano, she needs to feel kinesthetically the muscle tone of her arms, the pressure on her fingers, and the curvature of her palms in order to finesse her movements to achieve desired sounds. Without such an internal monitoring of one's bodily movements, muscle tensions, movements, and postures, it would be impossible to improve and master her instrument. In tandem with this kinesthetic awareness, learning an instrument requires a heightened and flexible sense of hearing; the student listens for different aspects of sound, including dynamics, length of notes, timbre, etc. at different times. At a particular moment the player may need to listen microscopically for a pitch decay, yet at another time one may need to listen macroscopically to the sound of the instrument in the ringing acoustics of a concert hall. In *Sheng*, I have exploited these notions of inward-looking kinesthesia and outwardly directed modes of listening by prompting the members of the audience to kinesthetically feel their own bodies and listen in different ways. Moreover, I have extended the concept of listening to include not only one's own sound production but also that of other participants (Figures 82.18- 82.20). The screen captures of instructions in the three different sections of the piece in Figures 82.18 to 82.20 show this exploration of feeling inwardly and listening outwardly and interactively. Contrasting kinds of listening are required of the four quadrants in the first section, Metal (Figure 82.18 right): while the red quadrant participants listen for "fluttering," (performed by the blue quadrant), the blue quadrant participants feel their own internal bodies (heartbeats). Both the green and yellow quadrants participants are prompted by instructions to listen to the whole sound in a macroscopic

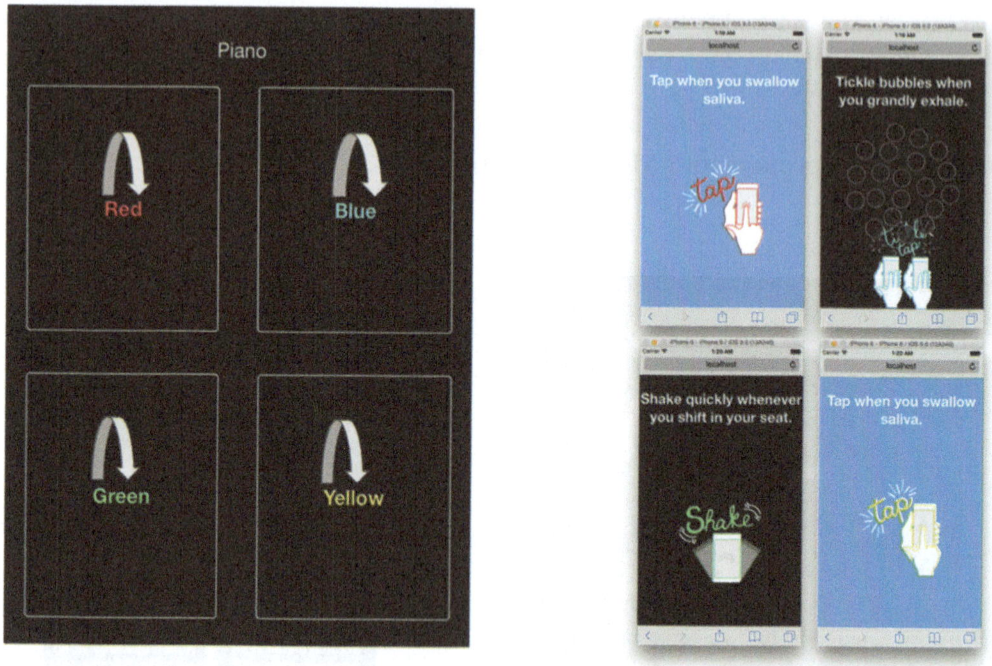

Fig. 82.19. Example of instructions (right) and audience dynamics (left) in second section, Underwater.

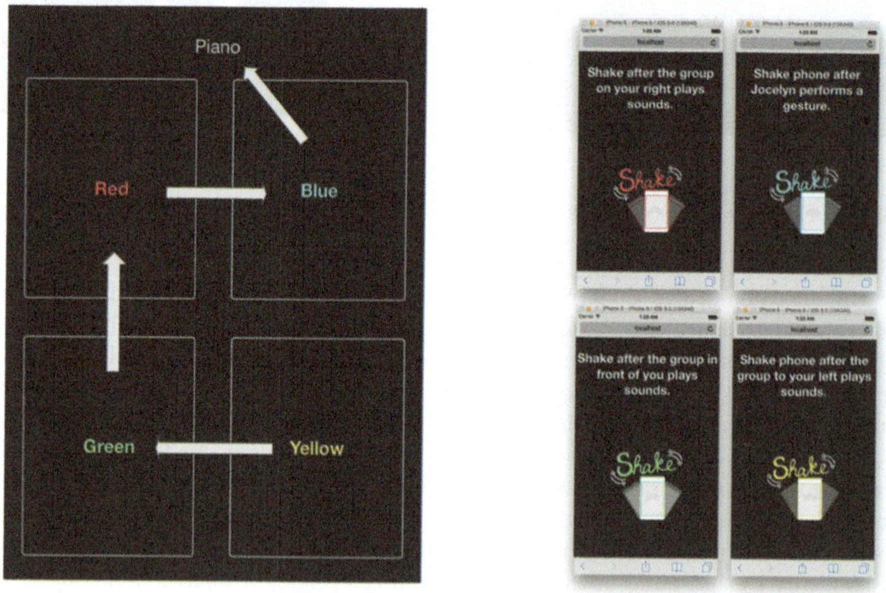

Fig. 82.20. Example of instructions (right) and audience dynamics (left) in third section, Air.

kind of listening. Thus, a social dynamic situation is created, where different participants respond, listen, feel, and act differently under the influence of different dispositions prompted by the instructions. The dynamics of the four quadrants are illustrated in the left portion of Figure 18, where the arrows indicate the direction of attention. As these arrows illustrate, the four quadrants are directed to listen inwardly (blue), interactively (red), and macroscopically (green and yellow).

Contrastingly, in the second section, Underwater (Figure 82.19), the four quadrants are asked to kinesthetically pay attention or "listen" inwards to their own bodies in a hyper-sensitive manner, much akin to when one is submerged underwater and can hear the sound of her own breath and swallowing. As the left portion of Figure 82.19 shows, the dynamics of all quadrants are solipsistic and introverted, in contrast to the active interaction between the participants in the first section. And in the final section, Air (Figure 82.20), the four quadrants are asked to listen to each other with a domino effect, much as in an aerial formation of a flock of birds, where one follows another in succession. The blue quadrant participants listen to and look at the pianist for their cue, red listens to blue, green listens to red, and yellow listens to green. At the very end of the piece, I play softer and softer sounds until the gestures become silent, as if I am ghosting the keys. Here, the blue quadrant's participants rely on their visual sense in order to play their gesture. As demonstrated in these three examples, all the instructions are tailored to elicit the duality of interiority and exteriority through internal kinesthesia and external listening, all the while adding an extra layer, asking participants to listen and respond to each other's sounds and gestures in an inter-corporeal engagement. The disposition suggested by the instruction, along with the gesture and its sound, constitutes the embodied musical gesture; the participants learn the gestures through an inter-sensorial enmeshment of feeling, listening, seeing and moving, all the while collectively engaging an inter-corporeal experience that is unique to their own bodies and their relationship to the other present bodies in the concert hall.

82.4 Conclusion: Foregrounding the Performer's Body

In the above two case studies, we encounter that which has been traditionally left by the wayside in the acts of analysis and composition: the performative body. Both, analytical and creative enquiries into Takemitsu's *Rain Tree Sketch II* and my composition *Sheng* respectively demonstrate that the musical experience is never divorced from the body that feels, moves, and resonates with those of others. It would be stating the obvious to say that without the body, music can neither be sounded nor experienced; yet traditionally analysts and composers disregard its importance by hiding it behind the score. In *Rain Tree Sketch II*, a performance-based analysis that takes the performer's body as a starting point reveals structural significances that cannot be derived from a score-based approach. It reveals that musical and bodily intentions are, in fact, inseparably unified. Although *Rain Tree Sketch II* is used as one particular test case for such a gestural approach, the efficacy of this methodology—analyzing from the performer's body—is neither limited to this piece nor to Takemitsu's music. Performance-based analysis is slowly and steadily gaining traction in mainstream musicology to counter the surfeit of analyses that seem entirely divorced from the act of performance, and consequently, from the experience of listening.

As a reciprocal process to analysis, the compositional process in *Sheng* is used as a mode of enquiry into the embodied musical gesture—its physical-affective-sonic constituency, cross-modal nature, and inter-corporeal potentiality. The preconceived idea of the composer who dreams up sound palettes and imagines sonic worlds and abstract structures is challenged by placing the performer's body at the front and center of the act of composition. Through the smartphone, I disrupt its functional usage and finger-visual association to convert it into a creative, full-bodied musical instrument. By placing performative gestures at the forefront, I draw on the cross-modality of gestures to create an intuitive and satisfying performance experience. Through their participation, the members of theaudience themselves are invited to explore different modes of listening and the concepts inter-corporeal resonances and inter-sensoriality. Composing with the performer's body as the focus of creative enquiry can lead to new physical-affective-sonic connections and provocative insights into instrument design with gesture technologies.

As the field of computer science enters an era when intuitive gesture technologies are at their peak, so must composers be at the frontiers to forge new gestural meanings and extend our artistic, corporeal experiences that are relevant to this age.

83

Gestural Analysis and Classification of a Conductor's Movements

Summary. Gestures can be studied as the connection of discrete points by continuous paths. In the gesture of the orchestral conductor, the points connected by the gestural path correspond to metric movements of time represented in space. Here, we will study the gestures of the conductor referring to some concept of homotopy theory. The basic metric gesture is a regular and symmetric spanning of the space between points. Musical interpretation modifies the form of these regular gestures, changing their time, velocity, energy, amplitude and directionality. Thus, the most important information for performance contained in the orchestral score can be described by gestures. The conductor can also, through his gesture, add elements not explicitly contained in the score. The conductor's gestures anticipate and continuously prepare the gestures of each musician in the orchestra in a hierarchical structure that corresponds to the structure of the score: from the general form to the articulation of each single note. In the first part of this chapter, we will discuss a case of study. In the second part, we will give some mathematical hints for a precise description of conducting gestures. In the third part, we will see an example of technology applied to conducting.

– Σ –

83.1 Gestures and Communication in Orchestral Conducting: A Case Study

To formally understand how music, from the pencil of composer to the brain of listener, can transmit its message, it is necessary to utilize methods from different branches of science: physics, mathematics, psychology, linguistics. Using concepts from all these disciplines we are able to construct the *musical ontology*: signs (expression, signification, content), realities (symbolic, psychologic, physical) and communication (poiesis, neutral level, aesthesis). Moreover, the concept of gesture, recently formalized [720], is a decisive connecting instrument between such analytical categories. In particular, orchestral conducting involves an intermediary step, the communication between conductor and musicians, which requires both specialized and common gestures. Here we present a case study, where the dimensions of signs and realities are used to analyze the gestural approach of Sergiu Celibidache during a rehearsal of Fauré's *Requiem*.

Something is *ineffable*—impossible to speak about—until one finds the appropriate words. Making a simple syllogism, if the appropriate words to describe nature are mathematical formulas, and the human being is a part of nature, then mathematical formalism can be applied to every human production and activity. Even science can describe art, and even art is present in science in the form of contemplation of beauty. In fact, the same word, *theory*, very common in scientific vocabulary, derives from Greek ϑεωρέω, from ὁράω, to see, and ϑέα (ϑέαμα), sight (spectacle). It is not the case that many artists have a double background, in science and in art. For example, the conductor Sergiu Celibidache, of which we will examine a rehearsal as a case study, was mathematician and musician. Among arts, we find that scientific concepts and computational tools have already been successfully applied in the field of music, for both analytic and creative purposes.

G. Mazzola et al., *The Topos of Music III: Gestures*, Computational Music Science,
https://doi.org/10.1007/978-3-319-64481-3_28

Musical ontology is given by three different layers, each of them having three components, see Chapter 2 for details.

- sign: expression, signification, content;
- realities: symbolic, psychological, physical;
- communication: poiesis, neutral level, aesthesis.

The communication strategy requires paths between points in this ontological scheme. The motor-entity that permits displacements among these points is the *gesture*. Gestures, present in all human activity (and not only human!), have been recently formalized [720], see Section 61.5. Playing music is effectively like emulating the movements of a dancer: he or she touches the stage in discrete points (the notes), when he or she moves continuously (gestures). The three-dimensional ontology has in fact been extended by a fourth dimension: embodiment, comprising gestures, too, see Section 57.1. Human gestures can be divided into two fundamental categories:

- specific gestures in specialized context: culturally invariant, known only by specialists in the field belonging to different cultural environments,
- gesture as support of oral communication, culturally dependent, known by all persons in a cultural environment; this kind of gesture implies a risk of misunderstandings among persons belonging to different cultures.

Orchestral conducting is a specific field of musical performance in which the two types of gestures are both present [150]. Conducting gestures are not directly finalized to the production of sound, but they suggest to orchestral musicians how to play, i.e., which gestures to perform. They are therefore kind of 'meta-gestures'. Although for many people orchestral conducting is still the kingdom either of ineffability or of uselessness, there are several treatises that explain the basilar techniques of this art [791, 941].

Let us analyze a specific case of study, represented by a chorus rehearsal [183] of the *Requiem in D Minor* op. 48 by Gabriel Fauré [311], conducted by Sergiu Celibidache, at the head of a semi-professional chorus, and the London Symphony Orchestra, recorded in 1983. This work is accessible via YouTube. The presence of music and text implies a double layer of meaning: musical content (not directly word-related) and textual content (precise reference to images and abstract concepts). In particular, the text refers to a mass for dead peoples' souls in a Catholic rite, but its fundamental concept is extensible to every human being: a person who is asking for forgiveness with remorses and guilt in the sadness of his last hour of life. For this reason, the interest of the text and its interaction with musical realities is not only related to the ambit of a specific rite, but covers a more general aspect of human life.

83.1.1 Problematics and Solving Methods

Here we will present some typical problematics of the work of the conductor, and the solutions found in this case of study. Generalizations will be discussed in the following section. In our case of study, the most relevant part is played by a chorus. Two layers are evident: the meaning of the text sung by chorus that—in this case—inspires the structure of the music, and the layer of the music itself. In some passages there are however some minor conflicts between them, and the conductor has to solve these problems case by case. We start with the primitive singing-gesture: *breath*. When performing the passage of Figure 83.1, Celibidache

Fig. 83.1. The beginning of the Soprano solo part in *Pie Jesu*. This and the following examples are excerpts from *Requiem* by G. Fauré.

says to the solo singer who was making a hurried—then wrong—take of breath: *You don't have to be afraid of the conductor. The breath, your breath is a part of expression! It is the dimension of your internal affective world. It is not a matter of being in time, because YOU create the time!* Then, he suggests making a change in the written score: *Don't sing as it is written, it is a crime!* referring to a more aesthetical anticipation of the consonant 'n' in the word 'donna'. Effectively, the mechanism of voice is not visible and external like the production of sound, for example by string instruments. The vocal chords are hidden, and the learning of vocal technique is still related to a complex ensemble of feelings and images. It is a field that needs more formalization and comprehension, see Chapter 86. Since the production of sounds by singers is strongly related to their visual and gestural imagination, the indications given by the conductor are strongly related to this field, too. Moreover, to augment the fascination of this field, there is a correspondence between two *topoi*: The pair inspiration-expiration and the pair raise-beat in symbolic scores, such as *arsis-thesis* in Greek prosody (Figure 83.3), that finds a natural realization in the primitive gesture of a conductor (Figure 83.2), as explained in [643]. Moreover, it is not the case that Celibidache was interested in Oriental spirituality and meditation, where breath and control of gestures plays a central role [184].

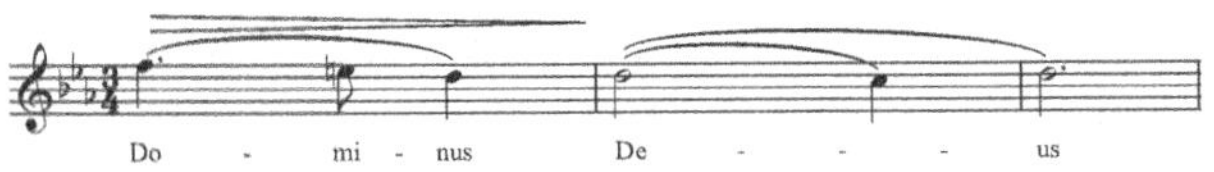

Fig. 83.2. and the corresponding *arsis—thesis* in breath: inspiration and expiration.

Fig. 83.3. The beat and the raise, a primitive gesture in conducting

In *Sanctus*, measure 32, the tenors are requested not to accent the third syllable of *Gloria*, following both the accent of the word and a soft sense of *diminuendo* in the *legato* phrase. Differently, in a later phrase, i.e., in measure 35, the shape of the melody would require an accent on the last part of the word (*Osanna*). In this case Celibidache proposes a compromise between the contrasting requirements of text and music. There is another change in measure 41 for sopranos: the last syllable of the word *excelsis* that has not to be emphasized; however it corresponds musically to a tonic accent: the first one in the measure. Referring to the repeated structure of *Osanna* in the score, the conductor uses the metaphor of 'taking in himself', in his own arms, to suggest the idea of catching the last sound, producing a *diminuendo* and a soft *staccato*. This particular gesture will be analyzed later in more detail, since it represents a point of connection between the world of specialized gestures used in conducting and the world of non-specialized gestures used by all persons in the same cultural context to express feelings. Celibidache explains that, for example, the text *Libera me domine* is not an imposition; it is rather a supplication. The concept of 'supplication' has to be translated into gesture and then into music (see our discusson of transmodal interpretations of gestural morphisms in Section 83.2). Listeners will recreate, in their minds, the original gesture of begging, then the abstract concept, and, finally, the corresponding mystical idea. To represent his intentions to musicians, the conductor uses also facial expressions: a *diminuendo* until silence is represented using whispering lips. A typically expressive role belongs to the left hand that does not in general express *tempo* indication in conducting. In a *Requiem* passage, Celibidache puts his left hand on his chest to underline a specific passage, a kind of gesture more similar to the non-specialized ones from daily life. The orchestra will not react to this gesture in the same way as to an other one. It is a more precise gesture that remarks the emerging of a loud theme or voice. One other gesture, with the open left hand, underlines the *pianissimo* in the *tutti* passage in the score.

According to Celibidache, the essence of creativity is a *primitive and direct act*. The reference to gesture, both in composition and in performance, is evident. What is the creative act in music? Of course, the composer creates. But the performer has also to give life to scores, unfreezing its gestures. In the specific case of Celibidache, according to a chorister, *he is a genius, because he recreated the piece, instead of limiting himself to reinterpreting it.* And, as the conductor himself says during an interview, *we have to find out what is not written in the score*, and then, quoting Gustav Mahler: *In the score there is everything but the essential.* Besides offering fascinating words, Mahler's statement reaches the central problem in performance.

In the score are indicated the points to reach, the notes, but not the essential instructions, how to play these notes, how to reach these points, i.e., which gestures to use. Musicians know specific gestures, but the role of the conductor is fundamental to making homogeneous in a central way the different personal interpretations and points of view. In the next section, we will discuss this problem from the general perspective of the musical ontology.

83.1.2 Results, Consequences, Applications

Performers start from the symbolic content of a score (its expression) to produce, using gestures, the physical phenomena that can convey a specific content to the listener. In our geometric frame of Figure 57.1, we start from expression/symbolic to reach content/psychological via the intermediate passage of content/physical [643]. It is a kind of transformation from gestures to musical facts, such as a rotation in the complex plane from the mental reality to the physical reality, see Section 59.2. The music via physical parameters sometimes has a dramatic effect on the psychological frame that can be described as an *allez-retour* from physical reality to psychological reality back to physical reality of the human body—a mechanism similar to the one present in psychosomatic diseases. The orchestral conductor's gestures realize an intermediary passage between expression/symbolic and content/physical, rather than expression/physical. As previously stated, the conductor uses specific gestures but also gestures derived from common experience shared by non-specialized persons in daily communication. Physical metaphors, visualized by hand or induced by words, suggest musical gestures that will produce in the listener the same content. Following the reference to the text of *Requiem*, the soul of the dead man is accepted in paradise as a 'hug' in heaven from spirits and angels. It is the case of the hug suggested by the conductor when singers in chorus sing the *Osanna*. Physical reality is, at the same time, an instrument to convey psychological content, but it is also the destination of the content. When music is artistically composed (good *poiesis*) and artistically played, its effect can affect also the physical side. Moreover, we talk of the *Stendhal syndrome* when the emotion provoked by a musical or visual artwork of extraordinary beauty causes tachycardia, dizziness, confusion, and, in some cases, even hallucinations or loss of consciousness.

What is, then, the role of a conductor? Of course, not only to give the same *tempo* for all musicians. He has to help orchestral musicians find the correct gesture to better express the content of a score, i.e., to transform symbolic reality into physical parameters that can produce a psychological effect on listeners. In order to accomplish this mission, the conductor utilizes, first of all, hand-arm gestures, and, next, verbal expressions. Verbal expressions often contain references to unexpressed gestures. If, for example, the conductor says "Here the soul of the man is received by angels in heaven," the underlying (gestural) image is an ensemble of angels with their arms toward the human, in a kind of 'welcome'. The 'welcome' is a concept understandable also by persons who do not believe in angels: it is a universal idea shared by all humans. However, also in cases not so strongly relied to visual situations, words can refer to gestures. If the conductor says "Here the altos have to stay behind tenors," he is implying a kind of spatial visualization of sound layers. It is not, of course, requiring them a change in the positions of the bodies of the singers; it is instead required to lower the loudness of voices of altos, to emphasize, in that specific time interval, the contribution of tenors. If we represent the symbolic content of a score by a simplified graph with only intensity, pitch and time (as noted using Maelzel's Metronome), the requirement of modifying intensity of a layer can be realized using a gesture that puts one layer behind, and displaces the other one to a frontal position. Therefore, in the conductor's activity, the words talk about gestures, and the hands, too, talk about gestures using gestures. The process of fruition of the artwork is a kind of signification between expression and content. What is the meaning of a *Requiem*? What is the meaning of this specific requiem we are talking about? Celibidache seems to be very conscious of this fact: For a soul to reach peace. But also musicians have to reach a sense of 'peace' during musical performance to communicate this feeling to listeners. It is interesting to note that the chorus is constituted by semi-professional singers, people who "come there only for the pleasure of making music."

83.1.3 Some Remarks

We have considered a collection of solutions proposed by a conductor to explain how to translate into real music the content of a score, also solving occasional conflicts between requirements of the text's meaning and musical structure. We have then tried to induce general relations among categories in the frame of musical ontology, described in Section 57.1, and graphically synthesized in Figure 57.2. We found that the starting point for performers, i.e., the score, corresponds to the symbolic reality and the semiotic expression. In order to reach the final state, i.e., the communication of the artwork to listeners, the intermediary steps are physical reality, and semiotic content. The physical reality affects the psychological reality, and sometimes, as in the aforementioned case of the Stendhal syndrome, affects again the physical reality in a sort of internal resonance. The mechanism that permits the evolution between two consecutive steps, constructing a *path* in the musical ontology's graphic visualization, is the gesture. Our method of analysis is in principle applicable to all music examples, also in non-Western contexts. The most striking result is that everything in art concerning feelings and emotions can be explained using precise terms, since gestures have been mathematically defined [720], and physical, physiological and psychological realities are described by science.

83.2 Hints for a Mathematical Description

The mathematical formalism developed in Chapter 78 can be applied also in the case of conducting. Here, the symbolic gesture is represented by the basic metric gesture, and the physical gesture by the real motion of the right hand of conductor. For each metric indication, there is an infinite variety of possible ways to realize the physical gesture, in the same way as we have seen for the pianist's hand movements.

Basic metrical structures for unary, binary, ternary, ternary in one movement, and tempo constitute the symbolic gesture. The number of movements in a measure is the number of points in the gesture's skeleton. If we think of a 2-dimensional skeleton, we can have a more clear idea of the starting point for conductor.

What is a symbolic gesture for the left hand? The left hand needs a separate description, because its role is not univocal. In general it carries the expressive content of the score and gives attack to performers. Expressive content of the score (*accelerando, ritardando, crescendo*, way of attack of instruments) also determines the deformation of the basic metric gesture. When the two hands move together they emphasize *gesture*. However, the risk of a mirror movement of right and left hands is accurately avoided by many professional conductors. Motions of both hands have been studied numerically, see Section 83.3.

Categories of unary, binary and ternary movements can be seen as homotopy categories. All deformations (i.e., physical realizations of symbolic metric indications) are homotopic to each other. When analyzing 5-, 6- and 7-metrical movements, and also compound meters, gestures can always be described as combinations of elements of these categories.

While referring to homotopy, we can build connections between conducting gestures and the gestures of other musicians (see also Figure 83.6), as well between music and other fields. In fact, if we define the *category of gestures*, whose morphisms are the morphisms between gestures, we can compare gestures from music (gestures of the orchestral players and of the conductor), between music and image (gestures of drawing and the motion of the bow or of the fingertip), and between music and emotion (the caressing gesture and gentle piano touch). However, we will not deal here with the complex topic of semiotics. We can call this concept *gestural similarity*, and it will be the subject of future research. These processes can be described in the diagram of Figure 83.5, if we identify $\overrightarrow{X}$ with music, $\overrightarrow{Y}$ with painting, and $\overrightarrow{f}$ with the change from the first to the second.

As suggested by global gesture theory (see Section 66.5), conducting gestures can also be constructed by gluing together simple gestures. Conducting didactics concentrates attention, during students' training, in such elementary movements, first in abstract examples, then in fragments of real scores, and finally in the context of complete orchestral pages. For details about the Russian conducting school of Ilya Musin, see [791].

More precisely, the conducting gesture can be defined in the following way. Let us consider the basic metric gesture C_{Im} (2-dimensional space, 1-dimensional symbolic time), with weights (parameters depending

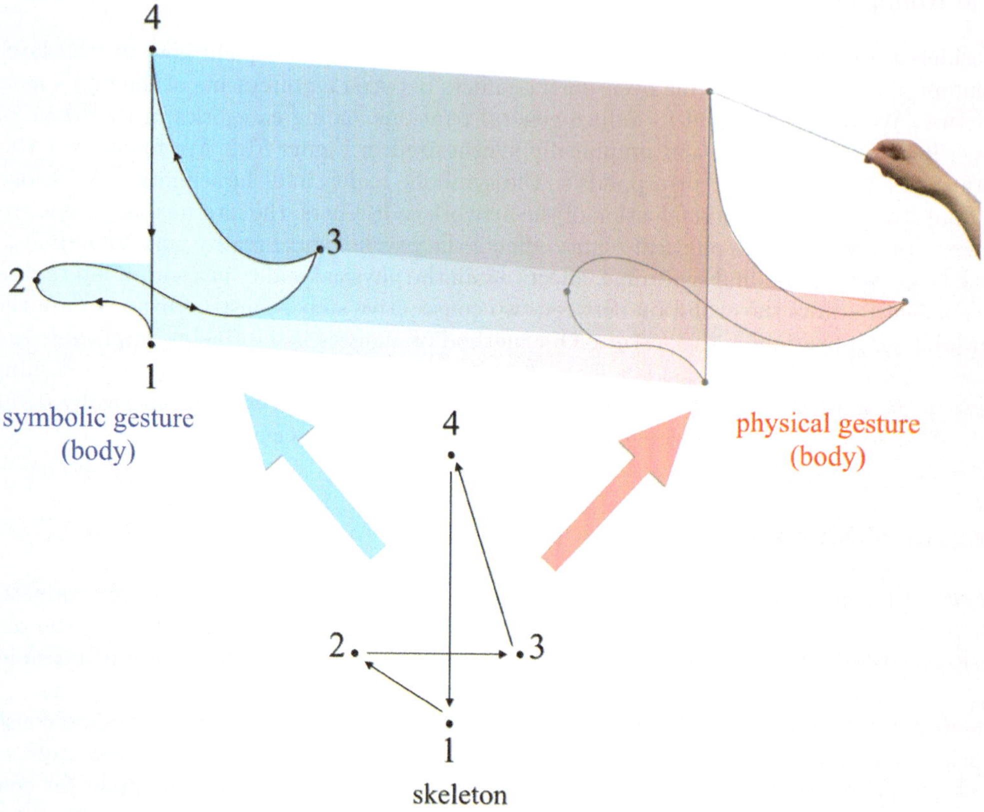

Fig. 83.4. The symbolic gesture is represented by the basic metric gesture (left), and the physical gesture (right) by the real motion of the right hand of conductor. They are connected by a world-sheet hypergesture.

on the content of the score). Let us also consider a performance operator $\mathcal{P}$ having these weights as arguments. The conducting physical gesture $\mathcal{C}_{Ph}$ is given by $\mathcal{P} \times \mathcal{C}_{Im} = \mathcal{C}_{Ph}$.

Moreover, we can observe that the full score, read by the conductor, is mathematically given by the colimit of fibers of the entire ensemble of pitches-durations-intensities, where each fiber corresponds to a set of these values labeled as *violin*, *cello*, *flute*, and so on. The opposite motion, from the full score to the separate parts, is in fact given by these fibers. It corresponds to the graphical appearance of an orchestral score, where the notes are distributed in groups for each instrument.

To end this section, we show in Figure 83.6 the coexistence of performance theory of notes, of gestures, and of conducting gestures.

83.3 Data Analysis

How is it possible to formally collect and analyze data of conducting gestures? The project PHENICX [380] answers this question. They have developed a technology of kinetic devices applied to the conductor's body, see Figure 83.7. They track the motion of articulations, and allow a precise description of gestures. This technique permits many different formal analyses. For example, movements of trained and not-trained musicians and conductors have been analyzed to describe the *instinctive* component for the conducting gesture [931]. The basis of this analysis is the schematization of the conductor's body using the position of torso and distances of points from its coordinates, normalized depending on the conductor's height. The

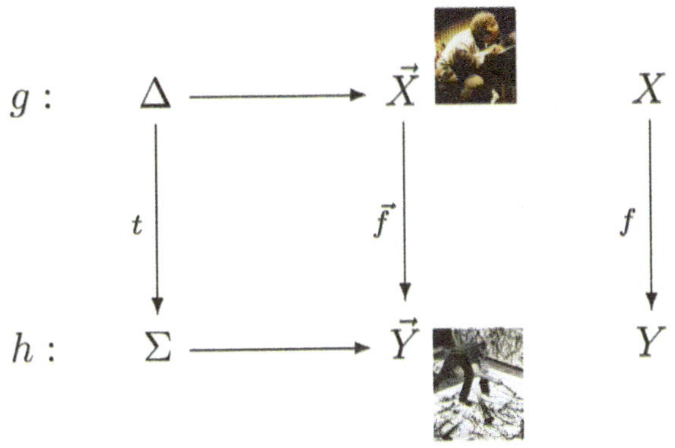

Fig. 83.5. We can build gestural relations between painting and music, in the formalism of category theory. Let $g : \Delta \to \vec{X}$ be a musical gesture and $h : \Sigma \to \vec{Y}$ a painting gesture, connected by a morphism f between X and Y ($\vec{f} : \vec{X} \to \vec{Y}$). The diagram must be commutative, which means $\mathbf{f}g = ht$. As two examples, we have on the bottom the painting gesture by Jackson Pollock and on the top the piano gesture by Cecil Taylor.

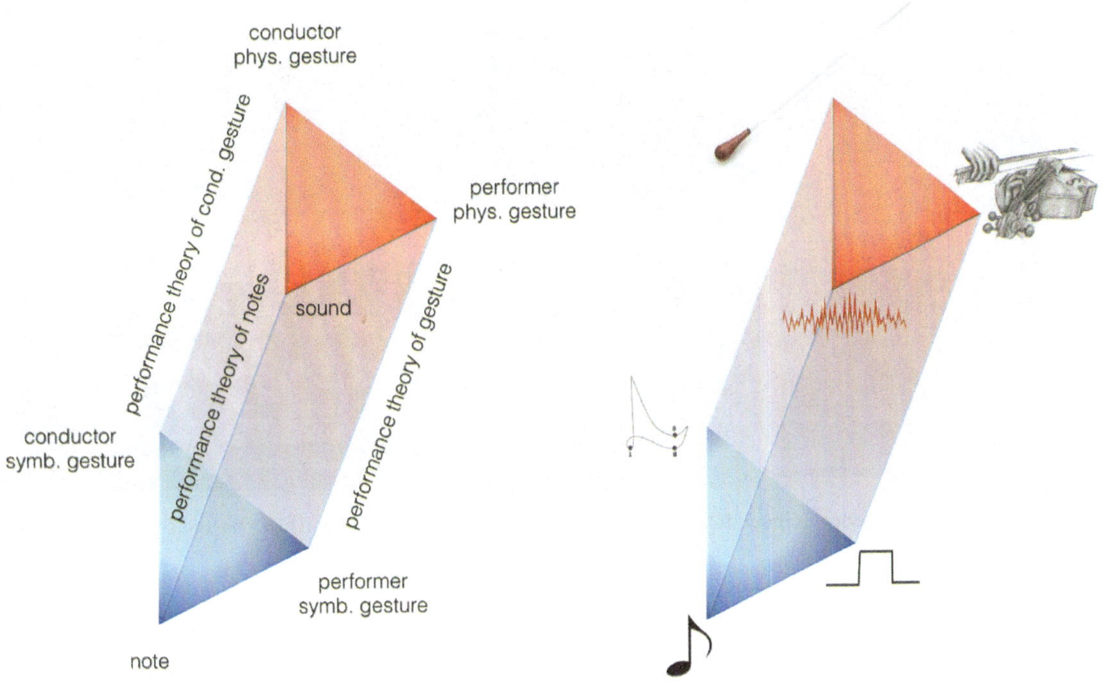

Fig. 83.6. The mechanism of conducting, between performance theory of notes, gestures, and conducting gestures.

empirically derived equations given in [931] are the following:

$$x_j^{tor} = \frac{(x_j - x_{torso})}{h}\frac{1}{1.8}, \; y_j^{tor} = \frac{(y_j - y_{torso})}{h}\frac{1}{1.8}, \; z_j^{tor} = \frac{(z_j - z_{torso})}{h}\frac{1}{1.4}, \tag{83.1}$$

where $h^2 = (x_{torso} - x_{head})^2 + (y_{torso} - y_{head})^2 + (z_{torso} - z_{head})^2$.

A particularly interesting example is the analysis of beat via gestural conducting tracking [931]. In this way, it is possible to understand the entity of deformation, along the time dimension, of intervals from basic metric gesture to real physical movement. In [931], the influence of the beat positions on conductors gestures has been quantitatively studied. There is a variety of gestures to anticipate the incoming beat, and their musical importance for communication to orchestral musicians is dramatic. It has been shown that acceleration along the trajectory is a potentially good candidate for beat extraction [621]. For every frame in which a musical composition has been previously divided, acceleration values have been calculated for all joints of the conductors' arms, by computing the second derivative via a second order polynomial fitting. Beat annotation has been accompanied by error distribution of the beat predictions.

Similarly to our study developed in Chapter 78, a preliminary study can focus just on the vertical dimension y, with elementary vertical up-down movements. In the case of conducting, as shown by observation on 3-dimensional models and stressed in [931], the most important information for beat is given by motion on the y axis. In fact, beat is often signaled by maximum upward-downward acceleration along the y axis.

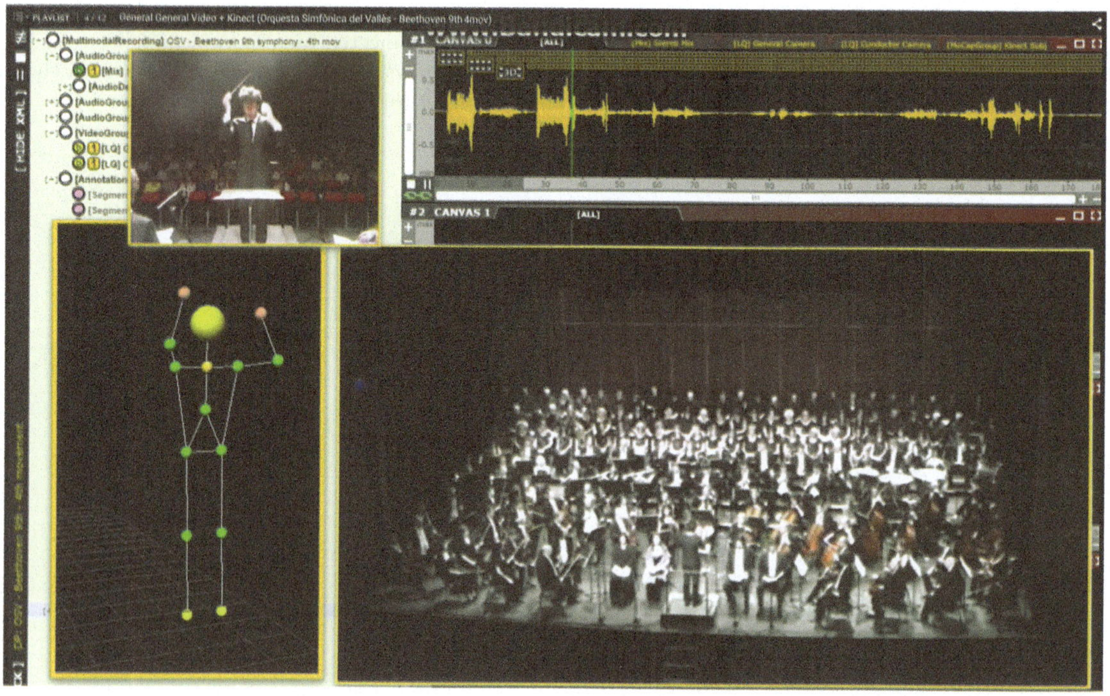

Fig. 83.7. The tracking device Kinect for conductor gestures, experimented within the context of the PHENICX project [380].

83.4 Conclusion

The Italian conducting school, whose important names include the well-known Arturo Toscanini, is very gestural-based. Franco Ferrara, famous for recordings of film music scores (such as Nino Rota's), and known to many conductors for his teaching activity at Accademia Musicale Chigiana di Siena, had a clear conception of gesture.

Franco Ferrara loved to "transfer into gesture all his musical thought, formed through an attentive reading of the score. This transfer operation is realized by modifying metric gestures (see Section 83.2). Mechanical gestures of 4/4, 3/4, 6/8 and so on, must be adapted to musical idea." [173] Bruno Aprea, former student of Ferrara, used to explain conducting with the same words. This is the idea at the basis of the Italian conducting school, but also applied by Herbert von Karajan in Austria. In general, the French conducting school presents a more developed idea of *concertazione*. In the Italian school, conductors are used to rarely stopping the orchestra, except in correspondence of passages containing elements that cannot be represented by the gesture [173]. Ferrara also used to sing to the orchestra: "If you sing, the orchestra will play in a good way." Also voice, in fact, contains hidden gestural indications, see Chapter 86.

It is interesting to know that Franco Ferrara after a stroke was not able to move his right arm. He conducted with his left arm and with his eyes. Metric and expressive content were so concentrated in his left hand. His students remember that his gesture, even after the stroke, was clear, effective, powerful. In fact, sometimes, observing the solutions found by persons with some physiological difficulties allows "normal" persons to improve their work. The two most important things in conducting are listening (to both mental and physical sound) and gesture. Absolute pitch always helps in fine-comprehension of orchestral sounds and mistakes. Sight helps establish a stronger communication with musicians, to better check their behavior, and, also, to quickly look at the score. However, can people conduct without looking at the orchestra? Yes, they can. It is the case with the young pianist Marco Orsini, blind from birth. He's able to conduct an orchestra through a perfect score memorizing (a practice followed by many professional conductors), and a precise study and control of gestures. For completeness, the first blind conductor is the Italian-Argentine Gabriel Francisco Bergogna. Another blind conductor is Luigi Mariani.

83.5 Addendum

We should not conclude this chapter without a hint at a recent Swiss publication: "DirigentenBilder—Musikalische Gesten — verkörperte Musik" [1012], which we cannot critically review here for time reasons, but it must be mentioned. The reason is that this book's approach is somewhat complemetary to the above. It refers to a number of scholars and musicians who mostly use the German language, with a few exceptions using other languages.

The first instance is a citation by Igor Stravinsky, where he complains that conductors are replacing the audible by the visual, and thereby also replacing the composer's central role by a kind of visual co-composer. Carl Dahlhaus completes this negative judgment by a positive point of view, attributing to the conductor the role of an aesthetic identifier, governor ("Statthalter" in German). He observes that this identifier is not transporting any semantic content, but simply a visual equivalent to the dynamics of sound. Dahlhaus identifies the conductor's gestures as a *presemiotic* phenomenon. In the writings of Friedrich von Hausegger the unity of visual and auditive utterance is only realized for a singer, and the instrumentalist has a split situation that the conductor must reconcile; he talks about[1] the "Orchester, als dessen Verkörperung der Dirigent erscheint." It is interesting that Helmuth Plessner's anthropology of music stresses the presemiotic character of the conductor's role: Pure music does not represent anything, tells us nothing, does not illustrate or symbolize anything. It only excites a psychological echo without any semantic charge. The psychologists Ernst Kurth and Alexander Truslit also stress the gestural character of music, which then is made accessible through the conductor's movements. Truslit even argues that the passage from neumes to modern notation is a step backwards to a less expressive notation. Compare Chapter 86 for such matters.

[1] Orchestra whose embodiment appears to be the conductor.

Check for updates

84

Reviewing Flow, Gesture, and Spaces in Free Jazz

Summary. Reviewing the production of the video *Imaginary Time*, we claim that the time that is created in free improvisation (let us take the purest type of improvisation here to deal with the unmixed phenomenon) is categorically different from score-generated time.

– Σ –

84.1 Improvisation: Defining Time

Improvisation is traditionally understood from its etymological meaning: playing "all'improvviso", in an unpredictable and unforeseen manner. It is a spontaneous production in the sense of a negative definition: improvisation as a negation of order, system, or rule-based behavior. This has been seen to be not only artistically invalid, it is also the opposite of what improvisers know.

But there is more than the concept of improvisation that we could summarize as an art that produces a flow which is generated by gestural communication in a collaborative space, an insight which we have gained in our free jazz book [721]. The difference with non-improvised music is not only produced by the absence of a written score; it is not merely an inner score (Siron's "partition intérieure" [979]) which controls the musical output much like an external material score. This approach to improvisation is driven by the same idea, namely that music, if it has an artistic quality, must be the result of a mental reflection, of a thought process similar to a mathematical result that is the output of a hidden formula.

The difference is in fact not only persistent on the poietic side, it is also an aesthesic difference: One cannot listen to improvised music as if it were the rendition of a predefined inner or outer score. When listening to written music, be it a well-known traditional composition from the first Viennese tradition or a complex New Music creation, such as György Ligeti's or Unsuk Chin's piano etudes, there is a categorical difference with a free jazz composition by, say, Cecil Taylor or John Coltrane. The difference is not in the complexity or character of these compositional structures, they could even be of comparable type. The difference is the creation of a radically different musical reality. The existential character of these musics is different, no score could have generated such a music. We claim that the time that is created in free improvisation (let us take the purest type of improvisation here to deal with the unmixed phenomenon) is categorically different from score-generated time.

84.2 Flow, Gestures, Imaginary Time and Spaces in the Music Movie *Imaginary Time*

The video *Imaginary Time* [361] was recorded at the famous jazz club *Yokohama Airegin* in Yokohama, Japan on the occasion of two concerts, on December 25 and 26, 2014. Our trio, Heinz Geisser (Switzerland) and Shiro Onuma (Japan) on drums and percussion, Guerino Mazzola (United States) on grand piano, had

G. Mazzola et al., *The Topos of Music III: Gestures*, Computational Music Science,
https://doi.org/10.1007/978-3-319-64481-3_29

collaborated since 2010, Geisser and Mazzola had collaborated since 1995, and Geisser with Onuma since 2005. Figure 84.1 shows the trio in action.

Fig. 84.1. The trio Geisser-Mazzola-Onuma in action at the *Yokohama Airegin* jazz club.

The video has been described in a short PR text as follows:

> *Imaginary Time* presents pianist *Guerino Mazzola* and drummers *Heinz Geisser* and *Shiro Onuma* performing 85 minutes of passionate e-motive music in swirling, telepathically attuned interplay.
> This film is suitable for all audiences: aficionados, experts, and children hungry for the vibrant vision of future music.
> In *Imaginary Time* what you hear is what you see, a deep connectivity of musical and gestural expression, exquisitely recorded by a Japanese team of audio and video technicians, and edited by a first class music film professional in Europe.
> This work is unique: performed, directed, and produced by creative virtuosi burning to transform musical imaginary time into fully enjoyable sensual reality.

The video's pieces are

1. Alter Space (14:33) (0:0:49-0:15:22)
2. Round About Midnight (4:00) (0:15:35-0:19:35)
3. One for Wang (15:40) (0:20:04-0:35:44)
4. Giant's Steps (15:12) (0:36:33-0:51:45)
5. Kanreki Onuma Gambare (17:00) (0:52:50-1:09:50)
6. ¡Ornette! (7:50) (1:10:35-1:18:25)
7. Suzuki's Delight (5:19) (1:18:40-1:23:59)

All compositions are by the trio, except the second, *Round About Midnight*, which is by Thelonious Monk. The fourth piece was also designed by Mazzola following a mathematical analysis of Coltrane's *Giant Steps* [731, Chapter 22]. In the following sections, we want to describe the characteristics of this music which in its style is derived from free jazz, but includes many other sources, cultural and scientific in nature.

84.2.1 The Compositional Character of the Pieces

As a general remark, we should make clear that no plan or compositional template was given in advance or fixed before a piece was played. Of course, everybody knew Monk's *Round About Midnight*, but even there, no structure was fixed, no repetition of choruses, or whatever scheme. The piece *Giant's Steps* had been designed by Mazzola, but not in the sense of a lead sheet structure. He had only invented a set of groups of chord changes or melodic units, but no syntax and even less a sequence of soli or similar musical roles.

That the music nevertheless sounds very much as if it had been composed is the result of two reasons: To begin with, the members of the trio have a huge repertory of structures and styles that have been accumulated through around four decades. It might suffice to hear or interpret a small extract of a known composition or rhythm or style to lead the trio in a corresponding direction. For example, at the end of *Kanreki Onuma Gambare*, the pianist cites the end of Duke Ellington's and Juan Tizol's *Caravan*, and everybody knows immediately that this initiates the composition's terminal sounds. Or in *¡Ornette!*, at time 1:17:28, the pianist starts a stride piano accompaniment with the left hand, a rhythmical figure that is immediately resonating in the drummers' beats. And in *Giant's Steps*, at time 0:38:20, Geisser had initiated a very delicate sound with his tiny bells, and thereby inspired Mazzola to make a short citation of *Silent Night*, just passing by and very soon leaving this reference (after a short, speeded-up repetition) for other sounds.

The second reason is that this music is the opposite of spontaneous or "all improviso", unprevisible. It is organized in what we now call *imaginary time*, a mental reality that is well-known to performing artists of all sorts. This layer of consciousness shapes the physical reality of performance, but it does so in a logical and knowledge-based way, virtually knowing everything in no *physical* time.

These two reasons add up to a performance that resembles the rendition of a composed musical creation.

Fig. 84.2. The strong agreement of the three musicians on large forms is manifested by their gestural expressions right after the piece's end.

84.2.2 Large Forms

The compositional character of this music is manifest in the shaping of large forms, too. The majority of the video's compositions are quite long, around 15 minutes in duration, except for the last piece, which was conceived as an encore. When one looks at the large sections of these pieces, they show prominent presence of even classical sections, such as the sonata form. For example, in *¡Ornette!*, it starts with an exposition, then adds a development with a strong rhythmical dominance driven by the drummers' dense interplay, then comes back in a recapitulation that includes the above mentioned stride piano reference, and then terminates with a short slow coda.

Fig. 84.3. Shiro Onuma's precise hit on the cymbal.

A strong sign of large formal organization is the magic of the pieces' ends. It never happens that some musician(s) is (are) terminating while others are still playing. It even happens as a rule that the trio's members stop in a well-organized musical way, and seemingly without any preconceived ending phase. All of a sudden, the piece ends in a logic that transcends communicative preparatory actions. One more proof of the imaginary time where creative logic is performed. The strong agreement of the three musicians is manifest by their gestural expressions right after the piece's end, as shown in Figure 84.2.

84.2.3 Precision of Attacks

The strong control of unfolding time is evident from the precision in time which is required to comply with this musical dynamics. Figure 84.3 shows Onuma's hit on the cymbal which must be performed in the interval of a 50th of a second to be correct.

The musicians are always aware of such small time intervals, not only for the percussive performance of the drummers, but also for the pianist's fast attacks due to his special interaction of left and right hand, as shown in Figure 84.4. The left hand is below the right one and is periodically hit by the right one, causing the left hand to move like a spring very fast between keys and the right hand's palm. The frequency of this periodic movement is around 20 Hz.

Fig. 84.4. The fast pianist's sound production is due to the gestural interaction of his hands.

84.2.4 Co-presence of Different Time Layers

The imaginary time presence is also effective in the simulateous perception and reification of different time layers. This music is very percussive in the sense that there is not only one big beat to be followed, but there are several, typically three, four or more beat pulsations that coexist. The musicians do not choose to play only one of them but several layers simultaneously. Focusing their rhythmical attention on one, then on another layer, according to the music's momentous configuration. This effect is evidently related to the flow, which generates a suspension of time in favor of gestural utterances per se, not as a slave of clocks.

Fig. 84.5. Geisser in extreme flow, catching a cymbal that was flying away after having been thrown onto the snare drum.

84.2.5 The Reality of Imaginary Time

Imaginary time might at first look like a mathematical artifact, but it is a musical reality of first order. This has been evidenced in the previous Sections 84.2.1, 84.2.2, 84.2.3, and 84.2.4. This reality cannot be proved entirely, but the individual experiences of musicians and their shared knowledge confirm this fact.

84.2.6 Measuring Flow

Flow usually cannot be measured as it is a kind of *qualia*. Nevertheless, the video proves in many instances that the musicians are in a flow state where they act in a thoroughly controlled way within very short time intervals, as shown, for example for Geisser's action (Figure 84.5) when catching a cymbal that wants to slip away. This is performed within a very short time, a 50th of a second, and Geisser maintains his absolute control over the rhythmical structure he is creating.

84.2.7 Explicit Perception of Gestures

Gestures can be observed during the entire concert. But sometimes they appear in a pure form, i.e., not producing any musical sounds through instrumental interaction, but simply as moving hands in the air in the sense of an understanding of resonance according to Cavaillès: catching the gesture and continuing. Here, in Figure 84.6, the pianist catches Onuma's percussive gestures and continues with his hands in order to prepare his own intervention.

Fig. 84.6. Mazzola drawing hand gestures in the air as a resonance to Onuma's drum play.

85

Gesture and Vocalization

Summary. The curves traced by a drummer's sticks, the various characteristic hand shapes adapted for various note clusters on a piano, the various ways that elbow and shoulder joints can support strokes on a violin all have sonic consequences. Indeed, if the previous chapters have taught us anything, it is that musicking is inherently (rather than incidentally) gestural. But there may yet be a lingering suspicion in some readers' minds (particularly those who are accustomed only to playing from notation) that the graceful arc of a pianist's hand is less like a dancer twirling across the stage and more like a blacksmith hammering a piece of metal into a horseshoe. The skeptical claim would be *that gesture is a necessary practical step in the production of a finished, pre-figured sonic product,* and no more.

We feel the preceding chapters have provided ample evidence against this claim. But, even granting its hypothetical possibility, how would it account for the extensive gestural movement of vocalists? Though it is certainly possible to willfully sing while sitting still, most of us nonetheless move our hands and arms as we sing (in addition to the elaborate internal motion of the laryngeal cartilages, the diaphragm, the tongue). Indeed, in many traditions, it is difficult not to. Yemenite cantors, Hindustani and Carnatic vocalists, flamenco singers, gospel soloists, and many other singers all move in elaborate ways as part of elaborate vocalization. This chapter will address this motion-unified action of the hands and the voice.

– Σ –

85.1 Vocal Gesture

Of course, the gesture that concerns us is of a very specific kind. We are not interested in every little muscular twitch that occurs during singing. Singers, like anyone else, straighten their clothing, sip coffee, adjust their posture, etc. We also, for the purposes of this book, specifically exclude encoding messages gesturally (the *semantic* level of gestures in Genevieve Calbris's terms; *emblems* in Adam Kendon's strict sense [530, 529])—pointing upward at the sound technician to get more voice in the monitors, nodding at an accompanist to encourage him to take a solo, drawing the index finger across the throat to cut it off. (We acknoweldge, of course, important ethnographic insights that accrue from attending to such signs [201], [525], etc.) We also exclude (without denigrating or dismissing) the elaborate, disciplined, conventional gestures of expression that often accompany theatrical singing—slowly extending the hand toward a distant beloved, tearing at one's hair, holding both hands over the heart [987]. These gestures certainly accomplish things in the broadest sense, and have their own elaborate rhetorical and kinesic structure—but they are easily detachable from vocalization, and even taught explicitly by vocal coaches; thus they are not spontaneously tied up in the production of melody in the sense we care about. We also are not considering the formalized and elaborate movements of *dance.* While singers do often dance while singing, there are several significant differences, though prior to the country's independence, Indian courtesans—among others—cultivated elaborate traditions of singing and dancing simultaneously. Dance ordinarily does not reflect melodic form the way that vocalization and song gesture do. Dancers (while occasionally sounding with their feet) dance to melody that

G. Mazzola et al., *The Topos of Music III: Gestures*, Computational Music Science,
https://doi.org/10.1007/978-3-319-64481-3_30

is produced by *others*, coordinated through a combination of choreography and improvisation. Vocalists, on the other hand, move and vocalize in a single coordinated action, and the two serve as vehicles for a single improvisational practice. Finally, and most importantly, dancers submit to years of training dedicated to the cultivation of disciplined movement. In contrast, singers' gestures are mostly implicit and unintentional.

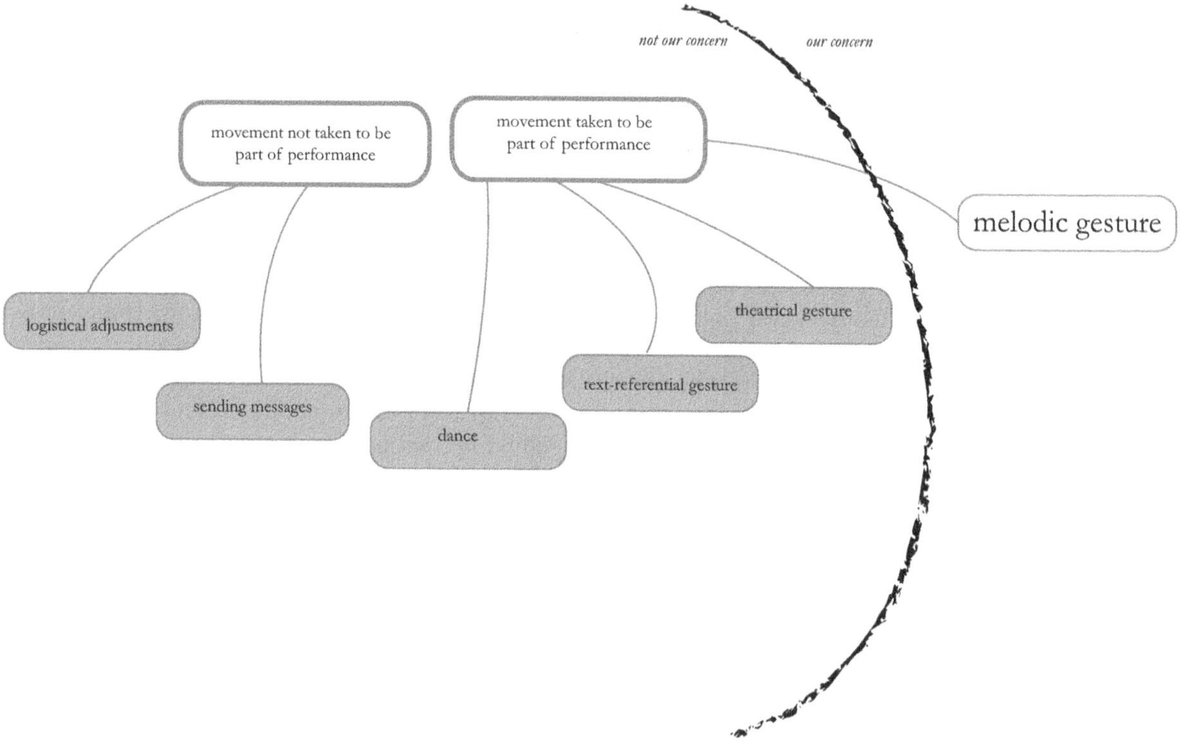

Fig. 85.1. Inclusions and exclusions.

We are specifically interested in movement that is inherently melodic, that is so closely co-timed with vocalization so as to form a single action. This we call *melodic gesture.* As Clayton [201] points out, these gestures are not "sound-producing" in the way that the arc of a mallet on a xylophone is. But, as we will see, hand movement and vocalization are somehow part of a single process, no more separable in practice than the tongue and the soft palette in articulating a dipthong. This is so much the case that gesture is hard to separate from sound, hard to thematize analytically. Gesture seems to be a natural, transparent background to vocalizing—in the most fundamental phenomenological sense, *invisible.*

As it turns out, this is equally true of speech vocalization. As gesture is rarely foregrounded analytically, it is often taken to be a secondary phenomenon, a mere ornament on the more essential substance of words-as-signs. And yet, the shapes we trace in the air in conversation are neither entirely arbitrary nor entirely determined by our speech. Gesture studies as a discipline has largely concerned itself with the exquisitely timed co-performance of speech and gesture, and in so doing it suggests a way beyond the usual boundaries of classical linguistics: beyond a system of discrete signs and lexical meanings.

Rendering gestures analytically thinkable does not produce evidence that gestures are interesting in and of themselves, or an argument for the universal beauty or goodness or meaningfulness of gestures. Indeed, attention to gesture has just as often led critics to conclude that it is an impediment to certain ethical goals of music. (For a consideration of questions of descriptive ethics, aesthetics, and politics with regard to song gestures, see [872] and [1110].) But attending to gesture as an integral part of vocalization can teach us a great deal. Seeing vocal performance as inherently gestural takes us beyond the apparent primacy of discrete notes and words as the key material of vocal action: toward the open-ended, creative, relational play of space, accomplishment, and self-fashioning.

85.2 Vocal and Manual Motion

The first thing to notice is that the motion of the hand and the motion of the voice are exquisitely co-timed. The hand does not follow the voice, nor does the voice follow the hand; they move as one, united toward a single purpose, like the shoulder and elbow joints of a tennis player.

This does not mean, however, that the shared goal of the voice or the hand is *articulating notes*. While this may sometimes be the case, it is most eminently not the case in spontaneous flows of improvised melody geared toward larger melodic sweeps. What follows is an excerpt from a performance by Mariah Carey, an exquisitely trained pop singer renowned for her melismatic improvisation. The excerpt here serves as a kind of cadenza at the end of the song *Vision of Love*, performed live on Good Morning America on July 20, 1990. The full performance is available on YouTube. The part analyzed here begins at 3:25.

We have split this four-second-long phrase into eight subphrases, labeled A-H (Figure 85.2). Subphrases A-G are nearly identical vocal gestures, swooping from a fifth scale degree (graced from below by a sharp four) up to a flat seven (articulated twice quickly, in the manner of a quick vibrato, but perceptible as a single note). Subphrase H jumps up to the high tonic and descends via a minor pentatonic run to the low tonic, looping around on 4.

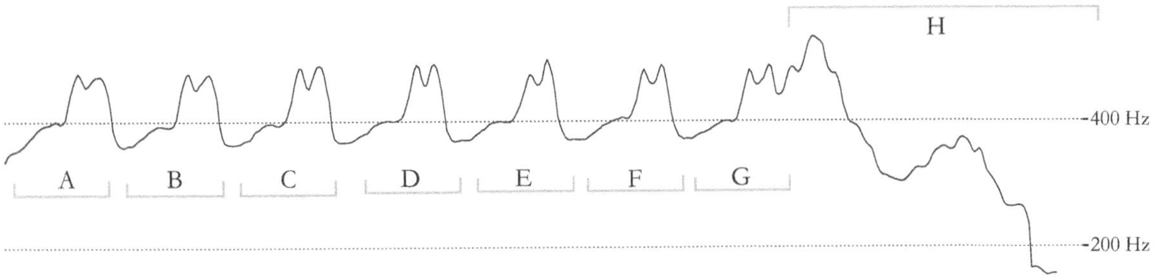

Fig. 85.2. Pitch vs. time trace of a cadenza.

While phrases A-G are sung vocally, the hand, loosely open with the palm facing forward, traces circling loops in the air along the vertical plane (situated just in front of the body like a plane of glass.) Figure 85.3 shows one such circle articulated by the hand, encoded in an undulating line that shows pitch movement on the vertical axis. The alternating black and grey tones indicate movement away from the body (black) and toward the body (grey). All seven circles follow this schema, though their microtimings of course vary. Phrase H, not pictured in Figure 85.2, follows an undulating course from high to low, more or less along with the pitch of the voice. Thus far, we might quite reasonably suspect that the hand is merely mapping pitch height onto the vertical axis. Certainly, phrase H would seem to confirm this, as the hand traces a looping descent from high to low. And in phrase A, the first manual zenith is linked with the first vocal zenith on 7b.

But as we look at the rest of the phrases, in Figure 85.4, it quickly becomes clear that the hand is not merely encoding pitch; at times, the hand descends when the voice ascends, and vice versa. Once the singer slips into a gestural groove (in phrases B-F), the manual zeniths fall into alignment with 5s in the voice, and the vocal loop up to the 7b is affiliated instead with manual descent *from* the zenith to the nadir. These gentle turns are not anchored to discrete points in space measured in centimeters along the y-axis

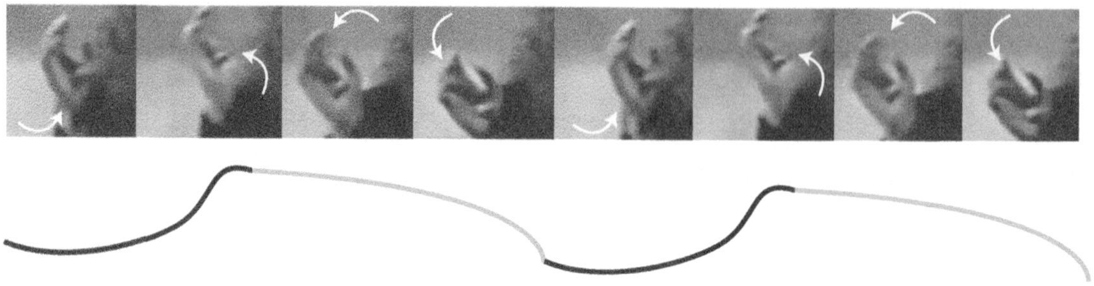

Fig. 85.3. Looping gesture.

or points in time measured in milliseconds along the t-axis. They rather have the temporality of a dancer twirling against the meter, a magician's flourish, a comedian indicating a particular texture of insomniac loopiness with a rotating finger in space. The melodic activity here (in both hand and voice) is not sequential repetition (i.e., not a repeatedly linear moving from point to point in what turn out to be identical phrases) but falling into a groove: a continuous rotation around a stable point. Spinning itself (rather than a sequence of absolute spatial positions of the flesh at particular time-points) is at the heart of this musical moment.

It is thus most parsimonious to analyze this segment as consisting not of twenty-odd notes or seven phrases, but of two melodic spaces: the looping space and the space of descending from the high tonic. These spaces, I will argue later, are not cartesian or Euclidean spaces, but spaces of potential bodily action, akin to the hypergestures discussed in Section 61.6.

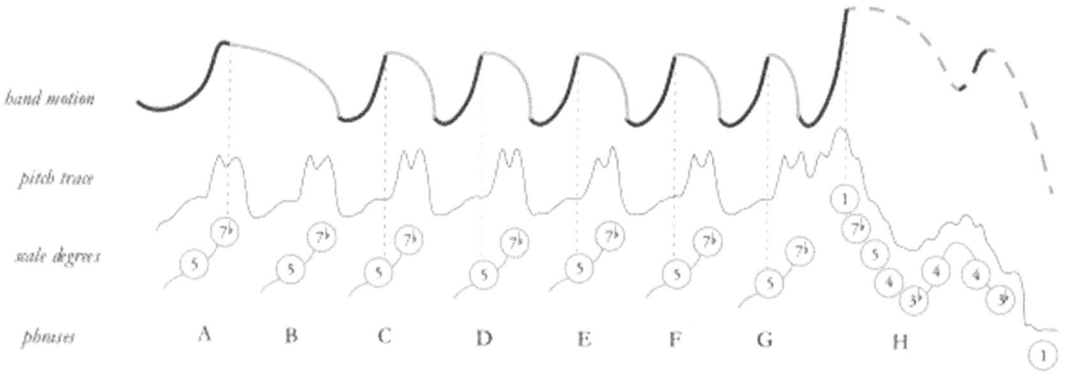

Fig. 85.4. Gestural and vocal analysis of looping segment.

85.3 Gait

On the way there, we will have recourse to an analogy with a familiar kinesthetic activity: walking. While we certainly may analyze walking strictly in terms of destination (as an illuminated line, for example, on Google Maps) the analytic loss would be great: we would miss much that is at the heart of walking. What would we make of strolling, strutting, *flânerie, passeggiata* [783]? What would we make of the very different lived experiences of trotting and galloping? Just so, to think of vocal gesture only terms of its notic content would ignore much of what is at the heart of vocal action.

A similar sense of melodic motion as a gait emerges from Katsman's study of Yemenite Torah recitation. Here, the consistent set of normative gestures used in teaching and recitation have no clear relationship to the ascent or descent of pitch. Instead, the gestures are linked to the melodic progress of the voice along the path of the text: they indicate walking, stopping, continuity, and breaking. The voice, text, and body become intimately linked, so that "the verse becomes a living, acting body" ([525, p.5]) rather than a sequence of words or pitches.

85.4 Hindustani Vocal Music

In the remainder of the chapter, we will focus largely on examples drawn from singers of North Indian raga music, the topic of *Musicking Bodies: Gesture and Voice in Hindustani Music* [872]. As most readers will no doubt be unfamiliar with raga music, the following background is intended to provide an orientation. But, like all orientations, it renders certain things especially visible: in this case, movement. Most accounts of Hindustani music fundamentals proceed by explicating grammatical structures. In principle, grammatical rules such as *there are two possible species of third scale degree,* or *phrases in Raga Hameer skip the fifth scale degree in ascent* or *phrases in Bageshri don't end in dangling fifth scale degrees* are understood to describe eternal structures that determine proper musical utterances, independent of their instantiation in bodies and voices. All of this is very useful—for teaching, for notation, for understanding the discursive history of music theory. But in contrast, the following account (largely a summary of [872]) puts bodily action and discipline at the heart of Hindustani music practice.

Of the several dozen vibrant song traditions in South Asia, traditions of raga music stand out for the great centrality of spontaneous melodic elaboration. No two performances of a raga, even when drawing on the same compositional material, consist of even remotely the same sequence of melodic moves; 80 to 90% of a performance is spontaneously generated according to the needs of the moment. This is part of why seeing a great Hindustani vocalist sing is viscerally thrilling even for newcomers to the tradition. The grace, power, flexibility, melodic imagination, and sheer speed in his disciplined vocal rendering is a display of spontaneous mastery akin to watching Michael Jordan playing basketball or Yip Man performing Chi Sao or Baryshnikov dancing. This may be why awestruck writers from V. H. Deshpande to Alain Danielou have tended to depict Hindustani musicians as inspired geniuses, effortlessly drawing on some transcendent source.

But undergirding this breathtaking kinetic freedom are years of grueling, extended, boringly repetitive training. To a romantic eye, spontaneity and mechanical repetition may seem to be at odds. But, as we have seen in Section 59.4 and in [721], the "freedom" of improvisation traffics in the passing on of gesture, and the awesome swoops and arcs of a master do not come from nowhere. One part of training consists of sitting with a great master, doing as as he, imbibing his musical disposition as an eaglet learns to hunt—this imbibing even, at times, resembles a kind of inspiration. But the great bulk of training consists of thousands of solitary hours grinding through drills: singing phrases, scales, and compositions over and over again. Even in the warm, nostalgic glow of childhood memories, adult musicians recall these interminable drills as tedious, bothersome [785]. But this highly unromantic training, carried out at the expense of what one *wants to do*, makes possible the magnificent kinesthetic-melodic grace and freedom evident in a master. Repetition generates durable habits of singing in tune, gliding smoothly between notes, generating clear timbre, and—most of all—graceful melodic motion.

A phrase such as 7134577655343, repeated hundreds of times, teaches a great deal about melodic motion quite apart from its notic content. It is not simply that one inductively determines the rule "skip 2 and 6 in ascent" and is ready to carry on; nor is it simply that one perfects the particular phrase and "has" it forever after. The repetitive drilling trains one to fall into a melodic groove, flowing along steadily at a wide range of speeds—simple on paper, but a far from trivial task in practice. But it allows building more specific melodic habits: to find one's footing along 77655 as a reliable path for descending motion to the 5, to feel 71345 as a reliable path for ascending to the 5, to feel 343 as a point of repose. One practices beyond the point of "perfection," or simply being able to sing note sequences accurately, to the point of being able to move effortlessly in the space of a raga [785]. A trained singer, after all, needn't think of every note that s/he sings any more than Pele would need to count the steps he takes before kicking a soccer ball. The capacity to sing freely in broad, graceful, tuneful melodic sweeps—rather than note by punctilious note—is a crucial sign of mastery in Hindustani music.

85.5 Notic Models and Kinetic Models

Singers are nonetheless perfectly capable of generating what I call notic models of melody: breaking their musical utterances into notes, constituting melody as a sequence of atomic units (as I did above: 7134577655343, for example.) This task is made easier by a strong, reliable tonal center, reinforced by a continuously sounded tonic drone, and a conventional system of note names in which 1 and 5 are never modified by accidentals.[1] Notic models are remarkably useful, providing a basis for remembering compositions, correcting mistakes, and analyzing scalar content.

But singers just as often analyze vocal action as motion—drawing on what I call *kinetic models*. Particularly at the level of large-scale melodic action, melody is constituted as motion: for example, we commonly hear of *aroh* (ascent), *avaroh* (descent), *vakr* (crooked) and *sapat* (straight) motion, *harkat* (subtle movements with no net motion), and *sukoon* (stillness). Kinetic models provide a robust way of understanding vocal action that complements a note-centric model.

But, coming to Hindustani music after 18 years of training in notic analysis, these kinetic models at first struck me as fanciful. Isn't kinetics just a sloppy, idiosyncratic way of describing the fundamental objective givens of melody, the atomic building blocks that we rigorously designate by note names? In the first place, there is nothing acoustically objective about thinking of melody as discrete stopping points. The voice, after all, is always moving. Consider the following pitch trace of a phrase in Raga Alhaiya Bilwal, sung by Ulhas Kashalkar, see Figure 85.5.

Though the note names given on the diagram represent a conventional analysis, they certainly are not given or obvious from the acoustic signal. Surely, there are broad correlations between pitch height and note name. But there are no straight horizontal lines in the pitch trace that would indicate perfect stillness of the vocal apparatus, or the unchanging thingness that we imagine when we hear notes. (It is this thingness that allows us to speak of a "fifth," for example, as a thing we hear.) The "notes" that arise correspond roughly to the peaks of continuous curves; even if we thought that notes had absolute pitches, each pitch would only be sounded for an infintessimal second (as, geometrically speaking, the moon is only full for an infintessimal fragment of clock-time.)

In the second phrase (from the same performance), we see that even prolonged tones are in motion, acoustically speaking. The sustained third scale degree which closes the phrase arises phenomenologically as a steady, unchanging pitch, a thing, exquisitely in tune. Its acoustic signal, however, is constantly in motion.

This is not a sign of a vocal shortcoming. Nor is our perception of notes a failure to apprehend what really is going on. This is simply the way vocal action and vocal perception work. Notic analysis is not a description of acoustical reality—it is an act of phenomenological constitution after the fact. The melodic utterance is hearable as notes only through a disciplined constitutive process (honed not through reading and examination, but by training the throat and ear with the repetitive vocal practice regimens noted above).

Nor is the sense that melody is in motion objectively given in the signal. (It certainly may be given in the movement of the vocal muscles and the hands, but we defer that until later.) The sense that melodies rise and fall, sweep and swell, pause and rush forward, is familiar even to listeners who haven't been trained to hear notes. Indeed, thinking in discrete notes can be a distraction from learning the fine kinesic details of rendering phrases.

The Hindi word commonly translated as "note" in cases like these, *svara*, is analytically ambiguous. In one sense, it can indeed refer to the seven atomic scale degrees, like solfege (sa re ga ma pa dha ni), but in another it refers to the voice that is already set into motion [757]. Were we looking for a parsimonious

[1] Though intonation is a fascinating and contentious field of debate (see [596], [498], [872]), for our purposes it is most parsimonious to accept the consensus of most theorists of raga and consider the gamut to consist of seven svaras, chosen from twelve possibilities roughly equivalent to the semitones on a piano.

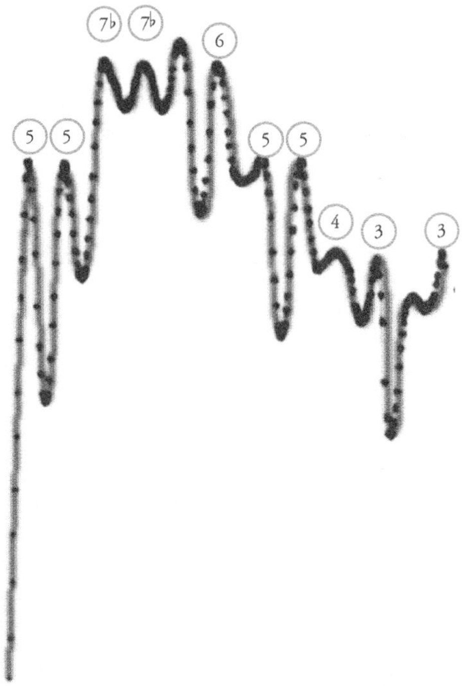

Fig. 85.5. Pitch trace, Ulhas Kashalkar, Alhaiya Bilawal.

description of the acoustic symbol, we would find it not in notes, but in this aspect of *svara*: swaying, sliding, bobbing, undulating.

Once we accept the kinetic model as no more or less fanciful than the notic model, it is no longer surprising that the hands, arms, head, and torso move along with the voice. The voice meanders slowly upward, and the hands trace a meandering path. The phrase repeats, this time with more elaboration, and the hands trace spirals on their ascent. Then, shooting straight upward, in a "straight run" (*sapat taan*) to the high major second, the hands grab and stretch an elastic substance suddenly available in front of the body, holding it steady as the high note rings out still and plain, letting go of it as I resolve to the high tonic. It is almost impossible not to move when one is improvising something new, furthest from what one has committed to memory; movement is less pronounced when singing a composition plainly that one already knows well.

Shifting our focus from melody-as-notes to melody-as-motion makes song gesture slightly less mysterious. However, before proceeding to analyze concrete examples of song gesture, we must first of all avoid two theoretical pitfalls in making sense of the relationship between the motion of the hand and the motion of the voice.

It is tempting, in fact, to think that the movement of the hands somehow encodes the movement of pitch up and down. As we will see, things turn out to be quite a bit more interesting.

85.6 The Realist Pitfall

Since the gestures of singers (unlike the gestures of drummers) do not, in themselves, produce sound, it is easy to imagine that they are mere *depictions* of sound. Drawing on the theoretical resources offered by classical mechanics, we might imagine that singing really just consists of air vibrations produced by the vocal mechanism. And so, when we see the voice and the hand ascend together, we might, from a realist point of

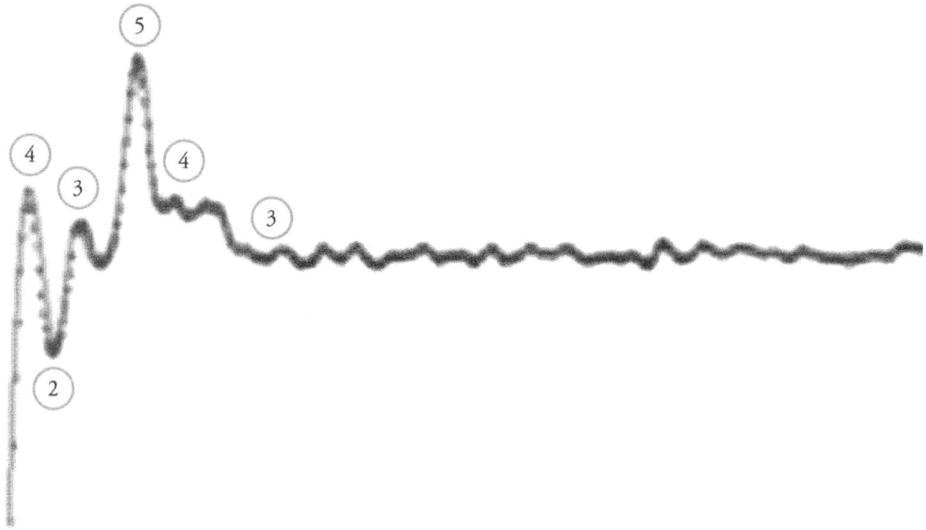

Fig. 85.6. Pitch trace of prolonged 3. Note the continuous, irregular acoustical movement even in a note that sounds steady.

view, imagine that the movement of the hand is merely a metaphoric *depiction* of what the sound is doing. Nothing in these vibrations, a realist might say, is actually exhibiting net motion—this is precisely why they are vibrations and not translations. Thus, even to speak of a melody "rising," "falling," "slowing down," or "circling around a note" is to indulge in a fanciful but misleading metaphor, however convincing it may seem. Such a metaphor, we might think, is merely a handy way to come to grips with the inherent slipperiness of sound, like speaking of "building" trust (as though it were an edifice) or prescribing a vacation as an "antidote" for heartbreak (as though one had been poisoned). From such a realist point of view, then, the ascending movement of the hand at best *encodes* the real, given acoustical essence of the ascending melodic fragment.

The sequence CFEG is a series of pitches with a net increase in frequency:

120 160 150 180

In this realist view, the sense that 180 Hz is "higher pitched" than 120 Hz is merely a metaphoric mapping of vibrational frequency onto the y-axis. As Lawrence Zbikowski has demonstrated, this particular mapping is widespread but by no means universal [1152, p.63]. Just as practices of love may be implicitly framed by metaphors of war, hunting, or travel, so too may changes in frequency may be implicitly understood as motion, ascent, or descent. The strictly realist view holds that these metaphors are mere fanciful ornamentation superimposed on what is truly the case out there in the real world.

But the graphic representation of pitch versus time can also be misleading. We do not hear motion quite like this, and, it turns out, we certainly do not move like this.

85.7 The Subjectivist Pitfall

[The claim to] unmediated experience signals a danger that is worse, if anything can be, than naive realism: its polar opposite, naive subjectivism.

Brian Massumi

Since the gestures of a singer do not make any sound, and are not beholden to any grammar, it is easy to imagine that they are utterly free expressions of personal fancy. In this view, the gestures of the hand express arbitrary imaginings of the singer's mind, going wherever and whenever they please. In this view,

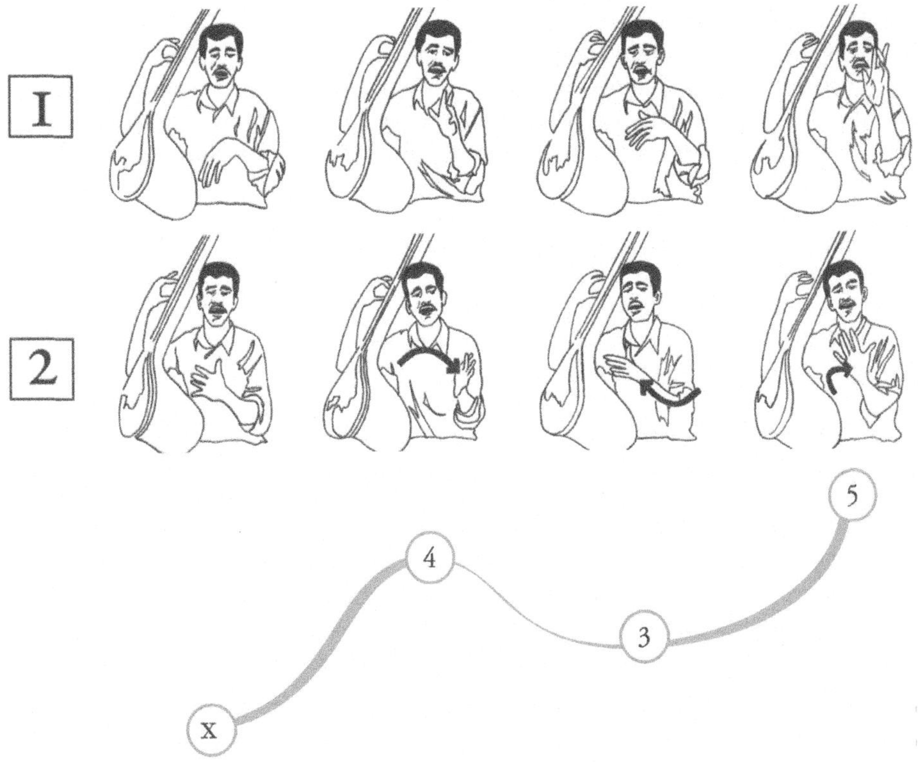

Fig. 85.7. Gestures accompanying two different articulations of a x-4-3-5 ascent.

gesture is a physical *expression* of a pure subjectivity free of purpose, free of responsibility, free of others. The mirror image of radical realism, then, is a radical subjectivism in which the only thing that exists is unmediated experience. Indeed, this idea of gestural freedom is closely akin to the idea of "free" jazz as mere honking, its merely doing as one pleases at the expense of an audience.

It is partially for this reason that many Indian music critics raised on gramophone recordings in the age of small, private gatherings were astonished, once large public concerts became common, to find that their favorite singers moved while they sang. They had grown accustomed to listening to records as a pursuit of detached, spiritual joys, and were uniformly disheartened by the physicality of singers on stage. They decried the "violent and spasmodic movements" of singers (Bailur in [872, p.30]), who moved conspicuously, as though the body were expected to remain as still as a phonograph speaker.

But musical gestures do not come from nowhere. They are "logical", and they traffic between bodies. Though not derived algorithmically from a sound product, they are neither willful nor arbitrary.

Surely, we do not (as the realist might think) simply depict vocal action with our hands. Nor do we (as the subjectivist might think) simply flap our hands around at will.

Improvisation in khyal unfolds within modal and metric frameworks (raga and tala, respectively). A quick summary follows. Talas are metrical cycles marked by strokes on a tabla or other drum, and provide a measured, periodic place of return for melodic elaboration. Ragas are complex modal entities that are less specific than individual songs, but more specific than scales. While allowing for infinite possibilities for improvisation, they indicate sets of notes, guidelines for their arrangement in phrases, and, most importantly, characteristic ways of moving between them. These melodic features are often associated with specific emotional states, and in some cases to times of the day or seasons. Ragas are also often spoken of as a space in which melodic action takes place ([872]). Khyal improvisation, then—like extemporaneous speech—takes place within structures even as it requires the creation of novel material at the moment of performance. For a body-centered take on raga and tala, see [872]; for a more conventional grammatical account, see [67].

One important context for this motion is the space in which melody moves. Both gesture and vocalization navigate through the specialized topographies of ragas. As melodic motives are developed over time, they solidify into features of a raga landscape, in both gesture and vocalization.

... and can also take on the status of virtual objects and materials, which are manipulated gesturally. Gripping, turning, pulling, and releasing these virtual objects are important modes of engagement with melodic ideas. A phenomenology of melodic performance emerges from these observations that includes motion, participation in flow, and manipulation of objects as an extension of the musicking body. This body includes the mechanics of vocal production and gesture as well as the virtual world of melody: both flesh and form.

[...]

We caution against this interpretation. In fact song gesture neither represents sound nor is purely arbitrary. Gestures beget "dynasties of gestures."

[...]

In Figures 85.8 and 85.9, we see two consecutive phrases (marked here as A and B) in a performance of Raga Rageshri by Amir Khan. This is an interesting, but hardly unique, counterexample to the oft-cited claim that the height of the hand is a map of pitch height. Indeed, the hand tends to be lower in space as the pitch rises, and vice versa. Nonetheless, it would seem at first glance as though pitch height is nonetheless being mapped linearly onto cartesian space. The first part of phrase A seems to fit this logic: the distance along an axis stretching diagonally downward from the chest to the ground in front of the singer's body seems to correspond to pitch height. This correlation is not robust across all of the melodic action, however. In particular, the crucial closing cadences in each phrase do not map onto a single axis.

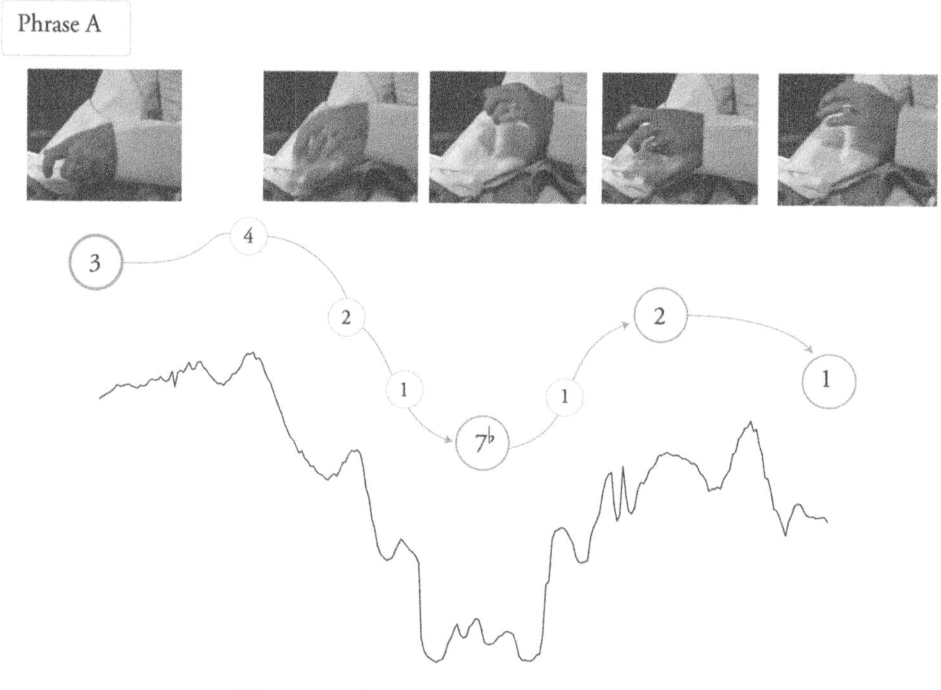

Fig. 85.8. Amir Khan, Rageshri, Phrase 1.

The situation becomes clearer if we attend to the shapes of the hands. Amir Khan, after all, is not randomly scuttling about along a single dimension of pitch. He is building a melodic world and gracefully exploring its architecture, playing with energetic tension and release. In subphrase A, he flirts with the common cadential formula in Rageshri: 3421. He begins with a tense handshape on the 3: the wrist joint flexes, deviates ulnarly, and grips a virtual object rigidly. We expect the release of this tension on the tonic:

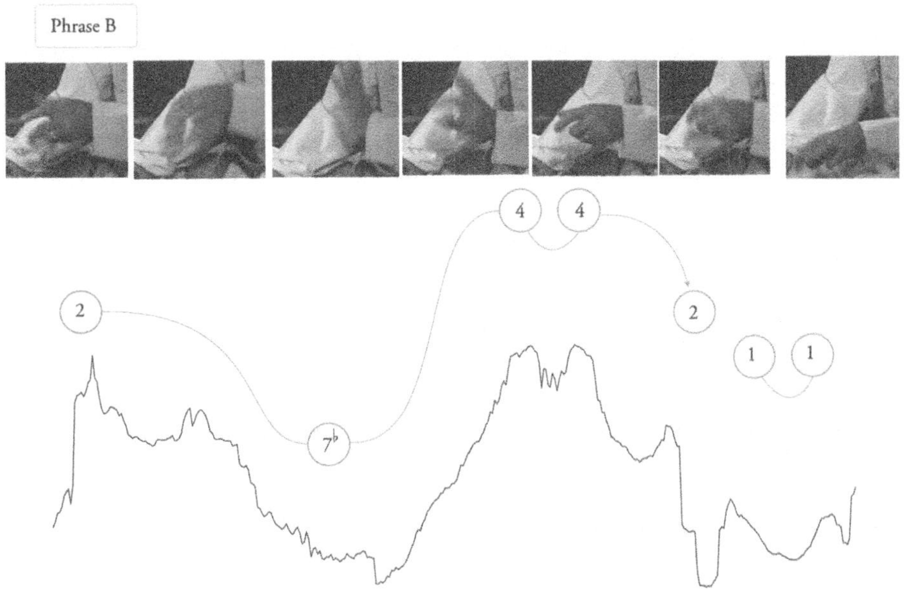

Fig. 85.9. Amir Khan, Rageshri, Phrase 2.

we imaginatively protend the joint unflexing, undeviating, and relaxing. But instead of ending on the tonic, the melodic gesture (both hand and voice) continues right on through to perch on the flat-7 for a moment, as the wrist—instead of relaxing—extends in the opposite direction. On the return to the 1, he again overshoots, landing on the 2 (with a more relaxed version of the first handshape, with more moderate flexion, deviation, and grip) before coming to partial rest on the 1. The 1 at the end of subphrase A is not a completion of the macrogesture, however. The wrist maintains a moderate extension, the handshape retains a slight, relaxed cupped grip, and the arm draws upward in preparation for subphrase B.

The 2 at the beginning of subphrase B has a nearly identical handshape to the 2 at the end of subphrase A: with moderate flexion, moderate ulnar deviation, moderate grip. But this time instead of the simple return to the tonic, there is a dramatic, continuous swoop down and up, touching the flat-7 (with maximum wrist extension and an open palm) before reaching the 4 (with one last flexion/deviation/grip handshape) Throughout, the play of melodic tension and release is marked as much manually as it is vocally.

Hierarchies of kinesic units, marked by various degrees of rest, correspond roughly to hierarchies of vocal phrasing. Phrasal hierarchies of gesture can be seen in rest positions ranging from fully engaged hands (shaping air, accelerating, manipulating virtual objects) to fully disengaged hands (buttoning shirt, resting in lap, drinking tea).

85.8 Speech Gesture

Now that we have a way of thinking through the gestural achievements of melody, we can turn to the other major kind of vocal activity: speech. Though speech is often construed as a way of getting a message across, there is of course much more happening.

Even commonsense descriptions of this unified vocal-gestural action often draw on semiotic metaphors ("body language," "speaking with the hands," etc.); the relation between hand and voice reveal that much more is going on than the mere transmission of signals.

The analysis of speech performance is fraught with theoretical pitfalls.

For one thing, because speech is so easily analyzed as words (and because we think that words are, first and foremost, signs) we are likely to assume that the sole purpose of speech is to convey information. To be sure, there are moments when we do speak primarily to convey information: a stranger asks us if this is Broadway, and we say "yes." A grandmother asks her granddaughter how old she is, and she says "three." In these odd, remarkable moments of information exchange, certain specialized gestural actions would work just as well—we could just as well nod and smile at the stranger; the child could just as well hold up three fingers. Such gestures are called emblems—quotable gestures, gestures that do their job by virtue of having, in principle, a pure, unambiguous semiotic content.

But most speech is not like this. Day-to-day exchanges mostly consist of formulaic greetings, subtle insults, attempts to impress each other, flirtation, one-upmanship, persuasion, requests, promises, scoldings, consolation. We call the meeting to order, we silence someone who is speaking out of turn, we encourage our timid colleague to speak up. We speak, that is, largely to achieve things.

When we shake hands with an old friend (or kiss his cheeks, or embrace, or merely make eye contact), and ask how he has been (Shu akhbarak? Kaise ho yaar? What's up?) we are not simply soliciting information—we are building rapport, strengthening bonds, inviting him into a shared interpersonal world of mutual care. When we are introduced to a new colleague and ask where he grew up, we are not only (or even primarily) gathering facts about thim—we are looking for some shared connection that will bring us closer together (I visited Damascus before the war!) or give us insight into his situation (as a refugee who has likely learned English in the last month or two, even this party may seem overwhelming.)

In Section 57.7, we focused on the ways in which the leading lights of the Anglo-Saxon tradition (Goffman [374], McNeill [741] and Kendon [529]) have appeared to treat speech gesture as semiotic, communicative action. But this tradition also treats speech-and-gesture as pragmatic and existential—as a kind of action that achieves things (Kendon [530, p.257, 351], and passim McNeill [741, p.19,100], and passim Goffman [374, p.34]) and calls into existence embodied, active modes of being-in-the-world; as McNeill says [741, p.99], "a gesture is not a representation, or is not only such: it is a form of being."

How long can we go on like this, trading unambiguous bits of information until one of these utterances becomes an issue of concern, before we begin celebrating, contesting, sympathizing? If an entire conversation consisted of a sequence of simple representations such as this, it would seem very odd to us indeed.

It is by now fairly common to speak about the "language of music" (in reference to the grammatical structures underlying note combinations) or the "music of language" (in reference to prosody). But the similarities between speech gesture and song gesture suggest that the improvisation of melody and speech may have more in common than is revealed in the structural and prosodic features of notes and words. Improvisation in khyal is guided by flexible frameworks such as raga and tala that operate like grammatical structures, and in this sense, even the vocal improvisation that occurs in the course of performance is akin to improvised speech. Of course, singing is not, in the classical linguistic sense, *language*.

But then again, neither is speaking. The performance of speech (the melodic and rhythmic gestures of the voice, the disciplined postures of the back and neck, the subtle play of the hands and arms) is not *language* (i.e., *langue*) in the classical linguistic sense. But nor do we suggest that vocal performance is exactly *parole*, a supplement to *langue* that merely embodies pre-existing syntactical structures, or prosodic ornamentation that merely lends emphasis or contextual specificity to a foundational, core, essential sequence of words. As we have seen, the spontaneous co-performance of gesture and vocalization need not be directed toward the communication of words or notes at all; but it is always directed instead toward conjuring and inhabiting gaits, worlds, ways of being—ways, in short, of chanting the world ([750, p.218]).

The "world" we chant is not typically the size of a planet. As the preceding examples have shown, the kinesthetic horizon of a gestural world tends to be roughly on the scale of a human body; the aural horizon tends to be roughly on the scale of a room. And yet the ethical, political, and spiritual horizons opened by inhabiting these gestural worlds may stretch well beyond our everyday life of eating, walking, waking, dreaming, and driving: well beyond ourselves.

86

Elements of a Future Vocal Gesture Theory

Summary. Gesture Theory has been first developed using the pianist's gesture as paradigm. However, the analytical techniques and the results found can be applied to other musical situations, once symbolic and physical gestures have been identified. For this reason it is possible to develop a gesture theory of voice. Voice teachers explain vocal technique also via gestures and references to imaginary movements of the voice through the resonant cavities of the body: we can think, as just a first example, of the passage from the *di petto* register to *di testa* register. For voice, there really are some *inner* movements, of larynx, vocal folds, tongue, as well of the diaphragm. Thinking of imaginary movements, the singer effectively changes the real shape of his or her phonatory system, obtaining the desired effect. These movements, connecting (imaginary once, and then embodied) points, are gestures. In fact, there are gestures that help one sing, but the singing is *itself* a gestural activity. Moreover, we can adapt to the modern gestural math-musical formalism a powerful instrument of the past, the *neumes*. The neumatic notation is the ancient way to notate the shape of the voice singing Gregorian melodies. This system successively evolved into a precise notation of pitches via points (square notes) in a four-line staff, and finally evolved to today's notation of (round) notes in the five-line staff. Explicit reference to gestures are also used in textbooks about the didactics of the Gregorian chant. We end the chapter with the proposal of a new neumatic notation for voice didactics and composition that can complete the information given by the musical score.

– *Σ* –

86.1 Why a Theory of Vocal Gestures?

The first reference to voice inside the gesture theory was a question of a reviewer: Interesting ideas and work, but what about the cases when gestures are not evident, even hidden, as for singing performance?[1]

Looking through the existing references of a so-called *vocal gesture*, it is possible to find, on one side, a reference to significant symbols about the association between a grapheme and a sound generated by the human voice, and on the other side, a not-well-defined general concept that would be an attempt, by voice specialists and singers and teachers, to define the ensemble of physical movements and positions of the phonatory system that allow the production and the control of the singing voice.

It therefore seems that the definition of 'vocal gesture' is first of all an exigency to avoid the vagaries of verbal expression in describing the singing activity, connecting what singers *think* and *do* when translating into sounds the information contained in the score. This is also the general philosophy of gesture theory. One of the main problems of singing is that the same mental images that help some singers to correctly use their phonatory system don't work for others. Vice versa, following other scholars [333], a knowledge just restricted to physiology could seduce the singers to a non-natural use or an over-stressing of their muscles. The use of abstract images reminds one of non-physiological gestures (more poetically than mathematically

[1] In Ancient Greek, *voice* is φωνή, translated as both *voice* and *sound*.

G. Mazzola et al., *The Topos of Music III: Gestures*, Computational Music Science,
https://doi.org/10.1007/978-3-319-64481-3_31

we can call them *imaginary* gestures) in real time, as opposed to real physiological gestures. The use of the first, combined with the knowledge of what really happens in their bodies, allows singers to have a deeper consciousness and mastery of their technique.

86.1.1 Studying the Voice Without the Singer?

Study of the voice is not limited to the analysis of the final result, *the sound*. As for all music, there cannot be music without sounds, but sounds are not all music. Moreover, the case of the singing voice is unique in the sense that the performer and the instrument are the same; it is an excellent case of musical embodiment. However, there are entire branches of research on the quality of the voice, also without singers, as the reconstruction of the voice via acoustical synthesis, as done at IRCAM in the context of the CHANT project [913]. Our approach is the opposite, because we are interested in the *human gesture* that produces the sound.

86.1.2 Parts of the Phonatory System and Their Functions

What are the parts of the human phonatory system, and how are they involved in motion?

According to [333] the human phonatory system is divided into three main parts with three main functions. First, the motor that sustains the sound production, second, the system that produces the sound, and third, the filter that acts on the sound, by enriching it with harmonics and amplifying it with resonances.

The *first part*, the motor of the entire process, is the diaphragm, that acts as a bellow. The diaphragm, shown in Figure 86.3, is a muscle constituted of three superposed folds. It is traversed by the aorta, by the superior vena cava and by the esophagus. It is connected to the central nervous system mainly by the vagus and frenic nerves.[2] When the diaphragm moves downward, the chest cavity expands, compressing the visceral organs. As the chest expands, the lungs are filled with air. When the diaphragm returns to its equilibrium position by moving upward, the lungs are compressed, letting the air flow out. The motion of the diaphragm can be described via parameters, such as the speed and the acceleration of the ascending motion, that influence the length and the energy of the vocal emission.

The *second part* is about the sound that is produced inside the larynx (see Figures 86.1 and 86.2). The larynx is an ensemble of cartilages (linked to the hyoid bone) and muscles that act on them. While the proper sound emission is produced at the level of the (true) vocal folds (we also have two *false* vocal folds) as a consequence of the repeated opening-closing, the entire motion of the larynx modifies some parameters of the vocal folds, influencing the produced sound. There are three main cartilages in the larynx, called thyroid, cricoid, and arythenoid.

A characteristic rotation of the larynx with respect to the trachea, called laryngeal rocking, affects the length and the thickness of the vocal folds. If we ask a singer to sing a C and then its upper octave, he/she will feel a movement upward of the larynx while singing the octave. This idea suggested a first, intuitive definition of vocal hypergesture, where we have a macro-gesture of the larynx containing micro-gestures of opening-closing of the vocal folds. Another fundamental gesture is, of course, the movement of diaphragm. In Section 86.1.6 we will however give a more precise and formally useful definition of vocal hypergesture using spectra. The vocal cords (often called *folds* for their layered structure), that open blowing up for the air pressure from the diaphragm, hurt regularly each other during the phonation, causing waves of rarefaction and compression of the air above them. These are the acoustical waves that will then be filtered by the resonant cavities (laryngopharynx, oropharynx, nasopharynx), the space immediately above the folds, to the external space.

Third part: The harmonic quality of the voice also depends on the characteristics of these resonant spaces. A great part of the difficulties in singing is due to the impossibility, for a singer, to directly control the inner gestures involved. There are several studies about the biomechanics of voice [558, 1055, 1056]; however we will not deal here with this approach, trying instead to create the basis for a general mathematical model to be inscribed in the frame of performance gesture theory.

[2] It is interesting to note that the Greek word φρήν is also used to indicate the *mind* as a physiological entity (we can think of words such as *phrenology*, and *schizophrenia*).

86.1.3 Imaginary Gestures in Real Time?

I'm giving here a small collection of some of the imaginary gestures that singers think of in order to shape in a desired way their phonatory system[3] [348]:

- Head voice. What is it? When singers think to *make the sound on the head* it means that the sound emission, for high pitches, has vibrations that can be perceived in the crane. As it is explained in [348], what really happens is that the cricothyroid muscle little by little starts to work, thickening the vocal folds, and the vibratory perceptions are directed toward the crane.
- Chest voice. The opposite happens for the chest voice, when the perception of vibration is at the level of the ribs. Also according to [348], the vocal muscle shortens the vocal folds, augmenting their thickness. The resulting vibrations are mainly directed toward the ribs.
- *Sostegno.* When singers try to *sustain* the breathing and the voice, they contract abdomen muscles to facilitate the upward motion of the diaphragm (as we will see, they return toward the equilibrium position). In this way, the singers can better control the pressure under glottis at each time of the exhalation. The final effect is of keeping constant the sound emission with a homogeneous intensity for the entire musical phrase.
- *Affondo.* What does it mean? By lowering the larynx to augment the space above vocal folds and with hypertonic lips, the voice is darkened.
- *Falsetto.* It is a kind of 'false' voice, much higher than normal. Cricothyroid muscle and larynx elevation have the effect of making more rigid and longer the vocal folds that vibrate only at their free border, with a shortened time of contact.
- "You should put metal in the voice" means: make the sound more brilliant and shrill. How? The truth is that it is necessary to find the good point of the resonators with the diaphragm pressure and the intervocalic sound.

86.1.4 Space of Voice Parameters Gestures

What are vocal gestures? To identify candidates for vocal gestures and hypergestures, it is necessary to study the physiology and the inner voluntary movements of the phonatory system. The first vocal gesture—a 'pre'-gesture—is the breathing. Breathing does not imply the phonation, but the phonation is impossible without breathing. We will define now the space of vocal gestures, the set of variables that contributes to the output parameters of sound emission and vowel specialization. The space of vocal gestures is defined by:

- air pressure from the lungs, also determined by the movements of the diaphragm (Figure 86.3),
- amount of air,
- vocal folds,
  - mutual position of the vocal folds as open or closed on their plane, angle of opening,
  - density, length,
- larynx: angle with respect to the trachea, detail of cartilages of larynx,
- time,
- filters: resonant cavities,
- vocal formants.

Candidates for vocal gestures can be envisaged in the movements of vocal folds, larynx and diaphragm. Although we don't move the vocal cords directly, we move the muscles that put them into vibration. We can define geometric parameters for the degrees of freedom associated to these gestures. We mainly have two angles, ϕ and θ, between the vocal cords,[4] and between larynx and trachea (see Figure 86.2), as well the length ρ of the vocal cords. In fact, the movement of the larynx implies their stretching. See Figure 86.1 for the details of muscles that act on vocal folds. Geometrically, the system can be described via modified

[3] Thanks to Salvatore Sutera, tenor, voice professor and writer, for the little collection and the reading suggestions.
[4] We should mention that we have four vocal folds, two true and two so-called false.

spherical coordinates—with a little change in the coordinate ϕ for the orientation. The diaphragm gesture, permitting the breathing, can be schematized as a vertical movement. It will be described later in more detail.

As previously mentioned, by taking these parameters as a domain, we can define a function that, as output, gives us the usual musical parameters such as pitch and loudness, for example, as well other parameters, linked to the vowel specialization. In fact, it is possible to sing different notes by using the same vowel. In the same way, it is also possible to change the vowel while singing the same note, this is a well-known exercise for vocal performers. The specialization into a vowel or into another one is given by the different vocal formants; it means, a harmonic of a note augmented by a resonance [507]. Formants are objects of deep study [175]. We can describe gestures of vocal folds as points in a high-dimensional space. Lines connecting them can be the movements of the larynx, as well the *legatura* given by the same breathing: more than one or two notes performed during the same exhalation. In this way, we obtain vocal hypergestures.

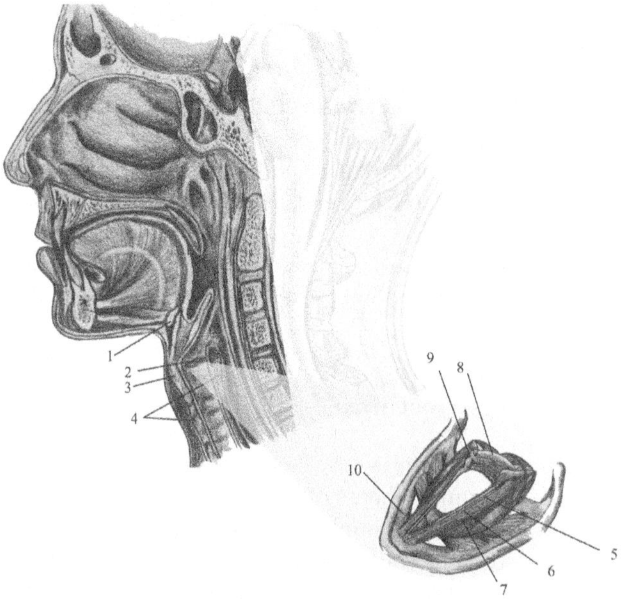

Fig. 86.1. Section of the human head showing the larynx. 1: hyoid bone; 2: vocal folds; 3: thyroid cartilage; 4: cricoid cartilage. Details with muscles of vocal folds. 5: vocal ligament; 6: vocalis muscle; 7: thyro-arytenoid muscle; 8: cricoid cartilage; 9: arytenoid cartilage; 10: thyroid cartilage. The phonation is produced by the movement (the gesture!) that alternately brings together and moves apart the vocal folds between themselves. Drawings by Maria Mannone. The parts of the human body responsible for the phonation are the vocal cords (sometimes called vocal folds due to their layered structure), moved by muscles. Vocal cords are part of the larynx, constituted by cartilages and muscles. The motion of the larynx stretches the vocal cords, changing their length and shape and changing the pitch. Moreover, to sustain high notes, a singer needs much more breathing: in this way, the change of breathing contributes in changing the sound emission.

86.1.5 About the Importance of Breathing and of Laryngeal Movements

It is true that, to change note or octave, the singers use gestures of the larynx. The idea is that, due to the rocking of the larynx, that is due to laryngeal tilt, the vocal folds are stretched and the sound is higher. However, it is also important, and even more for very high notes, to increase the amount of breath. And, because the breathing is itself a gesture, it is interesting and more complete to see their combinations. The increase of a necessary amount of air to reach high pitches reminds us of the technique employed for higher octaves on the flute.

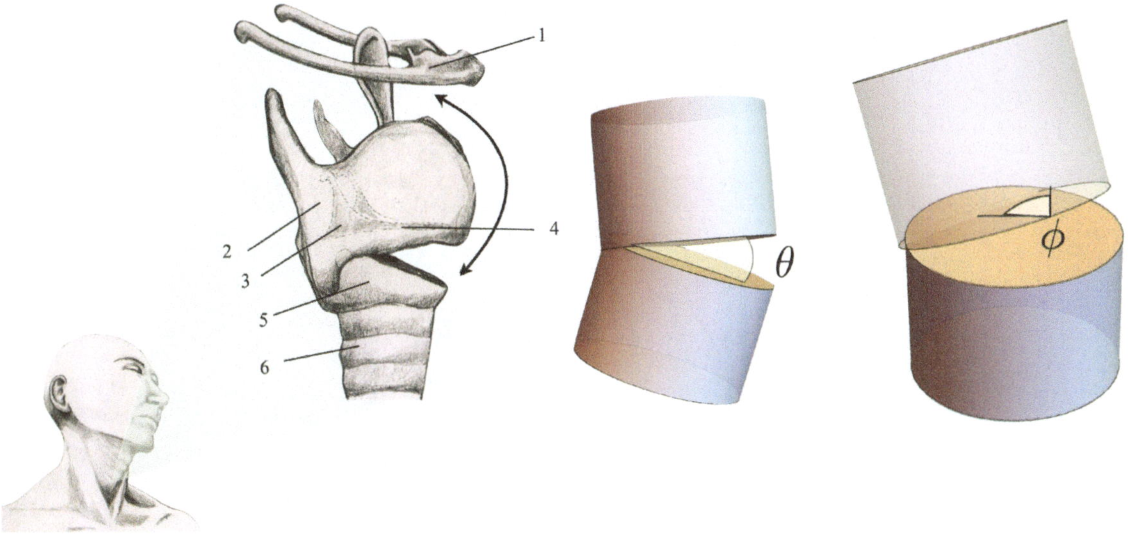

Fig. 86.2. Detail of larynx cartilages. Left, position of the larynx in the human neck. Details of the larynx: 1: hyoid bone; 2: thyroid cartilage; 3: arytenoid cartilage; 4: vocal folds; 5: cricoid cartilage; 6: trachea. Middle: angle θ of the larynx with respect to the trachea; and right: angle ϕ between the two vocal folds. It reminds us of spherical polar coordinates. A third coordinate ρ can be envisaged in the length of the vocal folds. Drawings by Maria Mannone.

In the specific case of singers, breathing is the necessary condition for the sound emission. For other performers, such as pianists, breathing is not directly related to the sound emission, but it is useful to control it, to ameliorate several parameters, such as the phrasing and the quality of touch. These parameters are influenced by the breathing in a sort of gestural analogy. Moreover, we can cite the case of musicians of a chamber music ensemble, who have to 'breathe together' in order to give coherence to the piece, coordinating the gestures, handling time. In orchestra there can be the same phenomenon, but in a much more complex way, due to the number of musicians, the complexity of music, and the unifying role of the conductor, in what we can poetically call 'leading the breathing.' This fundamental gesture is relevant also in the case of improvisation, because it can strongly determine the structure of the music as such.

We can geometrically describe the gesture of breathing from the point of view of the diaphragm: movement downward, enlargement of abdomen cavity, inspiration; movement upward, narrowing of abdomen cavity, exhalation. This gesture can also be combined with the movement of the larynx.

Regarding the larynx there is an interesting anecdote: Lowering of larynx permits a more brilliant sound but shortens the career of a singer. It is the case of Corelli, as opposed to the technique, without systematic lowering of larynx, used by Domingo, that allowed him an extraordinarily long career. These examples emphasize the practical application of these ideas also in voice didactics. There are several physiological and physical studies about the influence of laryngeal movements on vocal emission, see [1055, 1056] as references.

86.1.6 Mathematical Description of Vocal Gestures

Let A be the space of parameters related to larynx, diaphragm, and upper resonators of the head, as well as lip position—important for the vowel specialization of the sound. We focus on the parameters that depend on the technique of the singer, and so, on his/her voluntary and more or less direct choices. We won't take here into account the parameters that depend on the physiology of each person—we are going to consider them as fixed. Let X be the space of spectral values in $\mathbb{R}^2$, expressed in Hz and dB.

We can define an A-parametrized gesture as a continuous function from the cartesian product of the parameter space ∇ with the category A ($\nabla = [0, 1]$ as a topological category, see Section 62.1) with values

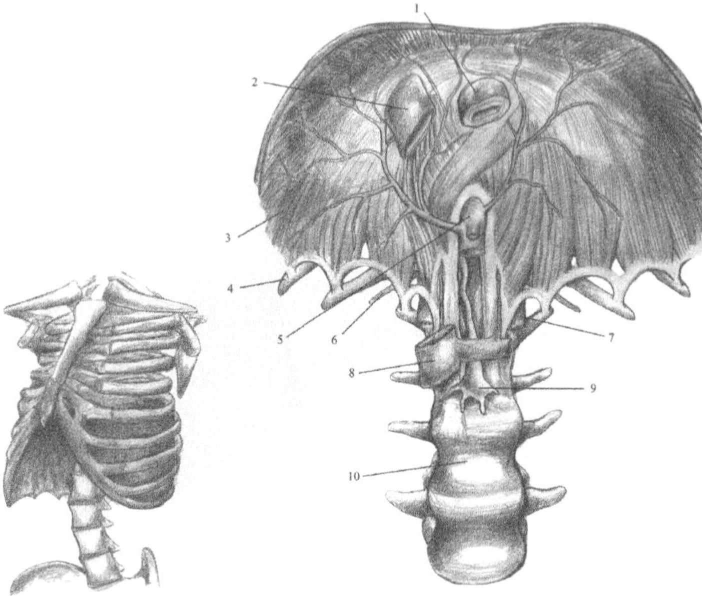

Fig. 86.3. On the left, the position of the muscle diaphragm among the ribs; on the right, the detail of the diaphragm, constituted by three crossing muscular folds, as the closer other parts of the human body. 1: esophagus, 2: inferior vena cava; 3: one of the muscular tissue; 4: ribs; 5: aorta; 6: anamostosis between azygos vein and vena cava; 7: right ascendent lombar vena; 8: inferior vena cava; 9: cisterna chyli; 10: spine. Drawings by Maria Mannone.

in the topological space X, that is, $\nabla \times A \to X$, see Section 62.6. According to what is called the currying theorem in informatics, that states, for a category C, that

$$C(X \times Y, Z) \xrightarrow{\sim} C(X, Z^Y),$$

where Z^Y are the curves from Y to Z, we can write that

$$q : \nabla \times A \to X \sim q : A \to \nabla @X,$$

for the gesture $q \in \uparrow @_A X$, where $\uparrow$ indicates a simple skeleton given by an arrow between two points. This second way of writing this, $q : A \to \nabla @X$, means that for every $a \in A$ we have a gesture $q(a)$ given by $q(a) : [0,1] \to X$, the curves from 0 to 1 mapped into the topological space X.

We define q as a *spectral gesture*, because X, as said before, is the space of spectra.

We can also define an *instrumental gesture* v_A, and we will see why. If we start from a skeleton Δ with, for example, two points connected by one arrow, we can map it into a curve in A that connects two points of A, i. e., two different configurations of A-parameters. We can indicate this gesture as an element of $\Delta \overrightarrow{@} A$, where the $\overrightarrow{@}$ indicates the topological category of all gestures from Δ in A.

There isn't any *time* information until now. We can add the time information by using the cartesian product $A \times T$:

$$v \in \Delta \overrightarrow{@} (A \times T) \xrightarrow{\sim} \Delta \overrightarrow{@} A \times \Delta \overrightarrow{@} T.$$

Time $T \xrightarrow{\sim} \mathbb{R}$ can be the physical time, if we are considering the final, real, vocal performance, as well the symbolic time, as happens for the singer who thinks of the positions and the parameter choices *before* singing. We can also symbolize by T complex time, comprehending both contributions. In the previous formula for v, we distinguish the parameter-dependent contribution v_A from the time-dependent contribution v_T, given by $\Delta \overrightarrow{@} A$ and $\Delta \overrightarrow{@} T$, respectively.

Now it is possible to put together the two gestures, the spectral one, q, and the instrumental-per-time one, v. In fact, q induces maps

$$\tilde{q} : \Delta\vec{@}A \times \Delta\vec{@}T \to \Delta\vec{@}(\nabla@X) \times \Delta\vec{@}T, \tag{86.1}$$

where we used $q : A \to \nabla\vec{@}X$, and the identity with respect to the time contribution. Conceptually, we are parametrizing points in space-time, getting configurations of curves, developed in the product $A \times T$.

To rewrite the mapping of Equation 86.1 in a more homogeneous way, we shall consider the embedding of time-points into time-curves. In fact, if we consider points as particular curves, we can write the injection

$$T \rightarrowtail \uparrow \vec{@}T.$$

Thus,

$$\Delta\vec{@}T \to \Delta\vec{@} \uparrow \vec{@}T,$$

and we can compose the instrumental gesture with the spectral one:

$$\tilde{q} : \Delta\vec{@}A \times \Delta\vec{@}T \to \Delta\vec{@}(\nabla@X) \times \Delta\vec{@} \uparrow \vec{@}T \tilde{\to} \Delta\tilde{@} \uparrow \vec{@}(X \times T),$$

and $\tilde{q}(v) \in \Delta\vec{@} \uparrow \vec{@}(X \times T)$. In this last passage we used

$$\nabla@X \tilde{\to} \uparrow \vec{@}X.$$

Returning to our problem, we finally get a hypergesture because there is a sequence of gestures (parametrized) into time. In the last writing,

$$\Delta\vec{@} \uparrow \vec{@}(X \times T),$$

there is no reference to A, which has been 'absorbed' inside the formalism. This fact corresponds to the reality of voice production: if we just listen to or watch a singer, we are not able to directly understand what is the parameter choice effectuated by the performer.

In fact, for the voice, at each instant of time we have a spectral gesture: $g(\alpha)$ for a choice α of parameters. A change in physiological parameters involves a passage $g(\alpha_1) \to g(\alpha_2)$, as shown in Figure 86.4. This passage can be interpreted as a hypergesture. Its skeleton Σ can represent the temporal connection between two instantaneous spectral gestures, corresponding to two parameters α_1 and α_2 in A, respectively. The corresponding space of hypergesture is denoted by $\Sigma@\Delta@(A@X)$.

86.1.6.1 Why Such a Formalism?

Intuitively, we could talk about three main voice gestures: those of the diaphragm, of the larynx, and of the vocal folds—inside the larynx. However, instead of analyzing the motion of these parts in terms of gestures, we characterized them as parameters inside the category A, used to defined the A-parametrized vocal spectral gestures. Why?

The motion of the diaphragm determines the pressure of the air that can be defined as a parameter. Vocal folds have some parameters independent of the larynx, and some others depending on it, such as the stretch that provokes a modification of the pitch. Geometrically, we have two angles, one between larynx and trachea, and the other between the two vocal folds, see Figure 86.2. The angle between the vocal folds actually varies regularly from 0 to ϕ and again to 0, and so on, during the sound emission. If we describe this motion in an onset-angle graph, we get a curve. We could describe this motion as a gesture. In the same way, we can describe the motion of the larynx as a gesture. To describe an octave-variation in singing, for example, we need two gestures, the one of the vocal folds, and the one of the larynx, as shown in Figure 86.5. In this way we could define a hypergesture, as in the classic example of the tube of circles, where we need two parameters: one for the point in the 'external' curve indicating a specific horizontal circle, and the other one to identify a point on that circle, see Figure 61.9.

However, this representation is not convenient. The final scope of this mathematical modeling is to connect the choice of physiological parameters to the final vocal result. If we deal with an onset-angle representation for the vocal folds-motion curve, we introduce a new parameter, the time of this regular motion, and moreover it is not clear where the curve ends. As an application of the currying theorem, we can see the problem from another perspective, the one of parameters. Given a choice of parameters, we get a spectral gesture, and so on as described. While dealing with parametrized gestures instead of gestures, we simplify the problem. Parametrized gestures are 'a kind of' hypergestures, but they are not real gestures, because the contributions can be added as cartesian products.

The details of vocal folds' gestures, as well those of the larynx, are difficult to control, but we can simplify the problem by using Fourier transformation and transforming these gestures into other ones, with the spectrum. With this formalism, we include all information in a complete and exhaustive way, and we finally get the sound result in terms of a spectrum. A last step is necessary to complete the theory. Voice production can in fact be specialized to different vowels. To reduce the complexity of this topic, we can think of a *formant* curve as being applied to the spectral curve. The formant curve acts as a filter, shaping the spectral curve into a specific vowel. The final result, a voice sound with a vowel specialization, is obtained as the product of these two curves.

86.1.6.2 Other Comments on Vocal Hypergestures

Let's consider a simple skeleton $\uparrow$. For every choice of parameter α inside the topological category A, as will be precisely discussed in the Appendix, there is an α-labeled curve $c(t) = (\alpha, t)$ in the space $A \times \nabla$. It means that each skeleton $\uparrow$ is α-labeled. And this curve will be mapped into a curve in X, yielding a spectrum. For every choice of parameter, we get a spectrum. If there is a transformation from α to α', there is a change of parameters, which leads to a change of spectra. Because we are here considering spectral gestures, the gesture that transforms a spectrum into another one is a spectral hypergesture. One could ask for an example of $\uparrow$. The skeleton is just an oriented graph, not already a *gesture*; it is just part of the definition of gesture. All the physiological parameters are inside the category A. However, to give *a first empirical idea*, not mathematically precise, we can consider $\uparrow$ as, for example, the motion of the diaphragm during exhalation, moving toward its equilibrium position. The motion of the diaphragm with a certain choice of laryngeal parameters will lead to a spectrum. Another choice of parameters will lead to another one. However, this is formally not precise because the motion of the diaphragm, described with its speed and acceleration, is part of the parameters' collection, and, as previously said, $\uparrow$ is just an oriented graph, not yet a *gesture*.

In every case, it is true that the voice first singing with a certain laryngeal configuration and then switching to another configuration is making a spectral hypergesture. See Figure 86.4 for the mechanism of construction. Another hypergesture can be realized when the laryngeal configuration is the same, but the parameter of the diaphragm's pressure is different: it is the case of a singer first singing a *piano* and then a *forte*. The *crescendo*, for example, is another case of hypergesture, because it implies a change of parameters of breathing. Because spectra are curves, the connection between them is given by a surface. Thus, the vocal hypergesture can be described as a surface connecting two or more spectral gestures.

This process can also be described in general terms as reparametrization. Although in physics the invariance for reparametrization is a frequent situation, here there isn't in general any invariance, because the final spectra will be different for different choices of physiological parameters. However, there can also be very similar spectra for different choices of parameters. It is the case when, for example, a voice student tries to imitate a particular sound, while forcing his phonatory system in a not-comfortable way, with an incorrect technique. This case is similar to the one of a pianist who is playing without the necessary muscular relaxation. After a while, these musicians can incur in muscular pain and, if the mistakes are reiterated, damage.

86.1.6.3 Branching

Melodic branching has been described in Chapter 78. Also in the case of voice it is possible to have melodic branching. It is the very rare case of Mongolian throat singers, able to produce multi-sound emissions at the same time.

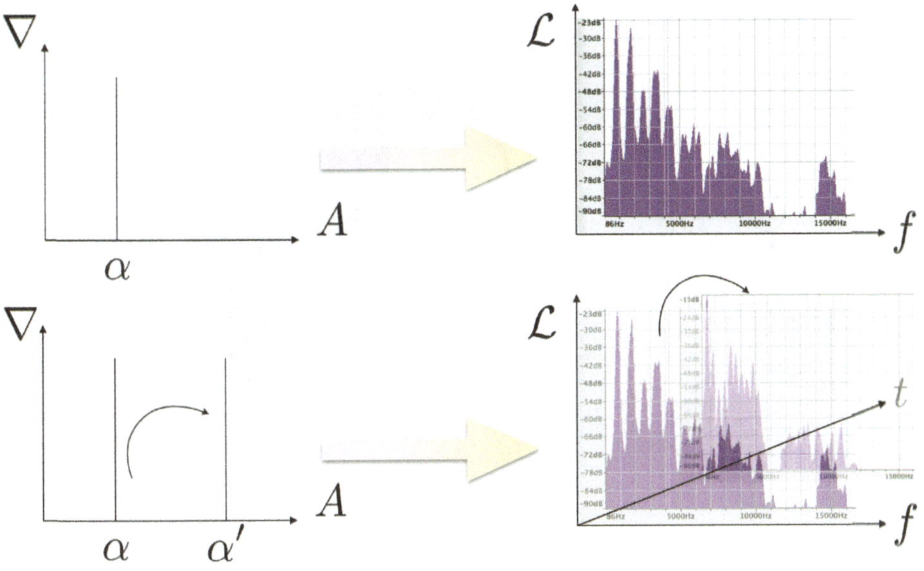

Fig. 86.4. Top: for a choice of laryngeal parameters we get a spectrum (loudness $\mathcal{L}$ in dB and frequency f in Hz); bottom: two different spectra that correspond to two different parametrizations. The first curve can be transformed into the second one via a surface. The third axis is labeled as the *time*, to represent the case when a spectrum follows the other temporally. The graphs of the spectra have been realized with Audacity software, while singing two notes in two different registers.

86.1.7 Gestures Thought by Singers

Singers, while teaching vocal techniques or singing themselves, are used to thinking of imaginary movements of the voice. These movements help the realization of real physical gestures. There is a sort of morphism between gestures in real time but in imaginary space, and gestures in real time and real space, mainly corresponding to the motion of vocal folds, larynx, and diaphragm.

86.1.7.1 Cultures of the Voice: An Example from Ethnomusicology

In the case of classically trained singers, a sustained tuned sound corresponds to a regular opening-closing of the vocal folds. The singing technique has, as its scope, the most projecting, loud or soft, and *vibrato* sound by a motion of the entire surface, without forcing any part of the vocal system. If we look at the shape of vocal folds of singers from other cultures, we can see that, in some cases, the vocal folds are forced to reach an 'unnatural' shape, by a progressive and dedicated training. Some singers of these cultures compare these changes in the natural shape of vocal folds to the callouses of violinists', harpists' and guitarists' fingers, as an example from traditional Korean *Han* singing. In recent time, more and more Koreans have been looking for a smoother *bel canto* style, toward what they call a *clean voice*, particularly in the context of Christian Lutheran singing. See [442] for more details.

86.1.7.2 Gregorian Chant and Gauls

The research of vocal qualities such as smoothness is a recurring theme. It is interesting to read the description by John the Deacon (9th century) of Gaul singers approaching the Gregorian chant (see Chapter VIII of [1106]): *These men from across the Alps seem unable to tone down to the supple delicacy of the Gregorian melodies, those stupendous noises which burst from their crude throats like claps of thunder.* Also in this

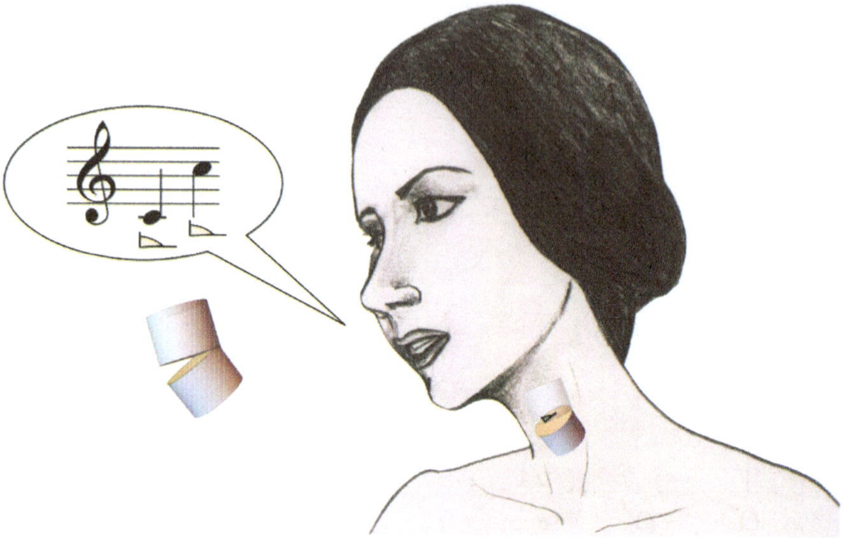

Fig. 86.5. A very first example of a vocal hypergesture (in Section 86.1.6 I give a more complete and mathematically manageable definition of a vocal hypergesture). The vocal folds open and close while the larynx is moving up and down. It can be used to realize a primitive melody with an octave: each note is a gesture (a repeated gesture) of vocal folds, and the jump of octave is a gesture of the larynx. As explained in the text, singers don't control directly the motion of vocal folds, but they control muscles that activate such movement. We use the score notation to recall the octave-movement gesture. Drawing by Maria Mannone.

case, the tendency toward a beautiful voice is seen as part of the liturgy: in fact, Pope Pius X wished to *pray in beauty*. It reminds us of Saint Augustine: *Who sings prays twice.*

This last example is also an introduction for the following topic. We will briefly describe a technique from the past that can be used as resource for the future.

86.2 A Powerful Tool from the Past for the Mathematical/Physical Theory of the Future: The Neumes of Gregorian Chant

Gregorian chant is the basis of Western music and Western music notation. We can recall, for example, the birth of the first harmony from counterpoint, when different chants were superposed, producing different simultaneous combinations of sounds that required new rules. We will examine here some characteristics of the training for Gregorian-chant singers, strongly related to the concept of gesture, and its ancient notation that is also derived from (vocal) gestures.

The first Christian liturgical music, the Gregorian chant with Latin texts from holy writings, was developed in the early Middle Ages and influenced by the Greek modes. It is a form of initially unaccompanied monophonic chant. The Western musical notational system is derived from Gregorian chant, whose origins, as we will see, stem from gestures. In 4th- and 5th-century Europe, there was no musical notation for melodies. Cantors in monasteries learned them by heart. Later, some signs called neumes (from the Greek word for sign) were added to words written in manuscripts. Initially, these neumes were positioned above syllables of the text, without any reference to a precise pitch, see Figure 86.10. Successively, neumes have been introduced in a four-line musical staff. the ancestor of our modern five-line staff. The use of a stave established the passage from non-diastematic to diastematic notation, where the precise pitch is shown by the vertical position on the musical stave [731].

A first attempt at a definition of new neumes is given in Figure 86.8.

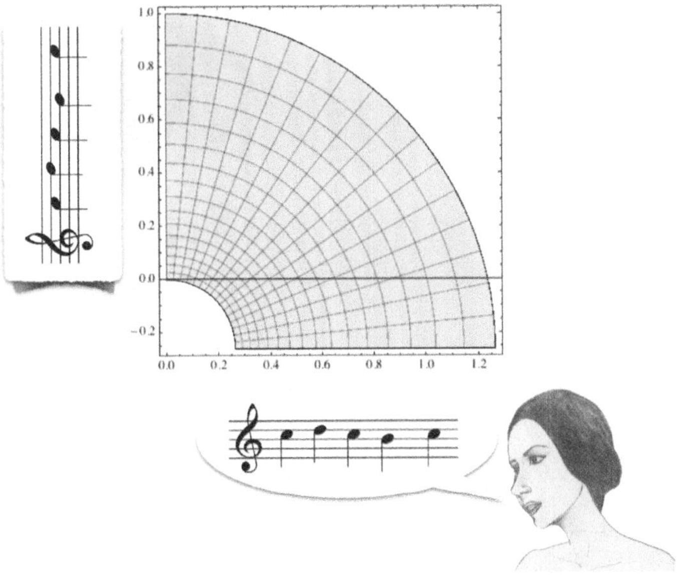

Fig. 86.6. Transition from imaginary time of a vocal score to real time of vocal performance.

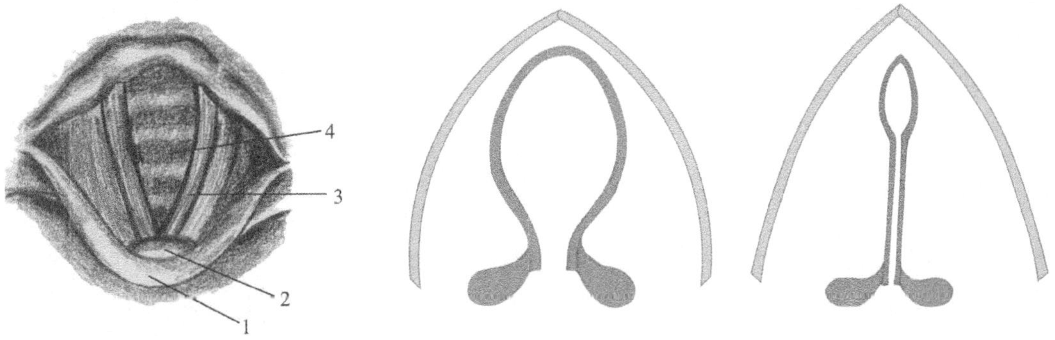

Fig. 86.7. Vocal folds. Left: in more detail, 1: epiglottis, 2: tubercle of epiglottis, 3: vocal folds, 4: trachea. Center: schematic representation of vocal folds while producing a lower sound, Right: the same scheme for a higher sound; see [347] for details. Drawings by Maria Mannone.

86.2.1 Gestures in Gregorian Chant Didactics

Here, we will describe some hints from a method of Gregorian chant dedicated to children [1106]. In this text, gestures of the entire body and of the voice have a fundamental role in learning this specific vocal technique and singing style.

86.2.2 Concept of Rhythm and Time

The connection between symbolic and physical gestures appears evident from the description of the approach to rhythm: *it cannot be merely mental. Rhythm is movement, and is acquired largely through the muscle sense.* The idea of smooth (vocal) gestures, a result of the gestural theory operators acting on non-smooth curves of symbolic gestures, is clearly present: *...a perfect legato, as on a stringed instrument, and never to sing as though by blows as on a piano*, putting the accent on the percussive character of piano playing (not of

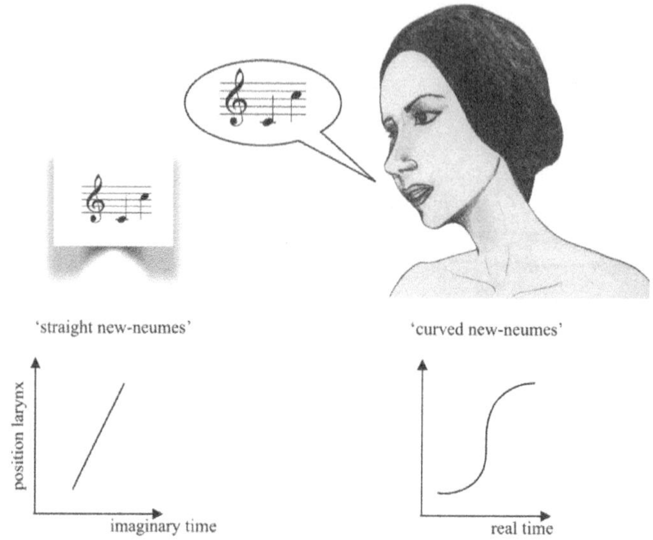

Fig. 86.8. One point: gesture of vocal folds. Curve: gesture up-down of the larynx connecting the points. We can define hypergesture also as including the gesture of breathing with the movements of diaphragm muscle, particularly relevant to reach very high notes that require an increased amount of breathing.

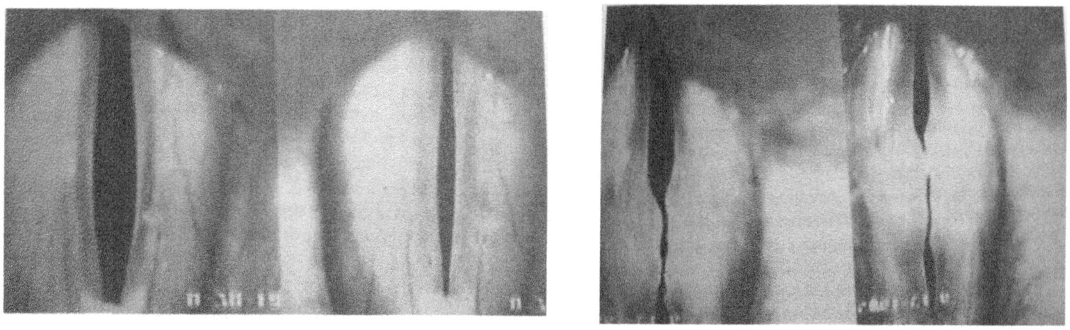

Fig. 86.9. Vocal folds of a classic trained *bel canto* singer (left), compared with respect to the vocal folds of a folkloristic Han Korean singer (right), from the book [442].

gestures). The goal of such a musical education is to develop a smooth voice, with a flexible and almost free (but not absent) rhythm. The steps to reach the mastery of the new rhythm comprehend gestural exercises to feel the movement as alternation of lift and weight, and the drawing of curves to reveal defect in the voice melody. There is, in fact, a kind of aesthetics of movement, it means, aesthetics of gestures: according to St. Augustine, *the rhythm is the science of beautiful movements.*

The first chapter of the treatise [1106] deals with the accent, corresponding to a slight raising of the melody of speech. The graphic tool used to indicate this movement is the wave shown in Figure 86.11 (a), where the first part signals the *arsis*, the second one the *thesis*.[5] It is also a notation for the hand: this kind of movement suggests the vocal reaction as shown in Figure 86.11 (b); the onset-pitch notation would

[5] We recall that the terms *arsis* and *thesis*, that come from Greek, are used to indicate the part of the movement of 'get off the ground' and of 'the down-beat'. They first appear in the Greek prosody. For the importance of arsis and thesis as musical paradigms, see Chapter 83 about conducting.

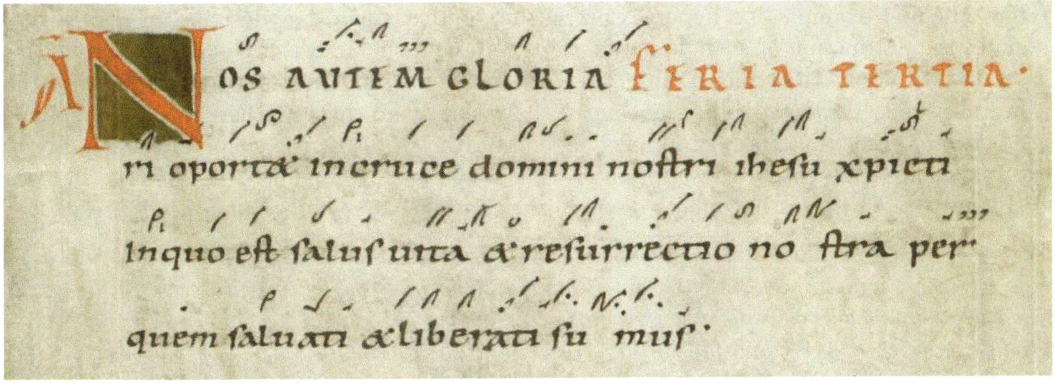

Fig. 86.10. An example of neumatic notation.

be used for the corresponding concept. This graph is close to the pick-up gesture in orchestral conducting, traditionally represented as shown in Figure 86.11 (c) and corresponding to the physical in 86.11 (d).

In Figure 86.11 (a), the smoothness on the apex is a clear signal for the required smoothness of the voice. The conductor's gesture is the mirrored version of other gestures: musicians looking at the conductor see a movement from the left to the right, i.e., in the 'natural' way. What is the arsis—thesis in vocal physiology? A first correspondence can be envisaged in the gesture of phonation. But let's proceed with this technique description.

In conducting, all the information for the pick-up is contained in its preparation; in the same way, as explained in [1106], the energy for both arsis and thesis is inside the arsis' gesture. We suggest a comparison with the throwing a ball over the head. The ball rises, and then falls back. The impulse is given by the initial gesture of throwing. The suggested image is then slightly modified: the Gregorian chant should 'drop' *more lightly than a snowflake.*

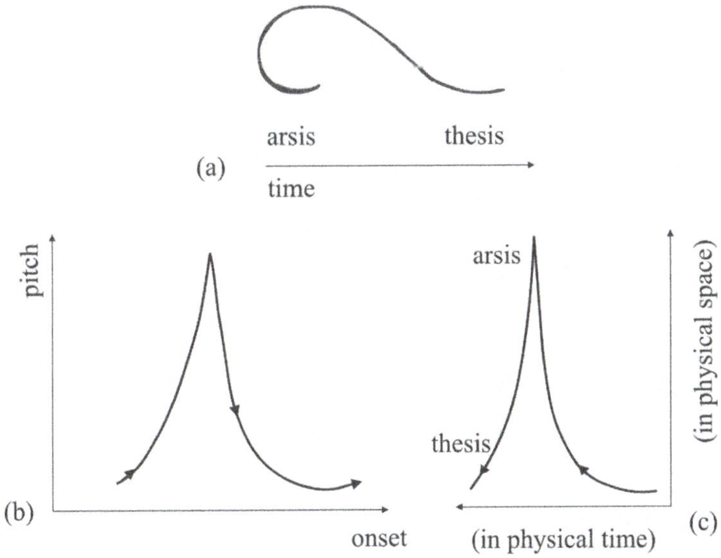

Fig. 86.11. (a) The wave shown in [1106] to represent arsis and thesis with the voice shown by a hand's gesture; (b) the corresponding vocal gesture; (c) the similar way to traditionally indicate the pick-up in orchestral conducting, and (d) its effect in onset-position graph.

We have the first example of a hypergesture in this context when they describe an arsis carrying more than one note (the production of a single note is a gesture, and the arsis is a gesture).

Let's talk now about time and rhythm. Gregorian chant is not measured music, because there is no division into measures. However, it does not mean that there is no rhythm: there are groupings of notes in sequences of two or three, but these groupings depend on the words or the shape of the melody itself, not on an 'external' time. After a short description of what we can call secondary arsis, realized in two different ways (see Figure 86.12), the author finally distinguishes between *time* and *rhythm.* He writes:

> *When we speak of* rhythm *we mean the great waves of sound that rise and fall. [...] When we speak of* time *we mean something different, something smaller which seems to pulsate within the rhythmic wave, just as our hearts pulsate in our bodies. Our hearts go on beating quietly, evenly, even when our bodies move to the rise and fall of the greater motions of rhythm.*

and again:

> *We should feel clearly [...] this steady, large, flowing movement of the rhythm, and then deep down, below the surface, the* time *moving, pulsating,* inside *the rhythm.*

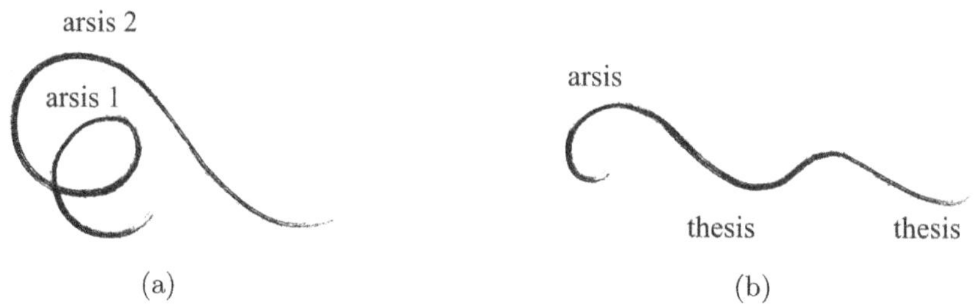

Fig. 86.12. Reproduction from [1106]; examples of nested arsis and thesis.

The concept of rhythmical pulsation is present also in other music worlds: classic symphony and jazz, for example. In the first case, it can be considered as a 'discrete' pulse, while in the second case the 'points' of pulsation are so short and so close that we can think of continuity. As an example of the first case, in the Symphony no. 40 by Wolfgang Amadeus Mozart (Figure 86.13), below the well-known melody there is the 'motor' of violas, that gives the vital pulse to the entire theme. The more or less hidden *tempo* can be modified, for example when a *fermata* is occurring. If we imagine the tempo as an elastic straight line (measure), the parameters of a fermata intervene as little *weights* that modify the shape of the line. In a two-dimensional example, we can also think of an elastic carpet, deformed by the gravitational field of some parameters. This is a common image used in popularization of space-time description in theoretical physics. The concept of *weight* comes from the performance theory (see Section 39.1), where the *fermata* is just an example.

In [1106] are described:

- the simple rhythm, when a rhythmic wave contains just one arsis and one thesis and not more,
- the simplex time, when, in each rhythmical wave, there is one note for an arsis and one note for a thesis,
- the complex time, when there is more than one note for an arsis, and more than one note for a thesis.

In Gregorian chant, the commonly used groupings are of two and four (duplex and triplex groups). As a final remark, it is recommended not marking 'loudly' the grouping of notes: *let the duplex and triplex time groups exist in* our mind only. The dimension of a mental, inner reality is implied, with its imaginary time, and it affects the flow of musical events inside the physical time.

Fig. 86.13. First page of Symphony no. 40 by Wolfgang Amadeus Mozart. The well-known melody played by violins is sustained by the pulse of violas (Leipzig: Breitkopf & Härtel, 1880. Plate W.A.M. 550).

86.2.2.1 The Chironomic Game

What in [1106] is called chironomic game is a compositive exercise where the teacher makes a hand gesture, and the students have to think about a melody that can fit the proposed gesture. This concept has striking similarities with the compositions inspired by gestures, recently proposed [649].

86.2.2.2 Voice in Imaginary Time, Silence in Physical Time?

It is suggested, before a certain exercise, to make a silent rhythmic wave. In this way, the student is thinking the melody, but not yet singing it (Figures 86.14, 86.15). The melody is still symbolic—completely imaginary, before its physical realization with the phonation. It can be called a gesture of silence, similar to the orchestral music in the first part of the pick-up—the attack!

86.2.3 The Neumes

The word *neume* originates from Greek *pneuma* (Greek letters), which means 'breathing.' Here we have a list of neumes, the square notation derived from them, and their evolution into modern musical notation [731].

We have these transitions:

Fig. 86.14. An image from [1106], representing the transition from the gestural indication for a firstly thought sound to a really sung one, for Gregorian chant learners. This passage reminds us to the transition from symbolic to physical gesture, and from imaginary to real time.

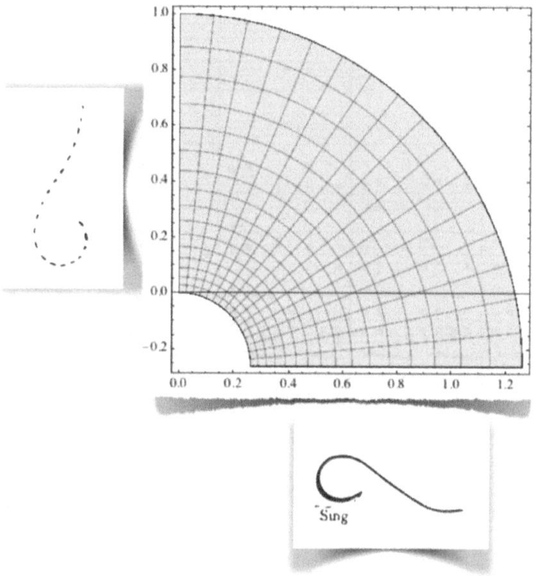

Fig. 86.15. The transition from symbolic to physical gesture, and from imaginary to real time.

1. from acute accent to virga
2. from grave accent to punctum (with the stem)
3. from circumflex accent to clivis
4. from anti-circumflex accent to pes or podatus.

Some of the Gregorian musical figures are the following (Figure 86.16):

- Punctum quadratum (simplest symbol, just a square note; see the first line of Figure 86.16, without stem)
- Punctum inclinatum (lozenge, used in a group of descending notes; see second line of Figure 86.16)
- Virga (like punctum quadratum, but with a stem)
- Scandicus, three ascending notes
- Quilisma (the modern mordente)
- The *torculus* is obtained as combination of three neumes, *porrectus*, *climacus*, and *scandicus*.

All of these musical figures have the same duration, and they are usually transcribed in modern notation using eighth notes. By combining these figures, it is possible to obtain more complex neumes. For example, by combining two ascending or descending notes, we respectively obtain the pes and the clivis, or the torculus/porrectus by using three notes. What is the origin of neumes? The neumes reproduced the ascending-descending movements used by the choral conductor to represent variations of the shape of the melody. These hand movements that indicate the shape of the melody constitute the cheironomic movements, from the Greek χέιρ, hand, and νόμος, law (in the sense of 'custom'). Due to its origin, we can state that the Western musical notation is derived from gestural indications. There is a transition from a continuous shape, characteristic of gestures, as we shall see in the following sections, to a discrete set, characteristic of the notes in the symbolic score. We argue that the idea of freezing gestures into simple signs is exactly what a composer does when he or she is writing a new score. Mathematically, that corresponds to a procedure of discretization.

Figure 86.16 [731] shows the derivation of some modern musical figures and combinations of figures from neumes first, and square notation then. Figure 86.17 shows an example of both neumatic and square notation.

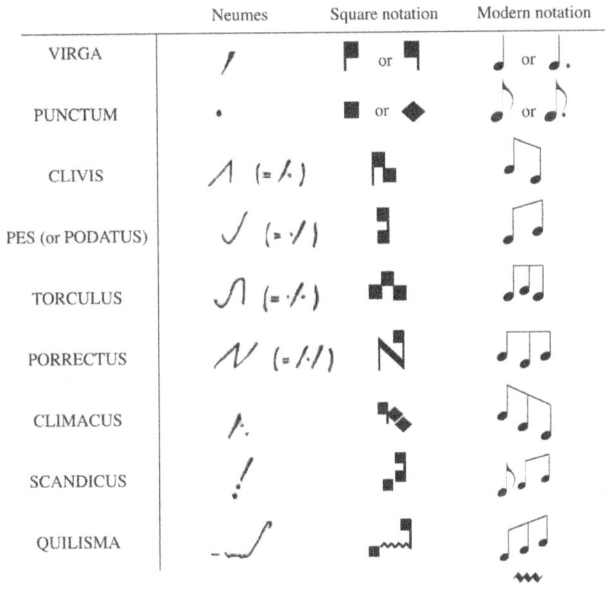

Fig. 86.16. The transition from neumes to square notation to modern musical notation.

86.3 Connecting Physiology, Gestures and Notation. Toward New Neumes?

86.3.0.1 A New Score

It is also possible to compose music starting on a vocal gesture. A fun example of such a piece is shown in the short original score of Figure 86.18. It contains the variation of loudness during the emission of a sound with the same pitch (from *pp* to *f* and back to *ff*), a continuous variation of pitch (*glissando*), and a discrete variation of pitch—a scale fragment, a variation on ascending three-note fragment, some *staccato* notes, and again a continuous sequence. The variation of loudness during the emission of the same note implies an augmented pressure of the diaphragm, keeping the same position of vocal folds. If the singer is also singing the same vowel, also the position of the mouth is kept unchanged. About the continuous change of pitch, short sequences of glissando in the form of gradual slides from one note to another closer one are frequently used in bel canto practice, although not written: they are examples of *portamento*. Continuous musical movements correspond to continuous movements of the larynx, as opposed to step-like notes as in the scale degrees. Highly staccato notes are usually made with blowing-like movements of the diaphragm.

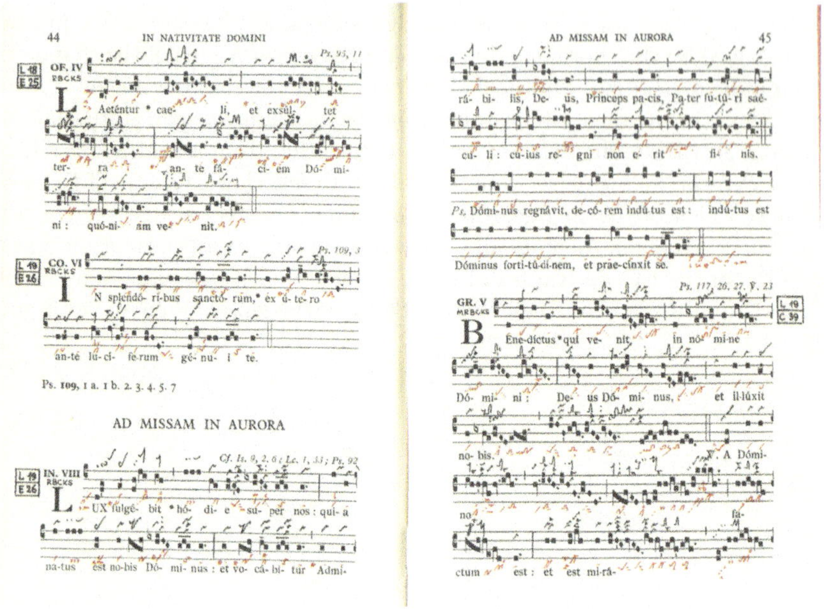

Fig. 86.17. An example of both neumatic and square notation.

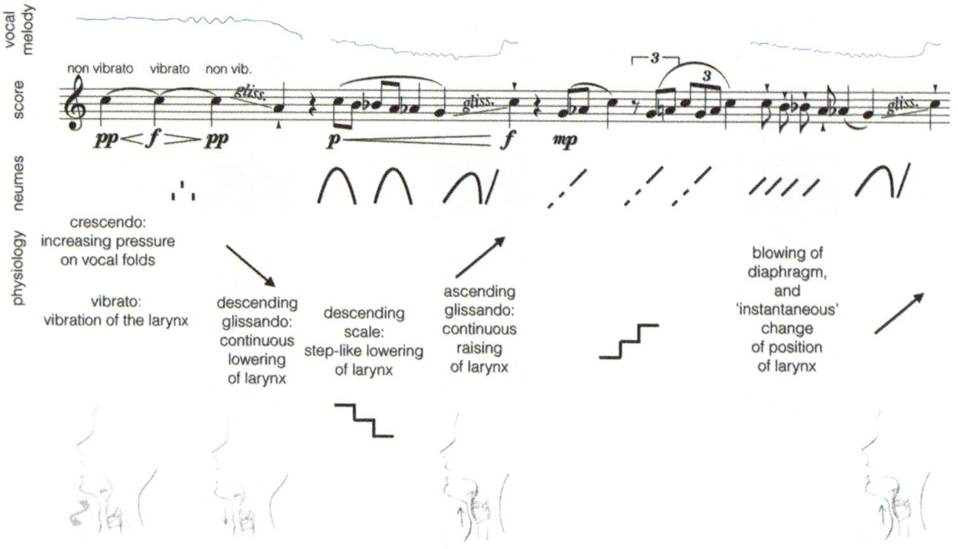

Fig. 86.18. Short original music fragments made by thinking of vocal gestures. Different notations are here used: From the top to the bottom, the vocal melody obtained with software Praat for vocal sound analysis; the musical 5-line staff, the Gregorian neumes—that don't take into account the pitches; synthetic physiological description accompanied by lines for *glissando* and 'stairs' for discrete scales; and, finally, schematic illustrations for such movements.

86.3.1 New Neumes

Symbolic and physical gestures, once defined as mathematical curves, can be included into the formalism of gesture theory as boundary conditions for the Poisson equation (see Section 78.2.1). The first candidates for symbolic and physical gestures are represented in Figure 86.8. The *vibrato* needs a separate description.

For example, it could be indicated via a couple of parameters, one for the speed, and the other for the pitch difference between higher and lower notes that constitute its boundary.

In fact, by our defining points in the space of vocal gestures, these points and the lines connecting them assume the character of new neumes, as gestural and hypergestural entities. In this way, an instrument from the past can give us the key for formally understanding the mathematics of vocal gestures, extending the gesture theory to the difficult case of inner gestures used in singing.

Part XXIV

References and Index

References

1. Abel V and Reiss P: MUTABOR II - Software Manual. Mutabor Soft, Darmstadt 1991
2. Abraham R and Marsden J: Foundations of Mechanics. Benjamin, New York et al. 1967
3. Abramowitz M and I A Stegun: Handbook of mathematical functions. With formulas, graphs and mathematical tables. Dover, 1965
4. Ackermann Ph: Computer und Musik. Springer, Wien and New York 1991
5. Aczel P: Non-well-founded Sets. No. 14 in CSLI Lecture Notes. Center for the Study of Language and Information, Stanford 1988
6. Ackermann Ph: Developing Object-Oriented Multimedia Software. dpunkt, Heidelberg 1996
7. Adorno Th W: Fragment über Musik und Sprache. Stuttgart, Jahresring 1956
8. Adorno Th W: Zu einer Theorie der musikalischen Reproduktion. Suhrkamp, Frankfurt/M. 2001
9. Adorno Th W: Der getreue Korrepetitor (1963). Gesammelte Schriften, Bd. 15, Suhrkamp, Frankfurt am Main 1976
10. Agawu V K: Playing with Signs. Princeton University Press, Princeton 1991
11. Agmon E: A Mathematical model of the diatonic system. JMT, 33, 1-25, 1989
12. Agmon E: Coherent Tone-Systems: a study in the theory of diatonicism. JMT, 40(1), 39-59, 1996
13. Agon A: OpenMusic: Un langage visuel pour la composition musicale assistée par ordinateur. PhD Dissertation, Université Paris VI, Paris 1998
14. Agustín-Aquino O A: Counterpoint in $2k$-tone equal temperament. Journal of Mathematics and Music, Vol 3, nr. 3, pp. 153-164, 2009
15. Agustín-Aquino O: Extensiones Microtonales de Contrapunto. PhD Thesis, Facultad de Ciencias, UNAM, Mexico City, 2012
16. Agustin-Aquino, O A, J Junod, G Mazzola: Computational Counterpoint Worlds. Springer Series Computational Music Science, Heidelberg 2014
17. Akmajian A et al.: Linguistics. MIT Press, Cambridge MA 1995
18. Alain: Alain Citation on Almada's painting in the Gulbenkian Foundation Center. Lisbon 1968
19. Albert R S and M A Runco: A History of Research on Creativity. In: R J Sternberg (Ed.): Handbook of Creativity. Cambridge University Press, New York 1999
20. Aleinikov A G: On Creative Pedagogy. Higher Education Bulletin, 12, 29-34, 1989
21. D'Alembert J Le Rond: Einleitung zur Enzyklopädie (1751). (German Translation) Fischer, Frankfurt/Main 1989
22. Alforhs L V: Complex Analysis. McGraw-Hill, New York et al. 1966
23. Alpaydin R and G Mazzola: A Harmonic Analysis Network. Musicae Scientiae, Vol. 19(2) 191–213, 2015
24. Alunni Ch: Diagrammes & catégories comme prolégomènes à la question: Qu'est-ce que s'orienter diagrammatiquement dans la pensée? In: Batt, N (éd): Penser par le diagramme. Presses Universitaires de Vincennes no.22-2004, Vincennes 2004
25. Alunni Ch: Relativités et puissances spectrales chez Gaston Bachelard. Revue de synthèse, 1999, 120-1, p. 73-110, 1999
26. Alunni Ch: Diagrammes et catégories comme prolégomènes à la question : qu'est-ce que s'orienter diagrammatiquement dans la pensée ? In: Théorie Littérature Enseignement no 22 - Penser par le diagramme de Gilles Deleuze à Gilles Châtelet, p. 83-93, 2004

G. Mazzola et al., *The Topos of Music III: Gestures*, Computational Music Science,
https://doi.org/10.1007/978-3-319-64481-3

27. Alunni Ch: Le Lemme de Yoneda : enjeux pour une conjecture philosophique ? (variations sous la forme pro-lemmatique, mais en prose), in Andreatta M, F Nicolas and C Alunni (Eds.): A la lumière des mathématiques et à l'ombre de la philosophie. Coll. Musique/Sciences, ENS, Delatour, Ircam, Centre Pompidou, p. 195-211, 2012
28. American Mathematical Society (AMS): Mathematical Subject Classification 2010, 47 pages `http://www.ams.org/msc/pdfs/classifications2010.pdf`
29. Amuedo J: Computational Description of Extended Tonality. Master's Thesis, U Southern California, Los Angeles 1995
30. Andreatta M: Group-theoretical Methods Applied to Music. Independent Study Dissertation, University of Sussex, Sussex 1997
31. Andreatta M: Méthodes algébriques en musique et musicologie du XXème siècle : aspects théoriques, analytiques et compositionnels. PhD thesis, Ircam/EHESS 2003
32. Andreatta M, A Ehresmann, R Guitart, G Mazzola: Towards a Categorical Theory of Creativity for Music, Discourse, and Cognition, in J. Yust, J. Wild, and J.A. Burgoyne (Eds.): MCM 2013, LNAI 7937, pp. 19-37, 2013
33. Andreatta M: La théorie mathématique de la musique de Guerino Mazzola et les canons rythmiques. Mémoire DEA, EHESS Paris IV & IRCAM, Paris 1999
34. Andreatta M and G Mazzola: From a Categorical Point of View: K-nets as Limit Denotators. Perspectives of New Music, volume 44, number 2 (Summer 2006)
35. `http://www.staps.uhp-nancy.fr/bernard`
36. `http://www.staps.uhp-nancy.fr/bernard/revues_corps.htm`
37. Ansermet E: Die Grundlagen der Musik im menschlichen Bewusstsein. 4th Ed. Piper, München and Zürich 1986
38. Apfel E: Diskant und Kontrapunkt in der Musiktheorie des 12. bis 15. Jahrhunderts. Heinrichshofen, Wilhelmshafen 1982
39. Apollonius: Les coniques d'Apollonius de Perge, trad. Paul Ver Eecke. Blanchard, Paris, 1959
40. Aristoteles: Topik (Organon V, 345 b.C.). Rolfes E (German transl.), Meiner, Hamburg 1992
41. Artin M: Algebra. Birkhäuser, Basel et al. 1993
42. Artin M, Grothendieck A, Verdier J L: Théorie des Topos et Cohomologie Etale des Schémas (Tomes 1,2,3). Springer Lecture Notes in Mathematics 269, 270, 305, Springer, New York et al. 1972-1973
43. Askenfelt A et al.: Musical Performance. A Synthesis-by-Rule Approach. Computer Music J. 7/1, 1983
44. Assayag G: Du calcul secret au calcul visuel. In: Delalande F and Vinet H (Eds.): Interface homme-machine et création musicale. Hermes, Paris 1999
45. Assayag G, G Bloch, M Chemillier: OMax-Ofon. Sound and Music Computing (SMC), Marseille 2006
46. Assayag G and S Dubnov: Using Factor Oracles for Machine Improvisation. Soft Computing 8, pp. 1432-7643, September 2004
47. Assmann J: Schöpfungsmythen und Kreativitätskonzepte im Alten Ägypten. In: R M Holm-Hadulla (Ed.): Kreativität. Springer, Heidelberg 2000
48. Auroux S: La sémiotique des encyclopédistes. Payot, Paris 1979
49. Babbitt M: Some Aspects of Twelve-Tone Composition. In: Hays W (Ed.): The Score and IMA Magazine, 12, 53-61, 1955 (reprinted in "Twentieth Century Views of Music History", 364-371, Scribner, New York 1972)
50. Babbitt M: Twelve-Tone Invariants as Compositional Determinants. Musical Quarterly, 46, 245-259, 1960
51. Babbitt M: Set Structure as a Compositional Determinant. JMT, 5(2), 72-94, 1961
52. Babbitt M: Twelve-Tone Rhythmic Structure and the Electronic Medium. PNM, 1(1), 49-79, 1962
53. Babbitt M: The Structure and Function of Music Theory. College Music Symposium, Vol.5, 1965 (reprinted in Boretz and Cone, 1972, 10-21)
54. Babbitt M: Words about Music. Dembski S and Straus J N (Eds.), University of Wisconsin Press, Madison 1987
55. Bach J S: Krebskanon. In: *Musikalisches Opfer*, 70. (BWV 1079), Neue Gesamtausg. sämtl. Werke Ser. VIII, Bd.1, Bärenreiter, Kassel 1978
56. Bach J S: Choral No.6, In: *Himmelfahrtsoratorium* (BWV 11), Neue Gesamtausg. sämtl. Werke Ser. II, Bd.8, Bärenreiter, Kassel 1978
57. Bachelard G: La Valeur inductive de la Relativité. 257 p., Vrin, Paris 1929
58. Bacon F: De dignitate et augmentis scientiarum. 1st Ed. 1605, improved 1623
59. Bacon F: Novum Organum, trad. Malherbe et Pousseur. coll. épiméthée, PUF, Paris, 1986
60. Badiou A: Court traité d'ontologie transitoire. Seuil, Paris, 1998

61. Badiou A, F Tarby: La philosophie et l'événement. Entretiens. Suivis d'une courte introduction à la philosophie d'Alain Badiou. Germina, 2010
62. Badiou A: La République de Platon. Pluriel, 2012
63. Badiou A: Le Séminaire – Parménide. L'être 1 - Figure ontologique. 1985-1986. Fayard, 2014
64. Badiou A: Le Séminaire – Heidegger. L'être 3 - Figure du retrait. 1986-1987. Fayard, 2015
65. Badiou A: Éloge des mathématiques. Flammarion, 2015
66. Baer J and J C Kaufman: Creativity Research in English-Speaking Countries. J C Kaufman and R J Sternberg (Eds.): The International Handbook of Creativity. Cambridge University Press, Cambridge, pp. 10-38, 2006
67. Bagchee S N: Understanding Raga Music. BPI, New Delhi 1996
68. Bahle J: Der musikalische Schaffensprozess. Psychologie der schöpferischen Erlebnis- und Antriebsformen. S Hirzel, Leipzig 1936
69. Baker J, Beach D, Bernard J: Music Theory in Concept and Practice. Eastman Studies in Music, University of Rochester Press, 1997
70. Bakhtin M: Toward a Philosophy of the Act. Translated by Vadim Liapunov. University of Texas Press, Austin 1993
71. Balzano G: The group-theoretic description of 12-fold and microtonal pitch systems. CMJ, 4, 66-84, 1980
72. Bandemer H and Gottwald H: Fuzzy Sets, Fuzzy Logic, Fuzzy Methods. Wiley, New York et al. 1995
73. Banter H: Akkord-Lexikon. Schott, Mainz 1982
74. Barbin E: Histoire et enseignement des mathématiques : pourquoi ? comment ?. Bulletin de l'Association Mathématique du Québec, XXXVII-1, 20-25
75. Barbin E: La révolution mathématique du XVIIème siècle. Ellipses, 2006
76. Barbin E et R Guitart: La pulsation entre les conceptions optiques, algébriques, articulées et projectives, des ovales cartésiennes, in: L'Océan indien au carrefour des mathématiques arabes, chinoises, européennes et indiennes. IUFM de La Réunion, 1998, pp. 359-394
77. Barbin E et R Guitart: Algèbre des fonctions elliptiques et géométrie des ovales cartésiennes. Revue d'histoire des mathématiques, SMF, n7, 2001, pp. 161-205
78. Barbin E: Gaston Tarry et la doctrine des combinaisons. Labyrinthes et carrés magiques, conférence au colloque *Les travaux combinatoires entre 1870 et 1914 et leur actualité*, Guéret, 30 Sept. - 2 Oct. 2015
79. Barbin E et J-P Cléro (Eds.): Les mathématiques et l'expérience : ce qu'en ont dit les philosophes et les mathématiciens. Hermann, Paris 2015
80. Barbin E: La révolution mathématique du XVIIème siècle. Ellipses, 2006
81. Bareil H: Analyse de "La pulsation mathématique", *A.P.M.E.P.* 427, March-April 2000
82. Barlow K: Über die Rationalisierung einer harmonisch irrationalen Tonhöhenmenge. Preprint, Köln 1985.
83. Barrow J and F Tipler: The Anthropic Cosmological Principle: Oxford University Press, New York 1986
84. Bars I: Extra Dimensions in Space and Time (Multiversal Journeys). Springer, Heidelberg 2009
85. Bartenieff I: The Roots of Laban Theory: Aesthetics and Beyond. In: Bartenieff et al.: Four Adaptations of Effort Theory. Dance Notation Bureau, New York 1970
86. Barthes R: Eléments de sémiologie. Communications 4/1964
87. Barwise J and L Moss: Hypersets. Mathematical Intelligencer, 13(4):3141, 1991
88. Basten A: Personal e-mail communciaction to G.M. November 22, 1996
89. Bastiani A (= A Ehresmann): Théorie des ensembles. CDU, 328 p., Paris, 1970
90. Bätschmann O: Einführung in die kunstgeschichtliche Hermeneutik. Wissenschaftliche Buchgemeinschaft Darmstadt, Darmstadt 1986
91. Batt N (éd): Penser par le diagramme. Presses Universitaires de Vincennes no.22-2004, Vincennes 2004
92. Bauer M: Die Lieder Franz Schuberts. Breitkopf und Härtel, Leipzig 1915
93. Bazelow A and F Brickel: A Partition Problem Posed by Milton Babbitt. PNM, 14(2), 15(1), 280-293, 1976
94. Becker M: Nineteenth Century Foundations of Creativity Research. Creativity Research Journal, Vol. 8, pp. 219-229, 1995
95. Bednarik R G: A Figurine from the African Acheulian. Current Anthropology, Vol. 44:3, pp. 405-413, 2003
96. Beethoven L van: *Grosse Sonate für das Hammerklavier* op.106 (1817-1818). Ed. Peters, Leipzig 1975
97. Békésy G von: Experiments in Hearing. McGraw-Hill, New York 1960
98. Bell J L: From absolute to local mathematics. *Synthese* 69, 409-426, 1986
99. Bellal A: Il faut sortir des mathématiques (humour) (see also [437]), `http/::environ-energie.org/category/humourpastiches/`
100. Benveniste É: La nature des pronoms. In: Benveniste É: Problèmes de linguistique générale, I. Gallimard, Paris 1966
101. Beran J: *Cirri.* Centaur Records, 1991

102. Beran J: Statistics in musicology. Chapman & Hall (CRC Press), New York 2003
103. Beran J and G Mazzola: *Immaculate Concept.* SToA music 1002.92, Zürich 1992
104. Beran J and G Mazzola G: Rational composition of performance. In Auhagen W, R Kopiez (Eds.): Regulation of creative processes in music, Staatliches Institut für Musikforschung (Berlin), pp. 37-67, Lang Verlag, Frankfurt/New York, 1997
105. Beran J and G Mazzola: Analyzing Musical Structure and Performance—a Statistical Approach. Statistical Science. Vol. 14, No. 1, 47-79, 1999
106. Beran J and G Mazzola: Visualizing the Relationship Between Two Time Series by Hierarchical Smoothing. Journal of Computational and Graphical Statistics, Vol. 8, No. 2, 213-238, 1999
107. Beran J and G Mazzola: Timing Microstructure in Schumann's "Träumerei" as an Expression of Harmony, Rhythm, and Motivic Structure in Music Performance. Computers and Mathematics with Applications, Vol. 39, Issue 5/6, 99-130, 2000
108. Beran J and G Mazzola: Musical composition and performance - statistical decomposition and interpretation. Student, Vol. 4, No.1, 13-42, 2001
109. Beran J: Maximum likelihood estimation of the differencing parameter for invertible short and long-memory ARIMA models. J.R. Statist. Soc. B, 57, No.4, 659-672, 1995
110. Beran J, R Goswitz, G Mazzola, P Mazzola: On the relationship between tempo and quantitative metric, melodic and harmonic information in Chopin's Prélude op. 28, No. 4: a statistical analysis of 30 performances. Journal of Mathematics and Music, Vol 8., No. 3, p. 225-248, 2014
111. Beranek L L: Acoustics, 1954 -1993, The Acoustical Society of America, ISBN: 0-88318-494-X
112. Berg A: Die musikalische Impotenz der Neuen Ästhetik Hans Pfitzners. Musikblätter des Anbruchs. Jg 02, S. 400, Wien 1920
113. Berger M: Geometry I, II. Springer, Berlin et al. 1987
114. Bernard J: Chord, Collection, and Set in Twentieth-Century Theory. In: Baker J et al. [69], 11-52
115. Béziau J-Y: La Théorie des Ensembles et la Théorie des Catégories : Présentation de deux Sœurs Ennemies du Point de Vue de leurs Relations avec les Fondements des mathématiques. Boletin de la Asociación Matemática Venezolana, Vol. IX, No 1, pp. 45-52, 2002
116. Bilkey D and K M Douglas: Amusia Is Associated with Deficits in Spatial Processing. Nature Neuroscience 10, pp. 915 - 921, 2007
117. Biok H-R: Zur Intonationsbeurteilung kontextbezogener sukzessiver Intervalle. Bosse, Regensburg 1975
118. Biot J-B: Essai de géométrie analytique, 8 ième éd. (1ère éd. 1802), Bachelier, Paris 1834
119. Birkhoff G: Universal Algebra. In: Rota G-C and J.S. Oliveira J S (Eds.): Collected papers of G. Birkhoff, pp. 111-115, Birkhäuser, Basel
120. Blind Boy Fuller: `ttp://www.allaboutbluesmusic.com/blind-boy-fuller/` 2015
121. Block I: Oscar Peterson Lays Down the Story of a Life Lived in the Swing of Things. The Gazette (Montreal), August 31, 2002
122. Blumenthal L M: Distance Geometry. Chelsea 1953
123. Boden M A: Conceptual Spaces. In [753], pp. 235-244
124. Boden M A: The Creative Mind: Myths and Mechanisms. Routledge, London 2004
125. Boden M A: Creativity & Art. Oxford University Press, Oxford 2010
126. Blumenthal L M: Distance Geometry. Chelsea 1953
127. Blood A, J Zatorre, P Bermudez, A C Evans: Emotional response to pleasant and unpleasant music correlates with activity in paralimbic brain regions. Nature Neuroscience, Vol. 2, No. 4, 382-387, April 1999
128. Boi L: Le problème mathématique de l'espace. Une quête de l'intelligible. Springer, 1995
129. Boirel R: Comment résoudre aisément les problèmes de mathématiques. Initiation au dynamisme opératoire mathématique. éd. Culture humaine, Sermaise, 1964
130. Boissière A: Geste, interprétation, invention selon Pierre Boulez. Revue DEMéter, Dec. 2002, Univ. Lille-3
131. Borceux F: Handbook of Categorical Algebra: Volume 1. Cambridge University Press, Cambridge 1994
132. Borceux, F: Handbook of Categorical Algebra: Volume 2. Cambridge University Press, Cambridge 1995
133. Borgo D: Sync or Swarm. Continuum International Publishing Group 2006
134. Boretz B and E T Cone: Perspectives on Contemporary Music Theory. W.W. Norton and Company, New York 1972
135. Bose R C and S S Shrikhande: On the construction of sets of mutually orthogonal Latin squares and the falsity of a conjecture of Euler. Trans. Amer. Math. Soc. 95, 191-209, 1960
136. Bose R C, S S Shrikhande and E T Parker: Further results on the construction of mutually orthogonal Latin squares and the falsity of Euler's conjecture. Canadian J. of Math. 12, 189-203, 1960

137. Bott J, J Crowley, J LaViola Jr: Exploring 3D Gestural Interfaces for Music Creation in Video Games. CFDG 2009 April 2630, 2009, Orlando 2009
138. Boulez P: Musikdenken heute I,II; Darmstädter Beiträge V, VI. Schott, Mainz 1963, 1985
139. Boulez P: Le timbre et l'écriture, le timbre et le langage. In: Bourgeois Chr (Ed.): Le timbre, métaphore pour la composition. IRCAM, Collection Musique/Passé/Présent, Paris 1991
140. Boulez P: Jalons (dix ans d'enseignement au Collège de France). Bourgeois, Paris 1989
141. Boulez P: L'Ecriture du geste. Bourgeois, Paris 2002
142. Boulez P: Structures, premier livre. UE, London 1953
143. Boulez P: Structures, deuxièxime livre. UE, London 1967
144. Bourgeois Chr (Ed.): Le timbre, métaphore pour la composition. IRCAM, Collection Musique/Passé/Présent, Paris 1991
145. Bourbaki N: Eléments de Mathématique, Algèbre, Ch.1-9. Hermann, Paris 1970-1973
146. Bourbaki N: Eléments de Mathématique, Algèbre Commutative, Ch.1-7. Hermann, Paris 1961-65
147. Bourbaki N: Eléments de Mathématique, Topologie Générale, Ch.1-4. Hermann, Paris 1971
148. Bourbaki N: Foundations of mathematics for the working mathematician. Journal of Symbolic Logic, 14, pp. 1-8, 1949
149. Boyer C B: Newton as an Originator of Polar Coordinates. American Mathematical Monthly 56: 73-78, 1949
150. Braem P B and T Braem. A Pilot Study of the Expressive Gestures Used by Classical Orchestra Conductors. In: Emmorey K and H Lane (Eds.): The signs of Language Revisited. Lawrence Erlbaum Associates, 2000
151. Brändle L: Die "Wesentlichen Manieren" (Ornamente in der Musik). Oesterreichischer Bundesverlag, Wien 1987
152. Brandt C and C Roemer: Standardized Chord Symbol Notation: Roerick Music, Sherman Oaks, CA 1976
153. Brecht B et al.: Conductor Follower. In: ICMA (Ed.): Proceedings of the ICMC 95, San Francisco 1995
154. Brockwell P J and R A Davis: Time Series: Theory and Methods. Springer, New York 1987
155. Brofeldt H: Piano Music for the Left Hand Alone. `http://www.left-hand-brofeldt.dk` Non-dated
156. Bruhn H: Harmonielehre als Grammatik der Musik. Psychologie Verlags Union, München et al. 1988
157. Bruter C-P: Comprendre les mathématiques. Les 10 notions fondamentales. Ed. Odile Jacob, Paris, 1996
158. Bulwer J: Chironomia. Thomas Harper, London 1644
159. Buteau Ch and G Mazzola: From Contour Similarity to Motivic Topologies. Musicae Scientiae 2, pp. 125-149, 2000
160. Buteau Ch and G Mazzola: From Contour Similarity to Motivic Topologies. Musicae Scientiae, Vol. IV, No. 2, 125-149, 2000.
161. Busch H R: Leonhard Eulers Beitrag zur Musiktheorie. Bosse, Regensburg 1970
162. Buser P and M Imbert: Audition. Hermann, Paris 1987
163. Buteau Ch: Motivic Topologies and their Signification in Musical Motivic Analysis. Master's Thesis, U Laval/Québec 1998
164. Buteau Ch: Reciprocity Between Presence and Content Functions on a Gestalt Composition Space. Tatra Mt. Math. Publ. 23, 17-45, 2001
165. Butsch C and H Fischer: Seashore-Test für musikalische Begabung. Huber, Bern/Stuttgart 1966
166. Byers W: How Mathematicians Think. Using Ambiguity, Contradiction and Paradox to Create Mathematics. Princeton U. Press, 2007
167. Cabral B and L Leedom: Imaging vector fields using line integral convolution. Computer Graphics 27 (SIGGRAPH'93 Proceedings), 263-272, 1993
168. Cadoz C and M M Wanderley: In: Gesture - Music. Trends in Gestural Control of Music, M M Wanderley and M Battier eds., Ircam - Centre Pompidou, 2000
169. Cadoz C: Le geste canal de communication homme-machine. La communication 'instrumentale'. Sciences Informatiques, numéro spécial: Interface homme-machine. 13(1): 31-61, 1994
170. Cage J: ASLSP. `http://www.john-cage.halberstadt.de`
171. Calvet O et al.: Modal synthesis: compilation of mechanical sub-structure and acoustical sub-systems. In: Arnold S, Hair G, ICMA (Eds.): Proceedings of the 1990 International Computer Music Conference. San Francisco 1990
172. Campos D G: Peirce on the Role of Poietic Creation in Mathematical reasoning, trans. Charles S. Peirce Society: A Quaterly J. in American Philosophy, Vol. 43, No 3, pp. 470-489, Summer 2007
173. Franco Ferrara's conducting technique, Personal conversation with maestro C. Caruso at State Music Conservatory of Palermo, 2014
174. Carnot L: Mémoire sur la Relation qui existe entre les distances respectives de cinq points quelconques dans l'espace, suivi d'un Essai sur la théorie des transversales. 1806

175. Castellengo M and N H Bernardoni: Interplay between harmonics and formants in singing: when vowels become music. Proceedings of ISMA, International Conference on Noise and Vibration Engineering 2014
176. Camuri A et al.: Toward a cognitive model for the representation and reasoning on music and multimedia knowledge. In: Haus G and Pighi I (Eds.): X Colloquio di Informatica Musicale. AIMI, LIM-Dsi, Milano 1993
177. Carey N and D Clampitt: Aspects of well-formed scales. MTS, 11, 187-206, 1989
178. Cardelli L and P Wegner: On Understanding types, data Abstraction and polymorphism. Computing Surveys, Vol. 17,4, 1985
179. Castagna G: Foundations of Object-oriented programming. ETAPS, Lisbon 1998
180. Castine P: Set Theory Objects. Lang, Frankfurt/Main et al. 1994
181. Cavaillès J: Méthode axiomatique et formalisme, p. 178, 1938. In: Oeuvres complètes de Philosophie des sciences. Hermann, 1994
182. Cayley A: On a Theorem in the Geometry of Position. Camb. Math. Jour., t. II, pp. 267-271, 1841
183. Celibidache Video: `http://www.youtube.com/watch?v=_BAsS5SmiNU`
184. Celibidache at Wikipedia: `http://en.wikipedia.org/wiki/Sergiu_Celibidache`
185. CERN. Internet Information via http://www.cern.ch 1995
186. Certeau M de: L'étranger ou l'union dans la différence. 1ère éd. 1969, Le Seuil, 2005
187. Châtelet G: Arithmétique et algèbre modernes, II, PUF, Paris 1956
188. Chastel A: Le geste dans l'art. Liana Levi, Paris 2001
189. Châtelet G: Les enjeux du mobile. Seuil, Paris 1993
190. Châtelet G: Figuring Space. Kluwer 2000 (translation of the French original *Les enjeux du mobile*, [189])
191. Chatterjee S and B Price: Regression Analysis by Example. Wiley, 2nd ed., New York 1995
192. Chauviré C: Peirce et la signification. PUF, Paris 1995
193. Chauviré C: L'œil mathématique. Essai sur la philosophie mathématique de Peirce. éd. Kimé, 2008
194. Cherlin M: Memory and Rhetorical Trope in Schoenberg's String Trio. Journal of the American Musicological Society, 51/3, pp. 559-602, fall 1998
195. Cherry D & E Blackwell: Mu (Parts I and II). Actuel 1969
196. Choi I: Cognitive Engineering of Gestural Primitives for Multi-Modal Interaction in a Virtual Environment. In: Proceedings of the 1998 IEEE International Conference on Systems, Man and Cybernetics (SMC'98), 1998
197. Chomsky N and M Halle: The Sound Pattern of English. Harper and Row, New York 1968
198. Chowning J: The Synthesis of Complex Audio Spectra by Means of Frequency Modulation. Journal of the Audio Engineering Society 21 (7), 1985
199. Chowning J: The synthesis of complex audio spectra by means of frequency modulation. Journal of the Audio Engineering Society, 21, 1973.
200. Clarke E: Imitating and Evaluating Real and Transformed Musical Performances. Music Perception 10/3, 317-341, 1993
201. Clayton M and G Fatone, L Leante, and M Rahaim: Imagery, Melody, and Gesture in Cross-Cultural Perspective. In: King E: New Perspectives on Music and Gesture (2016), Routledge 2011
202. Clough J and Myerson G: Variety and multiplicity in diatonic systems. JMT, 29, 249-270, 1985
203. Clough J and J Douthett: Maximally even sets. JMT, 35, 93-173, 1991
204. Clough J: Diatonic Interval Cycles and Hierarchical Structure. PNM, 32(1), 228-253, 1994
205. Clynes M: Sentics. The Touch of Emotions. Anchor Doubleday, New York 1977
206. Clynes M: Secrets of life in music. In: Analytica, Studies in the description and analysis of music in honour of Ingmar Bengtsson, Royal Swedish Academy of Music, No 47 pp. 3-15, 1985
207. Colbourn C J and J H Dinitz: Handbook of Combinatorial Designs. Chapman & Hall/CRC, 2007
208. Cont A, S Dubnov and G Assayag: On the Information Geometry of Audio Streams with Applications to Automatic Structure Discovery. IEEE Transactions on Audio, Speech and Language Processing 2009
209. Cohn R: Introduction to Neo-Riemannian Theory: A Survey and a Historical Perspective. JMT 42(2), 167-180, 1998
210. Conen H: Formel-Komposition. Schott, Mainz et al. 1991
211. Cook P: `http://www.cs.princeton.edu/ prc/SingingSynth.html`
212. Cook N: Playing God: Creativity, Analysis, and Aesthetic Inclusion. In: I Deliège and G A Wiggins (Eds.): Musical Creativity: Multidisciplinary Research in Theory and Practice. Psychology Press, Hove/New York, pp. 9-24, 2006
213. Coolidge J L: The origin of polar coordinates. American Mathematical Monthly, 59 : 78-85, 1952
214. Cooley J W and J W Tuckey: An Algorithm for the Machine Calculation of Complex Series. Math. Comput. 19, 297301 1965
215. Cope D: Virtual Music: Computer Synthesis of Musical Style. MIT Press, Cambridge, MA 2001

216. Cope D: Computer Models of Musical Creativity. MIT Press, Cambridge, MA 2005
217. Corbin A, J-J Courtine, G Viagarello (Eds.): Histoire du corps, T3, Les mutations du regard. Le XX siècle. Seuil, Paris 2006
218. Cox A: Hearing, Feeling, Grasping Gestures. In: Gritten A and E King (Eds.): Music and Gesture. Aldershot: Ashgate, 2006
219. Cooper K N D et al.: Handwritten Music-Manuscript Recognition. In: ICMA (Ed.): Proceedings of the ICMC 96, San Francisco 1996
220. Couasnon B et al.: Using a Grammar For a Reliable Full Score Recognition System. In: ICMA (Ed.): Proceedings of the ICMC 95, San Francisco 1995
221. Creutzfeldt O D: Cortex Cerebri. Springer, Berlin et al. 1983
222. Cros S C H: Théorie de l'homme intellectuel et moral, vol.2. Bachelier, 1842
223. Csíkszentmihályi M: Beyond Boredom and Anxiety. Jossey-Bass, San Francisco 1975
224. Csíkszentmihályi M: Society, Culture, and Person: A Systems View of Creativity. In: R J Sternberg (Ed.): The Nature of Creativity: Contemporary Psychological Perspectives. Cambridge University Press, New York, pp. 325-339, 1988
225. Csíkszentmihályi M: The Psychology of Optimal Experience. Harper and Row, New York 1990
226. Csíkszentmihályi M: Creativity: Flow and the Psychology of Discovery and Invention. HarperCollins, New York 1996
227. Cusick S: Feminist Theory, Music Theory, and the Mind/Body Problem. Perspectives of New Music 32/1 (Winter 1994): 8-27
228. Czerny C: Pianoforte Schule. 1840
229. Dacey J S and K Lennon: Understanding Creativity: The Interplay of Biological, Psychological, and Social Factors. Josey-Bass, San Francisco, 1998
230. Dahlhaus C: Zur Theorie des klassischen Kontrapunkts. Kirchenmusikalisches Jb 45, 1961
231. Dahlhaus C: Über den Begriff der tonalen Funktion. In: Vogel M (Ed.): Beiträge zur Musiktheorie des 19. Jahrhunderts. Bosse, Regensburg 1966
232. Dahlhaus C and H H Eggebrecht: Was ist Musik? Heinrichshofen, Wilhelmshaven et al. 1985
233. Dahlhaus C: Untersuchung über die Entstehung der harmonischen Tonalität. Bärenreiter, Kassel et al. 1967
234. Dahlhaus C et al.: Neues Handbuch der Musikwissenschaft, Bd. 1-13: Athenaion and Laaber, Laaber 1980-1993
235. Dahlhaus C and G Mayer: Musiksoziologische Reflexionen. In: Dahlhaus, C and de la Motte-Haber H (Eds.): Neues Handbuch der Musikwissenschaft, Bd. 10: Systematische Musikwissenschaft. Laaber, Laaber 1982
236. Dahlhaus C: Ludwig van Beethoven und seine Zeit. Laaber, Laaber 1987
237. D'Alembert J Le Rond: Einleitung zur 'Enzyklopädie'. Fischer, Frankfurt/Main 1989
238. D'Alembert, J Le Rond: Einleitung zur Enzyklopädie (1751). (German Translation) Fischer, Frankfurt/Main 1989
239. Dallos P: The active cochlea. J. Neurosci. Dec;12(12):4575-85, 1992
240. Dannenberg R B: An on-line algorithm for real-time accompaniment. In: ICMA (Ed.): Proceedings of the ICMC 84, San Francisco 1984
241. Dannenberg R B et al.: Automatic Ensemble Performance. In: ICMA (Ed.): Proceedings of the ICMC 94, San Francisco 1994
242. Alighieri, D: La Divina Commedia (1321). CreateSpace Independent Publishing Platform 2014
243. Danuser H et al. (Eds.): Neues Handbuch der Musikwissenschaft, Bd. 11: Interpretation. Laaber, Laaber 1992
244. Darwin Ch: The Origin of Species. London 1859
245. Daval R and G-T Guilbaud: Le raisonnement mathématique. PUF, 1945
246. de Bruijn N G: Pólya's Theory of Counting. In: Beckenbach E F (Ed.): Applied Combinatorial Mathematics, Ch.5. Wiley, New York 1964
247. de Bruijn N G: On the number of partition patterns of a set. Nederl. Akad. Wetensch. Proc. Ser. A 82 = Indag. Math. 41, 1979
248. Debussy C: *Préludes*, Livre I (1907-1910). Henle, München 1986
249. Dechelle F et al.: The Ircam Real-Time Platform and Applications. In: ICMA (Ed.): Proceedings of the ICMC 95, San Francisco 1995
250. Delalande F: La gestique de Gould. In: Guertin G (Ed.): Glenn Gould — pluriel. Corteau, Verdun 1988
251. Degazio B: A Computer-Based Editor For Lerdahl and Jackendoff's Rhythmic Structures. In: ICMA (Ed.): Proceedings of the ICMC 96, San Francisco 1996
252. de la Motte D: Harmonielehre. Bärenreiter/dtv, Kassel 1976
253. de la Motte-Haber H and H Emons: Filmmusik. Eine systematische Beschreibung. München, Hanser 1980
254. de la Motte-Haber H: Handbuch der Musikpsychologie. Laaber-Verlag, 2.Ed., Laaber 1996

255. de la Motte-Haber H: Rationalität und Affekt. In: Götze H and Wille R (Eds.): Musik und Mathematik. Springer, Berlin et al. 1985
256. de la Motte-Haber H: Musikalische Hermeneutik und empirische Forschung. In: Dahlhaus, C and de la Motte-Haber H (Eds.): Neues Handbuch der Musikwissenschaft, Bd. 10: Systematische Musikwissenschaft. Laaber, Laaber 1982
257. de la Motte-Haber H: Die Umwandlung der Interpretationsparameter in Struktureigenschaften. In: "Das Paradox musikalischer Interpretation", Symposion zum 80. Geburtstag von K. von Fischer, Univ. Zürich 1993
258. Deleuze G: Francis Bacon. La logique de la sensation. Éditions de la Différence, Paris 1981
259. Demazure M and P Gabriel: Groupes Algébriques. Masson & Cie./North-Holland, Paris/Amsterdam 1970
260. Dennett, D C: Quining qualia. In Marcel A and Bisiach E (Eds.): Consciousness in Contemporary Science. Oxford University Press, 1988
261. Desain P and Honing H: The Quantization of Musical Time: A Connectionist Approach. Computer Music Journal 13 (3), 56-66, 1989
262. Descartes R: Musicae Compendium. Herausgegeben und ins Deutsche übertragen als "Leitfaden der Musik" von J. Brockt, Wiss. Buchgesellschaft, Darmstadt 1978
263. Descartes R: Principia philosophiae (Principles of Philosophy). Translation with explanatory notes by Valentine Rodger and Reese P. Miller (Reprint ed.). Dordrecht: Reidel
264. Descartes R: Œuvres. Vrin, Paris, 1989. See also in English: The Geometry of René Descartes, transl. by E Smith and M L Latham, Dover Publ., 1954
265. Detienne M: Les Maîtres de Vérité dans la Grèce archaïque. 1ère éd. 1967, Le livre de poche/Références, no 611, Paris, 2012
266. De Tranquelléon: Sur l'intersection de deux cônes. Nouvelles Annales de mathématiques, 2ème série, tome III, pp. 539-540, 1864
267. Deyoung L: Pitch Order and Duration Order in Boulez' Structure 1a. PNM 16/2: 27-34, 1978
268. Dieudonné J: Foundations of Modern Analysis. Academic Press, New York et. al. 1960
269. Dieudonné J: La conception des mathématiques chez Valéry. In: Robinson-Valéry J: Fonctions de l'esprit. Treize savants redécouvrent Paul Valéry. Hermann, p. 183-191, Paris 1983
270. Dieudonné J: Mathematics – Music of Reason. Springer, 1992
271. Discogs: Horace Parlan. `http://www.discogs.com/artist/253065-Horace-Parlan` Not-dated
272. Döhl F: Webern - Weberns Beitrag zur Stilwende der Neuen Musik. Katzbichler, München et. al. 1976
273. Donald M: Mimesis and the Executive Suite: missing links in language evolution. In: J R Hurford, M Studdert-Kennedy, Ch Knight (Eds.): Approaches to the evolution of language, Cambridge U Press 1998
274. Dominguez M, D Clampitt and Th Noll: Well-formed Scales, Maximally Even Sets, and Christoffel Words. In: Proceedings of the MCM 2007, Springer, Heidelberg et al. 2009
275. Douady R: Jeux de cadres et dialectique outil-objet dans l'enseignement des mathématiques. Une réalisation dans le cursus primaire. Thèse, Paris 7, 1984
276. Douthett J and R Steinbach: Parsimonious Graphs: A Study in Parsimony, Contextual Transformations, and Modes of Limited Transposition. Journal of Music Theory 42.2, pp. 241-263, 1998
277. Doyle P, J Conway, J Gilman, B Thurston: Geometry and the Imagination in Minneapolis, version 0.91 dated 12 April 1994
278. Drake R M: Drake Musical Aptitude Tests. Science Research Associates, Chicago 1954
279. Dreiding A et al.: Classification of Mobile Molecules by Category Theory. In: Symmetries and Properties of Non-Rigid Molecules. Studies in Physical and Theoretical Chemistry, 23, 1983
280. Drossos C A: Sets, Categories and Structuralism. In: Sica, G. (Ed.): What is Category Theory?, Polimetrica, Monza 2005
281. Duffy M and S Tsuda: Shield Your Eyes Sonic Anarchy. The Sonic Alchemists, Minneapolis 2010
282. Dufourt H: Les difficultés d'une prise de conscience théorique. In: Le compositeur et l'ordinateur. 6-12, Ircam, Centre Georges Pompidou, Paris 1981
283. Duvignau K, B Gaume, J-L Nespoulous: Proximité sémantique et stratégies palliatives chez le jeune enfant et l'aphasique. Revue Parole, numéro spécial: 'Handicap langagier et recherches cognitives: Apports mutuels', J-L Nespoulous and J Virbel (Eds.), UMH Belgique, vol. 31-32: 219-255, 2004
284. Eberle G: "Absolute Harmonie" und "Ultrachromatik". In: Kolleritsch O (Ed.): Alexander Skrjabin. Universal Edition, Graz 1980
285. Ebisch S J H, M G Perrucci, A Ferretti, C Del Gratta, G Luca Romani, V Gallese: The Sense of Touch: Embodied Simulation in a Visuotactile Mirroring Mechanism for Observed Animate or Inanimate Touch. Journal of Cognitive Neuroscience 20:1-13, 2008
286. Eco U: Kunst und Schönheit im Mittelalter. Hanser, Wien 1991

287. Eco U: The Search for the Perfect Language. Wiley-Blackwell 1997
288. Eco U: Die Suche nach der vollkommenen Sprache. Beck, München 1994
289. Eco U: Le Signe. biblio essais n4159, éd. Labor 1988
290. Edelsbrunner H and J L Harer: Computational Topology. AMS 2010
291. Eggebrecht H H: Interpretation. In: "Das Paradox musikalischer Interpretation", Symposion zum 80. Geburtstag von K. von Fischer, Univ. Zürich 1993
292. Eggebrecht H H: Musik im Abendland. Piper, München and Zürich 1996
293. Ehrenfels Chr von: Über Gestaltqualitäten. Vierteljahresschrift für wissenschaftliche Philosophie XIV, 1890
294. Ehresmann C: Œuvres complètes et commentées. Part. IV-2 : esquisses et structures monoïdales fermées. ed. par Andrée Ch. Ehresmann, Amiens, 1983
295. Ehresmann A C, J-P Vanbremeersch: Memory Evolutive Systems. Hierarchy, Emergence, Cognition. Elsevier Amsterdam, 2007
296. Ehresmann A and J Gomez-Ramirez: Conciliating Neuroscience and Phenomenology via Category Theory. Progress in Biophysics and Molecular Biology, Vol. 19, issue 2, Elsevier, 2015
297. Eitz C: Das mathematisch reine Tonsystem. Leipzig 1891
298. Eimert H: Grundlagen der musikalischen Reihentechnik. Universal Edition, Wien 1964
299. Eisenbud D: Commutative Algebra with a View Toward Algebraic Geometry. Springer New York et al. 1996
300. Eisenbud D, and J Harris: The Geometry of Schemes. Springer, New York 2000
301. Engström B: Stereocilia of sensory cells in normal and hearing impaired ears. Scand. Audiol. Suppl. 19, 1-34, 1983
302. Engquist B and W Schmid (Eds.): Mathematics Unlimited — 2001 and Beyond. Springer, 2001
303. Ericson K: Heinz Holliger. Spurensuche eines Grenzgängers. Peter Lang, Bern 2004
304. Erlewine S: Movin' & Groovin' - Horace Parlan. `http://www.allmusic.com/album/movin-groovin-mw0000381525` 2015
305. Essl K: Strukturgeneratoren. Beiträge zur elektronischen Musik 5, IEM, Graz 1996
306. Euclid: Euclid's Elements. Translated with introduction and commentary by Sir Th L Heath, vol. 1, 2 and 3, Dover, 1956
307. Euler L: Tentamen novae theoriae musicae (1739). In: Opera Omnia, Ser.III, Vol.1 (Ed. Bernoulli, E et al.). Teubner, Stuttgart 1926
308. Euler L: Conjecture sur la raison de quelques dissonances générales reçues dans la musique (1764). In: Opera Omnia, Ser.III, Vol.1 (Ed. Bernoulli, E et al.). Teubner, Stuttgart 1926
309. Euler L: De harmoniae veris principiis per speculum musicum representatis (1773). In: Opera Omnia, Ser.III, Vol.1 (Ed. Bernoulli, E et al.). Teubner, Stuttgart 1926
310. Euler L: Recherches sur une nouvelle espèce de quarrés magiques. Opera Omnia, Ser. I, Vol. 7: 291-392, 1782
311. Fauré G: Requiem in D minor, Op. 48. Edited by P. Legge 2000
312. Feldman J et al.: Force Dynamics of Tempo Change in Music. Music Perception, *10*, 1992
313. Felver Chr: Cecil Taylor - All The Notes. A Chris Felver Movie, 2003
314. Ferretti R and G Mazzola: Algebraic Varieties of Musical Performances. Tatra Mt. Math. Publ. 23, 59-69, 2001
315. Feulner J et al.: MELONET: Neural Networks that Learn Harmony-Based Melodic Variations. In: ICMA (Ed.): Proceedings of the ICMC 94, San Francisco 1994
316. Fichte J G: Doctrine de la science. Nova methodo (1799), Livre de poche 4621, LGF, 2000
317. Fichtner R: Die verborgene Geometrie in Raffaels "Schule von Athen". Oldenburg 1984
318. Fink E: Grundlagen der Quantenmechanik. Akademische Verlagsgesellschaft, Leipzig 1968
319. Finke R A, Th B Ward, S M Smith: Creative Cognition: Theory, Research, and Applications. MIT Press, Cambridge, MA 1992
320. Finscher L: Studien zur Geschichte des Streichquartetts. Bärenreiter, Kassel 1974
321. Finsler P: Über die Grundlegung der Mengenlehre. Erster Teil. Die Mengen und ihre Axiome. Math. Z. 25, 683-713, 1926
322. Finsler P: Aufsätze zur Mengenlehre. Unger G (Ed.), Wiss. Buchgesellschaft, Darmstadt 1975
323. Fisher G and J Lochhead: Analyzing from the Body. In: Theory and Practice, Journal of the Music Theory Society of New York State, Vol. 27
324. Fleischer A: Eine Analyse theoretischer Konzepte der Harmonielehre mit Hilfe des Computers. Magisterarbeit, MWS, HU Berlin 1996
325. Fleischer A, G Mazzola, Th Noll: Zur Konzeption der Software RUBATO für musikalische Analyse und Performance. Musiktheorie, Heft 4, 314-325, 2000
326. Florida R: The Rise of the Creative Class. Basic Books, 2003

327. Floris E: La rupture cartésienne et la naissance d'une philosophie de la culture dans les œuvres juvéniles de J.-B. Vico. Thèse de 3ème cycle, dir. P. Ricœur, Université Paris X-Nanterre, 22 juin 1974
328. Flusser V: Gesten — Versuch einer Phänomenologie. Fischer, Frankfurt am Main 1994
329. Forster M: Technik modaler Komposition bei Olivier Messiaen. Hänssler, Neuhausen-Stuttgart 1976
330. Forte A: Structure of Atonal Music: Practical Aspects of a Computer-Oriented Research Project. In: Musicology and the Computer. Musicology 1966-2000. A Practical Program. Three Symposia. American Musicological Society, NY 1970
331. Forte A: Structure of Atonal Music. Yale University Press, New Haven 1973
332. Forte A: La Set-complex theory: Elevons les enjeux! Analyse musicale, 4e trimestre, 80-86, 1989
333. Foster W C: Singing Redefined. Recital Publications 1998
334. Frank H: RUBATO® Broadcast. ORF2: Modern Times, Jan. 10, 1997
335. Freedman D Z and P van Nieuwenhuizen: Supergravitation und die Einheit der Naturgesetze. In: Dosch H G (Ed.): Teilchen, Felder und Symmetrien. Spektrum der Wissenschaft, Heidelberg 1984
336. Friberg A: Generative Rules for Music Performance: A Formal Description of a Rule System. Computer Music Journal, Vol. 15, No. 2, 1991
337. Friberg A et al.: Performance Rules for Computer-Controlled Contemporary Keyboard Music. Computer Music Journal, Vol. 15, No. 2, 1991
338. Friberg A et al.: Recent Musical Performance Research at KTH. In: Sundberg J, (Ed.): Generative Grammars for Music Performance. KTH, Stockholm 1994
339. Friberg A: A Quantitative Rule System for Musical Performance. KTH PhD-Thesis, Stockholm 1995
340. Fripertinger H: Enumeration in Musical Theory. Beiträge zur elektronischen Musik 1, Hochschule für Musik und Darstellende Kunst, Graz 1991
341. Fripertinger H: Die Abzähltheorie von Pólya. Diplomarbeit, Univ. Graz 1991
342. Fripertinger H: Endliche Gruppenaktionen in Funktionenmengen—Das Lemma von Burnside—Repräsentantenkonstruktionen—Anwendungen in der Musiktheorie. Doctoral Thesis, Univ. Graz 1993
343. Fripertinger H: Untersuchungen über die Anzahl verschiedener Intervalle, Akkorde, Tonreihen und anderer musikalischer Objekte in n-Ton Musik. Magisterarbeit, Hochschule für Musik und Darstellende Kunst, Graz 1993
344. Fripertinger H: Anwendungen der Kombinatorik unter Gruppenaktionen zur Bestimmung der Anzahl "wesentlich" verschiedener Intervalle, Chorde, Tonreihen usw. Referat an der Univ. Innsbruck, Math. Institut d. Karl-Franzens-Univ., Graz 1996
345. Fripertinger H: Enumeration of Mosaics. Discrete Mathematics, 199, 49-60, 1999
346. Fripertinger H: Enumeration of Non-isomorphic Canons. Tatra Mt. Math. Publ. 23, 47-57, 2001
347. Frova A: Fisica nella musica. Zanichelli, Bologna 1999
348. Fussi F and S J Ghiotti: Il canto fra terminologia e didattica, 2015 `http://www.voceartistica.it/it-IT/index-/?Item=ghiotti`
349. Fux J J: Gradus ad Parnassum (1725). Dt. und kommentiert von L. Mitzler, Leipzig 1742
350. Gabriel P: Personal Communication. Zürich 1979
351. Gabriel P: Des catégorie abéliennes. Bull. Soc. Math. France 90, 1962
352. Gabriel P: Representations of Finite-Dimensional Algebras. In: Kostrikin, A.I.; Shafarevich, I.R. (Eds.): Encyclopaedia of Mathematical Sciences, Vol. 73, Springer Berlin 1992
353. Gabrielsson A: Music Performance. In: Deutsch D (Ed.): The Psychology of Music (2nd ed.). Academic Press, New York
354. Gabrielsson A: Expressive Intention and Performance. In: Steinberg R (Ed.): Music and the Mind Machine. Springer, Berlin et al. 1995
355. Gaither C C and A E Cavazos-Gaither: Mathematically Speaking. A Dictionary of Quotations. IOP Publishing Ltd, 1998
356. Galenson D W: Old Masters and Young Geniuses: The Two Life Cycles of Creativity. Princeton University Press, Princeton, NJ 2006
357. Gamer C: Some combinatorial resources of equal-tempered systems. JMT, 11, 32-59, 1967
358. Gamma E, R Helm, R Johnson, J Vlissides: Design Patterns, Elements of Reusable Object-Oriented Software. Addison-Wesley, Reading, MA et al., 1994
359. Gardner H: Kreative Intelligenz: Was wir mit Mozart, Freud, Woolf und Gandhi gemeinsam haben (Translated by Andreas Simon). Campus Verlag, Frankfurt/New York 1999
360. Gasquet S: L'illusion mathématique. Le malentendu des maths scolaires. Syros, 1997
361. Geisser H, G Mazzola, S Onuma: Imaginary Time. Video, Vimeo 2015

362. Gentilucci M & and M C Corballis: From manual gesture to speech: A gradual transition. Neuroscience & Biobehavioral Reviews Vol. 30, Issue 7, pp 949-960 2006
363. Gerwin Th: IDEAMA, Zentrum fur Kunst und Medientechnologie, Karlsruhe 1996
364. Geweke J: A comparison of tests of independence of two covariance-stationary time series. J. Am. Statist. Assoc., 76, 363-373, 1981
365. Giannitrapani D: The Electrophysiology of Intellectual Functions. Karger, Basel 1985
366. Gilson E: Introduction aux arts du beau. Vrin, Paris 1963
367. Ginzburg C: Leçon de Montaigne. Sud-Ouest, 17 octobre 2015
368. Ginzburg C: L'estrangement. In à distance, Neuf essais sur le point de vue en histoire. Traduction française par Pierre Antoine Fabre, éditions Gallimard NRF, Paris 2001
369. Giocanti S: L'art sceptique de l'estrangement dans les *Essais* de Montaigne, in *L'estrangement, Retour sur un thème de Carlo Ginzburg* [573, p. 19-35]
370. Godement R: Topologie algébrique et théorie des faisceaux. Hermann, Paris 1964
371. Godøy R I and M Leman (Eds.): Musical Gestures—Sound, Movement, and Meaning. Routledge, New York and London 2010
372. Goeller S: Object Oriented Rendering of Complex Abstract Data. PhD. thesis, Universität Zürich, 2004
373. Goethe J W von: Brief an Zelter. 9. Nov. 1829
374. Goffman E: The Presentation of Self in Everyday Life. University of Edinburgh, Edinburgh 1956
375. Goffman E: Behavior in Public Places. The Free Press, New York 1963
376. Goldblatt R: Topoi. North-Holland, Amsterdam et al. 1984
377. Goldin-Meadows S: Hearing Gesture: How Our Hands Help Us Think. Harvard University Press 2003
378. Goldstein J L: An Optimum Processor Theory for the Central Formation of the Pitch of Complex Tones. J. Acoust. Soc. Am. 54, 1496 1973
379. Gomes Teixeira F: Traité des courbes spéciales remarquables planes et gauches. 3 vol., (Coïmbre, Impr. de l'Université, 1908-1915), éd. Jacques Gabay, 1995. Traduction de *Tratado de las curvas especiales notables*, Ac. Sc. Madrid, 1899
380. Interview with Emilia Gómez: "PHENICX is a new way of experiencing a classical music concert", `http://www.upf.edu/enoticies/en/entrevistes/1232.html#.VbfU7s63QUU`
381. Gorenstein D: Classifying the finite simple groups. Bull. A.M.S. 14, 1-98, 1986
382. Gottschewski H: Tempohierarchien. Musiktheorie, Heft 2, 1993
383. Götze H and R Wille R (Eds.): Musik und Mathematik. Springer, Berlin et al. 1985
384. Gould G: The Glenn Gould Reader. Alfred A. Knopf, New York 1984
385. Goupillaud P, A Grossmann, J Morlet: Cycle-octave and related transforms in seismic signal analysis. Geoexploration, 23, 85-102, 1984-1985
386. Grabusow N: Vielfalt akustischer Grundlagen der Tonarten und Zusammenklänge - Theorie der Polybasiertheit. Musiksektion des Staatsverlags, Moskau 1929
387. Graeser W: Bachs "Kunst der Fuge". In: Bach-Jahrbuch 1924
388. Graeser W: Der Körpersinn. Beck, München 1927
389. Gramain A: Géométrie élémentaire. Hermann, 1997
390. Greimas A J: Les actants, les acteurs et les figures. In: Chabrol C (Ed.): Sémiotique narrative et textuelle. Larousse, Paris 1974
391. Greub W: Linear Algebra. Springer, Berlin et al. 1967
392. Grisey G: Tempus ex machina: a Composer's Reflections on Musical Time. Contemporary Music Review 2, no. 1:238-275, 1987
393. Grondin J: L'herméneutique. Coll. Que sais-je ? no 3758, PUF, Paris 2006
394. Gross D: A Set of Computer Programs to Aid in Music Analysis. Ph. Diss. Indiana Univ. 1975
395. Grothendieck A and J Dieudonné: Eléments de Géométrie Algébrique I. Springer, Berlin et al. 1971
396. Grothendieck A and J Dieudonné: Eléments de Géométrie Algébrique I-IV. Publ. Math IHES no. 4, 8, 11, 17, 20, 24, 28, 32, Bures-sur-Yvette 1960-1967
397. Grothendieck A: Correspondence with G. Mazzola. April 1, 1990
398. Grothendieck A: Récoltes et Semailles. Université Univ. Sci. et Tech. Languedoc et CNRS, Montpellier, 1st edn. 1985
399. Guevara R C L et al.: A Modal Distribution Approach to Piano Analysis and Synthesis. In: ICMA (Ed.): Proceedings of the ICMC 96, San Francisco 1996
400. Guillaume P: La psychologie de la forme. Journal de Psychologie, XXII, p. 768-800, 1925
401. Guillaume P: La psychologie de la forme. Flammarion, 1979

402. Guitart R: Sur l'ébauche des structures. Proc. 3d Congress of Bulgarian math., summaries part II, p. 354, Varna, Sept. 6-15, 1972
403. Guitart R: Sur le foncteur diagrammes, CTGD XIV-2, pp. 181-182, 1973
404. Guitart R: Remarques sur les machines et les structures, CTGD XV-2, pp. 113-144, 1974
405. Guitart R: Remarques sur les machines et les structures. CTGD XV-2, pp. 113-144, 1974
406. Guitart R and L Van den Bril: Décompositions et lax-complétions. CTGD XVIII-2, pp. 333-407, 1977
407. Guitart R: Des machines aux bimodules. polycopié, Univ. Paris 7, 24 p., 1978
408. Guitart R: Relations et carrés exacts. Ann. Sc. Math. Qué., vol. IV-2, p. 103-125, juillet 1980
409. Guitart R and C Lair: Calcul syntaxique des modèles et calcul des formules internes, *Diagrammes*, vol. 4, déc. 1980, 106 p.
410. Guitart R: On the geometry of computations, I and II. Cahiers Top. Géo. Diff. Cat. XXVII,4, 107-136, 1986; XXIX,4, p. 297-326, 1988
411. Guitart R: Mathématiques. 1ère année. Polycopié U. Paris 7; 27 chap., 432 p. 1992-1993
412. Guitart R: La pulsation mathématique. L'Harmattan, Paris, 1999
413. Guitart R: évidence et étrangeté. PUF, Paris, 2000
414. Guitart R: Sur les places du sujet et de l'objet dans la pulsation mathématique. Questions éducatives, revue du Centre de Recherche en éducation de l'Université Jean Monnet de Saint-Étienne, no. 22-23, pp. 49-81, décembre 2002
415. Guitart R: Toute théorie est algébrique et topologique. Cahiers Top. Géo. Diff. Cat. vol. XLIX-2 (2008), p. 83-128. 2008 (first partial version: "Toute théorie est algébrique", in Journée Mathématique en l'Honneur d'Albert Burroni, vendredi 20 septembre 2002, Université Paris 7, Institut de Mathématiques de Jussieu, Prépublication 368, p. 79-102, avril 2004
416. Guitart R and Ch Ehresmann: Au carrefour des structures locales et algébriques. Colloque International Ch Ehresmann : 100 ans, Amiens 7-9 octobre 2005; Cahiers Top. Géo. Diff. Cat., vol. XLVI-3, p. 172-175, 2005
417. Guitart R: L'évidemment des objets et le dehors comme substance. Colloque Le lemme de Yoneda, "Enjeux mathématiques et philosophie", ENS, Paris 18 juin 2007
418. Guitart R: An anabelian definition of abelian homology. Cahiers Top Géo Diff Cat, XXXXVIII, 4, pp. 261-269, 2007
419. Guitart R: Mathematical Pulsation at the root of invention. Anais do Congresso Htem 4, Rio de Janeiro, de 5 a 9 de maio de 2008
420. Guitart R: Sur la représentation de l'herméneutique comme commerce gratuit des interprétations, conférence au *CERCI*, Université de Nantes, 7 mars 2009
421. Guitart R: Cohomological Emergence of Sense in Discourses (As Living Systems Following Ehresmann and Vanbremeersch). Axiomathes, 19(3), 245-270, 2009
422. Guitart R: Les coordonnées curvilignes de Gabriel Lamé, représentations des situations physiques et nouveaux objets mathématiques, 12 p. In: Actes du Colloque International Gabriel Lamé. Les pérégrinations d'un ingénieur au XIXème siècle, 15-16-17 janvier 2009, Nantes. SABIX, no 44, p. 119-129, octobre 2009
423. Guitart R: Sense and Signs in the Mathematical Invention of Coordination (in the Perspective of Arrows, Relations and Sketches). Talk at the colloquium *Semiotic Approaches to Mathematics, the History of Mathematics and Mathematics Education (SemMHistEd)*, 3rd Meeting Aristotle University of Thessaloniki, July 16-17, 2009
424. Guitart R: Modélisation qualitative catégoricienne : modèles, signes et formes. In: Andreatta M, F Nicolas and Ch Alunni (Eds.): à la lumière des mathématiques et à l'ombre de la philosophie, Coll. Musique/Sciences, ENS, Delatour, Ircam, Centre Pompidou, pp. 133-147, Paris 2012
425. Guitart R: Bachelard et la pulsation mathématique. Revue de synthèse, tome 136, 6ème série, n. 1-2, p. 33-74, 2015
426. Guitart R: Note sur deux problèmes pour une épistémologie transitive des mathématiques, *Revue de synthèse*, tome 136, 6ème série, n. 1-2, p. 237-279, 2015
427. Guitart R: Nietzsche face à l'expérience mathématique, in: Barbin E and J-P Cléro éds, *Les mathématiques et l'expérience : ce qu'en ont dit les philosophes et les mathématiciens*, Hermann, p. 247-278, 2015
428. Guitart R: History in mathematics according to André Weil. ESU7, Copenhagen, July 2014
429. Guitart R: L'évidement et la nécessité de la contingence. Conférence au Colloque *Les fins de cure : la poursuite du désêtre*, 7/8, AECF Lille, novembre 2015
430. Gurlitt et al. (Eds.): Riemann Musiklexikon/Sachteil. Schott, Mainz 1967
431. Gusdorf G: La révolution galiléenne, tomes I et II. Payot, Paris, 1969
432. Hába A: Neue Harmonielehre. (Reprint from 1927) Universal Edition, Wien 1978
433. Habermann R: Elementary Applied PDEs, Prentice Hall 1983

434. Hadamard J: An Essay on the Psychology of Invention in the Mathematical Field. Princeton University Press, 1945, Dover, 1954
435. Halsey D and E Hewitt: Eine gruppentheoretische Methode in der Musiktheorie. Jahresber. d. Dt. Math.-Vereinigung 80, 1978
436. Handschin J: Der Toncharakter. Zürich 1948
437. Hansen P: Pourquoi il faut sortir des mathématiques. Mercredi 1er avril 2015 (see also [99]). `http://www.energie-crise.fr/spip.php?article176`
438. Hanslick E: Vom Musikalisch-Schönen. Breitkopf und Härtel (1854), Wiesbaden 1980
439. Hardy G H: A Mathematician's Apology. Cambridge U. Press, 1967
440. Hashimoto K: D-Brane. Springer, Heidelberg et al. 2012
441. Hardt M: Zur Zahlenpoetik Dantes. In: Baum R. and Hirdt W (Eds.): Dante Alighieri 1985. Stauffenburg, Tübingen 1985
442. Harkness R: Songs of Seoul. University of California Press 2014
443. Harris C and A R Brinkman: A unified set of software tools for computer-assisted set-theoretic and serial analysis of contemporary music. Proc. ICMC 1986, ICA, San Francisco 1986
444. Hartshorne R: Algebraic Geometry. Springer, New York et al. 1977
445. Hashimoto S and H Sawada: Musical Performance Control Using Gesture: Towards Kansei Technology for Art. In: Kopiez R and Auhagen W (Eds.): Controlling Creative Processes in Music. Peter Lang, Frankfurt am Main et al. 1998
446. Hatten R: Interpreting Musical Gestures, Topics, and Tropes. Indiana University Press 2004
447. Hawking S: A Brief History of Time: From the Big Bang to Black Holes. Bantam Books, New York 1988
448. Haugh L D: Checking for independence of two covariance stationary time series: a univariate residual cross correlation approach. J. Am. Statist. Assoc., 71, 378-385, 1976
449. Haus G et al.: Stazione di Lavoro Musicale Intelligente. In: Haus G and Pighi I (Eds.): X Colloquio di Informatica Musicale. AIMI, LIM-Dsi, Milano 1993
450. Hayes J R: Cognitive Processes in Creativity. In: J A Glover, R R Roning, C R Reynolds (Eds.): Handbook of Creativity. Plenum, New York 1989
451. Hebb D O: Essay on mind. Hillsdale, New Jersey, Lawrence Erlbaum Associates, 1980
452. Hegel G W F: Wissenschaft der Logik I (1812). Felix Meiner, Hamburg 1963
453. Hegel G W F: Wissenschaft der Logik, Erster Band, Erstes Buch. Schrag, Nürnberg 1812 Online `http://www.deutschestextarchiv.de/book/show/hegel_logik0101_1812` Modern German online version: `http://www.zeno.org/Philosophie/M/Hegel,+Georg+Wilhelm+Friedrich/Wissenschaft+der+Logik`
454. Hegel G W F: Hegel's Science of Logic. Online `http://www.marxists.org/reference/archive/hegel/works/hl/hlconten.htm`
455. Heiberg J L and H Menge (Eds.): Euclidis opera omnia. 8 vol. & supplement, in Greek. Teubner, Leipzig 1883-1916.
456. Heijink H, P Desain, H Honing, L Windsor: Make me a match: An evaluation of different approaches to score-performance matching. Computer Music Journal, 24(1), 43-56, 2000
457. Helmholtz H von: Die Lehre von den Tonempfindungen als physiologische Grundlage der Musik (1863). Nachdr. Darmstadt 1968
458. Henck H: Karlheinz Stockhausens Klavierstück IX. Verlag für systematische Musikwissenschaft, Bonn-Bad Godesberg 1978
459. Hentoff N: Liner notes to Coltrane's last album "Expression". Impulse AS-9120, New York 1967
460. Hecquet S and A Prokhoris: Fabriques de la danse. PUF, Paris 2007
461. Herbort H J: Keine Ausweispflicht für cis. Die Zeit Nr. 43, 21. Oktober 1988
462. Hermann J: Consideratio curvarum in punctum positione datum projectarum, et de affectionibus earum inde pendentibus. Commentarii Academiae Petropolitanae, IV, 37-46, 1729
463. Hesse H: Das Glasperlenspiel (1943). Fretz und Wasmuth, Zürich 1943
464. Hesse H: Materialien zu H. Hesses "Das Glasperlenspiel", Vol. I. : Vom Wesen und von der Herkunft des Glasperlenspiels (1934). Suhrkamp, Frankfurt am Main 1981
465. Heussenstamm G: Norton Manual of Music Notation. Norton & Comp., New York 1987
466. Hichert J: Verallgemeinerung des Kontrapunkttheorems für die Hierarchie aller starken Dichotomien in temperierter Stimmung. Diplomarbeit, TU Ilmenau 1993
467. Hindemith P: Unterweisung im Tonsatz. Schott, Mainz 1940
468. Hilbert D and S Cohn-Vossen: Geometry and the Imagination. AMS Chelsea Pub. (transl. by Paul Nemenyi from *Anschauliche Geometrie*, 1932), 1952

469. Hiller L and P Ruiz: Synthesizing sounds by solving the wave equation for vibration objects. J. of the Audio Engineering Soc. 19: 463-470, 542-551, 1971
470. Hilton P J and S Wylie: Homology Theory. Cambridge University Press, London et al. 1967
471. Hjelmslev L: La Stratification du Langage. Minuit, Paris 1954
472. Hjelmslev L: Prolégomènes à une théorie du langage. Minuit, Paris 1968-71
473. Hjelmslev L: Nouveaux essays. PUF, Paris 1985
474. Hofstädter D: Gödel, Escher, Bach. New York: Basic Books, New York 1979
475. Hofstadter D R: Staring Emmy Straight in the Eye—and Doing my Best not to Flinch. In: D Cope (Ed.): Virtual Music: Computer Synthesis of Musical Style. MIT Press, Cambridge, MA 2001
476. Hodgson J: Mastering Movement: the life and work of Rudolf Laban. Routledge, New York 2001
477. Holden S P: Right-Handed and Right On. The New York Times, No. 3, June 2, 1999
478. Holm-Hadulla R M: Kreativität. Konzept und Lebensstil. Vandenhoeck & Ruprecht, Göttingen 2005
479. Hong Y: Testing for independence between two covariance stationary time series. Biometrika, 83, No.3, 615-626, 1996
480. Honing H: Expresso, a strong and small editor for expression. In: ICMA (Ed.): Proceedings of the ICMC 92, San Francisco 1992
481. Hooft G 't: Symmetrien in der Physik der Elementarteilchen. In: Dosch H G (Ed.): Teilchen, Felder und Symmetrien. Spektrum der Wissenschaft, Heidelberg 1984
482. Hörnl D et al.: Learning Musical Structure and Style by Recognition, Prediction and Evolution. In: ICMA (Ed.): Proceedings of the ICMC 96, San Francisco 1996
483. Horry Y: A Graphical User Interface for MIDI Signal Generation and Sound Synthesis. In: ICMA (Ed.): Proceedings of the ICMC 94, San Francisco 1994
484. Hosken D: An Introduction to Music Technology. Routledge New York 2011
485. Howe H: Some combinatorial properties of Pitch-Structures. PNM, 4(1), 45-61, 1965
486. Howe M J A: Prodigies and Creativity. In: R J Sternberg (Ed.): Handbook of Creativity. Cambridge University Press, New York 1999
487. Hu S-T: Mathematical Theory of Switching Circuits and Automata. University of California Press, Berkeley and Los Angeles 1968
488. Hudak P et al.: Haskore Music Notation - An Algebra of Music. J. Functional Programming Vol. 6 (3) 1996
489. Hudspeth A J and D P Corey: Sensitivity, Polarity, and Conductance Change in the Response of Vertebrate Hair Cells to Controlled Mechanical Stimuli. Proc. Nat. Acad. Sci. Am. 74(6), 2407-2411, 1977
490. Hull P et al.: The Music Business and Recording Industry. Third Edition. Routledge, New York and London 2011
491. Humphreys J E: Introduction to Lie Algebras and Representation Theory. Springer, New York et al. 1972
492. Hung R et al.: The Analysis and Resynthesis of Sustained Musical Signals in the Time Domain. In: ICMA (Ed.): Proceedings of the ICMC 96, San Francisco 1996
493. Hunziker E and G Mazzola: Ansichten eines Hirns. Birkhäuser, Basel 1990
494. Husmann H: Einführung in die Musikwissenschaft. Heinrichshofen, Wilhelmshaven 1975
495. Ilchmann A: Kritik der Übergänge zu den ersten Kategorien in Hegels Wissenschaft der Logik. Hegel-Studien, Bd. 27, Bouvier, Bonn 1992
496. Jackendoff R and F Lerdahl: A Generative Theory of Tonal Music. MIT Press, Cambridge MA, 1983
497. Jackson J D: Classical Electrodynamics. Wiley, New York 1998
498. Jairazbhoy N A: What Happened to Indian Music Theory? Indo-Occidentalism? Ethnomusicology 52 (3): 34977, 2008
499. Jakobson R and M Halle: Fundamentals of Language. Mouton, Le Hague 1957
500. Jakobson R: Linguistics and Poetics. In: Seboek, TA (Ed.): Style in Language. Wiley, New York 1960
501. Jakobson R: Language in relation to other communication systems. In: Linguaggi nella società e nella tecnica. Edizioni die Communità, Milano 1960
502. Jakobson R: Six leçons sur le son et le sens. Ed. de Minuit, Paris 1976
503. Jakobson R: Hölderlin, Klee, Brecht. Suhrkamp, Frankfurt/Main 1976
504. Jakobson R and K Pomorska: Poesie und Grammatik. Dialoge. Suhrkamp, Frankfurt/Main 1982
505. Jakobson R: Semiotik - Ausgewählte Texte 1919-1982. Holenstein E (Ed.), Suhrkamp, Frankfurt/Main 1988
506. Jauss H R: Rückschau auf die Rezeptionstheorie—Ad usum Musicae Scientiae. In: Danuser H and Krummacher F (Hsg.): Rezeptionsästhetik und Rezeptionsgeschichte in der Musikwissenschaft. Laaber, Laaber 1991
507. Jeans J H: Science and Music. 1938, reprinted by Dover, 1968
508. Jeppesen K: Kontrapunkt. Breitkopf und Härtel, Wiesbaden 1952
509. John F: Partial Differential Equations. Springer, Heidelberg et al. 1978

510. Johnson T: See his web page at http://www.tom.johnson.org
511. Johnson-Laird P N: Freedom and Constraint in Creativity. In: R J Sternberg (Ed.): The Nature of Creativity: Contemporary Psychological Perspectives. Cambridge University Press, New York, pp. 202-219, 1988
512. Junod J: Counterpoint Worlds and Morphisms. Dissertation, University of Zurich, Zurich 2010
513. Julia G: Sur l'itération des fonctions rationnelles. J. de Math. Pure et Appl. 8, 1918
514. Jurek T: The Complete Blue Note Horace Parlan Sessions. AllMusic 2015. `http://www.allmusic.com/album/the-complete-blue-note-horace-parlan-sessions-mw0001007267`
515. Kagel M: Translation - Rotation. Die Reihe Bd.7, Universal Edition, Wien 1960
516. Kahle W: Taschenatlas der Anatomie, Bd.3: Nervensystem und Sinnesorgane. Thieme/dtv, Stuttgart 1979
517. Kaiser J: Beethovens 32 Klaviersonaten und ihre Interpreten. Fischer, Frankfurt/Main 1979
518. Kan D: Adjoint functors. Transactions of the American Mathematical Society, 87: 294-329, 1958
519. Kant I: Kritik der reinen Vernunft. Meiner, Hamburg, 2nd edn. 1957
520. Kantor J-M (Ed.): Ou en sont les mathématiques?. Vuibert/SMF, 2002
521. Karg-Elert S: Polaristische Klang- und Tonalitätslehre (1931). Out of print, cf.: Schenk P.: Karg-Elerts polaristische Harmonielehre. In: Vogel, M (Ed.) Beiträge zur Musiktheorie des 19. Jahrhunderts. Bosse, Regensburg 1966
522. Kasner E and J Newman: Mathematics and the imagination. Simon and Schuster, New York, 1940
523. Katayose H et al.: Demonstration of Gesture Sensors for the Shakuhachi. In: ICMA (Ed.): Proceedings of the ICMC 94, San Francisco 1994
524. Katsman R: Anthropoetic Gesture: A Key to Milorad Pavić's Poetics (Landscape Painted With Tea). Academic Electronic Journal in Slavic Studies No. 16, Univ. Toronto, 2006
525. Katsman R: Gestures Accompanying Torah Learning/recital among Yemenite Jews. Gesture 7 (1): 119, 2007
526. Kayser H: Akróasis. Die Lehre von der Harmonik der Welt. Schwabe, Basel 1946
527. Kelley J L: General Topology. Van Nostrand, Princeton et al. 1955
528. Kempe A B: A memoir on the theory of mathematical form. Philosophical Transactions of the Royal Society of London, 177, 1-70, 1886
529. Kendon A: How Gestures Can Become like Words. Crosscultural Perspectives in Nonverbal Communication. 1988
530. Kendon A: Gesture: Visible Action as Utterance. Cambridge University Press, Cambridge 2004
531. Keyser C J: The Meaning of Mathematics. Scripta Mathematica, 1933, p. 15-28
532. Kiczales G, J Rivieres, D G Bobrow: The Art of the Metaobject Protocol. The MIT Press, Boston 1991
533. Kinderman W: Artaria 195. University of Illinois Press, Urbana and Chicago 2003
534. King E: New Perspectives on Music and Gesture. Routledge 2016
535. Kirsch E: Wesen und Aufbau der Lehre von den harmonischen Funktionen. Leipzig 1928
536. Klemm M: Symmetrien von Ornamenten und Kristallen. Springer, Berlin et al. 1982
537. Klumpenhouwer H: Deep Structure in K-net Analysis with special reference to Webern's opus 16,4. Unpublished manuscript
538. Klumpenhouwer H: A Generalized Model of Voice-Leading for Atonal Music. Ph.D. Thesis, Harvard University 1991
539. Knapp J: Franz Liszt. Berlin 1909
540. Köhler E: Brief an Guerino Mazzola. Hamburg 1988
541. Kollmann A: An Essay on Musical Harmony. London 1796
542. Komparu K: The Noh Theatre. Weatherhill/Tankosha, New York et al. 1983
543. Koenig Th: Robert Schumanns *Kinderszenen* op.15. In: Metzger H-K und Riehn R (Hrg.): Robert Schumann II. edition text+kritik, München 1982
544. Kononenko N O: Ukrainian minstrels and the blind shall sing. Armonk, N.Y.: M E Sharpe, 1998
545. Kopiez R and J Langner: Entwurf einer neuen Methode der Performanceanalyse auf Grundlage einer Theorie oszillierender Systeme. In: Behne K-E and de la Motte H (Eds.): J.buch der D. Ges. für Musikpsychologie. **12** (1995), Wilhelmshaven 1996
546. Kopiez R: Aspekte der Performanceforschung. In: de la Motte H: Handbuch der Musikpsychologie. Laaber, Laaber 1996
547. Kopiez R: Mensch - Musik - Maschine. Musica, 50 (1), 1996
548. Kopiez R: "The most wanted song/The most unwanted song" – Klangfarbe als wahrnehmungsästhetische Kategorie. Musicology Conference, Halle 1998
549. Kopiez R, J Langner, Ch Stoffel: Realtime analysis of dynamic shaping. Talk at the 6th International Conference on Music Perception and Cognition (ICMPC), Keele, England, August 5-10, 2000

550. Kopiez R, J Langner and B Feiten: Perception ad Representation of Multiple Tempo Hierarchies in Musical Performance ad Composition: Perspectives from a New Theoretical Approach. In: Kopiez R and W Auhagen (Eds.): Controlling Creative Processes in Music. Peter Lang, Frankfurt am Main et al. 1998
551. Körner T W: Fourier Analysis. Cambridge University Press 1988, Cambridge
552. Kostelanetz R: John Cage. Praeger, New York 1968
553. Kouneiher J, D Flament, P Nabonnand, J-J Szczeciniarz (Eds.): Geometrie au XXème siècle, 1930-2000. Histoire et horizons. Hermann, Paris 2005
554. Kouzes R T et al.: Collaboratories: Doing Science on the Internet. Computer, August 1996
555. Kozbelt A, R A Beghetto, M A Runco: Theories of Creativity. In: J C Kaufman and R J Sternberg (Eds.): The Cambridge Handbook of Creativity. Cambridge University Press, New York, pp. 20-47, 2010
556. Krömer R: La théorie des catégories : un outil de musicologie scientifique aux yeux de la critique philosophique. In: Andreatta M, F Nicolas et C Alunni (Eds.): Á la lumière des mathématiques et à l'ombre de la philosophie. Coll. Musique/Sciences, ENS, Delatour, Ircam, Centre Pompidou, p. 213-222, Paris 2012
557. Kronland-Martinet R: The Wavelet Transform for Analysis, Synthesis, and Processing of Speech and Music Sounds. Computer Music Journal, 12 (4), 1988
558. Kreiman J and D Sidtis: Foundations of Voice Studies. Wiley-Blackwell 2013
559. Kronland-Martinet R: The Wavelet Transform for Analysis, Synthesis, and Processing of Speech and Music Sounds. Computer Music Journal, 12 (4), 1988
560. Kronman U and J Sundberg: Is the Musical Ritard an Allusion to Physical Motion? In: Gabrielsson A (Ed.): Action and Perception in Rhythm and Meter. Bibl. of the Royal Swedish Acad. of Sci. 55, Stockholm 1987
561. Kuratowski K and A Mostowski: Set Theory. North Holland, Amsterdam 1968
562. Kühner H: Virtual Table. CG TOPICS 3/97
563. Laban R: Language of Movement: A guidebook to Choreutics, ed. Lisa Ullmann, Plays, Boston 1974
564. Laban R: Die Welt des Tänzers. W. Seifert, Stuttgart 1920
565. Laban R: The Laban Sourcebook. McCaw D (Ed.), Routledge, New York 2011
566. Laban R: The Mastery of Movement, 4th ed. Macdonald & Evans, Plymouth 1980
567. Laban R: The Rhythm of Effort and Recovery I. LAMG Magazine, in Hodgson, Mastering Movement, November 1959
568. Lachenmann H: Mouvement (- vor der Erstarrung) für Ensemble. Breitkopf und Härtel, 1984
569. Lages Lima E: Fundamental Groups and Covering Spaces. Peters, Natick, MA 2003
570. Lakoff G and R Núñez: Where mathematics comes from: How the embodied mind brings mathematics into being. Basic Books, New York 2000
571. Lamé G: Examen des différentes méthodes employées pour résoudre les problèmes de géométrie (1903). Hermann, Paris 1918
572. Lamé G: Leçons sur les coordonnées curvilignes, et leurs diverses applications. Mallet-Bachelier, 1859
573. Landi S (Ed.): L'estrangement, Retour sur un thème de Carlo Ginzburg. Essais, revue interdisciplinaire d'humanités, No Hors série, école doctorale Montaigne-Humanités, Univ. Michel de Montaigne, Bordeaux 3, 2013
574. Landi S (Ed.): Avant-propos. In: L'estrangement, Retour sur un thème de Carlo Ginzburg [573, p. 9-17]
575. Landry, E and J-P Marquis: Categories in Context: Historical, Foundational, and Philosophical. Philosophia Mathematica 13.1: 1-43, 2005
576. Lang S: Introduction to Differentiable Manifolds. Interscience, New York et al. 1962
577. Lang S: Elliptic Functions. Addison-Wesley, Reading, Mass. 1973
578. Lang S: $SL_2(\mathbb{R})$. Addison-Wesley, Reading, Mass. et al. 1975
579. Langley P, H A Simon, G L Bradshaw, J M Zytkow: Scientific Discovery: Computational Explorations of the Creative Process. MIT Press, Cambridge, MA, 1987
580. Langner G: Evidence for Neuronal Periodicity Detection in the Auditory System of Guinea Fowl: Implications for Pitch Analysis in the Time Domain. Exp. Brain Res. 52, 333-355 1983
581. Langner G: Periodicity coding in the auditory system. Hear Res. Jul;60(2):115-42, 1992
582. Langer S: Feeling and Form, Routledge and Kegan Paul, London 1953
583. Lawson R A: The First Century of Blues: One Hundred Years of Hearing and Interpreting the Music and the Musicians. Southern Cultures, pp. 39-61, 2007
584. Langner J, R Kopiez, Ch Stoffel, M Wilz: Realtime analysis of dynamic shaping. In: Woods C et al. (Eds.): Proceedings of the Sixth International Conference on Music Perception and Cognition, Keele, UK, 2000
585. Lawvere F W: An elementary theory of the category of sets. Proc. Natl. Acad. Sci. 52, 1506-1511, 1964
586. Lawvere F W and R Rosebrugh: Sets for mathematics, Cambridge U. Press, 2003

587. Lazier R: Making Schoenberg Dance. In: Sound Moves, Proceedings of the Int. Conf. on Music and Dance, Roehampton University 2005
588. Leach J L: Towards a Universal Algorithmic System for Composition of Music and Audio-Visual Works. In: ICMA (Ed.): Proceedings of the ICMC 96, San Francisco 1996
589. Ledderose L: Kreativität und Schrift in China. In: R M Holm-Hadulla (Ed.): Kreativität. Springer, Heidelberg 2000
590. Lee, J M: Introduction to Topological Manifolds. Springer, Heidelberg et al. 2000
591. Legrand M: Revue critique de "La pulsation mathématique". Repères-IREM, 39, avril 2000
592. Le Guin E: Boccherini's Body. University of California Press, Berkeley and Los Angeles 2006
593. Leman M: Schema Theory. Springer, Berlin et al. 1996
594. Lerdahl F: Timbral Hierarchies. Contemporary Music Review, vol.2, no.1, 1987
595. Lerdahl F: Cognitive Constraints on Compositional Systems. In: ed. J Sloboda (Ed.): Generative Processes in Music. Oxford University Press, Oxford 1988. Reprinted in Contemporary Music Review 6, no. 2 (1992):97-121
596. Levy M: Intonation in North Indian Music. Biblia Impex, 1982
597. Lévi-Strauss C: Le cru et le cuit: Mythologies I. Plon, Paris 1964
598. Lewin D: Re: Intervallic Relations between Two Collections of Notes. JMT, 3(2), 298-301, 1959.
599. Lewin D: Klumpenhouwer Networks and Some Isographies that Involve Them. Music Theory Spectrum 12/1, 1990
600. Lewin D: The Intervallic Content of a Collection of Notes, Intervallic Relations between a Collection of Notes and Its Complement: An Application to Schoenberg's Hexachordal Pieces. JMT, 4(1), 98-101, 1960
601. Lewin D: Forte's Interval Vector, My Interval Function, and Regener's Common-Note Function. JMT, 21(2), 194-237, 1977
602. Lewin D: A Formal Theory of Generalized Tonal Functions. JMT 26(1), 32-60, 1982
603. Lewin D: On Formal Intervals between Time-Spans. Music Perception, 1(4), 414-423, 1984
604. Lewin D: Musical Form and Transformation: 4 Analytic Essays. Yale University Press, New Haven and London 1993
605. Lewin D: Generalized Musical Intervals and Transformations (1987). Cambridge University Press 2007
606. Lewis C I: Mind and Word Order. Dover, New York 1956
607. Leyton M: Symmetry, Causality, Mind. MIT Press, Cambridge/MA and London 1992
608. Leyton M: A Generative Theory of Shape. Springer, Berlin et al. 2001
609. Lichtenhahn E: Romantik: Aussen- und Innenseiten der Musik. In: "Das Paradox musikalischer Interpretation", Symposion zum 80. Geburtstag von K. von Fischer, Univ. Zürich 1993
610. Lidov D: Mind and Body in Music. Semiotica 66/1 (1987), pp. 77-78
611. Ligeti G: Pierre Boulez: Entscheidung und Automatik in der Structure Ia. Die Reihe IV, UE, Wien 1958
612. Ligeti G and G Neuweiler: Motorische Intelligenz. Wagenbach, Berlin 2007
613. Lippe E and A H M ter Hofstede: A Category Theory Approach to Conceptual Data Modeling. Informatique Theorique et Applications, Vol 30, 1, 31-79, 1996
614. Liszka J: A general introduction to the Semeiotic of Charles S. Peirce. Bloomington M.A. : Harvard University Press, 1996
615. Liszt F: Mephisto-Walzer I. Ed. Peters, No. 7203, (Edited by E. Von Sauer) 1913-17
616. Locke J: An essay concerning human understanding (first publication 1690). Pennsylvania State University 1999
617. Loomis L H and S Sternberg: Advanced Calculus. Addison-Wesley, Reading, MA et al. 1968
618. Loria G: Perfectionnements, Evolution, Métamorphoses du concept de "coordonnées" Contribution à l'histoire de la géométrie analytique. Osiris, Vol. 8, pp. 218-288, 1948
619. Loria G: A. L. Cauchy in the History of Analytic Geometry. Scripta Mathematica, p. 123-128, 1933
620. Losada C: Isography and Structure in the Music of Boulez. Journal of Mathematics and Music 2008, Vol. 2, no. 3, pp. 135-155, 2008
621. Luck G, P Toiviainen, M Thompson: Perception of Expression in Conductors' Gestures: A Continuous Response Study. Music Perception, 20(1), 47-57, 2010
622. Louis A K et al.: Wavelets. Teubner, Stuttgart 1994
623. Lubet A: (Paralyzed on One) Sideman: Disability Studies Meets Jazz, through the Hands of Horace Parlan. Critical Studies in Improvisation/Études critiques en improvisation 6.2. 2010 `http://www.criticalimprov.com/article/view/1268`
624. Lubet A: Music, Disability, and Society. Temple University Press, Philadelphia 2011
625. Lubet A: Social Confluence and Citizenship: A View from the Intersection of Music and Disability. In: Hirschmann N J, and B Linker (Eds.): Civil Disabilities: Citizenship, Membership, and Belonging. University of Pennsylvania Press, Philadelphia 2015

626. Lubet A: Oscar Peterson's piano prostheses: Strategies of performance and publicity in the post-stroke phase of his career. Jazz Research Journal, 7(2), pp. 151-182, 2015
627. Lubet A: A Case of the Blues (misc). Unpublished, 2014
628. Lucas E: Récréations mathématiques. Gauthier-Villars, Paris, 1882
629. Lüdi W: Fax to G Mazzola. Malans 01/23/1991
630. Ludwig H: Marin Mersenne und seine Musiklehre. Olms, Hildesheim 1971
631. Lussy M: Traité de l'expression musicale. Paris 1874
632. Lutz R: Le poète dans son oeuvre. Seminar Jung, Univ. Zürich, 1980
633. Lyttleton H: The Best of Jazz. Robson Books, London 1998
634. Machado A: Chant XXIX, Proverbios y cantarès, Campos de castilla, (1917), Ed. Cátedra, Madrid, 1989, 2015
635. Mac Lane S: Homology. Springer, Heidelberg et al. 1963
636. Mac Lane S: The development of mathematical ideas by collision: The case of Categories and Topos Theory. In: Categorical Topology and its relation to analysis, algebra and combinatorics, 1989
637. Mac Lane S: Categories for the Working Mathematician. Springer, New York et al. 1971
638. Mac Lane S: Mathematics, Form and Function. Springer, New York 1986
639. Mac Lane S and Moerdijk I: Sheaves in Geometry and Logic. Springer, New York et al. 1994
640. MacLean P D: The triune brain, emotion, and scientific bias. In: Schmitt F O (Ed.): The Neurosciences: Second Study Program, 336-348, Rockefeller Univ. Press, New York 1970
641. Maeder R: Programming Mathematica. Addison-Wesely, Reading, Mass. 1991
642. Manin Y: Interview in the Notices of the American Mathematical Society, November 2009, p. 1268
643. Mannone M and G Mazzola: Hegel's Conceptual Group Action on Creative Dynamics in Music. Gli spazi della musica, University of Torino 2014
644. Mannone M and G Compagno: Characterization of the degree of Musical non-Markovianity. Unpublished
645. Mannone M and S Lo Franco and G Compagno: Comparison of non-Markovianity criteria in a qubit system under random external fields. Phys. Scr. T, 2013
646. Mannone M: Segmentation des séries temporelles pour l'orchestration automatique. IRCAM - UPMC, Paris 2013 `http://www.atiam.ircam.fr/Archives/Stages1213/MANNONE_Maria.pdf`
647. Mannone M: Dalla Musica all'Immagine, dall'Immagine alla Musica - Relazioni matematiche fra composizione musicale e arte figurativa. Compostampa 2011
648. Mannone M and G Mazzola: Hypergestures in Complex Time: Creative Performance Between Symbolic and Physical Reality. Proceedings of the MCM 2015 conference, Springer, Heidelberg 2015
649. Mannone M and G Mazzola: Hypergestures in Complex Time: Creative Performance Between Symbolic and Physical Reality. Talk at MCM conference 2015, Queen Mary U, London 2015
650. Mansouri B: Normes internationales élargies et dépassement du calcul classique de la valeur. Thèse, Doctorat ès Sciences de gestion, Université de Bordeaux, 321 p., 3 juillet 2015
651. Marek C: Lehre des Klavierspiels. Atlantis 1977
652. Marino G, M-H Serra, J-M Raczinski: The UPIC system: Origins and innovations. Perspectives of New Music, 31(1):258–69, 1993
653. Marquis J-P: Category theory and the foundations of mathematics: philosophical excavations, *Synthese*, 103, pp. 229-254, 1995
654. Marrou H-I: De la connaissance historique. 1954, nouv. éd. coll. Points/Histoire, H21, Paris, 1975
655. Martinet A: Eléments de Linguistique Générale. Colin, Paris 1960
656. Martin B: CeDell Davis: The Last Man Standing. Huffington Post, 18 February 2015 `http://www.huffingtonpost.com/barrett-martin/cedell-davis-the-last-man_b_6696710.html`
657. Martino D: The Source Set and Its Aggregate Formations. JMT, 5(2), 224-273, 1961
658. De Martino E: La terra del rimorso. Bellaterra edizioni, 1999
659. Marx A B: Lehre von der musikalischen Komposition. Breitkopf & Härtel, Leipzig 1837-1847
660. Mason R M: Enumeration of Synthetic Musical Scales (...). J. of Music Theory 14, 1970
661. Massey W S: A Basic Course in Algebraic Topology. Springer, New York et al. 1991
662. Mathiesen Th J: Transmitting Text and Graphics in Online Database: The Thesaurus Musicarum Latinarum Model. In: Hewlett W B and Selfridge-Field E (Eds.): Computing in Musicology Vol.9, CCARH, Menlo Park 1993-94
663. Matisoff S: The legend of Semimaru, blind musician of Japan. Columbia University Press, New York 1978
664. Mattheson J: Der vollkommene Kapellmeister. Hamburg 1739
665. Maxwell H J: An Expert System for Harmonizing Analysis of Tonal Music. In: Balaban M et al. (Eds.): Understanding Music with AI: Perspectives on Music Cognition, MIT Press. Cambridge, MA 1992
666. May J P: A Concise Course in Algebraic Topology. The University of Chicago Press, Chicago 2007

667. Mazzola G: *Akroasis—Beethoven's Hammerklavier-Sonate in Drehung (für Cecil Taylor)*. LP, recorded August 30, 1979, WERGO/Schott, Mainz 1980
668. Mazzola G: Musique et théorie des groupes. Conference at Institut de Mathématique, U Genève, January 31, 1980
669. Mazzola G: Die gruppentheoretische Methode in der Musik. Lecture Notes, Notices by H. Gross, SS 1981, Mathematisches Institut der Universität, Zürich 1981
670. Mazzola G: Gruppen und Kategorien in der Musik. Heldermann, Berlin 1985
671. Mazzola G et al.: Rasterbild-Bildraster. Springer, Berlin et al. 1986
672. Mazzola G: Die Rolle des Symmetriedenkens für die Entwicklungsgeschichte der europäischen Musik. In: Symmetrie, Katalogband Vol.1 zur Symmetrieausstellung, Mathildenhöhe, Darmstadt, 1986
673. Mazzola G: Obertöne oder Symmetrie: Was ist Harmonie?. In: Herf F R (Ed.): Mikrotöne, Helbling, Innsbruck 1986
674. Mazzola G: Mathematische Betrachtungen in der Musik I,II. Lecture Notes, Univ. Zürich 1986/87
675. Mazzola G: Der Kontrapunkt und die K/D-Dichotomie. Manuscript, University of Zürich 1987
676. Mazzola G: Die Wahl der Zahl - eine systematische Betrachtung zum Streichquartett. In: Dissonanz 17, 1988
677. Mazzola G and G R Hofmann: Der Music Designer MD-Z71 - Hardware und Software für die Mathematische Musiktheorie. In: Petsche H (Ed.): Musik - Gehirn - Spiel, Beiträge zum 4. Herbert-von-Karajan-Symposion. Birkhäuser, Basel 1989
678. Mazzola G et al.: A Symmetry-Oriented Mathematical Model of Classical Counterpoint and Related Neurophysiological Investigations by Depth-EEG. In: Hargittai I (Ed.): Symmetry II, CAMWA, Pergamon, New York 1989
679. Mazzola G et al.: Hirnelektrische Vorgänge im limbischen System bei konsonanten und dissonanten Klängen. In: Petsche H (Ed.): Musik - Gehirn - Spiel, Beiträge zum 4. Herbert-von-Karajan-Symposion. Birkhäuser, Basel 1989
680. Mazzola G: *presto* Software Manual. SToA music, Zürich 1989-1994
681. Mazzola G: *Synthesis*. SToA 1001.90, Zürich 1990
682. Mazzola G: Geometrie der Töne. Birkhäuser, Basel 1990
683. Mazzola G and Muzzulini D: Tempo- und Stimmungsfelder: Perspektiven künftiger Musikcomputer. In: Hesse H P (Ed.): Mikrotöne III. Edition Helbling, Innsbruck 1990
684. Mazzola G and Muzzulini D: Deduktion des Quintparallelenverbots aus der Konsonanz-Dissonanz-Dichotomie. Accepted for publication in: Musiktheorie, Laaber 1990, but unpublished (!)
685. Mazzola G: Mathematische Musiktheorie: Status quo 1990. Jber. d.Dt. Math.-Verein. 93, 6-29, 1991
686. Mazzola G: Mathematical Music Theory—An Informal Survey. Edizioni Cerfim, Locarno 1993
687. Mazzola G: RUBATO at SMAC KTH, Stockholm 1993
688. Mazzola G and O Zahorka: Tempo Curves Revisited: Hierarchies of Performance Fields. Computer Music Journal **18**, No. 1, 1994
689. Mazzola G and O Zahorka: The RUBATO Performance Workstation on NeXTSTEP. In: ICMA (Ed.): Proceedings of the ICMC 94, San Francisco 1994
690. Mazzola G and O Zahorka: Geometry and Logic of Musical Performance I, II, III. SNSF Research Reports (469pp.), Universität Zürich, Zürich 1993-1995
691. Mazzola G et al.: Analysis and Performance of a Dream. In: Sundberg J (Ed.): Proceedings of the 1995 Symposium on Musical Performance. KTH, Stockholm 1995
692. Mazzola G et al.: The RUBATO Platform. In: Hewlett W B and E Selfridge-Field (Eds.): Computing in Musicology 10. CCARH, Menlo Park 1995
693. Mazzola G and O Zahorka: The PrediBase Data Base System of RUBATO on NEXTSTEP. In: Selfridge-Field E (Ed.): Handbook of Musical Codes. CCARH, Menlo Park 1995
694. Mazzola G: Inverse Performance Theory. In: ICMA (Ed.): Proceedings of the ICMC 95, San Francisco 1995
695. Mazzola G et al.: Neuronal Response in Limbic and Neocortical Structures During Perception of Consonances and Dissonances. In: Steinberg R (Ed.): Music and the Mind Machine. Springer, Berlin et al. 1995
696. Mazzola G and O Zahorka: RUBATO und der Einsatz von Hypermedien in der Musikforschung. Zeitschrift des Deutsches Bibliotheksinstituts, Berlin Jan. 1996
697. Mazzola G and O Zahorka: Topologien gestalteter Motive in Kompositionen. In: Auhagen W et al. (Eds.): Festschrift zum 65. Geburtstag J.P. Fricke. Köln 1995 `http://www.uni-koeln.de/phil-fak/muwi/publ/fs_fricke`
698. Mazzola G: Towards Big Science. Geometry and Logic of Music and its Technology. In: Enders B and Knolle N (Eds.): Symposionsband Klangart '95, Rasch, Osnabrück 1998

699. Mazzola G et al.: The RUBATO Homepage.
http://www.rubato.org, Univ. Zűrich, since 1996
700. Mazzola G: Objective C and Category Theory. Seminar Notes, Multimedia Lab, CS Department, U Zurich, Zurich 1996
701. Mazzola G et al.: *Orbit.* Music & Arts CD-1015, Berkeley 1997
702. Mazzola G and J Beran: Rational Composition of Performance. In: Kopiez R and W Auhagen (Eds.): Proceedings of the Conference "Controlling Creative Processes in Music". Lang, Frankfurt and New York 1998
703. Mazzola G: Semiotic Aspects of Musicology: Semiotics of Music. In: Posner R et al. (Eds.): A Handbook on the Sign-Theoretic Foundations of Nature and Culture. Walter de Gruyter, Berlin and New York 1998
704. Mazzola G: Semiotic Aspects of Musicology: Semiotics of Music. - Art. 152, In: Posner et al. (eds): Semiotics (Handbook, Vol.III). W. de Gruyter, Berlin 1999
705. Mazzola G: Music@EncycloSpace. In: Enders B and J Stange-Elbe (Eds.): Musik im virtuellen Raum (Proceedings of the Klangart congress'98). Rasch, Osnabrück 2000
706. Mazzola G: Humanities@EncycloSpace. FER-Studie Nr.XX. Schweizerischer Wissenschaftrat, Bern 1998
http://www.swr.ch
707. Mazzola G: Die schöne Gefangene – Metaphorik und Komplexität in der Musikwissenschaft. In: Kopiez R et al. (Eds.): Musikwissenschaft zwischen Kunst, Aesthetik und Experiment. Königshausen & Neumann, Würzburg 1998
708. Mazzola G: L'Essence du Bleu (sonate pour piano). Acanthus, Rüttenen 2002
709. Mazzola G: Les mathématiques de l'interprétation musicale. Séminaire MaMuX, IRCAM, Paris, May 18, 2002
710. Mazzola G, G Milmeister, J Weissmann: Comprehensive Mathematics for Computer Scientists, Vols. I, II; 700 pp. Springer, Heidelberg et al. 2004
711. Mazzola G: The Topos Geometry of Musical Logic. Proceedings of the Fourth Diderot Symposium of the European Math. Soc., Springer, Heidelberg et al. 2002
712. Mazzola G: Degenerative Theory of Tonal Music. In: Proceedings of the Klangart congress'99. Universität Osnabrück 2002
713. Mazzola G: Classifying Algebraic Schemes for Musical Manifolds. Tatra Mt. Math. Publ. 23, 71-90, 2001
714. Mazzola G et al.: The Topos of Music. Birkhäuser, Basel 2002
715. Mazzola G, G Milmeister, K Morsy, F Thalmann: Functors for music: The Rubato Composer System. In: Digital Art Weeks Proceedings, ETH Zürich, 2006
716. Mazzola G: Global Networks in Computer Science? Invited Talk at the Computer Science Department Colloquium, ETH Zurich, January 2006
717. Mazzola G: Elemente der Musikinformatik. Birkhäuser, Basel et al. 2006
718. Mazzola G: La vérité du beau dans la musique. Delatour/IRCAM, Paris 2007
719. Mazzola G and M Andreatta: From a Categorical Point of View: K-nets as Limit Denotators. Perspectives of New Music, Vol. 44, No. 2, 2006
720. Mazzola G and M Andreatta: Diagrams, Gestures, and Formulas in Music. Journal of Mathematics and Music 2007, Vol. 1, no. 1, 2007
721. Mazzola G and P B Cherlin: Flow, Gesture, and Spaces in Free Jazz—Towards a Theory of Collaboration. Springer Series Computational Music Science, Heidelberg et al. 2009
722. Mazzola G et al.: Topos Theory for a Creative Analysis of Boulez's Structures. In: Naimpally S, G Di Maio (Eds.): Quaderni di Matematica, Theory And Applications of Proximity, Nearness and Uniformity, Vol. 23, 2009
723. Mazzola G: Categorical Gestures, the Diamond Conjecture, Lewin's Question, and the Hammerklavier Sonata. Journal of Mathematics and Music Vol. 3, no. 1, 2009
724. Mazzola G, C Losada, F Thalmann, S Tsuda: Topos Theory for a Creative Analysis of Boulez's Structures. In: Somashekhar Naimpally, Giuseppe Di Maio (Eds.): Quaderni di Matematica, Theory And Applications of Proximity, Nearness and Uniformity, Vol 23, 2009
725. Mazzola G: Musical Performance. Springer, Heidelberg et al. 2011
726. Mazzola G, J Park, F Thalmann: Musical Creativity. Springer, Heidelberg 2011
727. Mazzola G: Singular Homology on Hypergestures. Journal of Mathematics and Music Vol. 6, Nr. 1, 2012
728. Mazzola G: Manifolds and Singular Homology of Compositions, Networks, and Gestures: Classification, Connecting Functors, Examples. Unpublished
729. Mazzola G and F Thalmann: The BigBang Rubette: Gestural Music Comoposition With Rubato Composer. Proceedings of the ICMC 2008, ICMA, Ann Arbor 2008
730. Mazzola G and F Thalmann: Musical Composition and Gestural Diagrams. In Agon C et al. (Eds.): Proceedings of the Third Conference on Mathematics and Computation in Music (MCM), Springer, Heidelberg, 2011

731. Mazzola G, M Mannone, Y Pang: Cool Math for Hot Music. Springer, Heidelberg 2016
732. Mesnage M: La Set-Complex Theory: de quels enjeux s'agit-il? Analyse musicale, 4e trimestre, 87-90, 1989
733. Messiaen O: Technique de mon langage musical. Leduc, Paris 1944
734. Meyer J: Akustik und musikalische Aufführungspraxis. Verlag Das Musikinstrument, Frankfurt/Main 1980
735. Meyer-Eppler W: Grundlagen und Anwendungen der Informationstheorie. Springer, Berlin 1959
736. Meyer-Eppler W: Zur Systematik der elektrischen Klangtransformationen. In: Darmstädter Beiträge III, Schott, Mainz 1960
737. Michels U: dtv-Atlas zur Musik I,II. dtv/Bärenreiter, München/Kassel 1977, 1985
738. Miklaszewski K: A case Study of a Pianist Preparing a Musical Performance. Psychology of Music, 17, 95-109, 1989
739. Milmeister G: The Rubato Composer Music Software: Component-Based Implementation of a Functorial Concept Architecture. Springer Series Computational Music Science, Heidelberg 2009
740. McKay B D and I M Wanless: On the number of Latin squares. arXiv:0909.2101v1 [math.CO] 11 Sep 2009
741. McNeill D: Gesture and Thought. University of Chicago Press, Chicago 2005
742. McPartland M: Marian McPartland's Piano Jazz. National Public Radio, 9 November 1997 `http://www.npr.org/player/v2/mediaPlayer.html?action=1&t=1&islist=false&id=17622315&m=90469170`
743. Mead V H: More than Mere Movement: Dalcroze Eurhythmics. Music Educators Journal, Feb. 1986
744. Mendes R S et al.: Universal patterns in sound amplitudes of songs and musical genres. Phys. Rev. E, Vol. 83, pp. 017101 ff., 2011
745. Menez M: L'expérience mathématique. In: Dorier J-L, S Coutat (Eds.): Enseignement des mathématiques et contrat social : enjeux et défis pour le 21ème siècle — Actes du Colloque EMF20123 (GT4, pp. 632-640)
746. Meredith D, G Wiggins, K Lemstrom: Pattern induction and matching in polyphonic music and other multidimensional datasets. In: Callaos N et al. (Eds.): Proceedings of the 5th World Multiconference on Systemics, Cybernetics and Informatics. Orlando 2001
747. Meredith D, K Lemstrom, G Wiggins: Algorithms for discovering repeated patterns in multidimensional representation of polyphonic music. Journal of New Music Research, Vol 4, 2002
748. Merker B H: Layered Constraints on the Multiple Creativities of Music. In: I Deliège and G A Wiggins (Eds.): Musical Creativity: Multidisciplinary Research in Theory and Practice. Psychology Press, Hove/New York, pp. 25-41, 2006
749. Merleau-Ponty M: The Phenomenology of Perception in The Body: Classic and Contemporary Readings, (ed. Donn Welton), Blackwell, Oxford 1999
750. Merleau-Ponty M: Phenomenology of Perception. Trans. Colin Smith, Routledge & Kegan Paul, London 1978 [1962]
751. Merleau-Ponty M: Phénoménologie de la perception. Gallimard 1945
752. Merleau Ponty M: The prose of the world, translation by J O'Neill. Northwestern University Press, Evanston 1973 (La prose du monde, Gallimard, 1969)
753. Meusburger P, J Funke, E Wunder (Eds.): Milieus of Creativity: An Interdisciplinary Approach to Spatiality of Creativity. Springer, 2009
754. Miller M: A Jazz Legend Handles his Own Legacy with Care. Globe and Mail (Canada), 23 February: C3. 1999
755. Misch C and Wille R: Stimmungslogiken auf MUTABOR: Eine Programmiersprache. In: Herf F R (Ed.): Mikrotöne II. Edition Helbling, Innsbruck 1988
756. `http://forumnet.ircam.fr/701.html`
757. Mokkapati V: Modal Action: An Indo-European Perpective on Gregorian Chant. Phd Thesis, Wesleyan University 1996
758. Molino J: Fait Musical et Sémiologie de la Musique. Musique en Jeu 17 1975
759. Monti G: Théorie des catégories et méthode conditionnelle appliquées aux comptes en normes internationales élargies. Thèse, Doctorat ès Sciences de gestion, Université de Bordeaux 2017
760. Montiel-Hernandez M: El denotador: Su estructura, construcción y papel en la teoría matemática de la musica. Master's thesis, UNAM, Mexico City, 1999
761. Moog R: MIDI, Musical Instrument Digital Interface. Audio Eng. Soc. 34, Nr.5, 1986
762. Moon T: A Master of Improv is Renewed by Melody. Philadelphia Inquirer, 29 September: H02, 2002
763. Moore C-L: The Harmonic Structure of Movement, Music and Dance According to Rudolf Laban. Edwin Meller Press, Lewiston 2009
764. Moritz R E: On Mathematics. A collection of witty, profound, amusing passages about mathematics and mathematicians. Copyright 1914 Robert Edouard Moritz, 410 p., Dover 1958
765. Morris R D: Composition with Pitch-Classes. Yale University Press, New Haven et al. 1987
766. Morris R D: Compositional Spaces and Other Territories. PNM 33, 328-358, 1995

767. Morris R D: K, Kh, and Beyond. In: Baker J et al. [69], 275-306
768. Morse Ph M and H Feshbach: Methods of Theoretical Physics, Part I. McGraw-Hill, New York et al. 1953
769. Mozart W A: *Die Zauberflöte.* (Klavierauszug) Ed. Peters 1932
770. Müller W: Darstellungstheorie von endlichen Gruppen. Teubner, Stuttgart 1980
771. Müller S: Pianist's Hands: Synthesis of Musical Gestures. PhD Thesis, University of Zurich, Zurich 2004
772. Müller S and G Mazzola: Constraint-based Shaping of Gestural Performance. Proceedings of the ICMC 03, ICMA, Ann Arbor 2003
773. Müller S and G Mazzola: The Extraction of Expressive Shaping in Performance. Computer Music Journal, Vol. 27(1), MIT Press, 2003
774. Mumford D: Lectures on Curves on an Algebraic Surface. Princeton University Press, Princeton 1966
775. Mumford D and K Suominen: Introduction to the theory of moduli. In: Oort F (Ed.): Algebraic Geometry Oslo 1970. Wolters-Noordhoff, Groningen 1972
776. Murenzi R: Wavelets. Combes J M, Grossman A, Tchmitchian P (Eds.), Springer Berlin et al., 1988
777. MusicKit Online-Documentation. Version 4.0 1994
778. Muzzulini D: Konsonanz und Dissonanz in Musiktheorie und Psychoakustik. Lizenziatsarbeit, MWS Univ. Zürich 1990
779. Muzzulini D: Musical Modulation by Symmetries. J. for Music Theory, Vol. 39, No. 2, pp. 311-328, 1995
780. Muzzulini D: Tempo Modifications and Spline Functions. NF-Report 1993, Univ. Zürich 1993
781. Narayan S S, Temchin A N, Recio A, Ruggero M A: Frequency tuning of basilar membrane and auditory nerve fibers in the same cochleae. Science. 282(5395):1882-4, Dec 4 1998
782. Nattiez J-J: Fondements d'une Sémiologie de la Musique. Edition 10/18 Paris 1975
783. Negro G Del: The Passeggiata and Popular Culture in an Italian Town: Folklore and the Performance of Modernity. Montréal; McGill-Queen's University Press 2004
784. Neumaier W: Was ist ein Tonsystem? Lang, Frankfurt/Main et al. 1986
785. Neuman D A: A House of Music: The Hindustani Musician and the Crafting of Traditions. PhD Thesis, Columbia University 2004
786. Neuwirth G: Josquin Desprez, "Erzählen von Zahlen". In: Musik-Konzepte 26/27, edition text+kritik, München 1982
787. Newlove J and J Dalby: Laban for All. Routledge, New York 2004
788. NeXTSTEP Online-Documentation; Version 3.3. NeXt Inc., Redwood City 1995
789. Nicolas F: René Guitart : la pulsation d'une intellectualité mathématique étrangement évidente. Conférence à la *Journée en l'honneur de René Guitart pour ses 65 ans*, 9 novembre 2012, Université Paris 7 Denis Diderot, 8 p, inédit. See a summary in: `http://www-lmpa.univ-littoral.fr/ sic`
790. Nicolas F: Le Monde-Musique III. Le musicien et son intellectualité musicale. Aedam Musicae, 384 p., juin 2015
791. Nicotra E: Introduzione alla tecnica della direzione d'orchestra. Edizioni Curci 2007
792. Nieberle R: IRCAM Signal Processing Workstation. Keyboards, Dec. 1992
793. `http://www.nime.org`
794. Noether E: Hyperkomplexe Grössen und Darstellungstheorie. Math. Zeitschr., Vol. XXX 1929
795. Noland C: Agency and Embodiment: Performing Gestures/Producing Culture. Harvard University Press, Cambridge 2009
796. Noll J: Musik-Programmierung. Addison-Wesley, Bonn 1994
797. Noll Th: Morphologische Grundlagen der abendländischen Harmonik. Doctoral Thesis, TU Berlin 1995
798. Noll Th: Fractal Depth Structure of Tonal Harmony. In: ICMA (Ed.): Proceedings of the ICMC 95, San Francisco 1995
799. Noll Th: `http://www.cs.tu-berlin.de/ noll/ChordDictionary.sea.hqx`, TU Berlin 1996
800. Noll Th: Harmonische Morpheme. Musikometrika 8, 7-32, 1997
801. Noll Th: The Consonance/Dissonance-Dichotomy Considered from a Morphological Point of View. In: Zannos I (Ed.): Music and Signs. ASCO Publ., Bratislava 1999
802. Noll Th: Morphologische Grundlagen der abendländischen Harmonik. Doctoral Thesis, TU Berlin 1995
803. Noll Th: Wie blättert man in einem hypermedialen Gebärdenlexikon? In: Schmauser C, Th Noll (Eds.): Körperbewegungen und ihre Bedeutungen. Berlin Verlag Arno Spitz, Berlin 1998
804. Noll Th et al.: Computer-aided transformational analysis with tone sieves. Contribution au congrés: SMC 2006
805. Nunn D et al.: Acoustic Quanta. In: ICMA (Ed.): Proceedings of the ICMC 96, San Francisco 1996
806. Oettingen A von: Das duale Harmoniesystem. Leipzig 1913
807. Okopenko A: Lexikon-Roman. Residenz Verlag, Salzburg 1970
808. Opcode: MAX. http://www.opcode.com/products/max/, Opcode Systems Inc. 1997

809. Marco Orsini conducting Le Nozze di Figaro 2014 https://www.youtube.com/watch?v=M7QuLXy_61U
810. Osgood C E et al.: The Nature and Measurement of Meaning. Psychological Bulletin 49, 1952
811. Osher S and R Fedkiw: Level Set Methods and Dynamic Implicit Surfaces. Springer, 2003
812. Owens Th: Bebop: The Music and the Players. Oxford University Press, New York 1995
813. Pakkan M and V Akman: Hypersets. Information Sciences: An International Journal, 85(1):4361, 1996
814. Pallasmaa J: The Thinking Hand. Wiley 2009
815. Palmer R: Liner notes, Oscar Peterson: A Night in Vienna. DVD. Verve 02498 6253-4, 2004
816. Pappas T: The Music of Reason: Experience the Beauty of Mathematics Through Quotations. Wide World Publishing/ Tetra, 1995
817. Parker W: Conversations. Rogueart, 2011
818. Parlan H: The Maestro. CD. SteepleChase SCCD 31167, 1995
819. Parlan H: Horace Parlan by Horace Parlan. DVD. Image ID1505R5DVD, 2000
820. Parlan H: In Copenhagen. CD. Storyville 101 8521, 2008
821. Parlan H: Musically Yours. CD SteepleChase SCS 31141, 2010
822. Parmentier M: Connaissances, vérités et idées mathématiques dans l'*Essai sur l'entendement humain* de Locke. In: Barbin E et J-P Cléro (éds): Les mathématiques et l'expérience : ce qu'en ont dit les philosophes et les mathématiciens, p. 95-114, Hermann, Paris 2015
823. Parncutt R: The Perception of Pulse in Musical Rhythm. In: Gabrielsson A (Ed.): Action and Perception in Rhythm and Music. Royal Swedish Adademy of Music, No.55
824. Parncutt R: Recording Piano Fingering in Live Performance. In: Enders B, Knolle N (Eds.): KlangArt-Kongress 1995, Rasch, Osnabrück 1998
825. Parncutt R et al.: Interdependence of Right and Left Hands in Sight-read, Written, and Rehearsed Fingerings. Proc. Euro. Sco. Cog. Sci. Music, Uppsala 1997
826. Parncutt R: Modeling Piano Performance: Physics and Cognition of a Virtual Pianist. In: ICMA (Ed.): Proceedings of the ICMC 97, San Francisco 1997
827. Patras F: La pensée mathématique contemporaine. PUF, Paris, 2001
828. Patras F: La possibilité des nombres. PUF, Paris, octobre 2014
829. Peck R: Klein-Bottle Tonnetze. MTO, Vol. 9, no. 3, August 2003
830. Penrose R: The Road to Reality. Vintage, London 2002
831. Perle G: Serial Composition and Atonality: An Introduction to the Music of Schoenberg, Berg and Webern. 5th ed., revised, University of California Press, Berkeley 1981
832. Peterson O: An Oscar Peterson Christmas. CD, Telarc 515956K, 1995
833. Peterson O: Oscar in Paris. Telarc 2CD-83414, 1997
834. Peterson O: A Tribute to Oscar Peterson: Live at the Town Hall. Telarc 2CD-83401
835. Peterson O: A Summer Night in Munich. CD. Telarc CD-83450, 1999
836. Peterson O: A Night in Vienna. DVD, Verve 02498 6253-4, 2004
837. Peterson O, R Brown, M Jackson: The Very Tall Band: Live at the Blue Note. CD. Telarc CD-83443, 1999
838. Peterson O, R Brown, M Jackson: What's Up? The Very Tall Band. CD. Telarc CD-83663, 2007
839. Peterson O and B Green: Oscar and Benny. CD. Telarc CD-83406, 1998
840. Peterson O, R Hargrove, R Moore: Oscar Peterson Meets Roy Hargrove and Ralph Moore. CD. Telarc, CD-83399, 1996
841. Peterson O, and M LeGrand: Trail of Dreams: A Canadian Suite. CD. Telarc CD-83500, 2000
842. Peterson O, and I Perlman: Side by Side. CD. Telarc, CD-83341, 1994
843. Petsche H et al.: EEG in Music Psychological Studies. In: Steinberg R (Ed.): Music and the Mind Machine. Springer, Berlin et al. 1995
844. Petsche H: Private correspondence. Vienna, March 2001
845. Piaget J: Morphismes et categories. Comparer et transformer, Delachaux et Niestlé, Neuchâtel, 1990
846. Pfeifer R and J Bongard: How the Body Shapes the Way We Think: A New View of Intelligence. MIT Press 2006
847. Peirce C S: Peirce, *Collected Papers*, vol. III, vol. IV The simplest mathematics, 1960
848. Pierce A: Deepening Musical Performance through Movement: The Theory and Practice of Embodied Interpretation (Musical Meaning and Interpretation). Indiana University Press, Bloomington 2010
849. Plato: Phaidron.
850. Plomp R and W Levelt: Tonal Consonance and Critical Bandwidth. J. Acoust. Soc. Am. 38, 548, 1965
851. Plotinus: The Six Enneads, in *Great Books of the Western World*, Vol. 17. Encyclopædia Britannica, Inc., Chicago, 1952
852. Poincaré H: La valeur de la science. Flammarion, Paris 1905

853. Pólya G: Mathematical Discovery. On Understanding, Learning and Teaching Problem Solving (Combined Edition). Wiley, New York, 1962, 1965, 1981
854. Pólya G: Comment poser et résoudre un problème. Dunod, 1965
855. Polyanin A D: Handbook of Linear Partial Differential Equations for Engineers and Scientists. Chapman & Hall/CRC, 2002; see also: `http://eqworld.ipmnet.ru`, section 3.2.
856. Pope S T: Music Notations and the Representation of Musical Structure and Knowledge. Perspectives of New Music, Spring-Summer 1986
857. Popelard M-D and A Wall: Des Faits et gestes. éditions Bréal, Rosny-sous-Bois 2003
858. Popper K R: Conjectures and Refutations. Routledge & Kegan Paul, London 1963
859. Popper K: The open society and its enemies. Vol. 2, Routledge & Kegan, London, 1945
860. Posner R: Strukturalismus in der Gedichtinterpretation. In: Blumensadth H (Ed.): Strukturalismus in der Literaturwissenschaft. Kiepenheuer & Witsch, Köln 1972
861. Prame E: Measurements of the Vibrato Rate of 10 Singers. In: Friberg A et al. (Eds.): Proceedings of the 1993 Stockholm Music Acoustic Conference. KTH, Stockholm 1994
862. Pribram K H: Brain and Perception. L Erlbaum, Hillsdale, NJ 1991
863. Priestley M B: Spectral Analysis of Time Series. Academic Press, London 1981
864. Prinz W: Modes of linkage between perception and action. In W Prinz & A-F Sanders (Eds.): cognition and motor processes (pp. 185–193), Springer, Berlin 1984
865. Promies W: Symmetrie in der Literatur. In: Symmetrie, Katalogband Vol.1 zur Symmetrieausstellung. Mathildenhöhe, Darmstadt, 1986
866. Promies W: Stolbergs Gedicht im Poesiegarten In: Symmetrie, Katalogband Vol.3 zur Symmetrieausstellung. Mathildenhöhe, Darmstadt, 1986
867. Puckette M and Lippe C: Score Following in Practice. In: ICMA (Ed.): Proceedings of the ICMC 92, San Francisco 1992
868. Pulkki V et al.: DSP Approach to Multichannel Audio Mixing. In: ICMA (Ed.): Proceedings of the ICMC 96, San Francisco 1996
869. Puttke M: Learning to Dance Means Learning to Think! In: Bläsing B, Puttke M, Schack Th. (Eds.): The Neurocognition of Dance. Psychology Press, Hove and New York 2010
870. Quinn I: Fuzzy Extensions to the Theory of Contour. Music Theory Spectrum. Vol. 19/2, 1997
871. Radl H: Versuch über die Modulationstheorie Mazzolas in reiner Stimmung. Diploma Thesis, U Augsburg, Augsburg 1998
872. Rahaim M: Musicking Bodies. Wesleyan U P, Middletown, CT 2012
873. Rahn J: Basic Atonal Theory. Longman, New York 1980
874. Rahn J: Review of D. Lewin's "Generalized Musical Intervals and Transformations". JMT, 31, 305-318, 1987
875. Raffman D: Language, Music, and Mind. MIT Press, Cambridge et al. 1993
876. Rameau J-Ph: Traité de l'Harmonie, Réduite à ses Principes Naturels. Paris 1722
877. Ramachandran V S: Mirror Neurons and imitation learning as the driving force behind the "great leap forward" in human evolution, 2015. See: `http://www.edge.org/3rd.culture/ramachandran/ramachandran p6.html`
878. Ramstein C: Analyse, Représentation et Traitement du Geste Instrumental. PhD thesis, Institut National Polytechnique de Grenoble, December 1991
879. Ratz E: Einführung in die musikalische Formenlehre. Universal Edition, Wien 1973
880. Rauchfleisch U: Musik schöpfen, Musik hören. Ein psychologischer Zugang. Vandenhoeck und Ruprecht, Göttingen and Zürich 1996
881. Rav Y: Why do we prove theorems?. Philosophia mathematica, 7 (1), p. 5-41, 1999
882. Read G: Music Notation. Crescendo Publ., Boston 1969
883. Reeves H: Patience dans l'azur. L'évolution cosmique. Seuil, Paris 1981
884. Reichardt J F: Vermischte Musikalien. Riga 1777
885. Reidmeister K: Elementare Begründung der Knotentheorie. Abh. Math. Sem. Univ. Hamburg, 5, 24-32, 1926
886. Repp B: Diversity and Commonality in Music Performance: An Analysis of Timing Microstructure in Schumann's "Träumerei". J. Acoustic Soc. Am. 92, 1992
887. Repp B: e-mail communication of tempo data. Haskins Laboratories, New Haven, June 2, 1995
888. Repp B: Patterns of note onset asynchronies in expressive piano performance. J. Acoustic Soc. Am. **100**, 1996
889. Repp B: Pedal Timing and Tempo in Expressive Piano Performance: A Preliminary Investigation. Psychology of Music **24**, 1996
890. Repp B: The Art of Inaccuracy: Why Pianists' Errors Are Difficult to Hear. Music Perception, *14*, 2 1997
891. Repp B: Expressive Timing in a Debussy Prélude: A Comparison of Student and Expert Pianists. Musicae Scientiae 1, 1997

892. Reti R: The Thematic Process in Music (1951). Greenwood Press, Westport, 2nd ed. 1978
893. Reti R, commented by Kopfermann M: Schumanns Kinderszenen: quasi Thema mit Variationen. In: Musik-Konzepte Sonderband Robert Schumann II, edition text + kritik, München 1982
894. Riguet J: Systèmes de coordonnées relationnels. *CRAS*, t. 236, p. 2369-2371, séance du 22 juin 1953
895. Riguet J: Systèmes de coordonnées relationnels II. Applications à la théorie des groupes de Kaloujnine. *CRAS* t. 238, p. 435-437, séance du 25 janvier 1954
896. Riguet J: Systèmes de coordonnées relationnels III. Fermetures et systèmes symétriques. *CRAS*, t. 238, p. 1763-1765, séance du 3 mai 1954
897. Rhode W S and L Robles: Evidence from Mössbauer Experiments for Nonlinear Vibration in the Cochlea. J. Acoust. Soc. Am. 48, 988, 1970
898. Richter P et al.: How Consistent are Changes in EEG Coherence Patterns Elicited by Music Perception? In: Steinberg R (Ed.): Music and the Mind Machine. Springer, Berlin et al. 1995
899. Riemann B: Über die Hypothesen, welche der Geometrie zugrunde liegen (1854). Gött. Abh. No.13 (published in 1867)
900. Riemann H: Musikalische Logik. Leipzig 1873
901. Riemann H: Musikalische Syntaxis. Leipzig 1877
902. Riemann H: Vereinfachte Harmonielehre oder die Lehre von den tonalen Funktionen der Akkorde. London 1893
903. Riemann H: Über Agogik. In: Präludien und Studien II. Leipzig 1900
904. Riemann H: System der musikalischen Rhythmik und Metrik. Breitkopf und Härtel, Leipzig 1903
905. Riemann H: Handbuch der Harmonielehre. Leipzig 6/1912
906. Riemann H: Grundriss der Musikwissenschaft. Leipzig 1928
907. Riemann H and H H Eggebrecht: Riemann Musiklexikon, Sachteil, 12. Auflage. Schott, Mainz 1967
908. Ries F: Biographische Notizen über L. van Beethoven (1838). New edition by Kalischer A Ch, 1906
909. Risset J-C: Real-time: Composition or performance? Reservations about real-time control in computer music and demonstration of a virtual piano partner. In: Kopiez R and W Auhagen (Eds.): Controlling creative processes in music. Peter Lang, Frankfurt/Main 1998
910. Rizzolati G and M Fabbri-Destro: Mirror neurons: from discovery to autism. Exp. Brain Res. 200:223-237, 2010
911. Roads C: The Computer Music Tutorial. MIT Press, Cambridge, Mass. and London 1998
912. Rodet X et al.: Xspect: a New X/Motif Signal Visualization, Analysis and Editing Program. In: ICMA (Ed.): Proceedings of the ICMC 96, San Francisco 1996
913. Rodet X: Synthesis and Processing of the Singing Voice. Proc.1st IEEE Benelux Workshop on Model based Processing and Coding of Audio (MPCA-2002), Leuven, Belgium, November 15, 2002
914. Roederer J G: Physikalische und psychoakustische Grundlagen der Musik. Springer, Berlin 1977
915. Rogers C R: Toward a theory of creativity, *ETC: A review of General Semantics*, Vol 11, 1954, 249-260
916. Rossi A: Die Geburt der modernen Wissenschaft in Europa. Beck, München 1997
917. Rostand F: Souci d'exactitude et scrupules des mathématiciens. Vrin, Paris 1960
918. Rüetschi U-J: Denotative Geographical Modelling—an attempt at modelling geographical information with the Denotator system. Diploma thesis, University of Zürich, 2001
919. Rufer J: Die Komposition mit zwölf Tönen. Bärenreiter, Kassel 1966
920. Runco M A and I Chand: Cognition and Creativity. Educational Psychology Review, Vol. 7:3, pp. 243-267, 1995
921. Ryan R A: Introduction to Tensor Products of Banach Spaces. Springer, Heidelberg 2002
922. Ruwet N: Langage, Musique, Poésie. Seuil, Paris 1972
923. *S-Plus.* MathSoft Inc., Seattle, Washington 1995
924. Sachs K-J: Der Contrapunctus im 14. und 15. Jahrhundert. AMW, Franz Steiner, Wiesbaden 1974
925. Sacks O W: Musicophilia: tales of music and the brain. (Rev. and expanded 1st ed.) Vintage Books, New York 2008
926. Saiber A and H S Turner (Eds.): Mathematics and the Imagination. Configurations, 17, 195 pp. 2009
927. Sandall R: Miraculously, Magically, the Years Fell Away. Daily Telegraph (London), 1 July: 21 2006
928. Sander E: Notre pensée progresse par analogies. La conscience. La Recherche Hors-série, no 15, pp. 54-57, octobre-novembre 2015
929. Sattler D E (Ed.): Friedrich Hölderlin – Einhundert Gedichte. Luchterhand, Frankfurt/Main 1989
930. Salzer F: Structural Hearing: Tonal Coherence in Music. (German: Strukturelles Hören. Noetzel, Wilhelmshaven 1960) Dover, New York 1962
931. Sarasúa A and E Guaus: Dynamics in Music Conducting: A computational Comparative Study Among Subjects. Proceedings of the International Conference on New Interfaces for Musical Expression, NIME'14, Goldsmiths London, UK, 2014

932. Sauvageot F: Pierre Deligne. *Images des mathématiques*, CNRS, 12 mai 2013 `http://images.math.cnrs.fr/Pierre-Deligne.html`
933. Saussure F de: Cours de Linguistique Générale. Payot, Paris 1922
934. Saussure F de: Linguistik und Semiotik - Notizen aus dem Nachlass. Fehr J (Ed.), Suhrkamp, Frankfurt/Main 1997
935. Savan D: An introduction to C.S. Peirce's Full System of Semeiotic. Toronto Semiotic Circle, Toronto 1988
936. Sawyer R K: Improvisation and the creative process: Dewey, Collingwood, and the aesthetics of spontaneity. The Journal of Aesthetics and Art Criticism, Vp;. 58, pp. 149-61, 2000
937. Sawyer R K: Group Creativity: Music, Theater, Collaboration. Lawrence Erlbaum Associates, Mahwah 2003
938. Scarborough D L et al.: Connectionist Models for Tonal Analysis. Computer Music Journal Vol. 13, 1989
939. Schenker H: Fünf Urlinien-Tafeln. Universal Edition, Wien 1932
940. Schenker H: Theorien und Phantasien III: Der freie Satz. Universal Edition, Wien 1935
941. Scherchen H: Manuale del direttore d'orchestra. Edizioni Curci 2009
942. Schewe K-D: Type concept in OODB Modelling and its Logical Implications. In Kawaguchi E et al. (Ed.), Information Modelling and Knowledge Bases XI, pp. 256–274. IOS Press, 2000
943. Schmalz R: Out of the mouth of mathematicians. A quotation book for philomaths. Mathematical Association of America, 1993
944. Schmidt J (Ed.): Celibicache. Book and Movie PARS (ISBN 3-9803265-1-9), München 1992
945. Schmidt-Biggemann W: Topica universalis. Meiner,Hamburg 1983
946. Schmitt, J-C: La raison des gestes dans l'occident médiéval. Gallimard, Paris 1990
947. Schneps L and P Lochak (Eds.): Geometric Galois Actions. Cambridge University Press, Cambridge, 1997
948. Schönberg A: Harmonielehre (1911). Universal Edition, Wien 1966
949. Schönberg A: Die Komposition mit zwölf Tönen. In: Style and Idea, New York 1950
950. Schönberg A: Fundamentals of Musical Composition. Edited by Gerald Strang, with an introduction by Leonard Stein. St. Martin's Press, New York 1967
951. Schopenhauer A: De la quadruple racine du principe de raison suffisante. éd. complète (1813-1847), Vrin, 1991
952. Schubert F: *Auf dem Wasser zu singen*, op. 72 (1823). Bärenreiter/Henle München/Kassel 1982
953. Schubert H: Kategorien I, II. Springer, Berlin et al. 1970
954. Schumann R: Kinderszenen, op. 15 (1839). Henle, München 1977
955. Schumann R: Dritter Quartett-Morgen. NZ f. Musik 1838
956. Schweitzer A: Johann Sebastian Bach. (1908) Breitkopf & Härtel, Wiesbaden 1979
957. Scriba C J and P Schreiber: 5000 Years of Geometry: Mathematics in History and Culture. Birkhäuser, 2015
958. Seashore C E: Psychology of Music. McGraw-Hill, New York 1938
959. Shepard R N and L A Cooper: Mental Images and Their Transformations. MIT Press 1982
960. Shepp A: The Geneva Concert. DVD, inakustic LC13921, 2006
961. Shepp A and H Parlan: Goin' Home. CD. SteepleChase (D) SCCD-31079, 1985
962. Shepp A and H Parlan: Trouble in Mind. CD. SteepleChase ? SCCD 31139, 1986
963. Shepp A and H Parlan: En Concert: First Set. CD. 52e Rue Est (F) RECD 015, 1987
964. Shepp A and H Parlan: En Concert: Second Set. CD. 52e Rue Est (F) RECD 016, 1987
965. Shepp A and H Parlan: Reunion. CD. L+R CDLR 45003, 1987
966. Shepp A and H Parlan: Goin' Home. CD. SteepleChase (D) SCCD-31079, 2010
967. Shepp A and H Parlan: Trouble in Mind. CD. SteepleChase ? SCCD 31139, 2010
968. Siebers T: Disability Aesthetics (pp. 2-3). University of Michigan Press, Ann Arbor 2010
969. Serre J-P: Faisceaux Algébriques Cohérents. Annals of Mathematics, 61 (2): 197-278, 1955
970. Serre J-P: Trees. Springer, New York 1980
971. Serres M (Ed.): Eléments d'histoire des sciences. Bordas, Paris 1989
972. Sessions R: Questions About Music. W. W. Norton and Company, 1971
973. Shaffer L H: Musical Performance as Interpretation. Psychology of Music, Vol. 23, 1995
974. `http://www.fau.edu/bodymindculture`
975. Shusterman R: Somaesthetics: A Disiplinary Proposa. The Journal of Aesthetics and Art Criticism. Vol.57, No.3, pp. 299-313, 1999
976. Simondon G: Du mode d'existence des objets techniques. Aubier, Paris 1989
977. Simonton D K: Thematic fame, melodic originality, and musical zeitgeist: A biographical and transhistorical content analysis. Journal of Personality and Social Psychology, Vol. 38, pp. 972-83, 1980
978. Simonton D K: Thesaurus of scales and melodic patterns. Coleman-Ross Company, New York 1947
979. Siron J: La partition intérieure. Outre Mesure, Paris 1992
980. Slawson W: The Musical Control of Sound Color. Canadian Univ. Music Review, No.3, 1982

981. Slawson W: Sound Color. U. Cal. Press, Berkeley 1985
982. Sloan D: From DARMS to SMDL, and back again. In: Haus G and I Pighi (Eds.): X Colloquio di Informatica Musicale. AIMI, LIM-Dsi, Milano 1993
983. Sloboda J: The Musical Mind: An Introduction to the Cognitive Psychology of Music. Calderon Press, Oxford 1985
984. Sloboda, A S and A C Lehmann: Tracking Performance Correlates of Changes in Perceived Intensity of Emotion During Different Interpretations of a Chopin Piano Prelude. Music Perception, Summer 2001, Vol 19, No. 1, pp. 87-120, 2001
985. Sloterdijk P: Selbstexperiment. Hanser, München & Wien 1996
986. Sluming V et al.: Broca's Area Supports Enhanced Visuospatial Cognition in Orchestral Musicians. The Journal of Neuroscience, Vol. 27(14), pp. 3799-3806, April 4, 2007
987. Smart M A: Mimomania: Music and Gesture in Nineteenth-Century Opera. New Ed edition, University of California Press, Berkeley 2005
988. SMDL committee: Standard Music Description Language Draft, Pittsford, NY 1995
989. Smith III J O: Synthesis of bowed strings. In: Strawn J, Blum T, ICMA: Proceedings of the International Computer Music Conference, San Francisco 1982
990. Smith III J O: Physical Modeling of Musical Instruments, part I. Computer Music Journal, 16 (4), 1992
991. Smith III J O: Physical Modeling Synthesis Update. Computer Music Journal, 20 (2), 1996
992. Sodomka A: KUNSTRADIO at documentaX. http://www.culture.net/orfkunstradio/BIOS/sodomkabio.html, Vienna 1997
993. Spanier E H: Algebraic Topology. McGraw-Hill, New York et al. 1966
994. Späth H: Eindimensionale Spline-Interpolations-Algorithmen. Oldenburg, München et al. 1990
995. Späth H: Zweidimensionale Spline-Interpolations-Algorithmen. Oldenburg, München et al. 1991
996. Spellman A B: Four Lives in the Bebop Business. Pantheon 1966
997. Sperry R W: Neurology and the mind-body problem. American Scientist, 40, 291-312, 1952
998. Spicker V: Das Abstraktmotiv bei Cecil Taylor. MusikTexte, 73/74 March 1998
999. Spivak M: Calculus on Manifolds. Benjamin, New York 1965
1000. Squire R L and Butters N (Eds.): Neuropsychology of Memory. D. Guildford Press, NewYork and London 1984
1001. Stahnke M: Struktur und Ästhetik bei Boulez. Wagner, Hamburg 1979
1002. Stange-Elbe J, G Mazzola: Cooking a Canon with RubatoPerformance Aspects of J.S. Bach's "Kunst der Fuge". In: ICMA (Ed.): Proceedings of the ICMC 98, 179-186, San Francisco 1998.
1003. Stange-Elbe J: Analyse- und Interpretationsperspektiven zu J.S. Bachs "Kunst der Fuge" mit Werkzeugen der objektorientierten Informationstechnologie. Habilitationsschrift, Osnabrück 2000
1004. Stange-Elbe J: Instrumentaltechnische Voraussetzungen für eine computergestützte Interpretation. In: Enders B (Ed.): Proceedings of the Klangart congress '01, Osnabrück
1005. Starr D and R Morris: A General Theory of Combinatoriality and the Aggregate. PNM, 16(1), 364-389, 16(2), 50-84, 1977-78
1006. Starr D: Sets, Invariance and Partitions. JMT, 22(1), 1-42, 1978
1007. Stein S and S Sandor: Algebra and Tiling: Homomorphisms in the Service of Geometry.: Math. Assoc. Amer., Washington, DC 1994
1008. Steinberg R (Ed.): Music and the Mind Machine. Springer, Heidelberg 1995
1009. Steinberg R and A Fairweather (Eds.): Blues-Philosophy for Everyone: Thinking Deep About Feeling Low. John Wiley 2012
1010. Sternberg R J and T I Lubart: The Concept of Creativity: Prospects and Paradigms. In: Sternberg R J (Ed.): Handbook of Creativity. Cambridge University Press, New York 1999
1011. Stinson D R: Combinatorial designs: constructions and analysis. Springer, 2004
1012. Stollberg A, J Weissenfeld, F H Besthorn (Eds.): DirigentenBilder—Musikalische Gesten — verkörperte Musik. Schwabe, Basel 2015
1013. Stravinsky I: The Poetics of Music. Harvard University Press 1947
1014. Steele G L: Common Lisp: The Language, 2nd Edition. Digital Press, 1990
1015. Stone P: Symbolic Composer. http:/www.xs4all.nl/~psto/
1016. Stopper B: Gleichstufig temperierte Skalen unter Berücksichtigung von Oktavstreckungen. In: Hosp I (Hrsg.): Bozener Treffen 1991 — Skalen und Harmonien. Südtiroler Kulturinstitut, Bozen 1992
1017. Straub H: Beiträge zur modultheoretischen Klassifikation musikalischer Motive. Diplomarbeit ETH-Zürich, Zürich 1989
1018. Straub H: Kadenzielle Mengen beliebiger Stufendefinitionen. Unpublished manuscript, Zürich 1999/2000
1019. Straus J N: Voice Leading in Atonal Music. In: Baker J et al. [69], 237-274

1020. Stumpf C: Tonpsychologie. Leipzig 1883-1890
1021. Sundberg J, A Askenfelt, L Frydén: Musical Performance: A Synthesis-by-Rule Approach, Computer Music Journal, 7, 37-43, 1983
1022. Sundberg J and V Verillo: On the Anatomy of the Retard: A Study of Timing in Music. J. Acoust. Soc. Am. 68, 772-779, 1980
1023. Sundberg J: Music Performance Research. An Overview. In: Sundberg J, L Nord, R Carlson (Eds.): Music, Language, Speech, and Brain. London 1991
1024. Sundberg J (Ed.): Generative Grammars for Music Performance. KTH, Stockholm 1994
1025. Sundin N-G: Musical Interpretation in Performance. Mirage, Stockholm 1984
1026. Suzuki T et al.: Musical Instrument Database with Multimedia. In: ICMA (Ed.): Proceedings of the ICMC 96, San Francisco 1996
1027. Suppan W: Zur Verwendung der Begriffe Gestalt, Struktur, Modell und Typus in der Musikethnologie. In: Stockmann D and Steszwski J (Eds.): Analyse und Klassifikation von Volksmelodien. PWM Edition Krakau 1973
1028. Swan, R G: Vector Bundles and Projective Modules, *Transactions of the American Mathematical Society* 105 (2): 264-277
1029. Sylvester D: Interviews with Francis Bacon. Pantheon Books, New York 1975
1030. Szczeciniarz J-J: Réflexion sur Riemann, philosophe mathématicien, in [79], p. 195-221
1031. Symmetrie. Katalogband Vol.1 zur Symmetrieausstellung, Mathildenhöhe, Darmstadt, 1986
1032. Takemitsu T: Confronting Silence. Fallen Leaf Press, Berkeley 1995
1033. Takhtajan L A: Quantum Mechanics for Mathematicians. American Mathematical Society, Providence 2008
1034. Teschl G: Mathematical Methods in Quantum Mechanics. American Mathematical Society, Providence 2009
1035. Tanaka A: Interaction, Experience, and the Future of Music. In: O'Hara K and B Brown (Eds.): Consuming Music Together. Springer, Dordrecht 2006
1036. Tarry G: Le problème des 36 officiers. AFAS, Compt 1900–1901
1037. Tatarkiewicz W: A History of Six Ideas: An Essay in Aesthetics. Polish Scientific Publishers, Warszawa 1980
1038. Taylor Y and H Barker: What's It Like to Be Blind? Faking It. `fakingit.typepad.com` 2007
1039. Taylor C: Burning Poles. VHS, Mystic Fire 1991
1040. Taylor C: Silent Tongues. Freedom 1974
1041. Thalmann F: Musical composition with Grid Diagrams of Transformations. Master's Thesis, University of Bern, Bern 2007
1042. Thalmann F: Gestural Composition with Arbitrary Musical Objects and Dynamic Transformation Networks. PhD thesis, University of Minnesota, 2014
1043. Thalmann F and G Mazzola: The Bigbang Rubette: Gestural music composition with Rubato Composer. In: Proceedings of the International Computer Music Conference, International Computer Music Association, Belfast 2008
1044. Thalmann F and G Mazzola: Affine musical transformations using multi-touch gestures. Ninad, 24:58–69, 2010
1045. Thalmann F and G Mazzola: Gestural shaping and transformation in a universal space of structure and sound. In: Proceedings of the International Computer Music Conference, International Computer Music Association, New York City 2010
1046. Thalmann F and G Mazzola: Poietical music scores: Facts, processes, and gestures. In: Proceedings of the Second International Symposium on Music and Sonic Art, MuSA, Baden-Baden 2011
1047. Thalmann F and G Mazzola: Using the creative process for sound design based on generic sound forms. In: MUME 2013 proceedings, AAAI Press, Boston 2013
1048. Thalmann F and G Mazzola: Visualization and transformation in general musical and music-theoretical spaces. In: Proceedings of the Music Encoding Conference 2013, MEI, Mainz 2013
1049. Tormoen D, F Thalmann, G Mazzola: The composing hand: Musical creation with leap motion and the Bigbang Rubette. In: Proceedings of 14th International Conference on New Interfaces for Musical Expression (NIME), London 2014
1050. Terhardt E: Zur Tonhöhenwahrnehmung von Klängen. II. Ein Funktionsschema. Acustica 26, 187-199, 1972
1051. Thomas d'Aquin: Première question disputée. La vérité (De Veritate). (teaching in the years 1256-1259), Paris, Vrin, 2002
1052. Thom R: Stabilité structurelle et morphogénèse. Benjamin, Reading, MA 1972
1053. Tinctoris J: Opera Theoretica, Vol. 2: Liber de arte contrapuncti. Seay A (Ed.): Amer. Institute of Musicology 1975. gopher:`//IUBVM.UCS.INDIANA.EDU/00/tml/15th/tincon2.text`
1054. Tittel E: Der neue Gradus. Doblinger, Wien-München 1959

1055. Titze I R and D T Talkin: A theoretical study of the effects of various laryngeal configurations on the acoustics of phonation. Acoustical Society of America Journal, Vol. 66, No. 60, 1979
1056. Titze I R: Speed, accuracy, and stability of laryngeal movement in singing. Acoustical Society of America Journal, Vol. 115, p. 2590 ff., 2004
1057. Todd N P M: A Model of Expressive Timing in Tonal Music. Music Percep. **3**, pp. 33-58, 1985
1058. Todd N P M: Towards a Cognitive Theory of Expression: The Performance and Perception of Rubato. Contemporary Music Review **4**, 1989
1059. Todd N P M: A Computational Model of Rubato. Contemporary Music Review **3**, 1989
1060. Todd N P M: The Dynamics of Dynamics: A Model of Musical Expression. J. Acoustic Soc. Am. **91**, pp. 3540-3550, 1992
1061. Todoroff T: A Real-Time Analysis and Resynthesis Instrument for Transformation of Sounds in the Frequency Domain. In: ICMA (Ed.): Proceedings of the ICMC 96, San Francisco 1996
1062. Tomasello M: Why Don't Apes Point? In: Enfield N J & S C Levinson (Eds.): Roots of Human Sociality: Culture, Cognition and Interaction. Berg, pp. 506-524, Oxford & New York
1063. Tomasello M, M Carpenter, U Liszkowski: A New Look at Infant Pointing. Child Development, Vol. 78, No. 3. pp. 705-722, May/June 207
1064. Tymoczko D: The Geometry of Musical Chords. Science, Vol 313, 2006
1065. Tymoczko D: Mazzola's Model of Fuxian Counterpoint. In: Agon, C, G. Assayag, E. Amiot, J. Bresson, J. Mandereau (Eds.): Mathematics and Computation in Music. LNAI 6726, Springer, Heidelberg 2011
1066. Uhde J: Beethovens Klaviermusik III. Reclam, Stuttgart 1974
1067. Uhde J and R Wieland: Forschendes Üben. Bärenreiter 2002
1068. Uhde J and R Wieland: Denken und Spielen. Bärenreiter, Kassel et al. 1988
1069. Ungvary T and R Vertegaal: The Sentograph: Input Devices and the Communication of Bodily Expression. ICMA 1995
1070. Utgoff P E and P B Kirlin: Detecting Motives and Recurring Patterns in Polyphonic Music. Proceedings of the International Computer Music Conference, ICMA 2006
1071. van Dalen D: Logic and Structure. Springer, Berlin et al. 1997
1072. Valéry P: Leçon Inaugurale du Cours de Poétique au Collège de France. Gallimard, Paris 1945
1073. Valéry P: Cahiers I-IV (1894-1914). Celeyrette-Pietri N and J Robinson-Valéry (Eds.), Gallimard, Paris 1987
1074. Valéry P: Philosophie de la danse. Oeuvres, tome I, Variété, "Théorie poétique et esthétique", pp. 1390-1403, Gallimard, 1956
1075. Valéry P: Analecta (1926), in *Tel quel*, tome II, Collection Idées, no 241, Gallimard, 1971
1076. Vallès J: "Penser c'est oublier" *Images des mathématiques*. CNRS, `http://images.math.cnrs.fr/Penser-c-est-oublier.html`, 10 juillet 2015
1077. Valtieri S: La Scuola d'Athene. Mitteilungen des Konsthistorischen Instituts in Florenz 16, 1972
1078. van der Waerden B L: Algebra I, II. Springer, Berlin et al. 1966
1079. van der Waerden B L: Algebra I. (English tranlsation) Springer, New York 2003
1080. van der Waerden B L: Die Pythagoreer. Artemis, Zürich 1979
1081. Varèse E: Erinnerungen und Gedanken. In: Darmstädter Beiträge III. Schott, Mainz 1960
1082. Varèse E: Rückblick auf die Zukunft. edition text+kritik, München 1978
1083. Vercoe B: The Synthetic Performer in the Context of Live Performance. In: ICMA (Ed.): Proceedings of the ICMC 84, San Francisco 1984
1084. Veyne P: Comment on écrit l'histoire, 1959. nouv. ed., Points/Histoire 226, Paris 1996
1085. Vico J-B: La méthode des études de notre temps. Présentation et traduction de A. Pons, (De nostri temporis studiorum ratione, 1708), online, site du Réseau Intelligence de la Complexité, 1981
1086. Vieru A: The Book of Modes. Editura Muzicală, Bucarest 1993
1087. Vinci A C: Die Notenschrift. Bärenreiter, Kassel 1988
1088. Violi P: Beyond the body: towards a full embodied semiosis. In: Dirven R, and R Frank (Eds.): Body, Language and Mind. Mouton de Gruyter, Berlin 2006
1089. Vogel M: Die Lehre von den Tonbeziehungen. Verlag für systematische Musikwissenschaft, Bonn-Bad Godesberg 1975
1090. Vogel M: Arthur v. Oettingen und der harmonische Dualismus. In: Vogel M (Ed.): Beiträge zu Musiktheorie des 19. Jahrhunderts. Bosse, Regensburg 1966
1091. Vogel M: Berechnung emmelischer und ekmelischer Mehrklänge. In: Herf F R (Ed.): Mikrotöne II. Helbling, Innsbruck 1988
1092. Vogt H: Neue Musik seit 1945. Reclam, Stuttgart 1972

1093. Vossen G: Datenbankmodelle, Datenbanksprachen und Datenbank-Management-Systeme. Addison-Wesley, Bonn et al. 1994
1094. Vuza D T: Sur le rythme périodique. Revue Roumaine de Linguistique—Cahiers de Linguistique Théorique et Appliquée 22, no. 1 1985
1095. Vuza D T: Some Mathematical Aspects of David Lewin's Book *Generalized Musical Intervals and Transformations*. Perspectives of New Music, vol. 26, no. 1, 1988
1096. Vuza D T: Supplementary Sets and Regular Complementary Unending Canons (Part One). Perspectives of New Music, vol. 29, no. 2, 1991
1097. Vuza D T: Supplementary Sets and Regular Complementary Unending Canons (Part Two). Perspectives of New Music, vol. 30, no. 1, 1992
1098. Vuza D T: Supplementary Sets and Regular Complementary Unending Canons (Part Three). Perspectives of New Music, vol. 30, no. 2, 1992
1099. Vuza D T: Supplementary Sets and Regular Complementary Unending Canons (Part Four). Perspectives of New Music, vol. 31, no. 1, 1993
1100. Vygotsky L S: Thought and Language (1934). Edited and translated by E Hanfmann and G Vakar, revised and edited by A Kozulin. MIT Press, Cambridge 1986
1101. Wanske H: Musiknotation. Schott, Mainz 1988
1102. Waldvogel M. et al.: *presto* source code. TRIMAX, Zürich 1986-1993
1103. Wallas G: The Art of Thought. Harcourt, New York 1926
1104. Wallis J: Tractatus de Sectionibus Conicis. 1655
1105. Walker J S: Fourier Analysis and Wavelet Analysis. Notices of the AMS, vol.44, No.6, July 1997
1106. Ward J: Gregorian Chant according to the Principles of André Mocquereau of Solesmes. The Catholic Education Press 1923 (Reprint)
1107. Waugh W A O'N: Music, probability, and statistics. In: Encyclopedia of Statistical Sciences, **6**, 134-137, 1985
1108. Webern A: *Variationen op.27* (mit Interpretationsanweisungen). UE 16845, Wien 1980
1109. Wegner P: Interactive Foundations of Object-Oriented Programming. Computer, IEEE Computer Soc., October 1995
1110. Weidman A: Singing the Classical, Voicing the Modern: The Postcolonial Politics of Music in South India. Duke University Press 2006
1111. Weil A: Œuvres scientifiques. Collected papers, Vol. I, Springer, New York, 1979
1112. Weisberg R W: Creativity and Knowledge: A Challenge to Theories. In: R J Sternberg (Ed.): Handbook of Creativity. Cambridge University Press, New York 1999
1113. Weyl H: Symmetrie. Birkhäuser, Basel 1955
1114. Weinberger N M: Musica Research Notes. Music and Science Information Computer Archive, Center for Neurobiology of Learning and Memory, U California Irvine, Vol. III, Issue II, Fall 1996
1115. Whiteside A: Indispensables of Piano Playing. Coleman-Ross, New York 1955
1116. Wicinski A A: Psichologyiceskii analiz processa raboty pianista-ispolnitiela nad muzykalnym proizviedieniem. Izviestia Akademii Piedagogiceskich Nauc Vyp., 25 [Moscow], 171-215, 1950
1117. Widmer G: Learning Expression at Multiple Structural Levels. In: ICMA (Ed.): Proceedings of the ICMC 94, San Francisco 1994
1118. Widmer G: Modeling the Rational Basis of Musical Expression. Computer Music J. 18, 1994
1119. Widmer G and W Goebl: Computational Models of Expressive Music Performance: The State of the Art. Journal of New Music Research 33(3), 203-216, 2004
1120. Wiedemann M and E Zimmermann: Kreativität und Form. Springer Vieweg, Berlin Heidelberg 2012
1121. Wieser H-G: Electroclinical features of the psychomotor seizure. Fischer, Stuttgart and Butterworth, London 1983
1122. Wieser H-G and G Mazzola: Musical consonances and dissonances: are they distinguished independently by the right and left hippocampi? Neuropsychologia 24 (6):805-812, 1986
1123. Wieser H-G and G Mazzola: EEG responses to music in limbic and auditory cortices. In: Engel J Jr, Ojemann G A, Lüders H O, Williamson P D (Eds.): Fundamental mechanisms of human function. Raven, New York 1987
1124. Wieser H-G: Musik und Gehirn. Revue Suisse Méd. 7, 153-162, 1987
1125. Wieser H-G and S Moser: Improved multipolar foramen ovale electrode monitoring. J Epilepsy 1: 13-22, 1986
1126. Wiggins G et al.: A Framework for Evaluation of Music Representation Systems. Computer Music Journal 17:3, 1993
1127. Wiil U K (Ed.): Computer Music Modeling and Retrieval (CMMR 2003). Springer, Heidelberg 2004
1128. Wiil U K (Ed.): Computer Music Modeling and Retrieval (CMMR 2004). Springer, Heidelberg 2005

1129. Wille R: Symmetrie — Versuch einer Begriffsbestimmung. In: Symmetrie, Katalogband Vol.1 zur Symmetrieausstellung, Mathildenhöhe, Darmstadt, 1986
1130. Wille R: Personal Communication. Darmstadt 1982
1131. Wille R: Musiktheorie und Mathematik. In: Götze H and R Wille (Eds.): Musik und Mathematik. Springer, Berlin et al. 1985
1132. Wille R: Bedeutungen von Begriffsverbänden. Preprint Nr. 1058, TH Darmstadt 1987
1133. Wille R: Personal Communication. Darmstadt 1985
1134. Winson I: Brain and Psyche. The Biology of the Unconscious. Anchor Press/Doubleday, Garden City, NY 1985
1135. Wilson F R: The Hand: How Its Use Shapes the Brain, Language, and Human culture. Pantheon Books, New York 1998
1136. Witten E: Magic, Mystery, and Matrix. Notices of the AMS, Vol. 45, Nr. 9, 1998
1137. Wittgenstein L: Tractatus Logico Philosophicus – Logisch-Philosophische Abhandlung, Wilhelm Ostwald (Ed.), Annalen der Naturphilosophie, 14, 1921
1138. Wittgenstein L: Tractatus Logico-Philosophicus (1918). Suhrkamp, Frankfurt/Main 1969
1139. Wolfram S: Software für Mathematik und Naturwissenschaften. In: Chaos und Fraktale. Spektrum der Wissenschaft Verlagsgesellschaft, Heidelberg 1989
1140. Xenakis I: Formalized Music. Indiana Univ. Press, Bloomington 1972
1141. Yaglom A M and I M Yaglom: Wahrscheinlichkeit und Information. Deutscher Verlag der Wissenschaften, Berlin, 1967
1142. Yoneda N: On the homology theory of modules. J. Fac. Sci. Univ. Tokyo, Sct. I,7, 1954
1143. Ystad S et al.: Of Parameters Corresponding to a Propagative Synthesis Model Through the Analysis of Real Sounds. In: ICMA (Ed.): Proceedings of the ICMC 96, San Francisco 1996
1144. Yusuke A et al.: Hyperscore: A Design of a Hypertext Model for Musical Expression and Structure. J. of New Music Research, Vol. 24, 1995
1145. Zahorka O: Versuch einer Charakterisierung des altrömischen Melodiestils. Seminar Lütolf, MWS Univ. Zürich 1993
1146. Zahorka O: From Sign to Sound—Analysis and Performance of Musical Scores on RUBATO. In: Enders B (Ed.): Symposionsband Klangart '95, Schott, Mainz 1997
1147. Zahorka O: PrediBase—Controlling Semantics of Symbolic Structures in Music. In: ICMA (Ed.): Proceedings of the ICMC 95, San Francisco 1995
1148. Zahorka O: RUBATO – Deep Blue in der Musik? Animato 97/3, 9-10, Zürich 1997
1149. Zalamea F: Synthetic Philosophy of Contemporary Mathematics. Urbanomic, Falmouth, UK, & Sequence Press, New York, USA 2011
1150. Zarca B: L'univers des mathématiciens. L'ethos professionnel des plus rigoureux des scientifiques. Presses Univ. Rennes, 2012
1151. Zarlino G: Istitutioni harmoniche. Venezia 1558
1152. Zbikowski Z: Conceptualizing Music: Cognitive Structure, Theory, and Analysis. Oxford University Press 2005
1153. Zbikowski Z: Music, Emotion, Analysis. Music Analysis 29/1-3(2010): 37-60
1154. Zekl G: Topos. Meiner, Hamburg 1990
1155. Zlotin B and A Susman: Creative Pedagogy. Journal of TRIZ. no.4, pp. 9-17 (in Russian) 1991
1156. Zweifel P F: Generalized Diatonic and Pentatonic Scales: A Group-theoretic Approach. PNM, 34(1), 140-161, 1996
1157. Zwicker E and H Fastl: Psychoacoustics. Facts and Models. Springer, Berlin, et al. 1999
1158. Zwicker E and M Zollner: Elektroakustik (2nd edition). Springer, New York et al. 1987
1159. Zwiebach B: A First Course in String Theory. Cambridge University Press, Cambridge et al. 2004

Index

G. Mazzola et al., *The Topos of Music III: Gestures*, Computational Music Science,
https://doi.org/10.1007/978-3-319-64481-3

A

E

J

N

O

Q

T

W

The manufacturer's authorised representative in the EU is Springer Nature Customer Service Centre GmbH, Europaplatz 3, 69115 Heidelberg, Germany. If you have any concerns regarding our products, please contact ProductSafety@springernature.com

Printed and bound by CPI Group (UK) Ltd, Croydon, CR0 4YY
01/10/2026
02225992-0003